Laser Scanning, Theory and Applications
Edited by Chau-Chang Wang

Published by InTech
Janeza Trdine 9, 51000 Rijeka, Croatia

Technical Editor Ivana Lorkovic

Cover Designer InTech Design Team

Additional hard copies can be obtained from orders@intechopen.com

Laser Scanning, Theory and Applications,
Edited by Chau-Chang Wang
p. cm.
ISBN 978-953-307-205-0

For EU product safety concerns:
IN TECH d.o.o., Prolaz Marije Krucifikse Kozulić 3, 51000 Rijeka, Croatia,
info@intechopen.com or visit our website at intechopen.com.

Meet the editor

He received the B.S degree in Mechanical Engineering from National Taiwan University, Taipei, Taiwan, in 1986 and the M.S. and Ph.D. degrees in mechanical engineering from University of Pennsylvania with a major in robotics in 1991 and 1995. After graduating from University of Pennsylvania, he worked as a postdoc fellow in the Department of Mechanical Engineering, National Cheng-Kung Univeristy, Tainan, Taiwan for one year. Since 1996 he has been an Associate Professor in the Institute of Undersea Technology, Kao-Hsiung, Taiwan. Further research interests include the development of ocean bottom seismometer, low power underwater mechatronic systems and underwater robotic systems.

Contents

Preface

Different laser-based methods have been developed to measure features or targets at various scales. It can be as large as landscape, or as small as cells. *Laser Scanning, Theory and Applications* collects up-to-date research on laser-based measurement. It covers topics ranging from theories, innovative applications to development of new instruments. Readers are urged to follow the chapter order to get familiar with the working principles (Chapter 1 to 6), then move on the various examples of medical, biological, archaeological and engineering and applications (Chapter 7 to 22.); at last, learn about the development of several laser-based measurement systems (Chapter 23 to 28.) Or just go directly to the part of interest if background knowledge is already equipped.

Chau-Chang Wang
Sun Yat-sen University
Kao-Hsiung, Taiwan

1

Quantitative Surface Characterisation Using Laser Scanning Confocal Microscopy

Steven Tomovich[1], Zhongxiao Peng[1], Chengqing Yuan[2] and Xinping Yan[2]

[1]*School of Engineering & Physical Sciences*
James Cook University, Queensland,
[2]*Reliability Engineering Institute, School of Energy and Power Engineering*
Wuhan University of Technology, Wuhan,
[1]*Australia*
[2]*China*

1. Introduction

This chapter reports the development of a new quantitative surface characterisation system based on laser scanning confocal microscopy (LSCM). This development involves: systemically studying and understanding the resolutions of a LSCM and its capacity; investigating appropriate hardware settings for acquiring suitable two-dimensional (2D) images; assess image processing techniques for eliminating noise and artefacts while constructing three-dimensional (3D) images for quantitative analysis; and developing quantitative image analysis techniques to characterise the required features of surfaces. This chapter will show examples of using a LSCM for image acquisition and the developed techniques for the surface characterisation of engineering surfaces and wear particles in engineering and bioengineering applications.

2. Importance

Quantitative surface characterisation is an essential and common practice in many applications. Two-dimensional surface measurement has been performed for over 50 years using traditional 2D surface roughness parameters including the arithmetic mean deviation (R_a), root mean squared deviation (R_q), surface skewness (R_{sk}) and surface kurtosis (R_{ku}), etc that are based on single line surface traces (Gadelmawla et al., 2002). The technique has been popularly used to measure large surfaces in 2D. Standard 2D measurements are commonly performed using the stylus profiler that traces a diamond stylus tip over the specimen surface, recording vertical displacement over distance travelled (DeGarmo, 1997). The resolution of the system is governed by the stylus tip size selected for testing surfaces of specific material type and surface finish. However, in operation the stylus unfortunately subjects the surface to damage (Bennett and Mattsson, 1989; Gjonnes, 1996; Conroy and Armstrong, 2005) and thus the technique cannot be used to perform surface measurement on soft materials. In addition, the minimum scan length for the stylus technique (ISO standards 4288, 1996) is not suited to small wear particle surfaces.

Due to the complexity of surface topography, the extraction of surface roughness information in 3D is often required for the accurate characterisation of surface features. Standard scanning electron microscopy (SEM) has been extensively used for characterising surface features. Scanning electron microscopy has outstanding 2D resolution and with specialised SEM operation can also provide 3D images (Podsiadlo and Stachowiak 1997; Bernabeu et al., 2001). However, in addition to the specialised preparation time required, the fact samples are modified during preparation and possibly during image acquisition are issues to consider when using SEM (Brown and Newton, 1994; Chescoe and Goodhew, 1990; Hanlon et al., 2001). Atomic force microscopy (AFM) is a technique often used for 3D surface measurement that is to a degree non-contacting or destructive, while also possessing sub nanometre resolution in 3D. The limitation with this technique is the maximum scan area of ≈100 μm^2 that is not sufficient for the study of some surfaces (Gauldie et al., 1994; Martin et al., 1994; Conroy and Armstrong, 2005).

A non-destructive and versatile technique for characterising surface information in 3D is the LSCM imaging system. This technique involves sequentially illuminating the focal plane with a diffraction-limited light spot point by point. Specimen information is transmitted back through the optics via a confocal aperture as a diffraction-limited spot to reduce light from above and below focal plane. The sharpened light spot is converted into an electrical signal by the photo multiplier tube (PMT) point detector and then digitised for storage or display. By sequentially scanning through the specimen field of view, a 2D digital image of the focal plane is formed. The significant benefit over standard optical microscopy is the LSCM ability to discriminate between in focus and out of focus light above and below the focal plane. In addition to imaging high-resolution 2D slices from the focal plane, another primary feature is the system's ability to capture 3D images. By compiling a series of adjacent 2D digital images captured at different heights in the specimen, constructed 3D images provide 3D surface and volume information by a non-contacting and un-intrusive method (Sheppard and Shotton, 1997; Pawley, 1995). Using specialised software, surface measurements can be readily performed on variations in pixel intensity and spatial locations contained in 3D images (Brown and Newton, 1994; Yuan et al., 2005).

By comparison the resolution of LSCM is possibly higher than a standard stylus profiler and can non-intrusively obtain surface information from large and small surfaces in 3D while the stylus profiler is restricted to measuring surface characteristics of large surfaces in 2D. The confocal system also provides a greater field of view and depth range than AFM, which is more suitable for industrial surfaces measurements. In addition, the minimal specimen preparation for LSCM imaging sets the technique to be a fast and versatile instrument for quantitative surface measurements (King and Delaney, 1994; Gjonnes, 1996; Bernabeu et al., 2001; Sheppard and Shotton, 1997; Haridoss et al., 1990). Past research using traditional 2D parameters for quantitative surface analysis found good correlation between roughness measurements of metal surfaces determined by LSCM, stylus and AFM surface measurement techniques (Gjonnes, 1996; Hanlon et al., 2001; Peng and Tomovich, 2008). Early research and development of surface measurement using LSCM found the technique to be viable for extracting surface information from 3D images in the form of numerical parameters, calculated from images generated using height encoded image (HEI) processing techniques (Peng and Kirk, 1998 & 1999). However, developments are still in their infant stages. The aim of this presented study is to develop a set of techniques so that LSCM can be used widely for quantitative surface characterisation in 3D for a range of studies and applications.

Specifically, this project involves the evaluation of the LSCM resolution and imaging system performance in relation to hardware parameter settings; the establishment of correct procedures for acquiring high quality images for quantitative 3D surface measurements and the development and validation of the 3D LSCM surface measurement technique. The details of the above developments are presented in the following section.

3. Development of 3D imaging, processing and analysis techniques for quantitative surface measurement

Surface measurement using any image acquisition systems involves three major steps, namely, data acquisition, processing and analysis. As this project is to develop 3D surface measurement techniques based on laser scanning confocal microscopy, the developments have been conducted in the following four phases. Phase 1 is to evaluate the LSCM system and to perform resolution testing to determine the optimal system settings for the acquisition of high quality images for numerical analysis. Phase 2 involves image processing to construct 3D images using acquired 2D image sets and to eliminate image artefacts and distortion. Phase 3 focuses on the development and selection of numerical parameters for surface measurement, and Phase 4 involves testing and validating the techniques using various samples. The section presents our research and developments on phases 1, 2 and 4.

3.1 Laser scanning confocal microscope resolution testing

A comprehensive study of the image acquisition system performance and resolutions has been conducted. To extract meaningful surface information from raw 3D images, the performance of the Radiance2000 LSCM imaging system and specimen influences on the image formation process must be fully understood to bring to light what is real morphological surface information or simply specimen and imaging system related artefact. It is for this reason that image acquisition tests focused on imaging metallic surfaces of known morphology and reflectivity.

For the tested objective lens, images of a front coated aluminium optical flat (Edmunds Optics 43402) were captured over several increments of scan rate, laser power, PMT gain, Kalman frame averaging and confocal aperture setting using zoom 999. Height encoding (Sheppard and Matthews, 1987; Pawley, 1995; Peng and Tomovich, 2008) the raw LSCM 3D images created HEI containing minimum and maximum z positions used to determine HEI depth. Plotted against increasing parameter settings, HEI depth highlighted the variance in depth discrimination (axial resolution) with various LSCM settings. Also useful in studying image quality were 2D maximum brightness images (MBI) constructed from raw 3D images with HEI. The coefficient of variance (CV) calculated from the standard deviation (SD) and mean pixel intensity (u) of MBI, provided a measure of image noise (Zucker and Price, 2001) that were plotted against LSCM settings. To visually assess the lateral resolution of lenses, chromium graticule and bar structures on the Gen III Richardson test slide were imaged using Nyquist zoom. Contrast measured (Oldenbourg et al., 1993; Pawley, 1995) across the structures and plotted against scan rate, frame averaging and aperture settings also highlighted the influence of these parameters on lateral resolution. For each objective lens, limiting HEI axial and lateral resolutions achieved for metrology applications were then determined from optical flat HEI depth tests and HEI of structures on the Richardson slide.

Scan Rate Testing

Using a Nikon LU Plan 100x lens, original tests for scan rate over the 166-750 Hz range produced HEI depths in Figure 1a that show the 600 Hz to provide the highest depth discrimination. Measured depths were expected to increase with image noise reflected by increasing CV in Figure 1b. When comparing Figures 1a original scan and Figure 1b, no clear relationship exists between increasing image noise and original HEI depths that were found to be impacted on by scanning distortions. However, depth discrimination was shown to reduce with increasing scan rate for the modified plot in Figure 1a on removing the distortions prior to height encoding.

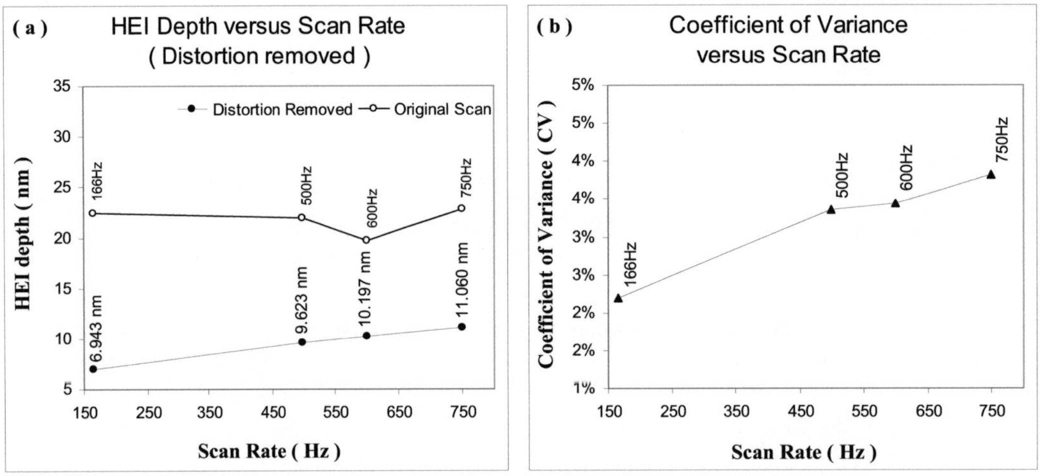

Fig. 1. Depth discrimination reduced with higher scan rates (a) as maximum brightness image noise increases (b).

Evidence of scanning distortion in Figures 2a-d show waves unique to each scan rate traversing left to right in sections from the full HEI representing a single optical flat location. The distortion potentially stems from mechanical instability in the galvanometer driving the left to right horizontal scanning process. In sections taken from the left side of the full HEI, Figures 3a-d show an increased waviness of scan lines due to the vertical scanning process traversing from top to bottom in the 500-750 Hz HEI. When comparing Figures 3a-d for image tilt, the 166 Hz HEI slope down towards the bottom of the image while faster scan rates show little tilt for the same optical flat location. This was present in all 166 Hz HEI indicating depths in Figure 1 for the 166 Hz original scan, included the effects of the stage drifting in the z direction with extended imaging time.

Comparing HEI over increasing scan rate in both Figures 2-3, there is a slightly higher level of graininess in HEI for the 500-750 Hz scan rates. To clearly associate increasing MBI noise reflected by CV in Figure 1b with increasing graininess of HEI and reduced depth discrimination, scanning distortions were therefore minimised in raw 3D data prior to height encoding. This was achieved by cropping images using the Radiance2000 LSCM operating system software LaserSharp2000, to provide five separate 25^2 xy pixel framed 3D image series. By reducing the sampling area the distortions traversing left to right were selectively filtered out and the tilt component substantially minimised, placing more significance on HEI depth associated with graininess. In a practical sense, the scale of

Fig. 2. Traversing left-right the (a) 166 Hz, (b) 500 Hz, (c) 600 Hz & (d) 750 Hz height encoded image distortions.

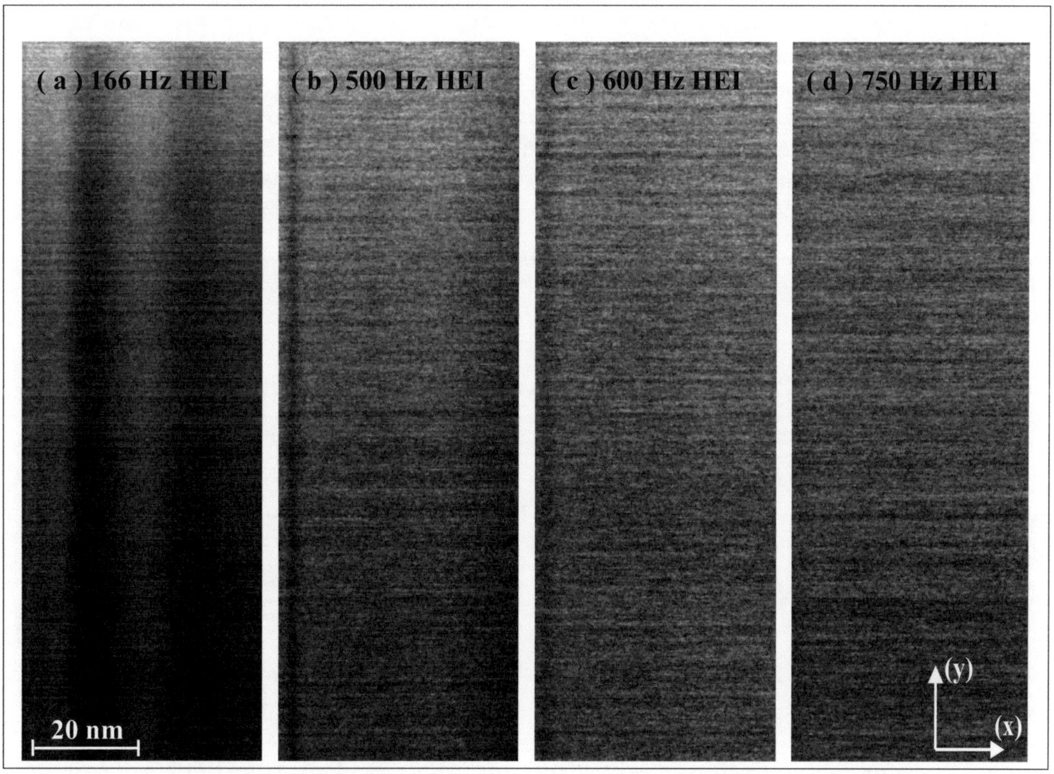

Fig. 3. Stage drift that increased the 166 Hz HEI depth (a) and scan line waviness increasing for the 500-750 Hz HEI (b-d).

distortions found in HEI are minimal when using Nyquist zoom and the cropping of images to remove their effect served only to assess parameter settings during preliminary tests. Although the modified HEI depths in Figure 1a now increase with CV in Figure 1b, the proportion image noise contributed to total HEI depth still remains unknown. Further, scan area geometric error and xy translational error of the stage will also introduce an unknown degree of shift in the optical flat surface that will impact on HEI depth. However, given these limitations the height encoding test did provide a more direct assessment of image quality and therefore remained a key indicator for LSCM testing.

To establish an ideal scan rate it was also necessary to assess the impact of scan rate on lateral resolution, defined as the minimum separating distance for two points to remain resolved with some level of contrast (Pawley, 1995). The standard test involves direct imaging of finely spaced graticule structures to visually assess limitations in lateral resolution. Extracting pixel intensities across the MBI of structures exampled in Figure 4a produced the intensity profile over x or y distance in microns. The contrast measured between the minimum and maximum profile intensities diminish with finer graticule spacing, or as in case with increasing image noise due to using higher scan rates. Imaging both vertical and horizontal 0.25 µm metal graticules on the Richardson test slide provided measured contrast versus scan rate in Figure 4b for the 100x lens. For the 166-750 Hz scan rates vertical contrast measured 10.34%, 11.32%, 15.50% and 25.17% more across horizontal graticules than across vertical graticules. The difference indicates a more symmetric lateral resolution was achieved using the 166-500 Hz scan rates. The reduced horizontal and increased vertical contrast for the 600-750 Hz scan rates, stems from the scanning and digitisation process that is continuous from left to right and discontinuous in the vertical. Scanning left to right, the continuous PMT output signal is digitised sequentially and appears to have generated a smearing of pixel intensities when scanning across passing differences in surface reflectivity. Circled in Figure 4c, the 750 Hz MBI of 2-4 µm horizontal graticules show smearing of pixel intensity across the non reflective substrate that is likely caused by some form of lag in PMT response. Graticules aligned perpendicular to the

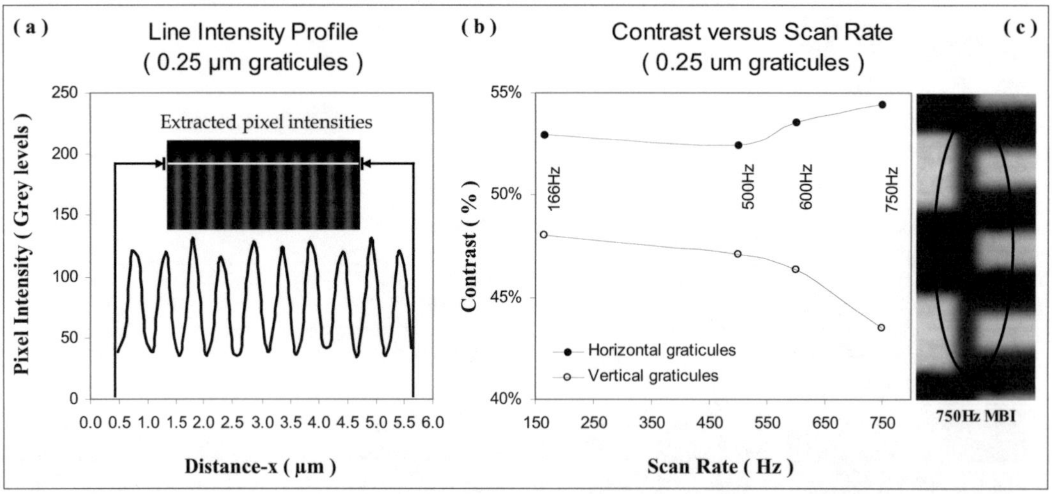

Fig. 4. Intensities data extracted across 0.25 µm vertical graticules (a), the impact of scan rate on measured contrast (b) and the carry over of reflection signal across the substrate regions when using the 750 Hz scan rate (c).

continuous scan direction, suffered reduced contrast and subsequent resolution due to the elevated signal from the substrate regions. Horizontally aligned graticules experienced a reinforcing of brightness for the reflective metal strips with no carry over signal across the substrate, overall increasing vertical contrast for the 600-750 Hz. Although the 166Hz scan rate provided the highest axial resolution with the most symmetric lateral resolution, engineering surfaces often have deep structures in addition to surface form and tilt that add significantly to the depth of a 3D image series. When imaging extended surfaces with the 166 Hz, acquisition time rapidly increases with the number of sections. The most practical scan rates for large image series are the 500-750 Hz that also eliminates stage drift in the z direction. For the 750Hz, pixel smearing and the excessive graininess of HEI need to be considered when imaging finer structures.

Laser Power and PMT Gain

Laser power and PMT gain were also investigated for their influence on HEI depth when imaging common industrial metals including aluminium (optical flat), polished stainless steel, brass and copper surfaces. Fixing PMT gain at 0.005% of the maximum setting while increasing laser power, several images of the aluminium optical flat were captured at a fixed location with the 100x lens using the 500 Hz scan rate. Generating MBI from each image provided in grey levels the mean pixel intensity (u) versus laser power % in Figure 5a. The intensity plot in Figure 5b was created when fixing laser power at 0.9% of the maximum and capturing images over varied PMT gain %.

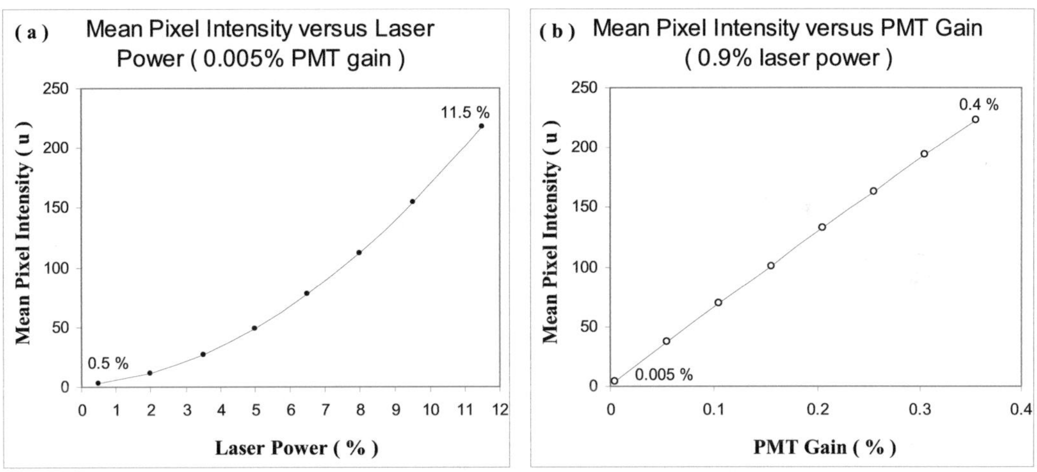

Fig. 5. Mean pixel intensity u versus laser power (a) and versus PMT gain (b), for the aluminium optical flat.

Shown in Figure 6a, the highest depth discrimination for both laser and PMT tests were realised above 50 grey levels. With increasing brightness, the constant HEI depth measurements also show both laser and PMT gain had no adverse effect on depth discrimination. The only constraint when increasing either of these parameters was to avoid image saturation above 256 grey levels. Below 50 grey levels and HEI depth measurements start to increase with reduced height encoding accuracy. An insufficient number of grey levels in the z axial intensity distribution cause the height encoding algorithm to behave more erratic. From varied laser and PMT tests, MBI also provided CV plots versus mean pixel intensity u in Figure 6b. Below 50 grey levels and the elevated CV reflects the poor suitability of images for height encoding. Above this level and CV remained stable with no upward trend to indicate increasing laser power or PMT gain were contributing an increasing noise component to images. For greater than 50 grey levels, taking the mean of all CV and HEI depth values show PMT results were consistently higher than laser results by an average 38% (3.04 nm) in Figure 6a and 49% (0.56%) in Figure 6b.

When imaging surfaces for the remaining metal types with the Nikon LU Plan Epi 50x lens, tests involved the same imaging procedure as the 100x lens except laser power was fixed at 0.5% for PMT tests and Kalman 4 frame averaging was used instead of direct scan. Figures 7a-b presents mean pixel intensity versus laser power and PMT gain, with varied PMT tests also including results when fixing laser power at 2.0% for comparison. Surfaces of different reflectivity not only verified laser and PMT results obtained with the 100x lens, but also provided a comparison for the influence of surface reflectivity on depth discrimination.

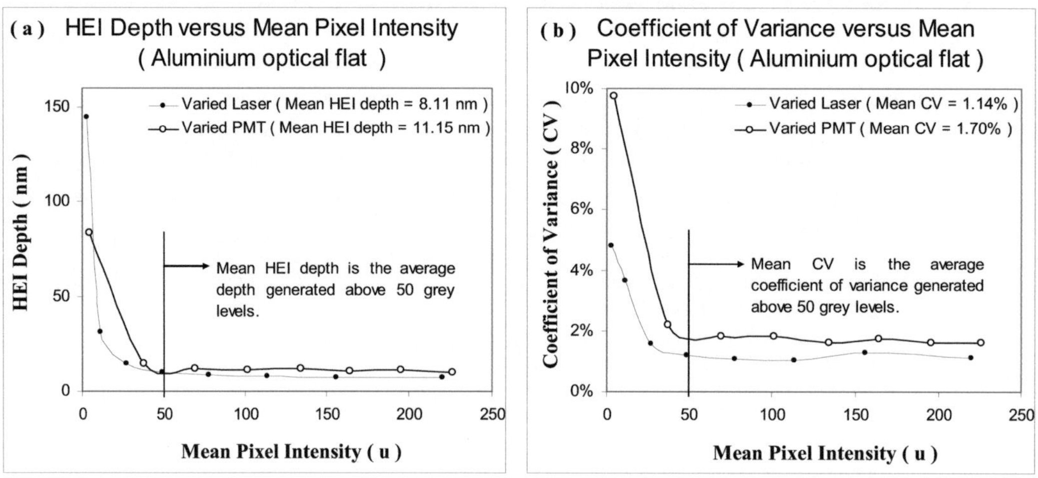

Fig. 6. HEI depth versus mean pixel intensity u (a) and versus coefficient of variance CV (b).

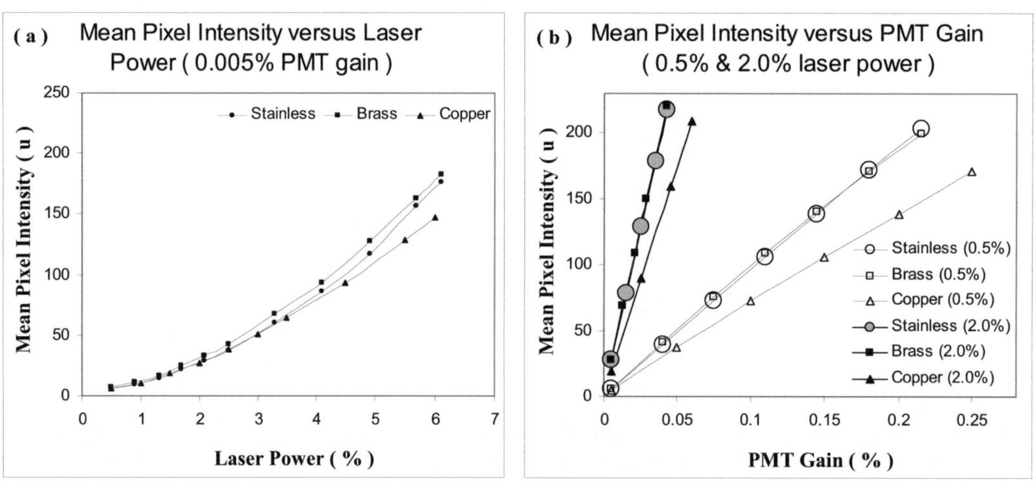

Fig. 7. Mean pixel intensity u versus laser power (a) and versus PMT gain (b) for stainless steel, brass and copper surfaces.

Height encoded results in Figures 8a-b show depth discrimination was constant for all metals after realising 50 grey levels, as was the case when imaging the aluminium optical flat with the 100x lens. In addition, the difference in reflectivity appears to have no impact on HEI depth. The strong reflection signal generated for each metal type required only a fractional shift in laser or PMT settings to compensate the effects of reflectivity. The HEI depth also indicates varied PMT results were once again consistently higher than laser results for over 50 grey levels, when comparing between Figure 8a and Figure 8b results. The margin of difference for the stainless steel, brass and copper surface were 10.4% (1.05 nm), 3.3% (0.36 nm) and 7.4% (0.81 nm). These margins are considerably less when compared to the difference between laser and PMT results obtained with the 100x lens. Comparing depth discrimination, Kalman frame averaging improved the 50x lens depth discrimination to nearly equal the 100x lens. To determine whether fixed laser power was impacting on PMT data, HEI depths were generated for varied PMT tests when fixing laser power at 2.0%. Figure 9 show the mean HEI depth for stainless steel, brass and copper surface tests were reduced to slightly less than varied laser HEI depths. For each surface when fixing laser power at 2.0%, varied PMT HEI depth averaged 1.40%, 5.65%, and 0.64% less than varied laser results. The small but apparent reduction in varied PMT depth was potentially the result of increased stability in laser illumination, reducing overall image noise at these low laser power settings.

Although Kalman frame averaging was used when imaging the stainless steel, brass and copper surfaces, the results were consistent with data captured using the 100x lens and direct scan. In addition to supporting the 100x test results, reflectivity tests have also shown depth discrimination (axial resolution) to improve with increased minimum laser power. Replicating this result when using the 100x lens and direct scan for the optical flat test, would confirm the benefit of avoiding minimum laser power. To maximise depth discrimination based on these findings, laser power should first be increased to provide a

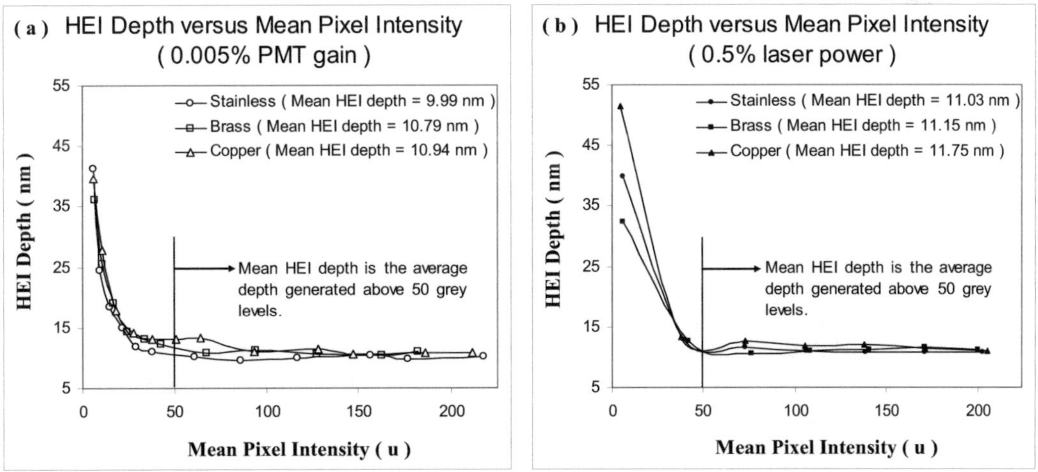

Fig. 8. HEI depth versus mean pixel intensity u for varied laser power using a 0.005% fixed PMT gain (a). With laser power fixed at 0.5% for varied PMT gain, HEI depth versus mean pixel intensity u (b).

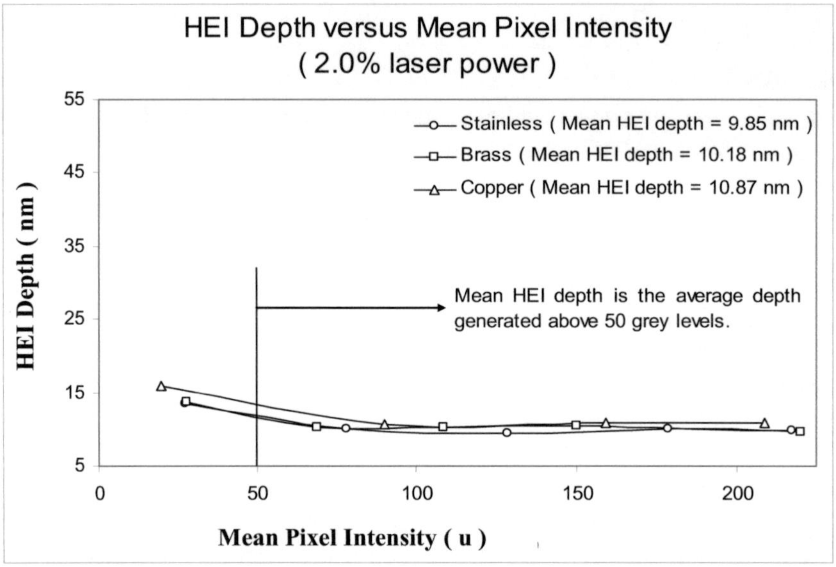

Fig. 9. HEI depth versus mean pixel intensity u for varied PMT gain using a 2.0% fixed laser power.

minimum of 50 grey levels with PMT gain fixed at 0.005%. Fine tuning image brightness should follow with increasing PMT gain to provide near image saturation. These adjustments are simplified by using the SETCOL function in the LSCM LaserSharp2000 software to highlight saturated pixels red and lower pixel intensities as green during scanning. However, difficulties striking a balance between saturation at highly reflective regions and dark regions falling below an intensity threshold may arise for surfaces containing multiple levels of reflectivity. To minimise dark regions that cause height encoding error, adjusting the SETCOL threshold to 50 grey levels would provide a good indicator during image acquisition while also highlighting the problem regions green prior to image processing.

For the 100x lens, several images of the optical flat were captured using Kalman 2-4 frame averaging and direct scan with the 500 Hz scan rate. The generated HEI depths in Figure 10a show Kalman 2 provided the most benefit, reducing HEI depth 38.66% in comparison to direct scan with Kalman 3 and 4 reducing HEI depth a further 6.72% and 0.46%. When comparing direct scan results in Figure 10a with previous depth discrimination results, the 100x lens delivered significantly greater depth discrimination in the previous tests. The loss of depth discrimination was caused by a failing laser power supply during Kalman tests. Although depth discrimination improved with frame averaging, no benefit to lateral resolution was achieved when imaging the 0.25 µm metal graticules at Nyquist zoom. Figure 10b presents the average contrast measured across the MBI of graticule structures captured using direct scan and Kalman 2-4 frame averages. Vertical contrast across horizontal graticules on average measured higher than horizontal contrast. The cause of this asymmetric contrast was in scan rate tests associated with continuous scanning in the x direction. In addition to limited improvements for lateral resolution, Kalman averaging becomes impractical for even the 500-750 Hz when imaging large 3D series due again to increased acquisition time.

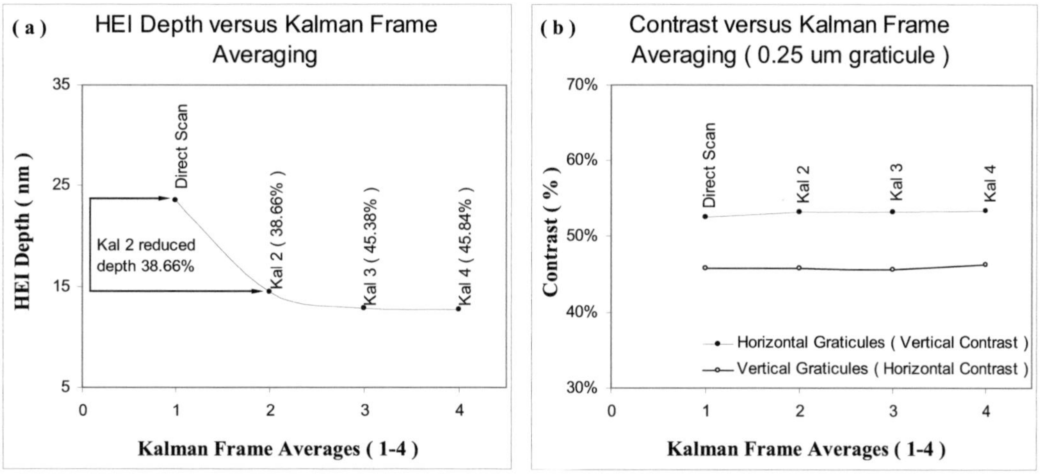

Fig. 10. HEI depth indicates frame averages improved depth discrimination (a) while no benefit to lateral resolution was gained with contrast remaining steady (b).

Confocal Aperture Settings

The impact of the confocal aperture settings on both depth discrimination and lateral resolution were also assessed using the aluminium optical flat and chromium structures on the Richardson slide. Since the confocal aperture is a critical parameter in maximising LSCM resolutions, varied aperture tests measured lateral contrast and measured HEI depth for the 10x, 20x, 50x and 100x objective lenses. Images of the optical flat were captured with each lens using zoom 999 for a 0.7, 1.0, 1.2, 1.7, 2.7 and 3.7 mm aperture setting. Height encoding generated the averaged HEI depth versus aperture size in Figure 11. On assessment, HEI

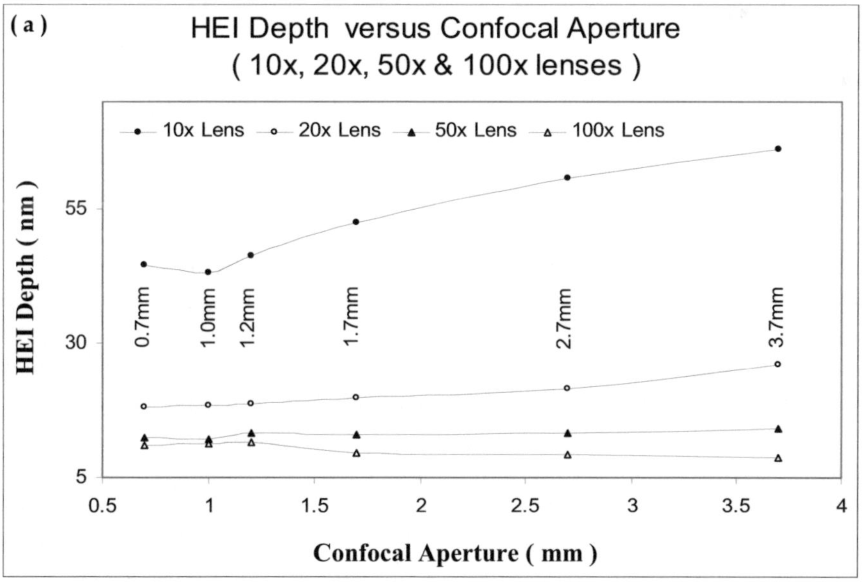

Fig. 11. The Influence of confocal aperture setting on depth discrimination for the 10x, 20x, 50x and 100x lenses.

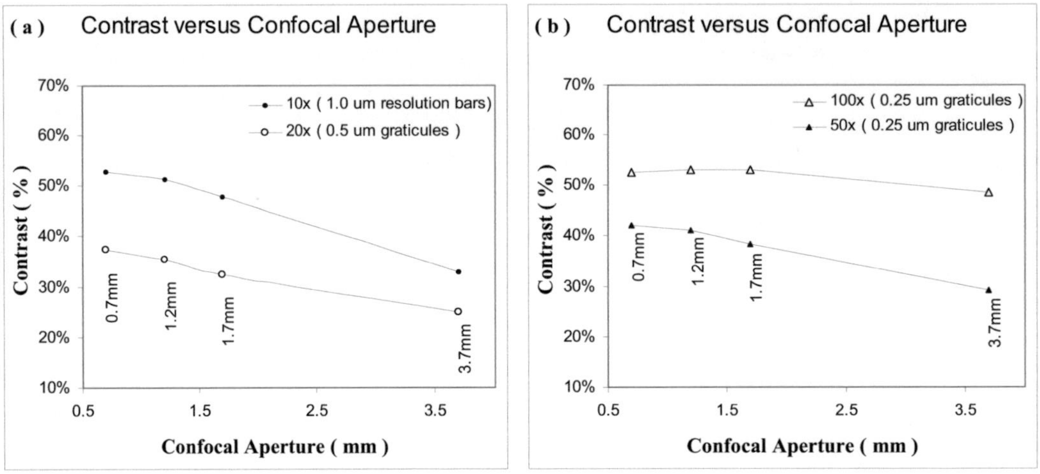

Fig. 12. Lateral contrast is reduced with increased aperture size for the 10x and 20x lenses (a), 50x and 100x lenses (b).

depth show each lens responded differently over the 0.7-1.2 mm range. Any mechanical effect related to opening the aperture would impact lenses equalling, implicating some form of optical effect (Pawley, 1995) as the likely cause. Further increases in the confocal aperture size and depth discrimination remained virtually unchanged for the 50x, while improving slightly for the 100x lens. The most significant reduction in depth discrimination beyond a 0.7-1.2 mm setting occurred for the 10x lens and to a lesser degree the 20x lens. When comparing the difference between 0.7 mm and 3.7 mm setting, a reduction in axial resolution measured 49.1%, 42.7%, 13% and -20.2% for the 10x, 20x, 50x and 100x lenses. To assess the impact of the same confocal aperture settings on lateral resolution, contrast was measured across resolved graticule and bar structures for each objective lens using their respective Nyquist zooming. A decreasing contrast plotted against increasing aperture size in Figures 12a-b indicated lateral resolution decreased with larger aperture selections. Figure 12a presents contrast decreasing with increased aperture size when imaging 1.0 µm bar structures with the 10x lens, and 0.5 µm graticules with the 20x lens. Figure 12b presents contrast decreasing with increased aperture size for the 50x and 100x lenses when imaging 0.25 µm graticules.

Over the tested aperture range, contrast losses for each lens measured 20%, 12.5%, 13% and 4% for the 10x, 20x, 50x and 100x lenses. Similar to HEI results, the 10x lens is presented with the highest sensitivity to increasing aperture size while the 20x, 50x and more specifically the 100x lateral resolutions were shown to be less affected. Consistent with all lenses, maximum lateral resolution was achieved using the smallest 0.7 mm confocal aperture setting. To better illustrate the importance of using a 0.7mm aperture for maximum HEI quality, MBI and HEI of the 600 line/mm ruled aluminium diffraction grate were captured and generated when imaging the surface with the 50x lens and Nyquist zoom. Figure 13a-c presents MBI and HEI of the surface captured using a 0.7 mm (a), 2.2 mm (b) and 3.7 mm (c) aperture settings. The optimal 2.2 mm aperture recommended for the 50x lens by LaserSharp2000 software, already demonstrates a loss of sharpness in HEI compared to using the 0.7mm aperture.

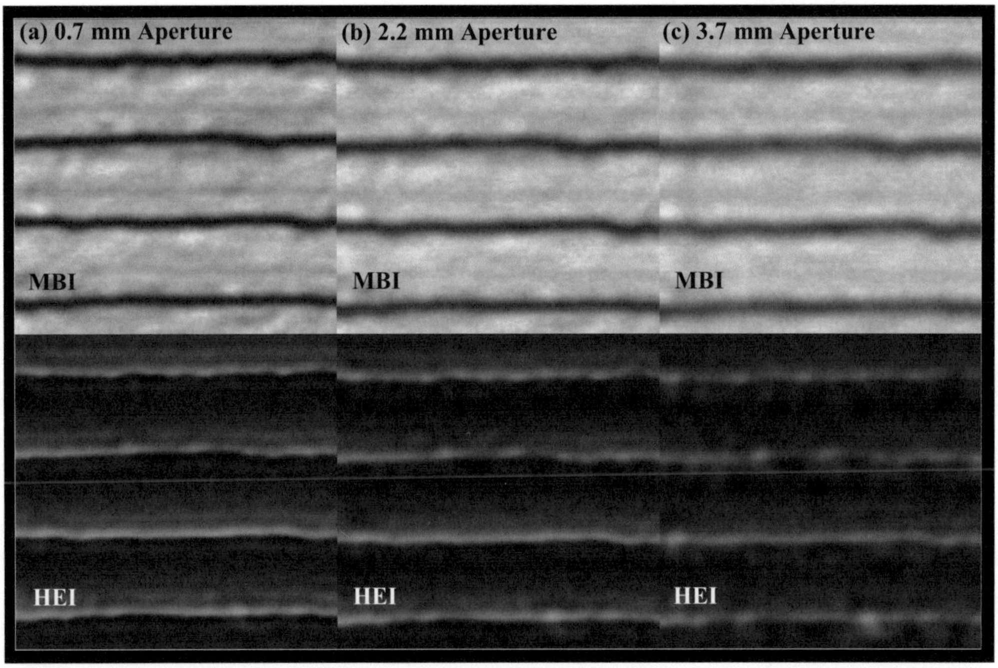

Fig. 13. Both HEI and MBI of the 600 line/mm ruled diffraction grate captured with a 0.7 mm (a), 2.2 mm (b) and 3.7 mm confocal aperture size.

Axial Resolution Tests

The Radiance2000 LSCM axial resolutions were measured from the through focal plane intensity distributions, applying the full width half maximum (FWHM) approximation often used by microscopist (Pawley, 1995; Webb, 1996). Performing xz scans of the optical flat for each objective lens with Nyquist zoom, through focal pixel intensities were then extracted and plotted against the z displacement of the microscope stage in Figure 14a using a 50 nm step size. Indicated on each intensity distribution are the measured FWHM approximations for the 10x, 20x, 50x and 100x lenses. Figure 14b illustrates the through focal xz scan captured with the 100x lens and the relationship of the intensity distribution to z displacement. Comparing measured and theoretical FWHM axial resolutions summarised in Table 1, measured axial resolutions were between 18-68% worse than FWHM Plane axial resolutions calculated for imaging a plane reflective surface by (Xiao and Kino, 1987).

It is well documented in literature that LSCM axial resolution is far greater when applying image processing techniques to extract height maps from the raw 3D image series. Throughout LSCM performance testing, HEI depth was calculated using the centre of mass image (CM) processing algorithm that delivered depth discrimination (axial resolution) many times higher than the measured or theoretical FWHM resolutions presented in Table 1. A measure of HEI axial resolution using the centre of mass technique was gained when performing earlier confocal aperture tests for the 10x, 20x, 50x and 100x lenses. In Figure 15, the same HEI depths generated for a 0.7 mm aperture setting highlights the difference in axial resolution for each objective lens NA. For comparison, reprocessing the images generated HEI axial resolutions in Figure 15 associated with the peak detection (PD) and the powers height encoding techniques. The peak detection algorithm determines surface height from the axial position containing the

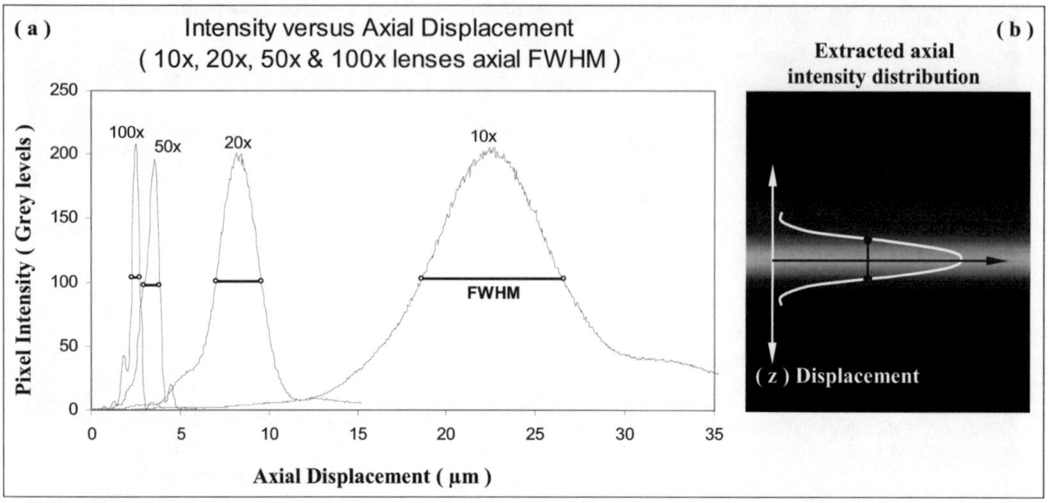

Fig. 14. Axial FWHM resolution approximations for the 10x, 20x, 50x and 100x lenses (a) and the typical axial intensity distribution extracted from a through focal plane xz scan with the 100x lens (b).

Objective Lens	NA	Measured FWHM	FWHM $_{Plane}$ (Xiao and Kino, 1987)	Difference in Measured From Theory
LU Plan 10x	0.30	8.02 µm	4.77 µm	68.1% more
LU Plan 20x	0.45	2.50 µm	2.12 µm	17.9% more
LU Plan 50x	0.80	0.90 µm	0.55 µm	63.6% more
LU Plan 100x	0.90	0.47 µm	0.39 µm	20.5% more

Table 1. Measured FWHM axial resolutions and theoretical FWHM differences.

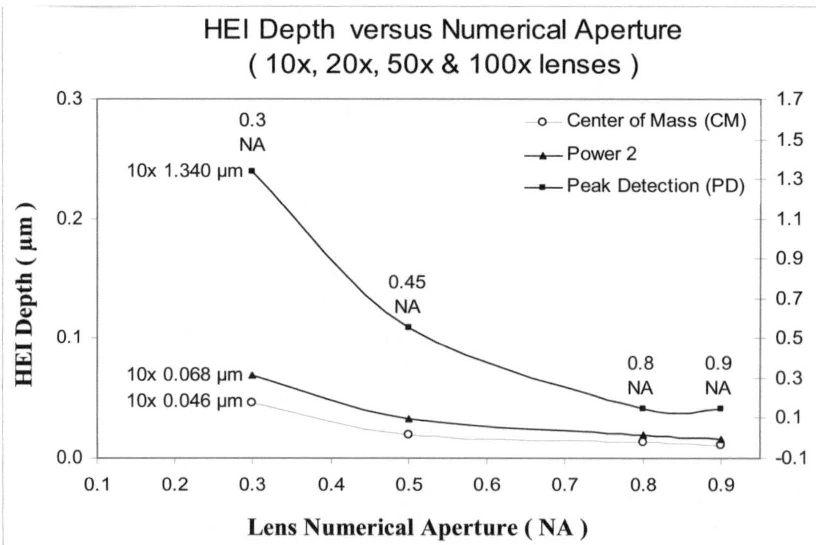

Fig. 15. Axial resolutions achieved in HEI using the centre of mass, powers and peak detection algorithms versus objective lens numerical aperture NA.

brightest pixel within the through focal plane intensity distribution (Sheppard and Matthews, 1987; Sheppard and Shotton 1997; Pawley 1995; Jordan et al., 1998). For the powers technique, through focal intensities are raised to the 2nd power before calculating the centre of mass surface position. Further increases in the power used and HEI depth approach those generated by peak detection as the weighting on brighter pixels increases.

Performing xz scans for the FWHM axial resolution approximations or capturing xyz series for generating HEI depths both involved using a 50-100 nm step size. Oversampling with finer steps ideally passes all xy surface points through the focal plane central region. Figure 16 illustrates a through focal intensity distribution for the 100x lens captured using a 50 nm step size. The peak intensity on the 50 nm stepped distribution coincides with a surface point passing through or near the focal plane central region. The two solid profile lines in Figure 16 illustrate when using a 0.5 µm step size, the surface may not pass through the ideal central region but rather above or below the central position. The location of the CM surface position using a 50 nm step size was 3.176 µm. The difference between CM surface positions generated for the two 0.5 µm stepped profiles is 0.06813 µm, which can vary depending on how the surface aligns with the focal plane. For peak detection identified by PD in Figure 16, the surface positions for the same 0.5 µm stepped profiles varied a much greater 0.3 µm. Table 2 summarises the centre of mass HEI depths, measured FWHM and theoretical FWHM axial resolutions. Also listed in Table 2 are the factors of improvement for HEI axial resolution over measured and theoretical values.

To assess the impact of step size on HEI depth and subsequent HEI axial resolutions, images of the optical flat were captured using oversampled to near Nyquist step size. Based on measured FWHM axial resolutions in Table 1 the approximate Nyquist step sizes for the 10x, 20x, 50x and 100x lenses are summarised in Table 3. Also summarised in Table 3 are the tested step sizes (1-3) in microns and as percentages of the approximated Nyquist steps. On capturing images for the tested step sizes using zoom 999 and the 500Hz scan rate, height encoding the images generated the HEI depth versus the percentage of Nyquist steps in Figure 17.

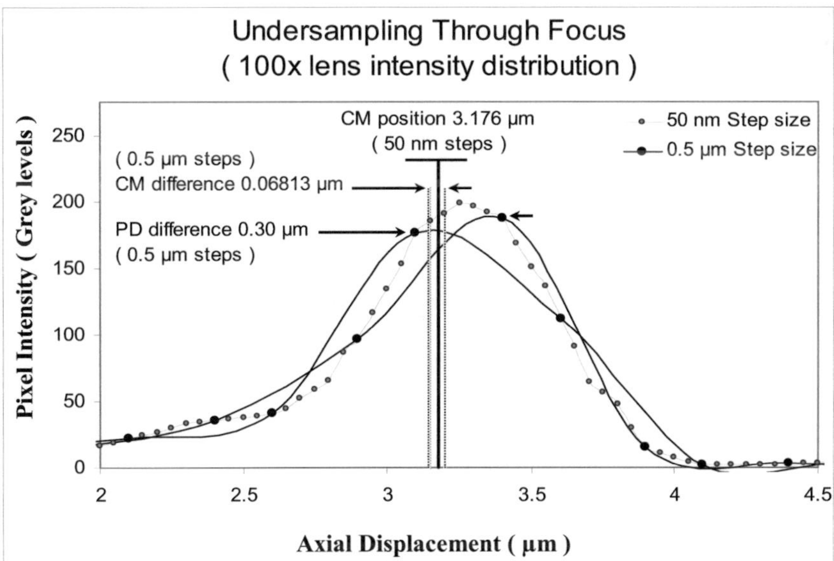

Fig. 16. Difference in accuracy between peak detection & centre of mass height encoding.

Objective Lens NA	Centre of Mass HEI Depth	Measured FWHM	Theoretical FWHM	Factor HEI Depth Improved on	
				Measured FWHM	Theoretical FWHM
0.30	44.4 nm	8.02 μm	4.77 μm	≈ 181 x	≈ 107 x
0.45	18.2 nm	2.50 μm	2.12 μm	≈ 137 x	≈ 116 x
0.80	12.4 nm	0.90 μm	0.55 μm	≈ 73 x	≈ 44 x
0.90	10.8 nm	0.47 μm	0.39 μm	≈ 44 x	≈ 36 x

Table 2. Comparison of measured and theoretical FWHM axial resolutions to HEI depths generated by the centre of mass height encoding algorithm.

Objective Lens NA	Nyquist Step size	Test step size (1) % of Nyquist Step Size	Test step size (2) % of Nyquist Step Size	Test step size (3) % of Nyquist Step Size
0.30	3.48 μm	0.10 μm (≈ 3%)	1.80 μm (≈ 52%)	3.60 μm (≈ 103%)
0.45	1.09 μm	0.05 μm (≈ 5%)	0.50 μm (≈ 46%)	1.00 μm (≈ 92%)
0.80	0.39 μm	0.05 μm (≈ 13%)	0.25 μm (≈ 64%)	0.50 μm (≈ 128%)
0.90	0.20 μm	0.05 μm (≈ 25%)	0.15 μm (≈ 75%)	0.30 μm (≈ 147%)

Table 3. Summary of tested step sizes for the 10x, 20x, 50x and 100x lens as percentages of Nyquist step size.

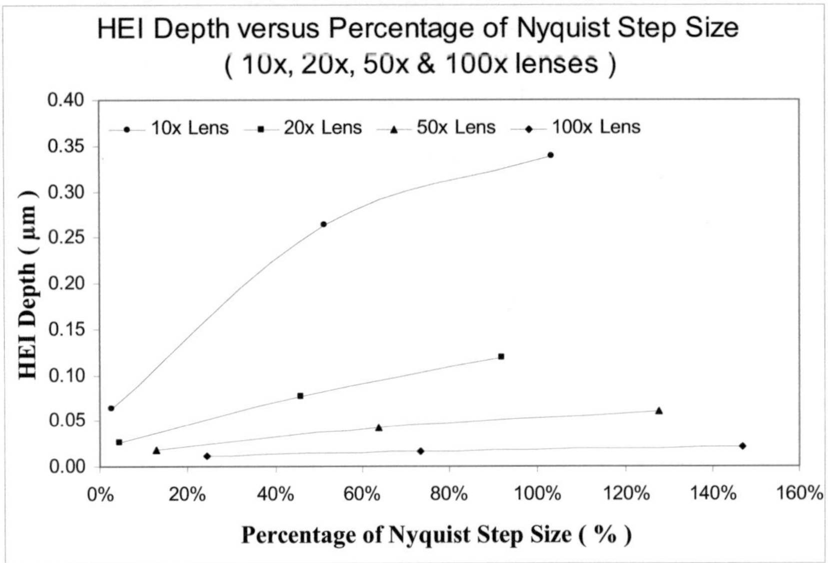

Fig. 17. HEI depth versus percentage of Nyquist step size for the 10x, 20x, 50x & 100x lenses.

Presented in Figure 17 and summarised in Table 4, tests show a marked increase in HEI depth and therefore reduced axial resolution when increasing step size to near Nyquist values. An insufficient sampling rate becomes obvious when attempting to image sloping surfaces. Figure 18a-c presents HEI of the silica Mahr PGN-3 roughness calibration standard captured with the 100x lens using 0.05 μm (a), 0.25 μm (b) and 0.5 μm step size (c). The Nyquist step size for the 100x lens is approximately 0.20 μm and using 0.25 μm steps in Figure 18b already show signs of missing surface information with Figure 18c clearly under sampled with 0.5 μm step size.

Objective Lens NA	Tested Step size (1)	HEI Depth	Tested Step size (2)	HEI Depth	Tested Step size (3)	HEI Depth
0.30	0.10 μm	64.0 nm	1.80 μm	263 nm	3.60 μm	339 nm
0.45	0.05 μm	26.0 nm	0.50 μm	76.0 nm	1.00 μm	119 nm
0.80	0.05 μm	18.0 nm	0.25 μm	42.0 nm	0.50 μm	60.0 nm
0.90	0.05 μm	11.0 nm	0.15 μm	17.0 nm	0.30 μm	21.0 nm

Table 4. Summary of HEI depths generated for the 10x, 20x, 50x and 100x objective lens tested step sizes.

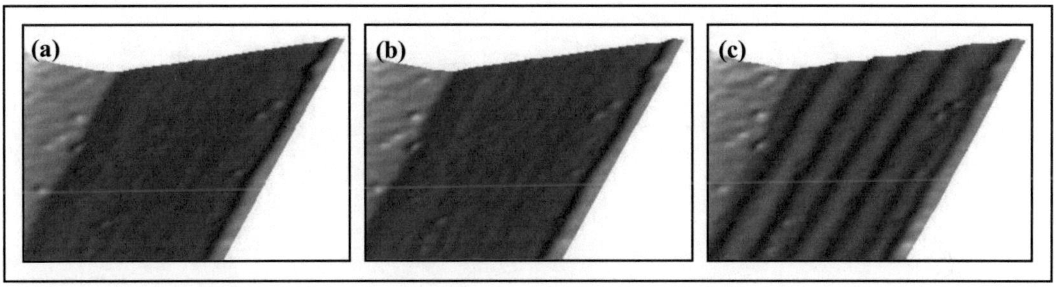

Fig. 18. Mahr PGN-3 roughness calibration standard HEI oversampled with 0.05 μm steps size (a), sample with near Nyquist 0.25 μm step size (b) and under sampled with a 0.5 μm step size (c).

When imaging the optical flat using zoom 999, HEI axial resolution (HEI depth) essentially provides a measure of the residual errors associated with LSCM image formation. Based on HEI depths presented in Table 4 and allowing for a 5% measurement error, the smallest measureable height between two real surface positions are listed in Table 5 for the tested step sizes. Important considerations when selecting step size include image acquisition and processing times that are traded with the level of measurement accuracy required. Data storage also places a limit on the number and size of captured image sections that subsequently affects the maximum vertical depth and scan area of a surface.

Objective Lens NA	Tested Step size (1)	Tested Step size (2)	Tested Step size (3)
0.30	1.28 +- 0.064 μm	5.26 +- 0.263 μm	6.78 +- 0.339 μm
0.45	0.52 +- 0.026 μm	1.52 +- 0.076 μm	2.38 +- 0.119 μm
0.80	0.36 +- 0.018 μm	0.84 +- 0.042 μm	1.20 +- 0.060 μm
0.90	0.22 +- 0.011 μm	0.34 +- 0.017 μm	0.42 +- 0.021 μm

Table 5. For the tested step sizes, the smallest vertical height measureable with 5% error.

Lateral Resolution Tests

Lateral resolutions for the Radiance2000 LSCM were measured directly using the high contrast metallic graticule and bar structures on the Richardson test slide. Captured MBI in Figure 19 presents resolved and unresolved horizontal and vertically aligned graticules for each objective lens. Figure 19a presents MBI of the 0.5 μm spaced graticules partially resolved by the 10x lens. Figure 19b presents the 20x lens having resolution between 0.25-0.5 μm, while Figures 19c-d show lateral resolutions for the 50x and 100x lenses falling between 0.125-0.25 μm.

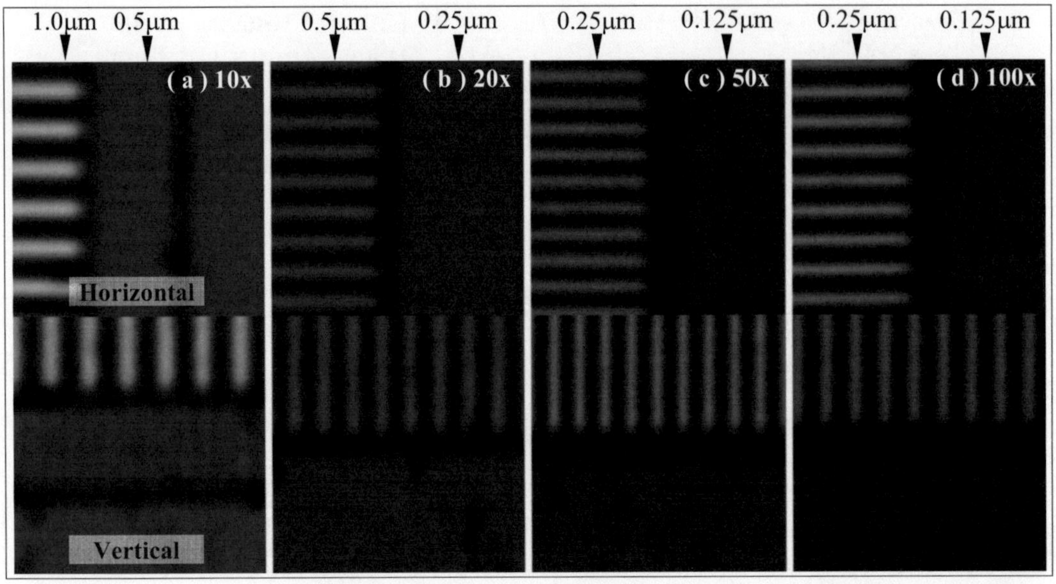

Fig. 19. Upper and lower binding limits for the 10x, 20x, 50x and 100x objective lens lateral resolution.

Contrast measured from MBI of the test structures were plotted against line spacing in Figure 20 for each lens. Also contained in Figure 20 are theoretical FWHM lateral resolutions at 33.3% contrast, corresponding to a FWHM separation distance predicted by (Brakenhoff, et al., 1979). Contrasts measured for the 20x, 50x and 100x lenses were limited by the degraded state of the finer bar structures found on the Richardson slide. Therefore the 0.125 µm graticules unresolved by the 50x and 100x lenses, and the 0.25 µm graticules unresolved

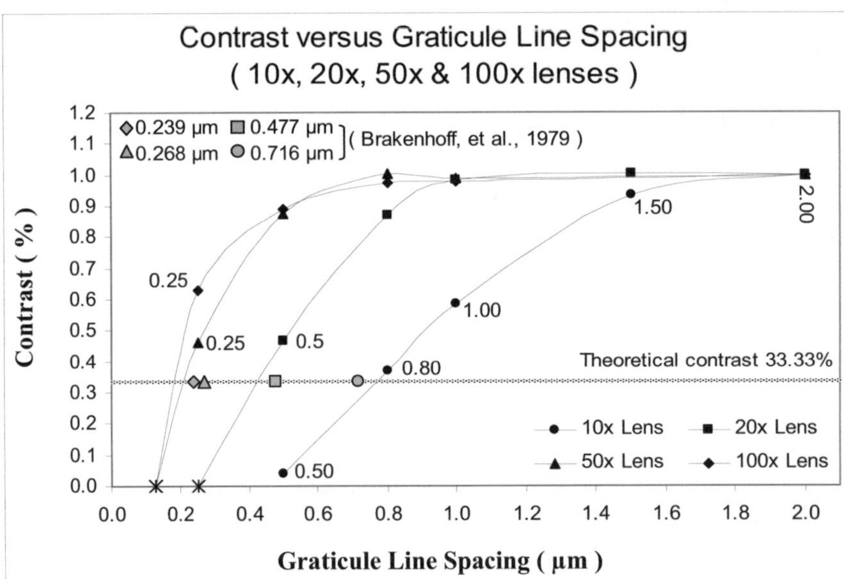

Fig. 20. Measured contrast versus graticule line spacing µm for the 10x, 20x, 50x, &100x lenses.

Objective Lens	NA	Measured FWHM or Limits	FWHM (Brakenhoff et al., 1979)	Difference in Measured From Theory
LU Plan 10x	0.30	0.77 μm	0.716 μm	7.5% more
LU Plan 20x	0.45	0.43 μm	0.477 μm	9.8% less
LU Plan 50x	0.80	0.22 μm	0.268 μm	17.9% less
LU Plan 100x	0.90	0.19 μm	0.239 μm	20.5% less

Table 6. Measured FWHM lateral resolutions, binding limits & theoretical FWHM.

by the 20x lens in Figure 19 were used to approximate zero contrast for these lenses in Figure 20. Interpolating a 33.3% contrast between 0.5-0.8 μm for the 10x, 0.25-0.5 μm for the 20x, and 0.125-0.25 μm for the 50x and 100x lenses estimated measured FWHM resolutions to be approximately 0.77, 0.43, 0.22 and 0.19 μm. It should be noted measured FWHM resolutions for the 20x, 50x and 100x lenses are slightly better than would usually be due to setting the resolution limit for these lenses at 0.125 μm or 0.25 μm, when in fact zero contrast will measure in at possibly higher line spacing. However, FWHM resolutions measured from plots in Figure 20 provide a fair comparison to FWHM calculated from theory (Brakenhoff et al., 1979). Table 6 summaries theoretical FWHM lateral resolutions, measured FWHM lateral resolutions and the percentile of difference for each objective lens.

In the LSCM reflection imaging mode, optical effects caused by discontinuous surface edges have significant influence on the formation of height encoded images. Light reflected from the edges of a surface relief constructively and destructively interfere (Juskaitis and Wilson, 1992), modifying the through focal plane intensity distribution. Both the centre of mass and peak detection height encoding algorithms break down in locating the surface position accurately in these regions. The centre of mass algorithm tends to over or undershoot the real surface position across such structures. To determine the limits of HEI lateral resolution for each objective lens based on limitations set by HEI artefact, images of a straight edge found on the Richardson slide were captured using twice Nyquist zoom and the 50 nm step sizes. Figure 21 present HEI of the edge structure captured with each objective lens to illustrate the lateral xy width (wavelength) of the image artefact unique to the 10x (a), 20x (b), 50x (c) and 100x (d) lenses. The artefact measured approximately 4.02 μm (a), 2.87 μm (b), 1.47 μm (c) and 1.14 μm (d) for the respective lens. Filter applications in TrueMap V4 image processing software established shortwave (λ_s) cut-offs limits of approximately 13.5 μm, 7 μm, 3.5 μm and 2.7 μm that completely removed the artefact from images prior to quantitative surface analysis.

Image Acquisition of Various Samples using Optimal Settings

Based on the LSCM resolution tests and assessment rules in the 4288 ISO standard for quantitative surface characterisation, it has been concluded that the confocal system can be used to measure surface roughness over a range of 0.05 microns to a few microns (Peng & Tomovich, 2008). The above studies on the Radiance2000 confocal microscope system performance and hardware settings have provided a guide to selecting the appropriate objective lens and system setting for quantitative surface characterisation. Images of engineering and biomaterial surfaces, metallic wear debris and collagen structures of cartilage samples extracted from sheep knee joints are shown in Figure 22 to demonstrate the capability of LSCM for various applications.

To acquire the images shown in Figure 22, different sample preparation methods were used. For the scanning of engineering surfaces shown in Figure 22a and biomaterial surface (dental ceramic) in Figure 22c right, the surfaces were cleaned and scanned directly. The

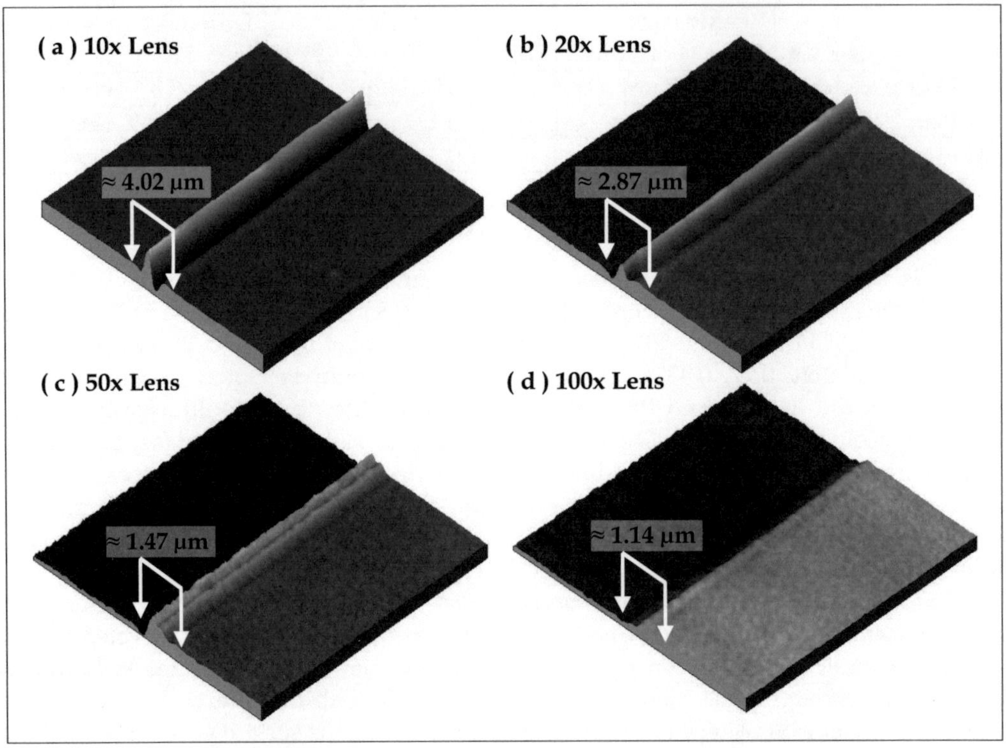

Fig. 21. HEI of the edge structure captured with each objective lens to illustrate the lateral xy width (wavelength) of the image artefact.

same method was applied when acquiring the metallic wear particle image shown in Figure 22b left. For the cartilage wear particle shown in Figure 22b right and cartilage surface in Figure 22c left, these samples required staining before imaging. The wear particle generated in a sheep's knee was collected from synovial fluid encapsulated within the knee joint. The extracted fluid was mixed with glutaraldehyde, agitated to mix the fluids and then stored in a refrigerator set at approximately 4 degrees Celsius for more than 24 hours. This process enabled the fixing of particles to prevent any further deterioration from occurring. Wear particles were then dyed and fixed onto glass slides for imaging. The cartilage image shown in Figure 22c left was acquired using the fluorescence imaging mode after the sample was stained via immunohistochemical (IHC) staining methods.

3.2 Image processing

Image processing is a three-step process before quantitative image analysis is performed using the developed system. The three steps are: (a) 3D image construction; (b) elimination of image distortion and artefacts; and (c) image stitching.

Before performing quantitative image analysis, a series of 2D images needs to be compiled into HEI format so that height information can be obtained for surface characterisation in 3D. Thus, HEI construction is the first step in the procedures used for image processing. Similar to other image acquisition systems, noise and distortion often associated with LSCM images are then reduced or eliminated with follow up HEI processing techniques. Finally, the stitching together of multiple HEI is a necessary step to form a large surface mosaic for

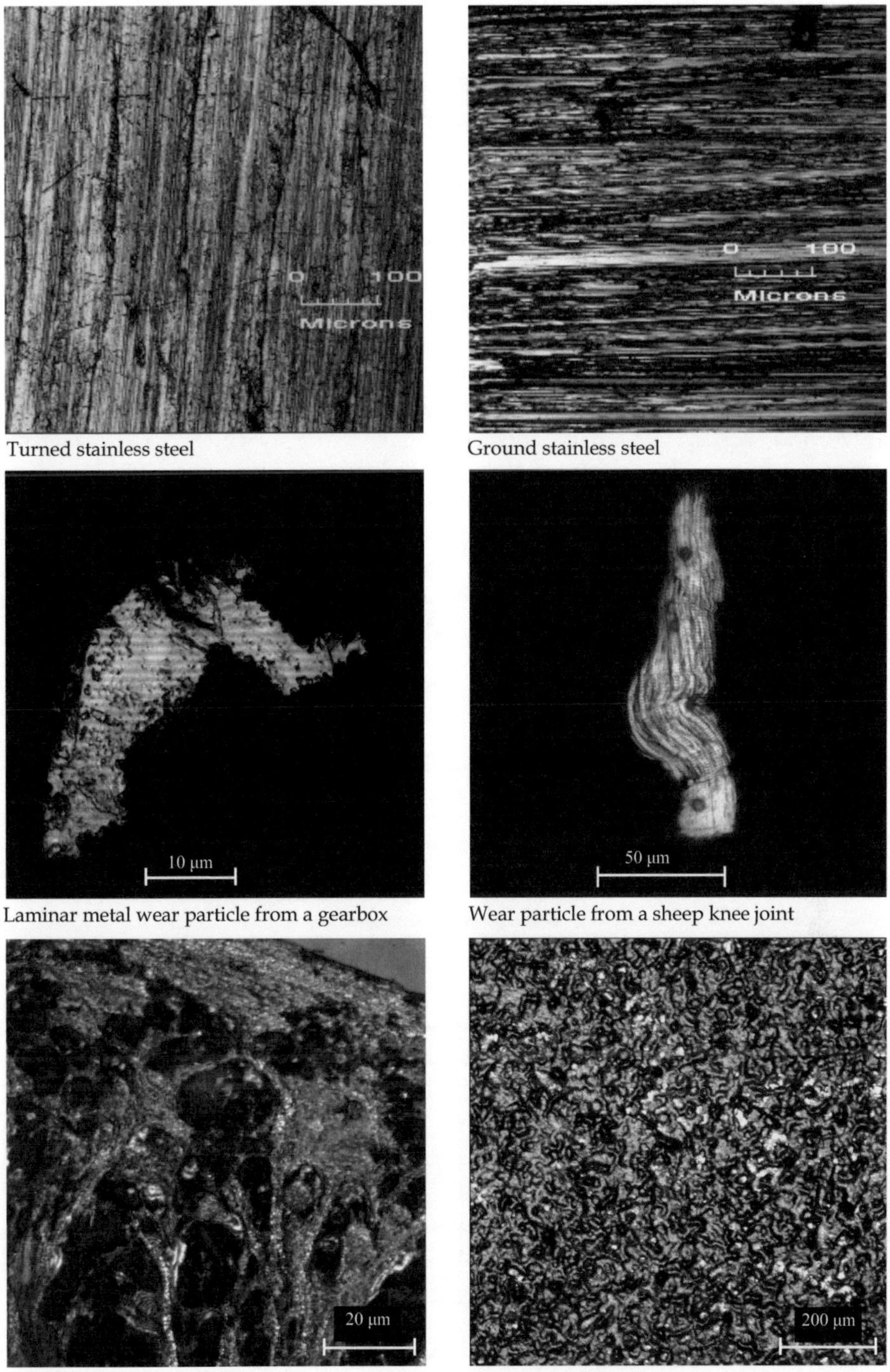

(a) Turned stainless steel — Ground stainless steel

(b) Laminar metal wear particle from a gearbox — Wear particle from a sheep knee joint

(c) Collagen fibres in cartilage — Dental ceramic empress

Fig. 22. Images of various surface acquired using the LSCM. (a) Engineering surfaces. (b) Laminar particle from a gearbox and a wear particle from sheep knee joint. (c) Cartilage surface with collagen fibre and biomaterial (dental ceramics).

ISO compliant measurements. so a large surface is generated for ISO compliant measurements. The details for the three phases of image processing are presented in as follows.

A series of 2D images is firstly obtained using the LSCM. To perform 3D surface characterisation, 3D information first needs to be extracted from the 2D image series. To construct 3D surface maps from raw LSCM data, a number of height encoding algorithms have been tested and evaluated. The tests also served to provide HEI cut off limits for developing a HEI reconstruction technique that removes sub resolution information such as edge artefact using wavelet filtering. Three surface detection algorithms including peak detection, centre of mass and centre of mass to a power have been tested for accuracy in locating the real surface. The established process can create the appropriate 3D images for surface characterisation. The procedures for image reconstruction can be found in Figure 23. Figure 24 shows a maximum brightness and height encoded image of an engineering surface complied from a set of 2D images acquired using the LSCM.

The second step in 3D image construction is to apply filters to separate surface form, waviness and roughness information for surface measurement. This step is necessary to analyse specific and meaningful surface information in different wave length for specific purposes. In this project, we have used the wavelet method to filter and separate original surface data into long, medium and short wave length, which corresponds to form, waviness and roughness information. Figures 25 and 26 show examples of the process conducted on an engineering surface and the surface of a metallic wear particle.

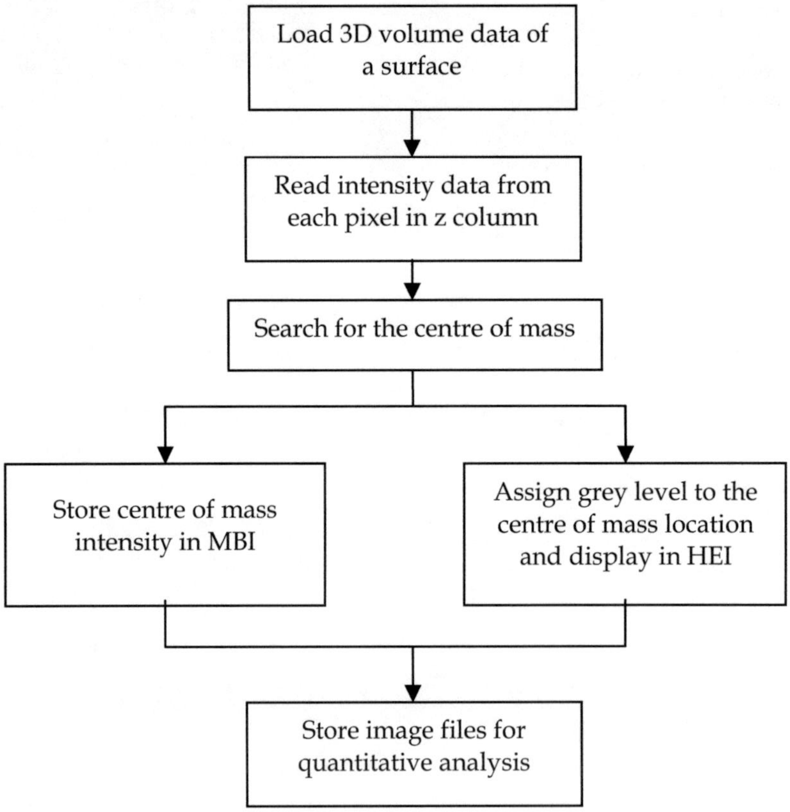

Fig. 23. Image reconstruction procedures using the centre of mass algorithm.

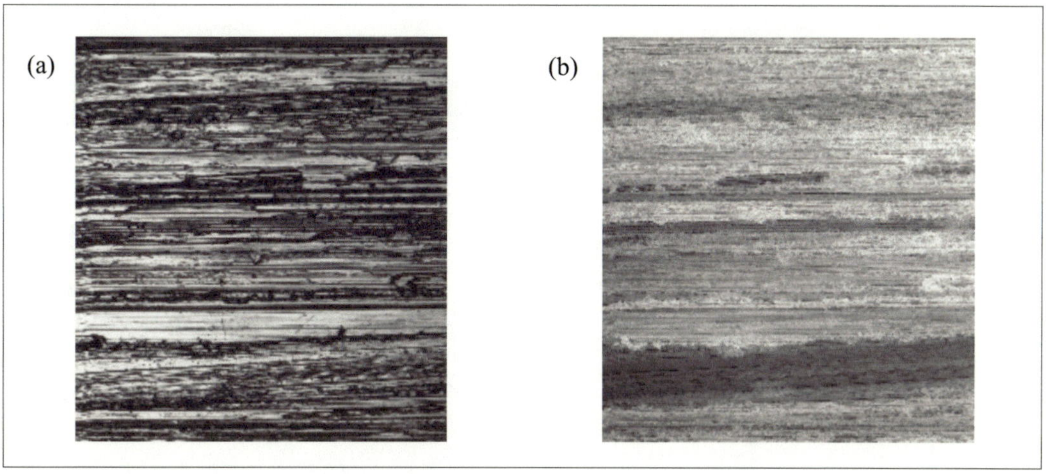

Fig. 24. Ground steel surface (a) maximum brightness image (MBI), and (b) height encoded image (HEI).

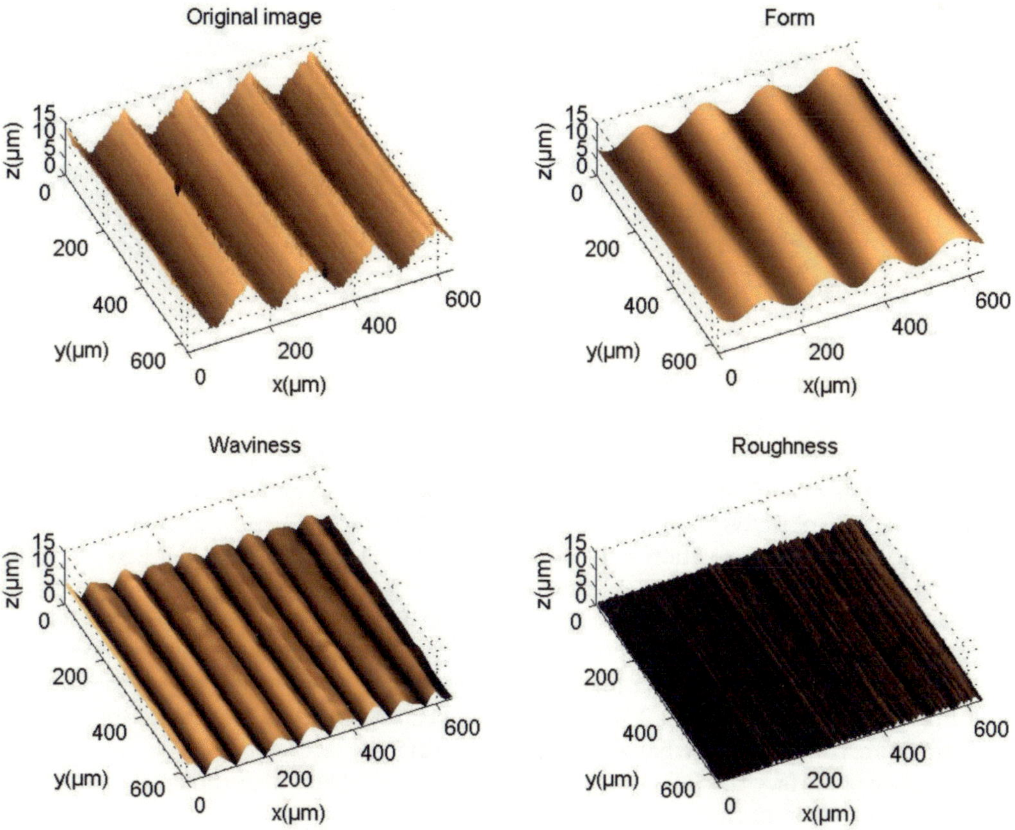

Fig. 25. Images of an engineering surface presented in the original image, Form, Waviness and Roughness.

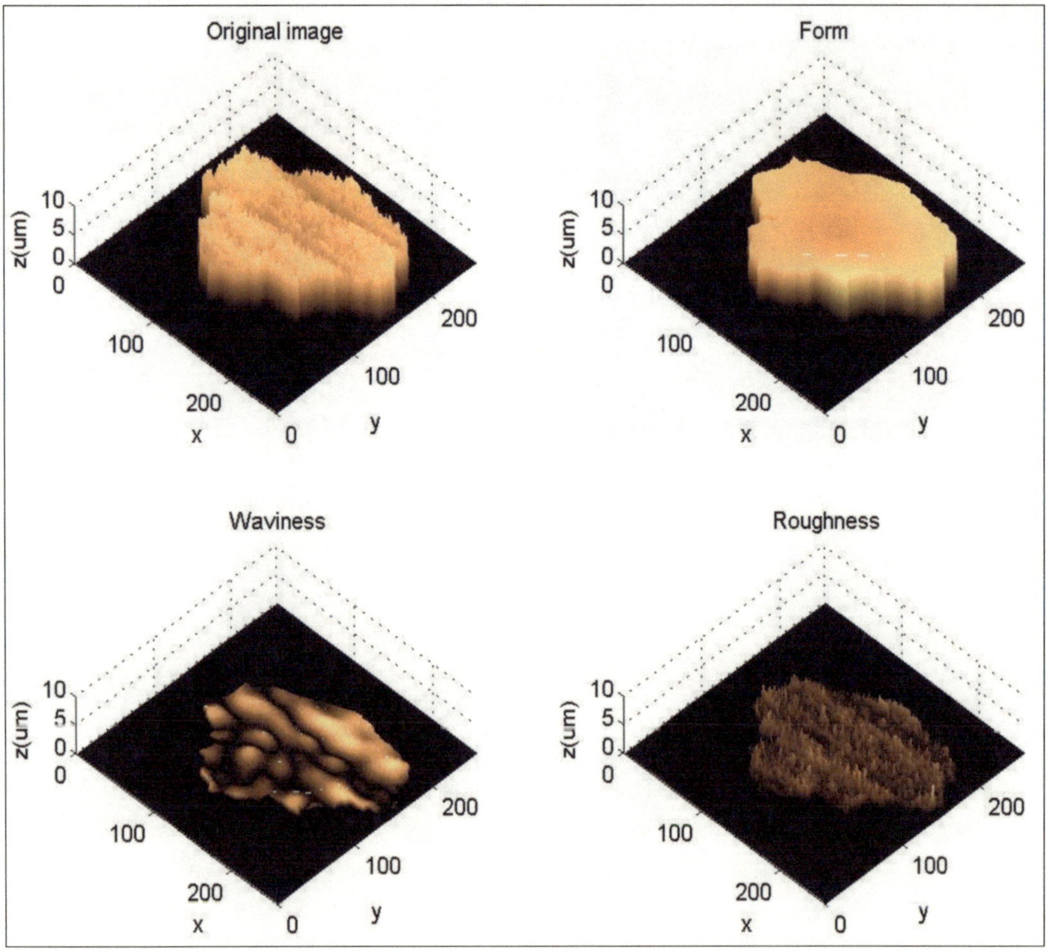

Fig. 26. Separated Form, Waviness and Roughness of a particle image.

Similar to any other imaging systems, image distortion exists and has to be eliminated to ensure the accuracy of surface measurements. A series of tests have been conducted to evaluate system related image distortions which may exist in generated 3D images. It has been found that objective lens field curvature and surface tilt are the two main issues affecting surface characterisation to be conducted later. Field curvature is a lens aberration that introduces distortion to most existing optical microscopy application. Although the 10x, 20x, 50x and 100x LU Plan objective lenses used in this project are all corrected for field curvature, noticeable curvature in HEI still remains when imaging with low zoom settings. Figure 27 present an example 3D height encoded surface view of the optical flat, imaged with the 10x lens using zoom 1.2. The z axis has been rescaled to accentuate the effects of field curvature on distorting what should be a flat surface.

The presence of surface tilt, field curvature and scanning system distortions all contribute to the total form error present in HEI. The application of least square curve fitting is one potential solution to removing field curvature and scanning distortion prior to more accurate industrial surfaces analysis. Another potential solution is to simply subtract known distortions obtained with the optical flat prior to industrial surface analysis. In our study, the removal process has been separated into two steps. The first step is to remove surface tilt resulting from a sloping stage or specimen surface. Surface tilt is removed by the application of a planar least square surface fit to both an optical flat HEI and the analysis surface HEI captured using identical LSCM settings. Field curvature present in the optical flat HEI is then simply subtracted from the levelled analysis surface HEI using the surface comparison utility in TrueMap V4 image analysis software. Figure 28 show two images before tilt correction (a) and after performing tilt correction (b). Once a 3D image has been constructed and major image distortions have been corrected, the next step is to generate a large surface area for numerical analysis. Since the frame size of one HEI is less than the ISO required measurement size, it is necessary to stitch a number of 3D images together so that surface measurements can be performed on a large area according to ISO requirements. A number of commercial stitching programs have been assessed. PanaVue Image Assembler was selected to stitch images together in this project. Figure 29 presents an example of a stitched image ready for numerical analysis.

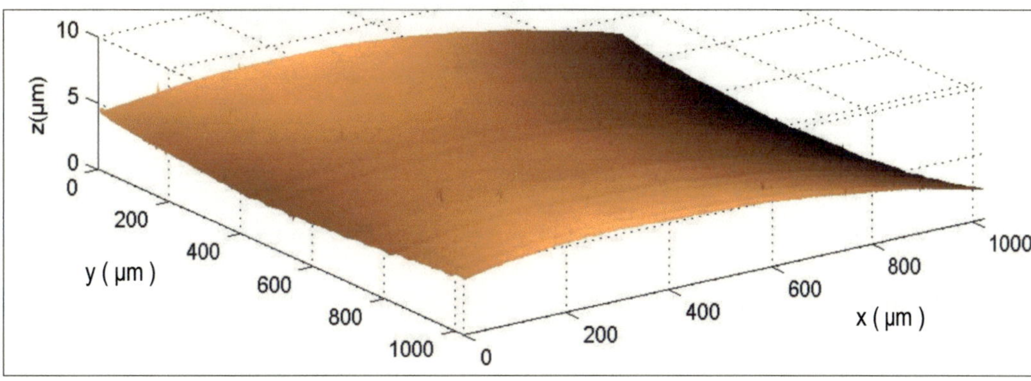

Fig. 27. Surface view of the optical flat surface captured with the 10x lens and zoom 1.2.

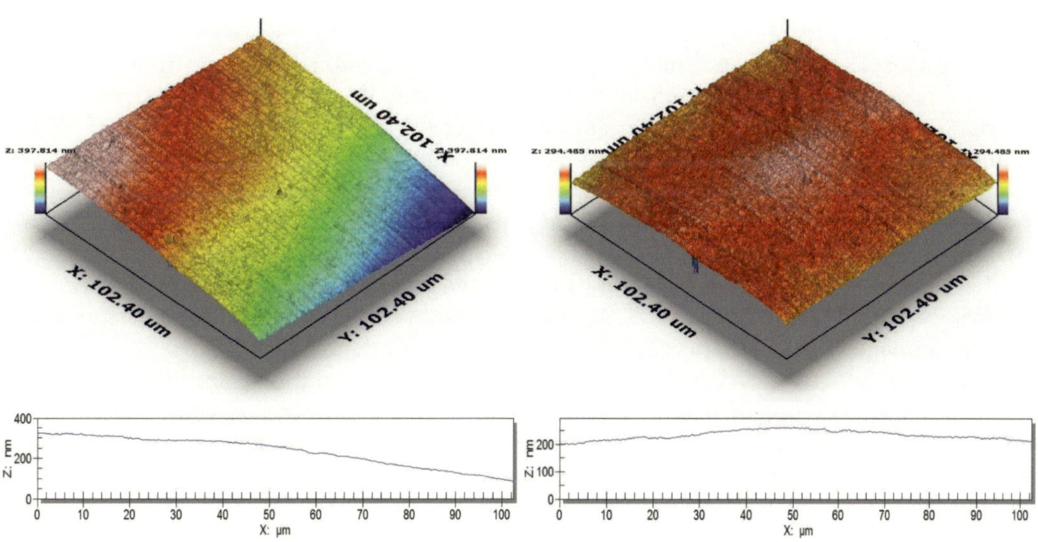

Fig. 28. Engineering surface (a) before and (b) after surface tilt correction.

Fig. 29. Stitched image after cross-correlation and blending for numerical analysis.

3.3 Quantitative image analysis

Quantitative image analysis involves using a set of numerical parameters to characterise surface morphology for various purposes. There are in general two sets of numerical parameters for surface characterisation. They are field parameters and feature parameters (Jiang et al., 2007; Scott, 2009).

Field parameters are often used to classify averages, deviations, extremes and specific features on a scale-limited continuous surface (Jiang et al., 2007). Surface roughness (S_a and S_q), skewness, kurtosis and spectral analysis are normally used as field parameters to describe how rough or smooth a surface is along with texture distribution. These parameters, also called profile surface texture parameters, have been commonly and traditionally used in many applications. Feature parameters characterise surface features and their relationship for functional diagnostics and prediction (Scott, 2009). The feature parameters are relatively new and have not been widely used for surface characterisation. For this reason, only common field parameters have been used in our study. In particular, the surface roughness parameter S_a is used to evaluate the performance of the above developed image acquisition, processing and analysis techniques.

A set of manufactured nickel specimens certified for various roughness specifications, and a front coated aluminium optical flat were used to test the above developments. These samples have a well defined surface structure and differences in surface roughness. The three selected nickel surfaces have average roughness values of 0.05 μm (ground surface), 1.60 μm and 3.20 μm respectively (vertically milled surface). The smooth front coated aluminium optical flat has a surface roughness of less than 5 nm. This sample was used to test the limits of the LSCM imaging system and approximate system measurement error. The surfaces were imaged using the system established in the section above. Image reconstruction, filtering and stitching processes were conducted as described. Table 7 presents the analysis results prior to corrective image processing, post corrective image processing and the associated percentile of improvement for S_a.

Sample (S_a)	Average S_a before correction (μm)	Average S_a after correction (μm)	Average S_a for stitched image (μm)	Percentage S_a being corrected (%)
0.005μm Optical Flat	0.114	0.009	0.009	92
0.05μm Ground Surface	0.591	0.054	0.054	91
1.60μm Milled Surface	1.511	1.516	1.624	7.5
3.20μm Milled Surface	4.100	3.059	3.200	22

Table 7. Average S_a for the tested samples before and after corrective image processing, and on stitching surface mosaics.

Prior to corrective imagine processing, the S_a values generated for raw images obtained with the LSCM did not match the reference values certified by the manufacturer. On processing images in TrueMap V4 to eliminate field curvature and surface tilt, S_a values for the corrected images matched or were closer to the manufacturer specified S_a values. In Table 7, the percentage S_a values were corrected by are calculated as follow:

$$\%S_{a\ corrected} = \frac{S_{a\ before\_correction} - S_{a\ after\_correction}}{S_{a\ before\_correction}}$$

Thereafter, the corrected images were stitched together forming a large image. Table 8 show the accuracy of S_a before and after the correction and stitching processes, highlighting the improvement in S_a measurements. This demonstrates that more accurate and precise topographical surface data has been extracted from large images.

Sample	Relative Error* of S_a before correction (%)	Relative Error* of S_a after correction and before stitching (%)	Accuracy of S_a after correction and stitching (%)
0.005μm Optical Flat	>100	80	20
0.05μm Ground Surface	>100	8	92
1.60μm Milled Surface	5.56	1.5	98.5
3.20μm Milled Surface	28.13	0.01	99.99

* The error is calculated as $\% = \frac{|measured\ S_a\ value - true\ S_a\ value|}{true\ S_a\ value} \times 100\%$

Table 8. The accuracy of S_a for test samples before and after corrective image processing, and on stitching surface mosaics.

4. Discussion

This project has clearly demonstrated that laser scanning confocal microscopy can be used for quantitative surface measurements. The technique is suited to acquiring surface roughness information within the range of approximately 0.05 microns to a few microns with reasonably high accuracy. This allows the use of the system for surface characterisation on a wide range of surfaces, particularly engineering surfaces and wear particles. The study has also examined optimal system settings and objective lens for quantitative surface characterisation. Although there is literature on the effects of LSCM settings in relation to image quality, this project has, for the first time, systematically investigated the issues associated with the measurement of engineering surfaces. From this study the recommended settings for scan rate, laser power, PMT gain, confocal aperture setting and step size provide the Radiance2000 LSCM user with a clear guide to obtaining images of suitable quality for quantitative analysis. The study has also identified the effects of field curvature and surface tilt on subsequent measurements, when selecting the appropriate objective lens and zoom to meet ISO evaluation area requirements. The results confirm that the removal of these distortions and artefacts during image reconstruction are important steps to ensuring images contain accurate surface information. Techniques have been tested, evaluated and selected for the removal of surface tilt and field curvature during 3D image reconstruction. However, artefact associated with the formation of HEI at edge structures require further assessment of appropriate techniques needed to minimise this effect and remove the residual artefact during image reconstruction. Part of this project was to research suitable image stitching algorithms. Due to the small field of view in confocal imaging, the stitching of multiple images into a large surface mosaic was a necessary procedure developed for performing measurements that satisfy ISO area of assessment rules.

The project used a range of samples with different surface roughness to evaluate the above developments. The results have shown the developed technique cannot accurately quantify the roughness of the very smooth optical flat, having a surface roughness of less than 5 nm. This was highlighted by the large error associated with the optical flat roughness measurement. The developed technique was able to measure surface roughness down to 0.05 microns with an accuracy of 90%. Based on measurement results obtained for the surfaces of known roughness, we have concluded the LSCM imaging technique is suited to measuring surfaces having roughness values ranging from 0.1 to approximately 5 microns. This roughness range is covered by the 250-2500 μm^2 assessment areas set out in the ISO 4288 measurement standards. However, further tests for surfaces having roughness values approaching 10 micron is required to confirm the 2500 μm^2 range is within practical limits of the LSCM imaging system. Important consideration when imaging for a 2500 μm^2 assessment area, include imaging time and storage space when capturing a sufficient number of images to form a surface mosaic.

5. Conclusion

This project, based on the studies of LSCM imaging fundamentals, has developed a reliable and accurate 3D quantitative surface measurement system using laser scanning confocal microscopy. The image acquisition technique and procedures have been developed for providing images with optimal quality, resolution and that are representative of real surface information. In addition, the project has investigated image processing techniques essential for accurate surface analysis. The project has made progress in the following aspects. (a) Quantitative determination of the LSCM resolutions and its capability in 3D surface measurement applications have been carried out in this study. (b) Comprehensive image acquisition and processing techniques have been developed for numerical analysis. In particular, reconstruction of 3D height encoded surface maps have been improved, and image processing methods have been investigated for the removal of field curvature, surface tilt and image artefact such as those generated by surface edges. (c) Based on the above developments, a quantitative surface measurement system using the Radiance2000 confocal microscope has been advanced for 3D surface characterisation. The reliability and accuracy of the system have been validated with the measurement of calibration standards, industrially machined surfaces and wear particles. We hope the work will extend the usage of the laser scanning confocal technique from visual inspection to quantitative analysis in many fields.

6. References

Bennett JM and Mattsson L (1989), Introduction to Surface Roughness and Scattering, Optical Society of America, Washington DC.

Bernabeu E, Sanchez-Brea LM, Siegmann P, Martinez-Anton JC, Gomez-Pedrero JA, Wilkening G, Koenders L, Muller F, Hildebrand M and Hermann H (2001), Classification of surface structures on fine metallic wires, Applied Surface Science, Vol. 180, pp 191-199.

Brakenhoff GJ, Blom P and Barends P (1979), Confocal scanning light microscopy with high aperture immersion lenses, J. Microsc., Vol. 117, pp 219-232.

Brown JM and Newton CJ (1994), Quantified three-dimensional imaging of pitted aluminium surfaces using laser scanning confocal microscopy, British Corrosion Journal, Vol 29 No 4, pp 261-269.

Chescoe D and Goodhew PJ (1990), in Royal Microscopical Society Microscopy Handbook 20, Oxford University Press, New York.

Conroy M and Armstrong J (2005), A comparison of surface metrology techniques, Journal of Physics: Conference Series 13, pp 458-165.

DeGarmo EP, Black JT and Kohser RA (1997), Materials and Processes in Manufacture, Eighth Edition, Prentice Hall International USA, pp 289-291.

Gadelmawla ES, Koura MM, Maksoud TMA, Elewa IM and Soliman HH (2002), Roughness Parameters, Journal of Materials Processing Technology, Vol. 123, pp 133-145.

Gauldie RW, Raina G, Sharma SK and Jane JL (1994), in Atomic Force Microscopy / Scanning Tunnelling Microscopy, Edited by al. SHC e, Plenum Press, New York, pp 85-90.

Gjonnes L (1996), Quantitative Characterisation of the Surface Topography of Rolled Sheets by Laser Scanning Microscopy and Fourier Transforms, Metallurgical And Materials Transactions A, 27A, pp 2338-2346.

Hanlon DN, Todd I, Peekstok E, Rainforth WM and Van Der Zwaag S (2001), The application of laser scanning confocal microscopy to tribological research, Wear, pp 1159-1168.

Haridoss S, Shinozaki DM and Cheng PC (1990), IEEE International Symposium on Electrical Insulation, Toronto, Canada, pp 392-397.

ISO standards 4288 (1996), Geometrical Product Specifications (GPS) - Surface texture: Profile method – Rules and procedures for the assessment of surface texture, ISO.

Jiang X, Scott PJ, Whitehouse DJ and Blunt L (2007), Paradigm shifts in surface metrology. Part II: The current shift, Proc. R. Soc. A, Vol. 463, pp. 2071-2099.

Jordan HJ, Wegner M and Tiziani H (1998), Highly accurate non-contact characterisation of engineering surfaces using confocal microscopy, Meas. Sci. Tecnol., UK, Vol. 9, pp 1142-1151.

Juskaitis R and Wilson T (1992), Surface Profiling with Scanning Optical Microscopes Using Two-Mode Optical Fibres, Applied Physics, Vol 31, No 22, pp 4569.

King RG and Delaney PM (1994), Confocal Microscopy, Materials Forum, Vol 18, pp 21-29.

Martin DC, Ojeda JR, Anderson JP and Pingali G (1994), in Atomic Force Microscopy / Scanning Tunnelling Microscopy, Edited by Cohen SH et al., Plenum Press, New York, pp 217-228.

Oldenbourg R, Terada H, Tiberio R and Inoue S (1993), Image sharpness and contrast transfer in coherent confocal microscopy, Journal of Microscopy, Vol. 172, Pt 1, pp. 31-39.

Pawley JB (1995), Handbook of Biological Confocal Microscopy, 2nd Edition. Plenum Press, New York.

Peng Z and Kirk TB (1998), Computer image analysis of wear particles in three-dimensions for machine condition monitoring, Wear, Vol. 223, pp 157-166.

Peng Z and Kirk TB (1999), The study of three-dimensional analysis techniques and automatic classification systems for wear particles, Journal of Tribology, Vol. 121, pp 169-176.

Peng Z and Tomovich S (2008), The development of a new image acquisition system for 3D surface measurements using confocal laser scanning microscopy, Advanced Materials Research, Vol. 32, pp. 173-179.

P. Podsiadlo and G.W. Stachowiak, Characterization of Surface Topography of Wear Particles by SEM Stereoscopy, Wear, Vol. 206, 1997, pp. 39-52.

Sheppard CJR and Matthews HJ (1988), The extended-focus, auto-focus and surface-profiling techniques of confocal microscopy, Journal Of Modern Optics, Vol. 35, No. 1, pp. 145-154.

Sheppard CJR and Shotton DM (1997), Confocal Laser Scanning Microscopy, BIOS Scientific Publishers, Oxford, pp 75.

Scott PJ (2009), Feature parameter, Wear, Vol. 266, pp. 548-551.

Webb RH (1996), Confocal optical microscopy, Rep. Prog. Phys., UK, Vol. 59, pp 427-471.

Xiao GQ and Kino GS (1987), A real-time confocal scanning optical microscope, Scanning Image Technology, Proc. SPIE Vol. 809, Wilson T and Balk L eds. pp 107-113.

Yuan C, Peng Z and Yan X (2005), Surface characterisation using wavelet theory and confocal laser scanning microscopy, Journal of Tribology, ASME, Vol. 127, pp. 394-404.

Zucker RM and Price O (2001), Evaluation of Confocal Microscopy System Performance, Cytometry, Wiley-Liss, Inc., Vol. 44, pp 273-294.

Zucker RM and Price OT (2001), Statistical Evaluation of Confocal Microscopy Images, Cytometry, Wiley-Liss, Inc., Vol. 44, pp 295-308.

2

Confocal Scanning Laser Microscopy: Applications for Imaging Dynamic Processes in Skin *In Vivo*

Melissa Chu[1], Nikiforos Kollias[1] and Michael A. Luedtke[2]
[1]*Johnson & Johnson Consumer Companies, Inc.*
[2]*Codman & Shurtleff, Inc.*
USA

1. Introduction

In vivo confocal scanning laser microscopy (CSLM) is a powerful imaging technique that enables physicians and researchers to evaluate skin dynamically and non-invasively at the cellular level. Traditionally, biopsy and histological processing are required to study the cellular details of skin. This involves tissue excision and staining, which may introduce artifacts to the sample. Staining and immunostaining have been used to identify specific proteins and organelles successfully. With CSLM, the examined areas are not damaged, removed, or processed, so imaging can be performed as often as desired. Studying dynamic changes in skin over an extended period of time or as a response to treatment is feasible. Unlike the vertical sections obtained by routine histology, *en face* (horizontal) images are captured with CSLM. During the last 20 years, advances in technology have resulted in images that are now near histological resolution and have enough detail for histological analysis (Pellacani et al., 2008; Scope et al., 2007; Rajadhyaksha et al., 1999).

CSLM has been used in dermatology to assess skin lesions and skin conditions without biopsy. Both benign and malignant melanocytic skin tumors have been described and diagnosed, and proliferative and inflammatory skin diseases have been characterized (Gonzalez & Gilaberte-Calzada, 2008; Calzavara-Pinton et al., 2008). Margins of lesions are also identifiable from non-lesional skin. For physicians, CSLM provides real-time diagnostics and the chance to follow the evolution of skin lesions over time, which would not be feasible otherwise.

Whereas the majority of CSLM skin research concentrates on dermatological conditions, this chapter focuses on capturing and studying non-diseased tissue, both healthy and wounded human tissue. We first assess the effect that the incident laser wavelength has on the images captured by comparing CSLM systems of three different wavelengths. Then, we specifically discuss two studies in which we use CSLM to study skin morphology and physiology: 1) a comparison of infant and adult skin morphology and 2) the wound healing response of micro-injuries in otherwise healthy human skin. The micro-injury study focuses on dynamic responses in skin over time. Ethically we are committed to minimize biopsy collection when studying healthy human skin. CSLM is our sole alternative to studying skin morphology and physiology at the cellular level non-invasively and to monitor biological processes such as blood flow at video rate, with the added benefit of investigating skin *in vivo*.

2. Structure of skin

2.1 Skin morphology

The skin is composed of two layers, the epidermis and dermis. The epidermis is the outermost layer of the skin. The top layer of the epidermis is known as the stratum corneum and functions as the barrier to the outside world. Its consists of corneocytes, or dead keratinocytes, which contain primarily keratin. Corneocytes are approximately 1 μm thick and 1000 μm^2 in area. The stratum corneum, approximately 10-15 μm thick, encounters various insults from the outside world, such as UV radiation, friction, scrapes and cuts, and irritants.

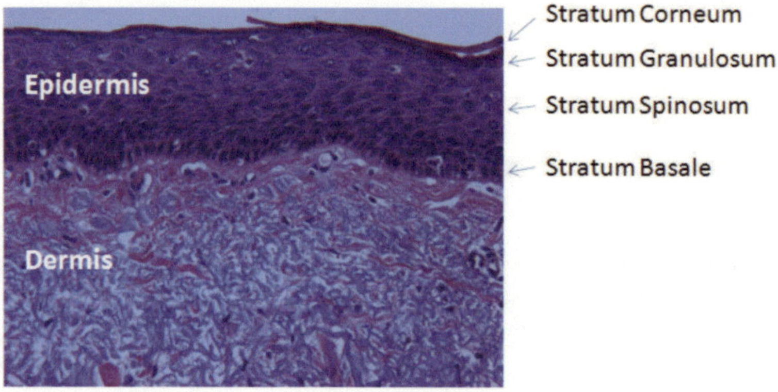

Fig. 1. Various layers of skin are noted in this histology slide of human skin. (hematoxylin and eosin stain, 20x)

Below the stratum corneum is the viable epidermis. The viable epidermis is composed of living skin cells, or keratinocytes, and is 50-100 μm thick. There are various layers noted within the epidermis: stratum granulosum (granular layer), stratum spinosum, and stratum basale. The stratum basale is a single layer of cells resting on the basal membrane that includes both keratinocytes and melanocytes. Keratinocytes are generated by mitosis in the stratum basale, and the cells differentiate and increase in size as they progress to the uppermost layer of living cells. At the highest layer, they then stop functioning as cells and exist as dead cell envelopes that compose the stratum corneum, which will eventually be shed. Melanin, pigment that contributes to the color of skin, is produced by melanocytes and is distributed to keratinocytes. The dermal-epidermal junction, where the epidermis meets the dermis, often consists of dermal papillae, which are small finger-like protrusions of the dermis into the epidermis. The dermis (1-3 mm) contains connective tissue, vasculature, and various appendages. The connective tissue, primarily collagen and elastin, provides cushion and elasticity to the skin. Blood and lymphatic vessels confined to the dermis bring nutrients to and remove waste from the keratinocytes.

2.2 Endogenous skin chromophores

Skin has different endogenous chromophores that have unique physical characteristics. Aside from optical scatterers found within skin, there are absorbers and fluorescent chromophores as well. In fluorescence, a particular stimulus at a given excitation wavelength elicits an emission at a higher wavelength due to the Stokes shift and basic principles of fluorescence. If coherent light of a specific wavelength is used in conjunction with appropriate filters and a detector sensitive enough to measure the fluorescence signal, endogenous (and exogenous) chromophores can be measured quantitatively from the skin surface.

In summary, skin is a complex organ with different appendages and structures that can scatter and absorb light. Table 1 lists some of these endogenous chromophores found within skin (Kollias, 1995).

Endogenous Skin Chromophore	**Characteristic**	**Principle wavelength (nm)**
Melanin	absorber and scatterer	Monotonic increase from 700-300 nm
Collagen crosslinks	fluorescent	X335 M380; X370 M460
Elastin crosslinks	fluorescent	X420 M500; X460 M540
Oxy Hemoglobin	absorber	412, 542, 577
Deoxy Hemoglobin	absorber	430, 555, 760

Table 1. Different endogenous chromophores within skin at wavelengths of interest. Excitation (X) and Emission (M) wavelengths are provided (Kollias, 1995).

3. Theory and technique of CSLM

3.1 Principles of CSLM

The confocal microscope is elementally comprised of a coherent scanning laser light source, condenser lens, objective lens, and detector. Details involving the intricacies of the optics, point spread function and other tissue-light interactions (Webb, 1996; Webb, 1999) are not included in this chapter.

A laser light point source illuminates a small region within the skin and is scanned across the skin. Light reflected from this focal point propagates back to the detector through a pinhole aperture after a single collision within the tissue, allowing us to observe reflections from single scattering events. Light from the in-focus plane is collected, as light emanating from above and below the focus plane minimally passes through the pinhole to the detector, resulting in images of high resolution and contrast. The term "confocal" stems from this design in which the pinhole in front of the light source and the pinhole in front of the detector are in optically conjugate focal planes. Figure 2 illustrates the principle of confocal microscopy.

For the research discussed in this chapter, a commercial *in vivo* confocal scanning laser microscope (Vivascope 1500, Lucid Inc., Rochester, NY) was used to non-invasively study skin at the cellular level. Unless otherwise stated, a 785 nm wavelength laser was used in the microscope.

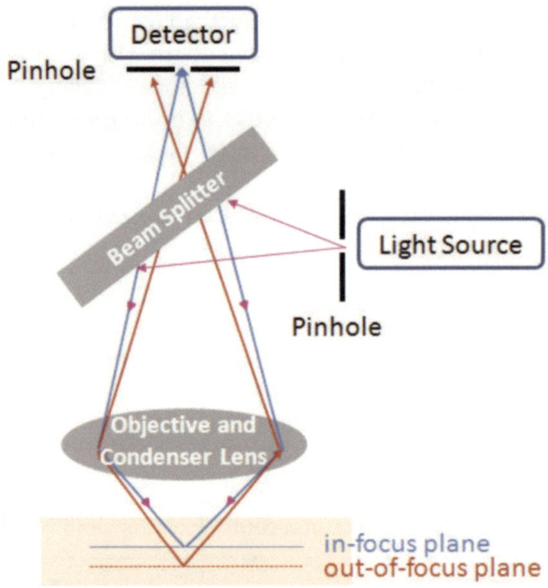

Fig. 2. Principle of CSLM. Note the same optical path is used for the detector and the source. Optics are used to direct the light towards the detector.

3.2 Image acquisition

To minimize lateral movement of the skin and motion artifacts, a metal tissue ring is adhered to the skin with a double-sided adhesive acrylic tissue window. The tissue ring, which holds the optical coupling (ultrasound gel), is magnetically attached to the microscope housing of this device.

Once the optical paths are fixed relative to the tissue imaged, a point source is rastered quickly within a conjugate focal plane. The pinholes are then used to reduce the amount of light that generates an image on the photodetector. This detector then produces the images as they are seen on screen.

There are various methods of generating a scanning laser effect (Webb, 1999). For example, a Nipkow disk (sometimes known as Nipkov disk) is a disk with a series of holes in it that spins rapidly. While resolution along a scanline is very high, scan rates are limited to the number of holes found within the disk. The images generated with these types of disks can cause a flickering and result in a less than video sampling rate.

Commercially available *in vivo* CSLM equipment typically uses a laser light point source that is scanned in a raster pattern by a multi-faceted polygon mirror and an oscillating galvanometric mirror. The polygon mirror rotates at a high speed, a few hundred revolutions per second, while the galvanometer oscillates at a slower speed and is driven by a saw tooth signal. The beam scans left to right along the horizontal axis, and after a single horizontal scan, the beam quickly returns to the starting point and shifts along the vertical axis to begin a new scan. Each horizontal scan line is generated by each facet of the quickly rotating multi-faceted polygonal mirror, and the vertical shift occurs during the oscillations of the galvanometer. Scanning is performed at a rate of 9 frames per second, and the resulting (1000 pixel x 1000 pixel) image is equivalent to an *en face* 0.5 mm x 0.5 mm horizontal microscopic section, at closer to full motion and flicker-free video rates.

The microscope can capture a series of focal planes at different depths by changing the focal length of the beam, and it has the capability of imaging 200-300 µm below the skin surface. A series of images is typically captured from the top of the stratum corneum (SC), through the epidermis and papillary dermis, and down into the superficial reticular dermis, forming a vertical image stack. This is known as "optical sectioning".

3.3 Visualization of cellular structures

Each two dimensional optical section is 500 µm x 500 µm in size and is obtained in grayscale relative to the positioning of the surface of the skin. Structures within tissue are visualized based on their optical properties, changes of index of refraction, and absorption within the tissue. There is good contrast within living tissue, allowing microscopic structures and cellular details in the epidermis and superficial dermis to be visualized. Contrast in images is provided by differences in refractive index of organelles and microstructures, such as keratin, melanin, and collagen, compared to the background. Melanin has one of the largest index of refraction found in skin, which makes it one of the strongest contrast agents in skin, and results in greater backscattering of light, causing pigmented structures to appear bright (Rajadhyaksha et al., 1995). Other structures such as inflammatory infiltrates, dendritic cells, and capillaries can also be imaged with CSLM. Total light is reflected back when structures appear white, while no reflection is represented by black.

3.4 Quantitative analysis

Both qualitative and quantitative analyses can be performed on the acquired CSLM images. From individual images, cellular size can be measured. The thickness of the SC and epidermis can also be measured when analyzing images captured from the image stacks. From the 8-bit grayscale image, changes of intensity can be normalized to the laser power for quantitative analysis of tissue structure within the images. This technique of quantifying a region of interest (ROI) or group of pixels found within an image is a powerful technique to measure changes in skin microstructure non-invasively and semi-automatically. With the appropriate software analysis routines, gigabytes of image data can be analyzed and meaningful measures can be made on data that was previously too cumbersome to analyze.

3.5 Imaging limitations

While CSLM is optimal for imaging optically thin tissues, certain appendages and other optically thick samples limit its capabilities. Since the microscope is a fluid immersion microscope, certain skin appendages and structures could change morphology while immersed in fluid. Bubbles, improperly adhered and inconsistent placement of the tissue ring are challenges that plague this imaging modality and increase the complexity of acquiring quality images. Due to the high magnification, the images acquired are prone to subject movement. Since *in vivo* CSLM measurements are made on living subjects who breathe, have a moving circulatory system and have involuntary movements, there will occasionally be added "noise" or movement in the acquired images. Great attention should be paid to acquiring quality images prior to arriving at conclusions. When imaging the same area over time, locating fiducial marks within tissue structures or carefully demarking the surface area is critical to finding the same cellular structures and to detect changes over time. Despite these various limitations, CSLM is one of the only methods to non-invasively image cellular morphology and is an extremely powerful tool if used correctly.

4. Incident wavelength effects on CSLM imaging

An important factor when imaging the skin with CSLM is to understand the effect that the incident light wavelength has on the images captured. Recognizing the various effects that the incident wavelength has on the images will help to properly understand and resolve different biological structures within skin. Near-infrared wavelengths are typically chosen for *in vivo* CSLM skin imaging because they can penetrate deeply into the skin. Shorter wavelengths and smaller pinholes can increase the axial and lateral resolution of reflectance imaging. The disadvantage of shorter incident wavelengths is shallower penetration depths, which may be attributed to the wavelength dependence of light scattering and absorption of light by endogenous skin tissue chromophores.

A study was performed to investigate the effect of three different wavelengths of incident light when imaging with CSLM: 405, 785, and 830 nm (Luedtke et al., 2009). These incident wavelengths were chosen because 1) 830 nm is the standard incident wavelength for commercial *in vivo* CSLM systems and is presumed to have the best penetration depth; 2) 785 nm could excite fluorescent dyes, e.g., indocyanine green, while also having good penetration depth; and 3) 405 nm could excite autofluorescence in skin and provide improved resolution in reflectance measurements, due to the shorter wavelength. Optical phantoms and human skin were imaged with these three incident wavelengths.

To determine differences attributed solely to the incident wavelength, the optics of the systems were made as comparable as possible. The optical design, incident laser wavelength, optical coatings and detectors are some of the factors that need to be considered. For the study, three microscopes were used: the commercially available Vivascope 1500 which uses an 830 nm laser, and two custom Vivascope 1500 microscopes (Lucid, Inc., Rochester, NY) to allow imaging with 405 nm and 785 nm lasers. The 785 nm and 830 nm CSLM systems were identical with regard to specifications. The exact same objectives were used for both microscopes, with the only major difference being the incident light source, to eliminate artifacts that may occur due to a change in optics. The 405 nm CSLM was designed to be comparable to the 830 nm system. Some adjustments were required to maintain a relative similarity despite the large wavelength difference, specifically changes in polygon mirror coatings, antireflective coatings on the telescope optics and objectives, pinhole size, and detector. These changes were essential to maintain equivalent optics, while delivering reflectance images that could be compared across the microscopes.

CSLM System	Polygon Mirror Coating	Telescope Design	Pinhole Diameter	Detector
785 and 830 nm systems	Lincoln Laser 01 protected gold	Optimized for 750-1400 nm	100 μm	Hamamatsu Avalanche Photodiode (C5460 Series) (Hamamatsu, Japan)
405 nm system	Lincoln Laser 04 enhanced aluminum	Optimized for 406-658 nm	50 μm	Hamamatsu Photomultiplier Tube (R7400U Series) (Hamamatsu, Japan)

Table 2. Differences in design of CSLM systems

In the tissue phantom study, an optical phantom that simulated the optical properties of human skin was evaluated to measure changes due to incident wavelength in a standard

frame of reference. The phantom was comprised of silicone, india ink, and titanium dioxide (Beckman Laser Institute, University of California - Irvine) and had absorption and scattering coefficients of μ_a= 0.08/mm and μ_s=2.53/mm at 600 nm. A 1 mm diameter steel wire was inserted into the phantom so that imaging at the same exact location was feasible. Six replicate image stacks were captured with each microscope, beginning at the surface of the phantom and progressing into the phantom until structures were no longer visible.
The normalized intensity profiles generated for each wavelength microscope show that the three wavelengths have similar rates of exponential decay down to depths of 175 µm (Figure 3). There are surface intensity differences, which may be due to sensitivity differences in the photodetector to each wavelength, or to possible fluorescence in the silicone phantom when excited at 405 nm. The high signal intensity observed in the 405 nm system may also be due to its photomultiplier tube (PMT) photodetector compared to the avalanche photodiode (APD) found in the near infrared wavelength systems.

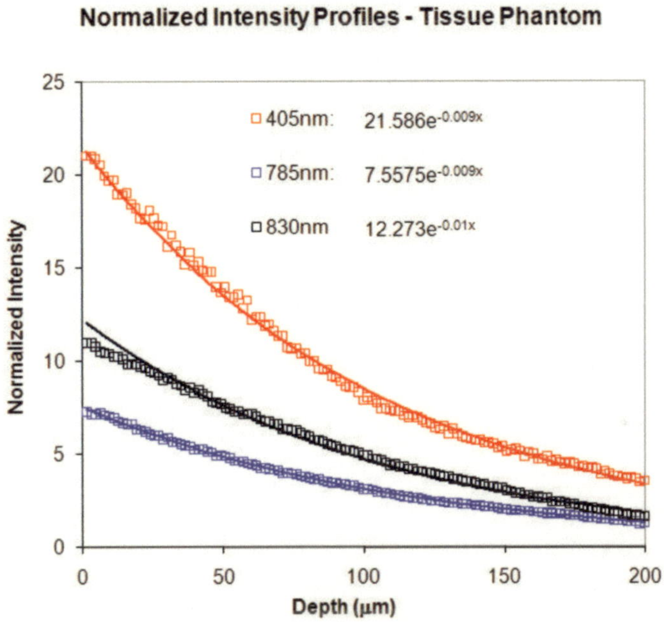

Fig. 3. Normalized intensity profiles of the tissue phantom, from different incident wavelength CSLM systems (405 nm, 785 nm, 830 nm). Normalized intensity refers to normalization of image intensity of a region of interest to the laser power used to capture the image.

In the human study, 8 female subjects, ages 25-45 years old, were evaluated on the upper volar arm after informed consent was obtained. After the tissue ring was applied on the arm, imaging with the three microscopes was performed, without removing the tissue ring. This helped minimize any artifacts that may be introduced when changing tissue windows or placing the tissue ring multiple times on the same site. The order in which microscopes were used was randomized for each subject. To image the same exact site, special attention was paid to identify fiducial marks (e.g., hair follicles and pigmented spots) in surface images to ensure imaging of exact locations. Image stacks were captured at 1.5 µm increments, until structures were no longer visible. For each subject, three image stacks were acquired at different locations with each microscope. When analyzing these images, the ROIs excluded areas of high contrast, e.g., hair follicles and surface microglyphics. When the three image stacks captured at the same site but with the different wavelength systems were analyzed, the size and shape of the regions were identical, and the average intensity of the ROI was normalized to the laser power used to capture the image (Figure 4).

Imaging with the 405 nm CSLM system was halted after 166 μm because structures were no longer visible due to the weak signal. The laser power of the 405 nm system cannot be greatly increased for better imaging in depth because of the strong absorption by melanin, which results in a stinging sensation for some subjects. Very little light penetration was expected, so imaging to a depth of 166 μm was unexpected. Due to strong endogenous skin tissue chromophore absorbers and increased scattering function, only the upper epidermal features were expected to be seen, but the entire epidermis and part of the superficial dermis were visible when using the 405 nm system. Two possible explanations for this surprising result are 1) the inhomogeneity of skin which may allow light to penetrate deeper than theorized or 2) the papillary dermal collagen reflectance observed in the images.

The intensity profiles were fit to exponential functions using least squares fitting. Data from the 405 nm system can be represented by a one-term exponential function, while data from the 785 and 830 nm systems require two-term exponential functions. When comparing the intensity profiles generated from the optical phantom and human skin, the complexity of the signals from human skin indicate inhomogeneities in skin structure.

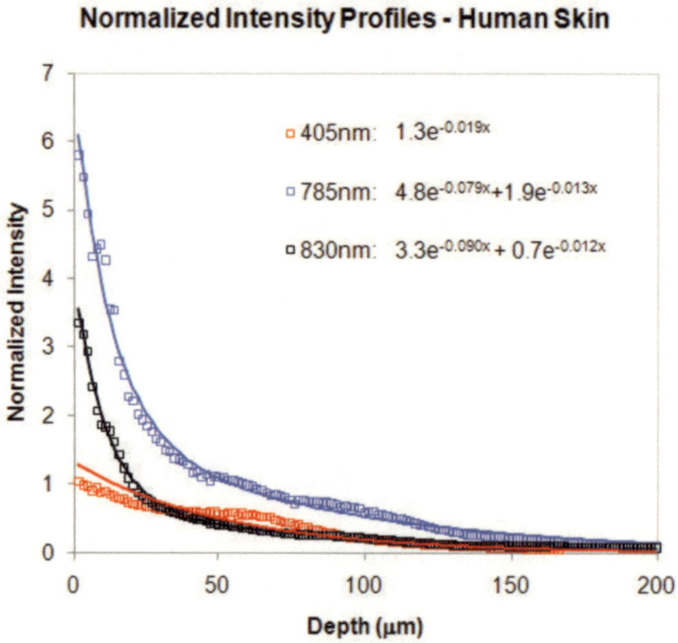

Fig. 4. Normalized intensity profiles of human skin, from different incident wavelength CSLM systems (405 nm, 785 nm, 830 nm)

The choice of wavelength may affect the ability of CSLM to visualize certain skin features, as differences in human skin intensity profiles at each wavelength are observed. The intensity profile generated by the 405 nm system shows an increase in signal intensity near 50 μm, which corresponds to the presence of dermal papillae near the dermal-epidermal junction. Although the dermal papillae are visible in the 785 and 830 nm images, they are most evident in the 405 nm images, which may indicate that using a 405 nm incident wavelength laser may be best when studying dermal papillae. There is a significant decrease in signal intensity from the 405 nm system, compared to the near-infrared wavelength systems, which is likely due to the absorption of light by collagen cross-links in the superficial dermis. The absorption is significantly reduced at near-infrared wavelengths, leading to strong signal intensities from the 785 and 830 nm systems, making them much better options when studying the superficial dermal tissue.

While the 405 nm system would theoretically generate images with improved resolution, it was not fully realized in this study. This may be due to using a 50 μm diameter pinhole, which may be too large, allowing light from adjacent layers to pass through. Imaging with a 35 μm diameter pinhole shows significantly improved resolution of cellular details.

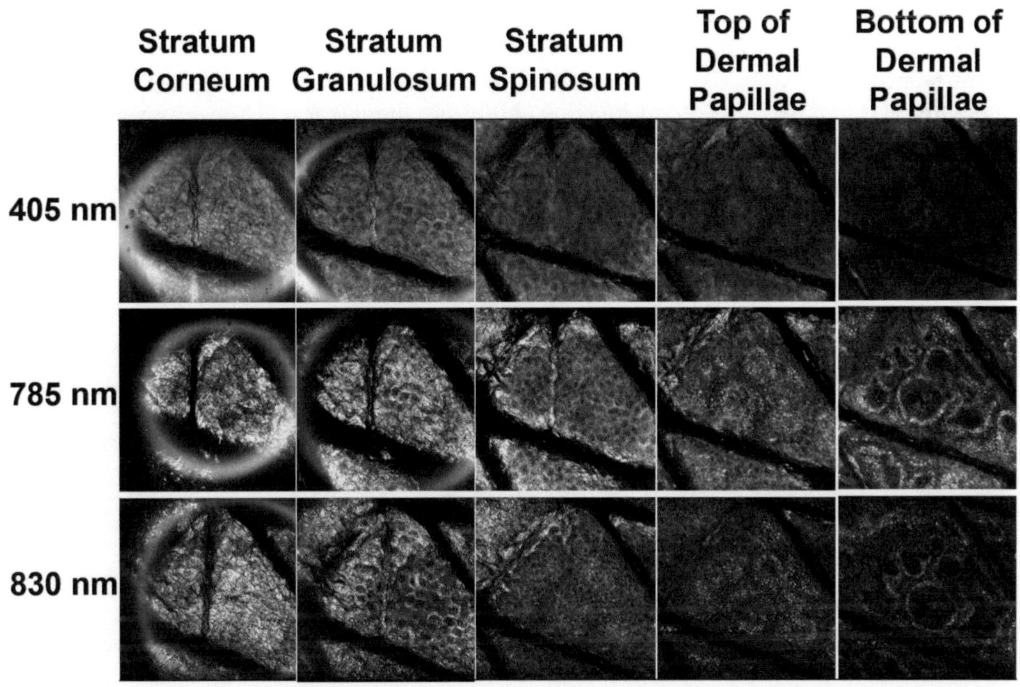

Fig. 5. Differences in image quality are observed with different incident wavelength lasers

It was anticipated that images obtained from the 785 and 830 nm systems would be nearly identical due to the small difference in wavelengths and the absence of skin chromophores that would absorb selectively at either 785 or 830 nm. The intensity values of skin at 785 nm, however, were consistently greater than those obtained at 830 nm. Enhanced contrast was also observed with the 785 nm, allowing better visualization of structures. Despite all efforts to create the microscope setups with nearly identical optics, differences in image detail were revealed at these wavelengths.

Overall, imaging at different wavelengths with CSLM systems can reveal different structural details in cells. Both qualitative and quantitative differences were observed when imaging the same location in healthy human skin with different incident wavelengths. The

405 nm CSLM system provides better spatial resolution than near-infrared wavelength systems, while still penetrating relatively deep into the skin, up to 166 μm. Despite the close proximity of wavelengths, the 785 and 830 nm CSLM systems show differences in signal intensity and structural contrast apparent in images. Optimization of the incident wavelength may improve CSLM performance depending on the application.

5. Comparison of infant and mother skin

Examples of such quantitative analyses of cell size, SC and epidermal thickness, and normalized intensities were used to measure age-related changes in a population of mothers and their biological infants (Stamatas et al., 2009). CSLM allows non-invasive visualization of different functional and microstructure differences between infant and adult skin.

In a clinical study comprised of 20 healthy mothers (25-43 years old) and their biological children (3-24 months old), the skin microstructure of the lower thigh area of infants and adults were examined and compared. Image stacks consisting of at least 32 images were acquired, with an axial section of 1.5 μm in step size. The lower thigh area was examined for ease of measuring infants, as it is difficult to keep them voluntarily still. The mothers held the infants in their laps throughout imaging. Although there was still some movement in the image stacks, every effort was taken to prevent movement of subjects during imaging. SC and epidermal thickness, as well as size of cells in the granular layer, were assessed from the confocal images. Sample images from various image depths are shown in Figure 6, comparing infant and mother skin.

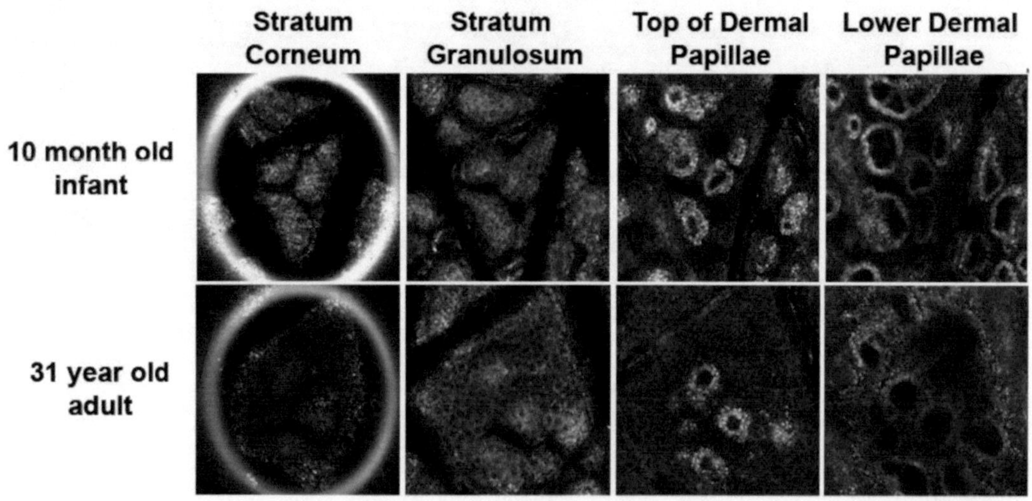

Fig. 6. CSLM images at various depths from sample image stacks of an infant and an adult. Both subjects are African-American. Pigmented keratinocytes are visible surrounding the dermal papillae.

Surface features evaluated showed differences in microglyphic density and surface area. Microglyphics appear as dark lines that separate "islands" of cells in CSLM images. For infant skin, a greater density of microglyphics is present. Differences in epidermal thickness and cell size were also documented. Infant SC and epidermis were found to be 30% and 20% thinner, respectively, than adult SC and epidermis. Infant corneocytes and granular cells were found to be 20% and 10% smaller, respectively, than adult cells indicating a more rapid cell turnover in infants (Figure 7).

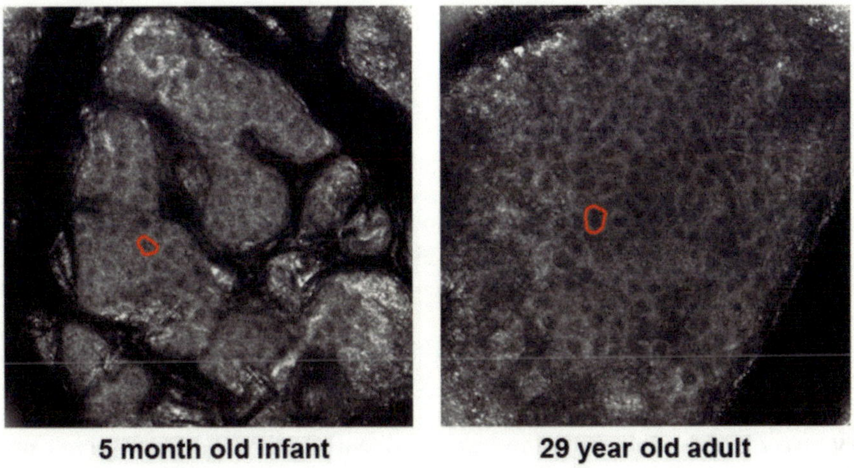

Fig. 7. A granular cell is outlined in infant and adult images.

Dermal papillae density and size distribution also differed. In infant skin, a distinct direct structural relationship between the SC morphology and dermal papillae was observed. A change in reflected signal intensity at ~100 µm, indicating the transition between papillary and reticular dermis, was evident only in adult skin (Figure 8). Upon performing software analysis of the optical sections, a normalized profile as a function of depth was captured.

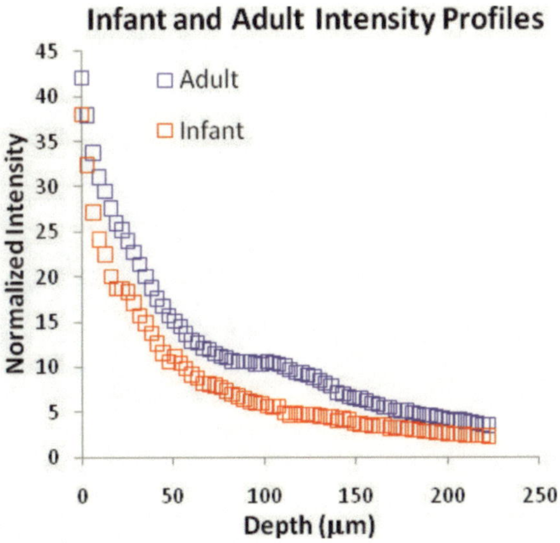

Fig. 8. Normalized intensity profiles as a function of depth, for infant and adult skin.

The centers of the images were analyzed to further account for movement within the images, and the averages were taken within the defined ROI. The ROI was averaged across the optical sections and the triplicate vertical image stacks acquired. These differences in skin microstructure may help explain some of the reported functional differences and assumptions made regarding skin softness and other aging related effects.

6. Wound healing response of micro-injuries

A common treatment for skin aging is the use of a laser device with infrared wavelength to induce an array of microscopic wounds in the skin. Micro-wounds are created when aqueous components within skin tissue absorb the light, and defined areas of photocoagulation stimulate a wound-healing response in the dermis, resulting in increased production of new collagen and improved skin tone and texture (Hantash & Mahmood, 2007; Manstein et al., 2004; Tannous, 2007). These micro-injuries quickly re-epithelialize by the surrounding undamaged tissue, and CSLM can be used to better understand the dynamic wound healing process as a response to micro-injuries by monitoring them non-invasively (Liu et al., 2009).

A single line of micro-injuries, 1 cm long with approximately 400 µm between centers of micro-injuries, were created with a fluence rate of 60 mJ on the volar forearms of 8 healthy subjects (Fitzpatrick skin type II-VI, ages 27-57 years old). Informed consent was obtained from all subjects. Micro-injuries were evaluated with both CSLM and video microscopy (Hirox, Japan) at various time points to study both short- and long-term effects of treatment (30 minutes, 2, 4, and 21 days post-treatment). Cross-polarized images were captured with video microscopy to highlight sub-surface features, including pigmentation and vasculature. For CSLM imaging, a series of 128 images were acquired, with an axial section of 1.5 µm in step size. To quantify the wound healing process at the microscopic level, the depth-dependent intensity profile from a defined ROI was obtained by calculating the average intensity within the ROI at each depth.

The micro-wounds appear as a line of dark brown spots in the video microscopy image, while they appear very bright in the corresponding CSLM image, captured approximately 80 µm below the SC surface, indicating strong scattering in the micro-wound area (Figure 9).

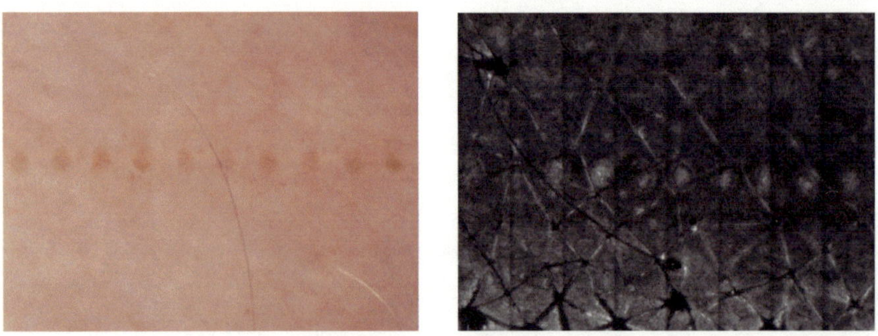

Fig. 9. Corresponding cross-polarized (left) and reflectance CSLM (right) images of a line of micro-wounds created by laser treatment, 1 day post-treatment, are shown. (4 mm x 3 mm)

To monitor the wound healing process of the micro-wounds, three ROIs were defined in each CSLM image stack and their intensity profiles were extracted. The individual treated micro-wound zone (1) and the surrounding collateral damage zone (2) were compared to a normal area away from the micro-wound (3). Their depth-dependent intensity profiles were extracted and analyzed to observe dynamic changes within skin as a response to the micro-wound. The depth-dependent intensity profiles of the 3 ROIs and a series of corresponding

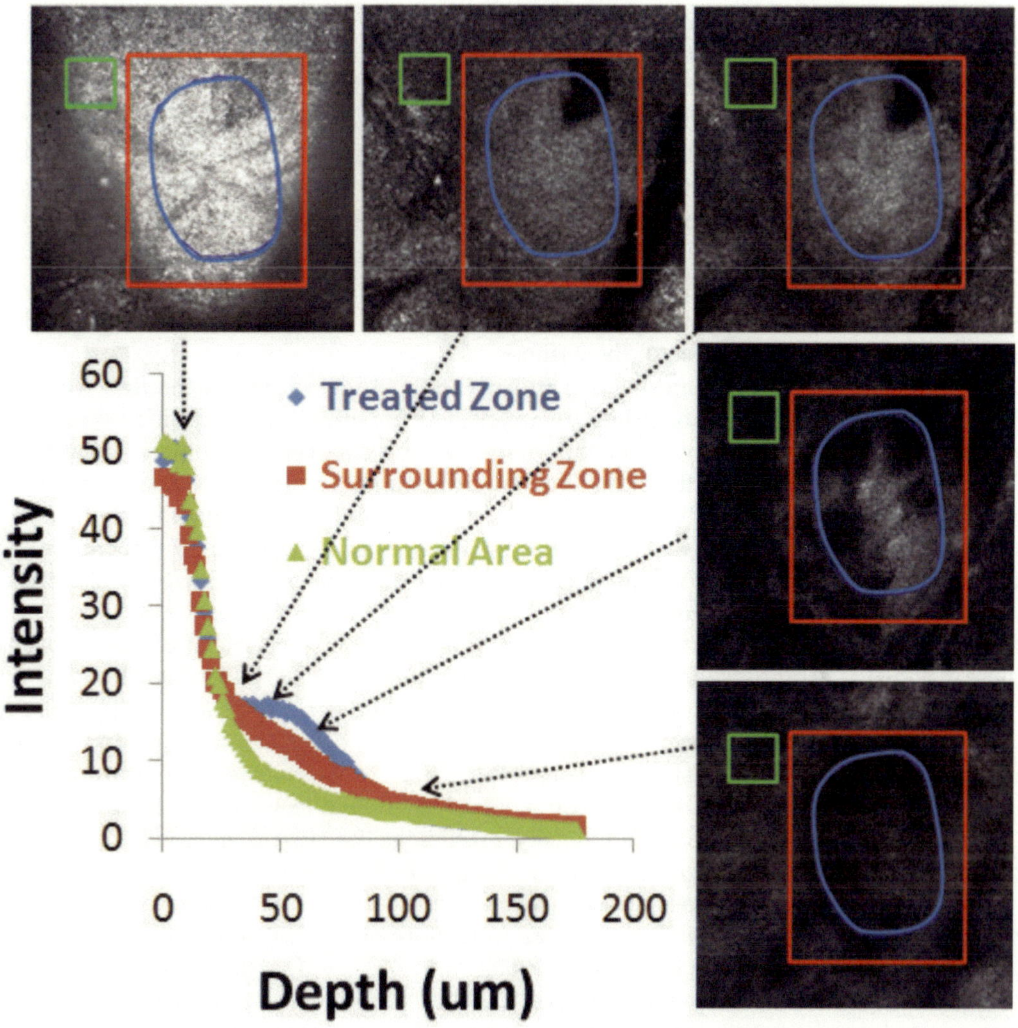

Fig. 10. The depth intensity profiles of the 3 regions are shown, along with confocal images at various depths from a sample image stack, 30 minutes post-treatment.

confocal images at various depths are shown in Figure 10 (data from the volar forearm of a 27 year-old Caucasian female 30 minutes post-treatment). In the confocal images, the region outlined in blue represents the treated micro-wound zone, and the region defined in red, excluding the micro-wound zone (blue region), represents the surrounding area. The normal "control" skin is defined by the green region.

As early as 30 minutes post-treatment, changes in reflectivity are observed in both the micro-wound and the surrounding tissue, indicating cellular changes are occurring both in and around the micro-wound as response to laser treatment. Within the superficial layers, 10-15 μm below the top of the SC, the intensity of reflection in the treated micro-wound zone is only slightly higher compared to the surrounding and normal areas, indicating that there is minimal damage to the surface of the skin. A significant increase in reflection is observed in the micro-wound zone 20-50 μm below the skin surface. This may be due to swelling, causing a mismatch in refractive index from the upper layers. The intensity of reflection slightly decreases within the micro-wound below 50 μm and into the dermis, indicating the dermal collagen may have been denatured from the laser treatment. This may also be due to the tissue becoming optically opaque, so as skin is sectioned deeper, light is lost to scattering and absorption of the tissue.

Dynamic changes of the wound-healing process were observed within the first week and then again 3 weeks after the one-time treatment. Using both CSLM and video microscopy together enables us to identify how changes on the cellular level contribute to what can be seen in magnified images of the skin. Cross-polarized images of the treated site show that micro-wounds are undetectable immediately after treatment, only showing slight edema and inflammation in the images. Within the first 48 hours post-treatment, the micro-wounds appear brown-colored, and as the micro-wounds healed over the 3 weeks, the micro-wounds became invisible (Figure 11).

The depth-dependent intensity profiles provide a quantitative measure of the associated cellular changes captured with CSLM and differentiate the 3 ROIs. Figure 11 also shows representative intensity profiles of the treated micro-wound zone, the surrounding collateral damage zone, and a normal area at various time points, and corresponding confocal images 50 and 120 μm below the surface. Throughout the study, normal skin shows similar intensity profiles, indicating that the treatment had minimal effects outside the treated area. Compared to the normal area, the intensity profiles of the treated micro-wound and the surrounding collateral damage zones show changes in the first 4 days post-treatment and return to profiles similar to normal skin 3 weeks after treatment. Although the surrounding collateral damage zone shows little change from normal skin 30 minutes post-treatment, there is a significant increase in intensity 2 days and 4 days after treatment. The intensity profile of the treated micro-wound, between 20-100 μm below the surface, progressively increases in intensity immediately after treatment until 4 days post-treatment.

To compensate for the intrinsic exponential decay in intensity with increasing depth in CSLM, the intensity profile of the treated micro-wound zone was normalized to the normal "control" area. Figure 12 shows a peak in the normalized intensity profile of the treated zone, and the peak position first shifts deeper before returning to shallower depths as the micro-wound heals. Images at intermediate time points between Days 4 and 21 were also captured for each subject. The transition time when the peak position shifts back to shallower depths appears to depend on age, with the peak of younger subjects shifting back to shallower depths sooner than they do for older subjects. This suggests that younger individuals heal faster than older individuals, which agrees with clinical observations. This quantitative method can be used to evaluate the wound-healing rate at the cellular level.

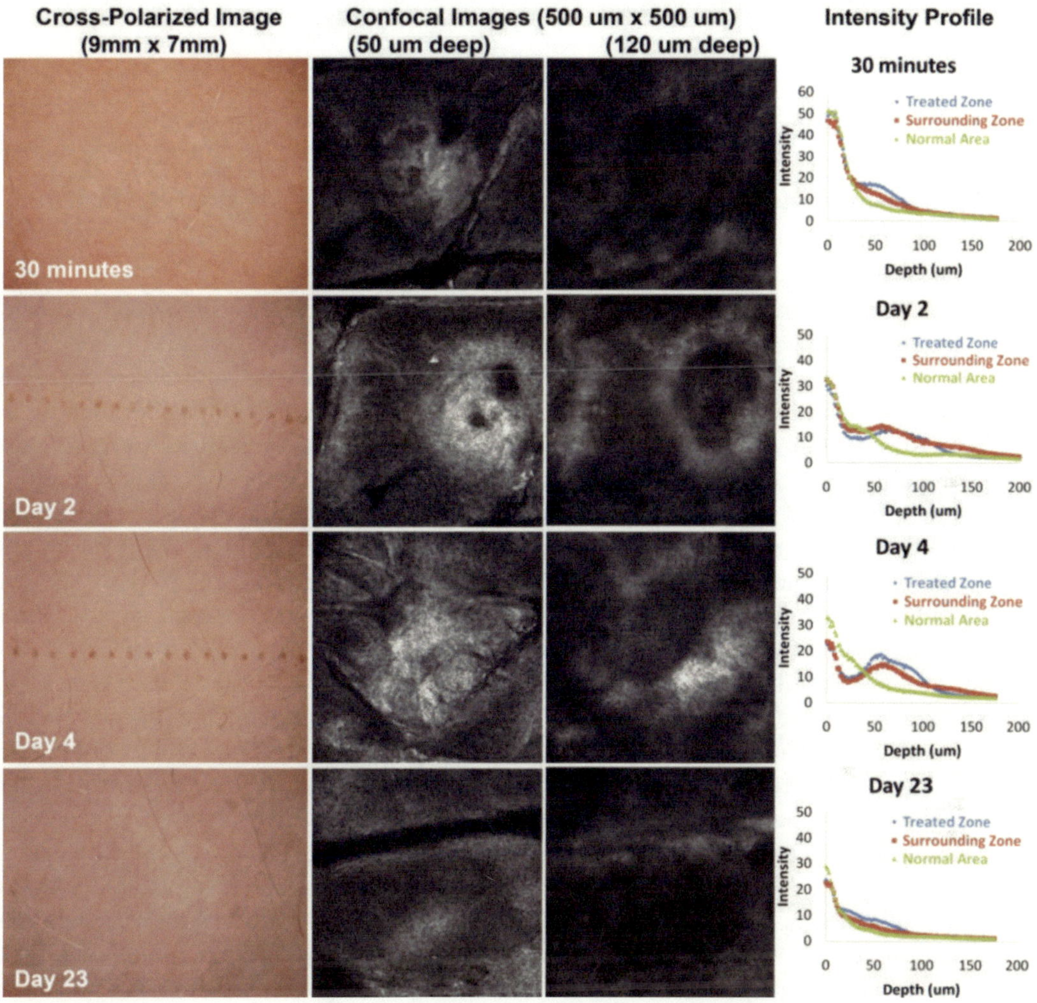

Fig. 11. Representative cross-polarized images of the micro-wounds at various post-treatment time points along with confocal images 50 and 120 µm below the top of the SC. Intensity profiles are also shown.

7. Summary and conclusions

Confocal laser scanning microscopy is a novel technique that enables us to non-invasively investigate skin at the cellular level and dynamic changes that may occur over a period of time. It was shown that the incident light wavelength in CSLM systems affects signal intensity and structural details at the cellular level. Systems equipped with shorter wavelengths may provide improved resolution in images, while those with longer

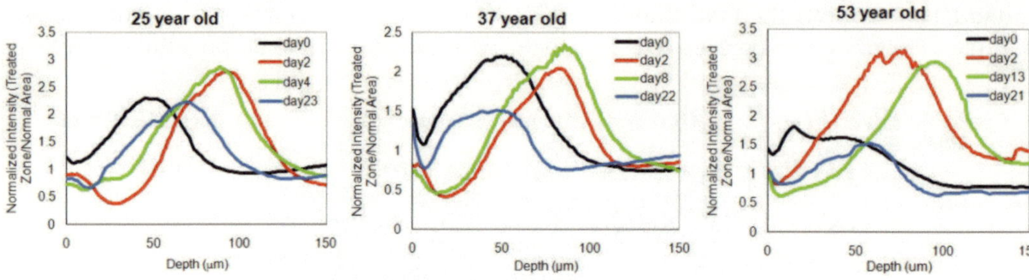

Fig. 12. Normalized intensity profiles within treated zone of 3 subjects at various time points post-treatment.

wavelengths can image deeper into the skin. When comparing systems of the same design and only differing in incident wavelength, the 785 nm system produces images with greater signal intensity and contrast in structural detail compared to the 830 nm system. It is peculiar that imaging human skin with 785 and 830 nm wavelengths produces intensity profiles that appear biphasic, while imaging at 405 nm result in intensity profiles that follow a simple exponential decay. The normalized intensity profiles generated from 405 nm images are also significantly less than those from 785 and 830 nm images at shallow depths.
The studies discussed show the potential of using CSLM to study skin and the cellular processes that occur over time and as a response to wounding. Infant skin is not fully developed and exhibits different characteristics compared to adult skin. With CSLM, we discovered morphological differences between infant and adult skin, which may contribute to the functional differences observed between the developing infant skin and fully-developed adult skin. Future work includes studying how and when infant skin transitions to being adult-like, morphologically. Since we can observe the wound-healing process, we can now evaluate various factors that may affect the wound-healing rate and evaluate how micro-wounds compare to larger skin wounds. We can also evaluate how the method used to create the micro-wound affects the resulting cellular damage.
Capturing cellular changes over time with CSLM is of great significance as it enables us to study the physiology of living skin tissue. Skin is the body's largest functioning organ, consisting of millions of cells that are in constant activity, whether it be proliferation, pigmentation, repair, etc. Capturing the same cells over time and observing their activity is powerful. By studying healthy skin tissue and understanding how it reacts and functions under average conditions, we can better characterize healthy tissue and differentiate between healthy and unhealthy tissue. Capturing physiological and morphological changes in skin due to aging, treatment, and wounding is only the beginning of our fundamental understanding of dynamic cellular changes using CSLM. Having a strong foundation in the knowledge of normal function and activity of healthy tissue is essential to the study and understanding of diseased tissue.

8. Acknowledgements

The authors would like to acknowledge Lucid, Inc. for their use of equipment, confocal microscopy expertise (Zachary M. Eastman) and accessibility to enable the research projects presented within this chapter. A special thanks to Gregory Payonk for the initial version of the normalized intensity analysis first used in the comparison of mother and infant skin investigation and to Yang Liu who contributed to the improvement of this analysis used in the wound healing response of micro-injuries. The authors would also like to thank Michael Neidrauer for his time and attention spent on data analysis and presentation of the different incident wavelength data.

9. References

Calzavara-Pinton, P., Longo, C., Venturini, M., Sala, R. & Pellacani, G. (2008). Reflectance confocal microscopy for in vivo skin imaging, *Photochemistry and Photobiology* Vol. 84(No. 6): 1421-30.

Gonzalez, S. & Gilaberte-Calzada, Y. (2008). In vivo reflectance-mode confocal microscopy in clinical dermatology and cosmetology, *International Journal of Cosmetic Science* Vol. 30(No. 1): 1-17.

Hantash, B. M. & Mahmood, M. B. (2007). Fractional photothermolysis: a novel aesthetic laser surgery modality, *Dermatologic Surgery* Vol. 33(No. 5): 525-34.

Kollias, N. (1995). The physical basis of skin color and its evaluation, *Clinics in Dermatology* Vol. 13(No. 4): 361-7.

Liu, Y., Bargo, P. & Kollias, N. (2009). Quantitative assessment of wound-healing process as a response to laser-induced micro-injuries, *Proceedings of SPIE*, SPIE, San Jose, California, pp. 71610W1-6.

Luedtke, M. A., Papazoglou, E., Neidrauer, M. & Kollias, N. (2009). Wavelength effects on contrast observed with reflectance in vivo confocal laser scanning microscopy, *Skin Research and Technology* Vol 15(No. 4): 482-8.

Manstein, D., Herron, G. S., Sink, R. K., Tanner, H. & Anderson, R. R. (2004). Fractional photothermolysis: a new concept for cutaneous remodeling using microscopic patterns of thermal injury, *Lasers in Surgery and Medicine* Vol. 34(No. 5): 426-38.

Pellacani, G., Longo, C., Malvehy, J., Puig, S., Carrera, C., Segura, S., Bassoli, S. & Seidenari, S. (2008). In vivo confocal microscopic and histopathologic correlations of dermoscopic features in 202 melanocytic lesions, *Archives of Dermatology* Vol. 144(No. 12): 1597-608.

Rajadhyaksha, M., Grossman, M., Esterowitz, D., Webb, R. H. & Anderson, R. R. (1995). In vivo confocal scanning laser microscopy of human skin: melanin provides strong contrast, *Journal of Investigative Dermatology* Vol. 104(No. 6): 946-52.

Rajadhyaksha, M., Gonzalez, S., Zavislan, J. M., Anderson, R. R. & Webb, R. H. (1999). In vivo confocal scanning laser microscopy of human skin II: advances in instrumentation and comparison with histology, *Journal of Investigative Dermatology* Vol. 113(No. 3): 293-303.

Scope, A., Benvenuto-Andrade, C., Agero, A. L., Malvehy, J., Puig, S., Rajadhyaksha, M., Busam, K. J., Marra, D. E., Torres, A., Propperova, I., Langley, R. G., Marghoob, A. A., Pellacani, G., Seidenari, S., Halpern, A. C. & Gonzalez, S. (2007). In vivo reflectance confocal microscopy imaging of melanocytic skin lesions: consensus terminology glossary and illustrative images, *Journal of American Academy of Dermatology* Vol. 57(No. 4): 644-58.

Stamatas, G. N., Nikolovski, J., Luedtke, M. A., Kollias, N. & Wiegand, B. C. (2009). Infant skin microstructure assessed in vivo differs from adult skin in organization and at the cellular level, *Pediatric Dermatology* Vol. 27(No. 2): 125-31.

Tannous, Z. (2007). Fractional resurfacing, *Clinics in Dermatology* Vol. 25(No. 5): 480-6.

Webb, R. H. (1996). Confocal optical microscopy, *Reports on Progress in Physics* Vol. 59(No. 3): 427-71.

Webb, R. H. (1999). Theoretical basis of confocal microscopy, In: *Methods in Enzymology Volume 307*, P. M. Conn (Ed.), 3-20, Academic Press, ISBN: 978-0-12-182208-8, New York.

3

Correcting Airborne Laser Scanning Intensity Data

Ants Vain[1] and Sanna Kaasalainen[2]
[1]Estonian University of Life Sciences
[2]Finnish Geodetic Institute
[1]Estonia
[2]Finland

1. Introduction

Airborne laser scanners (ALS) produce a point cloud where each point has X, Y and Z coordinates which are determined by the combination of GNSS (*Global Navigation Satellite System*), IMU (*Inertial Measurement Unit*) and laser range finder. Additionally to coordinates points have also an intensity value recoded. Intensity values are recorded in digital numbers and represent the number of photons that are impinging the detector (Coren & Sterzai, 2006). Intensity values are related to the energy coming back to the receiver. Since the ALS systems operate on the same physical basis as radar we can use the radar equation (Wagner, 2010):

$$P_r = \frac{P_t D_r^2}{4\pi R^4 \beta_t^2} \eta_{sys} \eta_{atm} \sigma \tag{1}$$

where P_r is the power that has been received, P_t is the power that was transmitted, D_r is the receiver aperture diameter, R is the range from sensor to the object, β_t is the laser beam width, η_{sys} is the systems transmission factor, η_{atm} is the atmospheric transmission factor and σ is the cross-section of the target. As we can see from equation (1) the intensity (or received power) is connected to the power that has been transmitted, the range from sensor to the object, how the beam is diverged, system and atmospheric transmission and the footprint size of the laser beam.
The cross-section σ can be described with a following equation (Höfle & Pfeifer, 2007):

$$\sigma = \frac{4\pi}{\Omega} \rho A_s \tag{2}$$

where Ω is the scattering solid angle of the target, ρ is the reflectance of the target and A_s the area illuminated by the laser beam. A_s can be written as a function of distance R and beam width β (Höfle & Pfeifer, 2007):

$$A_s = \frac{\pi R^2 \beta_t^2}{4} \tag{3}$$

When we substitute A_s in equation (3) to equation (2), we get:

$$\sigma = \pi \rho R^2 \beta_t^2 \cos\alpha \tag{4}$$

Substituting this into equation (1) we get a new radar equation:

$$P_r = \frac{P_t D_r^2 \rho}{4R^2} \eta_{sys} \eta_{atm} \cos\alpha \tag{5}$$

The η_{sys} is the optical transmission efficiency for all optical components of ALS system (Höfle & Pfeifer, 2007). It is a constant for a certain ALS system but it may vary with different sensors. The D_r component is also a constant when using one type of sensor. The α is an incidence angle, which is defined as an angle between incoming laser beam and surface normal.

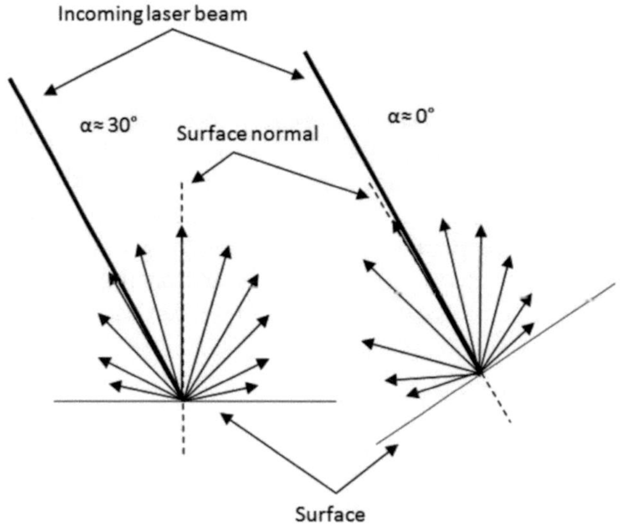

Fig. 1. Lambertian scattering of the surface. Topographic effect on incidence angle.

As we can see from Fig. 1 the cosα part in equation (5) depends on the surface roughness. When the measured area is flat then the incidence angle is the same as scan angle. The surfaces are assumed to have Lambertian backscattering properties. This means that the energy that is backscattered from the surfaces decreases with the increase of incidence angle. It is clear that those factors have to be taken into account when we want to use the intensity data.

2. Factors affecting the intensity

2.1 Atmospheric correction

When we are using ALS data it is clear that the atmospheric conditions are affecting the intensity values. Since most of the ALS systems use 1064 nm wavelength it is important to notice that scattering effect is much larger than absorbtion in that particular wavelength (Kim et al., 2001). It is also reported that the attenuation of horizontal propagation can be from 0.2 dB/km for very clear weather conditions to 3.9 dB/km for haze conditions (Höfle

& Pfeifer, 2007). If we look at the vertical path of laser beam then the atmospheric transmittance increases with higher altitudes (Höfle & Pfeifer, 2007). From the Beer-Lambert law the intensity I can be written as:

$$I = I_0 e^{-\alpha x} \tag{6}$$

where I_0 is the incident intensity, a is the atmospheric attenuation in dB/km and x is the path length (from the sensor to the object). The part e^{-ax} in the equation (6) is also called atmospheric transmittance (T). Knowing that we can write for equation (5) the η_{atm} as a function of atmospheric transmittance T:

$$\eta_{atm} = \frac{1}{T^2} \tag{7}$$

T describes the properties of atmosphere and can vary from 0 to 1 (1 referring to clear transmittance of atmosphere). The value of T depends on the weather conditions (e.g. visibility, air temperature, humidity, aerosol concentration) but also on the flying altitude. Most of the flight are made in clear weather but usually after the rain. This means that the humidity might be high. Since the lack of information about the weather conditions during the flights, we have to use programmes that calculate approximate values for T. The usual inputs for these programmes are: wavelenght, visibility, atmospheric model, path length, environment (rural or urban).

2.2 Range correction

When we look at the equation (5) we can notice that the recieved power (or intensity) has an inverse range square dependency ($1/R^2$). The $1/R^2$ dependency applies to the extended diffuse targets. This means that the entire footprint is reflected on one surface and it has Lambertian scattering characteristics (see Fig. 1). This is not always the case. Targets that are smaller than the laser footprint size (e.g. a small tree branch or a leaf) are range independent non-extended diffuse targets. And targets that have linear physical properties (e.g. wire) are linear range dependent (Höfle & Pfeifer, 2007). For a target smaller than laser footprint size the range dependency becomes $1/R^4$, and for linear targets $1/R^3$. In practical applications, it is very hard to determine weather the target was smaller than a footprint size or not.

To correct the intensity values for range dependency a reference altitude should be chosen. For this purpose we have to know the distance from sensor to the object for every individual laser point. Also, incdence angle correction should be applied. But in some cases it is hard to determine the incidence angle, e.g. for trees. Incidence angle correction can be left out if the scan angle does not exceed 20 degrees on the flat surface, because the error caused by the incidence angle is smaller than errors in measuremnts. Intensity values start to decrease significantly from 20 degrees (Kukko et al., 2008). The range correction or sometimes also called sperical loss can be written (Vain et al., 2009):

$$I_{Rcorr} = I_{obs} \frac{R_i^2}{R_{ref}^2} \tag{6}$$

where corrected intensity for sherical loss (I_{Rcorr}) is a function of observed intensity (I_{obs}), Ri (the slope distance from sensor to the object) and R_{ref} (the chosen reference distance). For example, if the flying altitude is around 500 m then the slope distance with 20 degree scan angle will be around 530 m. Then it is proper to choose 500 m as reference distance (R_{ref}).

2.3 Pulse energy correction

One of the critical parts in correcting ALS intensity data is the pulse energy correction. It is problematic because the transmitted pulse energy is usually unknown. There is an improvement with the newer pulse ALS systems and some pulse scanners can even record wavefroms of detected signal as well. Since the intensity data is strongly connected to the transmitted pulse energy level (see equation (5)) it has to be taken into consideration when using data from different flights that have different pulse energy settings. The average power (P_{av}) that is transmitted is a function of pulse energy and pulse rate (also called *pulse repetition frequency*, PRF) (Baltsavias, 1999):

$$P_{av} = EF = P_{peak} t_p F \tag{8}$$

The t_p is the pulse duration that is usualy couple of nanoseconds. Pulses with shorter duration time can carry more energy but are less stable than pulses with longer duration time. The P_{peak} is the peak power of the transmitted laser pulse. The pulse energy E is connected to the t_p and P_{peak} so that the energy level will decrease with the decrease of peak power P_{peak}. Peak powers for Optech scanners are reported by Chasmer et al., (2006). With the higher PRF setting the transmitted energy is lower (Chasmer et al., 2006; Baltsavias, 1999), which reduces the penetration rate of the laser pulse. This means less points under the canopy and also almost double error in height component when using 50 kHz or 100 kHz settings (Chasmer et al., 2006). Even though the higher PRF settings will give more points, they might not reach to the ground. The PRF settings should be selected according to the vegetation density and application purposes. For correcting intensity data for different energy levels, a simple formula is used (Vain et al., 2009):

$$I_{Ecorr} = I_{obs} \frac{E_{Tref}}{E_{Tj}} \tag{9}$$

The corrected intensity (I_{Ecorr}) is a function of observed intensity (I_{obs}), chosen transmitted reference pulse energy value (E_{Tref}), and the transmitted pulse energy value in current flight (E_{Tj}). The E_{Tref} is an energy value in one of the flights that are used, e.g. when you are comapring two flights with different PRF setting and you know the transmitted energy values, you choose one of the energy levels as reference.

2.4 Intensity correction formula

Summarizing the previous, we can write the correction formula as follows (Vain et al., 2009):

$$I_{corrected} = I_{original} \frac{R_i^2}{R_{ref}^2} \frac{1}{\cos\alpha} \frac{1}{T^2} \frac{E_{Tref}}{E_{Tj}} \tag{10}$$

To correct the intensity values we need to know several parameters. First, the atmospheric conditions during the flight campaign. Since the atmospheric conditions play quite important role in laser beam attenuation, weather conditions have to be taken into account as much as possible using atmospheric modelling programmes. The second important part is energy losses during the travelling time of laser pulse. The effect of different pulse properties might be large enough to be confused with temporal variation in vegetation structure (Chasmer et al., 2006). If the scan angle exceeds the 20 degee limit, it starts to affect the intensity values quite much (Kukko et al., 2008). But the correction proceedures so far are using incidence angle correction for flat surfaces only, because it is hard to determine the incidence angle for trees or other vegetation. For this reason, Wagner (2010) proposes the use of backscattering coefficient to be used as a calibration parameter for laser scanner intensity. The range differences has an effect on intensity data also. Therefore it has to be corrected using range squared dependency for large areas.

3. Calibrating the intensity data

3.1 Commercial tarps

When we want to compare the intensity data for the same area, but we have used different sensor types and settings, some kind of reference target is needed to unify the data sets. Brightness targets (see Fig. 2) have proven to be suitable for calibration purposes.

Fig. 2. Brightness targets placed on the ground during the campaign in August 2006, Espoonlahti, Finland.

These brightness targets (tarps) are made of polyester 1100 dtex with polyvinyl chloride (pvc) coating. The weight of that fabric is 600 g/m^2 and they are coated with titanium dioxide and carbon black paint mixing pigment. To decrease the non-Lambertian reflectance effect a delustring agent was added to the paint which gives a mat surface. However, the mat surface is quite sensitive to the dirt, which means that the tarps have to be cleaned often. The size of these tarps is 5x5 m, which makes them easy to transport but at the same time large enough to get reasonable number of laser points.

The tarps were measured at the Finnish Geodetic Institute (FGI) laboratory. For ALS application a 1064 nm laser (diode-pumped Nd:YAG) was used with the polarization ratio of 100:1. 1064 nm wavelength is usual for most of the ALS scanners. A basic laboratory measurement setup is shown on Fig. 3.

For the calibration purposes in the laser laboratory a 99% Spectralon (Labsphere, Inc.) plate was used. The tarps had nominal reflectances of 5%; 10%; 20%; 25%; 30%; 45%; 50% and 70%. Detailed information about the tarps is given in Kaasalainen et al., (2005) and Kaasalainen et al., (2008).

These tarps have proved to be good reference targets and have been used also in photogrammetric measurements in several studies (Honkavaara et al., 2008).

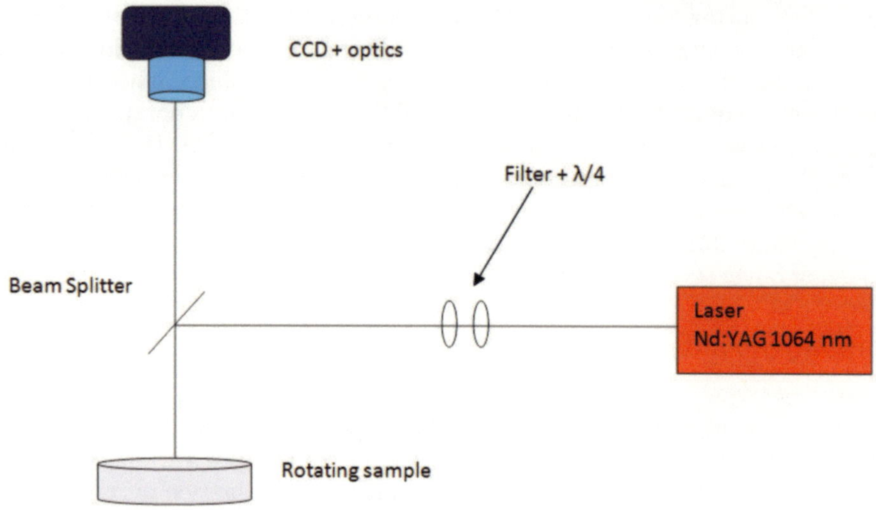

Fig. 3. Setup of the FGI laser laboratory. The sample is placed on a rotating base. CCD camera records the photons that backscatter from the sample. Filter was used to keep the signal in the linear range of the CCD camera. CCD camera is on top of the beamsplitter, 110 cm from the sample.The distance from the sample to the beam splitter is 15 cm. The quater-wave plate was used to scramble the polarization.

3.2 Using natural or semi-natural material to calibrate the intensity

The reference tarps are stable targets to use in calibration process. However, it is not always possible to use them. They have to be placed on the ground before the flight takes place and they must be cleaned once in a while to remove the dust. For this reason a possibility to use natural or semi-natural targets (in this case asphalt and concrete are referenced as semi-natural targets) was studied.

To study the stability of natural and semi-natural targets a data set from 2004-2007 for the test area in Espoonlahti, Finland was used. Sensors were Optech, Topeye and Leica with different flying altitudes and parameters. The intensity of the laser points were corrected according to the equation (10), leaving out the pulse difference correction, simply because of the lack of information about it. The samples were also collected and measured in the laser laboratory (see Fig. 3), and a digital camera solution was tested (see Fig. 6). The most promising and stable results were from the asphalt (Coren & Sterzai, (2006) have also used asphalt to calibrate the intensity data).

Fig. 4 shows that the asphalt is a quite stable calibrator. The differences between sensors are caused by the lack of information about the energy levels used. Also, ALS50-II system uses the AGC (see Ch. 4) and the correction formula for that is not applied in this study.

3.3 Moisture effect

One of the factors that affect the intensity values is the moisture level in the target. It is quite common that the airborne measurements are made after the rain. Fig. 5 shows the gravel road in East-Finland where one part of the road was made wet with water during the flight. The results show that the moisture has an effect on intensity data, making it lower than the dry road data.

Fig. 5 shows that the moisture reduces the intensity level and makes the intensity image „darker“. It means that the knowledge about the moisture level in the target is important. For this purposes a camera-based solution was introduced (Vain et al., 2009) with calibration frame. This helps to make *in-situ* measurements that are more realistic than laboratory measurements, because the conditions are more close to the actual ones. The camera uses NIR spectrum that is quite close to the ALS measurements (1064 nm). This kind of system helps to study the variations in one type of sample and also study the moisture effect on intensity.

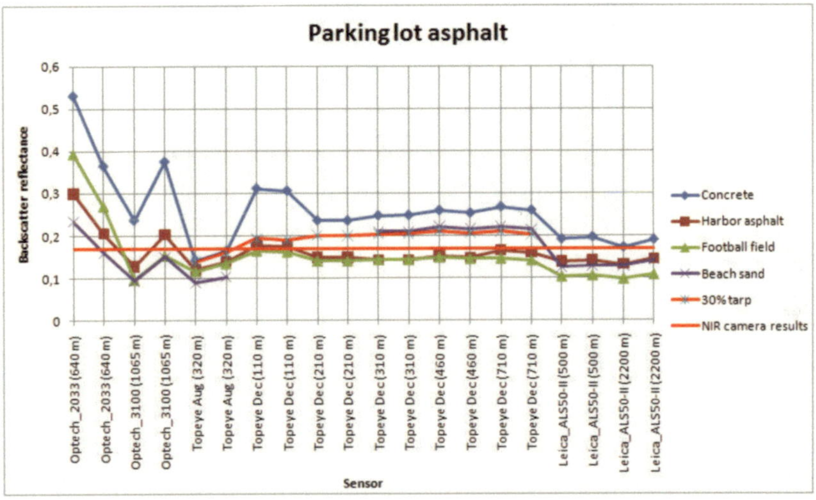

Fig. 4. Results of the parking lot asphalt. Calibrated intensity data from different sensors using parking lot asphalt sample as reference target. (Vain et al., 2009)

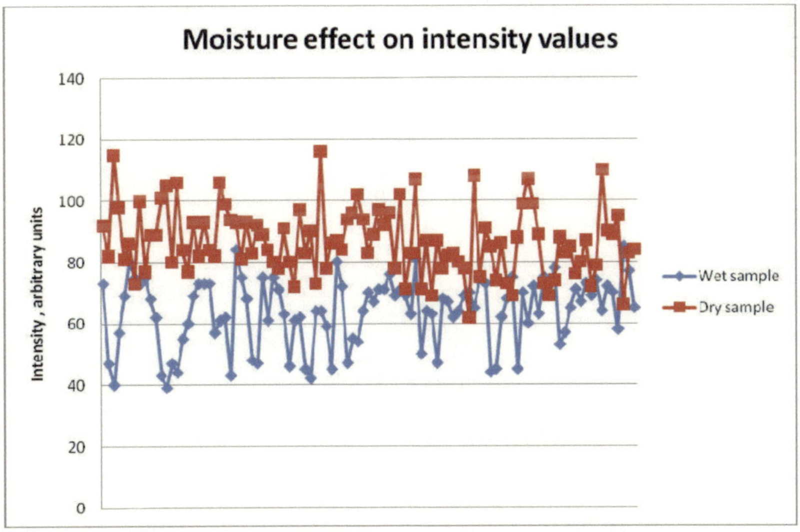

Fig. 5. The wet spot on the road during the flight campaign. Intensity values from the wet spot (blue line and dots) and the dry spot nearby (red). (Kaasalainen et al., 2010)

Fig. 6. Using digital camera and calibration frame to measure sample areas in Espoonlahti, Finland.

4. Automatic gain control

Automatic Gain Control (AGC) is one of the features that are used with the latest ALS systems. The purpose of AGC is to increase the number of received echoes and reduce oversaturation. Since the distance measurement in ALS is connected to the pulse width, the oversaturation (or ringing effect) has direct effect on range accuracy.

The ALS receiver has a lower and upper level. Any received signal that is weaker than the lower level is not recorded, and any signal that is stronger than upper level causes oversaturation which causes errors in time measurements (see Fig. 7).

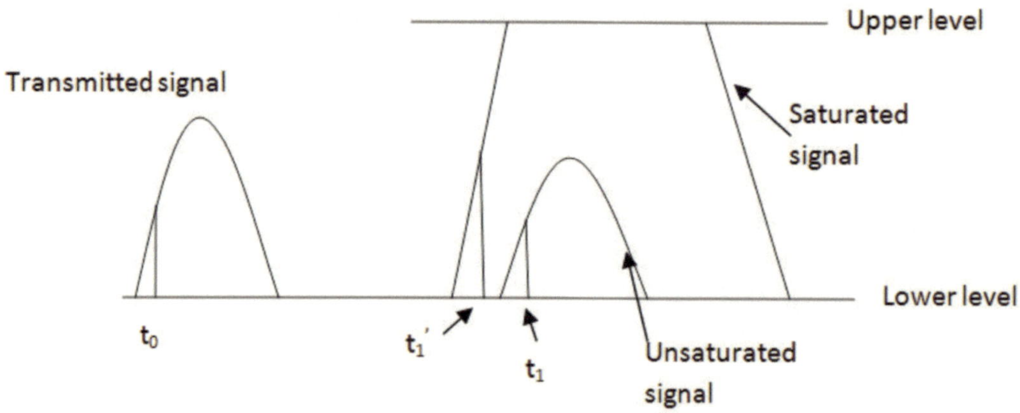

Fig. 7. The signals from very bright targets can cause received signals to oversaturate which in turn causes the error in the time measurement. The unsaturated "normal" signal does not cause error in time measurement.

Oversaturated signal in Fig. 7 causes the time measurement error (difference between t_1' and t_1) which means that the range is not right, because the range is determined by the measurements in time difference between transmitted and received signal (Baltsavias, 1999).
The task of AGC is to keep the signal within the upper and lower levels. If there is a signal that is weaker than the lower level, the gain will increase by 1 step. Or if the signal is stronger than the upper level, then it will reduce the gain by 1 step. The AGC in Leica ALS50-II system has 256 steps and the speed of increase or decrease can be adjusted. Two switches control the increase and decrease speed. For each speed (16 speeds) level there is a certain number of points assigned. For example, if the level is set to position 10, this means that there has to be 144 consecutive laser signals that are below the lower level to increase the gain by 1 step. These levels help to control the amount of how much the gain is fluctuating. If there are a lot of water object in the area, it means that the gain will ramp up if the laser shots are coming from the water and if the laser hits the ground or trees, the gain will start to ramp down again. With the AGC working properly the gain is fluctuating constantly.
The gain and intensity is correlated quite strongly. Therefore, if the gain is fluctuating, the intensity is changing also, even though the target area might be the same (e.g., asphalt road). On Fig. 8 the AGC effect on intensity is clearly visible. The test flight was made in January 2010 in North-West Estonia. The same area was flown twice at the altitude of 500 m. Once with AGC working properly and second time the AGC was set to a constant level (same as previous ALS systems). The AGC increases the number of returns which can be seen clearly on the water (top image have less points on the water than bottom image). And it also "messes up" the intensity image which makes the use on intensity data very difficult (e.g., in automatic classification of object based on intensity values).
The Fig. 9 shows that the intensity is changing according to the changes in AGC values. The sample data was collected in Hyytiälä, Finland 2008. The intensity, when AGC was turned on (blue line), is changing according to the changes in AGC (red line) values.
The intensity data has to be corrected for AGC effect. Vain et al., (2010) propose a correction model that is based on test flights made in Hyytiälä, Finland in 2008. Same area was flown twice: once with AGC working properly and second time when the AGC was set to a constant level. Using least square fitting a model was generated with $R^2 = 0.76$. The model itself is as follows (Vain et al., 2010):

$$I_{off} = a_1 + a_2 I_{on} + a_3 I_{on} AGC \tag{11}$$

where I_{off} is the intensity value when the AGC is turned off or set to a constant level, I_{on} is observed intensity value with AGC turned on and AGC is the automatic gain control value itself. The parameters are: $a_1=-8.093883$; $a_2=2.5250588$; $a_3=-0.0155656$. Using equation (11) to normalise the intensity for AGC effect the relation between observed $I_{off(obs)}$ (intensity values from the flight where AGC was set to constant) and calculated intensity I_{off} was determined. The root-mean-squared error between those two data sets was 5.65. Same model was used to correct the beach sand sample (see Fig. 10).
As we can see from Fig. 10 the model still follows the original (I_{off}) trend. That is the reason why AGC has to be studied further. More information about the AGC should be given by the manufacturers. There is also an other approach introduce by Korpela et al., (2010).

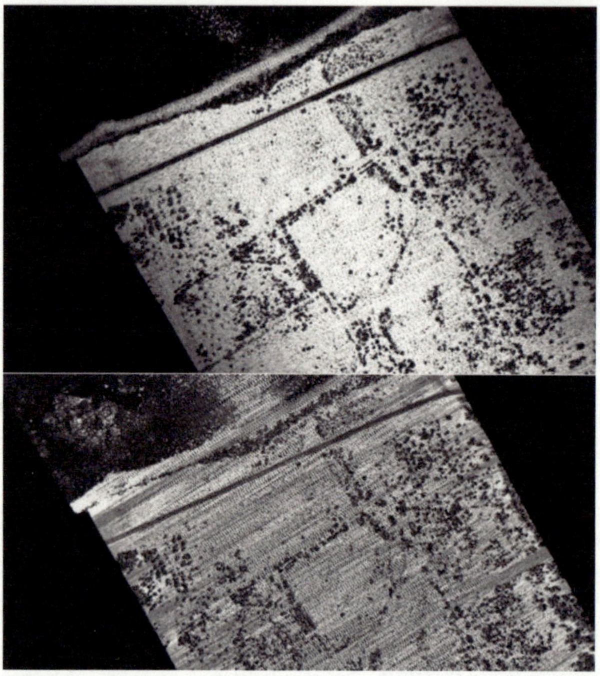

Fig. 8. **Top:** Test area flown with the gain set to a constant level. The intensity image is homogeneous and there are only few points on the water (top part of the picture).
Bottom: The same test area flown with gain fluctuating constantly. The intensity image is much more "messy" than with constant gain. Also there are much more point on the water.

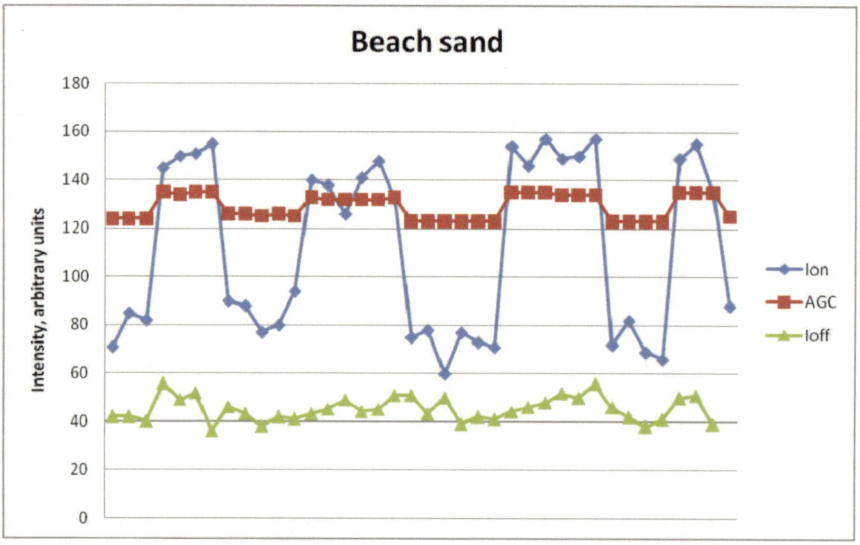

Fig. 9. AGC effect on intensity. Beach sand data collected from Hyytiälä, Finland. Blue line represent the intensity values when the AGC was turned on, red line is the AGC values itself and the green line is the intensity values when the AGC was turned off. (Vain et al., 2010)

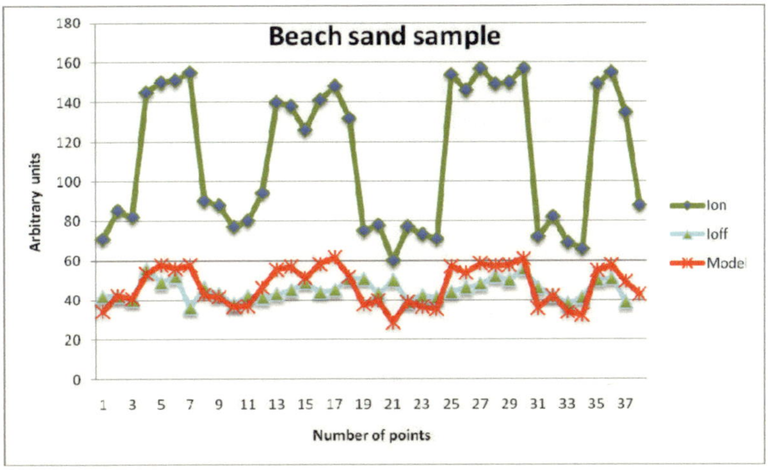

Fig. 10. Beach sand sample points corrected for AGC effect („Model" line, red). (Vain et al., 2010)

5. Conclusions

The intensity data can be used in several applications and therefore has a huge potential in the future studies of the environment. The correction and calibration of ALS intensity plays a crucial part in the usage of that data. This chapter introduces the problems that occur when we manage the intensity values. The atmosphere is an important factor that affects the results, also different flight settings (PRF, flying altitude, scan angle). The targets that can be used as calibrators were introduced (tarps and asphalt).

A new feature in the ALS systems is the AGC. It makes the range measurements more accurate and increases the point density. But if we look at the side of using intensity data with AGC, then there is a problem. AGC has a clear effect on the intensity data that was shown also in this chapter. More information from the manufactures is needed, and also more tests and studies to make the correction model as good as possible.

6. Acknowledgements

The authors would like to thank Prof. Kalev Sepp from the Estonian University of Life Sciences for supporting this work and Harri Kaartinen from the Finnish Geodetic Institute.

7. References

Baltsavias, E. P. (1999). Airborne laser scanning: basic relations and formulas. *ISPRS Journal of Photogrammetry & Remote Sensing*, Vol. 54, pp. 199-214, ISSN 0924-2716.

Chasmer, L., Hopkinson, C., Smith, B. & Treitz, P. (2006). Examining the influence of changing laser pulse repetition frequencies on conifer forest canopy returns. *Photogrammetric Engineering & Remote Sensing*, Vol. 72, No. 12 (December 2006), pp. 1359-1367. ISSN 0099-1112.

Coren, F. & Sterzai, P. (2006). Radiometric correction in laser. *International Journal of Remote Sensing*, Vol. 27, No. 15-16 (August 2006), pp. 3097-17, ISSN 0143-1161.

Honkavaara, E., Merkelin, L., Ahokas, E., Kuittinen, R. & Peltoniemi, J. (2008). Calibrating digital photogrammetric airborne imaging systems in a test field. *The International Archives of the Photogrammetry, Remote Sensing and Spatial Information Sciences,* Vol. 37/B1, pp. 555-560, ISSN 1682-1750.

Höfle, B. & Pfeifer, N. (2007). Correction of laser scanning intensity data : Data and model-driven approaches. *ISPRS Journal of Photogrammetry & Remote Sensing,* Vol. 62, pp. 415-433, ISSN 0924-2716.

Kaasalainen, S., Ahokas, E., Hyyppä, J. & Suomalainen, J. (2005). Study of surface brightness from backscattered laser intensity: calibration of laser data. *IEEE Geoscience and Remote Sensing Letters,* Vol. 2, No. 3 (July 2005), pp. 255-259. ISSN 1545-598X.

Kaasalainen, S., Kukko, A., Lindroos, T., Litkey, P., Kaartinen, H., Hyyppä, J. & Ahokas, E. (2008). Brightness measurements and calibration with airborne and terrestrial laser scanners. *IEEE Transactions on Geoscience and Remote Sensing,* Vol. 46, No. 2 (February 2008), pp. 528-534. ISSN 0196-2892.

Kaasalainen, S., Niittymäki, H., Krooks, A., Koch, K., Kaartinen, H., Vain, A, and Hyyppä, H. (2010). Effect of target moisture on laser scanner intensity. *IEEE Transactions on Geoscience and Remote Sensing,*Vol.48, No.4, pp. 2128–2136. ISSN 0196-2892.

Kim, I. I., McArthur, B. & Korevaar, E. (2001). Comparison of laser beam propagation at 785 nm and 1550 nm in fog and haze for optical wireless communication, In *Proceeding of SPIE, 2414*, pp. 26-37, Boston, November, 2000.

Korpela, I., Ørka, H. O., Hyyppä, J., Heikkinen, V. & Tokola, T. (2010). Range and AGC normalization in airborne discrete-return LiDAR intensity data for forest canopies. *ISPRS Journal of Photogrammetry and Remote Sensing,* Vol. 65, pp. 369-379. ISSN 0924-2716.

Kukko, A., Kaasalainen, S. & Litkey, P. (2008). Effect of incidence angle on laser scanner intensity and surface data. *Applied Optics,* Vol. 47, No. 7 (March 2008), pp. 986-992, ISSN 1559-128X.

Vain, A., Kaasalaine, S., Pyysalo, U., Krooks, A. & Litkey, P. (2009). Use of naturally available reference targets to calibrate airborne laser scanning intensity data. *Sensors,* Vol. 9, pp. 2780-2796 (April 2009), ISSN 1424-8220.

Vain, A., Yu, X., Kaasalainen, S. & Hyyppä, J. (2010). Correcting airborne laser scanning intensity data for automatic gain control effect. *IEEE Geoscience and Remote Sensing Letters,* Vol. 7, No. 3 (July 2010), pp. 511-514. ISSN 1545-598X.

Wagner, W. (2010). Radiometric calibration of small-footprint full-waveform airborne laser scanner measurements: Basic physical concept. *ISPRS Journal of Photogrammetry and Remote Sensing,* In Press.

4

LIDAR Remote Sensing Applications in Automated Urban Feature Extraction

Poonam S. Tiwari and Hina Pande
Indian Institute of Remote Sensing (NRSC)
P Box- 135, Dehradun -248001
India

1. Introduction

Terrain elevation is a key input in numerous environmental applications and its quality plays a critical role in understanding various earth surface processes (Kenward et al., 2000; Clarke and Burnett, 2003; Neeson et al., 2008). Laser altimetry, also called lidar (light detection and ranging), is a frontier remote sensing technology for mapping earth surface elevation with high vertical resolving ability (Sun et al., 2003; Lefsky et al., 2005; Hofton et al., 2006; Chen, 2007; Simard et al., 2008).In recent years LIDAR has emerged as an important mapping tool and is finding a niche in GIS. LIDAR is a relatively new technological tool that can be used effectively in geospatial decision making. LIDAR is an acronym for Light Detection And Ranging and in some literature it is referred to as laser altimetry. A LIDAR system is composed of a laser scanning system, global positioning system (GPS), and an inertial measuring unit (IMU).

Airborne *LiDAR* has established itself as a standard method for the acquisition of precise and reliable digital elevation data. The main application of airborne *LiDAR* technology is topographic surveying. Airborne LiDAR has found application in an increasing number of mapping and geo-data acquisition tasks. Apart from terrain information generation, applications such as automatic detection and modeling of objects like buildings, roads or vegetation for the generation of 3-D city models have been extensively explored. In the present chapter airborne Lidar data has been used for semiautomatic detection of buildings and roads.

2. Principle of LiDAR

2.1 Laser

Laser (Light Amplification by the Stimulated Emission of Radiation) light is highly monochromatic, coherent, directional, and can be sharply focused.

Laser Scanning is an active remote sensing technique which provides range measurements between the laser scanner and the illuminated object (most commonly earth topography). Laser altimeter produces short or continuous pulses of laser light, which is captured by a telescope after bouncing back when they are intercepted by a target. Laser system enables day and night observation and also range measurement in textureless areas. Observations from single direction are sufficient, unlike photogrammetric technique where at least two

perspective views are required. In addition it provides data with high vertical accuracy. While LIDAR is an active system that can be, theoretically, used 24 hours a day, it cannot be used above cloud cover or when fog, smoke, mist, rain, or snow storms are present. Additionally, high winds and turbulence will cause problems with the inertial system.
Mostly laser systems are Nd:YAG emitting in NIR (1064nm) wavelength in a narrow spectral width (0.1-0.5nm). Some systems emit at 810 nm (ScaLARS), 900 nm (FLI-MAP), 1540 nm (TopoSys, Riegl). Laser systems generally emit in one wavelength only however bathymetric lasers emit at1064 and 532 nm, to measure both water surface and water bottom.

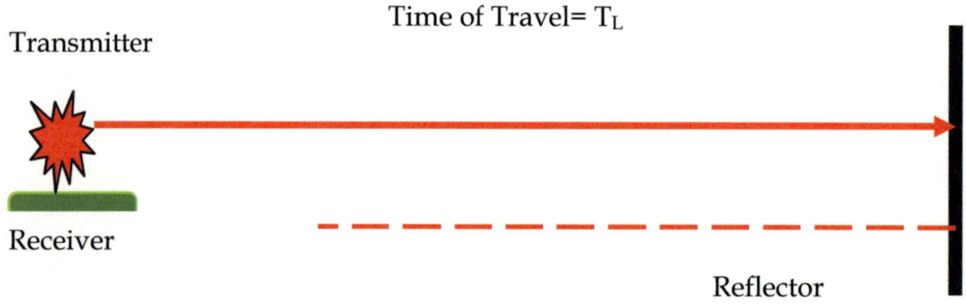

Fig. 1. Principle of Laser

2.2 Basic components and functioning

Typical system components include a position and orientation recording unit, a laser measurement system and a control and data recording unit.
The pulsed laser is optically coupled to a beam director which scans the laser pulses over a swath of terrain, usually centred on, and co-linear with, the flight path of the aircraft in which the system is mounted, the scan direction being orthogonal to the flight path. The round trip travel times of the laser pulses from the aircraft to the ground are measured with a precise interval timer and the time intervals are converted into range measurements. The position of the aircraft at the epoch of each measurement is determined by a phase difference kinematic GPS. Rotational positions of the beam director are combined with aircraft roll, pitch, and heading values determined with an inertial navigation system (INS), and with the range measurements, to obtain vectors from the aircraft to the ground points. When these vectors are added to the aircraft locations they yield accurate coordinates of points on the surface of the terrain.

2.3 Lidar system classification

Laser systems are categorized on the basis of:

1. Method of recording the return pulse:
   a. Discrete
   b. Full Waveform
2. Size/Area of illuminated spot
   a. Small footprint (few cm)
   b. Medium to large footprint (tens of m)
3. Sampling Rate/ Scanning pattern

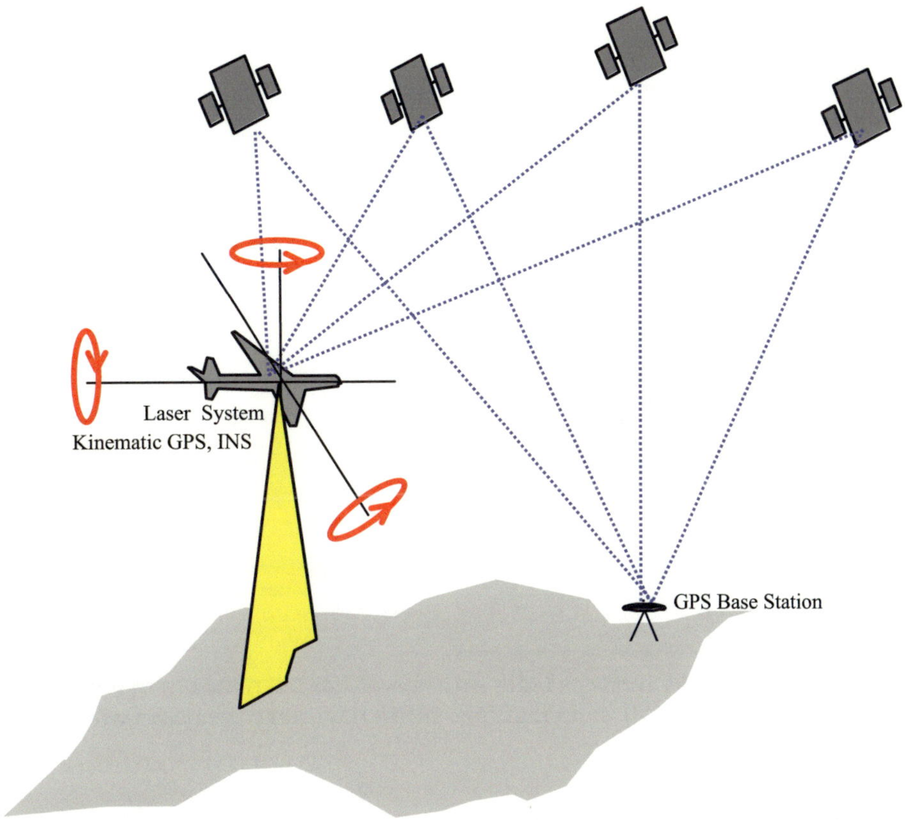

Fig. 2. Components of a Laser Altimeter

4. Pulsed and continuous wave
5. Platform Based
    a. Aerial
    b. Satellite
    c. Terrestrial

Discrete Lidar system measures the time of flight of Laser pulse until its first or last echo arrives. They can only differentiate a first, several intermediate and a last echo. Full waveform systems on the other hand sample the return signals very rapidly and store the entire echo waveform. Small footprint lidar illuminates a few cm area on the ground whereas the footprint size for medium to large footprint systems cover tens of meters. Small footprint systems provide high density data and an accurate altimetric description within the diffraction cone. However mapping large areas require extensive surveys. Besides small footprint systems often miss tree tops. Large footprint systems increase the probability to both hit ground and canopy top (Mallet & Bretar 2009). Pulsed systems measures the round trip time of the light pulse from the laser to the target and back to the receiver. Continuous wave systems use phase difference between transmitted and received signal for range measurements.

Apart from range, several other physical properties like width of backscattered echo, amplitude and cross section of backscatter are also measured.

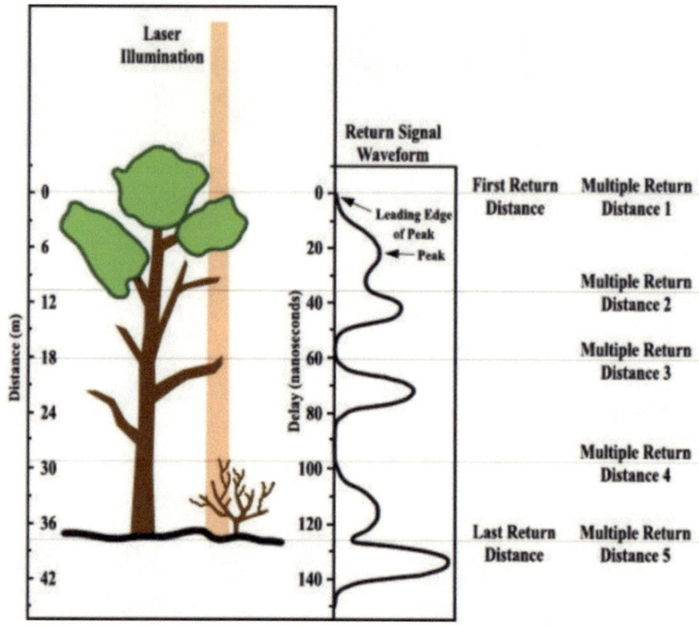

Fig. 3. **Discrete and Full Waveform Lidar** (In a waveform lidar, the entire return pulse is digitized and recorded. In a discrete multiple-return lidar, only the peaks would be recorded) Source: ASPRS

2.4 Lidar accuracy and error sources

The various sensor components fitted in the LiDAR instrument possess different precision. For example, in a typical sensor the range accuracy is 1-5 cm, the GPS accuracy 2-5 cm, scan angle measuring accuracy is 0.01°, INS accuracy for pitch/roll is <0.005° and for heading is < 0.008° with the beam divergence being 0.25 to 5 mrad. However, the final vertical and horizontal accuracies that are achieved in the data are of order of 5 to 15 cm and 15-50 cm at one sigma. The final data accuracy is affected by several sources in the process of LiDAR data capture (Lecture notes from International School on LiDAR Technology, IIT Kanpur, India2008). A few important sources are listed below:

1. Error due to sensor position due to error in GPS, INS and GPS-INS integration.
2. Error due to angles of laser travel as the laser instrument is not perfectly aligned with the aircraft's roll, pitch and yaw axis. There may be differential shaking of laser scanner and INS. Further, the measurement of scanner angle may have error.
3. The vector from GPS antenna to instrument in INS reference system is required in the geolocation process. This vector is observed physically and may have error in its observation. This could be variable from flight to flight and also within the beginning and end of the flight. This should be observed before and after the flight.
4. There may be error in the laser range measured due to time measurement error, wrong atmospheric correction and ambiguities in target surface which results in range walk.
5. Error is also introduced in LiDAR data due to complexity in object space, e.g., sloping surfaces leads to more uncertainty in X, Y and Z coordinates. Further, the accuracy of laser range varies with different types of terrain covers.

6. The divergence of laser results in a finite diameter footprint instead of a single point on the ground thus leading to uncertainty in coordinates.

3. Discrete airborne LIDAR for urban feature extraction

3.1 Road extraction

Automatic road extraction from remotely sensed images has been an active research in urban area during last few decades. But it is quite difficult in urban environment due to mix of natural and man-made features. An integrated approach of airborne laser scanning (ALS)/altimetry and high-resolution data has been used to extract road and differentiate them from flyovers. An integration of high-resolution satellite data and airborne laser scanning data offer exciting possibilities for automatically and accurate road extraction (Zhan et al., 2003). The urban area chosen for addressing the study problem is Amsterdam in The Netherlands. Aerial LiDAR and IKONOS PAN and XS dataset were utilized for road extraction.

Multiresolution segmentation allows the segmentation of an image into a network of homogeneous image regions at any chosen resolution. These image object primitives represent image information in an abstracted form serving as building blocks and information carriers for subsequent classification, beyond purely spectral information, image objects contain a lot of additional at-tributes which can be used for classification: shape, texture and - operating over the network - a whole set of relational / contextual information. Knowledge Base provides an intelligent and integrated knowledge solution that promotes easy image information extraction. Various rules were applied for classifying road segments. Spectral, shape, textural and contextual properties of the objects were utilized for formulating rules for road extraction (Fig 4a). Lidar Data was processed to generate Digital surface model. Lidar point cloud was filtered to classify ground and non ground points (Fig 4b). Height threshold was applied to extracted roads using lidar point cloud to separate out ground road and elevated roads (Fig 4c).

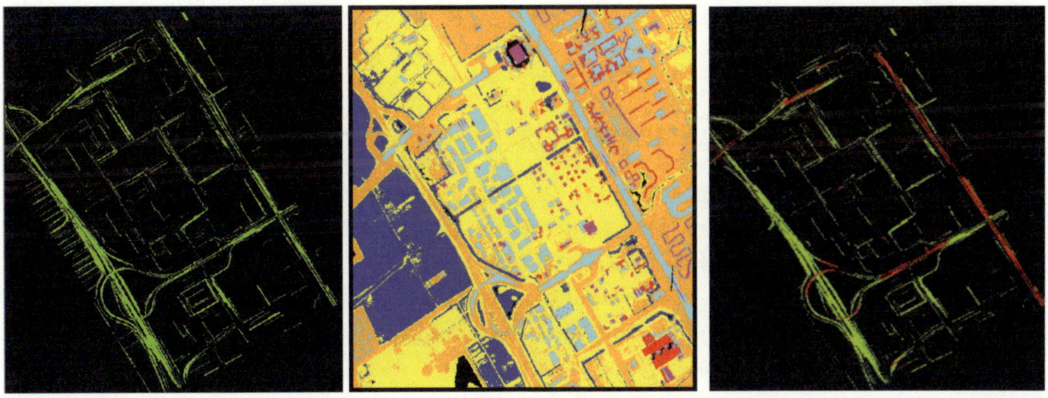

(a) (b) (c)

Fig. 4. a: Roads extracted with object oriented rule based classification
b: Filtered Lidar Data
c: Ground and Elevated Roads

To evaluate the success of the results of road extraction, the extracted roads were subjected to an accuracy assessment. Accuracy is the degree of conformity with a true reference. Wiedemann et al., (1998) have suggested that the quality of each road can be evaluated by contemplating the completeness and correctness of the extracted road network. The contiguity of the two road networks (reference and extracted road) may lack in coherence, hence the accuracy cannot be judged directly. Buffer method can be used for accuracy assessment of automatically extracted data with respect to reference data. For assessment of road extraction accuracy, a buffer of constant predefined width is constructed around the reference and extracted road data. (Fig 5).
The parameters are defined and calculated using these formulas:
Completeness is the ratio of the correctly extracted records to the total no of relevant records within the ground truth data, and can be calculated as-

$$\textbf{\textit{Completeness}} = \frac{\text{length of matched reference}}{\text{length of reference}} \tag{1}$$

$$\text{Completeness} \in \{0; 1\}$$

Correctness is the ratio of the number of relevant records extracted to the total number of the relevant and irrelevant record retrieval, and can be calculated as-

$$\textbf{\textit{Correctness}} = \frac{\text{length of matched extraction}}{\text{length of extraction}} \tag{2}$$

$$\text{Correctness} \in \{0; 1\}$$

Quality is the measure of final result combining completeness and correctness, and can be calculated as-

$$\text{Quality} = \frac{(\text{length of matched extraction})}{(\text{length of extraction} + \text{length of unmatched reference})} \tag{3}$$

$$\text{Quality} \in \{0; 1\}$$

The optimum value for completeness, correctness and quality is 1.
In the case of road, completeness of 76.26% and correctness of 50.78% was achieved. The overall accuracy observed for the road extraction was 43.85%. With this result it was evaluated that more than 3/4th part whole of the road network was completely extracted with the correctness of nearly half of the total road network.
While in the case of elevated road/ flyover, a completeness of 88.61% and correctness of 85.71%. The overall accuracy of the elevated road extraction was examined as 77.21%. The output result shows that most of the part of elevated road network was completely and correctly extracted to the total road network. In the study area, urban scene was highly complex and contained high-rise buildings, open-grounds, medium-rise buildings etc, which mainly affected the accuracy of road extraction. To overcome such problems an integrated approach of high-resolution satellite data and LiDAR data have been studied, which can extract urban road efficiently.

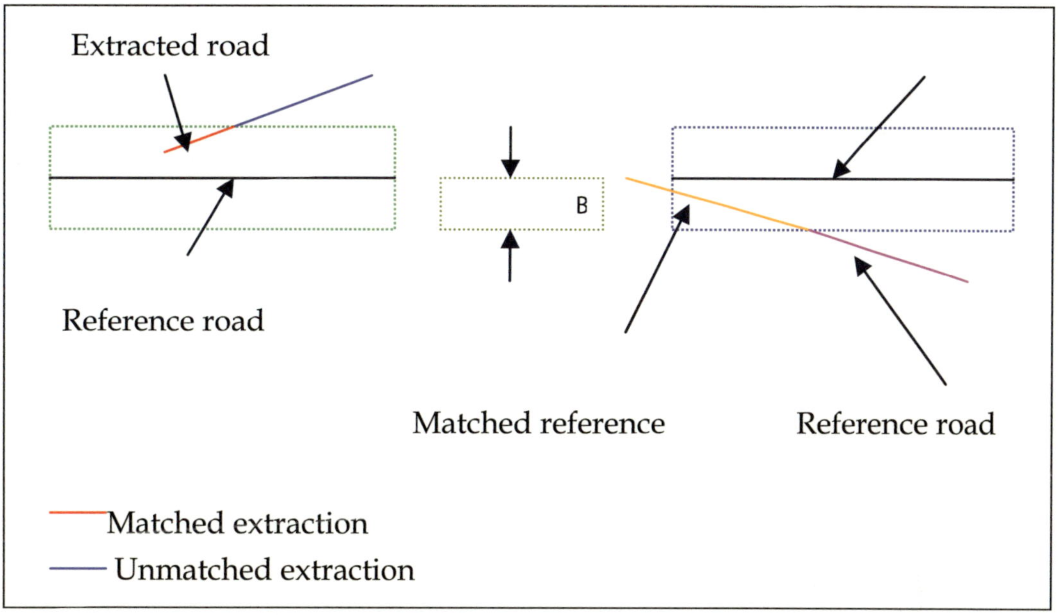

Fig. 5. Accuracy Assessment strategy

3.2 Building extraction

3.2.1 Integrated use of high resolution spatial data and aerial laser data

The automatic building extraction from remotely sensed imagery has attracted much attention during the last decade. Generally the building extraction is based on different kinds of knowledge of buildings: geometry, spatial, spectral, radiometry. Many computer-aided, GIS and remote sensing techniques are currently in use to allow a faster building extraction, development, updating, maintenance and urban planning. Different kinds of techniques are being used for building extraction. Most of them are pixel based, using multi source data alone.

The increasing availability of high spatial resolution satellite images has provided a new data source for building extraction. When compared with the aerial photographs, the high resolution satellite images provide several advantages that include the cost and the accessibility. In the recent studies the spectral reflectance values have been used to detect the buildings *[Lee, S., et al., 2003)*.

High resolution satellites image provides a good basis for reorganization and monitoring of structural changes to map urban details. Higher the resolution of the imagery, more man made features such as buildings, roads, moving objects, etc. can be identified. But even in high resolution image, the feature extraction suffers from the background objects. Airborne laser scanning is an emerging technique to identify the height objects. An integration of high-resolution satellite data and airborne laser scanning data offer exciting possibilities for automatically and accurate building extraction. *(Zhan, Q,. 2003)*. One single sensor technology seems unlikely to produce detailed and varying characteristics of building models. Combining the geometry, photometry, and other sensing sources can compensate for the shortcomings of each sensing technology and appears to be a promising methodology.

The study area selected is a very small part of Amsterdam, but having well planned buildings. When using images characterized by a higher spatial and spectral resolution, it is difficult to obtain satisfactory results using traditional classification methods. Object oriented approach utilizes group of pixels instead of individual pixels for classification process. Objects are created by segmentation which makes this step the most crucial in the methodology used. Segmentation results can be improved by iteratively selecting appropriate parameters like scale, smoothness and compactness to identify various objects in the image. In multi segmentation, different scale parameters were analysed to identify the scale value at which the image could be segmented. If the parameters are not optimally chosen, the features may merge producing complex polygons having parts of adjoining features merged with them hence producing erroneous results. Accuracy of extraction was analysed using the above mentioned parameters. IKONOS fused image and LIDAR data were segmented based on scale, shape, compactness and smoothness. An object-oriented knowledge base was generated to extract building (Fig 6).

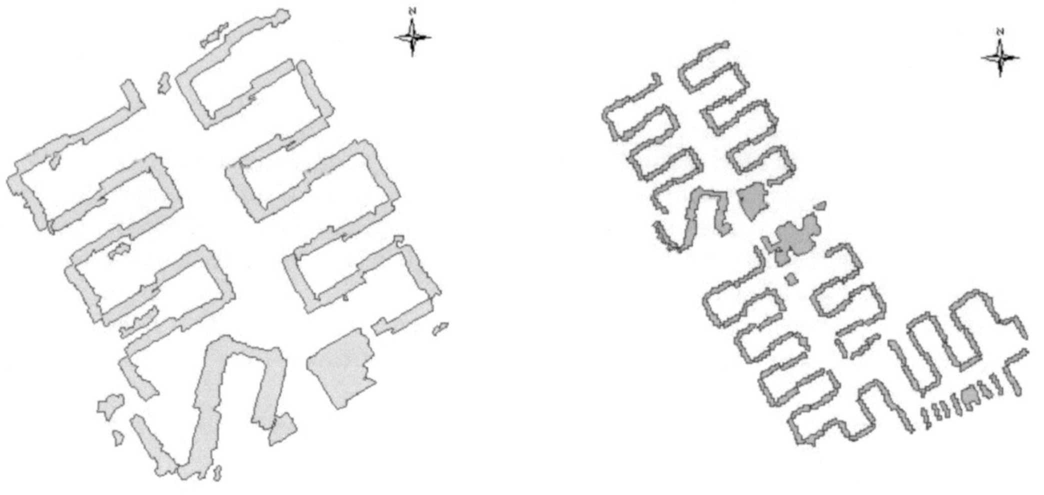

(a) (b)

Fig. 6. Buildings extracted from a: IKONOS b: IKONOS+LiDAR

After extracting the object of interest, it was found that the extracted objects did not show accurate and defined boundary. To remove these defects and regularize the building boundaries, mathematica5l morphological operators were applied (Fig 7).

In addition to completeness, correctness and quality, area and positional accuracy were calculated for the extracted buildings. The area accuracy using IKONOS is 95.05% and the area accuracy for the buildings extracted using IKONOS and LiDAR is 99.84%. In case of qualitative assessment, completeness correctness and quality of the extracted buildings was calculated. For IKONOS data the completeness was 84 %, correctness 100% and quality was found to be 86 %. For IKONOS and LiDAR data completeness is 90.9%, correctness is 100%

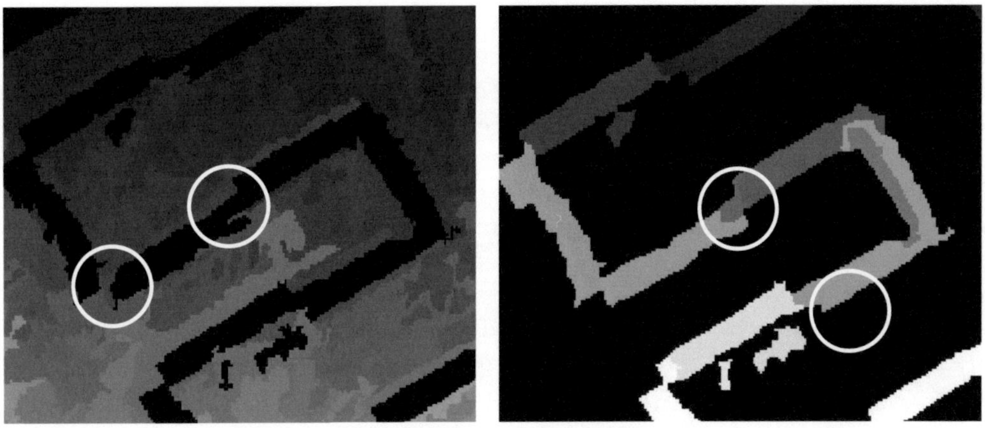

(a) (b)

Fig. 7. a Before using Closing Operator b. After using Closing Operator

and quality 90.9%. The buildings extracted using IKONOS data have well defined building boundaries so the error in X and Y direction are very less. While in LiDAR and IKONOS, the buildings extracted have not defined building boundaries. The edges and the corners of the extracted building boundaries using LiDAR and IKONOS are not matching. Thus, there is more positional error in X and Y direction in the extracted buildings. The results indicate a definite improvement in accuracy by utilizing complementary datasets. In high contrast areas buildings extracted using only high resolution intensity data gave satisfactory results, while in low contrast and shadow areas incorporating range data is necessary for proper boundary detection.

The present study indicates the usefulness of integrating LiDAR data for building extraction. It was found that in low contrast areas where IKONOS image was not giving good results, integration of LiDAR has improved the correctness and quality of the extracted data to a great extent. Though positional accuracy was higher with IKONOS dataset as it gives proper edges of the buildings.

3.2.2 Use of laser range and height texture clues

Detecting buildings directly from the raw LiDAR data is not a straightforward problem. This is due to the ambiguity of other vertically extended features which are not buildings in the raw data. (Alharthy & Bethel 2002). The raw LiDAR point cloud consists of a mixture of terrain, vegetation, buildings and other natural and man-made structures. Different types of objects require different methods for modeling, analyses and visualization. An inherent source of information of *LiDAR* point data is the analysis of height texture defined by local variations of height. Depending on the type of sensor used, objects to be classified show different behaviour in texture measures derived from these variations.

Most of the previous work in classification of aerial LiDAR data has concentrated on unsupervised clustering on a smaller number of classes often resulting in coarse classification while a few have attempted parametric classification with or without segmentation. [Arefi et al]

The aim of the work presented in the following is to classify raw laser scanner data using height texture measures. Texture is qualitatively and quantitatively defined by height, variation of height in local windows and measures. When a small-area patch has wide variation of gray level primitives, the dominant property is texture (Haralick and Shapiro, 1992). A number of texture measures will be defined and discussed. These texture measures are used as bands in a classification of the dataset.

Following texture measures for the purpose of this study were calculated:

Data range: It is the difference between minimum and maximum pixel value in a window around a pixel. It will be zero in homogenous area such as on flat roofs or streets, small on tilted roofs and large within trees.

Variance: The variance of height in a window around a pixel will show similar characteristics as the data range, with somewhat different behaviour concerning single outliers.

$$Variance = \sum \frac{(x_{ij} - M)^2}{n-1} \tag{4}$$

where: x_{ij} = DN value of pixel ij , n = number of pixels in a window, M= mean of moving window

Contrast: It is a measure of the amount of local variation in the Image. It increases exponentially as (i-j) increases. Contrast increases at the edges.

$$\sum_{i,j=0}^{N-1} P_{ij}(i-j)^2 \tag{5}$$

Dissimilarity: Similar to "Contrast", but increases linearly. High if the local region has a high contrast.

$$\sum_{i,j=0}^{N-1} P_{ij}(i-j) \tag{6}$$

i is row and j is column number, i,j is value in the cell i,j of matrix, P_{ij} is normalized value in cell i,j, N is number of rows or columns

Correlation: Measures the linear dependency of grey levels of neighbouring pixels.

$$\frac{\sum_{i,j=0}^{N-1} P_{ij}(i-\mu_i)(i-\mu_j)}{(\sigma_i^2)\sqrt{(\sigma_i^2)}} \tag{7}$$

σ is standard deviation

Variance image highlighted the building and tree boundaries. Data range, Contrast and Dissimilarity gave similar results. Correlation highlighted the building blobs and tree pixels. Since the study aimed at automatically extracting the building footprints, hence correlation was found useful. The original Height data, Variance and Correlation were stacked in form of bands to give a composite image. An analysis of the results obtained from the classification using different combinations of the texture measures discussed showed that good results are obtained when using these three bands, which were further processed to highlight the desired characteristics of buildings

Both unsupervised classification and supervised classification techniques were applied to these three bands. While unsupervised classification did not lead to a satisfactory segmentation, good results were obtained with maximum likelihood classification.

After training the ML algorithm quantitatively evaluates the variance and correlation of the class spectral response patterns when classifying all pixels in the image. The value of the discrimination function of a given pixel, being a member of a particular class, is computed. The pixel is then assigned to the class with the largest probability.

The classified results (Fig 8a) gave objects protruding from the terrain which included both buildings and tree pixels. The differentiation between building and tree classes was based on spectral values. NDVI was calculated and was used as an input for rule based expert classification (Fig 8 b).However few tree pixels were not removed as their spectral values were effected due to factors like shadows etc. The remaining scattered non building pixels were removed by applying morphological filters (Fig 8 c). The improvement achieved by post processing with opening and closing can be observed. The small holes are filled, few isolated pixels are removed or associated with other objects and the edges are smoother and closer to the shape of the buildings than before post processing. However in places few building pixels in close proximity to tree pixels have also been eliminated.

Visual interpretation of the IKONOS image was carried out for use as reference in error evaluation of the extraction process. Accuracy assessment was carried out and a confusion matrix was generated. The error matrix obtained is shown in Table1.

An overall accuracy of 91% was achieved. The error of commission for class buildings was found to be 9% that means 9% pixels which belong to other classes have been wrongly classified into buildings, whereas the error of omission was 6% which indicated that 6% pixels belonging to class buildings have been omitted in the final classification results.

The investigations in the present study have shown that the detection of buildings in laser range data is feasible and indicates a promising quality by just using standard remote sensing and image processing tools.

	BUILDINGS	*TREES*	*OTHERS*	*TOTAL*
BUILDINGS	32	3	0	35
TREES	2	9	1	12
OTHERS	0	1	30	31
TOTAL	34	13	31	78

Table 1. Error Matrix for knowledge based classification

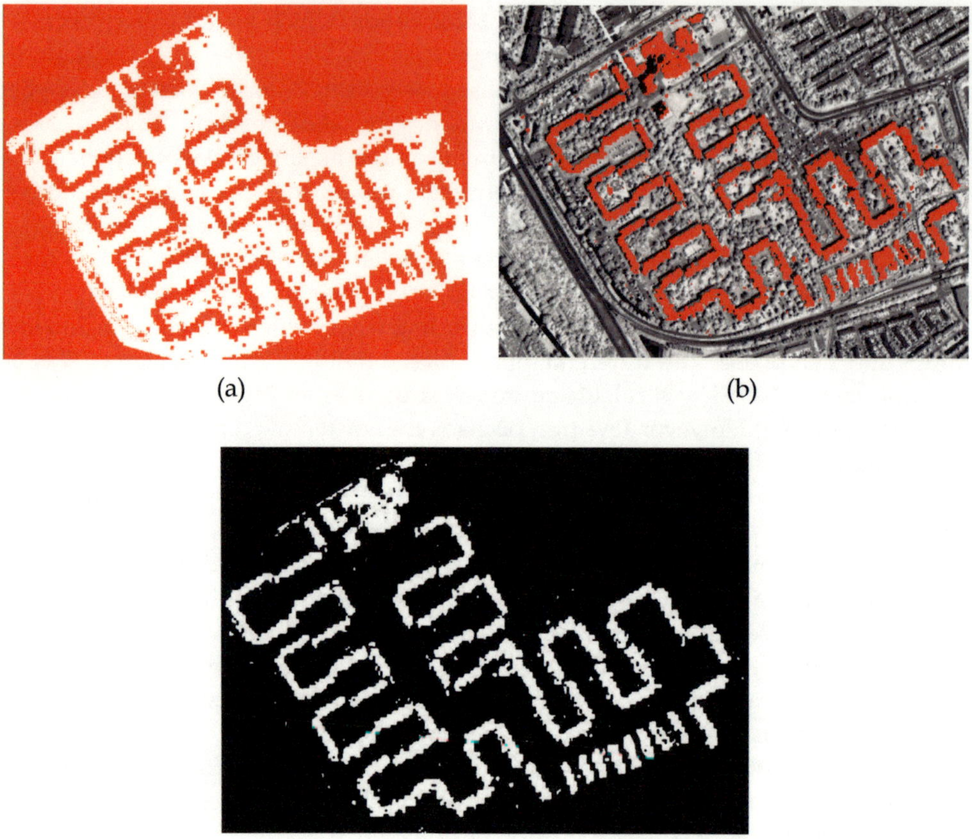

(a) (b) (c)

Fig. 8. a: Supervised classification results, b: Knowledge based classification c: After noise removal

4. Conclusions

In high resolution satellite data, urban features are no longer point objects but polygons with prominent shadows and other associated features. Shadows and surrounding objects give different manifestations during the feature extraction. Sometimes the roads parallel to the buildings and the trees mix up and pose a hindrance to the building extraction. Since numerous factors influence the brightness value of the image making it difficult to separate the desired information. To overcome this problem, multiple data sources have been exploited to compensate for these disadvantages. For this purpose LiDAR data has been used as an attractive supplement together with high resolution intensity data due to high vertical accuracy and high point density. LiDAR data have several advantages of feature localization and planer patch extraction compared to image dataset. On the contrary, high resolution imagery provides more accurate break lines information than LiDAR data. Moreover multispectral imagery is useful to identify and classify objects, such as building and vegetation. Thus, it is proposed to combine LiDAR data and high resolution satellite images for the urban feature identification and detection.

5. References

Alharthy. A, Bethel J., http://www.isprs.org/commission3/proceedings02/papers/ paper061.pdf accessed on 20-08-2006

Arefi. H., Hahn, M. and Lindenberger, J., http://www.igf.uni-osnabrueck.de/mitarbeiter/schiewe/papers/32.pdf accessed on 25-02-2005

C. Mallet and F. Bretar, "Full-waveform topographic lidar: State-of-the-art," *ISPRS Journal of Photogrammetry & Remote Sensing*, vol. 64, no. 1, pp. 1-16, 2009.

Chen, Q., 2007. Airborne lidar data processing and information extraction. Photogrammetric Engineering & Remote Sensing 73 (2), 109_112.

Clarke, S., Burnett, K., 2003. Comparison of digital elevation models for aquatic data development. Photogrammetric Engineering & Remote Sensing 69 (12), 1367_1375.

Haralick, R., Shapiro, L., 1992. Computer and Robot Vision, volume 1. Addison-Wesley Publishing Company.

Hofton, M., Dubayah, R., Blair, J.B., Rabine, D., 2006. Validation of SRTM elevations over vegetated and non-vegetated terrain using medium footprint lidar. Photogrammetric Engineering & Remote Sensing 72 (3), 279_285.

Kenward, T., Lettenmaier, D.P., Wood, E.F., Fielding, E., 2000. Effects of digital elevation model accuracy on hydrologic predictions. Remote Sensing of Environment 74 (3), 432_444.

Lee, S., Shan, J. and Bethel., J. S., 2003. Class-guided building extraction from IKONOS imagery. PE&RS, Vol. 69, No.2, pp. 143-150.

Lefsky, M.A., Harding, D.J., Keller, M., Cohen, W.B., Carabajal, C.C., Del Bom Espirito-Santo, F., Hunter, M.O., de Oliveira Jr., R., 2005. Estimates of forest canopy height and aboveground biomass using ICESat. Geophysical Research Letters 32, L22S02. doi:10.1029/2005GL023971.

Neeson, T.M., Gorman, A.M., Whiting, P.J., Koonce, J.F., 2008. Factors affecting accuracy of stream channel slope estimates derived from geographical information systems. North American Journal of Fisheries Management 28 (3), 722_732.

Simard, M., Rivera-Monroy, V.H., Ernesto Mancera-Pineda, J., Castañeda-Moya, E., Twilley, R.R., 2008. A systematic method for 3D mapping of mangrove forests based on shuttle radar topography mission elevation data, ICEsat/GLAS waveforms and field data: Application to Ciénaga Grande de Santa Marta, Colombia. Remote Sensing of Environment 112 (5), 2131_2144.

Sun, G., Ranson, K.J., Kharuk, V.I., Kovacs, K., 2003. Validation of surface height from shuttle radar topography mission using shuttle laser altimeter. Remote Sensing of Environment 88 (4), 401_411.

Wiedemann, C.; Heipke, C.; Mayer, H.; Jamet, O. (1998), Empirical evaluation of automatic extracted road axes. In Bowyer, K.; Philips, P. (Eds.), Empirical evaluation techniques in computer vision. IEEE Computer Society, Los Alamitos, pp.172-187.

Zhan, Q. (2003), A hierarchical object-based approache for urban land use classification from Remote Sensing data. PhD thesis, ITC.

5

The Basics of Confocal Microscopy

Vineeta Rai and Nrisingha Dey*
Institute of Life Sciences, Laboratory of Plant Biotechnology,
Dept. of Gene Function & Regulation,
Bhubaneswar (Orissa)
India

1. Introduction

Confocal microscopy is a powerful tool that creates sharp images of a specimen that would otherwise appear blurred when viewed under a conventional microscope. This is achieved by excluding most of the light from the specimen that is not from the microscope's focal plane. The image thus obtained has less haze and better contrast than that of a conventional microscope and represents a thin cross-section of the specimen (Diaspro, 2002; Hibbs, 2004; Matsumoto, 2002; Muller, 2002; Paddock, 1999; Pawley, 1995). Laser scanning confocal microscopy has become an invaluable tool for imaging thin optical sections in living and fixed specimens ranging in thickness up to 100 micrometers. In fact the confocal microscope is often capable of revealing the presence of single molecule (Peterman, Sosa et al., 2004). Modern confocal microscopes have kept the key elements of Minsky's design: the pinhole apertures and point by point illumination of the specimen. During the 1990s advances in optics and electronics afforded more stable and powerful lasers, high-efficiency scanning mirror units, high-throughput fiber optics, better thin film dielectric coatings, and detectors having reduced noise characteristics. In addition, fluorochromes that were more carefully matched to laser excitation lines were beginning to be synthesized (Mason, 1999). Coupled to the rapidly advancing computer processing speeds, enhanced displays, and large-volume storage technology emerging in the late 1990s, the stage was set for a virtual explosion in the number of applications that could be targeted with laser scanning confocal microscopy. Modern confocal microscopes can be considered as completely integrated electronic systems (Spring and Inoue, 1997) where the optical microscope plays a central role in a configuration that consists of one or more electronic detectors, a computer (for image display, processing, output, and storage), and several laser systems combined with wavelength selection devices and a beam scanning assembly. In most advanced systems, integration between the various components is so thorough that the entire confocal microscope is often collectively referred to as a digital or video imaging, capable of producing electronic images. These microscopes are now being employed for routine investigations on molecules, cells, and living tissues that were not possible just a few years ago (Goldman and Spector, 2005).
Modern instruments are equipped with **1)** 3-5 laser systems controlled by high-speed acousto-optic tunable filters (AOTFs), which allow very precise regulation of wavelength and excitation intensity; **2)** Photomultipliers that have high quantum efficiency in the near-ultraviolet, visible and near-infrared spectral regions, these microscopes are capable of

examining fluorescence emission ranging from 400 to 750 nanometers; **3)** Spectral imaging detection systems which further refine the technique by enabling the examination and resolution of fluorophores with overlapping spectra as well as providing the ability to compensate for autofluorescence; **4)** Spatial pinhole that eliminate out-of-focus light in specimens that are thicker than the focal plane thus increases optical resolution and contrast of a micrograph using point illumination. (Murphy, 2001; Wilson and Carlini, 1988; Pawley, 2006) Besides allowing better observation of fine details, it is also possible to build three-dimensional (3D) reconstructions of a volume of the specimen by assembling a series of thin slices taken along the vertical axis using confocal microscope (Al-Kofahi, Can et al., 2003).
In the recent years confocal microscopy has gained a tremendous popularity in the scientific and industrial communities owing to its great number of applications in life sciences, semiconductor inspection and material science (Conn, 1999; Diaspro, 2002; Gu, 1996; Hibbs, 2004; Mason, 1999; Masters, 1996; Matsumoto, 2002; Muller, 2002; Paddock, 1999; Pawley, 1995; Sheppard and Shotton, 1997; T.R.Corle and G.S.Kino, 1996; Wilson, 1990).

Principle

Current instruments are highly modified from the earliest versions, but the principle of confocal imaging that was developed by Marvin Minsky is employed in all modern confocal microscopes (Minsky 1961; 1988). The image in a confocal microscope is achieved by scanning one or more focused beams of light, usually from a laser or arc-discharge source, across the specimen. This point of illumination is brought to focus in the specimen by the objective lens, and laterally scanned using some form of scanning device under computer control. The sequences of points of light from the specimen are detected by a photomultiplier tube (PMT) through a pinhole (or in some cases, a slit) (Fellers and Davidson, 2007), and the output from the PMT is built into an image and displayed by the computer.
Although unstained specimens can be viewed using light reflected back from it, most of the times specimens are labeled with one or more fluorescent probes. A laser is used to provide the excitation light (in order to get very high intensities). The laser light reflects off a dichroic mirror hits two lenses mounted on motors. These mirrors scan the laser across the sample, dye in the sample fluoresces and the emitted light gets descanned by the same mirrors that are used to scan the excitation light from the laser. The emitted light passes through the mirror and is focused onto the pinhole. The light that passes through the pinhole is measured by a detector, i.e. a photomultiplier tube. Thus there is never a complete image of the sample at any given instant; only one point of the sample is viewed. The detector is attached to a computer which builds up the image, one pixel at a time.
Earlier, a 512x512 pixel image formation could be done probably 3 times a second due to the limitation in the scanning mirrors. Subsequently, to speed up scanning a special Acoustic Optical Deflector (AOD) was used in place of one of the mirrors. AOD uses a high-frequency sound wave in a special crystal to create a diffraction grating, which deflects the laser light. By varying the frequency of the sound wave, the AOD changes the angle of the diffracted light, thus allowing a quick scan leading to 512x480 pixel images 30 times per second. If one looks at a smaller field of view, then the process can be even faster (up to 480 frames per second).
The basic difference between the two techniques are as follows: (Lichtmann, 1994; Murray, 2005; Sandison and Webb, 1994; Swedlow, Hu et al., 2002; White, Amos et al., 1987; Wilson, 1989; Wright and Wright, 2002).

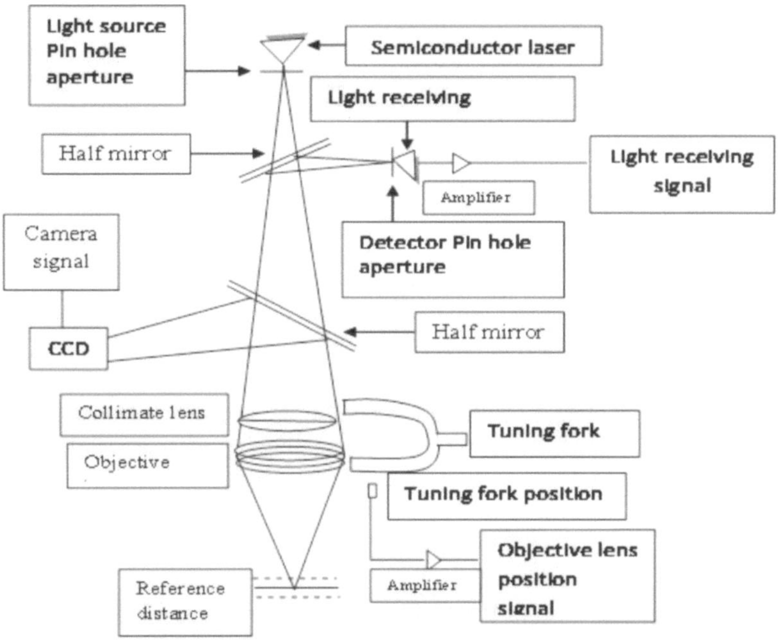

Fig. Schematic representation of basic principle of confocal microscope

Sl. No.	Characteristics	Widefield fluorescence microscope	Confocal microscope
1.	Fundamental setup	Comprises of fluorescence light Source, solid-state charge-coupled devices (**CCDs**) and fluorescence filter cube.	Comprises of laser, light source pinhole aperture, photomultiplier detector, detector pinhole aperture and dichrome mirror
2.	Principle	The entire specimen is bathed with light.	Only a single point is illuminated at a time to avoid unwanted scattering of light.
3.	Resolution	Low	High
4.	Regions of out-of-focus information	Blurred and large	Reduced
5.	Optical resolution in z	2-3 μm	0.5 μm
6.	Signals from optical sections	Many signals cannot be seen separately.	Improved z-resolution allow for more accurate signal discrimination.
7.	Acquisition of 3-D data	Not possible	Possible
8.	Cost	Relatively inexpensive	Expensive
9.	Hazards	No any	High intensity laser irradiation is hazardous.

Components of confocal microscope
1. Light source (laser system)
2. Filters
3. Acousto optical devices
4. Scanner
5. Detector (PMT, APD)
6. Pin hole

2. Laser systems

LASER- Light Amplification by the Stimulated Emission of Radiations can be defined as materials with altered distribution of atoms is such that there are more excited atoms, ready to emit energy. The process of altering the energy distribution of the atom so that most of them are in a high energy state, population inversion. Lasers did not become a reality, until methods to achieve population inversion become feasible in 1950s. Mercury and xenon light sources, the most common light sources for microscopes are found to be too weak for confocal system, hence it uses different type of lasers.

Laser types
Depending on the materials used as the medium lasers, are grouped into three categories:
- Gas lasers
- Solid state lasers
- Semi-conductor lasers

Laser safety
The two major concerns in safe laser operation are exposure to the beam and the electrical hazards associated with high voltages within the laser and its power supply. While there are no known cases of a laser beam contributing to a person's death, there have been several instances of deaths attributable to contact with high voltage laser-related components. Beams of sufficiently high power can burn the skin, or in some cases create a hazard by burning or damaging other materials. But the primary concern with regard to the laser beam is potential damage to the eyes, which are most sensitive to light.

3. Filtering devices

In fluorescence microscope filtering devices are used to separate light beams on the basis of their wavelengths. Four different types of filters are used to selectively transmit or block a desired range of wavelengths.
a. Short pass filters - they cut off wavelengths longer than a certain wavelength e.g. heat filters are used to exclude infra red light to reduce specimen heating by illumination.
b. Long pass filters -e.g. fluorescent filters that transmit light of longer than a certain wavelength.
c. Band pass filter- that transmit light only between a cut-on and cut-off wavelength, especially useful when one is trying to image signals from more than one fluorochrome simultaneously.
d. Dichrome mirrors –that separates the emitted light from the excited light.

Now-a-days these filters are rapidly replaced by active optical devices such as acousto-optical devices.

4. Acousto-optical devices

These devices work on the principle of acousto-optic effect. In general, acousto-optic effects are based on the change in refractive index of a medium due to the presence of sound waves in that medium. Sound waves produce a refractive index grating in the material and this grating is seen by the light wave. These variations in the refractive index, due to the pressure fluctuations, may be detected optically by refraction, diffraction, interference and reflection effects. The acousto-optic effect is extensively used in the measurement and study of ultrasonic waves. However, the growing principle area of interest is in acousto-optical devices for the deflection, modulation, signal processing and frequency shifting of light beams. This can be attributed to the increasing availability and performance of lasers, which have made the acousto-optic effect easier to observe and measure. Technical progresses in both crystal growth and high frequency piezoelectric transducers have brought valuable benefits to acousto-optic components' improvements. Some materials displaying acousto-optic effect include fused silica, arsenictrisulfide, tellurium dioxide and tellurite glasses, lead silicate, mercury(I) chloride, lead (II)bromide and other materials (Wachmann, 2000; Wachmann, Niu et al., 1996).

Various acousto-optic devices used now-a-days are as follows:

a. Acousto-Optic Tunable Filters (AOTFs)

The integration of opticoelectronic technology into confocal microscopy has provided a significant enhancement in the versatility of the spectral control for a wide variety of fluorescence investigations. AOFT is an adjustable quartz filter that works at frequencies as high as sound i.e. "Acousto". Light that passes the AOFT is diffracted depending on its own wavelength and the wavelength of the ultrasonic wave field. The ultrasonic wave field can be modulated so that the intensities of different laser lines can be changed between 0% to 100% by the software even during the scanning process (Chang, 1995). Several benefits of the AOTF combine to enhance the versatility of the latest generation of confocal instruments, and these devices are becoming ever-increasing popular for control of excitation wavelength ranges and intensity. The primary characteristic that facilitates nearly every advantage of the AOTF is its capability to allow the microscopist control the intensity and/or illumination wavelength on a pixel-by-pixel basis while maintaining a high scan rate. This single feature translates it into a wide variety of useful analytical microscopy

Fig. The AOTF enables you to select the wavelengths (laser lines on/off)

tools, which are even further enhanced in flexibility when laser illumination is employed. The rapid intensity and wavelength switching capability of the AOFT enables sequential line scanning of multiple laser lines in which each excitation wavelength can be assigned a different intensity in order to balance various signals for optimal imaging (Sharma, 2001). The AOFT speed and versatility controls the wavelength and intensity of light that simultaneously or sequentially scan each sample at sufficient speed to accurately monitor dynamic cellular processes (Chen, Mills et al., 2003). The AOFT functions allow the selection of small user-defined specimen areas termed as regions of interest. (ROI) can be illuminated with either greater or lesser intensity and at different wavelengths for precise control in photobleaching techniques, or spectroscopic measurements (Chen, Mills et al., 2003; Day, Periasamy et al., 2001; Wallrabe and Periasamy, 2005). The illumination intensity can only be increased in selected regions for controlled photobleaching experiments but can be attenuated in desired area in order to minimize unnecessary photobleaching. When the illumination area is under AOFT control, the laser exposure is restricted to the scanned area by default and the extremely rapid response of the device can be utilized to provide beam blanking during the fly-back interval of the galvanometer scanning mirror cycle there by limiting unnecessary specimen exposure (Claxton, Fellers et al.). When the AOFT is combined with multiple lasers and software that allow time course control of the sequential observations, time-lapse experiments (4-D imaging) can be designed to acquire data from several different areas in a single experiment, which might, for example, be defined to correspond to different cellular organelles.

Thus development of AOFT has ultimately provided substantial additional versatility to techniques such as FRAP (Lippincott-Schwartz, Altan-Bonnet et al., 2003; Klonis, Rug et al., 2002), FLIP (Phair and Misteli, 2002) as well as uncaging i.e. localized photoactivated fluorescence (Politz, 1999).

b. Acousto-optic Beam Splitter (AOBS)

The AOBS is actually an AOFT used in an imaginative manner to replace the dichoric mirror which is usually used to separate the illumination and detection paths in confocal microscope. In practice, AOBS is programmed so that it leaves most of the fluorescent light undetected and only deflects light at specific laser lens.

The advantages of AOBS over systems that use dichoric mirrors as the main beam splitter are its high efficiency, its lack of moving parts and its flexibility in terms of using new laser lines simply by changing the software. While proper operation of the AOBS depends on the light input being parallel as it passes through the crystal light originating above or below the plane of the focus, one may see slightly different transmission characteristics. Fortunately in confocal operation most of the out focus light is removed at the pin hole.

c. Acousto-optic modulator (AOM)

Optical wave can be modulated by varying the parameters like amplitude, phase, frequency and polarization etc. Besides the acousto-optic interaction also makes it possible to modulate the optical beam by both temporal and spatial modulation. A simple method of modulating the optical beam travelling through the acousto-optic device is by switching the acoustic field on and off. When off, the light beam is undiverted, the intensity of light directed at the Bragg diffraction angle is zero, when switched on Bragg diffraction occurs, the intensity at the Bragg angle increases. Hence, acousto-optic device modulates the output along the Bragg diffraction angle by switching it on and off. In confocal microscope the device acts as a modulator for keeping the acoustic wavelength (frequency) fixed and thereby varying the drive power to vary the amount of light in the deflected beam. There are several limitations associated with

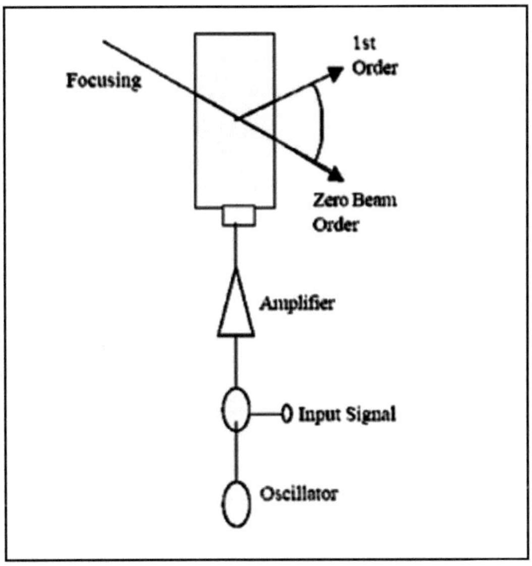

Fig. Schematic representation of AOM

the design and performance of acousto-optic modulators. The acousto-optic medium must be designed carefully to provide maximum light intensity in a single diffracted beam. The time taken for the acoustic wave to travel across the diameter of the light beam limits the switching speed and hence the modulation bandwidth. So to increase the bandwidth the light must be focused to a small diameter at the location of the acousto-optic interaction.

d. Acousto-optic deflectors (AOD)

AODs are essentially the same as acousto-optic modulators (AOMs). In an AOM, only the amplitude of the sound wave is modulated (to modulate the intensity of the diffracted laser beam), whereas in an AOD, both the amplitude and frequency are adjusted (Hosada, Seya et al.). AOD can scan a laser beam at up to 1000 KHz compared to about 500-1000Hz for linear galvanometer scanner and about 4-8 KHz for resonant galvanometer. However, since AOD produces deflection by diffraction and the scan angle depends on the wavelength of light beam, the longer wavelength fluorescent light will be deflected by a different amount on its way back through the crystals and will therefore fail to pass through the pin hole which

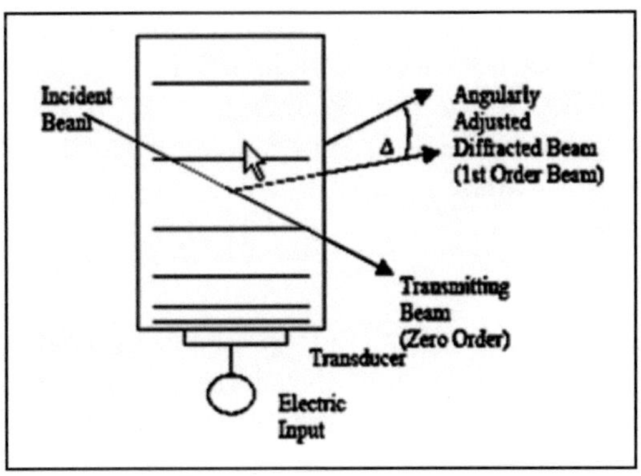

Fig. Schematic representation of AOD

restricts its use. Although this dispersion effect can be theoretically compensated by a chromatic corrections, but they are not practical in case of confocal. However, to overcome this limitation of AOD, a confocal microscope is designed to descan only along the vertical direction (controlled by slow galvanometer), followed by collection of light using a slit rather than a pin hole (Brakenhoff, Blom et al., 1979). The penalty is the reduction in the amount of optical sectioning and a very slight distortion in image caused due to loss of circular symmetry. Nevertheless, it is possible to obtain high quality image using slits (Keller, 1995).

5. Scanners

Confocal imaging relies upon the sequential collection of light from spatially filtered individual specimen points, followed by electronic signal processing and eventually the visual display as corresponding image points. The point-by-point signal collection process requires a mechanism for scanning the focused illuminating beam through the specimen volume under observation which is achieved by scanning the stage or the beam. Thus, the principle scanning variations commonly employed to produce confocal microscope images includes:

- Scanning a laterally translating specimen stage coupled to a stationary illuminating light beam (**stage scanning**)
- Scanning light beam with a stationary stage (**beam scanning).** In modern confocal microscope, two fundamentally different techniques for beam scanning have been developed.
    - Single beam scanning (Wilhelm, Grobler et al., 2003) uses a pair of computer controlled galvanometer mirror to scan the specimen at the rate of approximately 1 frame/sec.
    - Multiple beam scanning confocal microscope that are equipped with a spinning Nipkow disk containing an array of pinholes (Ichihara, Tanaami et al., 1996; Inoue and Inoue, 2002; Nokano, 2002), here the light source is arc-discharge lamp instead of lasers to reduce specimen damage and enhance detection of low fluorescence level during real time image collection. Apart from this multiple beam microscope can readily capture image with an array of detectors, such as CCD camera system, PMT etc. (Chong, Coates et al., 2004).

Scanning the stage has the advantage that the optical system is spatially invariant but beam scanning can be faster and avoids the problem of moving the sample. However, beam scanning has certain limitations as it is usually performed using a feedback controlled galvano-scanner mirror. So, in this case the light from the sample must be de-scanned so that it does not move relative to the pinhole (Murphy, 2001; Webb, 1996). Each technique has performance features that make it advantageous for specific confocal applications, but that limits their usefulness in others.

6. Detectors

As light emitted by fluorophores in the sample is mainly unpolarized, polarization sensitive beam splitters can only separate half of it from polarized excitation light. Instead, the emitted light is either sequentially split by a combination of short pass or long pass dichrome mirror before being passed through a band pass or long pass emission filters to detectors or it strikes a dispersive element and then can be projected to detector. In confocal microscopy fluorescence emission is directed through a pinhole aperture positioned near the image plane to exclude light from fluorescent structures located away from the objective focal plane, thus reducing the amount of light available for image formation. As a result, the exceedingly low light levels most often encountered in confocal microscopy necessitate the use of highly sensitive photon detectors that do not require spatial discrimination, but instead respond very quickly with a high level of sensitivity to a continuous flux of varying light intensity. Thus in confocal microscopy the collection and measurement of secondary emission gathered by the objective is accomplished by several classes of photosensitive detectors, including photomultipliers (PMT) and assigning pseudocolors detector (APD) (Spring, 2001).

a. **Photomultiplier detector:** It measures intensity without spectral information. It actually outputs an electrical impulse to the position logical circuits.
   Principle of signal amplification:
   - Conversion of photons into electrons
   - Multiply electron
   - Signal read out
b. **Assigning pseudo colors detector**: For multiple-channel acquisition it is helpful to assign indexed colors to different grey-scale-images. The detector signals are adjusted by gain and offset such that maximum number of grey level is included in the resulting image (output).
   - **Gain:** Amplifies the input signal by multiplication which results in a higher gray level value; bright features are brought closer to saturation, general image brightness is increased.
   - **Offset:** sets the gray level of a selected background to zero; adjust the darkest features in the image to black.

7. Pinhole

The optical sectioning (the rate at which the detected intensity drops off in the axial direction) capability of a confocal microscope depends on pinhole and its capability to reject out-of-focus light rays i.e. the strength of the optical sectioning depends strongly on the size of the pinhole. Thus one can assume that making the pinhole as small as possible is the best way to enhance optical sectioning. However, as the pinhole size is reduced a large number of photons that arrive at the detector from the specimen is blocked. This may lead to a reduced signal-to-noise ratio. To offset the weaker signal more fluorescence is needed from the specimen but this can be done to a limit (by raising the intensity of the excitation light) on the contrary high intensities can damage the specimen, and in the case of fluorescence, also degrade the fluorophore. Moreover, it has been shown that optical sectioning does not improve considerably with the pinhole size below a limit, which approximates the radius of the first zero of the Airy disk. Thus, a fine approximation is to make the pinhole about the size of the Airy disk. Confocal fluorescence imaging using the pinhole along with the earlier optics significantly reduced the intensity of the emission that reaches the detector. With the advancement in technology, the use of photo detectors for capturing light the sensitivity of confocal imaging is enhanced. Thus, the detector's sensitivity and noise behavior are vitally important. The sensitivity is characterized by the quantum efficiency. That is, the accuracy of the measurement is improved by increasing the number of photons arriving at the detector.

Confocal working

Confocal imaging is accomplished by using a two-step process. First, excitation light that is focused on the specimen by the objective is initially passed through a small aperture, often a slit or pinhole. Alternatively, a very narrow beam of laser light can be introduced into the system via an optical fiber. By conditioning the excitation light this way, the amount of fluorescence not in focus can be controlled or minimized. Second, fluorescence emissions that originate from above or below the plane of focus are blocked by a second aperture or slit in front of the detector. The smaller this second opening, the higher the rejection rates of out-of-focus light and the thinner the optical section. These thin optical sections have greatly improved contrast and axial resolution, but they are obtained at the expense of overall specimen brightness.

Fluorophores

The choice of fluorophores is the most important aspects of fluorescence, confocal microscopy. It is typically influenced by several factors:

- The fluorophores should tag the correct part of the specimen.
- It must be sensitive enough for the given excitation wavelength.
- For living specimens it should not significantly alter the dynamics of the organism
- The specimen should not affect the fluorophore as its chemical environment can affect the position of the peaks of the excitation and emission spectra.

A major problem with fluorophores is that they fade (irreversibly) when exposed to excitation light, a phenomenon known as photobleaching. Although this process is not completely understood, it is believed in some instances to occur when fluorophore molecules react with oxygen and/or oxygen radicals and become nonfluorescent. The reaction can take place after a fluorophore molecule transitions from the singlet excited state to the triplet excited state. Although the fraction of fluorophores that transitions to the triplet state is small, its lifetime is typically much longer than that of the singlet state. This can lead to a significant triplet state fluorophores population and thus to significant photobleaching. Several strategies have been developed to reduce the rate of photobleaching. One method is to merely reduce the amount of oxygen that would react with the triplet excited state. This can be done by displacing it using a different gas. Another method is by the use of free-radical scavengers to reduce the oxygen radicals. Shortening the long lifetime of the triplet excited state has also been shown to be effective. Still other ways include using a high numerical aperture lens to collect more fluorescence light and thus use less excitation light. While photobleaching makes fluorescence microscopy more difficult, it is not always undesirable. One technique that takes advantage of it is fluorescence photobleaching recovery (FPR) or fluorescence recovery after photobleaching (FRAP). It involves exposing a small region of the specimen to a short and intense laser beam, which destroys the local fluorescence, and then observing as the fluorescence is recovered by transport of other fluorophore molecules from the surrounding region. Quantities such as the diffusion co-efficient of the dyed structures can then be determined.

Fluorophores for Confocal Microscopy

Biological laser scanning confocal microscopy relies profoundly on fluorescence as an imaging mode due to its high degree of sensitivity, coupled with the ability to specifically target structural components and dynamic processes in chemically fixed as well as living cells and tissues. In confocal microscope this is achieved by fluorophores. Recent advances in

Sl. No.	Fluorophores	Excitation (nm)	Emission (nm)
1.	5-Hydroxytryptamine (HAD)	370-415	520-540
2.	Acridine yellow	470	550
3.	Acridine orange	500	530
4.	Alexa Fluor488	494	519
5.	Alexa Fluor532	530	555
6.	Alexa Fluor546	554	570
7.	Bodipy 500/510	508	515
8.	Bodipy 530/550	534	554
9.	Cascade Blue	375	410
10.	Coumarin	384	470
11.	CY2	489	506
12.	CY3	548	562
13.	CY5	650	670-700
14.	Dansyl	340	520
15.	DAPI	345	458
16.	DPH	354	430
17.	Erythrosin	529	554
18.	Ethidium Bromide	510	595
19.	FITC	494	518
20.	Fluorescein	495	517
21.	FURA-2	340/380	500/530
22.	GFP	395/489	509
23.	Hoechst 33258	365	480
24.	Hoechst33342	355	465
25.	Laurand	364	497
26.	Lucifer yellow CH	428	535
27.	Nile Red	485	525
28.	Oregon Green 488	493	520
29.	Oregon Green 500	503	522
30.	Oregon Green 514	511	530
31.	Prodan	361	498
32.	Pyrene	341	376
33.	Rhodamine 110	496	520
34.	Rhodamine 123	505	534
35.	Rhodamine 6G	525	555
36.	Rhodamine B	540	625
37.	SITS	336	438
38.	SNARF	480	600/650
39.	Stilbene SITS, SITA	365	460
40.	Texas Re	589	615
41.	TOTO-1	514	533
42.	YOYO-1	491	509
43.	YOYO-3	612	631

Table 1.

fluorophore design have led to superior synthetic and naturally probes, including fluorescent proteins and quantum dots, which exhibit a high level of photo stability and target specificity. Synthetic probes are constructed around synthetic aromatic organic chemicals designed to bind with a biological macromolecule or to localize within a specific structural region, such as the cytoskeleton, mitochondria, golgi apparatus, endoplasmic reticulum, and nucleus. Various other probes are employed for monitoring dynamic processes and localization of environmental variables, including concentrations of inorganic metallic ions, pH, reactive oxygen species, and membrane potential. Still others are useful in monitoring cellular integrity, endocytosis, exocytosis, membrane fluidity, protein trafficking, signal transduction, and enzymatic activity. Besides, fluorescent probes have been widely applied to genetic mapping and chromosome analysis in the field of molecular genetics.

Excitation and Emission Wavelengths of some well-known fluorophores

A list of excitation and emission wavelengths of probes utilized in fluorescence research and applications is listed in Table 1. The positions of the maxima depend upon the solvent utilized in the measurements. The table is only intended to provide an indication of the wavelength range covered by each fluorophores.

Choosing fluorophore combinations for confocal microscopy

During multiple label fluorescence staining protocols, be it be wide-field and confocal microscopy, the sensible choice of probes is major criteria for obtaining the best target signal while simultaneously minimizing bleed-through artifacts.

Spectral bleed-through artifacts

The spectral bleed-through of fluorescence emission (often termed **crossover** or **crosstalk**), that occurs due to very broad bandwidths and asymmetrical spectral profiles exhibited by many of the common fluorophores, is a fundamental problem that must be addressed in both wide-field and confocal microscopy. The phenomenon is usually manifested by the emission of one fluorophore being detected in the photomultiplier channel or through the filter combination reserved for a second fluorophore. Bleed-through artifacts often complicate the interpretation of experimental results while working with a combination of fluorophores, mainly if sub-cellular co-localization of fluorophores is under investigation or quantitative measurements are necessary, such as in resonance energy transfer (**FRET**) and photobleaching (**FRAP**) studies.

Signal-to-Noise considerations

Confocal microscope can also be used for quantitative estimation. For quantitative assessment of imaging modes utilizing digital microscopy techniques, the effect of signal sampling on contrast and resolution must be considered. The measured signal level values do not directly represent the number of photons emitted or scattered by the specimen, but are proportional to the number of photons. Furthermore, each individual sample of signal intensity is only an estimation of the number of collected photons, and will vary with repeated measurement. This variation is referred to as noise which imparts an uncertainty in the quantification of intensity and therefore in the contrast and resolution of the image data. Thus for better quantitative estimation signal-to-noise ratio must be taken into consideration.

Resolution

Confocal microscopy has inherent resolution limitations due to diffraction. Resolution can be defined as the ability to recognize two closely associated points as two distant points. As discussed in confocal microscope the point source (LASER) is used to produces a point of light on the specimen. In fact it appears in the focal plane as an Airy disk, whose size depends on the wavelength of the light source and the numerical aperture of the objective lens. The Airy disk limits the maximum resolution that can be attained with the confocal microscope, the best resolution is usually about 200nm. Ideally, the image of a point would just be a single intense point right at radius ¼ 0. However, the finite size of the Airy disk sets the scale for which details can be resolved. According to the Rayleigh criterion, the minimum separation between two Airy disks for which they are distinguishable is equal to their radius. This corresponds to the maximum of one Airy disk superimposed on the minimum of the other. Resolution along the optical axis is also limited by diffraction effects. As in the lateral direction there is a periodic, but elliptical distribution of intensity in the shape of an Airy disk.

Types of confocal microscope

Two types of confocal microscopes are commercially available:

- **Confocal laser scanning microscopes**
- **Spinning-disk (Nipkow disk) confocal microscopes**

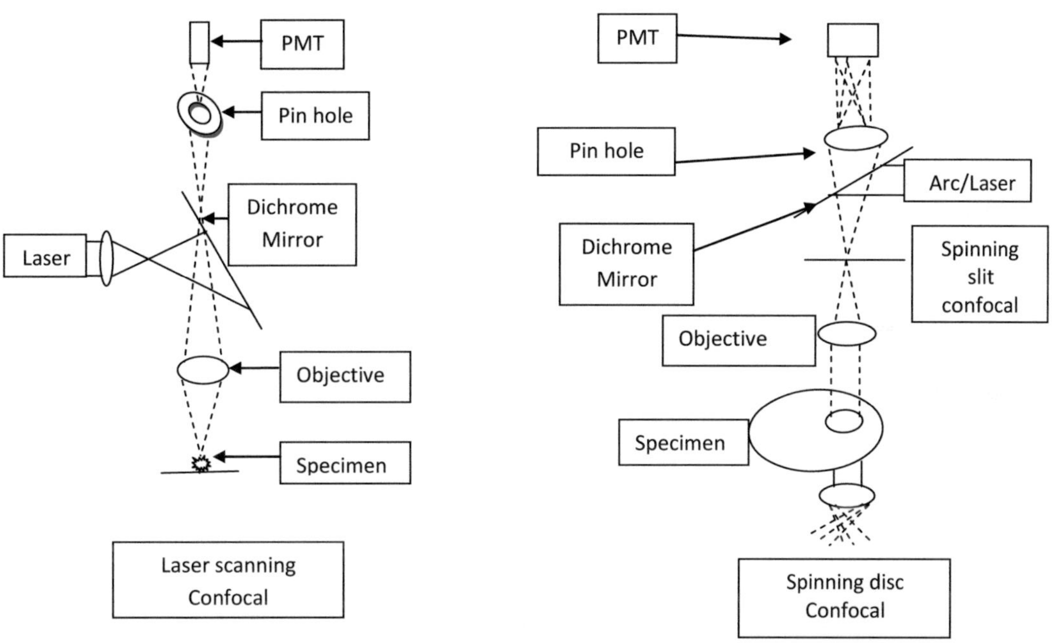

Fig. Schematic representation of the working principle of different types of confocal microscope

Each of these classes of confocal microscope has particular advantages and disadvantages. Most systems are either optimized for resolution or high recording speed (i.e. video capture). Confocal laser scanning microscopes can have a programmable sampling density while Nipkow use a fixed sampling density defined by the camera resolution. Imaging frame rates are typically very slow for laser scanning systems (e.g. less than 3 frames/second). Commercial spinning-disk confocal microscopes achieve frame rates of over 50 per second- a desirable feature for dynamic observations such as live cell imaging.

Cutting edge development in confocal laser scanning microscopy now allows better than video rate (60frames/second) imaging, by using multiple microelectronic mechanical systems based scanning mirrors. Confocal x-ray fluorescence imaging is a novel technique that allows control over depth, in addition to horizontal and vertical aiming. For example, when analyzing buried layers in a painting.

Difference between laser scanning and spinning disc confocal microscope:

Sl. No.	Characteristics	Laser Scanning confocal microscope	Spinning-disk confocal microscope
1.	Light source	Laser	Arc or Laser
2.	Pin hole size	Small	Large
3.	Mode of scanning	Point-by-point scanning	Full frame imaging
4.	Detector used	PMT	Full-frame CCD cameras are used for image formation.
5.	Dimension analysis	It is used for the re-construction of 3-D image	It is required for 4-D(time-lapse) imaging of living cells
6.	Types	Only one type: laser scanning confocal microscope.	Two types: Nipkow disk scanning confocal and DSU disk scanning confocal.
7.	Advantages	It provides highest level of confocality and ability to do thin optical sections.	It offers speed and great transmission.
8.	Disadvantages	Compromise detectible fluorescence in order to deliver high degree of confocality. Point-by-point acquisition of image is time consuming. Any specimen movement during scan can result in jagged edges in image or poor definition of intracellular details due to both time it takes to capture the image and damaging effects of the laser.	Less confocality.

Fast Confocal Microscope

Most confocal microscopes generate a single image in 0.1–1 sec. For many dynamic processes this rate may be too slow, particularly if 3D stacks of images are required. Even for a single 2D image, slow frame rates translate into long exposure times of the specimen to intense laser light, which may damage it or cause photobleaching. Two commonly used designs that can capture images at high speed are the Nipkow disk confocal microscope and a confocal microscope that uses an acousto-optic deflector (AOD) for steering the excitation light.

Applications of Confocal Microscope

The broad range of applications available to laser scanning confocal microscopy includes a wide variety of studies in neuroanatomy and neurophysiology, as well as morphological studies of a wide spectrum of cells and tissues. In addition, the growing use of new fluorescent proteins is rapidly expanding the number of original research reports coupling these useful tools to modern microscopic investigations. Other applications include resonance energy transfer, stem cell research, photobleaching studies, lifetime imaging, multiphoton microscopy, total internal reflection, DNA hybridization, membrane and ion probes, bioluminescent proteins, and epitope tagging. Many of these powerful techniques are described in these reviews.

CLSM is widely-used in numerous biological sciences discipline, from cell biology and genetics to microbiology and developmental biology.

FRAP (Fluorescence Recovery After Photo-bleaching)

FRAP is a technique designed to quantifying two dimensional lateral diffusion of a fluorescent probes tagged to biological samples like nucleotides and proteins in single cells or in molecularly thin film. In a FRAP experiment, a region with fluorescent molecules is irradiated or photo-bleached with laser light. This results in the fluorescent molecules inside that region to become non-fluorescent. The part recovered after a time lapse of this experiment is due to subsequent redistribution of fluorescent and bleached molecules throughout the volume. This gives information about their mobility. This technique is very useful in biological studies of the fluid mosaic model of cell membranes. It can also be used to determine if a protein is able to move within a membrane, or whether it is tethered to other structural components of the cell. FRAP allows quantitation of a number of 'dynamic parameters' such as diffusion coefficient, immobile fraction and binding or residence time.

- FRAP can be applied to one photon or multi-photon.
- Slow FRAP takes seconds and fast FRAP takes micro- to milli-seconds.
- The amount of light that recovers (returns) relative to the amount of light that was there before the photobleaching is the percent recovery.
- The speed with which the fluorescent molecules migrate back into the photobleached area is a measurement of the "diffusional mobility" which is usually called lateral mobility

FRET (Fluorescence Resonance Energy Transfer)

Fluorescence resonance energy transfer (FRET) is a distance-dependent interaction between the electronic excited states of two dye molecules in which excitation is transferred from a donor molecule to an acceptor molecule without emission of a photon. This mechanism is termed "Förster resonance energy transfer" and is named after the German scientist Theodor Förster. When both chromophores are fluorescent, the term "fluorescence resonance energy transfer" is often used instead Förster resonance energy transfer. Energy passes from the excited DONOR molecule to the ACCEPTOR molecule through nonradiative dipole–dipole coupling which then emits fluorescence. The fluorophores must be extremely close to one another for this to happen (< 0.1mm). Thus, if FRET occurs, the donor fluorescence decreases (quenching) and the acceptor fluorescence increases. So the ratio of acceptor/donor goes up. When the acceptor is photobleached, donor fluorescence increases. Thus the efficiency of FRET is dependent on following parameters:

- Donor and acceptor molecules must be in close proximity (typically 10-100Å).
- The absorption spectrum of the acceptor must overlap the fluorescence emission spectrum of the donor
- Donor and acceptor transition dipole orientations must be approximately parallel.

Some selected applications of FRET:

- Structural and conformation of proteins
- Spatial distribution and assembly of protein complexes
- Receptor/ligand interactions
- Immunoassays
- Prohibiting interactions of single molecules
- Real-time PCR assays and SNP detection
- Detection of nucleic acid hybridization
- Primer-extension assays for detecting mutations
- Automated DNA sequencing
- Distribution and transport of lipids
- Membrane fusion assays
- Fluorogenic protease substrates
- Indicators for cyclic AMP and zinc

Thus, FRET is an important technique for investigating a variety of biological phenomena that produce changes in molecular proximity.

FLIM (Fluorescence Lifetime Imaging)

The fluorescence lifetime is defined as the average time that a molecule remains in an excited state prior to returning to the ground state. The process of fluorescence emission by an excited molecule occurs like radioactivity with a characteristic half-life or decay lifetime of a few nanoseconds. The life time varies between different fluorescent molecules and for the same probe in a different environment. For example, the lifetimes of many fluorophores are altered by the presence of ions such as Ca^{2+}, Mg^{2+}, Cl^{-}, pH or K^{+}. This allows the researcher to follow environmentally induced changes. FLIM does not require wavelength-ratio metric probes. Fluorescence lifetime can give information about colocalisation of two molecules (e.g.FRET) and separation of two emissions with same spectrum.

An advantage of lifetime imaging is that the absolute values of lifetimes are independent of the probe concentration, photobleaching, light scattering and the amount of excitation intensity. Fluorescence lifetime imaging (FLIM) thus offers several opportunities to study dynamic events of living cells.

FLIP (Fluorescence loss in photobleaching)

A part of a cell at a distance from where there is an excess of fluorescent protein is bleached with a laser at low intensity. One then images the sample just before and at constant intervals after bleaching, allowing the bleached molecules to redistribute through the cell.

FLAP (Fluorescence localization after photobleaching)

The molecule to be located has 2 fluorophores, one to be bleached, and the other to act as a reference label. One can then track the distribution of the molecule after it is bleached. The

FLAP signal was obtained by subtracting the bleached signal from the unbleached one, allowing the tracking of the labeled molecule.

FISH (Fluorescence *in situ* hybridization)

FISH involves the preparation of short sequences of cDNA complementary to the DNA sequence of interest (Daims, Ramsing et al., 2001; Schmid, Twachtmann et al., 2000; Wagner, Horn et al., 2002). These cDNA probes are labeled with fluorescent tags, and once the probes hybridize to the complementary DNA, the area of DNA fluoresces. FISH can be performed on non-dividing cells as well as actively dividing cells.

Living Cells imaging

Confocal microscopy has been used effectively for the 3D study of dynamics in living cells. However, the imaging of living specimens adds the challenge of maintaining the life and normal function of the organism. There are of course difficulties involved in preparing the sample for viewing as is the case in conventional microscopy. But while performing the experiments effect of photo-damage on the specimen caused due to focused high intensity excitation light must be taken into account. This is complicated by the repeated exposure required for tracking the cellular dynamics—a problem that has worsen 3-D data collection. Fluorescence also introduces the problem of the fluorophore influencing the cell behavior as well as the risk that oxygen molecules reacting with fluorophores in triplet excited states may generate free radicals that damage the cell. Despite the challenges, a wide variety of sophisticated fluorophores have been developed to study different aspects of cell biology. They are designed to mark specific parts of the cell interior and often can simply be introduced to the cell wall. The fluorophores molecules make their way into the cell and attach to the intracellular structures of interest such as the mitochondria and the Golgi apparatus. This is not always the case, however, as some fluorophores must be injected directly into the cell. Labeling is even applied to the study of "nonphysical" structures of the cell, some fluorophores have been developed for the measurement of dynamic processes such as membrane potentials and ion concentrations. Multicolor fluorescence to distinguish between small features such as proteins within a cell it is useful to tag them with different fluorophores and image them as separate colors. There are two ways to do this **1)** fluorophores are selected to correspond with the wavelengths of a multiline laser and **2)** their response to the same excitation wavelength causes emission at different wavelengths. In both cases the resulting emission is separated with appropriate filters and directed to different detectors. However, there can be cross talk between channels of the emitted light. For most of the commonly used fluorophores there is usually some overlap between their emission spectra, making perfect channel separation impossible by filtering alone. To first order this can be corrected by determining the level of overlap of emission of each individual fluorophores into the channels of the other fluorophores and subtracting it out mathematically.

Thus, confocal microscopy can serve as a valuable tool for live cell imaging.

Clinically, CLSM is used in the evaluation of various eye diseases, and is particularly useful for imaging, qualitative analysis, and quantification of endothelial cells of the cornea (Patel and McGhee, 2007). It is used for localizing and identifying the presence of filamentary fungal elements in the corneal stroma in cases of keratomycosis, enabling rapid diagnosis and thereby early institution of definitive therapy. Research into CLSM techniques for endoscopic procedures is also showing promise (Hoffman, Goetz et al., 2006). In the pharmaceutical industry, it was recommended to follow the manufacturing process of thin film pharmaceutical forms, to control the quality and uniformity of the drug distribution (Le Person, Puigalli et al., 1998). CLSM is also used as the data retrieval mechanism in some 3D optical data storage systems and has helped determine the age of the Magdalen papyrus.

Advantages:

Confocal microscopy offers several distinct advantages over traditional widefield fluorescence microscopy-

- The primary advantage of laser scanning confocal microscopy is the ability to serially produce thin (0.5 to 1.5 micrometer) optical sections through fluorescent specimens that have a thickness ranging up to 50 micrometers or more.
- The ability to control depth of field.
- The most important feature of a confocal microscope is the capablity of isolating and collecting a plane of focus from within a sample, thus eliminating the out of focus "haze" normally seen with a fluorescent sample. Fine detail is often obscured by the haze and cannot be detected in a non-confocal, fluorescent microscope
- Elimination or reduction of background information away from the focal plane (that leads to image degradation) - The most important feature of a confocal microscope is the capablity of isolating and collecting a plane of focus from within a sample, thus eliminating the out of focus "haze" normally seen with a fluorescent sample. Fine detail is often obscured by the haze and cannot be detected in a non-confocal, fluorescent microscope.
- The capability to collect serial optical sections from thick specimens- this is due to the fact that with the confocal microscope, the z-resolution, or optical sectioning thickness, depends on a number of factors: the wavelength of the excitation/emission light, pinhole size, numerical aperture of the objective lens, refractive index of components in the light path and the alignment of the instrument.
- The confocal microscope has a stepper motor attached to the fine focus, enabling the collection of a series of images through a three dimensional object. These images can then be used for a two or three dimensional reconstruction.
- Double and triple labels can be collected with a confocal microscope. Since these images are collected from an optical plane within the sample, precise co-localizations can be performed.
- Confocal reflection microscopy can be utilized to gather additional information from a specimen with relatively little extra effort, since the technique requires minimum specimen preparation and instrument re-configuration. In addition, information from unstained tissues is readily available with confocal reflection microscopy, as is data from tissues labeled with probes that reflect light. The method can also be utilized in combination with more common classical fluorescence techniques. Examples of the latter application are detection of unlabeled cells in a population of fluorescently labeled cells and for imaging the interactions between fluorescently labeled cells growing on opaque, patterned substrata.
- Has the ability to compensate for autofluorescence.

The basic key to the above mentioned advantages is the use of spatial filtering techniques in confocal microscope to eliminate out-of-focus light or glare in specimens whose thickness exceeds the dimensions of the focal plane.

Disadvantages:

Co-localization of fluorophores in Confocal Microscope - Two or more fluorescence emission signals can often overlap in digital images recorded by confocal microscopy due to their close proximity within the specimen. This effect is known as **colocalization** and usually occurs when fluorescently labeled molecules bind to targets that lie in very close or identical spatial positions.

8. Summary

A confocal microscope provides a significant imaging improvement over conventional microscopes. It creates sharper, more detailed 2D images, and allows collection of data in three dimensions. In biological applications it is especially useful for measuring dynamic processes. A number of designs have been developed to achieve video-rate confocal microscopy, which enables the capture of short-timescale dynamics.

9. References

Minsky, M. (1961).Microscopy Apparatus.

Minsky, M. (1988). Memoir on inventing the Confocal Scanning Microscopy. *Scanning* 10 128-138.

Egger, M. D. & Petran, M. (1967). New reflected-light microscope for viewing unstained brain and ganglion cells. *Science* 157(786) 305-307.

Davidovitis, P. & Egger, M. D. (1973). Photomicrography of Corneal Endothelial Cells in vivo. *Nature* 244 366-367.

Amos, W. B. & White, J. G. (2003). How the confocal laser scanning microscope entered biological research. *Biol Cell* 95(6) 335-342.

Brakenhoff, G. J.; Blom, P., et al. (1979). Confocal Scanning light Microscopy with High Aperture immersion Lens. *Journal of Microscopy* 117 219-232.

Sheppard, C. J. R. & Wilson, T. (1979). Effects of Spherical Abbressions on the Imaging Properties of Scanning Optical Microscopes. *Applied optics* 18 1058.

Hamilton, D. K. & Wilson, T. (1986). Scanning Optical Microscopy by Objective Lens Scanning. *Journal of Physics E: Scientific Instruments* 19 52-54.

Diaspro, A. (Ed) (2002). *Confocal and two -Photon Microscopy: Foundations, Applications, and Advances.* New York: Wiley-Liss.

Hibbs, A. R. (Ed) (2004). *Confocal Microscopy for Biologists.* New York: kluwer Press.

Matsumoto, B. (Ed) (2002). *Cell Biology Applications of Confocal Microscopy, in methods in Cell Biology.* New York: Academic Press.

Muller, W. (Ed) (2002). *Introduction to Confocal Fluorescence Microscopy.* Maastrich, Neitherlands: Shaker.

Paddock, S. W. (1999). Confocal Microscopy : Methods and Protocols.

Pawley, J. B. (Ed) (1995). *Handbook of Biological Confocal Microscopy.* Plenum Press.

Peterman, E. J. G.; Sosa, H., et al. (2004). Single-Molecule Fluorescence Spectroscopy and Microscopy of Biomolecular Motors. *Ann. Rev. Phy. Chem.* 55 79-96.

Mason, W. T. (Ed) (1999). *Fluorescent and Luminescent Probes for Biological Activity.* New York: Academic Press.

Spring, K. R. & Inoue, S. (Eds) (1997). *Video Microscopy: The Fundamentals.* New York: Plenum Press.

Goldman, R. D. & Spector, D. L. (Eds) (2005). *Live Cell Imaging: A Laboratory Mannual.* New York: Cold Spring Harbour Press.

Murphy, D. B. (Ed) (2001). *Fundamentals of light Microscopy and Imaging.* New York: Wiley-Liss.

Wilson, J. & Carlini, A. R. (1988). Three dimensional imaging in confocal imaging system with finite-sized detectors. *J Microsc* 141 51-66.

Pawley, J. B. (Ed) (2006). Berlin: Springer.

Al-Kofahi, O.; Can, A., et al. (2003). Algorithms for Accurate 3D Registration of Neuranal Images Acquired by Confocal Laser Microscopy. *J Microsc* 211 8-18.

Conn, P. N. (Ed) (1999). *Confocal Microscopy in methods in Enzymology.* New York: Academic Press.

Gu, M. (Ed) (1996). *Principles of Three-Dimensional Imaging in Confocal Microscopes.* New Jersey: World Scientific.

Masters, B. R. (Ed) (1996). *Selected Papers on Confocal Microscopy,SPIE Milstone Series.* Bellingham, Washington: SPIE Optical Engineering Press.

Sheppard, C. J. R. & Shotton, D. M. (Eds) (1997). *Confocal Laser Scanning Microscopy.* Oxford, United Kingdom: BIOS Scientific Publishers.

T.R.Corle & G.S.Kino (Eds) (1996). *Confocal Scanning Optical Microscopy and Related Imaging Systems.* New York: Academic Press.

Wilson, T. (Ed) (1990). *Confocal Microscopy.* New York: Academic Press.

Fellers, T. J. & Davidson, M. W. (2007). Introduction to Confocal Microscopy. *Olympus Fluoview Resource Center.*

Lichtmann, J. W. (1994). Confocal microscopy. *Scientific American* 40-45.

Murray, J. (Ed) (2005). *live Cell Imaging: A Laboratory Manual.* New York: Cold Spring Harbour Press.

Sandison, D. & Webb, W. (1994). Background Rejection and Signal-to-noise Optimization in the Confocal and Alternative Fluorescence microscopes. *Applied optics* 33 603-610.

Swedlow, J. R.; Hu, K., et al. (2002). Measuring Tubulin Content in Toxoplasma gondii: A Comparision of Laser-Scanning Confocal and Wide-Field Fluorescence Microscopy. *Proc. Natl. Acad. Sci. USA* 99 2014-2019.

White, J. G.; Amos, W. B., et al. (1987). An Evaluation of Confocal versus Conventional Imaging of Biological Structures by Fluorescent Light Microscope. *J. Cell Biol.* 105 41-48.

Wilson, T. (1989). Optical Sectioning in Confocal Fluorescence Microscopes. *Journal of Microscopy* 154 143-156.

Wright, S. J. & Wright, D. J. (Eds) (2002). *Cell Biology Applications of Confocal microscopy.* New York: Academic Press.

Wachmann, E.S. (Ed) (2000). *Imaging Neurons: A Laboratory Mannual.* New York: Cold Spring Harbour Press.

Wachmann, E. S.; Niu, W., et al. (1996). Imaging Acousto-optic Tunable Filter with 0.35-Micrometer Spatial Resolution. *Applied optics* 35 5220-5226.

Chang, I.C. (Ed) (1995). *OpticsII: Fundamentals, Techniques, and Design.* New York: McGraw-Hill.

Sharma, D. (2001).Bio-Rad Technical Notes

In *The Use of an AOFT to Achieve High Quality Simultaneous Multiple Label Imaging*

San Francisco: Bio-Rad.

Chen, Y.; Mills, J. D., et al. (2003). Protein Localization in Living Cells and tissues using FRET and FLIM. *Differentiation* 71 528-541.

Day, R. N.; Periasamy, A., et al. (2001). Fluorescence Resonance Energy Transfer Microscopy of Localized Protein Interactions in the Living Cell Nucleus. *Methods* 25 4-18.

Wallrabe, H. & Periasamy, A. (2005). Imaging Protein molecules using FRET and FILM Microscopy. *Curr. Opin. Biotech.* 16 19-27.

Claxton, N. S.; Fellers, T. J., et al. Laser Scanning Confocal Microscopy. Florida: Dept. of Optical Microscopy and Digital Imaging, National High Magnetic Field Lab.,The Florida State University, Florida.

Lippincott-Schwartz, J.;Altan-Bonnet, N., et al. (2003). Photobleaching and Photoactivation: Following Protein Dynamics in Living Cells. *Nature Cell Biology* 5 S7-S14.

Klonis, N.; Rug, M., et al. (2002). Fluorescence Photobleaching Analysis for Study of Cellular Dynamics. *Eur. Biophys. J.* 31 36-51.

Phair, R. D. & Misteli, T. (2002). Kinetic Modelling Approaches to in vivo Imaging. *Nature Rev. Mol. Cell Bio.* 2 898-907.

Politz, J. C. (1999). Use of Caged Fluorophores to Track Macromolecular Movement in Living Cells. *Trends in Cell Biol.* 9 284-287.

Hosada, S.; Seya, E., et al. A High Speed Laser Beam Scanning using AOD. *Japanese Journal of Applied Physics* 26(7) 1026-1030.

Keller,H.E. (1995) Objective lenses for confocal microscopy. In *Handbook of Biological Confocal Microscopy.* 2nd Ed. Pawley, J.B., Ed. New York: Plenum Press. 111-126.

Wilhelm, S.; Grobler, B., et al. (2003).Confocal Laser Scanning Microscopy: Optical image formation and Electronic Signal Processing. Jena, Germany: Carl zeiss Advanced Imaging Microscopy.

Ichihara, A.; Tanaami, T., et al. (1996). High Speed Confocal Fluorescence Microscopy using a Nipkow Scanner with microlenses for 3-D Imaging of Fluorescent Molecules in Real-Time. *Bioimages* 4 57-62.

Inoue, S. & Inoue, T. (Eds) (2002). *Cell Biological Applications of Confocal Microscopy, in Methods in Cell Biology.* New York: Academic press.

Nokano, A. (2002). Spinning Disk Confocal Microscopy- A Cutting-Edge Tool for Imaging of Membrane Traffic. *Cell Struct Funct* 27 349-355.

Chong, F. K.; Coates, C. G., et al. (Eds) (2004). *Optimization of Spinning Disk Confocal Microscopy: Synchronization with the Ultra-Sensitive EMCCD*

Webb, R. H. (1996). Confocal optical Microscopy. *Rep. Prog. Phys* 59 427-471.

Spring, K. R. (Ed) (2001). *Methods in Cellular Imaging.* New York: Oxford University Press.

Daims, H.; Ramsing, N. B., et al. (2001). Cultivation-independent, semiautomatic determination of absolute bacterial cell numbers in environmental samples by fluorescence in situ hybridization. *Appl Environ Microbiol* 67(12) 5810-5818.

Schmid, M.; Twachtmann, U., et al. (2000). Molecular evidence for a genus-level diversity of bacteria capable of catalyzing anaerobic ammonium oxidation. System. . *Appl. Microbiol.* 23 93-106.

Wagner, M.; Horn, M., et al. (2002). Fluorescence in situ hybridization for the identification of prokaryotes. *Curr. Opinion Microbiol.* 6 302-309.

Patel, D. V. & McGhee, C. N. (2007). Contemporary in vivo confocal microscopy of the living human cornea using white light and laser scanning techniques: a major review. *Clin. Experiment. Ophthalmol.* 35(1).

Hoffman, A.; Goetz, M., et al. (2006). Confocal laser endomicroscopy: technical status and current indications *Endoscopy* 38 (12) 1275-1283.

Le Person, S.; Puigalli, J. R., et al. (1998). Near infrared drying of pharmaceutical thin film: experimental analysis of internal mass transport. *Chemical Engineering and Processing* 37 257-263

6

Rigid Confocal Endoscopy of the Larynx

Tino Just[1], Joachim Stave[2], Oliver Stachs[2], Gabriele Witt[3], Rudolf Guthoff[2] and Hans Wilhelm Pau[1]
[1]*Department of Otorhinolaryngology, Head and Neck Surgery, University of Rostock*
[2]*Department of Ophthalmology, University of Rostock*
[3]*Department of Phoniatry and Pedaudiology, University of Rostock*
Germany

1. Introduction

Early detection of pre-malignant and malignant lesions of the larynx may be the best method to improve patient quality of life and survival rates. Microlaryngoscopy, the Kleinsasser method (Kleinsasser, 1962), was introduced in 1962 as a new procedure to diagnose laryngeal diseases. Diagnosis of laryngeal disease depends on both the extent of sampling and the method used (McGuirtBrowne, 1991). To improve the accuracy of sampling during microlaryngeal surgery (MLS) and to maximize survival for patients with cancer of the larynx, the search for new in-vivo technologies is of great clinical relevance. The risk of local recurrence increases from 32% to 80% when positive margins are found (Batsakis, 1999). Several techniques have been introduced in laryngology to detect early stages of laryngeal cancer. Stroboscopy was thought to detect early laryngeal cancer, but when applied preoperatively, this method fails to differentiate clearly between invasive carcinoma of the true vocal cord and intracellular atypia and to determine the penetration depth of the laryngeal cancer (Colden et al., 2001). Several methods were introduced to improve the accuracy of sampling during microlaryngoscopy. Among these technologies, "optical" biopsy (Kothe et al., 2005), rigid endoscopy (Kawaida et al., 1998), contact endoscopy (Andrea et al., 1995a; Andrea et al., 1995b; Arens et al., 2003; Carriero et al., 2000; Cikojevic et al., 2008; Wardrop et al., 2000), autofluorescence endoscopy (Delank et al., 2000; Gillenwater et al., 1998; Harries et al., 1995; Malzahn et al., 2002), aminolevulinic acid-induced fluorescence (Arens et al., 2007; Csanady et al., 2004; Gillenwater et al., 1998) and optical coherence tomography (OCT) (Armstrong et al., 2006; Lüerßen et al., 2006; Wong et al., 2005) should be noted. Except for the latter method, all technologies mentioned mainly visualize macroscopically suspect lesions of the larynx. OCT is an optical imaging technique that clearly identifies basement membrane violation from laryngeal cancer, but fails to penetrate bulky exophytic tumors (Armstrong et al., 2006). Laryngeal tumors mostly appear as keratotic lesions. So far, there is no information on the correlation between penetration depth and thickness of keratotic lesions using OCT. Additionally, OCT fails to visualize nuclear abnormalities that will be important to differentiate between hyperplasia of the epithelium and dysplasia or early cancer (Armstrong et al., 2006). OCT images can be obtained at depths of up to 2 mm (Brezinski et al., 1996) and provide approximately 3–35 μm lateral and 15–30 μm axial resolution when applied to epithelial tissue (Schmitt, 1999).

OCT provides test outcomes for differentiation between benign laryngeal lesions and dysplasia/carcinoma in situ with sensitivity of 88%, specificity of 89%, and predictive accuracy of 88% (Just et al., 2010).

In contrast to OCT, laser scanning microscopy (LSM) provides cellular and sub-cellular information and achieves a resolution of 1 µm (Rajadhyaksha et al., 1995). The basic principle of LSM is that light from a laser passes through the illumination aperture, is reflected by a dichroic mirror, and is finally focused on one point in the specimen (Stave et al., 2002). Laser light that is backscattered from the specimen is captured through an aperture. Only the light from the in-focus plane is imaged. Out-of-focus light is eliminated. This gives a high-resolution image of cellular structures at the focal point. To obtain a two-dimensional image, the confocal microscope scans across the tissue at a single depth, recording images from a number of adjacent focal points. By moving the focal plane, different layers of the tissue are displayed. The imaging depth is up to 0.5 mm and allows identification of many intra-epithelial lesions (Collier et al., 2002; Rajadhyaksha et al., 1995; White et al., 1999). With special software, it is possible to 3D-reconstruct the volume scans in vivo. This procedure has been termed "virtual biopsy" (Amos & White, 2003; Kiesslich et al., 2004). The illumination of the tissue (penetration depth of the laser) depends primarily on the keratinization of the epithelium. It was hypothesized that increased levels of keratin diminish the ability for illumination of the deeper cell layer of the epithelium. The reason for this is the higher refractive index of keratin compared to cytoplasm (Brunsting & Mullaney, 1974). However, the penetration depth of LSM is limited due to two factors: background due to out-of-focus scattering and signal-to-noise ratio (Schmitt et al., 1997).

Recently, a confocal microscope has been miniaturized and integrated into a conventional flexible endoscope (Venkatesh et al., 2009). The diameter of the distal tip of the endoscope is 12.8 mm including the integrated miniaturized confocal microscope with a diameter of 5 mm (Venkatesh et al., 2009). The use of endoscopes with this large diameter has so far been limited to the gastrointestinal tract. This newly developed imaging system provides in-vivo histology during ongoing endoscopy and can be used to diagnose different diseases of the gastrointestinal tract (Kiesslich et al., 2005; Kiesslich et al., 2006; Kiesslich & Neurath, 2005; Kiesslich & Neurath, 2006; Venkatesh et al., 2009). There have been initial attempts to use the mentioned imaging technology for in-vivo assessment of the oropharyngeal mucosa. After topical application of acriflavine and intravenous fluorescein, cellular and sub-cellular structures of the anterior human oropharyngeal region were displayed (Haxel et al., 2010). Using a confocal microscope for use in ophthalmology, we obtained cellular and sub-cellular information in vivo without any staining (Just et al., 2006a; Just et al., 2007a; Just et al., 2005). Figure 1 demonstrates an in-vivo image of healthy human cheek mucosa.

Next, we commenced ex-vivo studies of laryngeal mucosa (Figures 2–7). A pre-clinical trial has been carried out to assess benign, pre-malignant and malignant lesions of the human larynx (Just et al., 2006b; Just et al., 2007b). Quantitative analysis of cellular and morphological features, e.g. nuclear size, nuclear density, number of nuclei per cell, nucleus/cytoplasm ratio, regularity of the cell layers, morphology of cells in a cell layer, and occurrence of cellular junctions, showed good agreement with histology, suggesting that LSM can identify pathologies in vivo (Just et al., 2006b). Confocal microscopy of the healthy human larynx revealed that the border between the superficial and basement cell layer is clearly visible, while the basement membrane itself cannot be detected (Fig. 2). In contrast to carcinoma (Fig. 3), dysplasia is diagnosed when the cellular junctions remain intact (Fig. 6). Figures 4 and 5 depict the typical findings of a papilloma (two papillae are indicated) and

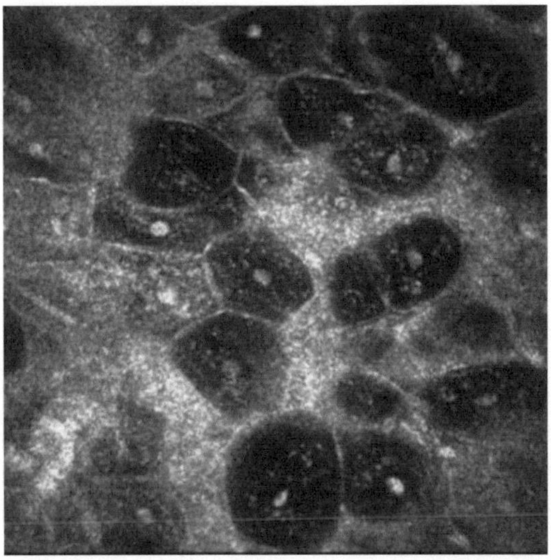

Fig. 1. Normal cheek mucosa. LSM displays a homogeneous cell architecture.

inflammation (koilocyte is marked). Visualization of the basement membrane is a precondition for in-vivo diagnosis of precancerous lesions. Figure 7 shows the infiltration of a carcinoma into the subepithelial space. Figures 2–7, obtained ex vivo, illustrate the hallmark differences between normal epithelium of the larynx and the epithelium of benign and pre-cancerous and cancerous lesions. The information from our ex-vivo studies of the human larynx is limited; we regard them as a necessary prelude to applying this technology in vivo during microlaryngoscopy.

To accomplish this, a rigid endoscope has been developed emerging as a potential tool for detecting pre-malignant lesions in vivo without the need for explorative excision. The

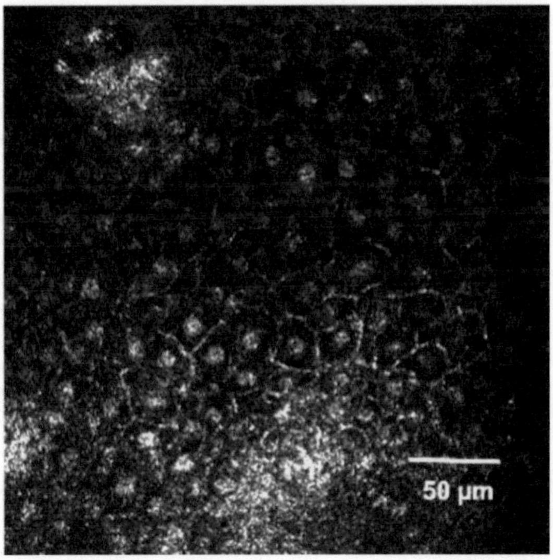

Fig. 2. Ex-vivo LSM image of healthy human true vocal cord epithelium at a depth of 5 µm. Regular cell architecture is apparent and both nuclei and cell membrane are clearly visible.

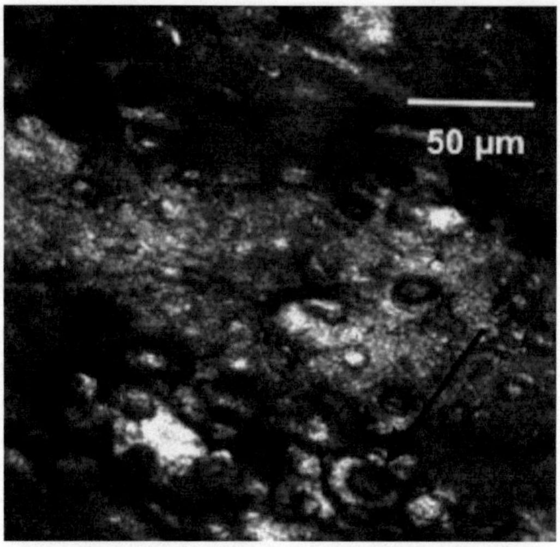

Fig. 3. Carcinoma. Ex-vivo LSM image presents all criteria of malignancy: enlarged nuclei of enlarged cells with variable shapes, clusters of cells, increased nucleus/cytoplasm ratio, irregular cell architecture and loss of cellular junctions.

experiments focus on the in-vivo application of LSM as a technology that enables visualization of cellular and sub-cellular information from the larynx epithelium. In our opinion, this technology has a high potential to significantly improve the detection rate, prevention and therapy of laryngeal cancer and its precursors. The application of this method intraoperatively is of scientific and clinical importance.

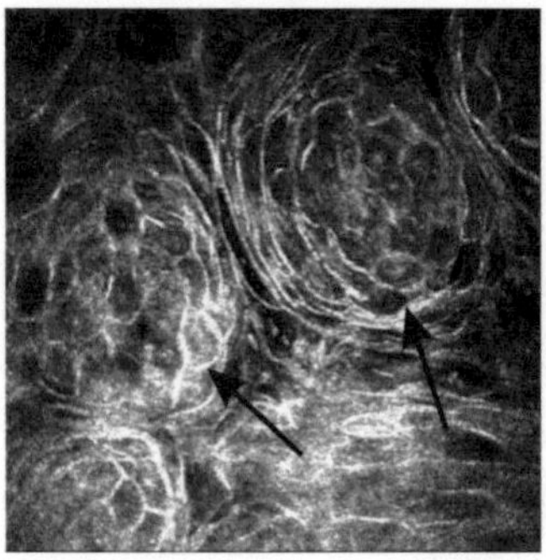

Fig. 4 Ex-vivo image of papilloma (arrows)

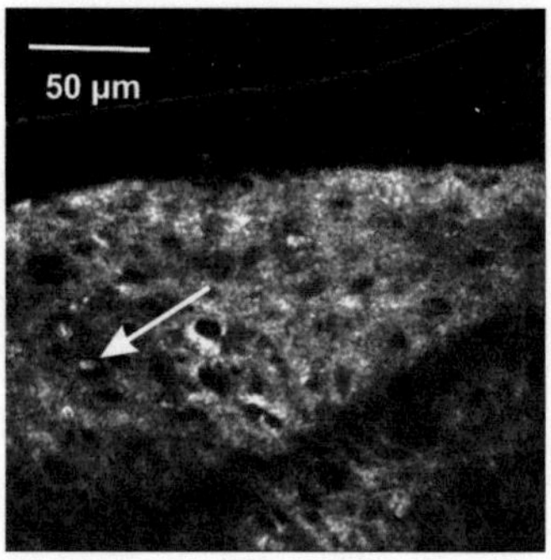

Fig. 5. Ex-vivo image of inflammation (koilocyte marked)

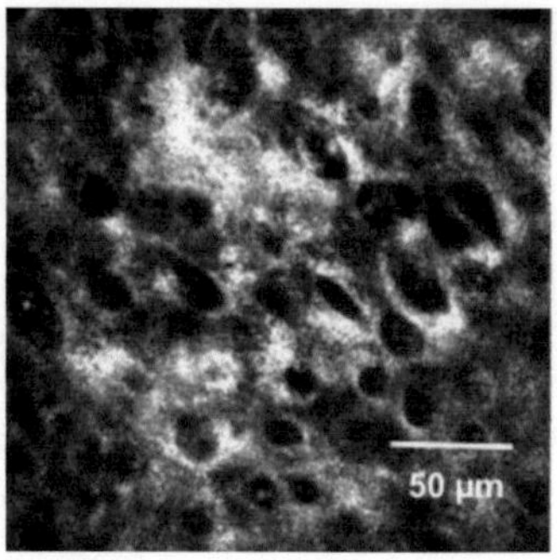

Fig. 6. Ex-vivo image of dysplastic cells

Next, we describe our approach to use confocal microscopy during microlaryngoscopy. Requirement was to construct rigid endoscopes rather than flexible probes. The potentially high impact of RCE is due to several unique properties and recent technical developments:

1. Imaging depths of 0.2 – 0.5 mm allow reliable identification of many intra-epithelial lesions (Collier et al., 2002; Rajadhyaksha et al., 1995; White et al., 1999).
2. LSM provides longitudinal-sections of the tissue and allows for characterization of normal laryngeal epithelium and of laryngeal lesions from different histogenetic origins (Just et al., 2006b; Just et al., 2007b).

3. Application of LSM gives the opportunity for rapid acquisition of high-quality images of the laryngeal tissue (Stave et al., 2002).
4. Good correlation between LSM images and conventional histology were achieved in normal tissue, but also in benign and cancerous lesions of the larynx (Just et al., 2006b; Just et al., 2007b) and oropharynx (Just et al., 2008; White et al., 1999).
5. Rigid endoscopes seem to be easier to handle during microlaryngoscopy than flexible endoscopic systems.

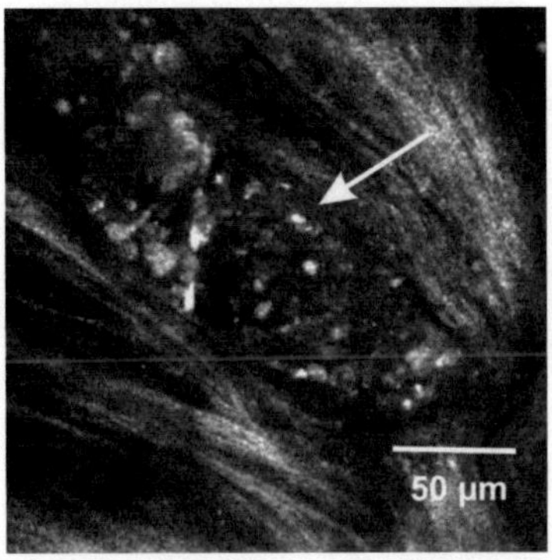

Fig. 7. Ex-vivo LSM image of infiltration of carcinoma into the subepithelial space (arrow)

2. Imaging system, analysis and application

A modified Heidelberg Retina Tomograph II (HRTII) (Heidelberg Engineering GmbH, Heidelberg, Germany) is used. The rigid endoscope is connected onto the scanner. Polarized light travels through a $\lambda/4$ plate. Reflections may occur at the water immersion objective and the endoscope. Illumination is delivered by a 670-nm diode laser. For endoscopic use, the $\lambda/4$ plate of the HRTII is removed to avoid strong reflections. This leads to a percentage reduction in the image quality. In order to improve the image quality, the laser power of the HRTII scanner was increased for animal experiments and for the first clinical applications.
Three different modes can be used for data sampling. In the volume mode or 3D imaging mode, 40 image planes are stored. The distance between two subsequent image planes is about 2 µm. Thus, the depth range is 80 µm per scan. Through resetting of the start plane, up to 4 volume scans can be made. The maximal penetration depth is about 300 µm and ranges from 100 to 450 µm. The acquisition time for a volume scan is about 6 s. In sequence acquisition mode, up to 100 images are stored. In section mode, single images can be stored.

2.1 Design of the rigid endoscope and the adapter

The prototype of the custom-made endoscope (length: 23 cm, diameter: 5 mm) consists of a steel endoscope shaft with integrated rod lenses and one microscope objective at the end-side of the endoscope (diameter 3.5 mm). Two different endoscopes have been developed. One endoscope has a flat microscope objective at the end of the endoscope. A second endoscope has a central hole perforation plate at the end-side of the endoscope. This modification avoids tissue compression in the field of vision. A flexible, transparent solid piece of cement is used to connect two adjacent rod lenses. This flexible material reduces the risk of damage to the rod lenses. The cement is non-scattering and does not impair the light transfer and image quality. The numerical aperture of the system is 0.9 and is limited by the endoscope. The endoscope is autoclavable and CE-certified. A connector has been developed (KARL STORZ GmbH & Co. KG; Tuttlingen, Germany) to connect the HRTII to the rigid endoscope (Fig. 8). With the connector, the starting plane can be set manually. To achieve optical sectioning of the tissue, the scanning mechanism of the HRTII needs to be activated. The devices (HRTII, adapter, and endoscope) supply images of 400 x 400 μm and reach average penetration depths of 100–300 μm (λ/4 plate of the HRTII was removed). The lateral and axial resolutions are about 1–2 μm and 2 μm, respectively.

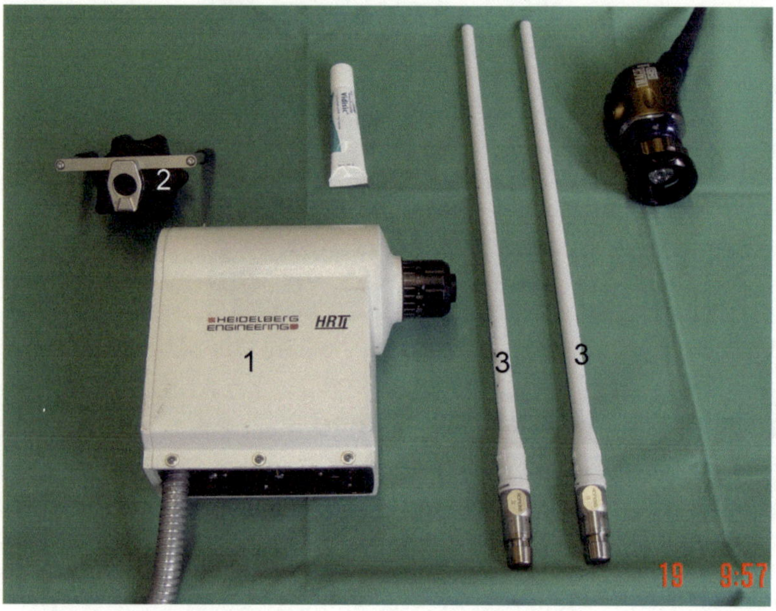

Fig. 8. Devices for in-vivo assessment of laryngeal mucosa.

1 – Scanner of the HRTII.
2 – Adapter to connect the endoscope to the scanner.
3 – Rigid endoscopes.

2.2 Image analysis and classification of squamous intra-epithelial lesions (SILs)

Confocal microscopic images are captured and stored in a data bank together with the microscopic images obtained during microlaryngology. The corresponding histopathological slices are reviewed separately. Each case is graded on a two-tiered scale as mild/moderate dysplasia (resembling low-grade squamous intra-epithelial lesion (SIL)), or severe dysplasia/carcinoma in situ (resembling high-grade SIL). Representative cases are discussed and arranged together with clinical and histological images in an image gallery of ascending severity SILs for synoptic viewing of lesions, thus enhancing reproducibility.

2.3 Intraoperative application

The larynx is exposed during microlaryngoscopy using laryngoscopes (Fig. 9). Suspicious lesions are identified with the operating microscope or with conventional endoscopes (0°, 30°, 70°).

For confocal endoscopy, the RCE is inserted through the laryngoscope (Fig. 10). The tip of the endoscope is gently placed on the lesion under macroscopic or endoscopic control. To avoid strong reflections, a contact gel (Vidisic®, Dr Mann Pharma, Berlin, Germany) is used. The key step in this procedure is to take a biopsy as precisely as possible from the same laryngeal region as for confocal endoscopy. This procedure enables us to compare the confocal-microscopic findings with histology at a later stage.

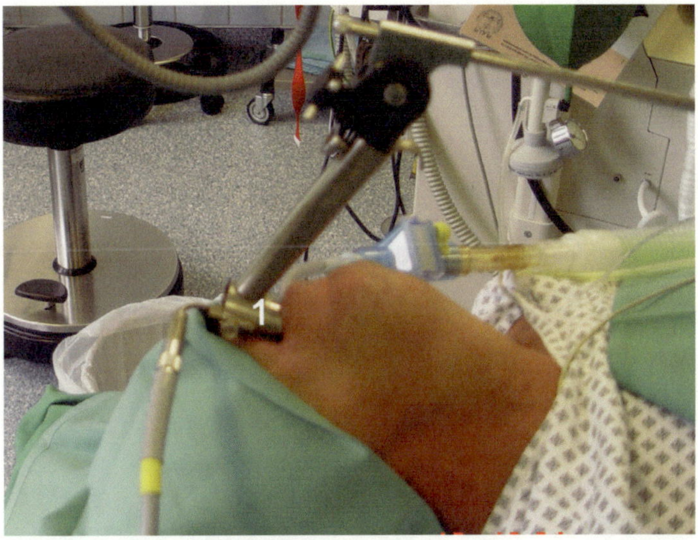

Fig. 9. Clinical setting during microlaryngoscopy.

1 - Laryngoscope was inserted into the oral cavity.

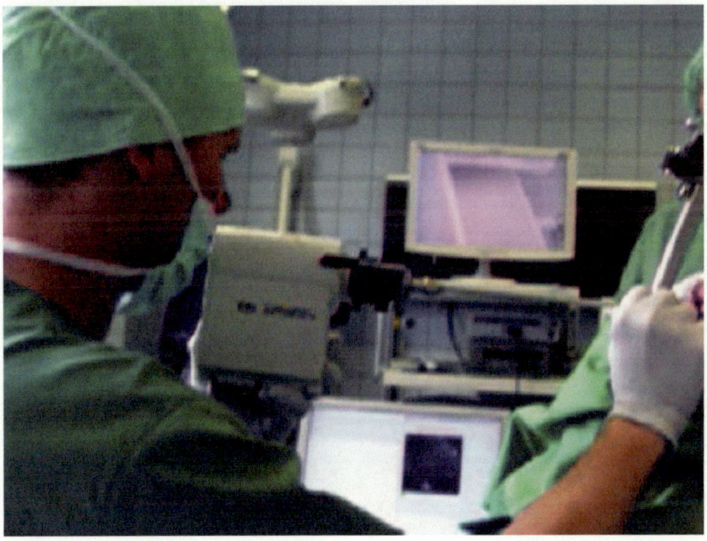

Fig. 10. Clinical setting. Rigid confocal endoscope needs to be inserted through the laryngoscope either under endoscopic or macroscopic control.

3. Human larynx findings with in-vivo-RCM

Figure 11 shows images of a volume scan through the epithelium of a healthy true vocal cord. Epithelial cell layers are visible up to a depth of 80 μm (first series). The second volume scan of the same vocal cord demonstrates the subepithelial fibers underneath the epithelium (Fig. 12).

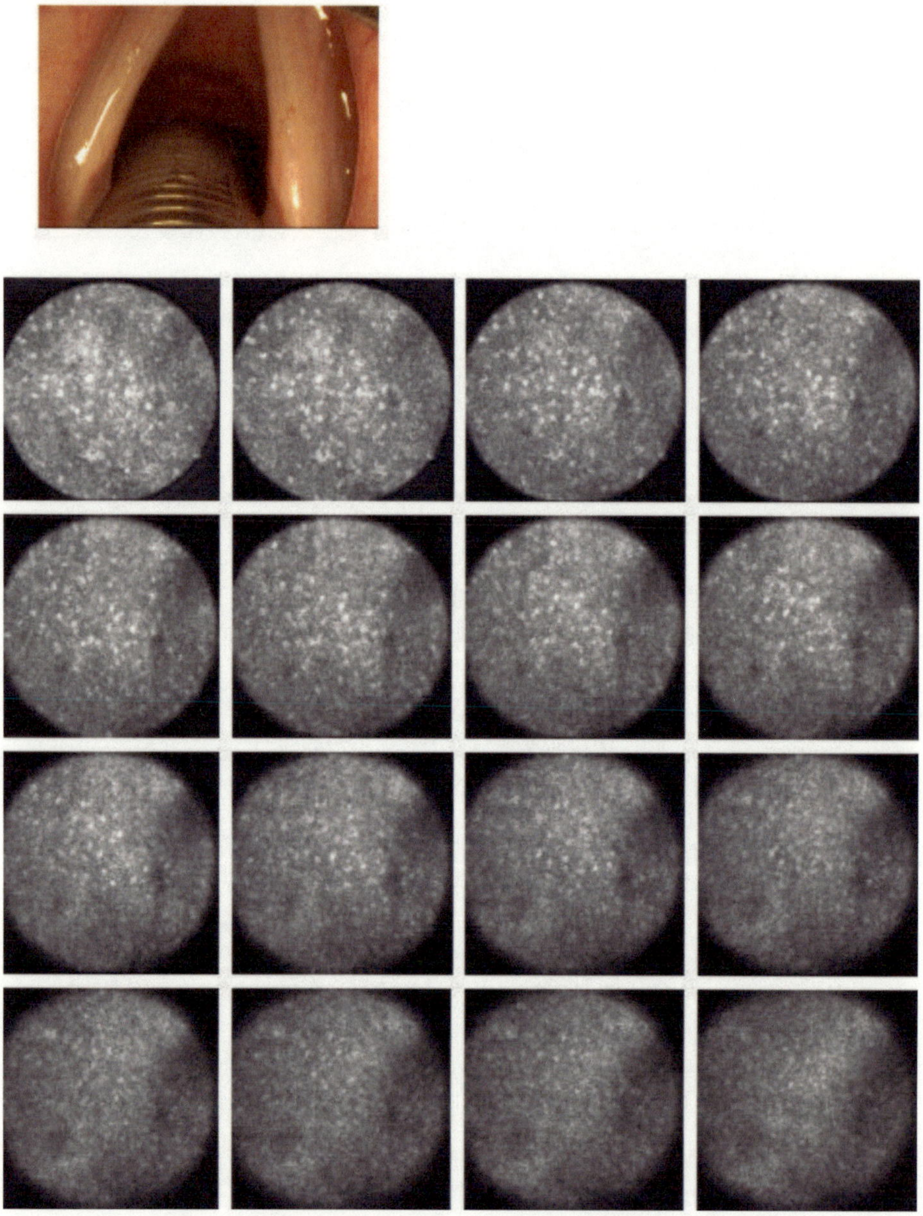

Fig. 11. Volume scan through a healthy true vocal cord. Cellular structures are visible up to 80 μm (stepwise volume scan started from the surface of the epithelium).

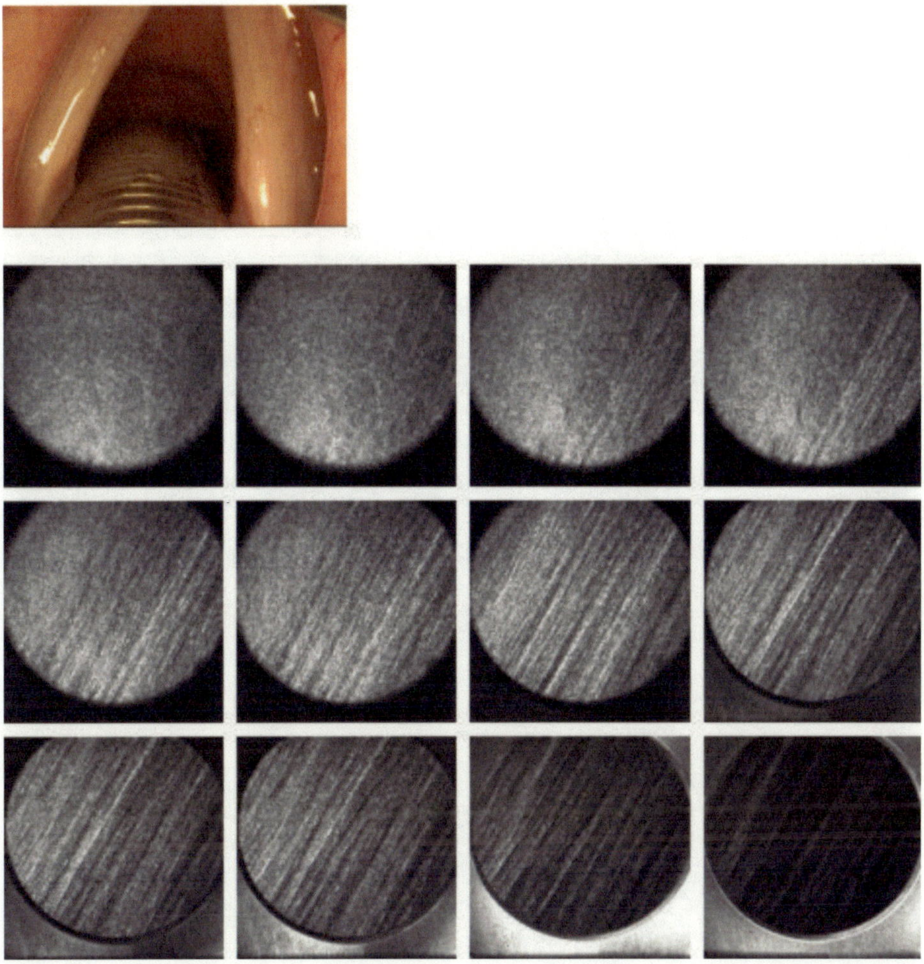

Fig. 12. Volume scan through a healthy true vocal cord. Subepithelial fibers are visible (stepwise volume scan started from basal cell layer).

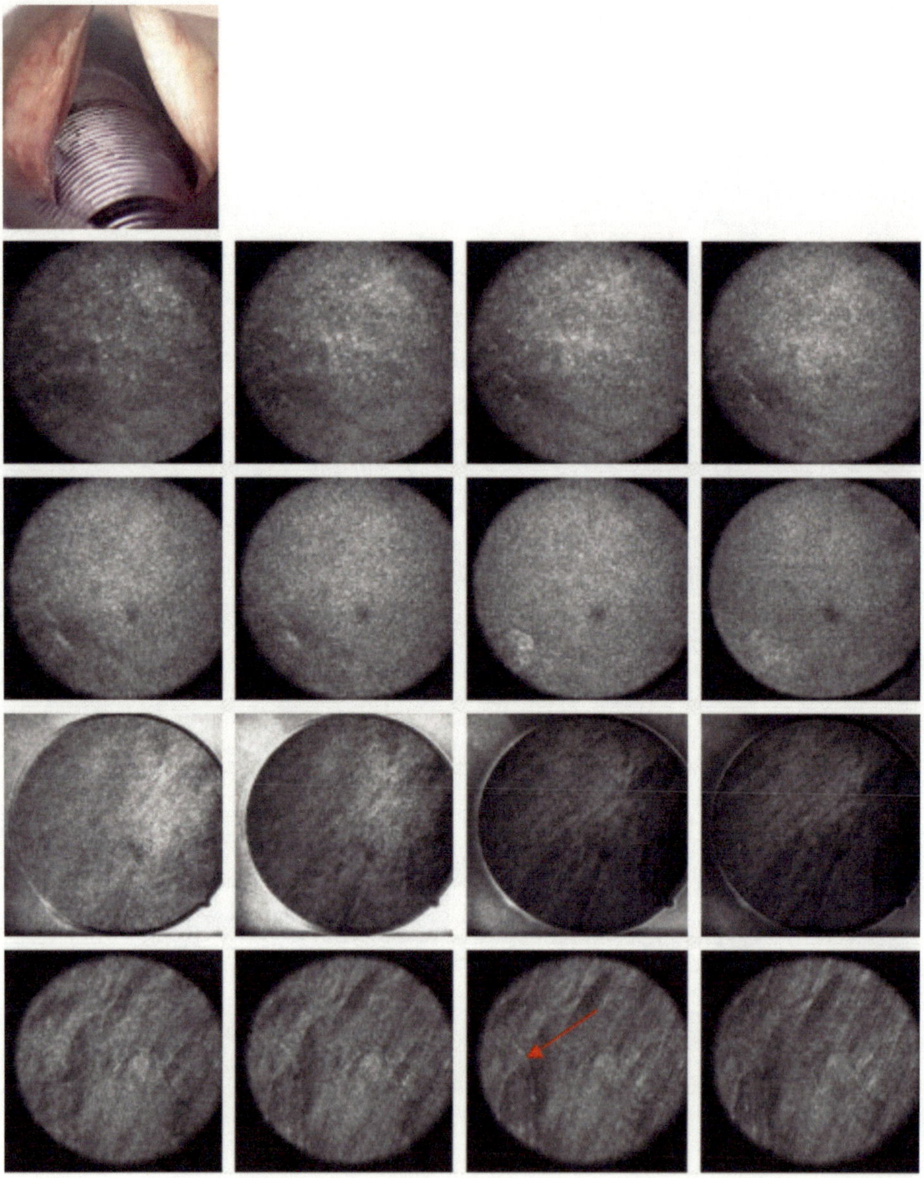

Fig. 13. Three months after left-sided cordectomy due to T1 laryngeal cancer. The images through the epithelium display no dysplastic cells. Vessel loops are visible in the subepithelial space. Histology revealed inflammation, but no dysplasia.

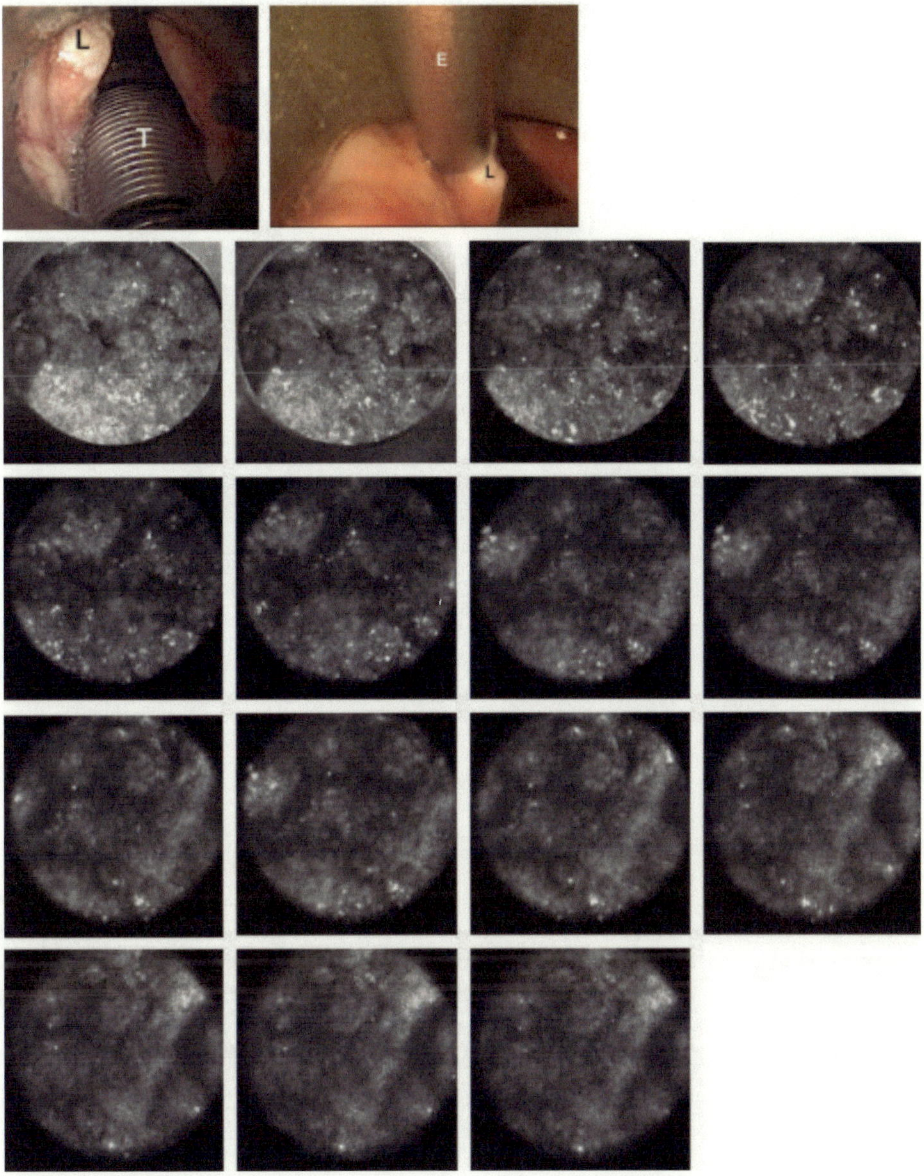

Fig. 14. Keratotic lesion of the left true vocal cord. LSM detected epithelial changes (irregular cell architecture and dysplastic cells). Histology revealed microinvasive carcinoma. L, lesion; T, tube; E, rigid confocal endoscope.

RCE findings are shown for a 56-year-old man 3 months after CO_2-laser assisted left-sided cordectomy (Fig. 13). The thin epithelial layer appears homogeneous and regular. No cellular abnormalities could be detected with confocal microscopy. Histology reveals inflammation but no dysplasia. Vessel loops are visible underneath the epithelium (arrow).
Figure 14 demonstrates a keratotic lesion of the posterior third of the left true vocal cord of a 59-year-old man who had undergone laser surgery of both sides several years earlier. Epithelial changes can be detected with confocal microscopy. Representative RCE images of a volume scan demonstrate enlarged nuclei of enlarged cells with variable shapes, clusters of cells, increased nucleus/cytoplasm ratio, and irregular cell architecture (Fig. 15). These are criteria for dysplasia. Histology revealed severe dysplasia.

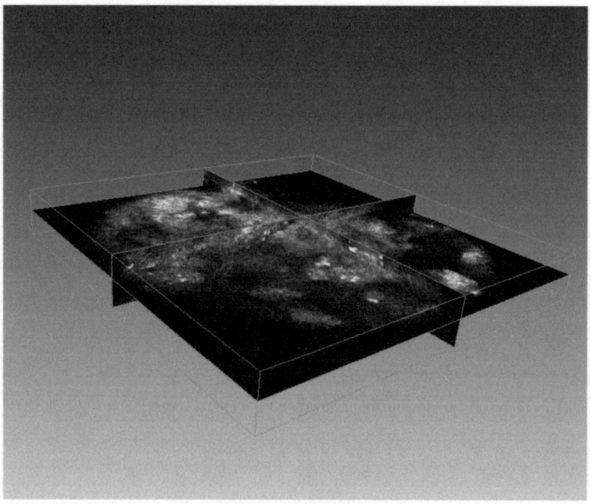

Fig. 15. 3D reconstruction of a volume scan (image size 400 x 400 x 80 μm) (carcinoma, compare with Fig. 14).

4. Future research and developments

To detect early laryngeal cancer, a precise biopsy is of great importance. Confocal microscopy has great potential for early diagnosis in laryngeal cancer. Initially, confocal microscopy was applied outside the larynx in ex-vivo studies (Just et al., 2006b; Just et al., 2007b) and later in vivo for detection of intraoral lesions (Just et al., 2007a). These studies demonstrated the proof of principle of this new method in the field of otorhinolaryngology. With the development of rigid endoscopes and coupling onto the confocal microscope, a new application field has opened.
Improvement of the confocal microscope is an ongoing process. Since the first application of this new technology in animal experiments, several improvements to the technique have been accomplished. Among these, manual setting of the starting plane, automatic volume scans and improvement in image quality should be mentioned (Farahati et al., 2010). A further improvement includes on-line mapping during investigation. Image sizes of 400 x 400 μm only allow off-line mapping of defined tissue regions. Large-scale maps may help to reduce the time required for re-locating tissue regions previously scanned. This method was first described for on-line mapping of corneal structures (Zhivov et al., 2010; Zhivov et al., 2009). Figure 16 demonstrates an on-line scan of a human tongue base.

Combination of RCE with other new technologies, for example Optical Coherence Tomography, is another aspect. OCT is a relatively new technology in laryngology. This technique has two major advantages over confocal microscopy. Firstly, OCT allows imaging of tissue layers, including the basal membrane. Optical cross-sections of the tissue (about 2 mm) can be displayed. The images can later be compared with the histopathology. Assessment of the integrity of the basal membrane is important for diagnosis of laryngeal lesions. Secondly, new OCT devices are coupled onto an operating microscope (Just et al., 2009). In contrast to previously published data on OCT devices, which mostly required special probes (Armstrong et al., 2006; Wong et al., 2005), OCT with an operating microscope allows a non-contact application. The optical system of the microscope can be used in its complete range without changing the center of field of vision, allowing very precise correlations between the location of the B-scan and the biopsy taken for histopathology (Just et al., 2010). But, cross sections of the tissue may also be a disadvantage of OCT compared to confocal microscopy. The latter method supplies en-face images and displays cellular and sub-cellular information on the tissue.
Intraoperative OCT measurements also allow the use of a rigid endoscope for confocal microscopy. With regard to OCT, RCE and other new technologies, primary research is required to validate the new technologies and to develop their clinical application for early detection of laryngeal cancer and its precursors (Hughes et al., 2010). The method demonstrated here will be explored in a multicenter study in patients with laryngeal cancer and its precursors. A new and non-invasive technology for diagnosis of laryngeal diseases should improve the accuracy of biopsies and reduce non-diagnostic biopsies.

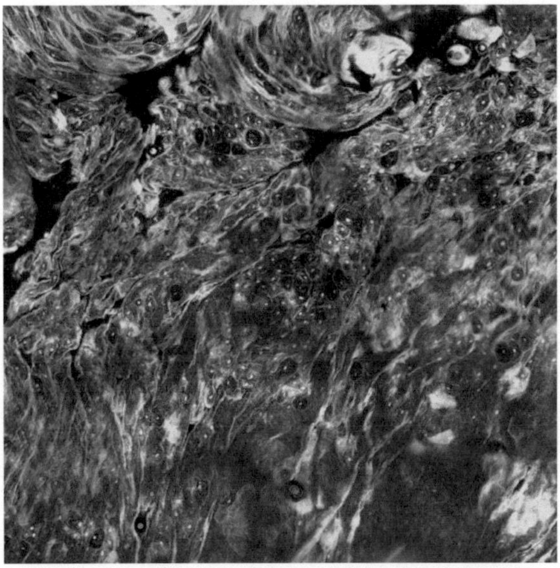

Fig. 16. On-line mapping of the human tongue edge (image size 3.2 x 3.2 mm).

5. References

Amos, W.B. & White, J.G. (2003). How the confocal laser scanning microscope entered biological research. *Biol Cell*, 95, 6, 335-42

Andrea, M.; Dias, O. & Santos, A. (1995a). Contact endoscopy during microlaryngeal surgery: a new technique for endoscopic examination of the larynx. *Ann Otol Rhinol Laryngol,* 104, 5, 333-9

Andrea, M.; Dias, O. & Santos, A. (1995b). Contact endoscopy of the vocal cord: normal and pathological patterns. *Acta Otolaryngol,* 115, 2, 314-6

Arens, C.; Glanz, H.; Dreyer, T. & Malzahn, K. (2003). Compact endoscopy of the larynx. *Ann Otol Rhinol Laryngol,* 112, 2, 113-9

Arens, C.; Reussner, D.; Woenkhaus, J.; Leunig, A.; Betz, C.S. & Glanz, H. (2007). Indirect fluorescence laryngoscopy in the diagnosis of precancerous and cancerous laryngeal lesions. *Eur Arch Otorhinolaryngol,* 264, 6, 621-6

Armstrong, W.B.; Ridgway, J.M.; Vokes, D.E.; Guo, S.; Perez, J.; Jackson, R.P.; Gu, M.; Su, J.; Crumley, R.L.; Shibuya, T.Y.; Mahmood, U.; Chen, Z. & Wong, B.J. (2006). Optical coherence tomography of laryngeal cancer. *Laryngoscope,* 116, 7, 1107-13

Batsakis, J.G. (1999). Surgical excision margins: a pathologist's perspective. *Adv Anat Pathol,* 6, 3, 140-8

Brezinski, M.E.; Tearney, G.J.; Bouma, B.E.; Izatt, J.A.; Hee, M.R.; Swanson, E.A.; Southern, J.F. & Fujimoto, J.G. (1996). Optical coherence tomography for optical biopsy. Properties and demonstration of vascular pathology. *Circulation,* 93, 6, 1206-13

Brunsting, A. & Mullaney, P.F. (1974). Differential light scattering from spherical mammalian cells. *Biophys J,* 14, 6, 439-53

Carriero, E.; Galli, J.; Fadda, G.; Di Girolamo, S.; Ottaviani, F. & Paludetti, G. (2000). Preliminary experiences with contact endoscopy of the larynx. *Eur Arch Otorhinolaryngol,* 257, 2, 68 71

Cikojevic, D.; Gluncic, I. & Pesutic-Pisac, V. (2008). Comparison of contact endoscopy and frozen section histopathology in the intra-operative diagnosis of laryngeal pathology. *J Laryngol Otol,* 122, 8, 836-9

Colden, D.; Zeitels, S.M.; Hillman, R.E.; Jarboe, J.; Bunting, G. & Spanou, K. (2001). Stroboscopic assessment of vocal fold keratosis and glottic cancer. *Ann Otol Rhinol Laryngol,* 110, 4, 293-8

Collier, T.; Lacy, A.; Richards-Kortum, R.; Malpica, A. & Follen, M. (2002). Near real-time confocal microscopy of amelanotic tissue: detection of dysplasia in ex vivo cervical tissue. *Acad Radiol,* 9, 5, 504-12

Csanady, M.; Kiss, J.G.; Ivan, L.; Jori, J. & Czigner, J. (2004). ALA (5-aminolevulinic acid)-induced protoporphyrin IX fluorescence in the endoscopic diagnostic and control of pharyngo-laryngeal cancer. *Eur Arch Otorhinolaryngol,* 261, 5, 262-6

Delank, W.; Khanavkar, B.; Nakhosteen, J.A. & Stoll, W. (2000). A pilot study of autofluorescent endoscopy for the in vivo detection of laryngeal cancer. *Laryngoscope,* 110, 3 Pt 1, 368-73

Farahati, B.; Stachs, O.; Prall, F.; Stave, J.; Guthoff, R.; Pau, H.W. & Just, T. (2010). Rigid confocal endoscopy for in vivo imaging of experimnetal oral squamous intra-epithelial lesions. *J Oral Pathol Med,* 39, 4, 318-327

Gillenwater, A.; Jacob, R. & Richards-Kortum, R. (1998). Fluorescence spectroscopy: a technique with potential to improve the early detection of aerodigestive tract neoplasia. *Head Neck,* 20, 6, 556-62

Harries, M.L.; Lam, S.; MacAulay, C.; Qu, J. & Palcic, B. (1995). Diagnostic imaging of the larynx: autofluorescence of laryngeal tumours using the helium-cadmium laser. *J Laryngol Otol,* 109, 2, 108-10

Haxel, B.R.; Goetz, M.; Kiesslich, R. & Gosepath, J. (2010). Confocal endomicroscopy: a novel application for imaging of oral and oropharyngeal mucosa in human. *Eur Arch Otorhinolaryngol,* 267, 3, 443-8

Hughes, O.R.; Stone, N.; Kraft, M.; Arens, C. & Birchall, M.A. (2010). Optical and molecular techniques to identify tumor margins within the larynx. *Head Neck; in press,*

Just, T.; Bombor, I.; Pau, H.W.; Klautke, G. & Fietkau, R. (2006a). Paraneoplastic changes of oropharyngeal mucosa in breast cancer. *Strahlenther Onkol,* 182, 2, 112-5

Just, T.; Lankenau, E.; Hüttmann, G. & Pau, H.W. (2009). Intra-operative application of optical coherence tomography with an operating microscope. *J Laryngol Otol,* 123, 9, 1027-30

Just, T.; Lankenau, E.; Prall, F.; Hüttmann, G.; Pau, H.W. & Sommer, K. (2010). Optical coherence tomography allows for the reliable identification of laryngeal epithelial dysplasia and for precise biopsy: a clinicopathological study of 61 patients undergoing microlaryngoscopy. *Laryngoscope,* in press,

Just, T.; Stave, J.; Boltze, C.; Wree, A.; Kramp, B.; Guthoff, R.F. & Pau, H.W. (2006b). Laser scanning microscopy of the human larynx mucosa: a preliminary, ex vivo study. *Laryngoscope,* 116, 7, 1136-41

Just, T.; Stave, J.; Bombor, I.; Kreutzer, H.-J.; Guthoff, R. & Pau, H.W. (2007a). In vivo diagnosis of epithelial changes of the oropharynx using confocal microscopy. *Laryngorhinootologie,* 87, 3, 174-80 (in German)

Just, T.; Stave, J.; Bombor, I.; Kreutzer, H.J.; Guthoff, R. & Pau, H.W. (2008). In vivo diagnosis of epithelial changes of the oropharynx using confocal microscopy. *Laryngorhinootologie,* 87, 3, 174-80 (in German)

Just, T.; Stave, J.; Kreutzer, H.J.; Guthoff, R. & Pau, H.W. (2007b). Confocal microscopic evaluation of epithelia of the larynx. *Laryngorhinootologie,* 86, 9, 644-8 (in German)

Just, T.; Stave, J.; Pau, H.W. & Guthoff, R. (2005). In-vivo observation of papillae of the human tongue using confocal laser scanning microscopy. *ORL J Otorhinolaryngol Relat Spec,* 67, 4, 207-212

Kawaida, M.; Fukuda, H. & Kohno, N. (1998). Multidirectional observations of the larynx using transurethral rigid endoscopes during direct laryngoscopy. *J Laryngol Otol,* 112, 5, 464-6

Kiesslich, R.; Burg, J.; Vieth, M.; Gnaendiger, J.; Enders, M.; Delaney, P.; Polglase, A.; McLaren, W.; Janell, D.; Thomas, S.; Nafe, B.; Galle, P.R. & Neurath, M.F. (2004). Confocal laser endoscopy for diagnosing intraepithelial neoplasias and colorectal cancer in vivo. *Gastroenterology,* 127, 3, 706-13

Kiesslich, R.; Goetz, M.; Vieth, M.; Galle, P.R. & Neurath, M.F. (2005). Confocal laser endomicroscopy. *Gastrointest Endosc Clin N Am,* 15, 4, 715-31

Kiesslich, R.; Hoffman, A. & Neurath, M.F. (2006). Colonoscopy, tumors, and inflammatory bowel disease - new diagnostic methods. *Endoscopy,* 38, 1, 5-10

Kiesslich, R. & Neurath, M.F. (2005). Endoscopic detection of early lower gastrointestinal cancer. *Best Pract Res Clin Gastroenterol,* 19, 6, 941-61

Kiesslich, R. & Neurath, M.F. (2006). Magnifying chromoendoscopy for the detection of premalignant gastrointestinal lesions. *Best Pract Res Clin Gastroenterol,* 20, 1, 59-78

Kleinsasser, O. (1962). Die Laryngomikroskopie (Lupenlaryngoskopie) und ihre Bedeutung für die Erkennung der Vorerkrankungen und Frühformen des Stimmlippenkarzinoms. *Arch Ohren Nasen Kehlkopfheilkd,* 180, 724-7

Kothe, C.; Münzenmayer, C.; Wittenberg, T. & Hess, M. (2005). Erfahrungen mit optischen Stimmlippenbiopsien bei Leukoplakien. *Laryngorhinootologie,* 84, 2, 92-5

Lüerßen, K.; Lubatschowski, H.; Ursinus, K.; Gasse, H.; Koch, R. & Ptok, M. (2006). Charakterisierung von Stimmlippen mittels optischer Kohärenztomographie. *HNO,* 54, 8, 611-615

Malzahn, K.; Dreyer, T.; Glanz, H. & Arens, C. (2002). Autofluorescence endoscopy in the diagnosis of early laryngeal cancer and its precursor lesions. *Laryngoscope,* 112, 3, 488-93

McGuirt, W.F. & Browne, J.D. (1991). Management decisions in laryngeal carcinoma in situ. *Laryngoscope,* 101, 2, 125-9

Rajadhyaksha, M.; Grossman, M.; Esterowitz, D.; Webb, R.H. & Anderson, R.R. (1995). In vivo confocal scanning laser microscopy of human skin: melanin provides strong contrast. *J Invest Dermatol,* 104, 6, 946-52

Schmitt, J.M. (1999). Optican coherance tomography (OCT): a review. *IEEE J Sel Top Quan,* 5, 1205-1215

Schmitt, J.M.; Lee, S.L. & Yung, K.M. (1997). An optical coherence microscope with enhanced resolving power in thick tissue. *Opt Common,* 142, 203-207

Stave, J.; Zinser, G.; Grummer, G. & Guthoff, R. (2002). Modified Heidelberg Retinal Tomograph HRT. Initial results of in vivo presentation of corneal structures. *Ophthalmologe,* 99, 4, 276-80 (in German)

Venkatesh, K.; Cohen, M.; Evans, C.; Delaney, P.; Thomas, S.; Taylor, C.; Abou-Taleb, A.; Kiesslich, R. & Thomson, M. (2009). Feasibility of confocal endomicroscopy in the diagnosis of pediatric gastrointestinal disorders. *World J Gastroenterol,* 15, 18, 2214-9

Wardrop, P.J.; Sim, S. & McLaren, K. (2000). Contact endoscopy of the larynx: a quantitative study. *J Laryngol Otol,* 114, 6, 437-40

White, W.M.; Rajadhyaksha, M.; Gonzalez, S.; Fabian, R.L. & Anderson, R.R. (1999). Noninvasive imaging of human oral mucosa in vivo by confocal reflectance microscopy. *Laryngoscope,* 109, 10, 1709-17

Wong, B.J.; Jackson, R.P.; Guo, S.; Ridgway, J.M.; Mahmood, U.; Su, J.; Shibuya, T.Y.; Crumley, R.L.; Gu, M.; Armstrong, W.B. & Chen, Z. (2005). In vivo optical coherence tomography of the human larynx: normative and benign pathology in 82 patients. *Laryngoscope,* 115, 11, 1904-11

Zhivov, A.; Blum, M.; Guthoff, R. & Stachs, O. (2010). Real-time mapping of the subepithelial nerve plexus by in vivo confocal laser scanning microscopy. *Br J Ophthalmol,* 94, 9, 1133-5

Zhivov, A.; Guthoff, R. & Stachs, O. (2009). On-line mapping of corneal structures with in vivo laser scanning microscopy. *Klin Monbl Augenheilkd,* 226, 12, 980-3 (in German)

7

High Precision and Fast Functional Mapping of Brain Circuitry through Laser Scanning Photostimulation and Fast Dye Imaging

Xiangmin Xu
The Department of Anatomy and Neurobiology, the University of California, Irvine
The United States of America

1. Introduction

The development of modern neuroscience tools has a significant impact on the progress in the field of neuroscience (Kandel, 1982). New tools has greatly faciliated the neuroscience resreach, and can be critical for studies of brain circuit organization and function. Although many approaches are useful by themselves, it is desirable to combine existing powerful techniques to harness each technique's advantages and compensate for limitations.

In many brain areas, neuronal circuits are segregated into anatomically discrete areas such as specific lamina and modules or compartments (Mountcastle, 1997). Functional imaging of these brain areas is particularly useful in characterizing circuit properties. Fast voltage-sensitive dye (VSD) imaging, which detects neuronal membrane potential changes via shifts in the dye absorption /or fluorescence emission in response to varying membrane potentials, offers a great means of simultaneous monitoring neuronal activities from many locations with high spatial and temporal resolutions. With new dyes and modern imaging apparatus, VSD imaging has been widely used to study spatiotemporal dynamics of population neuronal activity in cortical tissue both in vivo and in vitro (Grinvald & Hildesheim, 2004). Particularly, for in vitro brain slices, fast VSD imaging is important for mapping circuit organization and response dynamics, and more recently has been used to probe functional abnormalities in models of neurological and psychiatric disorders (Ang et al., 2006; Airan et al., 2007). One major limitation of most in vitro VSD imaging studies, however, lies in that the imaged neuronal responses are either spontaneous seizure activities through pharmacological manipulations or induced by electric stimulations (Petersen & Sakmann, 2001; Huang *et al.*, 2004; Ang *et al.*, 2006). Significant disadvantages of electric stimulation include indiscriminate activation of axons of passage, slow and inefficient placement of multiple stimulation locations, and tissue damage. In comparison, optical stimulation including laser scanning photostimulation (LSPS) either by glutamate uncaging or direct activation of light-sensitive channels (e.g., channelrhodopsin-2) enables rapid and noninvasive photoactivation of neurons with great convenience and superior spatial resolution in practical experiments (Callaway & Katz, 1993; Boyden *et al.*, 2005; Petreanu *et al.*, 2009). Combining whole-cell recordings from single neurons with photostimulation of clusters of presynaptic neurons permits extensive mapping of local functional inputs to individually recorded neurons (Schubert *et al.*, 2003; Shepherd & Svoboda, 2005; Xu & Callaway, 2009).

While both methods of VSD imaging and LSPS are already well established, the combination can offer new and potentially valuable advantages in studies of spatiotemporal dynamics and functional connectivity of populations of neurons in cortical circuits. LSPS does not activate the axons of passage and avoids the difficulty of restricted and unambiguous stimulation by electrode-based methods; VSD imaging affords the ability of reporting the information of population responses that is largely missed when the combination of photostimulation with single cell recordings is used for circuit input mapping. Although it is of great advantage to combine both VSD imaing and LSPS for circuit studies, the combination had never achieved due to technical challenges until recently my research group succeeded in this implemention. We have developed a new mapping technique that uses LSPS to provide precise and strong stimulation of neural circuit components, and allows for functional connectivity and spatiotemporal dynamics of population activity in large cortical areas to be examined by fast VSD imaging (Xu et al., 2010). This novel approach is an all-optical approach for brain circuit mapping, and it enables fast mapping and high-precision evaluation of cortical organization and function.
In this book chapter, I will report on this novel technique in detail, and introduce the system design and new technical improvement. Then I will describe effective applications of the technique in mapping brain circuitry.

2. General system design, calibration, VSD response characterization

2.1 The system design for combining LSPS with VSD imaging

Our overall system is illustrated in Figure 1, consisting of LSPS, VSD imaging, and electrophysiological recording systems. We adopted the design of the LSPS system described previously (Shepherd *et al.*, 2003; Shepherd & Svoboda, 2005). A laser unit (model 3501; DPSS Lasers, Santa Clara, CA) was used to generate 355 nm UV laser for glutamate uncaging. The laser beam was 1.5 mm in diameter and directed through the optical path of our system. Short durations of laser flashes (e.g., 1- 3 ms) were controlled by using an electro-optical modulator (i.e., pockels cell) (Conoptics, Danbury, CT) and a mechanical shutter (Uniblitz; Vincent Associates, Rochester, NY). Laser beam power was modulated by a neutral density gradient wheel and monitored by diverting a small fraction of the laser beam with a glass coverslip to a photodiode. The laser scanning system included an X-Y pair of scan mirrors, the scan lens, the tube lens, and the objective lens (Figure 1). The mirrors delivered the laser beam through a scan lens; then the beam entered the microscope (BX51WI; Olympus) via a dichroic mirror (351DRLP; Chroma Technology, Brattleboro, VT) and was focused by a custom-made UV-transmitting tube lens. The beam underfilled the back aperture of the microscope objective to provide a more columnar (as opposed to conical) illuminating beam, keeping the mapping as two-dimensional as possible by reducing the axial resolution. Various laser stimulation positions could be achieved through galvanometers-driven XY scanning mirrors (Cambridge Technology, Cambridge, MA), as the mirrors and the back aperture of the objective were in conjugate planes, translating mirror positions into different scanning locations at the objective lens focal plane. During uncaging, a variable number of patterned sites that covered the mapping field were stimulated sequentially in a nonraster, nonrandom sequence, following a "shifting-X" pattern designed to avoid revisiting the vicinity of recently stimulated sites (Shepherd *et al.*, 2003; Shepherd & Svoboda, 2005). A modified version of Ephus software (Ephus, available at https://openwiki.janelia.org/) was used to control photostimulation protocols and acquire photostimulation data.

As indicated in Figure 1, a dual camera port was used to couple the Q-imaging CCD camera (Retiga 2000, Q-imaging Inc, Austin, TX) and the LSPS system to a MiCAM02 fast imaging system (SciMedia USA Ltd, Costa Mesa, CA) for voltage sensitive dye imaging. The VSD imaging computer (B) was independent from the computer (A), but its image acquisitions were triggered and synchronized with computer A through output TTL pulses from computer A. Upon triggering, optical recording of VSD signals was performed by the MiCAM02 system with a sampling rate of 2.2 ms per frame (frame resolution 88 (w) x 60 (h) pixels). The photostimulation and imaging systems were aligned and uncaging tests were visualized by exciting caged fluorescein dextran on a glass slide (Xu et al., 2010). Under the 4x objective, the laser beam formed uncaging spots, each approximating a Gaussian profile with the estimated full width at the half maximal fluorescence intensity of about 150 μm laterally at the focal plane; and the laser beam caused uncaging of fluorescent dextran at ~100 μm in depth. It is important to note that the physical laser excitation size in the glass slide does not directly translate into the effective spatial resolution of physiological uncaging in brain slices (see below). In comparison, under the 60x objective, the UV laser induced uncaging was focused to a smaller spot with the estimated full width at the half maximal intensity of about 2 μm. While photostimulation through the 60x objective is useful for exciting smaller numbers of neurons or possibly even single cell stimulation, our current work focused on both imaging and excitation through the 4x objective to map propagation of activity within local cortical circuits.

Electrophysiological recordings, photostimulation, and imaging of the slice preparations were done in a slice perfusion chamber mounted on a motorized stage of the microscope. Slices were visualized with an upright microscope (BW51X; Olympus, Tokyo, Japan) with infrared differential interference contrast optics. At low magnification (4x objective lens, 0.16 NA; UPlanApo; Olympus), laminar and cytoarchitectonic features of brain slices were visualized under infrared bright-field transillumination; and the slice images were acquired by the Q-imaging camera. Digitized images from the camera were used for guiding and registering photostimulation sites in brain slices. During experiments, simultaneous electrophysiological recordings (whole cell recordings, loose-seal patchings or local field potential recordings) were conducted to monitor laser stimulation and correlate VSD signals with the electrical activity. To perform patch recording, cells were visualized at high magnification (60x objective, 0.9 NA; LUMPlanFl/IR; Olympus). Neurons were patched with borosilicate electrodes and recorded at room temperature in the whole-cell or loose-seal mode. The patch pipettes (4-6 MΩ resistance) were filled with an internal solution containing 126 mM K-gluconate, 4 mM KCl, 10 mM HEPES, 4 mM ATP-Mg, 0.3 mM GTP-Na, 10 mM phosphocreatine (pH 7.2, 300 mOsm). The electrodes also contained 0.5-1% biocytin for cell labeling and morphological identification. Resting membrane potentials were measured immediately after electrodes broke into the cells following formation of a gigaohm seal, and current pulses were injected to examine each cell's basic electrophysiological properties. Data were acquired with a Multiclamp 700B amplifier (Molecular Devices, Sunnyvale, CA), data acquisition boards (models PCI MIO 16E-4 and 6713; National Instruments, Austin, TX), and Ephus software. Data were filtered at 2 kHz using a Bessel filter and digitized at 10 kHz and stored on a computer. Once stable whole-cell recordings were achieved with good access resistance (usually <20 MΩ), the microscope objective was switched from 60x to 4x for laser scanning photostimulation. The same low-power objective lens was used for delivering the UV flash stimuli.

Stock solution of MNI-caged-l-glutamate (4-methoxy-7-nitroindolinyl-caged l-glutamate, Tocris Bioscience, Ellisville, MO) was added to 20-25 ml of circulating ACSF for a concentration of 0.2 mM caged glutamate. After 5-6 hours of experimentation, the bath solution and MNI-glutamate was refreshed. Care was taken to ensure a constant fluid level in the chamber of ~ 2.0-2.5 mm above the slice to avoid small fluctuations in UV attenuation. For some experiments, the addition of 10 μM CNQX (6-Cyano-7-nitroquinoxaline-2,3-dione disodium, Tocris Bioscience) and 10 μM CPP (3-(2-Carboxypiperazin-4-yl)-propyl-1-phosphonic acid, Tocris Bioscience) in solutions were used to block ionotropic glutamate receptors. In addition, modified ACSF containing 0.2 mM Ca2+, 4 mM Mg2+ was used to block synaptic transmission.
To prepare living brain slices, wild type C57/B6 mice (postnatal day 17–23) were deeply anesthetized with Nembutal (>100 mg/kg, i.p.), rapidly decapitated, and their brains removed. Visual cortical or hippocampal sections were cut 400 μm thick with a vibratome (VT1200S; Leica Systems, Germany) in sucrose-containing artificial cerebrospinal fluid (CSF) (in mM: 85 NaCl, 75 sucrose, 2.5 KCl, 25 glucose, 1.25 NaH2PO4, 4 MgCl2, 0.5 CaCl2, and 24 NaHCO3). Slices were first incubated in sucrose-containing ACSF for 30 minutes to 1 hour at 32°C, and after the initial incubation period, transferred to recording ACSF (in mM: 126 NaCl, 2.5 KCl, 26 NaHCO3, 2 CaCl2, 2 MgCl2, 1.25 NaH2PO4, and 10 glucose) for the dye staining at room temperature. The slices were stained for 1-2 hours in a staining chamber containing ACSF with 0.02 mg/ml of an oxonol dye, NK3630 (available from Nippon Kankoh-Shikiso Kenkyusho Co., Ltd., Japan), and then maintained in regular ACSF before use. The NK3630 dye has been characterized in previous studies and has its peak signal-to-noise ratio centered around 705 nm (Jin et al., 2002). Throughout the incubation, staining and recording, the slices were continuously bubbled with 95% O2-5% CO2.

2.2 LSPS system calibration and VSD imaging tests

The effective physiological resolution of photostimulation in our system was assessed by measuring neuronal excitation profiles, which quantify the spatial distribution of uncaging sites that produce action potentials in individually recorded neurons; similar approaches to assess the distribution of evoked neuronal excitability have been used by other groups (Shepherd *et al.*, 2003; Shepherd & Svoboda, 2005; Weiler *et al.*, 2008). Specifically, the spatial resolution of effective photostimulation, R, was quantitatively estimated as the mean distance between the soma and the spike-generating sites weighted by the number of spikes per site [i.e., $R = \Sigma(r \times n)/\Sigma n$, where for each site, r is the distance to the soma and n is the number of spikes]. Under our normal photostimulation conditions through the 4x objective (power level: 20–35 mW; pulse duration: 1-2 ms), the average R across excitatory neurons in primary visual cortex (V1) was about 90 μm across different V1 layers (Xu et al., 2010). This is illustrated in Figure 2A-B, as LSPS had a spatial resolution of 50-100 μm in evoking suprathreshold spikes from the recorded neuron in a V1 slice. Therefore, photostimulation in our experiments could provide spatially restricted neuronal activation, and offered a sufficient resolution for V1 laminar circuit mapping. However, under stronger photostimulation conditions (power level: 30-35 mW; pulse duration: 3 ms or above), the spatial resolution of photostimulation was lowered as spikes could be evoked from cells located 100-300 μm away from the stimulation sites. The stronger laser stimulation evoked neuronal spikes from the cells located far away from the photostimulation sites possibly either through direct activation of distant dendrites or through strong synaptic drive

(Dantzker & Callaway, 2000; Weiler *et al.*, 2008). Hence, in hippocampal circuit mapping experiments that we have conducted, strong laser stimulation was used to further assess trans-synaptic spread of activity.

During experiments, a 705 nm light trans-illuminated brain slices, and voltage-dependent changes in the light absorbance of the dye were captured by the MiCAM02 camera (Figure 1). Under the 4x objective, the imaging field covered the area of 1.28 x 1.07 mm2 with a spatial resolution of 14.6 x 17.9 μm2/pixel. With the 60x objective, the imaging field covered the area of 84.5 x 70.8 μm2 with a resolution of 0.96 x 1.18 μm2/pixel. The trials were obtained every 8 seconds and the recording periods were 1000 frames for each trial. VSD images were smoothed by convolving images with a Gaussian spatial filter (kernel size: 3x3 pixels; δ size: 1x1 pixel) and a Gaussian temporal filter (kernel size: 3 frames; δ size: 1 frame). VSD signals were originally measured by the percent change in pixel light intensity [ΔI/I %; the % change in the intensity (ΔI) at each pixel relative to the initial intensity (I)]. In addition, signal amplitudes were expressed as standard deviations (SD) above the mean baseline signal for display and quantification.

Because photostimulation and VSD imaging are combined, separation of the laser excitation and the VSD signal is important. Given the UV laser and the VSD absorption light have differing wavelengths (355 nm vs ~705 nm), the laser and its excitation artifact had been significantly reduced using a band-pass filter centered at 705 nm right before the imaging camera (See Figure 1). Furthermore, when using short laser pulses (i.e., 1-3 ms), the laser artifact signal only existed in the initial 2 image frames (2.2 ms per frame). Compared to the VSD signal reflecting neural activity (0.1-0.5%), the artifact signal was large, up to 2% change from baseline (Figure 3B1-B4), however did not interfere with true VSD signals, considering detectable VSD responses occurred approximately 5-6 frames after the laser onset.

Successful combination of VSD imaging with laser photostimulation was exemplified in Figure 2C and D through the 4x and 60x objectives, respectively. In Figure 2C, the image data were from a coronal slice of mouse primary visual cortex (V1) with photostimulation and imaging through the 4x objective. In response to the spatially restricted laser stimulation, the evoked neural activity reflected by VSDsignals was initiated at the stimulation site shortly after the laser onset, and gradually propagated and spread to other regions. As shown in Figure 2D, VSD imaging and photostimulation was also achieved through the 60X objective; the VSD response was the strongest at the stimulation location with weaker activation in the surroundings. In the present study, single-trial photostimulation-evoked VSD responses were sufficiently strong and could be discerned from background noise; hence no averaging over multiple trials was necessary.

2.3 Characterization of photostimulation-evoked VSD responses

In V1 imaging experiments, we characterized photostimulation-evoked VSD responses. The VSD signal results from the summed signal of neuronal ensembles; but the observed properties of photostimulation evoked VSD signals were closely related to membrane potential depolarization of individual neurons (Figure 3B1- B4). Spikes responded faster to photostimulation than VSD signals, as the average spike peak time in response to photostimulation was 11.8 ± 3 ms, and the average VSD signal peak time was 29.9 ± 6.3 ms (Xu et al., 2010). In these experiments, the average VSD response latency across different V1 layers was 15.4 ± 0.7 ms.

As exemplified in Figure 3A-E, the photostimulation-evoked VSD responses were mediated by glutamate and its receptors, as neuronal spiking and VSD responses to glutamate uncaging were essentially abolished by the ionotropic glutamate receptor antagonists (CPP and CNQX). There was no VSD signal without glutamate uncaging; laser flashes alone did not activate neurons or induce VSD responses. There was no detectable signal in brain slices without VSD staining in response to photostimulation, indicating that photostimulation-evoked VSD responses were not contaminated by activity-related tissue autofluorescence signals (Shibuki et al., 2003; Llano et al., 2009). Furthermore, reducing synaptic transmission from stimulated neurons using low Ca2+ and high Mg2+ ACSF solution restricted VSD changes to the region near the stimulation site. This indicates that most activity far away from the stimulation site reflected synaptic spread of activity to postsynaptic neurons rather than activity in the axons and distant dendrites of directly stimulated cells (Figure 3F-H). This is similar to the result observed in TTX, which blocks both synaptic spread and conduction of activity with the axons of stimulated cells.

3. Visual cortical circuit mapping

Compared to previous studies of V1 circuits that used VSD imaging with electrical stimulation or puffed glutamate at different cortical locations (Nelson & Katz, 1995; Yuste *et al.*, 1997), our new technique allows us to evaluate laminar propagation of evoked excitation and probe functional connectivity of mouse V1 laminar circuits, with a much improved precision and speed.

The new technique has been used to map activity propagation and layer-specific output patterns in V1 local circuits. Under the stimulation conditions used, LSPS (1ms, 20-35 mW) offered spatially restricted neuronal activation in a specific cortical layer so that we were able to map direct projections from the stimulated layer to its targeted layer(s) by VSD imaging of evoked activation. That is, photostimulation-evoked action potentials were restricted to neurons with cell bodies at or close to the stimulation site; activity then propagated through the axons of the stimulated neurons and generated postsynaptic responses in the neurons that were connected to the stimulated cells. The measured VSD signals reflected the combined contributions of these sources, but responses distant from the stimulation site were dominated by postsynaptic changes, as evidenced by control experiments in which synaptic transmission was blocked by using low Ca2+ and high Mg2+ ACSF (see above). Figure 4A, B and C show VSD image frames in response to laser photostimulation at cortical layers 2/3, 4 or 5 in a V1 coronal slice, respectively, while Figure 4D shows VSD image frames in response to electrical stimulation at V1 layer 4. In comparison, layer 4 photostimulation was effective as electrical stimulation in evoking population neuronal responses, but its activation was more restricted and specific (compare Figure 4B6 and Figure 4D6). Unlike electrical stimulation, a significant advantage of photostimulation is such that uncaged glutamate does not activate passing axon fibers (Xu et al., 2010), thus ensuring spatial specificity of stimulation and avoiding antidromic presynaptic activation. During photostimulation and VSD imaging experiments, stimulation in V1 cortical layers initiated excitation which resulted in VSD signals first localized to the stimulation site at around 10-20 ms after laser exposure; excitation then propagated to functionally connected cortical regions. For layer 2/3 stimulation, the activation in layer 2/3 was mostly localized in layer 2/3, however, relatively small but clear activation propagated from layer 2/3 to layer 5, bypassing most of layer 4 (Figure 4A1-A6). There was essentially no excitation from layer 2/3 propagating to layer 6. Stimulation in layer 4 caused excitatory activity to spread vertically to layers 2/3 and 5, but with little excitation in layer 6 (Figure 4B1-B6). Strong activation in layers 2/3 and 5 due to layer 4 stimulation indicates these layers receive strong direct projections from layer 4. Layer 5 stimulation resulted in distinct foci of activation in layer 2/3 (Figure 4C1-C6), and some activation spread into layers 6 and 4. Layer 6 stimulation resulted in mostly localized responses, but some activation could spread into upper layers with stronger phostimulation.

This new technique has also been used to investigate spatiotemporal patterns of activity in tangential slices of visual cortex. As shown in Figure 5, photostimulation evoked VSD responses propagated largely in horizontal directions, and their response patterns evolved into extended clusters of distinct activation domains. The pattern of population activity reflected anatomical horizontal connections to the stimulation sites. There were spatial shifts in the distribution of activation domains that resulted from different photostimulation sites. The clustered activation domains seen in the tangential slices were similar to that found in ferret visual cortex, but at a much finer scale. This is perhaps due to the spatially restricted neuronal activation by photostimulation instead of electrical stimulation used in previous studies (Tucker & Katz, 2003).

In addition, as LSPS activates many different sites with very short time intervals (e.g., see Figure 2A-B), our technique enables rapid mapping of neuronal circuitry by patterned photostimulation and VSD imaging of multiple V1 locations. This approach is much faster and more efficient than mapping with electrical stimulation at different locations, considering the time required for proper electrode placement on the order of minutes. As illustrated in Figure 6, an array of 4 x 4 photostimulation sites (cyan stars, in Figure 6A1) covered the V1 cortical area at different laminar locations. During this experiment, the laser stimulation was a 1ms laser flash at 35 mW with a stimulation interval of 8 seconds; the whole duration of VSD mapping for all 16 locations only took 128 seconds. Figure 6C1-C16 shows the overall laminar profile of photostimulation-evoked VSD responses by presenting peak activation frames at each of the 16 stimulation sites covering V1 from cortical layer 2/3 to layer 6. LSPS resulted in laminar specific patterns of excitatory output detected by VSD imaging.

Based upon high-resolution mapping data, we can quantify patterns of evoked activation in individual cortical layers to analyze interlaminar functional connectivity of local V1 circuits. Provided that VSD activation outside the stimulation site mostly reflects synaptic spread of activity to postsynaptic neurons rather than activity in the axons and distant dendrites of directly stimulated cells in the photostimulated layer, we can construct a functional connectivity diagram for mouse V1 local circuits based upon quantitative analysis of propagation activity from multiple laminar locations (Xu et al., 2010). Moreover, our data allows for quantification of functional interlaminar projection strength. Specifically, based upon the excitation profile data, it is estimated that each photostimulation activated approximately 250 neurons within ~90 μm of the laser uncaging center, as N spiking = ρ Vexc , where the number of spiking neurons, N spiking is determined by the mouse cortical neuronal density, ρ and the volume of excited neurons, Vexc [the product of photostimulation-evoked spiking area (π R2) and the photostimulation axial penetration depth in the brain slices (~ 50-100 um; an average of 75 um is used for calculation)]. The neuronal density in mouse visual cortical layers 2/3, 4, 5 and 6 has been previously quantified (Schuz & Palm, 1989). Therefore, the average number of spiking neurons (N spiking) per photostimulation under normal experimental conditions can be calculated. As the number of presynaptic spiking neurons by each photostimulation is known, the projection strength of these neurons could be quantitatively defined with the measurement of evoked activation by VSD imaging in their functionally connected regions (Olivas et al., submitted). Taken together, our new technique yields precise and quantitative mapping data for functional circuit analysis.

4. Hippocampal circuit mapping

We have extended our technique to investigation of spatiotemporal dynamics of hippocampal circuit activity. We are particularly interested in understanding whether or how a restricted population of dentate neurons (e.g., granule cells) can engage the entire trisynaptic circuit. We photo-stimulated different locations in the dentate gyrus (DG) such as the molecular layer, the granule cell layer and the hilus, and monitored how evoked responses initiated and propagated throughout the hippocampal circuitry.
For the illustration purpose, we present example data of stimulation in the DG molecular layer as shown in Figure 7. In this experiment, restricted photostimulation occurred in the superior blade of DG molecular layer (indicated by small cyan star in Figure 7A). This uncaging mimicked the perforant pathway stimulation, and induced serial excitatory propagation in DG, CA3 and CA1 (Figure 7C). As shown in Figure 7B, we simultaneously monitored optical recordings with field potential recordings at CA3, and the optical signal from the ROI in CA3 correlated well with electrical activity reflected by the field potential at that location. Our technique clearly revealed the information flow along the trisynaptic circuitry in hippocampus. After a short delay of the photostimulation in the molecular layer, the VSD response first initiated in the granule cell layer, traversed through polymorphic layer (hilus), and reached CA3 and then CA1 (Figure 7C). This stimulation caused long-lasting excitation in DG, CA3 and CA1. The excitation in CA3 and CA1 was extensive; the amplitudes of the optical signals in pyramidal cell layer were stronger than those in stratum radiatum or stratum oriens. In contrast, clearly the CA2 region did not have strong excitation (see the white arrows in Figure 7C), which supports evidence that CA2 lacks of the granule cell input (Swanson et al., 1978; Nakagami et al., 1997). The data illustrate a localized DG neuronal population effectively engaged in the excitatory flow of information throughout DG and the hippocampus proper collectively, thus providing a comprehensive perspective of the functional circuit organization and dynamics of the hippocampal pathway.

5. Further technical development

As an alternative method of photostimulation via glutamate uncaging, optogenetical approaches can be used to make neurons directly photoactivated via channelrhodopsin-2 (ChR2) (Boyden et al., 2005; Zhang et al., 2006; Gradinaru et al., 2010). ChR2 is a light-sensitive nonselective cation channel isolated from the green alga *Chlamydomonas reinhardtii,* which can transduce light energy into neural activity when expressed in neurons (Boyden et al., 2005). Therefore, in our new technique, glutamate uncaging can be replaced with photoactivation via channelrhodopsin or other genetically encoded photosensitive molecules expressed in cortical neurons (Boyden *et al.*, 2005; Kuhlman & Huang, 2008). Genetically encoded ChR2 can also enhance the ability of photostimulation in targeting define cell types, as glutamate uncaging indiscriminately stimulates all neurons expressing glutamate receptors. Figure 8 exemplifies our ability of VSD imaging with ChR2 photoactivation. We take advantage of the transgenic mouse line expressing ChR2 in subsets of excitatory pyramidal neurons under the control of the Thy1 promoter (Figure 8A), and have mapped local cortical circuitry using the blue laser photostimulation (475 nm laser, 2mW, 1ms) without the aid of caged glutamate. As shown in Figure 8 C, compared to photostimulation via glutamate uncaging, ChR2 activation was much faster, with an imaging response latency of less than 1 image frame (2.2 ms). In addition, the population neuronal activity induced by ChR2 photoactivation was quite strong and the activity propagation pattern was similar to that evoked by glutamate uncaging (Figure 8B and C). A caveat is that ChR2 activation has an issue of spatial precision of photostimulation as action potentials can be elicited by photostimulating the soma and dendrites as well as the axon (Petreanu et al., 2007). Regardless, this new development makes it possible to target not only specific cortical regions but also specific subset of neurons within their participating circuits through laser scanning photostimulation.

Our technique can be generally extended to other fast dye imaging. For example, when calcium indicators are used, LSPS can be combined with calcium imaging. As the optical signals of VSD and calcium indicators differentially report subthreshold postsynaptic potentials and local action potential firing, and as simultaneous VSD and calcium imaging using epifluorescence optics has been reported to measure the spatiotemporal dynamics of activity in cortical circuits in vitro and in vivo (Berger et al., 2007), we plan to combine both VSD and calcium imaging with laser scanning photostimulation, which will allow us to monitor both the synaptic drive and the spiking activity of a given circuit location at the same time. Compared to VSD measurements, the drawbacks of calcium measurements include the concern that some cells or cell types that produce action potentials may not produce measurable Ca2+ transients, and the inability of calcium dyes to detect inhibitory synaptic activity (Knopfel *et al.*, 2006).

Although we have successfully used synthetic voltage-sensitive dyes in our technical implementation, the conventional dyes have certain limitations such as non-specificity of cell staining and side effects including toxicity and dye interference with neuronal membrane potentials and excitability (Mennerick et al., 2010). However, emerging optogenetics methods have started to overcome the problems of the conventional dyes by developing genetically encoded sensors and expressing such sensors in defined cell populations in vivo, thus avoiding background noise and enabling a rigorous assignment of optical response signals to specific cellular sources (Akemann et al., 2010). Clearly, this is an important consideration for our future improvement.

6. Discussion and conclusions

As described in this chapter, we have developed a new photostimulation-based technique through the integration of VSD imaging and LSPS for high precision and rapid mapping of in vitro functional circuits at the neuronal population level. The incorporation of LSPS has greatly enhanced the ability of assessment of evoked network activity by fast VSD imaging. The "dream-team" methods of stimulation and imaging make this new technique quite effective in mapping neuronal circuit dynamics and organization.

Photostimulation based mapping techniques have been widely applied for analyzing cortical circuits. LSPS combined with whole cell recordings is an effective method for mapping local circuit inputs to single neurons, as the simultaneous recording from a

postsynaptic neuron with photostimulation of clusters of presynaptic neurons at many different locations provides quantitative measures of spatial distribution of excitatory and inhibitory inputs impinging onto single recorded neurons (Callaway & Katz, 1993; Schubert *et al.*, 2003; Shepherd & Svoboda, 2005; Xu & Callaway, 2009). LSPS has also been combined two-photon calcium imaging to generate detailed functional maps of inputs to individual cells with single-cell and three-dimensional precision (Nikolenko *et al.*, 2007). Different from the aforementioned approaches, our newly developed technique is intended to assess circuit activation and network connectivity at the neuronal population level through fast VSD imaging and photostimulation. As our technique is often performed with simultaneous electrophysiological recordings of single neurons, the method is readily combined with whole-cell patch-clamp measurements of electrical signals in brain slices so that local cortical circuits can be examined in the same brain slice at both single cell and population levels. Note that laser photostimulation has been combined with intrinsic flavoprotein auto fluorescence imaging to map long range neural connections in the thalamocortical slice (Llano et al., 2009), but the imaging of intrinsic signals generally lacks the temporal and spatial resolution required for studies of local circuit connections and dynamics.
Our technique can have important applications in the field of cortical circuitry as demonstrated in our studies of mouse V1 and hippocampal circuits. As the mouse is an important model system for cortical circuit studies and mouse V1 circuits are less well understood, we imaged and probed V1 with spatially restricted photostimulation, and mapped interlaminar functional connectivity and circuit dynamics. This technique enabled direct visualization of interlaminar functional connections in V1 circuits at a previously unattainable precision (Burkhalter, 1989; Yuste et al., 1997; Callaway, 1998). Overall, our mouse V1 data fit with and extend previous anatomical and physiological observations of laminar patterns of axonal projections in rodent V1 local circuits (Burkhalter, 1989; Yuste et al., 1997), and is generally consistent with proposed V1 laminar operational schemes (Gilbert, 1983; Callaway, 1998). Although our V1 results are as expected from known anatomy and connectivity in the rodent visual cortex (Burkhalter, 1989; Yuste et al., 1997), this method would be valuable for assaying brain areas, species, or genetically modified mouse lines in which these data are not presently available. In addition, the application of this technique to the hippocampal circuitry mapping further validated its technical power and effectiveness. The circuit dynamics and functional connection in the hippocampal circuitry were mapped through spatially restricted activation of a subset of DG neuronal population. With strong photostimulation, we were able to examine the perspective of the polysynaptic spread of activity in detail.
As this is all-optical technique to image and evoke circuit activity, our new method can be further developed as a methodology for identification and monitoring of real-time responses in vitro (e.g., cell cultures and slice preparations) to drugs, therapeutic or genetic interventions. The immediate application can be extended for fast and effective screening of circuit alterations in transgenic animal models recapitulating specific neurological diseases at a large scale.

7. Acknowledgements

I would like to thank my group members for their contributions in our new technical development. This work was funded by the US National Institutes of Health Grants DA-023700 and DA-023700-04S1.

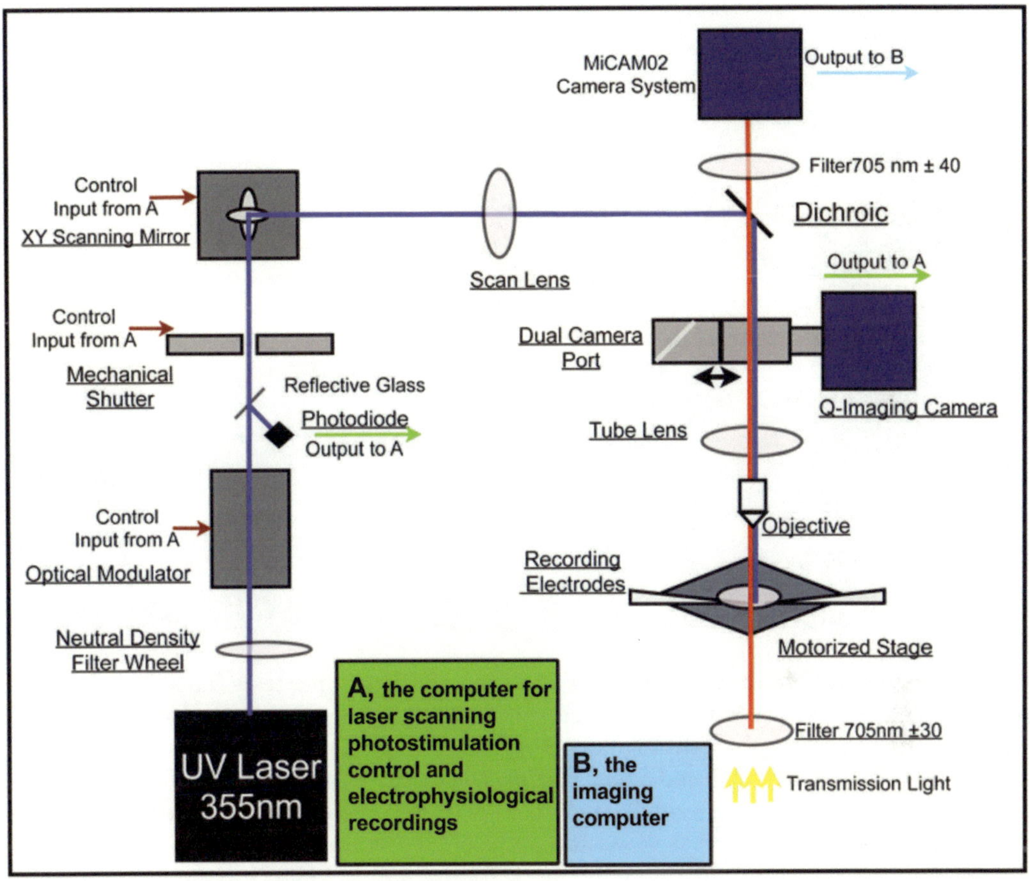

Fig. 1. **The general system design for combining laser scanning photostimulation (LSPS) with voltage sensitive dye (VSD) imaging.** The diagram shows the whole system consisting of laser scanning photostimulation, VSD imaging, and electrophysiological recording systems. This figure was modified from Xu et al. (2010) with permission of the American Physiological Society.

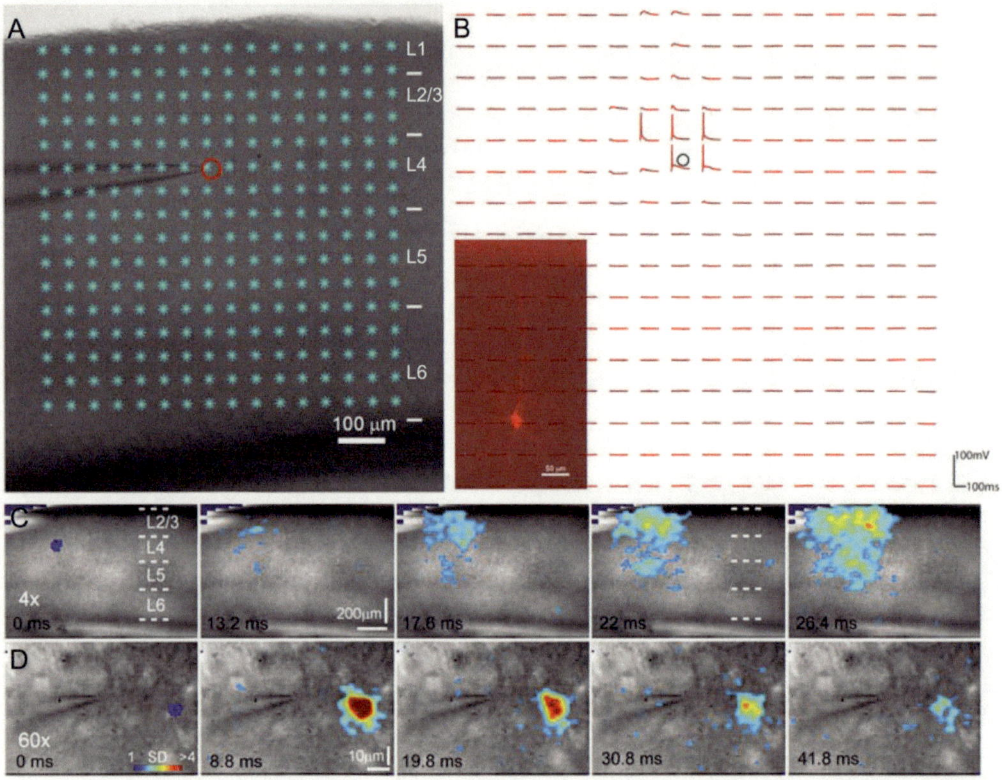

Fig. 2. **Physiological calibration of the LSPS system and example tests for combining photostimulation with VSD imaging.** A shows a mouse coronal V1 cortical slice image with the superimposed photostimulation sites (16 x 16 cyan stars, spaced at 50 µm x 50 µm) across cortical layers 1, 2, 3, 5 and 6 (i.e., L1-L6). The glass electrode was recording from an excitatory pyramidal neuron (with its morphology shown in the inset in B), with its location indicated by a small red circle. B shows the traces of membrane potential depolarization of the recorded neuron at the current clamp mode in response to LSPS (1ms, 30 mW, interstimulus interval, 400 ms) at the locations shown in A. Although subthreshold responses are seen at many locations, suprathreshold spikes only occur around the perisomatic area of the recorded neuron (indicated by the small black circle). C and D are sequences of VSD image frames in response to photostimulation through 4x and 60x objectives, respectively. Laser stimulation (32 mW) was 1 ms and 100 µs for C and D, respectively. The site of photostimulation can be identified by the laser excitation artifact in the initial frames of the sequences. Time progresses from left to right, and VSD signal amplitudes expressed as standard deviations (SD) above the mean baseline signal are color coded. Warmer colors indicate greater excitation. The map pixels with amplitudes ≥ 1 SD are plotted. Note that the CCD camera images have a slightly different aspect ratio. Under the 4x objective, the camera covers an area of 1.28 (w) x 1.07 (h) mm2 with a spatial resolution of 14.6 (w) x 17.9 (h) µm/pixel. The following figures use the same conventions.

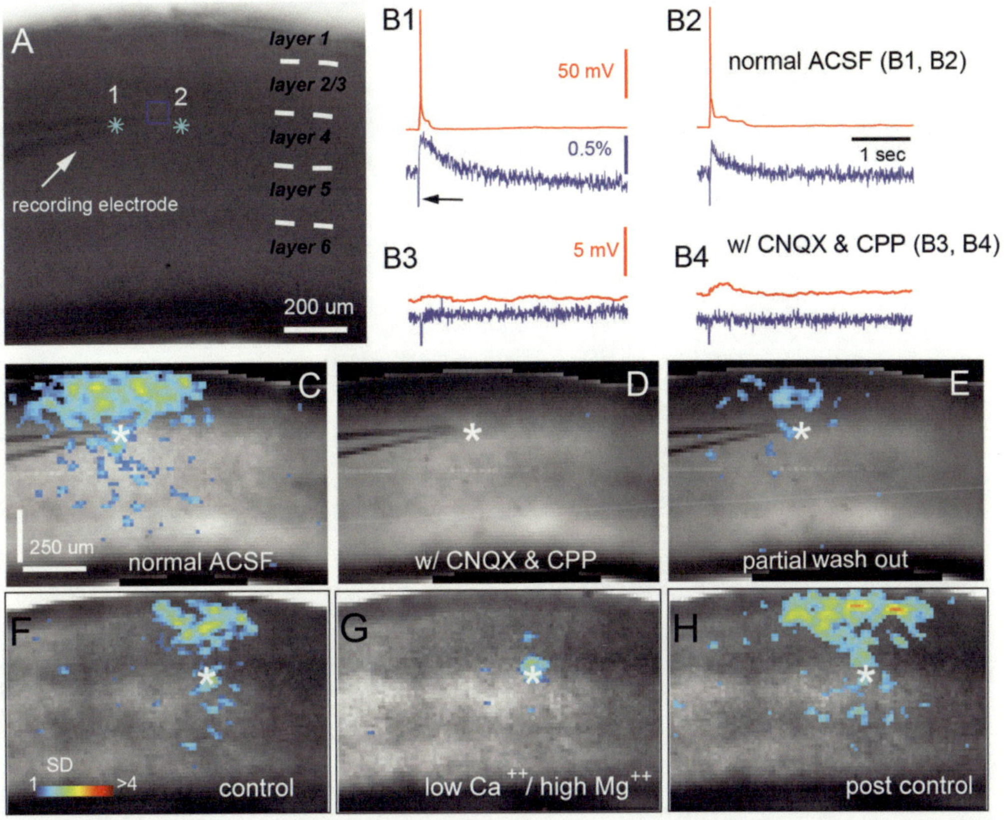

Fig. 3. **Characterization of photostimulation-evoked VSD responses.** A shows a gray scale image of V1 slice stained with NK3630, photostimulated (laser: 1 ms, 35 mW) at two sites (indicated by cyan stars) close to the recording electrode placed at the boundary of layers 3 and 4. The recorded neuron was identified as an excitatory pyramidal neuron. B1 and B2 are data traces of simultaneous whole-cell recording and VSD imaging in response to photostimulation at sites of 1 and 2 in normal artificial cerebrospinal fluid (ACSF), respectively. Red traces represent membrane potentials of the recorded neuron, and blue traces represent VSD signals that were measured from the region of interest (ROI) - marked by a blue square around the electrode tip shown in A. The black arrow in B1 points to the artifact signal of laser excitation in the VSD signal trace. B3 and B4 show data traces of simultaneous whole-cell recording and VSD imaging in response to photostimulation at the same sites as B1 and B2 but in ACSF with 10 μM CNQX and 10 μM CPP. C, D and E show VSD image frames of peak activation after glutamate uncaging at site 1 (indicated by the white star) before, after bath application and after wash out of CNQX and CPP, respectively. VSD signal amplitudes expressed as standard deviations (SD) above the mean baseline signal are color coded. F-H are peak activation frames of a different V1 slice in response to laser photostimulation in a layer 4 site (indicated by the white star; laser: 1ms, 35 mW) with perfusion of normal ACSF (control), low Ca2+ and high Mg2+ ACSF (containing 0.2 mM Ca2+, 4 mM Mg2+) and post-control normal ACSF, respectively. The data illustrate that most VSD responses outside the photostimulation site are predominantly postsynaptic responses which are blocked in the low Ca2+ and high Mg2+ solution. This figure is reproduced from Xu et al. (2010) with permission of the American Physiological Society.

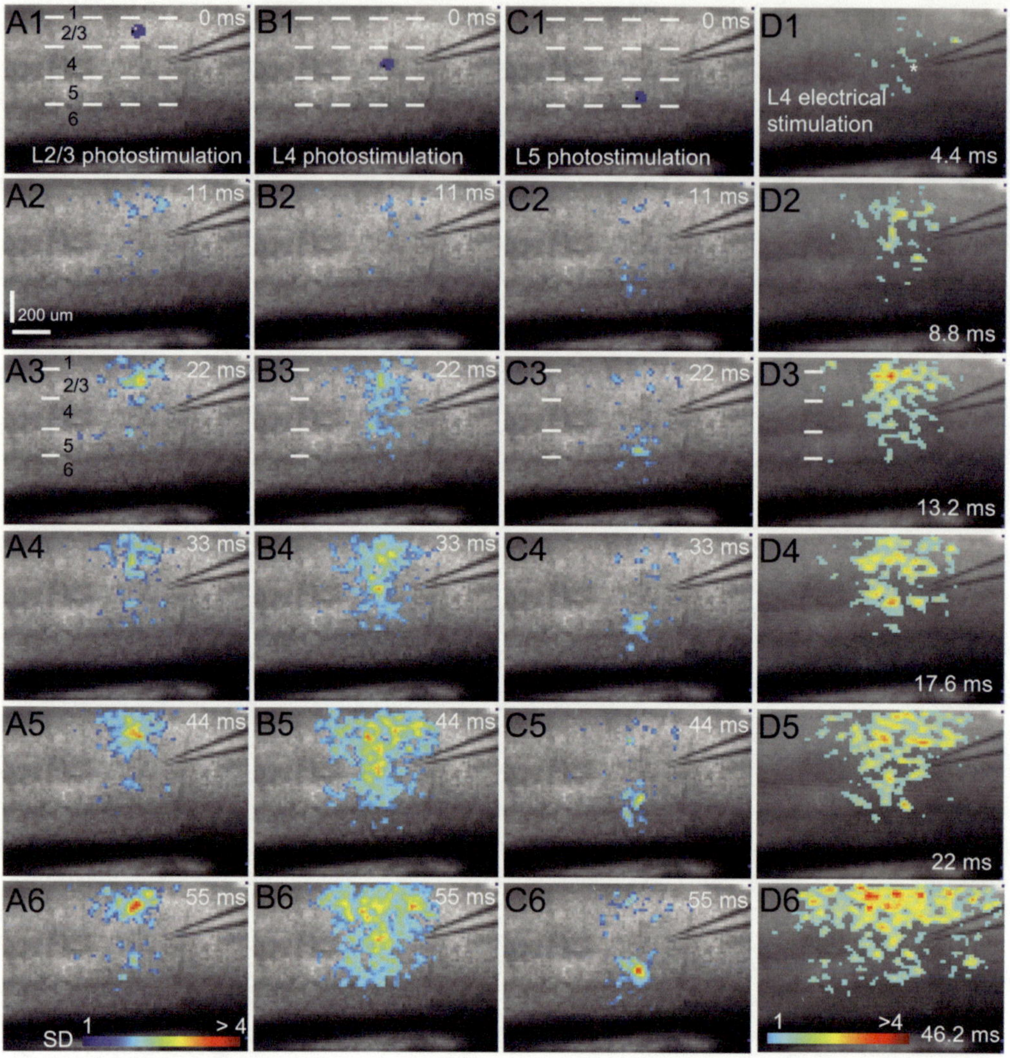

Fig. 4. **Spatially restricted neuronal activation via laser photostimulation enables high resolution mapping of interlaminar connections in mouse V1 local circuits.** A, B and C are sequences of VSD image frames in response to photostimulation (laser duration: 1ms; power: 32 mW) at cortical layers 2/3, 4 or 5 in a V1 coronal slice, respectively, while D shows VSD image frames in response to electrical stimulation (1ms, 50 μA current injection) through a microelectrode placed at V1 layer 4. VSD images were acquired at the rate of 2.2 ms/frame during the experiment, and are displayed at specific time points. Time progresses from top to bottom in the column, and color code is used to indicate VSD signal amplitudes expressed as standard deviations (SD) above the mean baseline. The map pixels with amplitudes ≥ 1 SD are plotted and included for further quantification. Warmer colors indicate greater excitation. The site of photostimulation can be identified by the laser excitation artifact (the blue spot) in the initial frames of the sequences. The short dashed white lines in the first image of A, B and C denote the laminar boundaries of V1 layers 1,

2/3, 4, 5 and 6. VSD images in D1-D6 are color-coded differently from A-C. In D1, the white star indicates the tip of the glass pipette for electric stimulation. This figure is reproduced from Xu et al. (2010) with permission of the American Physiological Society.

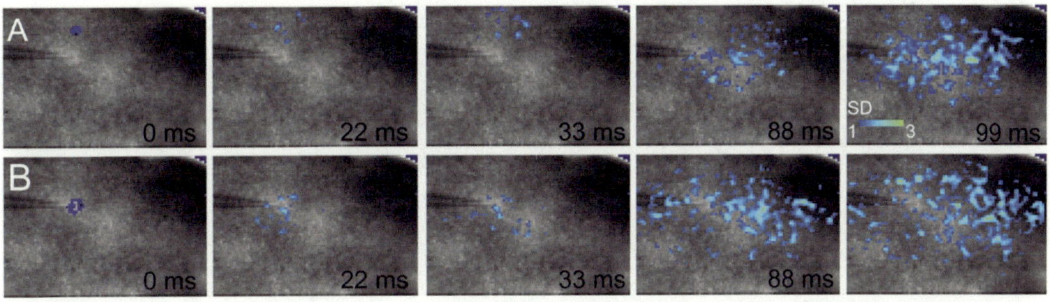

Fig. 5. **Photostimulation-evoked horizontal excitatory propagation in a V1 tangential slice.** A and B are time frame series of VSD responses to photostimulation (30 mW, 2 ms) in two different locations, respectively, in the tangential slice of layer 2/3 in mouse visual cortex. The resulting patterns of VSD responses are characterized by an extended zone of activation domains from the stimulation site.

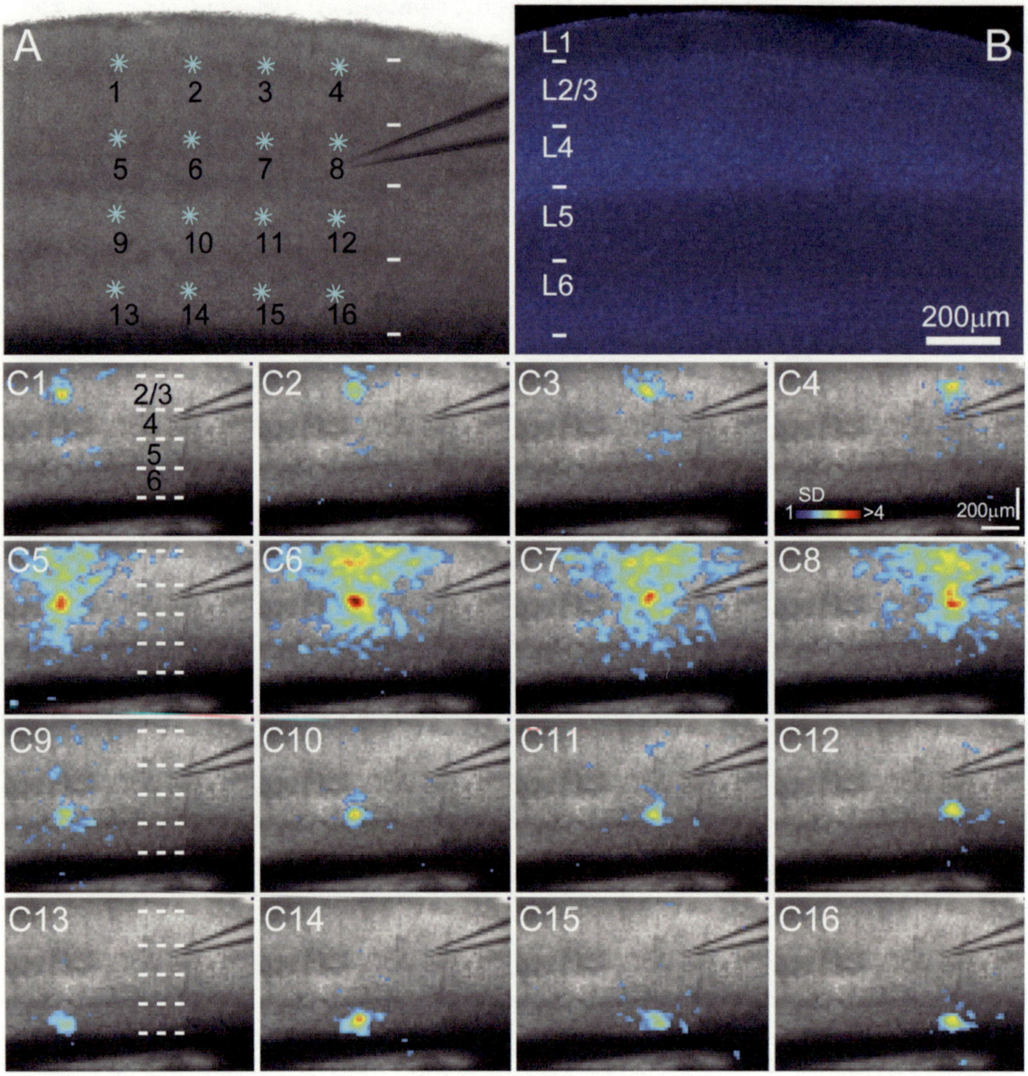

Fig. 6. **Rapid mapping of V1 local circuits through combination of VSD imaging and LSPS at multiple sites.** A shows the slice image with cyan stars indicating a 4x4 stimulus pattern covering V1 from cortical layer 2/3 to layer 6. The patch pipette was placed in the middle of layer 4 for extracellular electrical stimulation, while in most other experiments, a single pyramidal neuron was recorded during photostimulation and imaging for monitoring the effectiveness and spatial precision of laser photostimulation and correlating single-cell activity with the VSD imaged population response. B shows the DAPI-stained image of the same V1 slice as in A for laminar boundary designation. The short white lines in A, B, and C1, C5, C9 and C13 denote the laminar boundaries of layers 1, 2/3, 4, 5 and 6. C1-C16 show peak activation frames of the VSD map sequences corresponding to the 16 stimulation sites indicated in A.

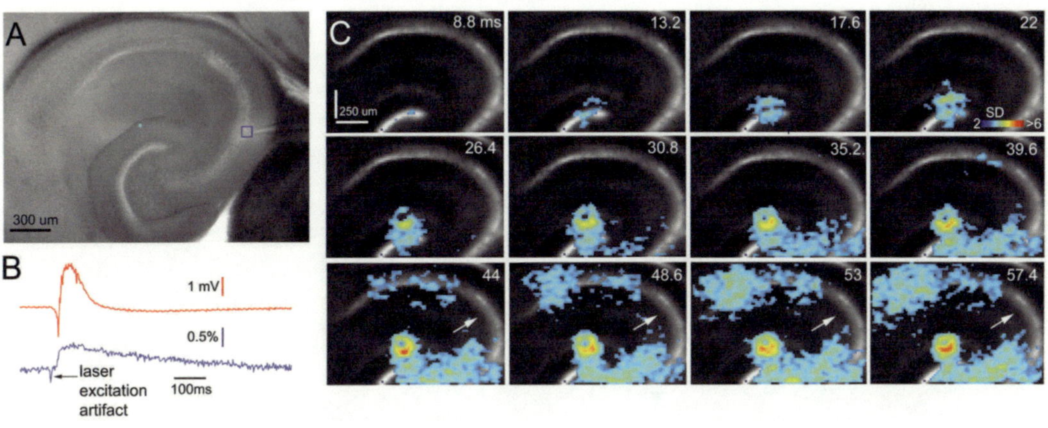

Fig. 7. **Hippocampal circuit mapping.** A shows a gray scale image of the hippocampus slice with the photostimulation site marked by a cyan star (in DG) and a blue square marking the ROI around the electrode tip in CA3. The data traces of the local field potential (in red) and the VSD signal (in blue) of the ROI shown in A are aligned in B to compare their temporal relationship. The VSD signal scale is in the percent change in light intensity [ΔI/I %]. C shows time frame series of the VSD response after photostimulation, and demonstrates a clear excitation flow through the hippocampal trisynaptic pathway. The white arrows in C point to the CA2 region which has little activation.

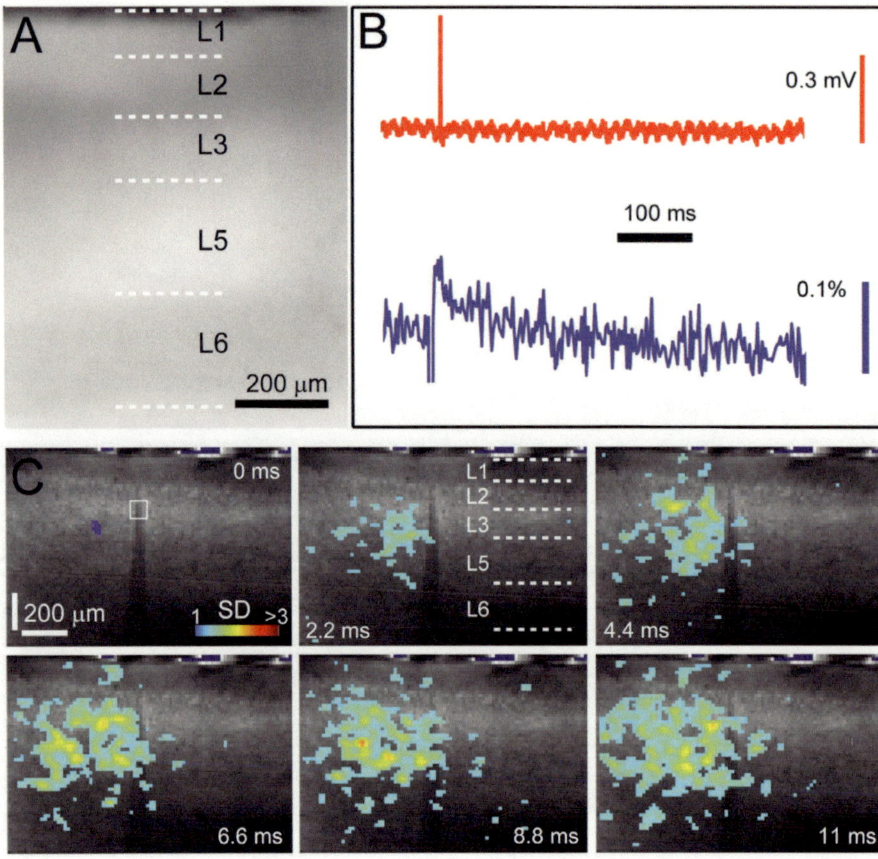

Fig. 8. **VSD imaging with channelrhodopsin-2 (ChR2) photoactivation.** A shows a yellow fluorescent proteins (YFP) fluorescent image of a medial prefrontal cortical slice from the transgenic mouse line expressing YFP- fused ChR2 mostly in subsets of excitatory pyramidal neurons under the control of the Thy1 promoter. This mouse line has relatively weak ChR2 in cortical layers 2 and 6, as reflected by the YFP fluorescence level. Note that medial prefrontal cortex lacks a granular layer 4 as seen in primary sensory cortices such as V1. B shows the aligned data traces of the local field potential (in red) and the VSD signal (in blue) of the ROI shown in C to compare their temporal relationship. The VSD signal scale is in the percent change in light intensity [ΔI/I %]. C shows time frame series of the VSD response to blue laser photostimulation (475 nm, 2 mW, 1 ms) in cortical layer 3 of the coronal slice, with both vertical and horizontal activity propagation.

8. References

Airan RD, Meltzer LA, Roy M, Gong Y, Chen H & Deisseroth K. (2007). High-speed imaging reveals neurophysiological links to behavior in an animal model of depression. *Science* 317, 819-823.

Akemann W, Mutoh H, Perron A, Rossier J & Knopfel T. (2010). Imaging brain electric signals with genetically targeted voltage-sensitive fluorescent proteins. *Nat Methods* 7, 643-649.

Ang CW, Carlson GC & Coulter DA. (2006). Massive and specific dysregulation of direct cortical input to the hippocampus in temporal lobe epilepsy. *J Neurosci* 26, 11850-11856.

Berger T, Borgdorff A, Crochet S, Neubauer FB, Lefort S, Fauvet B, Ferezou I, Carleton A, Luscher HR & Petersen CC. (2007). Combined voltage and calcium epifluorescence imaging in vitro and in vivo reveals subthreshold and suprathreshold dynamics of mouse barrel cortex. *J Neurophysiol* 97, 3751-3762.

Boyden ES, Zhang F, Bamberg E, Nagel G & Deisseroth K. (2005). Millisecond-timescale, genetically targeted optical control of neural activity. *Nat Neurosci* 8, 1263-1268.

Burkhalter A. (1989). Intrinsic connections of rat primary visual cortex: laminar organization of axonal projections. *J Comp Neurol* 279, 171-186.

Callaway EM. (1998). Local circuits in primary visual cortex of the macaque monkey. *Annu Rev Neurosci* 21, 47-74.

Callaway EM & Katz LC. (1993). Photostimulation using caged glutamate reveals functional circuitry in living brain slices. *Proc Natl Acad Sci U S A* 90, 7661-7665.

Dantzker JL & Callaway EM. (2000). Laminar sources of synaptic input to cortical inhibitory interneurons and pyramidal neurons. *Nat Neurosci* 3, 701-707.

Gilbert CD. (1983). Microcircuitry of the visual cortex. *Annu Rev Neurosci* 6, 217-247.

Gradinaru V, Zhang F, Ramakrishnan C, Mattis J, Prakash R, Diester I, Goshen I, Thompson KR & Deisseroth K. (2010). Molecular and cellular approaches for diversifying and extending optogenetics. *Cell* 141, 154-165.

Grinvald A & Hildesheim R. (2004). VSDI: a new era in functional imaging of cortical dynamics. *Nature reviews* 5, 874-885.

Huang X, Troy WC, Yang Q, Ma H, Laing CR, Schiff SJ & Wu JY. (2004). Spiral waves in disinhibited mammalian neocortex. *J Neurosci* 24, 9897-9902.

Jin W, Zhang RJ & Wu JY. (2002). Voltage-sensitive dye imaging of population neuronal activity in cortical tissue. *J Neurosci Methods* 115, 13-27.

Kandel E. (1982). The origins of modern neuroscience. *Annu Rev Neurosci* 5, 299-303.

Knopfel T, Diez-Garcia J & Akemann W. (2006). Optical probing of neuronal circuit dynamics: genetically encoded versus classical fluorescent sensors. *Trends Neurosci* 29, 160-166.

Kuhlman SJ & Huang ZJ. (2008). High-resolution labeling and functional manipulation of specific neuron types in mouse brain by Cre-activated viral gene expression. *PLoS ONE* 3, e2005.

Llano DA, Theyel BB, Mallik AK, Sherman SM & Issa NP. (2009). Rapid and sensitive mapping of long-range connections in vitro using flavoprotein autofluorescence imaging combined with laser photostimulation. *J Neurophysiol* 101, 3325-3340.

Mennerick S, Chisari M, Shu HJ, Taylor A, Vasek M, Eisenman LN & Zorumski CF. (2010). Diverse voltage-sensitive dyes modulate GABAA receptor function. *J Neurosci* 30, 2871-2879.

Nakagami Y, Saito H & Matsuki N. (1997). Optical recording of trisynaptic pathway in rat hippocampal slices with a voltage-sensitive dye. *Neuroscience* 81, 1-8.

Nelson DA & Katz LC. (1995). Emergence of functional circuits in ferret visual cortex visualized by optical imaging. *Neuron* 15, 23-34.

Nikolenko V, Poskanzer KE & Yuste R. (2007). Two-photon photostimulation and imaging of neural circuits. *Nat Methods* 4, 943-950.

Petersen CC & Sakmann B. (2001). Functionally independent columns of rat somatosensory barrel cortex revealed with voltage-sensitive dye imaging. *J Neurosci* 21, 8435-8446.

Petreanu L, Huber D, Sobczyk A & Svoboda K. (2007). Channelrhodopsin-2-assisted circuit mapping of long-range callosal projections. *Nat Neurosci* 10, 663-668.

Petreanu L, Mao T, Sternson SM & Svoboda K. (2009). The subcellular organization of neocortical excitatory connections. *Nature* 457, 1142-1145.

Schubert D, Kotter R, Zilles K, Luhmann HJ & Staiger JF. (2003). Cell type-specific circuits of cortical layer IV spiny neurons. *J Neurosci* 23, 2961-2970.

Schuz A & Palm G. (1989). Density of neurons and synapses in the cerebral cortex of the mouse. *J Comp Neurol* 286, 442-455.

Shepherd GM, Pologruto TA & Svoboda K. (2003). Circuit analysis of experience-dependent plasticity in the developing rat barrel cortex. *Neuron* 38, 277-289.

Shepherd GM & Svoboda K. (2005). Laminar and columnar organization of ascending excitatory projections to layer 2/3 pyramidal neurons in rat barrel cortex. *J Neurosci* 25, 5670-5679.

Shibuki K, Hishida R, Murakami H, Kudoh M, Kawaguchi T, Watanabe M, Watanabe S, Kouuchi T & Tanaka R. (2003). Dynamic imaging of somatosensory cortical activity in the rat visualized by flavoprotein autofluorescence. *J Physiol* 549, 919-927.

Swanson LW, Wyss JM & Cowan WM. (1978). An autoradiographic study of the organization of intrahippocampal association pathways in the rat. *J Comp Neurol* 181, 681-715.

Tucker TR & Katz LC. (2003). Spatiotemporal patterns of excitation and inhibition evoked by the horizontal network in layer 2/3 of ferret visual cortex. *J Neurophysiol* 89, 488-500.

Weiler N, Wood L, Yu J, Solla SA & Shepherd GM. (2008). Top-down laminar organization of the excitatory network in motor cortex. *Nat Neurosci* 11, 360-366.

Xu X & Callaway EM. (2009). Laminar specificity of functional input to distinct types of inhibitory cortical neurons. *J Neurosci* 29, 70-85.

Xu X, Olivas ND, Levi R, Ikrar T & Nenadic Z. (2010). High precision and fast functional mapping of cortical circuitry through a combination of voltage sensitive dye imaging and laser scanning photostimulation. J Neurophysiol 103, 2301–2312.

Yuste R, Tank DW & Kleinfeld D. (1997). Functional study of the rat cortical microcircuitry with voltage-sensitive dye imaging of neocortical slices. *Cereb Cortex* 7, 546-558.

Zhang F, Wang LP, Boyden ES & Deisseroth K. (2006). Channelrhodopsin-2 and optical control of excitable cells. *Nat Methods* 3, 785-792.

8

Confocal Laser Scanning Microscopy in Dermatology: Manual and Automated Diagnosis of Skin Tumours

Wiltgen Marco
*Institute for Medical Informatics, Statistics and Documentation, Medical University of Graz
Austria*

1. Introduction

The number of melanocytic tumours has increased in the last decades, whereby the frequency of melanoma doubles every 20 years. At present there is a risk of 1:100 to falling sick with a melanoma. According to a WHO report, in the year 2006, about 48,000 melanoma related deaths occur worldwide per year (Lucas et al., 2006). The malignant melanoma is one of the less common types of skin cancer but causes the majority (75%) of skin cancer related deaths. Most melanomas are brown to black looking lesions (Friedman et al., 1985). Warning signs that might indicate a malignant melanoma include change in size, shape and color. Early signs of melanoma are changes to the shape or color of existing moles (ABCD rule). Skin cancer has many potential causes, including: overexposure to UV-radiation (extreme sun exposure during sun-bathing), which may cause skin cancer either via the direct DNA damage or via the indirect DNA damage mechanism; smoking tobacco can double the risk of skin cancer; chronic non-healing wounds; genetic predisposition and the human papilloma virus (HPV), which is often associated with squamous cell carcinoma of the genitals, anus, mouth and fingers (Oliveria et al., 2006) The detection of malignant changes of skin tissue in the early stages will augment the success of the therapy (Marcovic et al., 2009). Metastasis of the melanoma in a progressive stage may spread out to the lymph nodes or even more distant places like: lungs; brain; bone and liver. Such metastatic melanoma may cause general symptoms like: fatigue; vomiting and loss of appetite. The greatest chance of cure is in the early surgical resection of thin melanomas. Therefore, a periodical screening of risk persons is necessary. In the fight against skin cancer, researchers have high hopes in improved and fast provisional screening methods for the clinical routine.
Confocal laser scanning microscopy (CLSM) is a novel imaging device enabling the non-invasive examination of skin lesions in real-time (Rajadhyaksha et al., 1999). Therefore, CLSM is very suitable for routine screening and early recognition of skin tumours. The CLSM technique allows the viewing of micro-anatomic structures and individual cells. In contrast to the conventional examination, where suspicious skin tumours have to be excised, embedded in paraffin and stained, this method is much more agreeable for the patient and faster. However, training and experience is necessary for a successful and accurate diagnosis in this new and powerful imaging technique. To diminish the need for training and to improve diagnostic accuracy, computer aided diagnostic systems are required by the derma pathologists.

Computer aided diagnosis means that the harmless (nevi) and malignant cases are discriminated by automatic analysis on a computer, providing optimised preventive medical checkups and accurate and reliable detection of skin tumours. Automated diagnostic systems need no input by the clinician but rather report a likely diagnosis based on computer algorithms. A main task in automated image analysis is the selection of appropriate features for a "computer friendly" description of the tissue. The choice and development of the features is driven by the diagnostic guidelines of the derma pathologists. In the diagnosis of CLSM views of skin lesions, architectural structures at different scales play a crucial role. The images of benign common nevi show pronounced architectural structures, such as: arrangements of nevi cells around basal structures and tumour cell nests. The images of malign melanoma show melanoma cells and connective tissue with few or no architectural structures. Features based on the wavelet transform have been shown to be particularly suitable for the automatic analysis of CLSM images because they enable an exploration of images at different scales (Wiltgen et al., 2008). A further task in automated analysis is the choice of the machine learning algorithm for classification, which enables it, after a previous training, to predict the class of a lesion (nevi or malignant melanoma). For medical diagnosis, the algorithm should duplicate the automated diagnostic process by making it understandable for the human interpreter. By the CART (Classification and Regression Tree) algorithm, the inferring rules are automatically generated during the training of the algorithm. The generated rules have a syntax that is understandable for the human interpreter and they can be discussed and explicitly used as diagnostic rules. The classification results are relocated to the images by use of the inferring rules as diagnostic aid. The regions enabling a high discrimination power are highlighted in the images, showing tissue with features in good accordance with typical diagnostic CLSM features.
In this paper, we introduce the basic principles of automated diagnosis of CLSM images of skin lesions. Special attention is given to the wavelet transform for the description of the tissues in a way that conforms with the guidelines of the human interpreter. Further, the machine learning algorithm for the class prediction and its diagnostic rules is discussed in some detail. The application and performance of the discussed methods is demonstrated by a selected study.
The procedure for image analysis presented in this paper, was developed with the "Interactive Data Language" software tool IDL, which is a computing environment for image analysis, data visualization, and software application development (IDL 7.1, ITT Visual Information Solutions (ITT VIS), formerly known as Research Systems Inc. (RSI), http://www.ittvis.com/). IDL belongs to the Fourth Generation Languages (4GL) and includes image processing procedures and tools for rapid prototyping and rapid application development. All software runs on a PC under Windows and supports the development of applications with graphical user interface (GUI). The CART (Classification and Regression Trees) analysis was done with the software from Salford Systems, San Diego, USA.

2. Confocal laser scanning microscopy

The principle of confocal microscopy was developed by Marvin Minsky in 1957. After the development of lasers, confocal laser scanning microscopy became a standard technique toward the end of the 1980s (Patel et al., 2007). Confocal laser scanning microscopy (CLSM) is a technique for obtaining high-resolution optical images with depth selectivity (Paoli et al., 2009). This technique enables the acquisition of in-focus images from selected depths

(Pellacani et al., 2008). Therefore, for non-opaque specimens, such as biological tissue, interior structures can be imaged. The images are acquired by a point-by-point scanning and reconstructed using a computer. For interior imaging, the quality of the image is greatly enhanced over conventional microscopy because the image information from multiple depths in the specimen is not superimposed.

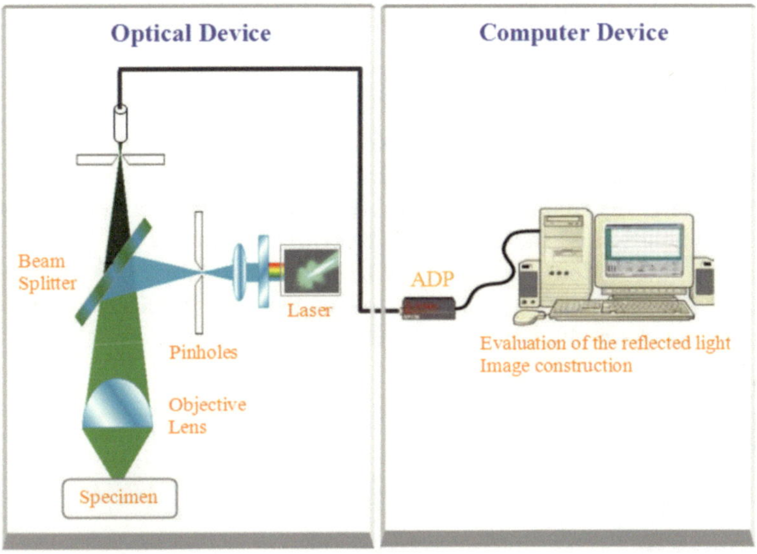

Fig. 1. Principle of the confocal laser scanning microscope

In a confocal laser scanning microscope, a laser beam passes trough a pinhole and is directed by a beam splitter (dichroic mirror) to the objective lens where it is focused into a small focal volume at a layer within the biological specimen (Fig. 1). The scattered and reflected laser light from the illuminated spot is then re-collected by the objective lens. The beam splitter separates it from the incident light and deflects it to the detector. After passing through a further pinhole, the light intensity is detected by the photon detection device (usually a photomultiplier tube or avalanche photodiode), transforming the intensity of the reflected light signal into an electrical one that is recorded by a computer. The depth of the layer (its vertical position) inside the specimen is controlled by the pinhole at the laser source. Each layer shows horizontal sections of the lesions. The electric signal, which is obtained from the intensity of the reflected light out of the illuminated volume element at a layer within the specimen, represents one pixel in the resulting image (Fig. 1). As the laser scans over the plane of interest, a whole image is obtained pixel-by-pixel and line-by-line. The brightness of a resulting image pixel corresponds to the relative intensity of the reflected light. The contrast in the images results from variations in the refractive index of microstructures within the biological specimen.

In this paper, the confocal laser scanning microscopy is performed with a near-infrared reflectance convocal microscope (Vivascope 1000, Lucid Inc., Rochester, NY, USA, http://www.lucid-tech.com/). The microscope uses a diode laser at 830 nm wavelength and a power of <35mW at tissue level. A x30 water-immersion objective lens with a numerical aperture of 0.9 is used with water (refractive index 1.33) as an immersion medium. The spatial resolution is 0.5-1.0 μm in the lateral and 3-5 μm in the axial dimension. The images contain a field-of-view of 0.5x0.5 mm on the skin tissue providing insights into cellular structures. Up to 16 layers per lesion can be scanned. Usually an examination depth of 350 μm can be reached (then the reflected light is absorbed by the preceding layers). All images, stored in BMP file format, are monochrome images with a spatial resolution of 640x480 pixels and a gray level resolution of 8 bits per pixel. (For the image analysis, the image matrix is rescaled to 512x512 pixels). The images are taken from the centre of the tumours representing the dermo-epidermal junction.

3. Automated image analysis

Prior to the automated image analysis, the CLSM views of the lesions are dissected into square elements. The size of the square elements ranges from 128x128 pixels to 512x512 pixels (whole image). This enables the analysis of different parts of the images which are of interest for the evaluation of the diagnostic relevant regions in the CLSM views. For an automated, or computer aided, diagnosis of CLSM views of skin lesions, the following steps are necessary (Wiltgen et al., 2003). First, the images must be appropriately analysed to enable a formulation of suitable features (Fig. 2).

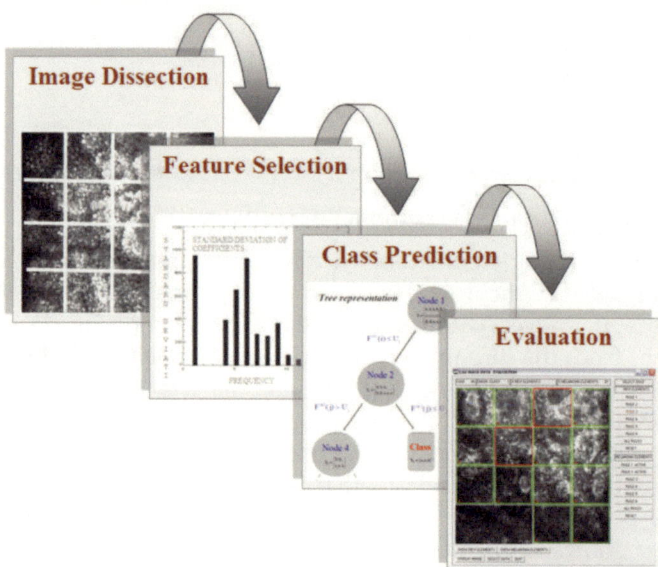

Fig. 2. Steps in the image analysis system

The formulation of appropriate features follows the derma pathologists´ diagnostic guidelines for CLSM views. Second, the images must be categorized according to their features, enabling a class prediction for every CLSM view of skin lesions. For medical purposes, it is important that the knowledge acquired during the training phase of the machine learning algorithm is represented in an understandable and readable form. Knowledge representation by rules, with a syntax that permits their use as diagnostic rules, is suitable. Third, the class prediction performance is evaluated. To this purpose, the single square elements are superimposed (relocated) onto the corresponding images by use of the diagnostic rules generated by the machine learning algorithm. The categorized square elements are highlighted in the images, enabling an identification of diagnostically highly relevant regions and an evaluation of typical diagnostic CLSM features by comparing them with the diagnostic decisions by the human observer.

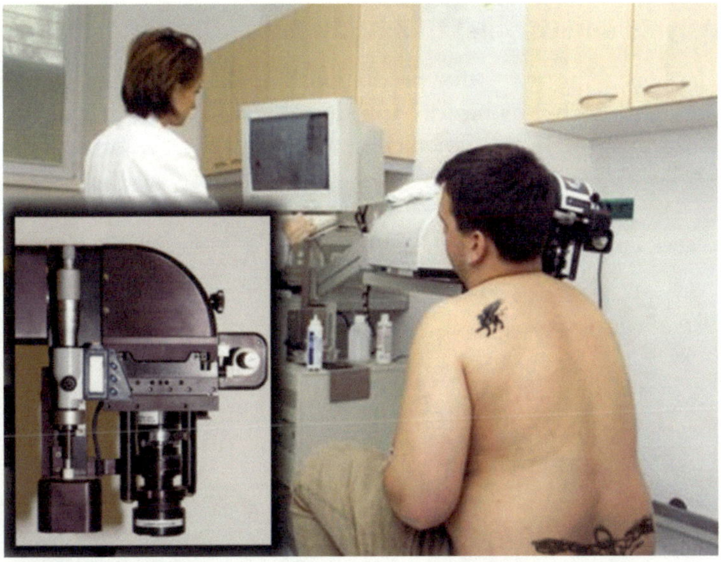

Fig. 3. Preventive medical skin checkups are of special importance in dermatology

4. Diagnosis of CLSM images

CLSM is a non-invasively method which takes 3-5 minutes per lesion whereby the microscope optical unit is positioned on the patient's skin (Fig. 3). An important step for a successful automated image analysis of histological tissue is the choice of the appropriate texture features. The selection of the features can be made according to the diagnostic guidelines used by the derma pathologist. For the diagnosis of the corresponding CLSM views, architectural structures such as: micro-anatomic structures; cell nests etc. play an important role in the diagnosis (Fig. 4).

Melanocytic cytomorphology and architecture and keratinocyte cell borders are taken into account for diagnostic decisions by the derma pathologist (Pellacani et al., 2008). Due to the high refraction of melanin, basal keratinocytes appear very intensive. The images of benign common nevi show, beside the nevi cells, pronounced architectural structures, whereas images of malign melanoma show melanoma cells and connective tissue with little or no architectural structures (Fig. 4). Therefore the information at different scales (from coarse structures to details) plays a crucial role in the diagnosis of CLSM images of skin lesions (Scope et al., 2007). This can be compared with the opinion of a human observer at different distances from an image. When the observer is close enough to the image he can study details. The greater the distance between the observer and the image, the better he can study the image as whole without being overwhelmed by the details (the details are smoothed out). In other words, he studies the images at different scales. This procedure is mathematically reflected by the wavelet transform. Therefore, features based on the properties of the wavelet transform enable an exploration of architectural structures, of different sizes, at different spatial scales. This makes the wavelet transform suitable for the automatic analysis of CLSM views of skin lesions (Wiltgen et al., 2008). The diagnostic guide lines are reflected by different groups and properties of wavelet coefficients.

5. Wavelet analysis and wavelet transform

A wavelet is a fast-decaying wave-like oscillation that starts with an amplitude of a given (non-zero) amount and decreases rapidly to zero. This means that a wavelet is zero-valued outside of a defined interval. It can typically be visualized as a "brief oscillation".

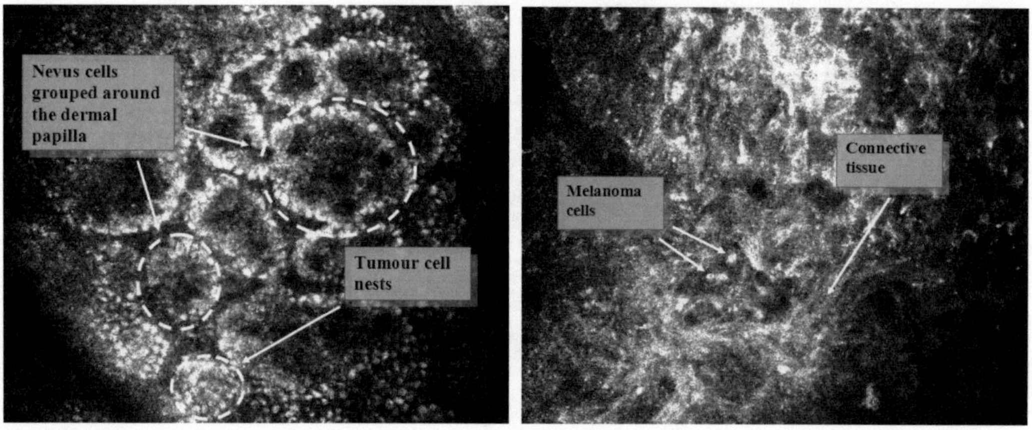

Fig. 4. CLSM images of common benign nevi (left) and malignant melanoma (right)

The wavelet transform is the representation of a given function by wavelets (Press et al., 1992). The basis wavelets are scaled and translated copies ("daughter wavelets") of the finite-length waveform ("mother wavelet"). The goal of the transformation is to represent data in another way, one which is more suitable for the analysis and interpretation (Prasad et al., 1997). Similar to the Fourier transformation, the wavelet transformation checks a given function for conformities with the basis functions and calculates the transformation from it. But in contrast to the Fourier transformation, where the basis functions (sine and cosines) are non-local in the spatial domain (and stretch out to infinity) and global properties of the analysed function are localized in prominent peaks of the frequency spectrum, the wavelet basis functions are localized in both: the spatial and the frequency domain. Wavelet transforms have advantages over the traditional Fourier transforms for accurate decomposition and reconstructing of finite, non-periodic functions. This is a crucial advantage for the analysis of the local properties of a function or a texture (Chui, 1992). By use of the wavelet basis functions with increasing spatial extension, the function or texture can be analysed at different scales showing detail (local) to coarse (global) properties (Burrus et al., 1988). This feature gives the wavelet transform a multi resolution property, enabling the study of functions and textures at varying resolutions (Daubechies, 1992).
The wavelet transforms can be divided into three classes: continuous transforms (CWT), discrete transforms (DWT) and multi-resolution-analysis (MRA) based transforms. The continuous wavelet transforms operate over every possible scale and translation whereas the discrete wavelet transforms use a discrete subset of scale and translation values. Whereas the CWT´s are mainly used in the mathematically analysis, the DWT´s have many practical applications, such as image processing. The MRA represents a design method of most of the relevant discrete wavelet transforms.

5.1 The continuous wavelet transforms

The wavelet transform (integral transform) is a linear and often invertible operation that defines a relationship between the spatial and the frequency domain. If the wavelet transform is invertible then, beside the analysis of a function, the synthesis of the function is also possible. The relationship between the spatial and the frequency domain is done by the wavelet function ("mother wavelet") Φ . The wavelet function provides the source function for the generation of the wavelets ("daughter wavelets"), which are the translated and scaled versions of the wavelet function:

$$\Phi_{s,l}(x) = \frac{1}{\sqrt{s}}\Phi(\frac{x-l}{s}) \tag{1}$$

The scale index s defines the width of the wavelet and the location index l its position. In this way, wavelets are able to localize behaviour in space (via translation) and at a characteristic scale (via dilation/contraction). The "wavelet family" is then given by:

$$\left\{\Phi_{s,l}(x) = \frac{1}{\sqrt{s}}\Phi\left(\tfrac{x-l}{s}\right) \middle| (s,l) \in \mathbb{R}_{+} \times \mathbb{R} \right\} \tag{2}$$

There exists a variety of different types of wavelet functions; such the Hermitian wavelet, the Shanon wavelet, the Morlet wavelet, the Mexican-hat wavelet, the Beta wavelets etc. The different wavelet types have different properties. For illustration, the Mexican-hat wavelet is presented (Fig. 5). By the choice of different values for s and l, the wavelet basis functions have different widths and are located at different spatial positions (Fig. 5). In continuous wavelet transforms, a given function is projected on a continuous family of frequency bands. The integral wavelet transform of a function $f(x) \in L^2(\mathbb{R})$, at a scale $s \in \mathbb{R}_{+}$ and translational value $l \in \mathbb{R}$, is expressed as convolution of the function with the wavelet:

$$\left(W_{\Phi}f\right)(s,l) = \int_{-\infty}^{+\infty} f(x)\Phi^{*}_{s,l}(x)\,dx \tag{3}$$

Whereby, $\Phi^{*}_{s,l}$ is the complex conjugate of the wavelet function. $\left(W_{\Phi}f\right)(s,l)$ is the continuous wavelet transform of *f*(*x*) at position *l* and scale *s*. It is the coefficient that measures the strength of the features at scale *s* and position *l*. In order to satisfy analytical requirements for the continuous wavelet transform and for the discrete wavelet transform, the wavelet functions are chosen from a subspace of the space: $\Phi(x) \in L^1(\mathbb{R}) \cap L^2(\mathbb{R})$. This is the space of measurable functions that are absolutely integrable ($L^1(\mathbb{R})$) and square-integrable ($L^2(\mathbb{R})$):

$$\int_{-\infty}^{+\infty} |\Phi(x)|\,dx < \infty \;;\; \int_{-\infty}^{+\infty} |\Phi(x)|^2\,dx < \infty \tag{4}$$

The functions in this space guaranty the admissibility requirement and the square norm. In principle, every finite-length function or fast-decaying oscillating waveform can be used as a wavelet function. But if the wavelet transform is required to be invertible, not every function can be used. For the wavelet transform to be invertible, the wavelet must satisfy the admissibility criterion:

$$0 < 2\pi \int_{-\infty}^{+\infty} \frac{\left|\hat{\Phi}(\omega)\right|^2}{|\omega|} d\omega < +\infty \tag{5}$$

To determine the admissibility criterion for a given wavelet $\Phi(x)$, first its Fourier transform $\hat{\Phi}(\omega)$ must be calculated. If a wavelet function fulfils the admissibility criterion, it is called an admissible wavelet. Because the integrand in the admissibility criterion has a null in the denominator for $\omega = 0$, the Fourier transform $\hat{\Phi}(0)$ must be zero. Then the admissibility criterion can be simplified to:

$$\hat{\Phi}(0) = \int_{-\infty}^{+\infty} \Phi(x)\,dx = 0 \tag{6}$$

Thereby the admissibility criterion can be reconsidered with the original wavelet $\Phi(x)$. By the integral wavelet transform, the wavelet functions, with given translation and scaling values, are compared successively with different sections of the function (Fig. 6). The wavelet transform coefficients are determined by scanning the function with the wavelets at different scales. The coefficients describe the conformity of the function at every position along the function with the analyzing wavelets.

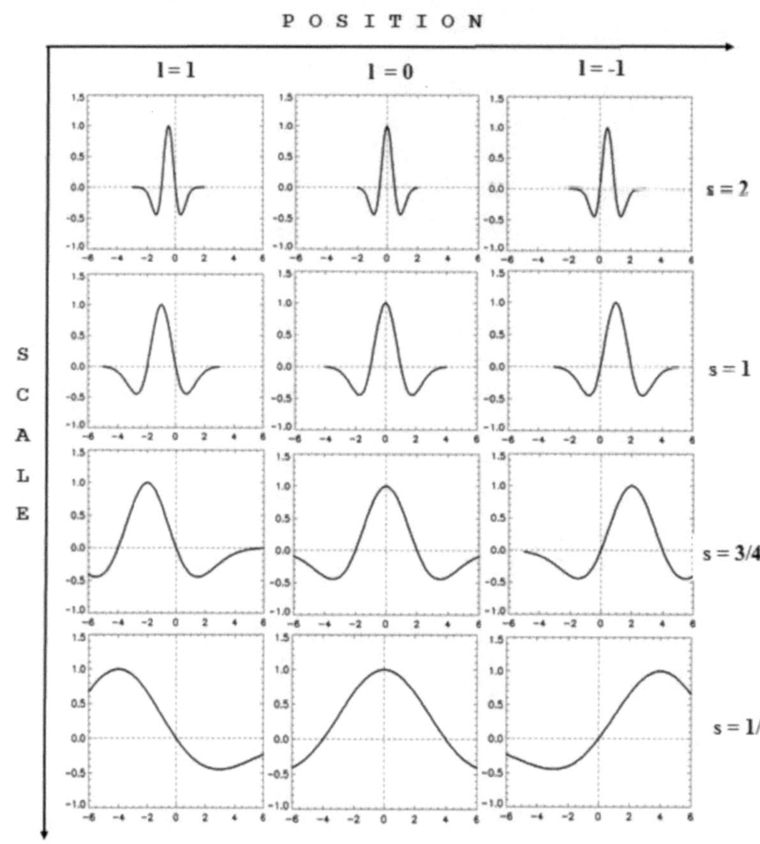

Fig. 5. Starting from the "mother function", the Mexican-hat basis functions are generated by selecting values for l and s

By analyzing a function with differently scaled wavelets, the degree of conformity is determined. A small wavelet (lower scale) at a certain position is well localized and delivers detailed information at this part of the function. An expanded wavelet (larger scale) has a lower spatial resolution and delivers coarse (details are smoothed out) information of the function. By the translation of the wavelets, the function is then analyzed at every position (Fig. 6). The wavelet transform coefficients reflect the degree of correspondence between the wavelet (at a certain position and scale) and the function. The product of a wavelet with a function section gives a new function and the enclosed area (by integrating out) is the corresponding wavelet coefficient. By use of wavelets with different widths, the function is analyzed at different scales (spatial resolution). With the wavelet transform coefficients together with the corresponding wavelets, the original function $f(x)$ can be recovered by the inverse continuous wavelet transform:

$$f(x) = \int_{0}^{+\infty} \int_{-\infty}^{+\infty} \frac{1}{s^2} (W_{\Phi} f)(s,l) \phi_{s,l}(x) \, dl \, ds \tag{7}$$

$\phi_{s,l}(x)$ is the dual function of $\Phi_{s,l}(x)$ satisfying the condition:

$$\int_{0}^{+\infty} \int_{-\infty}^{+\infty} \frac{1}{|s^3|} \Phi_{s,l}(x') \phi_{s,l}(x) \, dl \, ds = \delta(x - x') \tag{8}$$

The function is synthesized by the superposition of the analyzing wavelets, weighted by their coefficients.

5.2 The discrete wavelet transforms

The discrete wavelet transform is given by extracting a discrete subset of the set of wavelet functions in the continuous wavelet transform. In other words: only wavelet functions with discrete scaling and translation parameters are used. That means the wavelet functions are still continuous, but they exist only at discrete positions with discrete scales (Daubechies, 1988). The basis wavelet functions are translated and dilated by whole numbers $m \in \mathbb{Z}_+$ and $n \in \mathbb{Z}$.

$$\Phi_{m,n}(x) = 2^{-\frac{m}{2}} \Phi(2^{-m} x - n) \tag{9}$$

In spit of the reduction to a discrete subset, it can be shown that the information remains completely preserved. For many applications of the discrete wavelet transform, orthonormal basis wavelets are used.

A function $\Phi \in L^2(\mathbb{R})$ is called an orthonormal wavelet if it can be used to define a Hilbert basis (an orthonormal and complete basis) for the Hilbert space $L^2(\mathbb{R})$ of square integrable functions. The Hilbert basis is constructed as the family of basis functions:

$$\left\{ \Phi_{n,m}(x) = 2^{-\frac{m}{2}} \Phi(2^{-m} x - n) \,\middle|\, (m,n) \in \mathbb{Z}_+ \times \mathbb{Z} \right\} \tag{10}$$

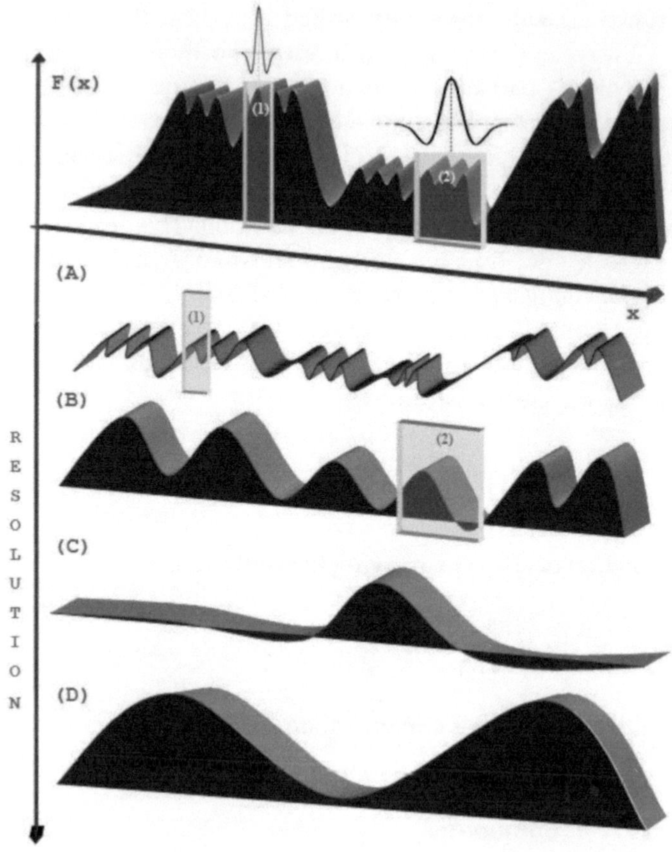

Fig. 6. In wavelet analysis, the wavelet basis functions are compared successively with different sections of the function

by means of dyadic translations and dilations of the wavelet "mother function" Φ . Or in other words: the mother wavelet is the generating function of an affine function system which constitute a Hilbert basis in $L^2(\mathbb{R})$. The family of wavelet basis functions is an orthonormal system if it is orthonormal under the inner product:

$$\left\langle \Phi_{m,n}, \Phi_{m',n'} \right\rangle = \delta_{mm'} \delta_{nn'} \tag{11}$$

Thereby, δ_{nm} is the Kronecker delta and $\langle \Phi, \Phi' \rangle$ is the standard inner product on $L^2(\mathbb{R})$:

$$\left\langle \Phi_{m,n}, \Phi_{m',n'} \right\rangle = \int_{-\infty}^{+\infty} \overline{\Phi}_{m,n}(x) \cdot \Phi_{m',n'}(x)\, dx \tag{12}$$

The requirement of completeness is that every given function $f \in L^2(\mathbb{R})$ can be decomposed into a linear combination of the basis wavelet functions weighted by wavelet coefficients:

$$f(x) = \sum_{n,m=-\infty}^{+\infty} c_{n,m} \Phi_{n,m}(x) \tag{13}$$

The convergence of the series is understood to be convergence in norm. Such a representation of a function $f(x)$ is known as a wavelet series and the basis functions are the building stones of the function (an orthonormal wavelet is self-dual). Then, $L^2(\mathbb{R})$ is spanned by the discrete set of basis wavelets $\{\Phi_{m,n} | (m,n) \in \mathbb{Z}_+ \times \mathbb{Z}\}$, and the discrete wavelet coefficients $\{c_{m,n} = \langle f, \Phi_{m,n} \rangle | (m,n) \in \mathbb{R}_+ \times \mathbb{R}\}$ of a function $f \in L^2(\mathbb{R})$ are determined by the discrete wavelet transform:

$$\langle f, \Phi_{m,n} \rangle = \int_{-\infty}^{+\infty} f(x) \bar{\Phi}_{m,n}(x) dx \tag{14}$$

Any function $f \in L^2(\mathbb{R})$ is therefore characterized by the set of its discrete wavelet coefficients $\{\langle f, \Phi_{m,n} \rangle | (m,n) \in \mathbb{Z}_+ \times \mathbb{Z}\}$ and it is possible to recover any function from its discrete wavelet coefficients in a numerically stable procedure. (Numerically stable means that small perturbations in the wavelet coefficients correspond to small perturbations of the function).

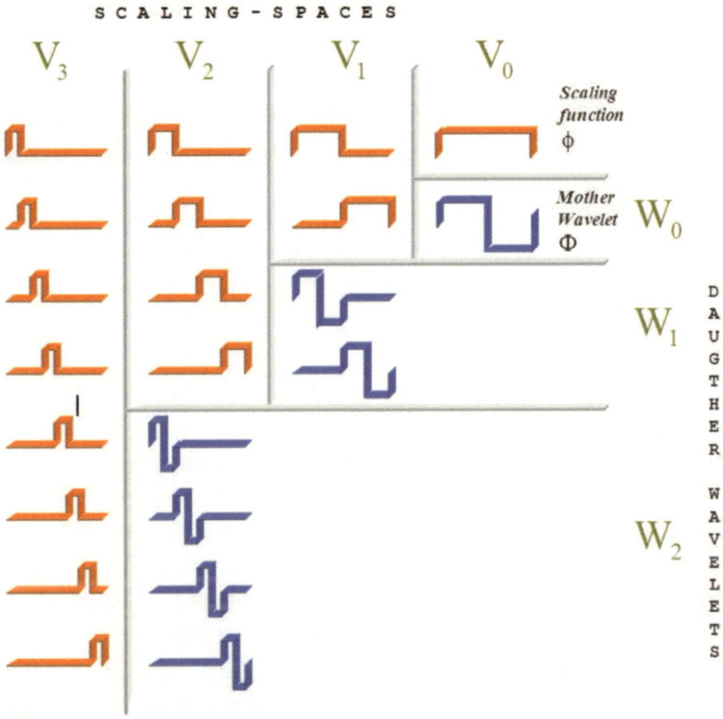

Fig. 7. The subspaces V_j represent the space of the scaling function and the W_j are spanned up by the wavelets

For any admissible wavelet, spanning up a basis $\Phi_{m,n}$ for $f \in L^2(\mathbb{R})$, the following condition holds:

$$\sum_{m,n} |< f, \Phi_{m,n} >|^2 \leq C \|f\|_2^2 \tag{15}$$

Thereby, C is a given constant value. Then, the discrete wavelet coefficients belong to:

$$l^2(\mathbb{Z}_+ \times \mathbb{Z}) = \left\{ c_{m,n} \mid \sum_{m,n} | c_{m,n} |^2 < \infty \right\} \tag{16}$$

Therefore the discrete wavelet transform $DWT : f \rightarrow \{\langle f, \Phi_{m,n} \rangle\}$ is a mapping from $L^2(\mathbb{R})$ (the space of square integrable functions) into $l^2(\mathbb{Z}_+ \times \mathbb{Z})$ (the space of square-summable, infinite sequences of numbers). For the wavelet coefficients, the following conditions are valid:

$$\forall_{m,n \in \mathbb{Z}_+ \times \mathbb{Z}} : \langle f, \Phi_{m,n} \rangle = \langle g, \Phi_{m,n} \rangle \Leftrightarrow f = g ; \quad \forall_{m,n \in \mathbb{Z}_+ \times \mathbb{Z}} : \langle f, \Phi_{m,n} \rangle = 0 \Leftrightarrow f = 0 \tag{17}$$

Then, numerical stability is defined by: given two functions $f \in L^2(\mathbb{R})$ and $g \in L^2(\mathbb{R})$, the characterization of the functions by their wavelet coefficients in $l^2(\mathbb{Z}^2)$ is stable if: whenever two sequences of wavelet coefficients in $l^2(\mathbb{Z}^2)$ are close, the corresponding functions are close in $L^2(\mathbb{R})$. As in the case of continuous wavelets, there also exists a variety of discrete wavelets; the Coiflet wavelets, the Legendre wavelets, the Haar wavelets, etc. For illustration, the Haar wavelet is presented (Fig. 7). It is now recognized as the first known wavelet and proposed 1909 by A. Haar (Alfréd Haar was a Hungarian mathematician). The Haar wavelet is also the simplest possible wavelet. The Haar wavelets give an example of a countable orthonormal system for the space of square-integrable functions on the real line. A large subclass of wavelets arises from special structures imposed on $L^2(\mathbb{R})$, known as multi-resolution-analysis (MRA). MRA is a framework for understanding and constructing wavelet bases, which generates discrete wavelet families of dilations and translations that are orthonormal bases in $L^2(\mathbb{R})$.

5.3 Multi-resolution analysis

Multi-resolution analysis (MRA) is a design method to generate a lot of practically relevant discrete wavelet transforms (DWT) and is the mathematical basis of the fast wavelet transform (FWT). The MRA satisfies certain self-similarity relations in spatial and the scale domain as well as completeness and regularity relations (Daubechies & Lagarias, 1991). The MRA describes the approximation properties of the discrete wavelet transform. The generated wavelet transformation is iterative and the analyzed function is split in successive "smoother" versions, which contain by progressive iterations successively poorer information. Of special importance in the multi-resolution analysis are the so called two scale equations for a scaling function and a wavelet function. The MRA is defined as follows: A MRA of the space $L^2(\mathbb{R})$ consists of a sequence of nested closed subspaces: $\{V_n \mid n \in \mathbb{Z}\}$ with the following properties:

a. Nesting property: $\{0\} ... \subset V_0 \subset ... V_n \subset V_{n+1} \subset ... \subset L^2(\mathbb{R})$
b. Invariance under translation: Self-similarity in the spatial domain requires that each subspace V_k is invariant under shifts by integer multiples of 2^{-k}:

$$f(x) \in V_k \Leftrightarrow f(x + m2^{-k}) \in V_k ; \ (m \in \mathbb{Z})$$

c. Scaling property: Self-similarity in scale requires that all nested subspaces $V_k \subset V_l$ with $k < l$ are scaled versions of each other, with the scaling factor 2^{-k}:

$$f(x) \in V_k \Leftrightarrow f(2^{l-k}x) \in V_l$$

d. Existence of a scaling function: This requires the existence of an orthogonal basis $\{\phi_{0n} | n \in \mathbb{Z}\}$ for the subspace $V_0 \subset L^2(\mathbb{R})$, where:

$$\phi_{jk}(x) = 2^{-j/2}\phi(2^{-j}x - k)$$

e. Completeness requires that the nested subspaces V_m fill the whole space, i.e., their union should be dense in $L^2(\mathbb{R})$:

$$\bigcup_{m\in\mathbb{Z}} : V_m = L^2(\mathbb{R})$$

f. The lack of redundancy requires that the intersection of the nested subspaces V_m should only contain the zero element:

$$\bigcap_{m\in\mathbb{Z}} V_m = \{0\}$$

The generating functions ϕ are known as scaling functions or "father wavelet". The structure of the MRA allows several conclusions for the construction of wavelet bases for practical applications. The orthonormal basis of each subspace, their scaling properties and resolutions follow from the MRA definition. The scaling property (c) implies that each subspace V_j is a scaled version of the central subspace V_0 (Fig. 7). Together with the invariance under translation (b) and the existence of a scaling function (d) it implies that $\{\phi_{jn} | n \in \mathbb{Z}\}$ is an orthonormal Hilbert basis of the subspace V_j:

$$\langle \phi_{j,k}, \phi_{j,k'} \rangle = \delta_{kk'} \tag{18}$$

Then, the sequence of scaling subspaces can be defined by spaces that are spanned off by a proprietary orthonormal basis system:

$$V_i = span\{\phi_{i,k} | k \in \mathbb{Z})\} \tag{19}$$

Further, the scaling property (c) implies that the resolution of the l-the subspace V_l is higher than the resolution of the k-the subspace V_k ($\forall_{k<l} : V_k \subset V_l$). In the case that there exists only one scaling function ϕ in the MRA which generates a Hilbert basis in V_0, the scaling function satisfies the two scale equation (or refinement equation):

$$\phi(x) = \sum_{k=-N}^{N} a_k \phi(2x - k) \tag{20}$$

Because of the nesting property $V_0 \subset V_1$, there exists a finite sequence of coefficients $a_k = 2\langle \phi(x), \phi(2x - k) \rangle$, for $|k| \leq N$ and $a_k = 0$ for $|k| > N$ |. This sequence of coefficients (real numbers) is called scaling sequence or filter (scaling) mask and is given by:

$$\{..., 0, a_{-N}, ..., a_0, ..., a_N, 0, ...\} \tag{21}$$

To fulfil filter properties, several conditions must be imposed on the coefficients of the scaling sequence. To demonstrate this, first the Fourier transform $\hat{\phi}$ of the scaling function ϕ is calculated:

$$\begin{aligned}\hat{\phi}(\omega) &= \frac{1}{2}\sum_k a_k e^{-i\frac{1}{2}\omega k} \int \phi(2x-k)e^{-i\omega x}dx \\ &= \frac{1}{2}\sum_k a_k e^{-i\frac{1}{2}\omega k}\hat{\phi}(\tfrac{\omega}{2})\end{aligned} \tag{22}$$

With: $\hat{a}(\omega) = \frac{1}{2}\sum_k a_k e^{-i\omega k}$ and $\hat{\phi}(\omega) = \int \phi(2x-k)e^{-i\omega x}dx$, the resulting equation can be formulated as follows:

$$\hat{\phi}(\omega) = \hat{a}(\frac{\omega}{2})\hat{\phi}(\frac{\omega}{2}) \tag{23}$$

The requirements for a low-pass filter are that the Fourier series must have the value 1 at the zero point $\omega = 0$ ($\hat{a}(0) = 1$) and the value 0 at $\omega = \pi$ ($\hat{a}(\pi) = 0$). From these requirements follows immediately the following restrictions for the coefficients of the scaling sequence:

$$\sum_{k=-N}^{N} a_k = 2 \ ; \ \sum_{k=-N}^{N} (-1)^k a_k = 0 \tag{24}$$

One task of the wavelet design is to impose several conditions on the coefficients a_k in order to obtain the desired properties of the scaling function ϕ. For example: if ϕ is required to be orthogonal to all dilations of itself, the coefficients of the scaling sequence must fulfil the following conditions:

$$\forall_{m \in \mathbb{Z}\setminus\{0\}} : \sum_{k=-N}^{N} a_k a_{k+2m} = 0 \ ; \ \sum_{k=-N}^{N} a_k^2 = 2 \tag{25}$$

This can easily be seen, since: $\langle \phi, \phi \rangle = 1$ and $\langle \phi_{j,k}, \phi_{j,k'} \rangle = \delta_{kk'}$. The requirement that the scaling sequence is orthogonal to any shifts of it by an even number of coefficients is the necessary condition for the orthogonality of the wavelets. In terms of the Fourier transform the orthogonality is given by the relation:

$$|\hat{a}(\omega)|^2 + |\hat{a}(\omega + \pi)|^2 = 1 \tag{26}$$

The trigonometric polynomial $\hat{a}(\omega)$ (filter function or transfer function) plays an important role in wavelet theory.
The "mother wavelet" is defined by a similar two scale equation as follows:

$$\Phi(x) = \sum_{k=-N}^{N} b_k \phi(2x-k) \tag{27}$$

Where the sequence of coefficients, called a wavelet sequence or wavelet mask, is given by:

$$\{\ldots,0,b_{-N},\ldots,b_0,\ldots,b_N,0,\ldots\} \text{ with: } b_k=(-1)^k a_{K-k} \tag{28}$$

Whereby, $K \in \mathbb{Z}$ is an arbitrary odd number. The corresponding wavelet subspaces W_i are spanned up by:

$$W_j = span\{\Phi_{j,k}(x)=2^{-j/2}\Phi(2^{-j}x-k)|k\in\mathbb{Z}\} \tag{29}$$

The space $W_0 \subset V_1$ is defined as the linear hull of the mother wavelets integer shifts. W_0 is the orthogonal complement to V_0 inside V_1 (that means: V_1 is the orthogonal sum of W_0 and V_0). By successive application of the orthogonal sum, the orthogonal dissection of the scaling spaces is obtained:

$$V_2 = V_1 \oplus W_1 = V_0 \oplus W_0 \oplus W_1 \tag{30}$$

(Explicitly, this formal notation is given by: $\forall_{u\in V_{i+1}} : u = v + w$ with: $v \in V_i$ and $w \in W_i$). This is illustrated for the Haar wavelet in Figure 7. By self-similarity, there exist scaled versions W_k of W_0 and by completeness one has:

$$L^2(\mathbb{R}) = closure\,of \bigoplus_{k\in\mathbb{Z}} W_k \tag{31}$$

This is the basic analytical requirement for an MRA. Starting from the two scale equation, respectively the equation for the "mother wavelet", the multi-resolution decomposition of $f \in L^2(\mathbb{R})$ follows by calculating the wavelet coefficients.

5.4 Multi-resolution wavelet decomposition of functions

The next task is to compute the wavelet coefficients $\langle f,\Phi_{lk}\rangle$ of the discrete wavelet transform. Starting point is the two scale equation for the "mother wavelet" (formula 27). Then the "daughter wavelets" $\Phi_{j,k}$ are generated from the "mother wavelet" Φ and inserted into the equation (by use of $x = 2^{-j}x - k$ in $\phi(2x-k)$):

$$\begin{aligned}\Phi_{j,k} &= 2^{-j/2}\Phi(2^{-j}x-k)\\ &= 2^{-j/2}\sum_n b_n 2^{1/2}\phi(2^{-j+1}x-2k-n)\end{aligned} \tag{32}$$

The expression can be rewritten as:

$$\Phi_{j,k}(x) = \sum_n b_{n-2k}\phi_{j-1,n} \tag{33}$$

By use of the inner product, the wavelet coefficients are then calculated immediately by:

$$\langle f,\Phi_{j,k}\rangle = \sum_n b_{n-2k}\langle f,\phi_{j-1,n}\rangle \tag{34}$$

In other words, if the sequence of coefficients $\{\langle f,\phi_{j-1,n}\rangle | n\in\mathbb{Z}\}$ is known, then the sequence $\{\langle f,\Phi_{j,k}\rangle | k\in\mathbb{Z}\}$ can be obtained by convolution of $\{\langle f,\phi_{j-1,n}\rangle | n\in\mathbb{Z}\}$ with the sequence $\{b_{-n} | n\in\mathbb{Z}\}$ where only the even indexed terms of the resulting sequence are retained. The same procedure can be done for the scaling function by use of its two scale equation (formula 20). As in the case of the "mother wavelet", results for the scaling function ("father wavelet"):

$$\phi_{j,k}(x)=\sum_n a_{n-2k}\,\phi_{j-1,n}(x) \tag{35}$$

And the wavelet coefficients are given by the convolution:

$$\langle f,\phi_{j,k}\rangle=\sum_n a_{n-2k}\langle f,\phi_{j-1,n}\rangle \tag{36}$$

The connection between both procedures is illustrated by the following:

$$\begin{array}{cccccc} \{\langle f,\phi_{0,n}\rangle | n\in\mathbb{Z}\} & \rightarrow \{\langle f,\phi_{1,n}\rangle | n\in\mathbb{Z}\} & \rightarrow \{\langle f,\phi_{2,n}\rangle | n\in\mathbb{Z}\} & \rightarrow \ldots \\ \searrow & \searrow & \searrow & \\ \{\langle f,\Phi_{1,n}\rangle | n\in\mathbb{Z}\} & \{\langle f,\Phi_{2,n}\rangle | n\in\mathbb{Z}\} & \{\langle f,\Phi_{3,n}\rangle | n\in\mathbb{Z}\} & \ldots \end{array} \tag{37}$$

Thus, successively coarser approximations of the function $f\in L^2(\mathbb{R})$ are computed along with the difference in information between successive levels of approximation. The wavelet decomposition can be considered as an orthonormal basis transformation:

$$\{\phi_{j,n} | n\in\mathbb{Z}\}\rightarrow\left(\{\phi_{j+1,n} | n\in\mathbb{Z}\},\{\Phi_{j+1,n} | n\in\mathbb{Z}\}\right) \tag{38}$$

on the coefficients of the projections. f^j is the projection of f onto the subspace V_j ($f^j=\sum_k\langle f,\phi_{j,k}\rangle\phi_{j,k}$), g^j is the projection of f onto the subspace W_j ($g^j=\sum_k\langle f,\Phi_{j,k}\rangle\Phi_{j,k}$). Then it results for the projections:

$$f^{j-1}=f^j+g^j \tag{39}$$

The wavelet decomposition of a function f means that f is successively projected onto the subspaces V_j and W_j. It is a fine-to-coarse decomposition. By use of $s^j=\{s_n^j=\langle f,\phi_{j,n}\rangle\}$ and $d^j=\{d_n^j=\langle f,\Phi_{j,n}\rangle\}$, the equations for the wavelet coefficients can be rewritten as:

$$s_k^j=\sum_n(a_{n-2k})s_n^{j-1}\ ;\ d_k^j=\sum_n(b_{n-2k})s_n^{j-1} \tag{40}$$

The d^j are the coefficients of the projection g^j onto the subspace W_j. d^j is the sequence of wavelet coefficients representing the difference in information between two consecutive levels of resolution of the function f. The coefficients d^j represent detail (fine) information about the function f. s^j is the sequence of wavelet coefficients of the projection of f onto V_j. The coefficients s^j represent smoothed (coarse) information of the function. From the consecutive decompositions of the function f results, after a finite number of iterations, a finite sequence of coefficients:

$$s^0 \rightarrow \left(d^1, d^2, d^3, ..., d^j, ..., d^N, s^N\right) \tag{41}$$

s^0 is the coefficient sequence of the function f^0 (projection in V_0). The procedures discussed above define wavelets abstractly by their properties. They enable a systematical construction of wavelet bases with certain desired properties. The abstract formulation can be illustrated by a wavelet decomposition of a function with filter cascades. This enables also the implementation of the fast wavelet transform. The fast wavelet transform is an algorithm, which implements the discrete wavelet transform by use of the multi scale analysis. By this procedure, the calculation of the inner product of the function with every basis wavelet is replaced by a successive dissection of the frequency bands. This is realized by a sequence of filters. The next step is to determine the scaling sequence $\{a_k\}$ and the wavelet sequence $\{b_k\}$.

5.5 Construction of wavelets

The Z-transform enables it to express the decomposition procedure in a form that is more suitable for the discussion in terms of sub-band filtering. The Z-transform converts a discrete function (sequence of numbers), into a special complex representation. Given a discrete sequence $\{a_n | n \in \mathbb{Z}\}$, the Z-transform is defined by:

$$a(Z) = \sum_n a_n Z^n \tag{42}$$

Z is in general a complex number ($Z = Ae^{i\varphi}$, where A is the magnitude of Z, and φ is the complex argument). In terms of the Z-transform, the transfer function $\hat{a}(\omega)$ is expressed as:

$$\hat{a}(\omega) = a(e^{-i\omega}) \tag{43}$$

Then the conditions for the coefficients of the scaling sequence to fulfil a low-pass filtering are (formula 24):

$$a(1) = 2 \quad \text{and} \quad a(-1) = 0 \tag{44}$$

We demonstrate the construction of orthogonal wavelets on hand of the Daubechies wavelets (named after Ingrid Daubechies, a Belgian physicist and mathematician).
In other words, the scaling sequence $\{a_k\}$ and in consequence the wavelet sequence $\{b_k\}$ are determined (Fig. 8). The Daubechies wavelet transform can be easy implemented for practical purposes by use of the fast wavelet transform. Ingrid Daubechies assumes that $\hat{a}(\omega)$ has an A-fold zero at the values $\omega = \pm\pi$. Therefore she formulated the following expression for $\hat{a}(\omega)$:

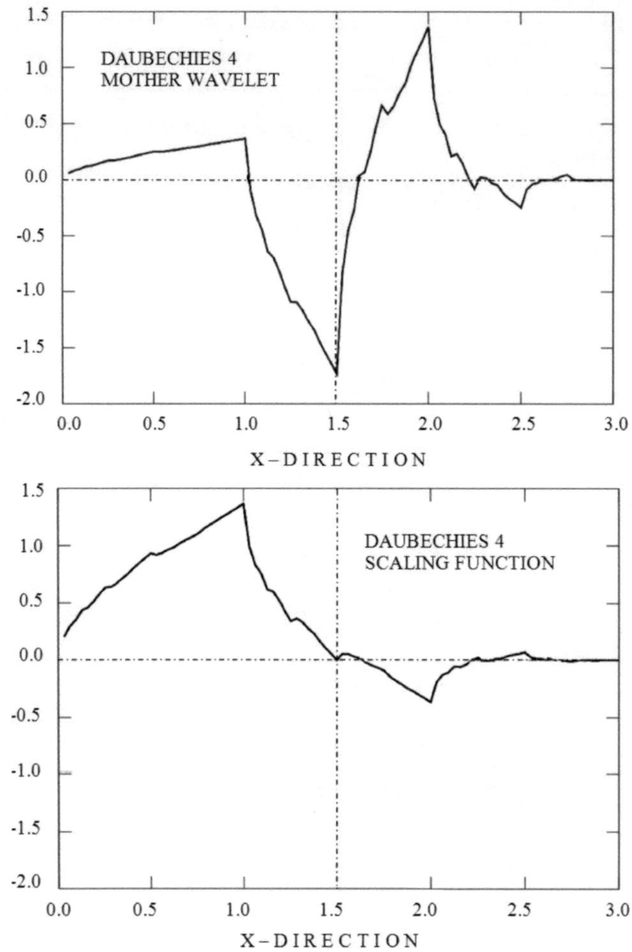

Fig. 8. Daubechies wavelets can not be represented analytically; they are generated by an iteration procedure

$$\hat{a}(\omega) = \left(\frac{1+e^{-i\omega}}{2} \right)^A \sum_{n=0}^{A-1} q_n e^{-in\omega} \tag{45}$$

This is called the factorization of the scaling sequence. As consequence, there is an A-fold zero of $\hat{a}(\omega+\pi)$ and $\hat{a}(-\omega-\pi)$ at $\omega=0$. The general representation for a scaling sequence of an orthogonal discrete wavelet transform with approximation order A, is given by the following factorization of the scaling sequence:

$$a(Z) = 2^{-A}(1+Z)^A p(Z) \tag{46}$$

$p(Z)$ is a polynomial in Z with degree A-1 and $p(1)=1$.

$$p(Z) = \sum_{n=0}^{A-1} q_n Z^n \tag{47}$$

The equation for $\hat{a}(\omega)$ results by setting $Z = e^{-i\omega}$. The Daubechies wavelets are characterized by a maximal number of vanishing moments (A) for some given support. The maximal potency A, enabling $(1+Z)^A$ to be a factor of $a(Z)$ is called the polynomial approximation degree. This degree reflects the ability of the scaling function ϕ to represent polynomial until the degree A-1 as a linear combination of translations (by whole numbers) of the scaling function. Then, by use of the Z-transform, the orthogonality condition (formula 25) can be written as:

$$a(Z)a(Z^{-1}) + a(-Z)a(-Z^{-1}) = 4 \quad (48)$$

The index number of a Daubechies DN wavelet refers to the number N of coefficients. Each wavelet has a number of vanishing moments (zero moments) equal to half the number of coefficients. (N=2A). In the case of Daubechies 4 (A=2), the polynomial p(Z) is linear and the following function can be used:

$$a(Z) = \frac{1}{4}(1+Z)^2\left((1+Z)+c(1-Z)\right) \quad (49)$$

By use of the factorization of the scaling sequence, the orthogonality condition can be formulated as Laurent polynomial:

$$(1-u)^A p(Z)p(Z^{-1}) + u^A\left(p(-Z)p(-Z^{-1})\right) = 1 \quad (50)$$

Whereby u is defined as: $u := \frac{1}{4}\left(2 - Z - Z^{-1}\right)$. By use of: $P_A(u) = p(Z)p(Z^{-1})$, the equation can be rewritten as:

$$(1-u)^A P_A(u) + u^A P_A(1-u) = 1 \quad (51)$$

A polynomial solution of this equation is the following binominal expansion:

$$P_A(u) = \sum_{k=0}^{A-1} \binom{A+k-1}{k} u^k \quad (52)$$

This polynomial plays an important role in the construction of the Daubechies wavelets. There exists a close connection between the zeros of $P_A(u)$ and the N scaling coefficients (filter coefficients) a_k of the Daubechies wavelet DN. When $P_A(u)$ is known, the product $p(Z)p(Z^{-1})$ can be calculated and the coefficients q_n be determined. In the case of N=4 (A=2, Daubechies D4), the polynomial till degree A-1 is given by $P_2(u) = 1 + 2u$ and the following equation is valid (with $P_2(u) = p(Z)p(Z^{-1})$):

$$1 + 2u = (q_0 + q_1)^2 - 4uq_0q_1 \quad (53)$$

This equation can be solved for q_0 and q_1. By use of $\hat{a}(0) = 1$ results: $q_0 + q_1 = 1$ and furthermore, by use of $\hat{a}(\omega)$ and the factorization of the scaling sequence (till A=2) results: $q_0 - q_1 = \sqrt{3}$. Then q_0 and q_1 are given by adding and subtracting the two relations:

$$q_0 = \frac{1+\sqrt{3}}{2}; \; q_1 = \frac{1-\sqrt{3}}{2} \tag{54}$$

The scaling coefficients are then calculated by equating $\hat{a}(\omega)$ and the factorization of the scaling sequence:

$$(a_0 + a_1 Z^{-1} + a_2 Z^{-2} + a_3 Z^{-3}) = \left(\frac{1+Z^{-1}}{2}\right)^2 (q_0 + q_1 Z^{-1}) \tag{55}$$

This gives the coefficients of the scaling sequence for Daubechies D4:

$$a_0 = \frac{1+\sqrt{3}}{4}, a_1 = \frac{3+\sqrt{3}}{4}, a_2 = \frac{3-\sqrt{3}}{4}, a_3 = \frac{1-\sqrt{3}}{4} \tag{56}$$

The wavelet coefficients b_k follow from the scaling coefficients a_k by use of $b_k = (-1)^k a_{K-k}$ (formula 28). In the case of Daubechies 4, the index K has the value 3. In the next chapter, the Daubechies 4 scaling and wavelet coefficients are used as filter coefficients to realize the wavelet transform as filter operations.

5.6 The relation between the discrete wavelet transform and digital filters

By use of discrete (digitized) data, the function is represented by a data vector with N components:

$$f = (f(1), f(2), \ldots, f(N-1), f(N))^{\dagger} \tag{57}$$

N is the dimension of the data vector or the number of the discrete data points. There exists a relation between the wavelet theory and the digital filter theory (Strang & Nguyen, 1996). Like the wavelet transform, digital filters extract details (high-pass filter) or smooth out details (low-pass filter). This correspondence between wavelets and digital filters enables the realization of a wavelet transform operation without an explicit formulation of basis wavelet functions. Just the scaling coefficients $\{a_n\}$ and the wavelet coefficients $\{b_n\}$ are necessary and in addition the transformation of the coefficients $s_n^j = \langle f, \phi_{j,n} \rangle$ and $d_n^j = \langle f, \Phi_{j,n} \rangle$, resulting from the two scale equations (formula 20 and formula 27). The transformation can be realized by definition and use of adequate filter operations. High-pass filters, which detect details (variations) in the data vector, are realized by subtraction operations. Along the data vector, the differences between the values of neighbour data components are considered: if there are great differences, the subtraction operation equals a finite number, if the values are nearby equal, the subtraction operation equals zero. Low-pass filters are realized by taking the mean value of neighbour data components: variations of the values are smoothed out. This corresponds with the dissection of a function f in a high-pass part in the subspace W_i and a low-pass part in the subspace V_i (beginning with the subspace V_{i+1}). The task of the scaling function is the generation of mean values of the data components (smoothing). It defines the global level of operation. The wavelet function detects the details of the data components.

The discrete wavelet transform can therefore be implemented and performed as convolution of the data vector with a filter bank. To realize a discrete wavelet transform by a filter operation, first a transformation matrix with appropriate filter properties is defined. In terms of the matrix operation, the scaling and wavelet functions apply to matrix rows (row basis). The transformation matrix consists of a set of filter coefficients that define special filter characteristics along the matrix rows. The matrix structure is determined by these filter coefficients, which are ordered using two dominant patterns: one works as a smoothing filter, the other works as a sharpening filter to bring out the data detail information. The associated wavelets are represented by the filter coefficients. Alternating, the even rows act as a high-pass filter and the odd rows act as a low-pass filter. The filter coefficients are associated with the corresponding "mother wavelet" and the scaling function, which determine the number of filter coefficients. The filter matrix for Daubechies 4 is defined as:

$$WT = \begin{pmatrix} a_0 & a_1 & a_2 & a_3 & & & & & & & \\ b_0 & b_1 & b_2 & b_3 & & & & & & & \\ & & a_0 & a_1 & a_2 & a_3 & & & & & \\ & & b_0 & b_1 & b_2 & b_3 & & & & & \\ & & & & & & \cdot & & & & \\ & & & & & & & a_0 & a_1 & a_2 & a_3 \\ & & & & & & & b_0 & b_1 & b_2 & b_3 \\ a_2 & a_3 & & & & & & & & a_0 & a_1 \\ b_2 & b_3 & & & & & & & & b_0 & b_1 \end{pmatrix} \quad (58)$$

The filter coefficients $[a_0, a_1, a_2, a_3]$ define the scaling function (low-pass filter) and the filter coefficients $[b_0, b_1, b_2, b_3]$ define the wavelet function (high-pass filter). Therefore, a data vector is decomposed into parts containing "smooth" information and parts with "detail" information. The filter coefficients are related according formula (28): $b_k = (-1)^k a_{3-k}$ and the transformation is orthogonal if the filter coefficients satisfy formula (25): $\sum_k a_k^2 = 1$ and $\sum_k a_k a_{k+2} = 0$. The transformation is then an image decomposition through the filter bank, in which the low-passed (smooth) and the high-passed (detail) parts serve as input for the next level (Fig. 9). This is called a pyramidal algorithm and the multi-resolution analysis is the theoretical foundation of the filter bank. This is an iterative procedure where the resulting output data vectors are used successively as input vectors (i) for the matrix operation in the next level (i+1):

$$f^{i+1} = WT \cdot f^i \quad (59)$$

The dimension of the transformation matrix is defined by the length of the input data vector. It should be noticed that by the first iteration, the data vector, with the analyzing data, is used as the input vector and the output vector contains the smooth and detail parts of the data components.

5.7 The 1-dimensional and 2-dimensional discrete wavelet transform

The matrix is applied to the data and produces alternating "smooth" (even rows) and "detail" (odd rows) components:

$$c^i(n) = \begin{cases} d^i(n);\ High-pass\ filtering\ with: [b_0, b_1, b_2, b_3] \\ s^i(n);\ Low-pass\ filtering\ with: [a_0, a_1, a_2, a_3] \end{cases} \tag{60}$$

The resulting output data vector is subsequently arranged into a part with "detailed" components and a part with "smooth" components (Fig. 9). Then the matrix is applied on the "smooth" part (halved vector, down sampling) of the output vector, then again on the "smooth-smooth" part of the resulting new output vector and so on until a trivial number of "smooth-…-smooth" components remain.

$$\begin{pmatrix} f(1)\\ f(2)\\ f(3)\\ f(4)\\ f(5)\\ f(6)\\ f(7)\\ f(8)\\ f(9)\\ f(10)\\ f(11)\\ f(12) \end{pmatrix} \xrightarrow[WT]{\text{LEVEL 1 } Multipl.} \begin{pmatrix} s^1(1)\\ d^1(1)\\ s^1(2)\\ d^1(2)\\ s^1(3)\\ d^1(3)\\ s^1(4)\\ d^1(4)\\ s^1(5)\\ d^1(5)\\ s^1(6)\\ d^1(6) \end{pmatrix} \xrightarrow[Coeff.]{Sort.} \begin{pmatrix} s^1(1)\\ s^1(2)\\ s^1(3)\\ s^1(4)\\ s^1(5)\\ s^1(6)\\ \hline d^1(1)\\ d^1(2)\\ d^1(3)\\ d^1(4)\\ d^1(5)\\ d^1(6) \end{pmatrix} \xrightarrow[WT]{\text{LEVEL 2 } Multipl.} \begin{pmatrix} s^2(1)\\ d^2(1)\\ s^2(2)\\ d^2(2)\\ s^2(3)\\ d^2(3)\\ \hline d^1(1)\\ d^1(2)\\ d^1(3)\\ d^1(4)\\ d^1(5)\\ d^1(6) \end{pmatrix} \xrightarrow[Coeff.]{Sort.} \begin{pmatrix} s^2(1)\\ s^2(2)\\ s^2(3)\\ \hline d^2(1)\\ d^2(2)\\ d^2(3)\\ \hline d^1(1)\\ d^1(2)\\ d^1(3)\\ d^1(4)\\ d^1(5)\\ d^1(6) \end{pmatrix} \xrightarrow[WT]{\text{LEVEL 3 } Multipl.} \bullet\bullet\bullet \tag{61}$$

The "detailed" parts obtained always remain conserved at the next operation levels. During the iterative procedure, the "smooth" parts of the resulting output vectors are successively filtered and their components rearranged in smooth ($s^i(n)$) and detail ($d^i(n)$) parts in each operation level. The output of the wavelet transform consists of the remaining "smooth-…-smooth" components and all the accumulated "detailed" components. The application of the filter bank to the smooth part, successively decimated by factor 2, corresponds to a successive stretching of the scaling function. In other words: the data vector is analyzed at successively larger scales (from high resolution to low resolution). At every operation level, the input data vectors are reduced to the half resolution (sub sampling). By the halving of the resolution at every level, the respective output vector always has the same number of components (wavelet coefficients) as the respective input vector. There is no loss of information during the procedure.

The 2-dimensional wavelet transforms works similarly to the 1-dimensional case. In the first step (or level), the rows of the input image are filtered by a high-pass filter and independently by a low-pass filter (Fig. 10). From both operations two images (sub-bands) result: one shows details and the other is smoothed out. Every second column is then

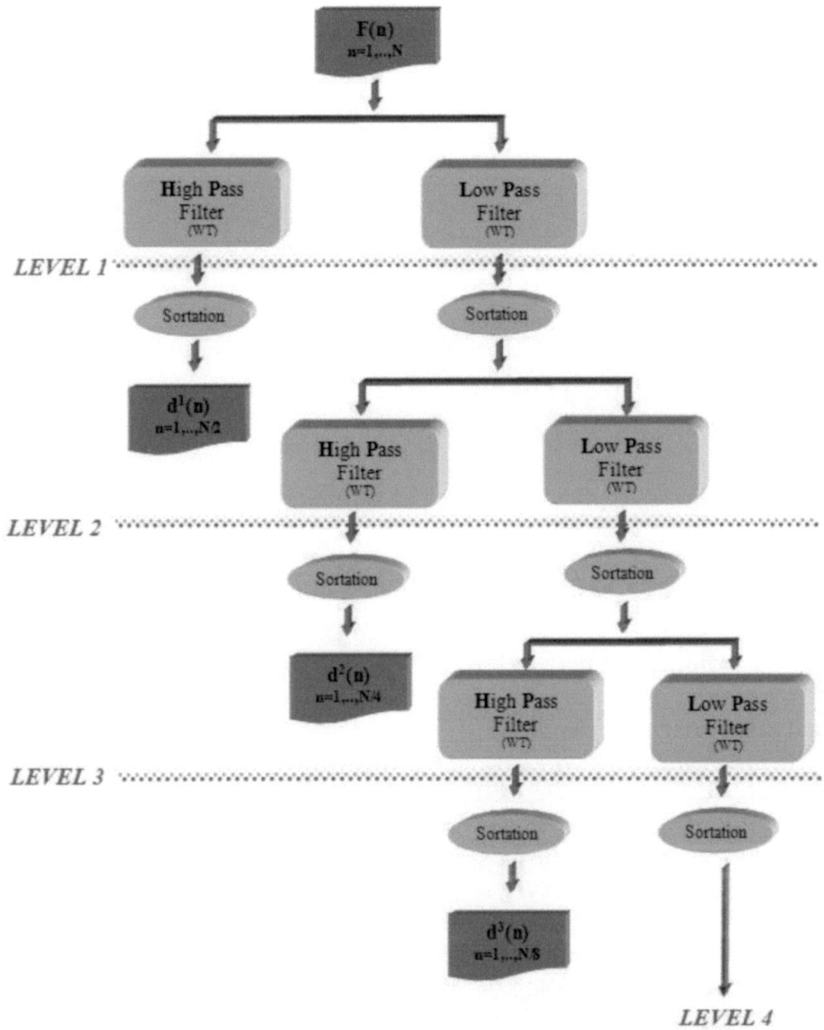

Fig. 9. The 1-dimensional wavelet transformation as filter operation

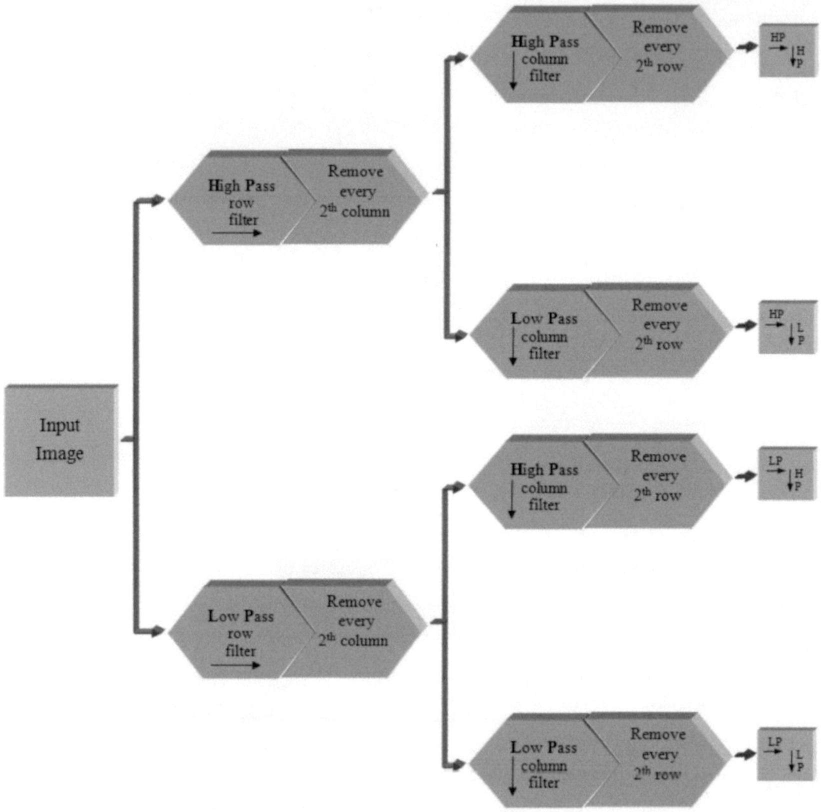

Fig. 10. The 2-dimensional wavelet transformation as filter operation

removed in both sub-bands, which corresponds to a sub sampling (half resolution). Subsequently the columns of both sub-bands are high-pass filtered and low-pass filtered. Again two sub-bands result from each of the two input sub-bands. A total of four sub-bands, which differ by the kind of filtering, are obtained in the first step. At the end of the first step, every second row is removed from each of the four sub-bands. Only the double low-passed sub-band is used in the next step, the other three sub-bands remain conserved for the next operation steps. During the next steps, the same procedure is used again and again. At each step the double high-passed sub-band, which results from the previous step, is used as an input sub-band. At every step, the resulting sub-bands are reduced to the half resolution. The sub-bands with higher spatial resolution contain the detail information (high frequencies) whereas the sub-bands with the low resolution represent the large scale coarse information (low frequencies). After the dissection of the quadratic sub-bands, they are usually arranged in a quadratic configuration where the sub-bands of the first level fill 3/4 of the square; the sub-bands of the second level fill 3/16 of the square; etc. (Fig. 11).

As in the case of the 2-dimensional function, for the 2-dimensional image array, the wavelet transform is computed along each dimension, e.g.: the array is first filtered on its first index (row), then on its second (column) and so on. The multi resolution analysis takes scale information into consideration and successively decomposes the original image into approximations (smooth parts) and details. That means, by the wavelet transformation, the two dimensional image array is split up into several frequency bands (containing various numbers of wavelet coefficients), which represent information at different scales. At each scale the original image is approximated with more or fewer details.

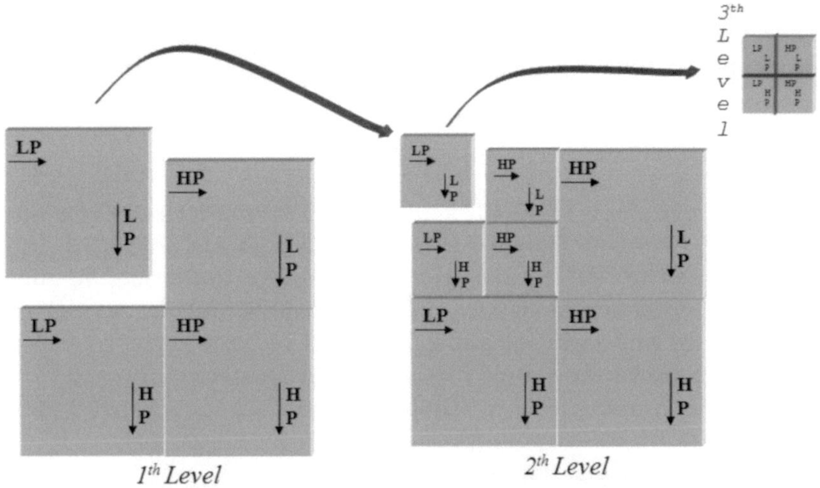

Fig. 11. Dissection of the sub-bands in the 2-dimensional wavelet transformation

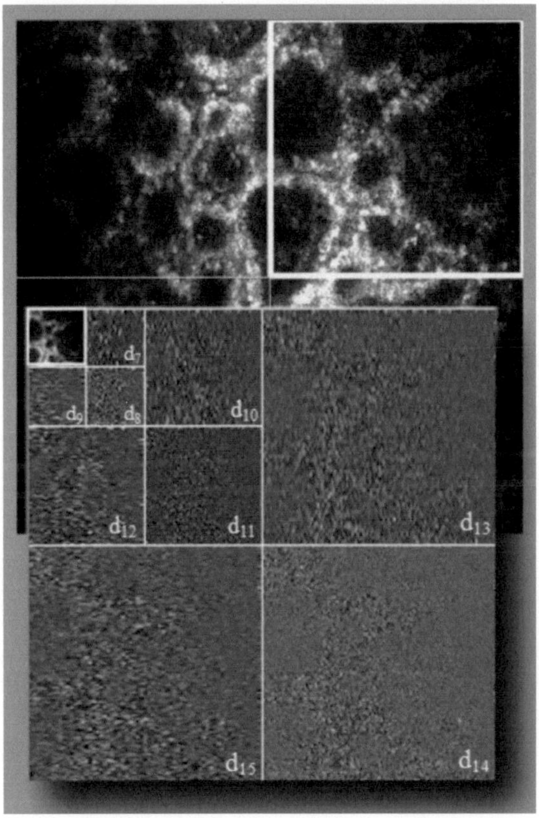

Fig. 12. Wavelet frequency band structure in the 2-dimensional wavelet transformation

The output of the last low-pass filtering is the mean gray level of the image. The frequency bands, representing information at large scale, are labelled with low indices and the frequency bands representing successively decreasing scales are labelled with higher indices (Fig. 12). Then the architectural structure information is accumulated along the energy bands (from course to fine), enabling the analysis of a given texture by its frequency components:

$$d_i = (c_i(k,l)) \tag{62}$$

where the subset of coefficients ($(c_i(k,l))$) are the coefficients contained in the i-th frequency band. At increasing frequency bands the detail structures are smoothed out and large architectural structures of the texture are analysed. At decreasing frequency bands, successively finer structures and details are registered. In wavelet texture analysis, the features are mostly derived from statistical properties of the wavelet coefficients. For the texture analysis, we choose the Daubechies 4 wavelets transform, because this has good localization properties in space and is computationally efficient (fast wavelet transform).

5.8 Wavelet texture analysis and texture features

After the wavelet transformation of the CLSM images, it is necessary to define texture features which reflect and describe the tissue properties. These features must be defined so that they allow a unique distinction between common benign nevi and malign melanoma tissues in the square elements. The texture features are based on the variations of the wavelet coefficients within the frequency bands and the distribution of the energy of the frequency bands in the power spectrum. As features, the standard deviations of the coefficients inside the frequency bands and the energy and entropy of the different frequency bands are used. The standard deviations of the coefficients in the frequency bands are calculated by:

$$F_{STD}(i) = \sqrt{\frac{1}{n_i}\sum_k \sum_l (m_i - c_i(k,l))^2} \tag{63}$$

where n_i is the number of coefficients in the i-th frequency band and $m_i = \frac{1}{n_i}\sum_k \sum_l c_i(k,l)$ is the mean value within the band. The energy of the i-th frequency band is determined by:

$$F_E(i) = \frac{1}{n_i}\sum_k \sum_l (c_i(k,l))^2 \tag{64}$$

The entropy of the i-th frequency band is calculated as:

$$F_{ENT}(i) = (-1)\sum_k \sum_l w_i(k,l)\log_2 w_i(k,l) \tag{65}$$

with the normalized values: $w_i(k,l) = \dfrac{c_i(k,l)^2}{\sum_k \sum_l (c_i(k,l))^2}$. The features in the different frequency bands reflect architectural structures and cell structures at different scales (Fig. 12). In total,

39 different features are calculated, representing large scale and low scale information. The features are calculated for 16 frequency bands (labelled from 0 to 15). The mean value is calculated from the 4 first frequency bands, therefore 13 values result for each of the 3 features (Fig. 12). The highest frequency bands contain only information about very fine grey level variations, such as noise, and are therefore not considered for the image analysis. The single square elements are represented by a feature vector:

$$x_n = \left(F_{STD}^{(n)}(0), F_{STD}^{(n)}(1), ..., F_{ENT}^{(n)}(15)\right) \tag{66}$$

The index n refers to the n-th square element. The next step is the splitting of the square elements, on hand of the features, in square elements with common benign nevi tissue and square elements with malign melanoma tissue.

6. Classification and class prediction

By the classification procedure, the inhomogeneous set of square elements (referring to the feature values) is split into homogeneous sub-sets, which are assigned to one of the two classes (common benign nevi or malignant melanoma). A homogeneous subset means that it contains only square elements with similar feature values. In the following, we will name the square elements as instances. (The input to a machine learning scheme is, in its general form, a set of instances. The instances are the things to be classified). In the context of machine learning the features are generally called attributes. The thing to be learned, in our case the discrimination of lesion tissues, is called the concept. We will use this terminology in the following. Classification is done by machine learning algorithms (Witten & Frank, 2005). After a training phase, these algorithms enable the automated prediction of the class of a new instance. Machine learning can be defined operationally as: machine learning algorithms learn when they change their behaviour in a way that makes them perform better in the future. In other words, the algorithm learns on hand of a training set how to assign the instances to given classes. Then, in future, the algorithm can apply the gained knowledge to predict the class of unknown instances. The task of machine learning algorithms is in general the search of patterns in big amounts of data, known as data mining (Van Rijsbergen, 1979). The set of square elements is used as instances:

$$I = \{x_1, ..., x_N\} \tag{67}$$

To simplify the expressions, the features (attributes) in the feature vector are rewritten as:

$$x_n = (a_0^{(n)}, a_1^{(n)}, ..., a_{L-1}^{(n)}) \tag{68}$$

Here $a_i^{(l)}$ is the i-th attribute of the l-th instance (L is the number of attributes). As an example, the attribute $a_0^{(n)}$ refers to $F_{STD}^{(n)}(i)$. The set of square elements is split into more or less homogeneous subsets, which are assigned to classes. The partition of the set of instances into g classes is then given by:

$$P = \{C_1, ..., C_g\} \tag{69}$$

For every class C_l the following conditions should be satisfied:

$$\text{i)} \bigcup_{k=1}^{g} C_k = I \quad \text{ii)} \ \forall_{i \neq j} : C_i \cap C_j = \varnothing \tag{70}$$

The first condition (i) means that every instance is assigned to a specific class. All the instances are classified, and then the union of all classes equals the set of instances. Or in other words, all the instances are assigned to a class, no one is left over. The second condition (ii) means that no two different classes contain the same instances (the classes are disjunct). Or in other words, no instance belongs to two or more classes. If the attributes have numerical values (as in our case) then for an optimal partition, the Euclidian distance of every feature vector of a class k to the class mean value $\overline{x}_k$ is less or equal to the mean values of every other class (j=1,…,g):

$$\left\| x_n - \overline{x}_k \right\|^2 \le \left\| x_n - \overline{x}_j \right\|^2 \tag{71}$$

Whereby the mean value of a class k containing K instances is $\overline{x} = \frac{1}{K}\sqrt{\sum_{i=1}^{K}(x_i)}$ and $\left\| x_i - x_j \right\| = \sqrt{\sum_{l=1}^{L}\left(a_l^{(i)} - a_l^{(j)}\right)^2}$ is the Euclidian distance. The first step for a successful use of machine learning algorithms is the representation of the knowledge. In computers, knowledge can be represented in different ways, for example: by numeric values, trees, rules or instance based (Brachman & Levesque, 1985). For medical diagnosis purposes it is important to duplicate the automated diagnostic process. Therefore, the rules that the algorithm uses to predict the class of an instance should be understandable for the human interpreter. Here we chose the CART (Classification and Regression Trees) algorithm for classification and class prediction, which uses a tree representation but where the inferring rules are automatically generated out of the tree.

6.1 Classification and Regression Trees (CART)

The CART algorithm was published for the first time in the year 1984 by Leo Breiman et al. CART is an algorithm which is used for optimal decision finding (Breiman et al., 1993). It is based on the generation of so called binary decision trees. The choice of the attributes is guided by the optimization of the information measure at each decision step. An important feature of the CART algorithm is that for every decision in a given node of the generated tree, the node is split only in two sub nodes (every node has maximally two child nodes). This is called a binary decision and the generated tree is called a binary tree. The central task of the algorithm is to determine the optimal binary decision for separation of the attributes at each step. Therefore, the special merit of the CART algorithm is the ability for the optimal separation of the data relating to classification purposes. Furthermore, the CART algorithm has the ability to capture the decision structure explicitly. This means that the decision rules generated from the tree are intelligible in that way that they can be understood discussed and explicitly used as diagnostic rules. A binary tree has a root node (starting point), several leaves (end points) and inner nodes which are branched into two succeeding nodes (Fig. 13). For a given node, the preceding node is the parent node and its successors are the child nodes. The root node of a tree is the node with no parents. There is, at most, one root node

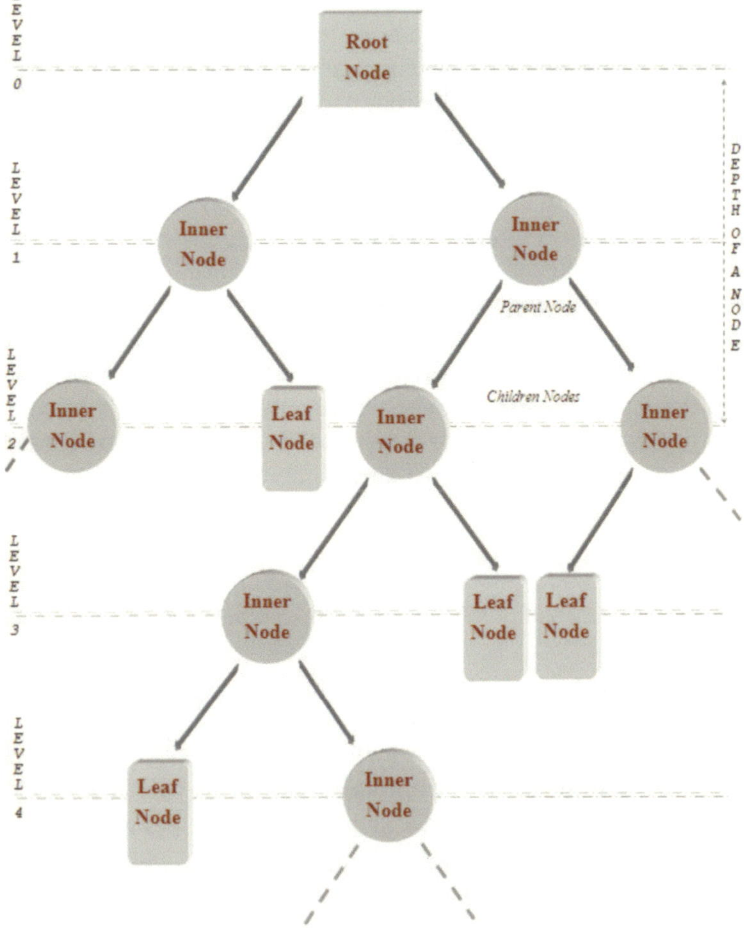

Fig. 13. In computers, knowledge can be represented as binary trees, consisting of a root node, inner nodes and leaves

in a rooted tree. A leaf node has no children. The depth of a node in the tree is the length of the path from the root node to the considered node. The set of all nodes at a given depth is called a level of the tree. The height of a tree is the length of the path from the root to the deepest leaf node in the tree. To classify an unknown instance, it is routed down the tree according to the values of the different features (Steinberg & Colla, 1995).

6.2 Generation of the decision tree

Divide-and-conquer (lat. divide et impera) algorithms are used to produce decision trees in a recursive way. A divide and conquer algorithm works by recursively breaking down a set of different objects (distinguishable by their specific feature values) into two (or more) sub-sets of similar kinds of objects, until the sub-sets contain only objects of the same kind. A node in the decision tree involves the testing of a particular feature. The root node, as the first node in a decision tree, contains all instances (square elements). A terminal (leaf) node contains instances that belong to the same class. The divide-and-conquer algorithm for constructing a decision tree consists in principal of three parts:

1. The determination of the optimal splitting at every node;
2. The decision whether the node is a leaf node or a inner node;
3. The assignment of a leaf node to a specific class.

First a feature (attribute) is selected at the root node and a branch is made for each possible value. Through this operation the set of the instances is split up into subsets. Then this process is repeated recursively for every branch, at an inner node, with the instances in the corresponding subset (Fig. 14). If the nodes cannot be divided anymore, the process is stopped and all the instances at a terminal node belong to the same class. The features that produce the purest daughter nodes are used for an optimal splitting. In the case of numerical features, the splitting is usually done by use of numeric thresholds that divide the range of the feature values. Trees with averaged numerical values are called regression trees. Trees with tests involving more than one feature at a time are called multivariate decision trees (in contrast to univariate trees, where only one feature is used).

Determining the optimal splitting: Each step in the tree construction consists of a binary decision leading to two daughter nodes. At every inner node, the actual instance set (parent set: t) is split into two children sets (t_R and t_L).

$$t \xrightarrow{s} \begin{cases} t_R = \{x \mid F_{STD}^{(n)}(i) \le U_i\} \\ t_L = \{x \mid F_{STD}^{(n)}(i) > U_i\} \end{cases} \tag{72}$$

The set of splits $S = \{s_i\}$ is generated by a set of tests $Q = \{q_i\}$. The tests $q_i \in Q$ consist of binary decisions, where each split $s_i \in S$ depends on the value of a single feature ($F_{STD}^{(n)}(l)$). The splitting operation is done by use of numerical threshold values U_i, which are selected for every attribute. The decisions are made by comparing the feature values with the threshold values separating the range of the features into two parts. The decision thresholds result from the optimization of the information measure. The fundamental principle for the construction of the decision tree is to select the split of a set in such a way that the instances of the children sets are purer than the instances of the parent set. The information measure $I(t)$ is commonly used for calculating the purity or impurity of the nodes. According to Shannon, the information measure of a set of objects $S = \{z_1, z_2, ..., z_n\}$ is defined by:

$$I = -\sum_{z \in S} [p_z \log_2(p_z)] \tag{73}$$

Whereby the probability of occurrence of an element in the set is $p_z = \frac{N(z)}{N}$ and $N(z)$ is the number of the elements z in the set with N elements. The sum of the probabilities is one ($\sum_z p_z = 1$). The information measures the amount of uncertainty associated with each element (object) in a given discrete set, depending on the probability of occurrence in the set. For a pure node (only elements of one kind), the information measure has a minimum (=0) value. For an impure node with equal distribution of the elements, the information measure has a maximum value.

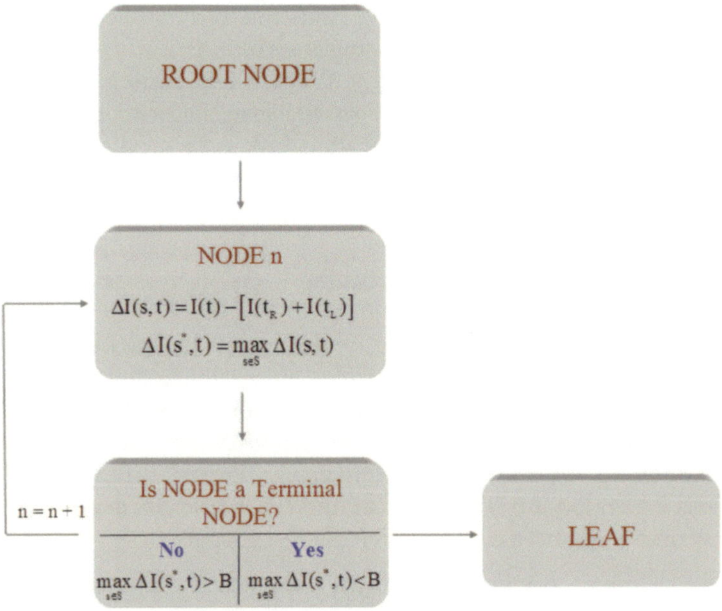

Fig. 14. The decision tree is generated recursively

For sets with different distributions of the elements, the value of the information measure lies between these two extremes. If the information measure is calculated by use of the binary logarithm, which is closely connected to the binary numeral system, the units of the corresponding values are expressed in bits. The information measure of an attribute is considered to be high when it enables a set splitting in such a way that a classification can be done with a high score. The higher the information measure of an attribute is, in relation to the classification values, the higher in the tree the attribute is selected. The difference between the impurities of the parent and the children sets is given by:

$$\Delta I(s,t) = I(t) - [I(t_R) + I(t_L)] \tag{74}$$

Then the parent set splits into children sets, by use of a splitting rule s^* generated by a test $q^* \in Q$, in such a way that the difference between the impurities of the parents and the children sets has the maximum value:

$$\Delta I(s^*,t) = \max_{s \in S} \Delta I(s,t) \tag{75}$$

This means that the set of parent square elements is split into children sets so that the feature values for the square elements in each of the children sets are as similar as possible. At every node, a feature is interrogated and a decision is made about the selection of the successor nodes . This procedure is repeated until a leaf is reached.

Decision whether the node is a leaf node. The decision whether the node is a terminal node or an inner node depends on the purity of a children node and is given by the stop rule:

$$\max_{s \in S} \Delta I(s^*,t) < B \tag{76}$$

The threshold B is a measure of how homogenous the terminal nodes must be. Ideally, the information measure for a set of equal elements is zero.
Assignment of a leaf node to a specific class. When a leaf node is reached, the instance is classified according to the class assigned to the leaf. Because the texture features are represented by numeric values, the classification task is to predict a numeric quantity by which the instance is assigned to the averaged numeric value of a class.

6.3 Inferring rules

The divide-and-conquer algorithm for generating decision trees enables a good separation of the instances. But the knowledge representation by binary trees is not as suitable for the interpretation of the classification process. In contrast, the knowledge representation by rules is very suitable in medicine because rules are easily understandable for the human interpreter. By the CART algorithm, the knowledge is extracted from the dataset in the form of inferring rules, which can easily be implemented as diagnostic rules. The inferring rules result from the splitting rules and are automatically derived directly off the decision tree, where one rule is generated for each terminal node. The rules have a syntax consisting of an antecedent (precondition) and a consequent (conclusion) part:

$$\forall_{x \in I} : P(x) \rightarrow C(x) \tag{77}$$

For all instances, which fulfil the precondition $P(x)$, follow the conclusion $C(x)$. The rule, represented by the formula, is called "Modus Ponens" meaning that: from $P(x)$ infer $C(x)$ or in other words "if $P(x)$ then $C(x)$". This rule is the formal description of the diagnostic process. The precondition may include several parts (single sub-conditions, for example clinical symptoms) $P(x) = p_1(x) \wedge p_2(x) \wedge ... \wedge p_N(x)$. If all the single sub-conditions ($p_n(x)$) are true then the precondition $P(x)$ is valid. The antecedent of a rule includes a condition for every node on the path from the root node to a specific terminal node. The consequence of the rule is the class assigned to the terminal node. For the implementation on a computer they are generally expressed as "IF-THEN" rules (IF $\{\text{precondition}\}$ THEN $\{\text{conclusion}\}$). These rules represent knowledge in a form that is easily understandable for the human observer. It enables him to understand why a specific square element is assigned to a benign common nevi or malignant melanoma. Such inferring rules are used as diagnostic rules. The rules can be compared with the diagnostic guidelines of the derma pathologist. Furthermore they enable the computer scientist to validate the gained knowledge and to combine it eventually with previously known facts. An example of such a (simplified) inferring rule is:

$$\text{IF} \left\{ \begin{array}{l} \text{Standard deviation in frequency band (5)} > \text{a} \\ \text{.and.} \\ \text{Energy of highest frequency band (0,1,2,3)} \geq \text{b} \end{array} \right\} \text{THEN} \{\text{tissue} := \text{nevus tissue}\} \tag{78}$$

According to the guidelines for the diagnosis, the rule can be translated in: if the tissue contains large and medium structures, for example: nevi cells grouped around basal structures, then it is nevus tissue. Generally, such rules are far more complex than necessary and therefore they are usually pruned to remove redundant tests. Pruning is a technique in machine learning that reduces the size of decision trees by removing sections of the tree that provide little power to classify instances. The goal of pruning is to reduce complexity of the final classifier and to improve the classification accuracy by the reduction of over fitting and removal of sections of a classifier that may be based on erroneous data. Pruning should reduce the size of a learning tree without reducing predictive accuracy as measured by a test set or using cross-validation.

7. Study

The discrimination power of texture features based on the wavelet transform and the performance of the CART algorithm are demonstrated in the following study (square size 512x512). Overall, 857 images of benign common nevi (408 images) and malignant melanoma (449 images) are used as study set. To get more insights into the classification performance, a percentage split was performed by using 66% of the dataset for training and the remaining instances (34%) as the test set. The classification results of 572 cases (276 benign common nevi, 296 malignant melanomas) in the training set and 285 cases (132 benign common nevi, 153 malignant melanomas) in the test set are shown in Table 1.

CART	Training Set			Test Set		
	% Correct	NZ	MM	% Correct	NZ	MM
NZ	96.6	267	9	78.0	103	29
MM	98.0	6	290	84.1	24	129

Table 1. Classification result for malignant melanomas (MM) and benign common nevi (NZ) in the training and test set

8. Discussion

Delayed recognition of skin malignancies puts the patient at risk of destructive growth and death from disease once the tumour has progressed to competence for metastasis. Therefore, preventive and periodical skin checkups are of special importance. Technological advancements in imaging systems have led to the development of convocal laser scanning microscopy (Ericson et al., 2008). This technique enables the examination of skin lesions in vivo and significantly higher prediction success than reported for dermoscopic examination can be achieved for the diagnosis of melanoma (Rajadhyaksha, 2009). However, due to the fact that the CLSM method is relatively new, there is still a lack of experiences with the diagnostic features and an intensive training is necessary for the clinician.

The study demonstrates the applicability of the automated diagnosis system for the discrimination of CLSM views of skin lesions (Table 1). The image analysis, based on the wavelet transform, together with tree based machine learning algorithms provide a powerful tool for automated diagnosis of CLSM images of skin lesions. For the diagnosis of the CLSM views, architectural structures such as: micro-anatomic structures; cell nests etc., are used as guidelines by the derma pathologist. Therefore features based on the spectral properties of the wavelet transform, enabling an exploration of architectural structures at different spatial scales are suitable for the automatic analysis. The images of benign common nevi show pronounced architectural structures whereas images of malign melanoma show melanoma cells and connective tissue with few or no architectural structures. These guide lines are reflected by the wavelet coefficients inside the different frequency bands. The standard deviations of the wavelet coefficients in the lower and medium frequency bands show higher values for the benign common nevi than for malignant melanoma tissue, indicating more pronounced structures at different orders of magnitude (Fig. 15). The energy of the lowest frequency band (large scale architectural structures) is higher for the CLSM views of benign common nevi than for malignant melanoma. The CART algorithm has the ability to capture the decision structure explicitly. This means that the, from the tree, generated decision rules are intelligible in that way that they can be understood, discussed and explicitly used as diagnostic rules.

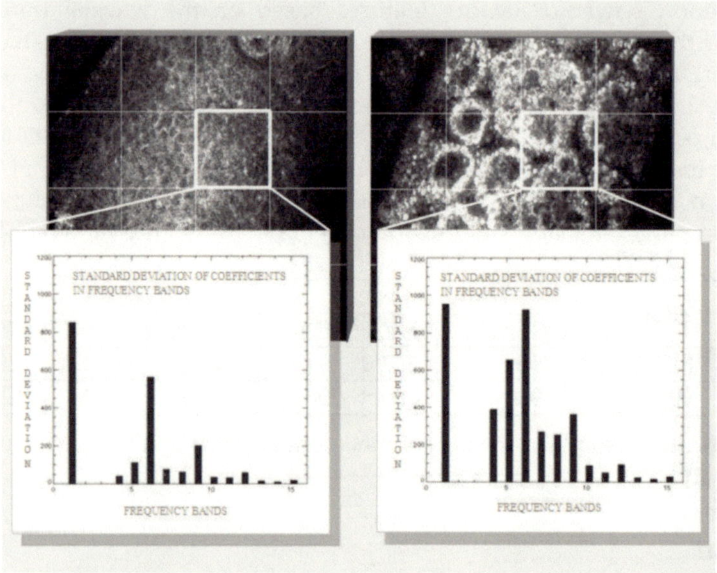

Fig. 15. The standard deviations of coefficients in the wavelet frequency bands for melanoma (left) and nevi (right)

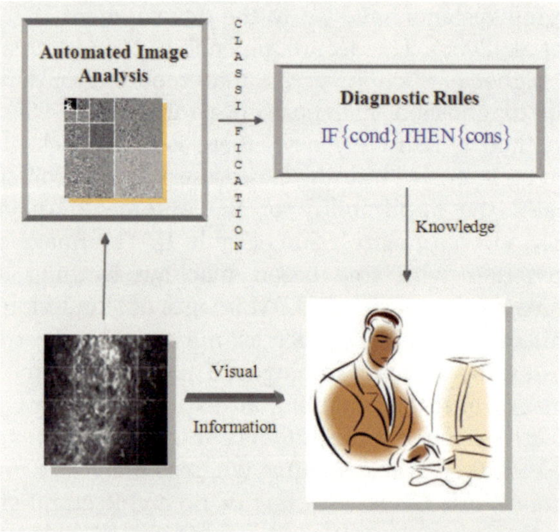

Fig. 16. From the CLSM image the features, based on wavelet transform, are extracted and automatically analyzed

Computer aided diagnosis, providing automated decisions, can be used as an expert second opinion or help and assist the non-experienced physician in the diagnostic procedure (Fig. 16). Automated diagnosis is the principal performance of medical expert systems. Although the handcrafted diagnostic rules in the inference machine of expert systems perform well in medical applications, machine learning has the advantage that the rules are generated automatically for systems where the producing of manual rules is too labour intensive and there is a lack of human expert knowledge and experience. The generating of manual rules requires expert knowledge whereas rules generated by machine learning algorithms represent knowledge that can be used to analyse and refine the diagnostic process.

To verify the performance of the method and to interpret the diagnostic process, the automatically generated inferring rules are implemented in specially developed viewer software (Fig. 17). By this viewer the classified square elements are indicated in the corresponding CLSM image in order to judge the performance of the analysis. For these purposes square elements of size 128x128 are used, because they enable a good localization of the different regions into the images and the texture features have still enough discrimination power (sensitivity and specificity of 88.12% for malignant melanoma and 84.75% for benign common nevi). The procedure is illustrated in the case of malignant melanoma tissue (Fig. 17). Square elements resulting from terminal nodes, with 100% of discrimination power, are taken as highly significant and are drawn with red margins at the graphical user interface of the viewer and labelled with the number 1. Square elements with discrimination power of 80-99% are drawn with green margins and labelled with the number 2. The relocated elements mainly show polymorphic tumour cells with structural

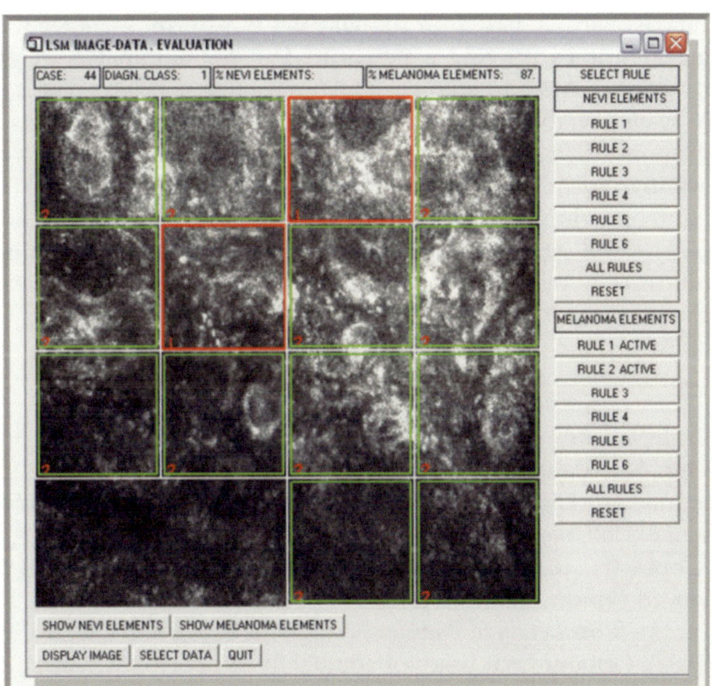

Fig. 17. Square elements containing diagnostic significant regions in a CLSM image of a malignant melanoma

disarray and are in good accordance with previously published diagnostic CLSM features. Due to the fact that the method of confocal laser scanning microscopy is relatively new, the diagnostic features for the clinician have not yet been completely investigated. Therefore the automated image analysis is a fundamental and important step toward the general assessment of the CLSM images by the clinician. The viewer software is further used for the interaction of the clinician with the automated diagnostic system. Then he can use it as computer aided tool for a second opinion or a non-experienced user can consult the system, learn from the system and ameliorate his diagnostic skills. The results of the inferring mechanism (the classification) are visualized by the graphical user interface enabling an interpretation and evaluation of the diagnostic process.

In this review, we demonstrated how diagnostic guidelines and experiences are represented by mathematical structures and implemented on a computer. To enable the human interpreter an interpretation of the results of the automated diagnostic, the output is visualized in an appropriate manner. Since the highlighted regions in the CLSM images show tissue structures which are in good accordance with the already known diagnostic guidelines, no explicit formulation of the diagnostic rules is necessary. Due to the fact that the tissue features are based on the wavelet transform, reflecting the guidelines of the derma pathologists, the highlighted regions are self explaining. Although it should be noted that, beside the visual interpretation, it seems that features not accessible to the human eye contribute to the automated diagnostic process.

9. Conclusion

In conclusion, image analysis, based on the wavelet transform, together with a tree based machine learning algorithm provide a powerful tool for automated diagnosis of CLSM images of skin lesions. Already known, but subjective CLSM criteria are objectively reproduced. The system enables the identification of highly significant parts in CLSM views of malignant melanoma. In a clinical application, the system can be used as a screening tool to improve preventive medical checkups and the early recognition of skin tumours. The automated decisions provided can be used as an expert second opinion and as a training system for inexperienced or student derma pathologists. In another clinical onset, the system may automatically pre-select the cases in such a way that the critical cases are first interpreted by the clinician.

10. References

Brachman, R.J.; Levesque, H.J. (1985). *Readings in knowledge representation*. Morgan Kaufmann, San Francisco

Breiman, L.; Friedman, J.; Olshen, R.A.; Stone, C.F. (1993). *Classification and Regression Trees*. Chapman & Hall, New York, London

Burrus, C.S.; Gopinath, R.A.; Guo, H. (1988). *Introduction to Wavelets and Wavelet Transforms:A Primer*. Prentice-Hall

Chui, C. K. (1992). *An Introduction to Wavelets*, Academic Press, San Diego

Daubechies I. (1988). Orthonormal bases of compactly supported wavelets, *Comm. Pure and Appl. Math.* 41

Daubechies, I.; Lagarias, J. (1991). Two-scale equations: Existence and global regularity of solutions. *SIAM J. Math. Anal*, 22

Daubechies, I. (1992). *Ten Lectures on Wavelets*. SIAM

Ericson, M.B.; Simonsson, C.; Guldbrand, S.; Ljungblad, C.; J. Paoli, J.; Smedh, M. (2008) Two-photon laser- scanning fluorescence microscopy applied for studies of human skin. *J Biophotonics*, 1, 4, 320-330

Friedman, R.; Rigel, D.; Kopf, A. (1985). Early detection of malignant melanoma: the role of physician examination and self-examination of the skin. *CA Cancer J Clin*, 35, 3, 130–151

Lucas, R. (2006). Global Burden of Disease of Solar Ultraviolet Radiation. *Environmental Burden of Disease Series*, July 25, 13. News release, World Health Organization

Markovic, S.N.; Erickson, L.A.; Flotte, T.J.; Kottschade, L.A. (2009). Metastatic malignant melanoma. *G Ital Dermatol Venereol*. 144(1) 1-26.

Oliveria, S.; Saraiya, M.; Geller, A.; Heneghan, M.; Jorgensen, C. (2006). Sun exposure and risk of melanoma. *Arch Dis Child*, 91, 2, 131–138

Paoli , J.; Smedh, M.; Ericson, M.B. (2009). Multiphoton laser scanning microscopy--a novel diagnostic method for superficial skin cancers. *Semin Cutan Med Surg*, 28, 3, 190-195

Patel, D.V.; McGhee, C.N. (2007). Contemporary in vivo confocal microscopy of the living human cornea using white light and laser scanning techniques: a major review. *Clin. Experiment. Ophthalmol*, 35, 1, 71–88

Pellacani, G.; Longo, C.; Malvehy, J.; Puig, S.; Carrera, C.; Segura, S.; Bassoli, S.; Seidenari, S. (2008). In vivo confocal microscopic and histopathologic correlations of dermoscopic features in 202 melanocytic lesions. *Arch Dermatol*. 144, 12, 1597-608

Pellacani, G.; Vinceti, M.; Bassoli, S.; Braun, R.; Gonzalez, S.; Guitera, P.; Longo, C.; Marghoob, A.A.; Menzies, S.W.; Puig, S.; Scope, A.; Seidenari, S.; Malvehy, J. (2009). Reflectance confocal microscopy and features of melanocytic lesions: an internet-based study of the reproducibility of terminology. *Arch Dermatol*, 145, 10, 1137-1143

Prasad, L.; Iyengar, S.S. (1997).*Wavelet analysis with applications to image processing*. CRC, Press Boca Raton, Boston, London, New York, Washington D.C.

Press, W.H.; Teukolsky, S.A.; Vetterling, W.T.; Flannery, B.P. (1992). *Numerical Recipes in C: The Art of Scientific Computing*. 2nd ed. Cambridge University Press, 591-606

Rajadhyaksha, M.; Conzales, S.; Zavislan, J.M.; Anderson, R.R.; Webb, R.R. (1999). In vivo confocal scanning laser microscopy of human skin: advances in instrumentation and comparision with histology. *J Invest Dermatol*, 113, 293-303

Rajadhyaksha, M. (2009). Confocal microscopy of skin cancers: Translational advances toward clinical utility. *Conf Proc IEEE Eng Med Biol Soc*. 1, 3231-3233

Scope, A.; Venute-Andrade, C.; Agero A.L. (2007). In vivo reflectance confocal microscopy imaging of melanocytic skin lesions: consensus terminology and illustrative images. *J Am Acad Dermatol*, 57, 644-658

Steinberg, D.; Colla, P. (1995). *CART: Tree-structured non-parametric data analysis*. Salford Systems, San Diego

Strang, G.; Nguyen, T. (1996). *Wavelets and Filterbanks*. Wellesley-Cambridge Press

Van Rijsbergen, C.A. (1979). *Information retrieval*. Butterworths, London

Wiltgen, M.; Gerger, A.; Smolle, J. (2003). Tissue counter analysis of benign common nevi and malignant melanoma. *International Journal of Medical Informatics*, 69, 17-28

Wiltgen, M.; Gerger, A.; Wagner, C.; Bergthaler, P.; Smolle, J. (2003). Discrimination of benign common nevi and malignant melanoma lesions by use of features based on spectral properties of the wavelet transform. *Anal Quant Cytol Histol,* 25, (5), 243-253

Wiltgen, M.; Gerger, A.; Wagner, C.; Smolle, J. (2008). Automatic identification of diagnostic significant regions in confocal laser scanning microscopy of melanocytic skin tumours. *Methods Inf Med,* 47, 15-25

Witten, I.H.; Frank, E. (2005). *Data Mining Practical Machine Learning Tools and Technique.* Elsevier, Morgan Kaufmann Publishers

9

Application of in vivo Laser Scanning Microscopy to Visualise the Penetration of a Fluorescent Dye in Solution and in Liposomes into the Skin after Pre-Treatment with Microneedles

Meinke Martina C., Kruithof Annelieke C., Bal Suzanne M., Bouwstra Joke A. and Lademann Jürgen
[1]*Charité - Universitätsmedizin Berlin, Department of Dermatology, Venerology and Allergology*
[2]*Division of Drug Delivery Technology, Leiden/Amsterdam Center for Drug Research, Leiden University, Leiden*
[1]*Germany*
[2]*The Netherlands*

1. Introduction

One of the most important functions of our skin is its barrier function. The skin keeps the water inside and protects us against environmental noxins such as light, pollutions or germs. Therefore, it is not easy to transmit substances into the skin even when it is desired. The uppermost layer of the skin, the horny layer (stratum corneum), acts as a physical barrier. Only small moderate lipophilic drugs can pass this barrier to achieve a therapeutic effect. For this reason the pharmaceutical industry has invested a high amount of effort in designing formulations or delivery methods to overcome the skin barrier and to enhance the penetration of active substances across the stratum corneum. One important novel area of research is the development of vaccines, which can be applied directly onto the skin without any pain sensation. However, the active part of the vaccine, the antigen, is a highly molecular weight substance, which makes it difficult to deliver by the dermal route using the more conventional methods.

One of the novel methods for the delivery of highly molecular weight substances through the skin barrier is the use of microneedles. Microneedles were introduced in the field of transdermal and dermal drug delivery more than a decade ago (Henry et al., 1998). Microneedles are shorter than hypodermal needles, but long enough to breach the stratum corneum barrier, the upper 20 µm of the skin. The application of 0.5 mm microneedles is described as painless (Kaushik et al., 2001). They have been developed as a replacement for the traditional needle and syringe and to facilitate the transport across the skin barrier without any pain sensation. When using solid microneedle arrays, small conduits are created in the skin, which are large enough to allow the penetration of high molecular

weight compounds across the stratum corneum (Prausnitz, 2004). Furthermore, compared to piercing with a conventional needle, there is only a minimal microbial infiltration through the conduits, and microorganisms do not reach the dermis (Donnelly et al., 2009). More recently, microneedles have also been used for vaccination (Ding et al., 2009, Kim et al., 2009, Koutsonanos et al., 2009, Matriano et al., 2002, Widera et al., 2006). Several studies have shown that the use of microneedles increases the efficiency of vaccination (Ding et al., 2009, Matriano et al., 2002, Van et al., 2009). In addition, microneedles can be used in combination with nanoparticles that may increase the transport of antigens or low molecular weight substances across the conduits (Bal et al., 2010a). However, in the development of these formulations, it is important to demonstrate the efficiency of transport by visualisation methods, preferably in vivo in humans.
The question is: How do we monitor the penetration? In this study, our aim was to estimate the size of the microneedle-induced conduits, their time behaviour and the diffusion of the active substances. In addition, it is of interest to monitor the efficiency of transport when the compound is encapsulated in liposomes.
There are several methods to study the skin penetration of substances, most of which are performed in vitro or invasively (Benfeldt, 1999, Gregoire et al., 2008, Rouse et al., 2007, Scheuplein, 1967, Schmook et al., 2001, Teichmann et al., 2005, Toll et al., 2004, Vogt et al., 2003, Weigmann et al., 2005). One non-invasive in vivo method is confocal laser scanning microscopy (CLSM) which can be used to visualise cell structures and to monitor the penetration of fluorescent dyes into the skin (Ardigo et al., 2010, Dietterle et al., 2008, Eichert et al., 2010, Lademann et al., 2003, Lange-Asschenfeldt et al., 2009, Martschick et al., 2007, Meyer et al., 2007). To show the ability of CSLM in vivo, a fluorescent dye (fluorescein) was injected into the epidermis using a syringe (Meyer et al., 2006). The dye is mostly distributed in the intercellular layers providing a strong contrast of the cellular structure in the CLSM images, as is shown in figure 1, in which three horizontal skin layers are shown. The cell structure of the corneocytes can be recognised (Fig 1a) and the reduction of cell diameter with increasing depth (Fig 1b). Furthermore, the papillary structure in the stratum basale can be observed (Fig 1c).

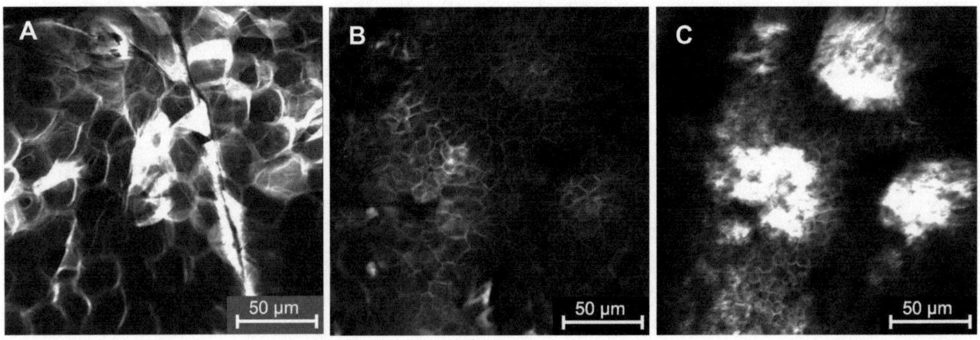

Fig. 1. CLSM images of A) stratum corneum at the surface, B) stratum spinosum at 70 µm, C) stratum basale with papillary structures at 110µm.

If the dye is applied on the surface and hair follicles or sweat glands are available, the dye will penetrate into the hair follicles and can be monitored (Meyer et al., 2007). Fig. 2 shows an example of a hair follicle and of a sweat gland at a depth of approx. 20 µm, again in in vivo settings.

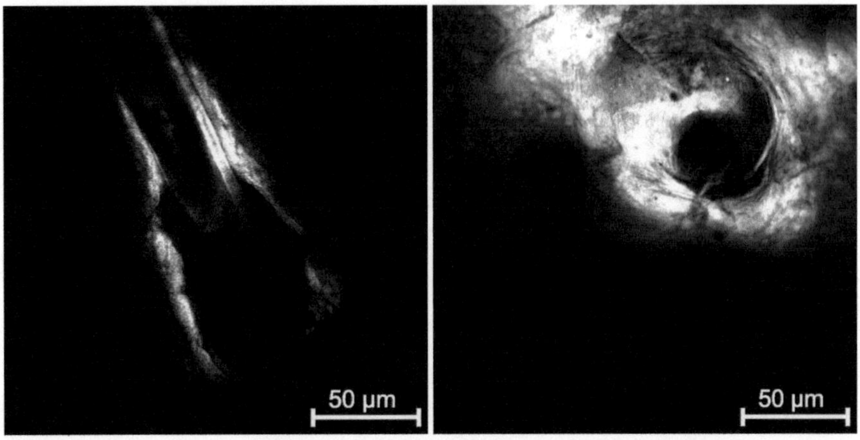

Fig. 2. CLSM images of a hair follicle (on the left) and a sweat gland (right).

These images illustrate that penetration into orifices can be monitored using CLSM. Therefore, it is highly probable that the conduits formed by microneedles can be monitored over time using CLSM.

The information obtained from CLSM studies of skin penetration can be used to optimise formulations for delivery of substances into the skin in general or into special parts of the skin, for example, transport through hair follicles or sebaceous glands (Grams et al., 2005, Lademann et al., 2009, van den Bergh et al., 1999). Often, these investigations were performed in vitro on porcine ear skin or excised human skin. In vitro, kinetic measurements can be performed, which can provide very useful information. However, measurements in the in vivo settings are more relevant, but suffer from the fact that only a limited number of dyes can be used.

CLSM can be applied both in vitro and in vivo. For example, Meyer et al. imaged Malassezia yeasts on living human cutaneous tissue using sodium fluorescein to label both the skin surface and the micro-flora structures with this method (Meyer et al., 2005). As well as visualising the penetration of substances into the skin, the effect of encapsulation of a dye into particles on the penetration pathways can be studied (Ossadnik et al., 2006), and CLSM can be used to study morphological changes of the skin (Dietterle et al., 2008, Eichert et al., 2010, Meyer et al., 2007).

Recently, we showed that by using CLSM it is possible to visualise the conduits made by solid microneedle treatment in human volunteers (Bal et al., 2010b) and to compare different microneedle systems (Bal et al., 2010c). One of these microneedle arrays was selected to compare the penetration of a fluorescent dye applied in solution with that applied in liposomes.

The microneedle arrays that were used in the present study have already been proven to be painless (Bal et al., 2008) and were shown to be efficient in enhancing the immune response when applying Diphtheria Toxoid as an antigen (Ding et al., 2009). In the present study, the model drug fluorescein was used.

2. Methods and materials

2.1 Volunteers

Six healthy volunteers (5 females and 1 male), aged between 20 and 58 years (mean 33 years) with no pre-existing skin conditions participated in the study with the fluorescent dye. The study had been approved by the Ethics Committee of the University Hospital Charité (Berlin, Germany) in accordance with the Rules of Helsinki. As a pilot study, the penetration of the fluorescent dye encapsulated in liposomes was visualised in one volunteer.

2.2 Microneedle system

Stainless steel microneedles were prepared by electrical discharge machining (300ED). The microneedles have a square base of 250 x 250 µm and a length of 300 µm and are also positioned in a 4x4 pattern with a pitch of 1.25 mm. The shape of the tip is defined by a diagonal plane, which runs from the top of one side of the square pillar to the opposed bottom, in this way forming an angle of approximately 40° relative to the bottom surface (figure 3).

Fig. 3. Array (left) and the applied microneedle more detailed (right).

To apply the microneedles in a controlled manner, an electrical applicator was used, as described previously (Verbaan et al., 2007). With this applicator the microneedles are applied onto the skin at a speed of 3 m/s. To visualise the conduits, 0.2% solution of sodium fluorescein was used (Alcon Pharma GmbH, Freiburg, Germany).

2.3 Preparation of liposomes

Fluorescently labelled liposomes were prepared from soybean phosphatidylcholine (Lipoid GmbH, Ludwigshafen, Germany) using the film hydration method followed by extrusion. A thin lipid film was formed at the bottom of a circular flask using a rotary evaporator. This film was hydrated in a 10 mM phosphate buffer pH 7.4 (7.7 mM Na_2HPO_4 and 2.3 mM NaH_2PO_4) containing 2 mg/ml fluorescein. The dispersion was shaken in the presence of glass beads at 200 RPM for 2 h at room temperature and subsequently extruded 4 times (LIPEX™ extruder, Northern Lipids Inc., Canada) through a carbonate filter with a pore size of 400 nm and 4 times through a filter with a pore size of 200 nm (Nucleopore Millipore, Amsterdam, The Netherlands). Unincorporated label was removed by filtration using a Vivaspin 2 centrifugal concentrator (PES membrane, MWCO 30 kDa, Sartorius Stedim, Nieuwegein, The Netherlands). The final lipid-to-fluorescein ratio was 1:2 (w/w). In this manner, liposomes of 143 ± 9 nm in size were obtained.

2.4 Set up of in vivo confocal laser scanning spectrometer

For the investigations, a CLSM Stratum® (Optiscan Ltd., Melbourne, Australia) was used to visualise the epidermal tissue in vivo. Stratum® has a hand-held optical scanner which is connected by optical fibres with the photo-detector of the basic system. The optical window of the hand-held device is placed directly on the skin. Fig. 4 shows a schematic diagram of the optical principle of Stratum®.

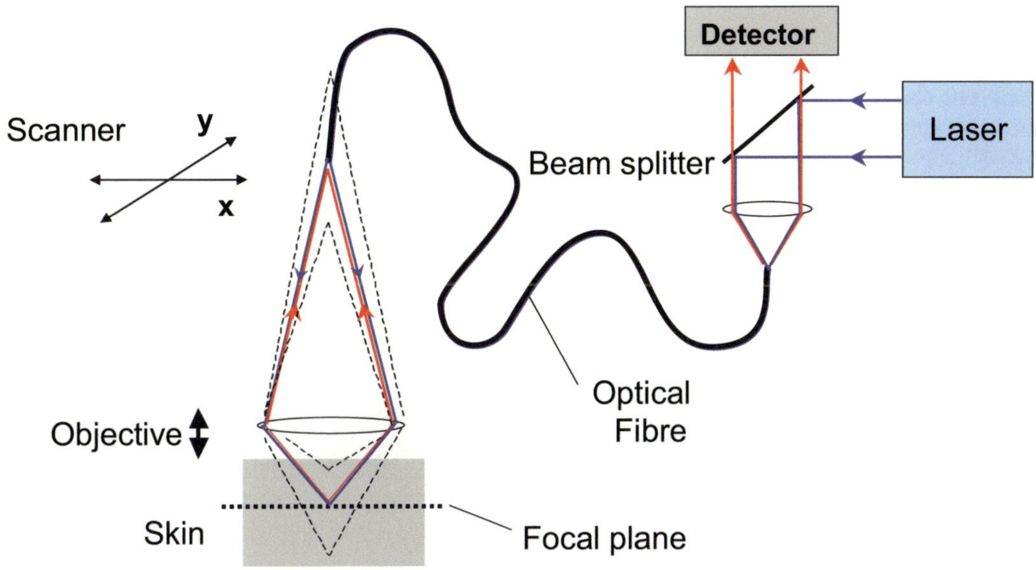

Fig. 4. Schematic diagram of the fibre based in vivo CLSM. Only the fluorescent light of the focal plane can enter the entrance of the fibre.

The illumination source of the Stratum® system is a blue 488 nm single-line argon ion laser. The scanned field of view is 235 by 235 μm^2 and the magnification is approximately 1000x. The obtained images are parallel to the skin surface resulting in horizontal or en face sections. The confocal imaging is restricted to the epidermis and superficial papillary dermis (Suihko et al., 2005). The imaging depth is varied by adjusting the focal plane manually with the focus tuning unit on the hand-held device.

2.5 Experimental procedure

The microneedles and the formulations were applied onto the ventral forearms of the volunteers. The skin was disinfected before the application of the formulations and the microneedle. The experiments were performed in triplicate on each volunteer. The fluorescein solution (50 µl / 0.8 cm^2) was applied after the skin had been treated with the microneedles. The dye was applied for 1 minute and removed with a tissue paper. Afterwards, the conduit was visualised with CLSM; 5, 10 and 15 min after application images were taken at different depths. At least 5 images were taken to monitor the dye at the surface, the lateral and vertical distribution and the maximal penetration depth. Between the measurements the laser was set out of focus to avoid bleaching. As a control, a drop of fluorescein was applied to untreated skin. This control experiment showed that at the concentration used in this study no bleaching of the fluorescein occurred.

Investigations with fluorescein encapsulated liposomes were performed on the forearm of 1 volunteer. One drop of fluorescein encapsulated liposomes as described above was applied after piercing with microneedles. The liposomes were removed using soft tissues after 1 minute of application. The CLSM evaluation was performed in the same way as in the in vivo study with fluorescein. The measurements were performed in triplicate.

2.6 Data analysis

The images were analysed with respect to fluorescence pixel intensity and area using Image J (National Institutes of Health, USA). The pixel intensity was categorised into three different classes: the class with the highest pixel intensity was set between 230 and 255 AU and the signal was referred to as "high intensity fluorescence" (HIF); the class with pixel intensity between 230 and 14 AU is referred to as "low intensity fluorescence" (LIF) and the class with pixel intensity values below 14 AU is regarded as background. The autofluorescence of the skin was always below 14 AU.

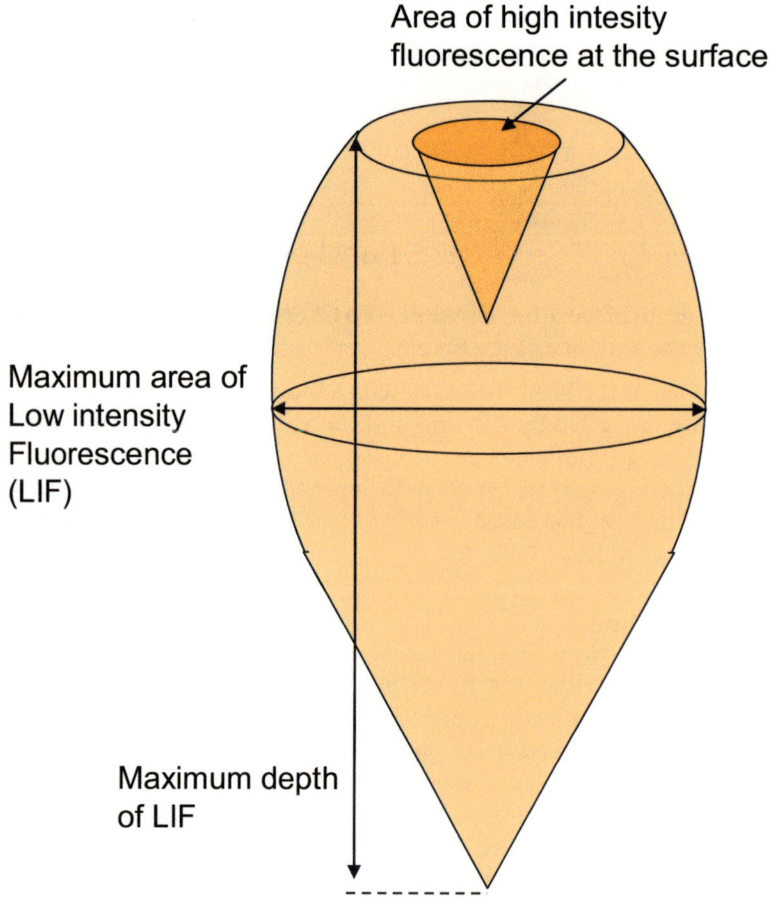

Fig. 5. Schematic diagram of the conduit formed by microneedles and the analysed parameters.

These thresholds were selected based on analysing 31 random images of different depths taken from two volunteers. The thresholds were selected in such a way that at least 90% of the test pixels were inside the described classes. The fluorescent signal from other skin structures such as furrows or hair follicles was removed manually. The area of either HIF or LIF was calculated by the number of pixels in the specified intensity areas. The following parameters were analysed: the area of HIF at the surface, the maximum area of LIF in the skin and the maximum depth where LIF can be detected. The parameters are further explained in figure 5.

3. Results

In Fig. 6, representative CLSM images of the fluorescent dye in a conduit formed by microneedles at different time points and at different depths are shown.
The HIF is clearly visible at the surface 5 min after removal of the formulation, thus the stratum corneum can easily be identified. The fluorescent dye preferably diffuses through the lipid regions surrounding the corneocytes, outlining the cells. The HIF at the surface decreases with time and after 15 min it has vanished. At the deeper skin layers a more diffuse LIF signal is present. From these images it appears that the largest area of HIF was present at the skin surface at all time points. However, this was not always the case. In

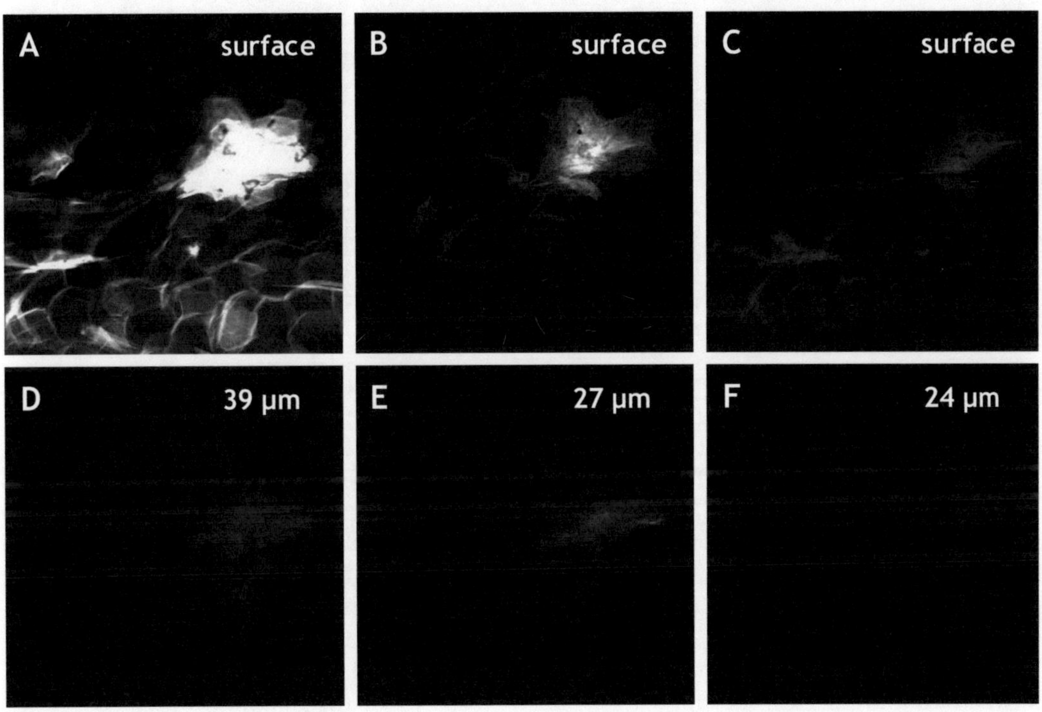

Fig. 6. Representative CLSM images of conduit closing over time in vivo after piercing and application of fluorescein in solution. Conduit at the surface at A) 5 min, B) 10 min and C) 15 min. At the indicated depths, the maximal diffusion area at D) 5 min, E) 10 min and F) 15 min was found. Field of view: 235 by 235 μm^2.

approximately 35% of the images, the largest area of HIF was observed below the surface, but within the upper 20 μm of the skin, in the stratum corneum.
In Fig. 7, representative CLSM images of conduits over time after piercing and application of fluorescein encapsulated liposomes are shown. The HIF is clearly visible at all time points at the surface and additionally up to a depth of 94 μm after 5 min. The LIF was observed in larger areas compared to the images obtained when fluorescein in solution was applied.
The changes in the area of HIF versus time are shown in Fig 8. for both formulations.

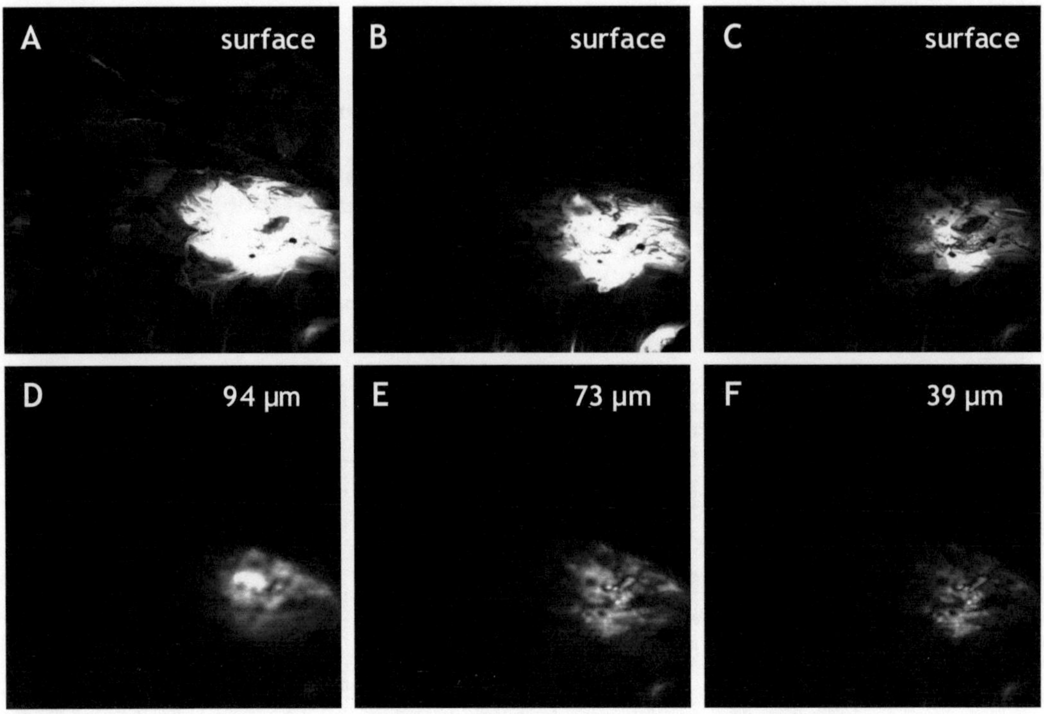

Fig. 7. Representative CLSM images of conduit closing over time after piercing and application of the fluorescein encapsulated liposomes. Conduit at the surface at A) 5 min, B) 10 min and C) 15 min. At the indicated depths, the maximal diffusion area at D) 5 min, E) 10 min and F) 15 min was found. Field of view: 235 by 235 μm^2.

At all 3 time points the area of HIF at the surface was elevated after liposome application compared to a fluorescein solution. Nevertheless, both showed a similar decay in HIF in time.
In Fig. 9, the maximum area of LIF is shown for both formulations as function of time.
In contrast to fluorescein encapsulated liposomes, the maximal diffusion area for the fluorescein solution decreased substantially between 5 and 10 min. In the time interval between 10 and 15 min, the opposite trend was observed; the maximum area of LIF for fluorescein encapsulated liposomes decreased substantially more than that of the fluorescein solution. Importantly, the maximum areas of LIF of fluorescein encapsulated liposomes were constantly higher compared to a fluorescein solution.
The time dependence of the maximum depth of LIF is shown in Fig. 10 for both formulations.

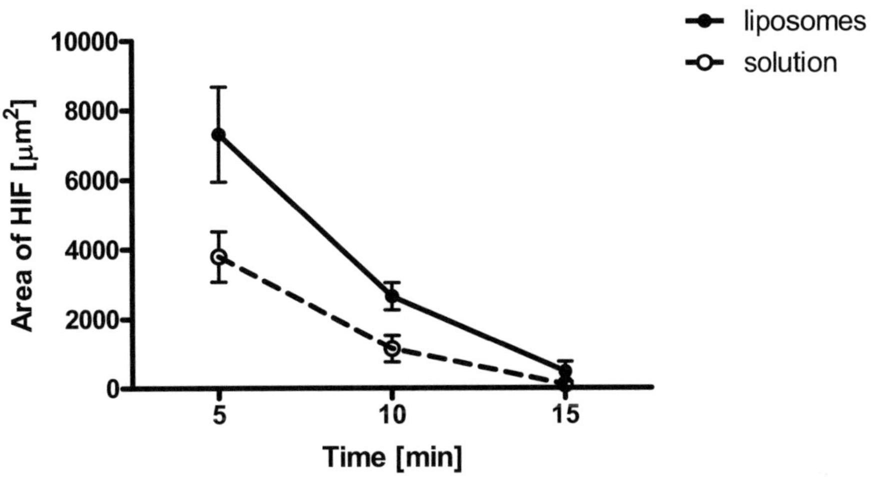

Fig. 8. Time dependency of the HIF area at the surface (μm²) for applying fluorescein encapsulated liposomes and fluorescein solution after microneedle piercing. Data are presented as mean value ± SE, n=3 for particles and n=18 for solution.

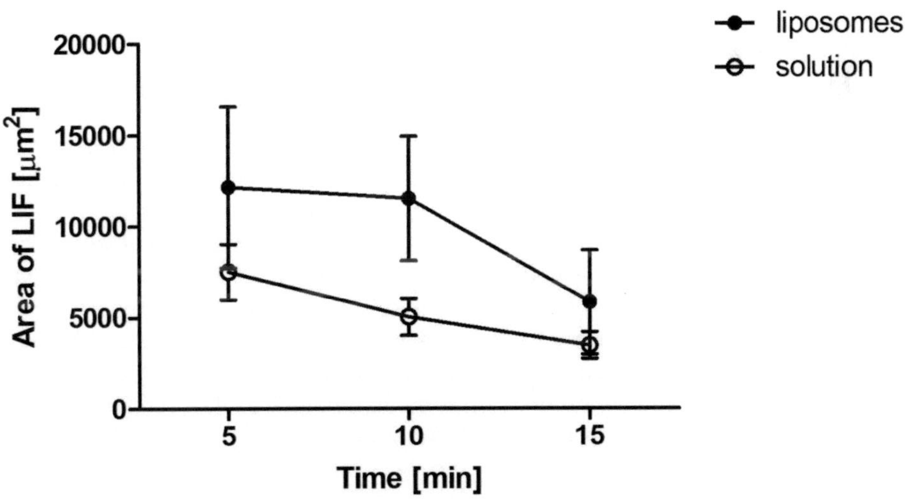

Fig. 9. Time dependence of LIF area for applying fluorescein encapsulated liposomes and fluorescein solution after piercing. Data are presented as mean value ± SE, n=3 for particles and n=18 for solution.

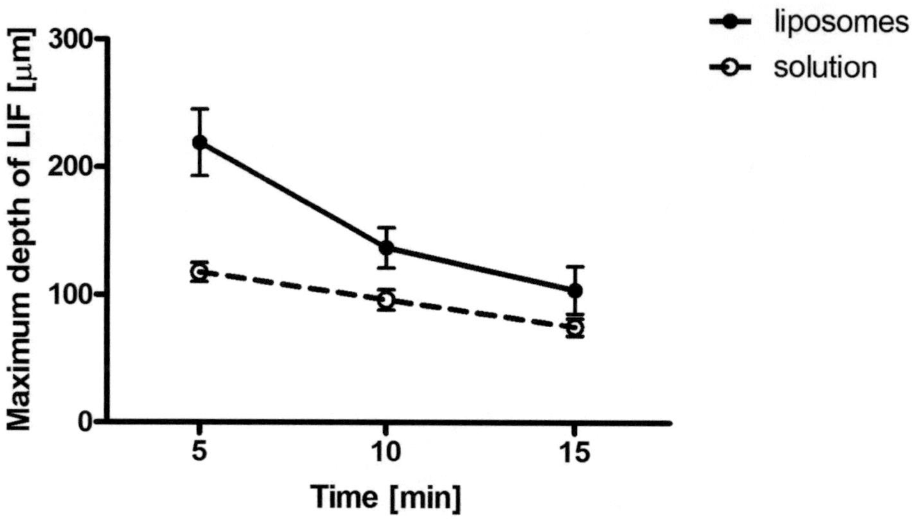

Fig. 10. Time dependence of the mean maximum depth of LIF (μm) for applying fluorescein encapsulated liposomes and fluorescein solution after piercing. Data are presented as mean value ± SE, n=3 for particles and n=18 for solution.

The maximum depths of the LIF of the solution were less compared to the liposome formulation. Moreover, the changes in time of the maximum LIF area were less distinct, in particular the change between 5 and 10 min.
These studies emphasise the potential of microneedles for cutaneous vaccine delivery and of using CLMS to monitor fluorescent species in microneedle conduits in vivo over time.

4. Discussion

The ultimate purpose of microneedles is to deliver active substances more rapidly and in sufficient amounts across the skin barrier to achieve a therapeutic effect. The results clearly demonstrated that the CLSM is able to monitor a fluorescent penetration applied either in solution or encapsulated in liposomes.
The present pilot in vivo study with fluorescein encapsulated liposomes was carried out to evaluate the penetration of fluorescein applied in liposomes in comparison to an aqueous fluorescein solution. A statistical analysis could not be performed, because the liposome data were obtained from only one volunteer, in triplicate. The preliminary results clearly showed that the diffusion area of fluorescein applied in liposomes was increased compared to fluorescein applied in a solution. In particular at 5 min the maximum depth of LIF, and the HIF area were increased when applied in liposomes. This may be due to an increased accumulation of the liposomes in the conduits compared to fluorescein only. This is consistent with the findings of Verbaan et al., who observed a considerably lower passive transport efficiency across human skin after microneedle array treatment for a large complex of FITC coupled Dextran (MW 72 kDa) compared to the smaller compounds Cascade Blue (MW 538) and Dextran - Cascade Blue (MW 10 kDa). This was observed for all examined needle lengths: 550, 700 and 900 μm. Nevertheless, all 3 compounds were significantly increased compared to the transport rate across the skin, when the skin was not pre-treated by microneedles (Verbaan et al., 2007). Most probably, during the transport

along the conduits, the dye is already partially released from the liposomes. However, because the time points were only 5, 10 and 15 min after application, the dye in the conduits may, at least partially, still be encapsulated in the liposomes. We cannot exclude that the difference in penetration of the fluorescent label is also due to a difference in activity of fluorescein in the two formulations. Fluorescein was applied at equal concentration and not at equal thermodynamic activity.

In contrast to the maximum observed area of LIF and the maximum depth of LIF, the time dependence of the decrease in area of HIF was remarkably comparable for both studies. Nevertheless, the preliminary results of the liposomes should be validated by further investigations.

5. Conclusion

In these studies, we have shown that CLSM is an excellent method to monitor the fluorescent distribution in time of a free fluorescence label, and the label incorporated in liposomes along conduits in human skin, in vivo. The pilot study shows an increased area of HIF at the skin surface and an increased maximum area of LIF within the tissue, as well as an increased depth of the fluorescent signal when liposomes were used.

6. References

Ardigo, M.; Cameli, N.; Berardesca, E. & Gonzalez, S.(2010). Characterization and evaluation of pigment distribution and response to therapy in melasma using in vivo reflectance confocal microscopy: a preliminary study. *J Eur Acad Dermatol Venereol*, 24., 1296 - 1303

Bal, S.M.; Caussin, J.; Pavel, S. & Bouwstra, J.A.(2008). In vivo assessment of safety of microneedle arrays in human skin. *Eur.J Pharm Sci.*, 35., 193 - 202

Bal, S.M.; Ding, Z.; van, R.E.; Jiskoot, W. & Bouwstra, J.A.(2010a). Advances in transcutaneous vaccine delivery: Do all ways lead to Rome? *J Control Release*, 148(3)., 266-282

Bal, S.M.; Kruithof, A.C.; Liebl, H.; Tomerius, M.; Bouwstra, J.; Lademann J & Meinke, M.(2010b). In vivo visualisation of microneedle conduits in human skin using laser scanning microscopy. *Laser Phys Lett*, 7., 242 - 246

Bal, S.M.; Kruithof, A.C.; Zwier, R.; Dietz, E.; Bouwstra, J.A.; Lademann, J. & Meinke, M.C.(2010c). Influence of microneedle shape on the transport of a fluorescent dye into human skin in vivo. *J Control Release*, 147(2)., 218-224

Benfeldt, E.(1999). In vivo microdialysis for the investigation of drug levels in the dermis and the effect of barrier perturbation on cutaneous drug penetration. Studies in hairless rats and human subjects. *Acta Derm Venereol.Suppl (Stockh)*, 206., 1 - 59

Dietterle, S.; Lademann, J.; Rowert-Huber, H.J.; Stockfleth, E.; Antoniou, C.; Sterry, W. & Astner, S.(2008). In-vivo diagnosis and non-inasive monitoring of imiquimod 5% cream for non-melanoma skin cancer using confocal laser scanning microscopy. *Laser Physics Letters*, 5., 752 - 759

Ding, Z.; Verbaan, F.J.; Bivas-Benita, M.; Bungener, L.; Huckriede, A.; van den Berg, D.J.; Kersten, G. & Bouwstra, J.A.(2009). Microneedle arrays for the transcutaneous immunization of diphtheria and influenza in BALB/c mice. *J Control Release*, 136., 71 - 78

Donnelly, R.F.; Singh, T.R.; Tunney, M.M.; Morrow, D.I.; McCarron, P.A.; O'Mahony, C. & Woolfson, A.D.(2009). Microneedle arrays allow lower microbial penetration than hypodermic needles in vitro. *Pharm Res.*, 26., 2513 - 2522

Eichert, S.; Mohrle, M.; Breuninger, H.; Rocken, M.; Garbe, C. & Bauer, J.(2010). Diagnosis of cutaneous tumors with in vivo confocal laser scanning microscopy. *J Dtsch.Dermatol Ges.*, 8., 400 - 410

Grams, Y.Y.; Whitehead, L.; Lamers, G.; Sturmann, N. & Bouwstra, J.A.(2005). On-line diffusion profile of a lipophilic model dye in different depths of a hair follicle in human scalp skin. *J Invest Dermatol.*, 125., 775 - 782

Gregoire, S.; Patouillet, C.; Noe, C.; Fossa, I.; Benech, K.F. & Ribaud, C.(2008). Improvement of the experimental setup for skin absorption screening studies with reconstructed skin EPISKIN. *Skin Pharmacol.Physiol*, 21., 89 - 97

Henry, S.; McAllister, D.V.; Allen, M.G. & Prausnitz, M.R.(1998). Microfabricated microneedles: a novel approach to transdermal drug delivery. *J Pharm Sci.*, 87., 922 - 925

Kaushik, S.; Hord, A.H.; Denson, D.D.; McAllister, D.V.; Smitra, S.; Allen, M.G. & Prausnitz, M.R.(2001). Lack of pain associated with microfabricated microneedles. *Anesth.Analg.*, 92., 502 - 504

Kim, Y.C.; Quan, F.S.; Yoo, D.G.; Compans, R.W.; Kang, S.M. & Prausnitz, M.R.(2009). Improved influenza vaccination in the skin using vaccine coated microneedles. *Vaccine*, 27., 6932 - 6938

Koutsonanos, D.G.; del Pilar, M.M.; Zarnitsyn, V.G.; Sullivan, S.P.; Compans, R.W.; Prausnitz, M.R. & Skountzou, I.(2009). Transdermal influenza immunization with vaccine-coated microneedle arrays. *PLoS One.*, 4., e4773 -

Lademann, J.; Patzelt, A.; Worm, M.; Richter, H.; Sterry, W. & Meinke, M.(2009). Analysis of in vivo penetration of textile dyes causing allergic reactions. *Laser Physics Letters*, 6., 759 - 763

Lademann, J.; Richter, H.; Otberg, N.; Lawrenz, F.; Blume-Peytavi, U. & Sterry, W.(2003). Application of a dermatological laser scanning confocal microscope for investigation in skin physiology. *Laser Physics*, 13., 756 - 760

Lange-Asschenfeldt, B.; Alborova, A.; Kruger-Corcoran, D.; Patzelt, A.; Richter, H.; Sterry, W.; Kramer, A.; Stockfleth, E. & Lademann, J.(2009). Effects of a topically applied wound ointment on epidermal wound healing studied by in vivo fluorescence laser scanning microscopy analysis. *Journal of Biomedical Optics*, 14.,

Martschick, A.; Teichmann, A.; Richter, H.; Schanzer, S.; Antoniou, C.; Sterry, W. & Lademann, J.(2007). Analysis of the penetration profiles of topically applied substances by laser scanning microscopy. *Laser Physics Letters*, 4., 395 - 398

Matriano, J.A.; Cormier, M.; Johnson, J.; Young, W.A.; Buttery, M.; Nyam, K. & Daddona, P.E.(2002). Macroflux microprojection array patch technology: a new and efficient approach for intracutaneous immunization. *Pharm Res.*, 19., 63 - 70

Meyer, L. & Lademann, J.(2007). Application of laser spectroscopic methods for in vivo diagnostic in dermatology. *Laser Phys.Lett.*, 4 (10)., 754 -

Meyer, L.; Otberg, N.; Richter, H.; Sterry, W. & Lademann, J.(2006). New Prospects in Dermatology: Fiber-Based Confocal Scanning Laser Microscopy. *Laser Phys Lett*, 15., 758 - 764

Meyer, L.; Otberg, N.; Tietz, H.J.; Sterry, W. & Lademann, J.(2005). In vivo imaging of Malassezia yeasts on human skin using confocal laser scanning microscopy. *Laser Physics Letters*, 2., 148 - 152

Ossadnik, M.; Richter, H.; Teichmann, A.; Koch, S.; Schafer, U.; Wepf, R.; Sterry, W. & Lademann, J.(2006). Investigation of differences in follicular penetration of particle- and nonparticle-containing emulsions by laser scanning microscopy. *Laser Physics*, 16., 747 - 750

Prausnitz, M.R.(2004). Microneedles for transdermal drug delivery. *Adv.Drug Deliv.Rev.*, 56., 581 - 587

Rouse, J.G.; Yang, J.; Ryman-Rasmussen, J.P.; Barron, A.R. & Monteiro-Riviere, N.A.(2007). Effects of mechanical flexion on the penetration of fullerene amino acid-derivatized peptide nanoparticles through skin. *Nano.Lett.*, 7., 155 - 160

Scheuplein, R.J.(1967). Mechanism of percutaneous absorption. II. Transient diffusion and the relative importance of various routes of skin penetration. *J Invest Dermatol.*, 48., 79 - 88

Schmook, F.P.; Meingassner, J.G. & Billich, A.(2001). Comparison of human skin or epidermis models with human and animal skin in in-vitro percutaneous absorption. *Int J Pharm*, 215., 51 – 56

Suihko, C. Swindle L.D., Thomas S.G., Serup J.(2005). Fluorescence fibre-optic confocal microscopy of skin in vivo: microscope and fluorophores. *Skin. Res. Technol.*, 11(4)., 254-267

Teichmann, A.; Jacobi, U.; Ossadnik, M.; Richter, H.; Koch, S.; Sterry, W. & Lademann, J.(2005). Differential stripping: determination of the amount of topically applied substances penetrated into the hair follicles. *J Invest Dermatol.*, 125., 264 - 269

Toll, R.; Jacobi, U.; Richter, H.; Lademann, J.; Schaefer, H. & Blume-Peytavi, U.(2004). Penetration profile of microspheres in follicular targeting of terminal hair follicles. *J Invest Dermatol.*, 123., 168 - 176

van den Bergh, B.A.; Vroom, J.; Gerritsen, H.; Junginger, H.E. & Bouwstra, J.A.(1999). Interactions of elastic and rigid vesicles with human skin in vitro: electron microscopy and two-photon excitation microscopy. *Biochim.Biophys.Acta*, 1461., 155 - 173

Van, D.P.; Oosterhuis-Kafeja, F.; Van der, W.M.; Almagor, Y.; Sharon, O. & Levin, Y.(2009). Safety and efficacy of a novel microneedle device for dose sparing intradermal influenza vaccination in healthy adults. *Vaccine*, 27., 454 - 459

Verbaan, F.J.; Bal, S.M.; van den Berg, D.J.; Groenink, W.H.; Verpoorten, H.; Luttge, R. & Bouwstra, J.A.(2007). Assembled microneedle arrays enhance the transport of compounds varying over a large range of molecular weight across human dermatomed skin. *J Control Release*, 117., 238 - 245

Vogt, A. & Blume-Peytavi, U.(2003). [Biology of the human hair follicle. New knowledge and the clinical significance]. *Hautarzt*, 54., 692 - 698

Weigmann, H.J.; Jacobi, U.; Antoniou, C.; Tsikrikas, G.N.; Wendel, V.; Rapp, C.; Gers-Barlag, H.; Sterry, W. & Lademann, J.(2005). Determination of penetration profiles of topically applied substances by means of tape stripping and optical spectroscopy: UV filter substance in sunscreens. *J Biomed.Opt.*, 10., 14009-1 - 14009-7

Widera, G.; Johnson, J.; Kim, L.; Libiran, L.; Nyam, K.; Daddona, P.E. & Cormier, M.(2006). Effect of delivery parameters on immunization to ovalbumin following intracutaneous administration by a coated microneedle array patch system. *Vaccine*, 24., 1653 - 1664

10

Application of Laser Scanning Cytometry to Clinical and Laboratory Hematology

Takayuki Tsujioka and Kaoru Tohyama
Department of Laboratory Medicine, Kawasaki Medical School, Okayama
Japan

1. Introduction

Morphological diagnosis by May-Gruenwald-Giemsa (MGG) staining is an essential method in clinical and laboratory hematology. In addition, the detection of chromosomal aberration by G-band methods, the phenotyping of surface markers by Flowcytometry (FCM) or genetic diagnosis by reverse transcriptase-polymerase chain reaction (RT-PCR), Fluorescence in situ hybridization (FISH) and Southern blotting are important approaches to facilitate the establishment of a definitive diagnosis. These methods are found to be much more useful for the diagnosis of specific diseases showing characteristic abnormalities on cytogenetics including acute myeloid leukemia (AML) with *RUNX1-RUNX1T1* or *PML-RARA* and chronic myelogenous leukemia (CML), *BCR-ABL1* positivity and other such diseases etc (Arber et al, 2008, Vardiman et al, 2008). However, there are few methods other than morphological diagnosis that are described as effective for heterogenous populations comprising the majority of myelodysplastic syndrome (MDS), the detection of minimal residual diseases and the discrimination of cases showing slight dysplasia. Therefore, it is necessary to establish an effective technique of collecting other information.

2. The diagnosis of blood pathological cell using flow cytometry

The measurement of DNA content in the hematopoietic cells by flow cytometry (FCM) has been useful for the characterization of pathological blood cells, detection of residual diseases and prediction of disease prognosis in the hematological disorders. Trueworthy et al. (Trueworthy et al., 1992) reported that ploidy of lymphoblasts was the strongest predictor of treatment outcome in B-progenitor cell acute lymphoblastic leukemia of childhood. Almeida et al. (Almeida et al., 1999) showed that the combined use of immunophenotyping and DNA ploidy studies by FCM is a suitable approach for minimal residural disease (MRD) investigation in multiple myeloma patients, based on their applicability and sensitivity. Though FCM is an objective and reliable method to identify DNA ploidy of hematological cells, it is difficult to make inquiries about morphology and DNA ploidy in each cell (Sasaki et al., 1996).

3. The structure and advantage of laser-scanning cytometry

A laser-scanning cytometry (LSC) was developed by Kamentsky et al. (Kamentsky, LA & Kamentsky,LD, 1991) in 1991. The LSC (CompuCyte Corporation, Cambridge, MA) is an instrument for being integrated image processing, flow cytometry, and automated digital microscopy, a rapidly high content analysis of cellular, tissue and other specimens. As cells pass through the scanning beams, the fluorescent light returns along the same path as the laser beams and is collected in the wavelength-specific photomultiplier. A photodiode scatter sensor detects scattered laser light for segmenting or brightfield imaging with Nomarski-like three-dimensional details. Specimens are observed with the microscope that is a component of the whole instrument and the morphological data are captured through the microscopic light path, and the LSC excites fluorescent-stained samples with up to 3 different wavelength, and acquires stoichiometric fluorescent and morphological data in as many as 5 detectors per laser.

LSC enables us to measure the cell population fixed on the slide glass by rapid scanning and acquire the fluorescent information of each cell like as FCM, and still better, LSC makes it possible to recall the cells of interest and evaluate their morphology with the microscope by switching the status between routine morphology observation and fluorescence condition (Martin-reay et al. 1994). LSC has recently been used for immunophenotyping of hematological malignancies in smear specimens or needle aspiration biopsy. Clatch et al. analyzed a total of 71 specimens including peripheral blood, tissue biopsy, bone marrow aspirates and lymphoid tissue, and concluded that LSC analysis provides advantages over FCM in point of using cells on the slide, examining microscopically at any time, low cost, easy for preparation (Clatch, 2001, Clatch & Walloch, 1997, Clatch & Foreman, 1998).

Whereas FCM yields useful information on the expression pattern and the amount of cellular molecules, LSC provides the morphologic information on the precise location of the molecules in the cell and makes it possible to retrieve the data file to examine the fluorescent features and morphology from a small number of the cells fixed onto the slide glass (Sasaki et al., 1996, Martin-Reay et al., 1994).

4. Comparative cell cycle analysis with LSC and FCM

A human myelocytic leukemia cell line, HL60 cells were fixed in 5ml cold ethanol (70%) for 30 minutes at 4°C, and after washing with PBS, the cells were resuspended in 500μl PBS and incubated with 2 mg/ml ribonuclease A (Sigma) at 37°C for 30 minutes, followed by the addition of 50μg/ml propidium iodide (PI, Sigma, ST. Louis, Missouri, USA) at room temperature for 20 minutes for DNA staining. HL60 cells were measured with FACS Calibur flow cytometer (Becton Dickinson Immunocytometry Systems, Franklin Lakes, NJ, USA). PI fluorescence was measured at FL2 channel. Up to 10,000 events for each sample was measured by Cellquest software (BD Bioscience, Mansfield, MA, USA) and the cell cycle was analyzed with ModFit LT™ (Verity Software House, Inc, Topsham, ME, USA).

LSC is the apparatus for analyzing the fluorescence of the cells fixed onto the slide glass and is capable of evaluating the cell cycle by integrating the data of the cells whose nuclei are stained with PI. As the cell cycle analysis has been established and standardized by FCM, we examined the consistency of the data by LSC with that of FCM using HL60 cells in the proliferating phase. Figure 1 shows the comparative cell cycle analysis with FCM and LSC. The percentages of G0/G1, S, G2/M and subG1 phases of HL60 cell population measured with FCM were 35.7%, 58.2%, 6.0% and 0.5%, respectively (Fig. 1a.), whereas those analyzed with LSC were 55.7%, 25.8%, 20.2% and 1.0%, respectively (Fig. 1b.) (Tsujioka et al.,2008). Collectively, cell cycle analysis by the use of LSC did not always coincide with that of FCM because of different analytic methods, but both histogram patterns seemed similar to each other. The measurement of DNA content of blood cells by FCM has been widely established and its accuracy control is excellently maintained. In contrast, the accuracy control of LSC method might be inferior to that of FCM. As shown in Figure 1, the peak pattern of the histogram demonstrating G0/G1 and G2/M phases is rather broad in LSC than that of FCM. It was reported that the coefficient of variation (CV) is almost in the range of 5-6% by LSC as compared with 3-7% in the case of FCM, but structural changes in the cells and nuclei associated with cell deposition on the solid surface of the slide glass may result in a broadening of the CV (Martin-Reay et al., 1994, Tarnok & Gerstner AOH, 2002, Kamada et al., 1997). In LSC, adjacent cells are often falsely recognized as a single giant cell with multiple nuclei, and the background fluorescence is not negligible particularly in the case of bone marrow specimens. To avoid such disadvantages, a relatively small number of the cells (less than five thousand cells per acquisition area) should be treated for cytospin preparation and the fluorescence-stained slide glass should be sufficiently washed (Pozarowski et al., 2006). Despite of some adverse points mentioned above, LSC exerts an excellent function that it enables us to evaluate both the morphology and the fluorescence information on each cell, and still better, it enables us to retrieve the individual cell and its information at any time as long as the fluorescence is sustained (Rodenburg et al., 1987, Chen TL et al., 1995, Rew et al., 1998, Kamiya et al., 1998).

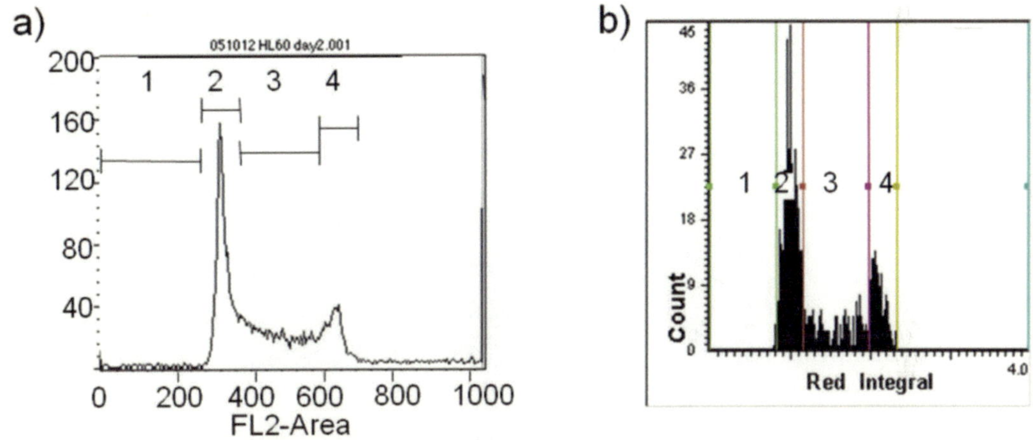

Fig. 1. Comparative cell cycle analysis with FCM and LSC (Tsujioka et al.,2008)
a) HL-60 cells were stained by PI as described in Materials and methods, then the cell cycle analysis was performed using FCM. 1:subG1 phase, 2: G0/G1 phase, 3: S phase, 4: G2/M phase
b) HL-60 cells were stained by PI as described in Materials and methods, then the cell cycle analysis was performed using LSC. 1: subG1 phase, 2: G0/G1 phase, 3: S phase, 4: G2/M phase

5. The model study on diagnosis of cases with megaloblastic anemia using LSC

Megaloblastic anemia is a common hematological disorder which is characterized by ineffective hematopoiesis due to impaired DNA synthesis of the bone marrow cells. This disease is usually caused by the disturbance of vitamin B12 absorption in pernicious anemia or after gastrectomy, folic acid deficiency, or the adverse effect of some cytotoxic drugs. The prominent morphological features of megaloblastic anemia are megaloblasts in the erythroid lineage and neutrophils with hypersegmented nucleus or with larger cell size such as giant metamyelocytes and giant band cells in the myeloid lineage (Shojania, 1980, Parry, 1980, Beck, 1991). In the present study, we assumed that the dysplastic bone marrow cells found in megaloblastic anemia are the adequate candidates for DNA ploidy analysis. We performed the measurement of DNA ploidy with morphological analysis of a single cell by LSC.

6. DNA ploidy analysis on megaloblastic anemia using LSC

Bone marrow smears were obtained by bone marrow aspirations of 6 patients with megaloblastic anemia including 5 cases with pernicious anemia and 1 case with after gastrectomy. Normal bone marrow as a control was obtained from 3 volunteers with non-hematological diseases for routine hematological examination after informed consent. Normal peripheral blood cells were obtained from 9 healthy volunteers. Alternatively, we used MDS92 cell line which was established from the bone marrow of an MDS patient (Tohyama et al., 1994).
After the sample smears were stained with May-Gruenwald-Giemsa, the cells were morphologically assessed, captured visually with a CCD camera and their coordinates (x- and y-) on the slide were recorded (Fig.2a-g, upper stand.) (Tsujioka et al., 2008). Next the sample smears were destained with ethanol/acetate, treated with 2mg/ml of ribonuclease A for 1 hour at 37°C, and 100μg/ml of PI was added to the specimen slides (Fig. 2a-g, lower

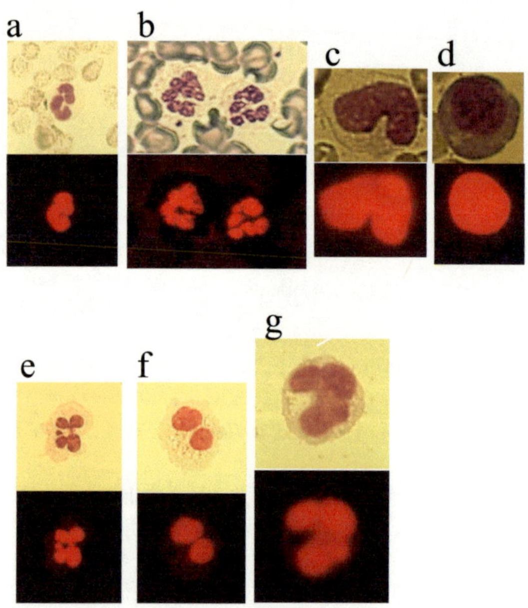

Fig. 2. Morphologic features of the cells with megaloblastic anemia and MDS92 cells (MDS cell line). Upper stand: May-Gruenwald-Giemsa staining, Lower stand: PI staining
a) normal neutrophil, b) hypersegmented neutrophil, c) giant metamyelocyte,
d) polychromatic megaloblast, e) neutrophil in MDS92 cell line, f) hyposegmented mature neutrophil in MDS92 cell line, g) giant metamyelocyte in MDS92 cell line (Original magnification 400X)

stand.) (Tsujioka et al.,2008). The slides were washed, scanned with the objective mode of LSC and the fluorescence data of more than 5,000 cells per slide were acquired. The nuclear stain PI was excited by a 488 nm wave-length argon-ion laser, and the red fluorescence emission was measured by appropriately filtered photomultiplier tubes. The mode value of DNA content was adjusted to 2.0 (this value implies a diploid cell at the G0/G1 phase). The data of each cell were retrieved corresponding to the identical cells formerly stained by May-Gruenwald-Giemsa according to their coordinates.

7. DNA ploidy analysis in abnormal neutrophils of the patients with megaloblastic anemia

Hypersegmented neutrophils, giant metamyelocytes, giant band cells or giant neutrophils are often detected in the peripheral blood or in the bone marrow of the patients with megaloblastic anemia. It has been one of the simple questions whether or not these morphologically abnormal neutrophils are ordinary diploid cells or perhaps hyperploid cells. Hence we selected these cells in May-Gruenwald-Giemsa-stained smears and subsequently determined their DNA ploidy after PI staining. We compared the DNA content of abnormal neutrophils found in the bone marrow of the patients with megaloblastic anemia with normal counterparts in the bone marrow of non-hematological disease. The average values±SD of DNA ploidy in normal neutrophils in the peripheral blood of healthy volunteers and hypersegmented neutrophils found in megaloblastic anemia were 2.0±0.1 (n=181) and 2.0±0.2 (n=18), respectively (data not shown). The average values±SD of DNA ploidy in normal neutrophils and hypersegmented neutrophils in the bone marrow were 2.0±0.1 (n=38) and 1.9±0.1 (n=36), respectively(data not shown). These data represented no significant difference of DNA content of normal neutrophils and hypersegmented neutrophils in both the peripheral blood and the bone marrow. Therefore it was suggested that DNA ploidy of hypersegmented neutrophils is not significantly increased as compared with that of normal neutrophils.

Giant metamyelocytes, giant band cells and giant neutrophils are extraordinary huge ones often detected in the bone marrow of the patients with megaloblastic anemia. The average DNA ploidy in giant metamyelocytes and giant neutrophils were 2.9±0.1 (n=23, $p<0.01$) and 3.5±0.2 (n=20, $p<0.01$), respectively. These data demonstrated that giant neutrophilic series have significantly higher DNA ploidy as compared with that of normal bone marrow counterparts (data not shown) (Tsujioka et al.,2008).

8. DNA ploidy analysis in the erythroid cells representing the megaloblastic change

The erythroid cells in the megaloblastic anemia represent morphological features of the comparatively large size and the delay of chromatin condensation as compared with cytoplasmic maturation. We measured DNA content of the abnormal erythroblasts including basophilic and polychromatic megaloblasts of the patients with megaloblastic anemia and compared them with normal erythroblasts of the bone marrow of non-hematological diseases

Table 1 show the distribution of DNA content in normal erythroblasts and megaloblasts, respectively. The cell cycle was divided into three phases by the measured values of DNA content: G0/G1 phase (1.8-2.4), S phase (2.5-3.8) and G2/M phase (3.9-4.2), and the

distribution pattern of three phases was compared between normal erythroblasts and megaloblasts (Tsujioka et al.,2008). Table 1 shows that the cell fraction in S phase was increased in the case of megaloblasts (Tsujioka et al.,2008). Wickramasinghe et al. (Wickramasinghe et al., 1982) previously reported the cell cycle distribution of normal erythroblasts in the bone marrow using a technique of combined Feulgen microspectrophotometry and ^{3}H-thymidine autoradiography and showed that the distribution of cell cycle was 32% in G0/G1 phase, 62% in S phase, 6% in G2/M phase in basophilic erythroblasts, and 15% in G0/G1 phase, 77% in S phase, 7% in G2/M phase in polychromatic erythroblasts. Fibach E et al. (Fibach & Rachmilewitz EA, 1993) reported that peripheral blood mononuclear cells were cultured with erythropoietin (2.0U/ml) and the cell cycle distribution of the erythroblasts was 47% in G0/G1 phase, 40 % in S phase and 13 % in G2/M phase. Considering Fibach's study and our data, the report by Wickramasinghe et al. might include overestimated proportion of S phase in normal erythroblasts.
On the contrary, the average DNA content of the cells in S phase were 2.7±0.1 in normal erythroblasts (n=27, range: 2.2-3.8) and 2.9±0.1 in megaloblasts (n=37, range: 2.3~3.8), and these values showed no significant difference (Tsujioka et al., 2008).

	Percentage		
	G0/G1	S	G2M
Megaloblastic anemia (6cases)	32.3% (n=20)	59.7% (n=37)	8.1% (n=5)
Normal control (3 cases)	66.1% (n=72)	24.8% (n=27)	9.2% (n=10)

Table 1. Proportion of the cells in three phases of cell cycle in normal erythroblasts and megaloblasts

The data of normal erythroblasts (n=109) were obtained from the bone marrow of 3 volunteers and those of megaloblasts (n=62) from the bone marrow of 6 patients with megaloblastic anemia. The specimens were stained by PI and up to 5,000 events for each sample were collected and analyzed by LSC. Normal samples and those with megaloblastic anemia were divided to G0/G1 (1.8-2.4), S (2.5-3.8), G2/M (3.9-4.2) phase. The G0/G1, S, G2/M fractions of the normal control were 66.1% (n=72), 24.8% (n=27), 9.2% (n=10), respectively. The G0/G1, S, G2/M fractions of those with megaloblastic anemia were 32.3% (n=20), 59.7% (n=37), 8.1% (n=5), respectively.

9. DNA ploidy analysis in normal megakaryocytes

In this study, we could not exactly estimate the analysis of DNA ploidy in megakaryocytes (MGK) with dysplasia, because bone marrow with megaloblastic anemia frequently shows hypercellular marrow and there are many other cells around MGK. Figure 3 shows DNA ploidy of MGK in normal bone marrow. DNA ploidy of MGK is much greater than that of erythroid cells or myeloid cells. As hematological diseases often show the characteristics of

MGK, such as micro-MGK and mononuclear-MGK in MDS or small-MGK in chronic myelogenous leukemia (Brunning et al, 2008, Vardiman et al, 2008), it would be interesting to measure the DNA ploidy of these cells using LSC in the future.

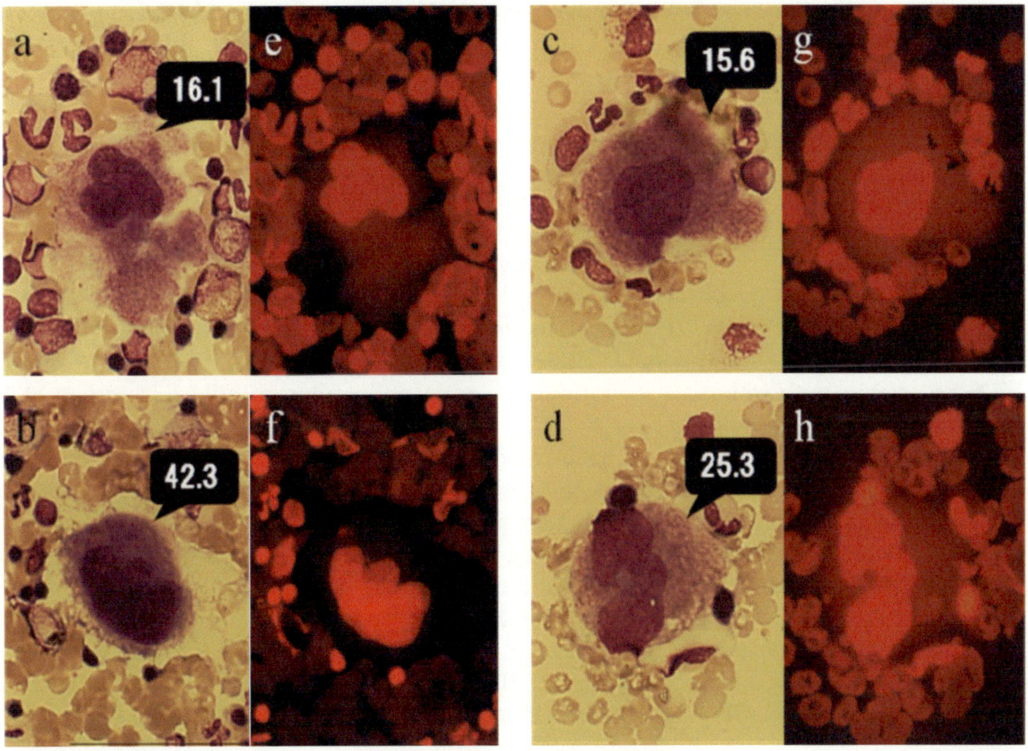

Fig. 3. DNA ploidy of megakaryocyte in normal bone marrow
a-d: May-Gruenwald-Giemsa staining
e-h: PI staining
a-h: megakaryocyte (Original magnification 400X)
The numerical value in the figure indicates DNA ploidy.

10. Overview of DNA ploidy analysis on megaloblastic anemia using laser scanning cytometry

In the present study, we have investigated the DNA ploidy of the erythroid /myeloid cells of the patients with megaloblastic anemia, because this anemia is characterized by dysplastic and often extraordinarily large erythroid/myeloid cells whose DNA ploidy is worthy to be examined. We actually confirmed for the first time that the DNA ploidy of giant metamyelocytes, giant band cells and giant neutrophils often found in the specimens of megaloblastic anemia is significantly increased over the diploid pattern of normal myeloid counterparts. As such maturing myeloid cells are considered to lose a capacity of cell division, the DNA ploidy of giant myeloid cells is obviously abnormal. On the contrary, the DNA ploidy of hypersegmented neutrophils is not significantly increased over normal diploid pattern.

As the studies dealing with clinical samples, DNA staining of the cells from normal or pathological tissues will be the best application for LSC. Our data about the DNA ploidy of the erythroid /myeloid cells of the patients with megaloblastic anemia demonstrate the impaired nuclear division although the underlying molecular mechanisms leading to the dysplasia remain unknown. From our data, we suspect that a larger number of megaloblasts are in the way of S-phase as a result of impaired DNA synthesis although it is not determined whether cell cycle process is retarded or arrested.

11. The future of LSC in clinical and laboratory hematology

The advantage of the 4th edition of the WHO classification in 2008 is the classification by genetic criteria newly defined for some diseases (eg: *JAK2/V*617F, *MPLW*515, *c-Kit* D816V) particularly among the myeloproliferative neoplasms (Vardiman et al, 2008). Myelodysplastic syndromes (MDS) are a group of clonal hematopoietic stem cell diseases characterized by cytopenia, dysplasia, ineffective hematopoiesis and increased risk of developing acute myeloid leukemia. These are highly heterogenous populations except in MDS associated with isolated del (5q) defined by genetic criterion only (Hasserjian et al, 2008). Therefore, morphological diagnosis of MDS by MGG staining has been the main method of diagnosis lagging behind the progress of genetic analysis made in other fields. However it is very difficult for us to distinguish between hypoplastic MDS without chromosomal abnormality and aplastic anemia with slight dysplasia by morphological diagnosis alone. DNA ploidy analysis using LSC is useful as an adjunct to the diagnosis of these confusing cells. In this study, though we performed the measurement of DNA ploidy in dysplastic cells with megaloblastic anemia as a previous step, we further intend to measure the DNA ploidy of several other hematopoietic cells in the near future in order to clarify the relation of DNA ploidy to cell dysplasia in bone marrow disorders such as neutrophils with nuclear hypolobation and micro-megakaryocyte in MDS using LSC.

12. References

Almeida, J, et al.(1999). High-sensitive immunophenotyping and DNA ploidy studies for the investigation of minimal residual disease in multiple myeloma.*Br J Haematol*, 107., 121-131, ISSN: 0007-1048.

Arber, DA, et al. (2008). Acute myeloid leukaemia with recurrent genetic abnormalities. In: *WHO Classification of tumours of Haemopoietic and Lymphoid tissues*, Sweldlow, SH, (ed.), 109-123, IARC Press, Lyon, ISBN: 978-92-832-2431-0 .

Beck, WS. (1991). Diagnosis of megaloblastic anemia. *Annu Rev of Med*, 42., 311-322, ISSN: 0066-4219.

Brunning, RD, et al. (2008).Myelodysplastic syndromes/ neoplasms, overview. In: *WHO Classification of tumours of Haemopoietic and Lymphoid tissues*, Sweldlow, SH, (ed.), 88-93, IARC Press, Lyon, ISBN: 978-92-832-2431-0.

Chen, TL, et al. (1995). Comparison of flow and image cytometry for DNA content analysis of fresh and formalin-fixed, paraffin-embedded tissue in breast carcinoma. *Cytometry*, 22., 181-189, ISSN: 1552-4949.

Clatch, RJ, Walloch, JL. (1997). Multiparameter immunophenotypic analysis of fine needle aspiration biopsies and other hematologic specimens by laser scanning cytometry. *Acta Cytol*, 41., 109-122,ISSN: 0001-5547.

Clatch, RJ, Foreman, JR. (1998). Five-color immunophenotyping plus DNA content analysis by laser scanning cytometry. *Cytometry*, 15., 34., 36-8, ISSN: 1552-4949.

Clatch, RJ. (2001). Immunophenotyping of hematological malignancies by laser scanning cytometry. *Methods Cell Biol*, 64., 313-342, ISSN: 0091-679x.

Fibach, E, Rachmilewitz EA.(1993). Stimulation of erythroid progenitors by high concentrations of erythropoietin results in normoblasts arrested in G2 phase of the cell cycle. *Exp Hematol*, 21., 184-188, ISBN: 0301-472x.

Hasserjian, RP, et al. (2008).Myelodysplastic syndrome with isolated del(5q). In: *WHO Classification of tumours of Haemopoietic and Lymphoid tissues*, Sweldlow, SH, (ed.), 102, IARC Press, Lyon, ISBN: 978-92-832-2431-0.

Kamada, T, et al. (1997). Sample preparation from paraffin-embedded tissue specimens for laser scanningcytometric DNA analysis. *Cytometry*, 27., 290-294, ISSN: 1552-4949.

Kamentsky, LA, Kamentsky, LD. (1991). Microscope-based multiparameter laser scanning cytometer yielding data comparable to flow cytometry data. *Cytometry*,12., 381-387, ISSN: 1552-4949 .

Kamiya, N, et al.(1998). Assesment of DNA content in formalin-fixed, paraffin-embedded tissue of lung cancer by laser scanning cytometer. *Pathol Int* , 49., 695-701, ISSN: 1320-5463.

Martin-reay, DG, et al. (1994). Evaluation of a new slide-based laser scanning cytometer for DNA analysis of tumors. *Am J Clin Pathol*, 102., 432-438, ISSN: 0002-9173.

Parry, TE. (1980). The diagnosis of megaloblastic anemia. *Clin Lab Haematol*, 2., 89-109, ISSN: 0141-9854.

Pozarowski, P, et al. (2006). Laser scanning cytometry: principles and applications. *Methods Mol Biol*, 319., 165-192, ISSN: 1064-3745.

Rew, DA, et al. (1998). Comparison of flow and laser scanning cytometry for the assay of cell proliferation in human solid tumors. *Cytometry* 33., 355-361, ISSN: 1552-4949,.

Rodenburg, CJ, et al. (1987). Tumor ploidy as a major prognostic factor in advanced ovarian cancer. *Cancer*, 59., 317-323, ISSN: 0008-543x.

Sasaki, K, et al. (1996). DNA ploidy analysis by laser scanning cytometry (LSC) in colorectal cancers and comparison with flow cytometry. *Cytometry*, 23., 106-109, ISSN: 1552-4949.

Shojania, AM.(1980). Problem in the diagnosis and investigation of megaloblastic anemia. *Can Med Assoc J*, 122., 999-1004, ISSN: 0820-3946.

Tarnok, A, Gerstner AOH. (2002). Clinical applications of laser scanning cytometry. *Cytometry*, 50., 133-143, ISSN: 1552-4949.

Tohyama, K, et al. (1994). Establishment and characterization of a novel myeloid cell line from the bone marrow of a patient with the myelodysplastic syndrome. Br J Haematol, 87., 235-242, ISSN: 0007-1048.

Trueworthy, R, et al.(1992). Ploidy of lymphoblasts is the strongest predictor of treatment outcome in B-progenitor cell acute lymphoblastic leukemia of childhood: a Pediatric Oncology Group study. *J Clin Oncol*, 10., 606-613, ISSN: 0732-183x.

Tsujioka, T, et al. (2008).DNA ploidy and cell cycle analysis in the bone marrow cells of patients with megaloblastic anemia using laser scanning cytometry. *Cytom part B-clin CY*, 74B., 104-109, ISSN: 1552-4949.

Vardiman, JW, et al. (2008). Chronic myelogenous leukemia, BCR-ABL1 positive. In: *WHO Classification of tumours of Haemopoietic and Lymphoid tissues*, Sweldlow, SH, (ed.), 32-37, IARC Press, Lyon, ISBN: 978-92-832-2431-0.

Wickramasinghe, SN, et al. (1982). Cell cycle distribution of erythroblasts in P. falciparum malaria. *Scand J Haematol*, 29., 83-88, ISSN:0036-553x.

11

Fluorescence Immunohistochemistry by Confocal Laser-Scanning Microscopy for Studies of Semi-Ultrathin Specimens of Epoxy Resin-Embedded Samples

Shin-ichi Iwasaki and Hidekazu Aoyagi
The Nippon Dental University School of Life Dentistry at Niigata
Japan

1. Introduction

It is difficult to visualize histological details on semi-ultrathin sections by light microscopy after immunohistochemical labeling because the histological structures in such sections cannot be distinguished by standard counter-staining. To solve this problem and to visualize the immunoreactivity specific for various antigens, we have developed a technique that involves a combination of immunofluorescent staining of semi-ultrathin sections of epoxy resin-embedded samples and either the corresponding differential interference contrast (DIC) images or the corresponding images in transmission mode obtained by confocal laser-scanning microscopy (LSM), providing detailed information about the immuno-localization of histological and cellular structures.

Haraguchi and Yokota (2002) were the first to describe a similar method, including treatment of sections with 10% sodium ethoxide to remove epoxy resin (Litwin et al., 1984) and reaction with fluorescence-labeled second antibodies. In our method, we use 10% sodium methoxide to remove epoxy resin from sections (Mayor et al., 1961). There is no significant difference between the two treatments, but Haraguchi & Yokota (2002) used fluorescence-labeled second antibodies while we use a combination of biotin-conjugated second antibodies and streptavidin fluorescence. Our preliminary experiments indicated that the latter system has greater sensitivity than the former in the case of semi-ultrathin sections. Using our above-mentioned method, we are now easily able to detect immunofluorescence on semi-ultrathin sections of epoxy resin-embedded specimens.

To demonstrate the effectiveness of our method, in the present study we examined the immunofluorescence of immuno-stained keratins and collagens and the images obtained, in transmission mode, of the lingual mucosa of the filiform and circumvallate papillary areas of tongues of fetus and juvenile rats by LSM. By combining images, we were able to define clearly the histological locations of keratin 13 (K13) and keratin 14 (K14) in the lingual epithelium (Iwasaki et al., 2006a, 2011a) and those of collagen II (CII) and collagen III (CIII) in the lingual connective tissue (Asami et al., 2008, Iwasaki et al., 2008, 2011b). Furthermore, we demonstrated that our newly developed technique for localization of pairs of antigens should be useful for investigations of very small specimens, such as fetal tissues and organs (Aoyagi et al., 2008).

2. Experimental methods

2.1 Experimental animals

Sprague-Dawley (SPF) rats were used for all observations. Tongues were removed from fetal and juvenile rats after they had been killed by an intraperitoneal overdose of sodium pentobarbital.

2.2 Preparation of tissue

Rat lingual tissues were fixed in 4% formaldehyde titrated from paraformaldehyde (Table 1-A) in 0.1 M phosphate buffer (PBS; pH 7.4; Table 1-B) at 4 °C for 5 h. After rinsing in 0.1 M PBS, all samples were transferred to an ascending ethanol series (60%, 70%, 80%, 90% and 99% ethanol; 10 min each) and then to absolute ethanol (twice for 15 min each) for dehydration. After immersion in propylene oxide (twice for 15 min each), each specimen was transferred to a mixture of propylene oxide and epoxy resin (1:1, v/v), and embedded in epoxy resin (Table 1-C), which was allowed to polymerize overnight at 60 °C. Then the epoxy resin-embedded blocks were sectioned at 500-nm thickness with a diamond knife on an ultramicrotome (MT-XL; RMC, Tucson, AZ, USA). The resultant semi-ultrathin sections were mounted on glass slides (Matsunami Glass Inc., Osaka, Japan) and incubated in 10% sodium methoxide (Table 1-D) for 2 min at room temperature to remove the epoxy resin (Mayor et al., 1961). After passage through an acetone series, which consisted of absolute acetone (two times) and 50% acetone (once), sections were transferred to PBS (pH 7.4). Each specimen was passed rapidly through each solvent, that is to say, with immersion for 30 sec or so.

2.3 Immunofluorescence staining

After retrieval of antigens (Shi et al., 1991) by heating in a microwave oven at 500 Watts for 2 min in 10 mM sodium citrate buffer (pH 6.0; Table 1-E), sections on slides were allowed to cool for 8 min and were then transferred to PBS (pH 7.4) at room temperature. Sections were then incubated with primary antibodies overnight at 4 °C. The primary antibodies that we used in this study are shown in Table 2. To determine the optimum working dilutions of each preparation of antibodies, we tested dilutions from 1:25 to 1:800. After washing in PBS, sections were incubated with biotin-conjugated antibodies raised in rabbit against mouse IgG, IgA and IgM (Nichirei Biosciences, Tokyo, Japan) or with biotin-conjugated antibodies raised in goat against rabbit IgG (Nichirei Biosciences) for 30 min at room temperature. Secondary antibodies are also shown in Table 2. Sections were incubated with streptavidin-Alexa Fluor 488 or 633 (Molecular Probes, Eugene, OR, USA) for 30 min at room temperature. Fluorescent reagents are shown in Table 3. Sections were then mounted with FluoroGuard™ antifade reagent (Bio-Rad Laboratories, Hercules, CA, USA) after washing in PBS. Each specimen was covered with a glass coverslip (Matsunami Glass Inc) with clear nail varnish as adhesive. The main steps for immunofluorescence staining are summarized in Table4. The specificity of immunoreactions was checked by preparation of the following controls: a control without primary antibodies; a control incubated with normal mouse serum instead of primary antibodies; and controls incubated with antibodies that had been incubated for 24 h at 4 °C with the corresponding antigen at 10 to 100 μg/ml. No immunolabeling of the lingual mucosa of fetal and juvenile rats was seen in any of the negative controls.

	Reagent	Preparation
A	4% formaldehyde titrated from paraformaldehyde	(1) 25 ml of distilled water (DW) are heated almost to boiling point. (2) 2 g of paraformaldehyde (fine granules; TAAB) are added to the hot DW and are dissolved by gentle shaking. (3) Upon addition of one to three drops of 1 N NaOH, the 2 g of paraformaldehyde are completely dissolved in DW. (4) Finally, a small amount of DW is added to give bring the total volume back to 25 ml. (5) Phosphate-buffered saline (PBS) is prepared as a 10x stock solution that contains 1.37 M NaCl, 27 mM KCl, 100 mM Na_2HPO_4 and 18 mM KH_2PO_4 (adjusted to pH 7.4 with HCl, if necessary) and is stored at 4 °C. (6) A working solution is prepared by dilution of one part stock solution with four parts distilled and deionized water (DDW). (7) Mix 25 ml of the solution of paraformaldehyde with 25 ml of 2x PBS just before use.
B	0.1 M phosphate buffer (pH 7.4)	(1) Prepare a 10x stock solution of PBS with 1.37 M NaCl, 27 mM KCl, 100 mM Na_2HPO_4 and 18 mM KH_2PO_4 (adjust to pH 7.4 with HCl, if necessary) and store at 4 °C. (2) Prepare a working solution by dilution of one part stock solution with nine parts DDW.
C	Epoxy resin	(1) Solution A: 62 ml of Epon 812 and 100 ml of dodecenyl succinic anhydride (DDSA; TAAB). (2) Solution B: 100 ml of Epon 812 and 89 ml of methyl nadic anhydride (MNA). (3) Mix equal volumes of solutions A and B, adding 1.5 to 1.8% tri-dimethylaminomethyl phenol (DMP-30; TAAB) for polymerization.
D	10% sodium methoxide	(1) Add 20 g of sodium metal to 200 ml of absolute methanol in a 1,000-ml flask; bubbling will occur for approximately 20 min. (2) Add methanol to give a volume of 200 ml, to compensate for losses due to evaporation. (3) After addition of 200 ml of benzene, add 50 to 100 ml of methanol until complete fusion of methanol and benzene has occurred.
E	10 mM sodium citrate buffer (pH 6.0)	(1) 10 mM sodium citrate buffer (pH 6.0): prepare a 20x stock solution with 20 ml of 200 mM citric acid and 80 ml of 200 mM sodium citrate. (2) Prepare the working solution by dilution of one part stock solution with nineteen parts DDW.

Table 1. Preparation of reagents for fixation of tissues and immunofluorescence staining.

Antibodies	Origin	Host	Type	Dilution	Company
Anti-K13	Human	Mouse	Monoclonal	1:50-1:100	Progen Biotech GmbH, Germany
Anti-K14	Human	Mouse	Monoclonal	1:50-1:100	Cymbus Bioscience Ltd., UK
Anti-CII	Rat	Rabbit	Polyclonal	1:50-1:100	Chemicon International, Inc., USA
Anti-CIII	Rat	Rabbit	Polyclonal	1:50-1:100	Chemicon International, Inc., USA
Biotin-conjugated anti-mouse	---	Rabbit	---	10 μg/ml	Nichirei Biosciences, Japan
Biotin-conjugated anti-rabbit	---	Goat	---	10 μg/ml	Nichirei Biosciences, Japan

Table 2. Primary and secondary antibodies used in this study.

Label	Dilution	Company
Streptavidin-Alexa Fluor 488	10 μg/ml	Molecular Probes, USA
Streptavidin-Alexa Fluor 633	10 μg/ml	Molecular Probes, USA

Table 3. Fluorescent reagents used in this study.

Step	Procedure
1	Retrieval of antigens by heating in a microwave oven (500 W) for 2 min in 10 mM sodium citrate buffer (pH 6.0)
2	Cooling for 8 min at room temperature
3	Incubation with the primary monoclonal or polyclonal antibodies overnight at 4℃
4	Washing in PBS
5	Incubation with biotin-conjugated rabbit antibodies against mouse IgG, IgA and IgM or with biotin-conjugated goat antibodies against rabbit IgG for 30 min at room temperature
6	Washing in PBS
7	Incubation with streptavidin-Alexa Fluor 448 or 633 for 30 min at room temperature
8	Mounting with FluoroGuard™ antifade reagent

Table 4. A summary of methods, showing the main steps for immunofluorescence staining.

Using our new technique, we were able easily to detect and localize immunofluorescence in the tongues of fetal and juvenile rats. In the present study, we used DIC images and images in transmission mode obtained by LSM to examine the same specimens as those in which we monitored the fluorescence of Alexa Fluor 488 or 633. We were able to define the histological location of K13, K14, CII and CIII by combining the respective images.

2.4 Confocal laser-scanning microscopy (LSM)

All specimens were examined with a confocal laser-scanning microscope (LSM510 or LSM710; Carl Zeiss, Jena, Germany) that was equipped with an argon laser or a helium-neon (HeNe) laser. The dimensions of all images displayed on the monitor were 1024 x 1024 pixels. For single scanning for detection of the fluorescence of Alexa Fluor 488 (Molecular Probes), we used a 488-nm laser wavelength filter; a 488-nm primary dichroic beam-splitter; and a 505- to 530-nm band-pass filter. These specimens were examined with a confocal laser-scanning microscope that was fitted with an argon laser. For single scanning for detection of the fluorescence of Alexa Fluor 633, we used a 633-nm laser wavelength filter; a 514/633-nm primary dichroic beam-splitter; and a 650-nm low-pass filter. These specimens were examined with a confocal laser-scanning microscope that was fitted with a HeNe laser. Combinations of pixel sizes from 0.12 μm x 0.12 μm to 0.17 μm x 0.17 μm and a 40x objective with a numerical aperture (NA) of 0.75 were used for observations. We also examined DIC images that revealed the histology and morphology of cells on the same semi-ultrathin sections. We stacked the immunofluorescence images and the corresponding DIC images by computer, as summarized in Table 5. Furthermore, after staining of specimens with 0.2% toluidine blue (Waldeck GmbH & Co., Münster, Germany) in 2.5% Na_2CO_3, we examined the corresponding images by LSM in the transmission mode. Finally, two images, showing the immunoreactivity of a specific antigen and the histology, recorded in transmission mode, were stacked on top of one another, by computer, for analysis.

Step	Procedure
1*	Fluorescent immunostaining
2*	Immunofluorescent images obtained by LSM
3	DIC images obtained by LSM
4	Toluidine blue staining
5	Images in transmission mode obtained by LSM
6	Combination of immunofluorescent and DIC images
7	Combination of immunofluorescent and transmission-mode images

Table 5. A summary of methods for the stacking of images of immunoreactivity and DIC images or images obtained by LSM in transmission mode.
* Steps 1 and 2 should be repeated using adjacent semi-ultrathin sections and antibodies with two different specificities to show the localization of pairs of antigens.

To reveal the combined localization of the pair of antigens of K13 and of K14, we stacked three images obtained by LSM, which showed the immunoreactivity of each antigen and the histology, recorded in transmission mode, on top of one another by computer and analyzed the results after examining the same respective DIC images and images obtained in transmission mode.

3. Experimental results and analyses

3.1 Localization of immunofluorescence specific for K13 and K14 in the lingual epithelium during morphogenesis of non-gustatory papillae

We have not yet succeeded in the double immunolabeling of samples prepared by the epoxy-resin method. Therefore, instead of double immunolabeling, we immunolabeled K13 and K14, separately, in adjacent sections. We then stacked the two resultant images together with an image obtained by LSM, in transmission mode, using a computer (Aoyagi et al., 2008). The combination of laser-scanning micrographs, which show the immunolabeling of K13 and K14 in the fetal rat tongue, and images obtained by LSM in transmission mode is shown in Figure 1.

Localization of immunofluorescence specific for K13 and K14 in the lingual epithelium during morphogenesis of non-gustatory papillae, as revealed by the above-mentioned double immunolabeling, is shown in Figure 2. Filiform papillae, which are non-gustatory lingual papillae, were compactly distributed on the dorsal surface of the lingual body of rats after birth, and these papillae developed rapidly just before birth (Iwasaki et al., 1997). Morphogenesis of filiform papillae is closely related to the keratinization of the dorsal lingual epithelium (Iwasaki et al., 1999). Immunoreactivity specific for K13 and for K14 was not detected on the lingual epithelium of fetuses on day 15 after conception (E15), at which time the lingual epithelium was composed of a few layers of cuboidal cells. Immunoreactivity specific for K13 and K14 was distinct at all postnatal stages examined. Although the respective patterns of K13-specific and K14-specific immunoreactivity differed as the filiform papillae developed, immunoreactivity specific for K13 was generally evident in the suprabasal cells of the interpapillary cell columns, where keratinization was weaker than in the papillary cell columns. Immunoreactivity specific for K14 was detected in the basal and suprabasal cells, which were mitotically active, of the papillary and interpapillary cell columns. Immunoreactivity specific for K13 in immunopositive cells in the interpapillary cell columns was densely distributed in the cytoplasm exclusively and non was evident in the nuclei. The same was true of the immunoreactivity specific for K14. Furthermore, immunoreactivity specific for K13 in the suprabasal cells of the interpapillary cell columns was more densely distributed than it was in the suprabasal cells of the papillary cell columns. By contrast, immunoreactivity specific for K14 in the basal and suprabasal cells of the papillary cell columns was more distinctively distributed than it was in the interpapillary cell columns. The corresponding images in transmission mode clearly revealed that the lingual epithelium was composed of stratified squamous cells and, in addition, that rounded rudiments of filiform papillae were arranged at equal intervals, for the most part, just before and just after birth. The sizes of basal cells in the papillar and interpapillar regions were almost same as those of cells in the dorsal epithelium of the tongue. At this stage, the connective tissue was beginning to penetrate into the central part of each filiform papilla and, as a result, the epithelial-connective tissue border was undulated (Iwasaki et al., 2006a).

3.2 Localization of immunofluorescence specific for K13 and K14 in the lingual epithelium during the morphogenesis of gustatory papillae

Localization of immunofluorescence specific for K13 and K14 in the epithelium during the morphogenesis of circumvallate papillae, which are gustatory papillae, in fetal and juvenile Sprague-Dawley rats is shown in Figures 3, 4 and 5. Only a single circumvallate papilla, one type of gustatory lingual papilla, is located medially on the dorsal surface at the end of the lingual body (Iwasaki et al., 1997). We used fluorescence immunohistochemistry, analysis of DIC images and LSM in the transmission mode, after staining specimens with toluidine blue, to examine the localization of K13 and K14 in the lingual epithelium of rats during the prenatal and postnatal morphogenesis of the circumvallate papillae. No immunoreactivity

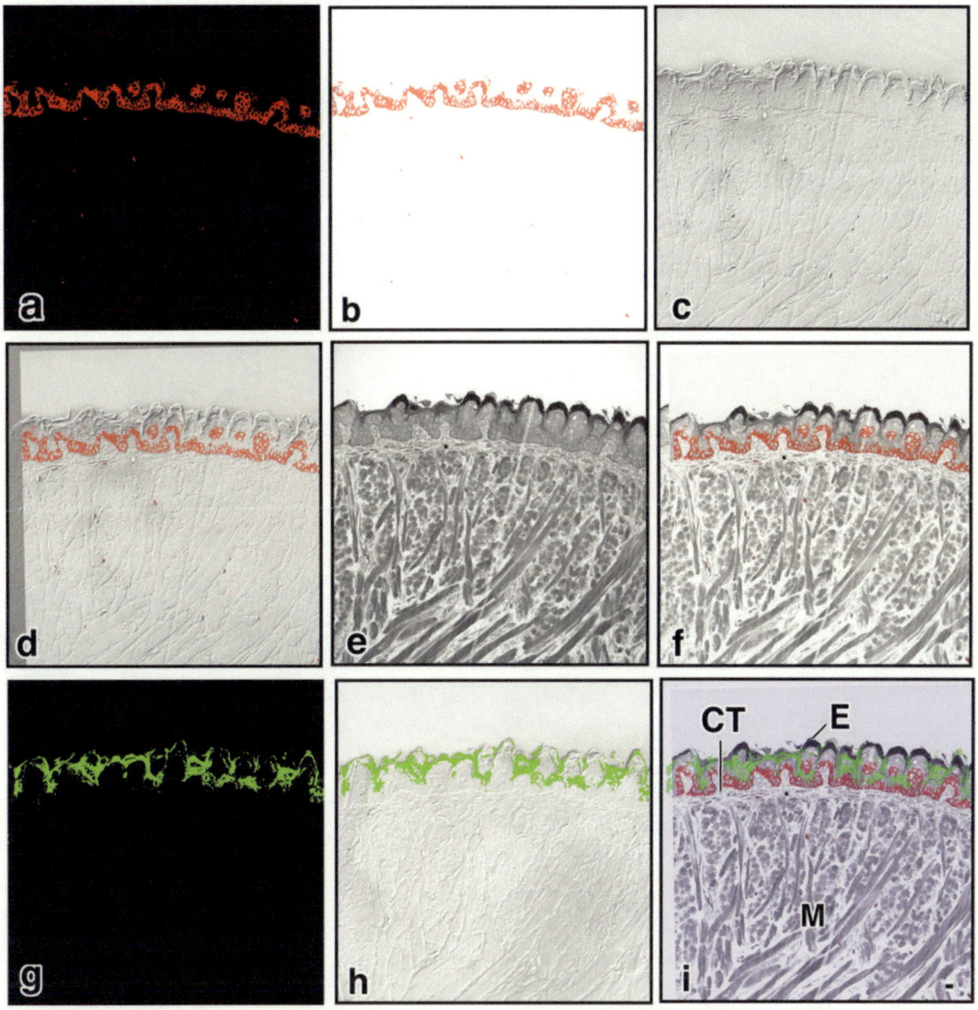

Fig. 1. The stacking of images of immunofluorescence with DIC images and with images obtained by LSM in transmission mode. Micrographs of two sagittally adjacent sections on P0, which show the immunolabeling of K13 (Alexa Fluor 633; green) and of K14 (Alexa Fluor 488; red) in the lingual epithelium of rats on P0 and the images obtained by LSM in the transmission mode, are stacked as follows.
(a) Immunofluorescence of K14 obtained by LSM;
(b) removal of the background from image in (a);
(c) DIC image of the same section as in (a), obtained by LSM;
(d) stacking of images in (b) and (c);
(e) image obtained by LSM in transmission mode after toluidine blue staining;
(f) stacking of images in (b) and (e);
(g) immunofluorescence of K14 obtained by LSM;
(h) stacking of image in (g), without background, and in (c); and
(i) stacking of images in (b), in (g) without background, and in (e).
E, Dorsal lingual epithelium; CT, connective tissue; and M, muscle. Bars = 10 μm.

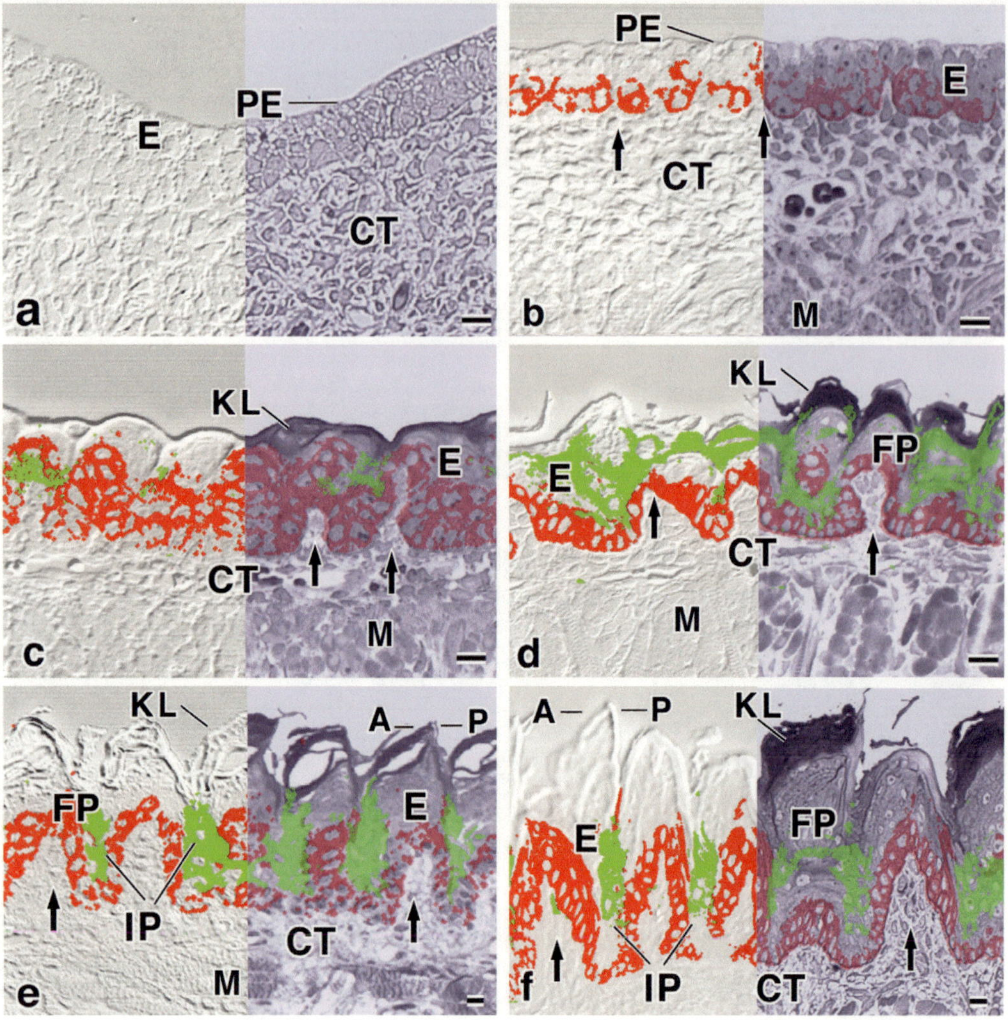

Fig. 2. Combination of laser-scanning micrographs, which show the immunolabeling of K13 (Alexa Fluor 633; green) and of K14 (Alexa Fluor 488; red) in the fetal rat tongue, and DIC images (left half of each micrograph) or images obtained by LSM in transmission mode (right half of each micrograph) that show the histology and cellular morphology of semi-ultrathin sections.

(a) Sagittal section from a fetus on E15;

(b) sagittal section from a fetus on E17;

(c) sagittal section from a fetus on E19;

(d) sagittal section from a juvenile on P0;

(e) sagittal section from a juvenile on P7; and

(f) sagittal section from a juvenile on P14.

E, Dorsal lingual epithelium; PE, periderm; CT, connective tissue; FP, filiform papillae; KL, keratinized cell layer; M, muscle; IP, interpapillary cell column; A, anterior cell column of filiform papilla; P, posterior cell column of filiform papilla; and arrows, connective tissue papillae. Bars = 10 μm.

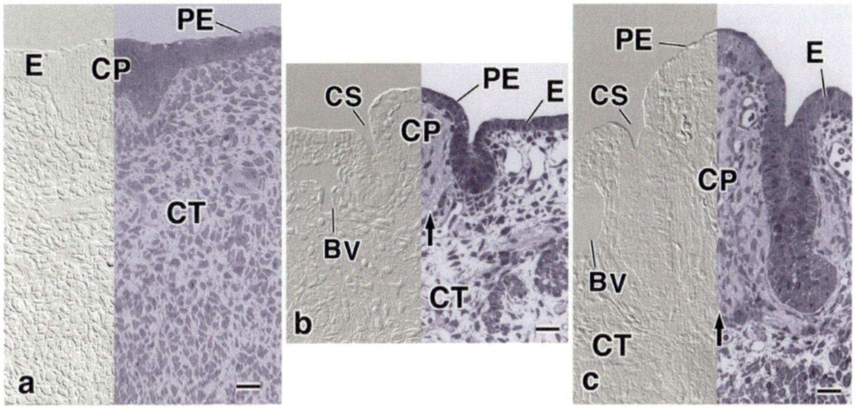

Fig. 3. Combination of laser-scanning micrographs, which show the immunolabeling of K13 in the fetal rat tongue, and DIC images (left half of each micrograph) or images obtained by LSM in transmission mode (right half of each micrograph) that show the histology and cellular morphology of semi-ultrathin sections. No immunoreactivity was recognizable.
(a) Transverse section from a fetus on E15;
(b) transverse section from a fetus on E17; and
(c) transverse section from a fetus on E19.
E, Dorsal lingual epithelium; PE, periderm; CT, connective tissue; CP, circumvallate papilla; CS, circular sulcus; M, muscle; BV, blood vessel; and arrows, connective tissue papillae. Bars = 20 μm.

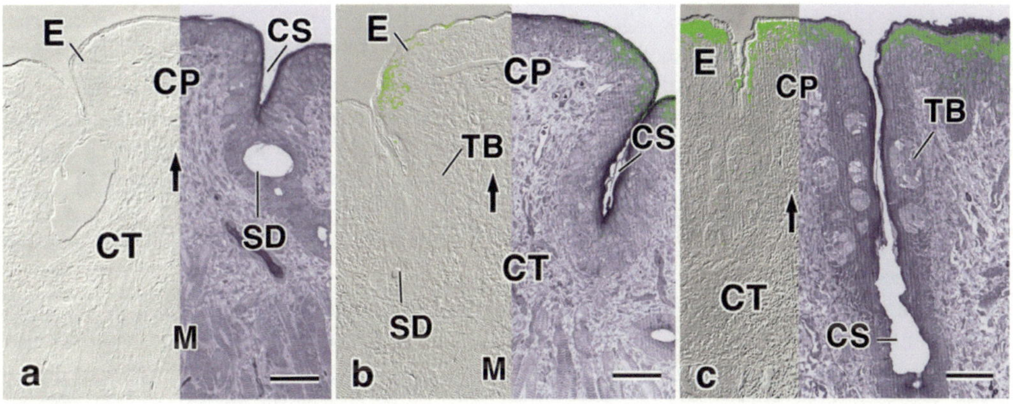

Fig. 4. Combination of laser-scanning micrographs, which show the immunolabeling of K13 (Alexa Fluor 633; green) in the juvenile rat tongue, and DIC images (left half of each micrograph) or images obtained by LSM in transmission mode (right half of each micrograph) that show the histology and cellular morphology of semi-ultrathin sections.
(a) Transverse section from a juvenile on P0;
(b) transverse section from a juvenile on P7; and
(c) transverse section from a juvenile on P14.
E, Dorsal lingual epithelium; CT, connective tissue; CP, circumvallate papilla; CS, circular sulcus; M, muscle; TB, taste bud; SD, secretory duct of von Ebner's gland; and arrows, connective tissue papillae. Bars = 50 μm.

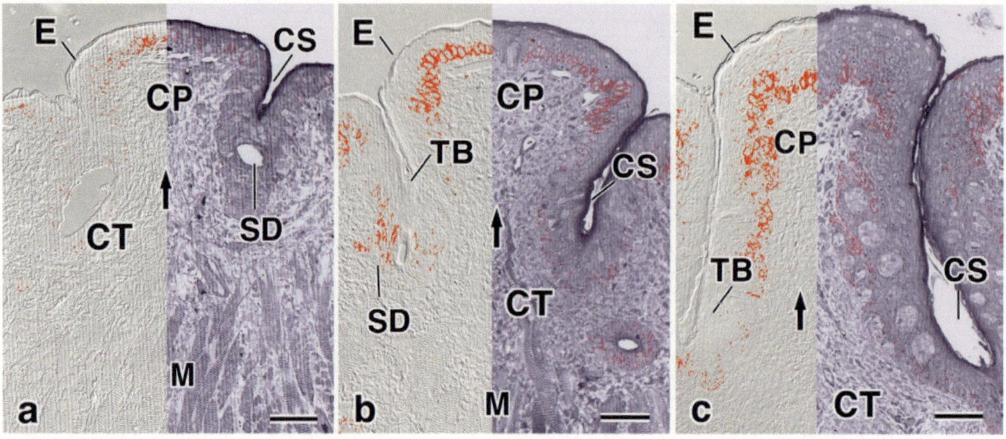

Fig. 5. Combination of laser-scanning micrographs, which show the immunolabeling of K14 (Alexa Fluor 488; red) in the juvenile rat tongue, and DIC images (left half of each micrograph) or images obtained by LSM in transmission mode (right half of each micrograph) that show the histology and cellular morphology of semi-ultrathin sections.
(a) Transverse section from a juvenile on P0;
(b) transverse section from a juvenile on P7; and
(c) transverse section from a juvenile on P14.
E, Dorsal lingual epithelium; CT, connective tissue; CP, circumvallate papilla; CS, circular sulcus; M, muscle; TB, taste bud; SD, secretory duct of von Ebner's gland; and arrows, connective tissue papillae. Bars = 50 μm.

specific for K13 and K14 was detected in the lingual epithelium of fetuses on E15, at which time the circumvallate papillary placode, the primitive rudiment of the circumvallate papilla, was detectable as the thickening of several layers of cuboidal epithelial cells. On E17 and E19, the developing circumvallate papillae were clearly recognizable, each onsisting of a central papilla and the surrounding sulcus. No immunoreactivity specific for K13 and K14 was evident in the lingual epithelium around these structures at this time. K14-specific immunoreactivity was first detected in the basal layer of the epithelium of the circumvallate papillae on P0 (the day of birth) and K13-specific immunoreactivity was detected on P7. Morphogenesis of the circumvallate papillae progressed significantly from P0 to P14, and immunoreactivity specific for K13 and K14 was clearly recognizable after P7. The respective patterns of K13-specific and K14-specific immunoreactivity differed during the development of the circumvallate papillae. K13-specific immunoreactivity was generally evident in cells in the intermediate layer of the epithelium, while K14-specific immunoreactivity was detected in cells in the basal and suprabasal layers.

The immunostaining method used in the present study, with removal of epoxy-resin and antigen retrieval by microwaving, was fundamentally the same as that used in previous studies (Aoyagi et al., 2008; Asami et al., 2008; Iwasaki et al., 2008). Therefore, the observed significant differences between the respective patterns of distribution of immunoreactive K13 and K14 in filiform papillary regions and circumvallate papillary regions are unlikely to be based on differences in methodology. The main reason for the differences in the respective patterns of distribution of K13 and K14 between the two areas might be the differences in the patterns of stratification and keratinization of the epithelium between the two areas. In particular, the difference in the timing of the appearance of K14 seems to be related to the difference in the timing of initiation of the stratification of the epithelium. It is possible that the appearance of K14 in basal cells of the lingual epithelium might induce changes in the lingual dorsal epithelium, namely, morphogenesis of filiform papillae, keratinization of the epithelium of the papillary cell column, and so on (Aoyagi et al., 2008). The initiation of stratification in the filiform papillary area clearly occurs earlier than that in the circumvallate papillary area, and K14-specific immunoreactivity in the filiform papillary area appears earlier than that in the circumvallate papillary area. By contrast, the appearance of K13-specific immunoreactivity seems to be related both to the stratification and to the keratinization of the epithelium. As indicated by Aoyagi et al. (2008), K13 is widely distributed in the suprabasal layer of the entire lingual epithelium before the initiation of the hard keratinization of the filiform papillae. However, the region of which K13 appears is restricted to the suprabasal and intermediate layers of the interpapillary cell columns, in which hard keratinization does not develop. After P7, K13-specific immunoreactivity was evident on the surface and upper intermediate layers of the epithelium in the circumvallate papillary area, but no hard keratinization was evident in this region. Thus, the pattern of keratinization of the epithelium clearly differs between the filiform papillary area and the circumvallate papillary area (Iwasaki et al., 2011a).

3.3 Localization of immunofluorescence specific for CII and CIII in the lingual mucosa during morphogenesis of non-gustatory papillae

Localization of immunofluorescence specific for CII and CIII in the connective tissue of the mucosa during morphogenesis of the circumvallate papillae, which are gustatory papillae, of fetal and juvenile rats is shown in Figures 6 and 7.

As shown in Figure 6, immunoreactivity specific for CII was scattered in the extracellular matrix over a wide area of the mesenchymal connective tissue of the fetal tongue on E15, when the lingual epithelium was composed of one or two layers of cuboidal cells. Immunoreactivity became more and more significant in the connective tissue of the lamina propria as morphogenesis of the filiform papillae advanced at birth. In addition, immunoreactivity was widely distributed in the connective tissue around the lingual muscle, as myogenesis in the tongue proceeded. The lingual epithelium was composed of stratified squamous cells, and keratinization of the lingual epithelium increased gradually as morphogenesis of filiform papillae continued during postnatal development. The present observations indicate that expression of CII might be related to the development of the endomysium and perimysium after myogenesis of the tongue is complete on P0. Rahkonen and Savontaus (2003) reported that CII is expressed in the epithelial-mesenchymal area of the developing heart and participates in the morphogenesis of cardiac valves and septa.

Thus, CII might also be widely expressed during morphogenesis of connective-tissue components after myogenesis of striated muscle. Some common mechanism might be involved in both phenomena. However, in the present study, we failed to define the stages at which the expression of CII begins and when it ends because the period during which animals were collected began too long after conception and ended too soon after birth (Asami et al., 2008).

We also examined, in semi-ultrathin sections of epoxy resin-embedded samples, the distribution of immunostained CIII, using images obtained in transmission mode, after toluidine blue staining, by LSM, during the morphogenesis of the filiform papillae, the keratinization of the lingual epithelium and the myogenesis of the tongue (Fig. 7).

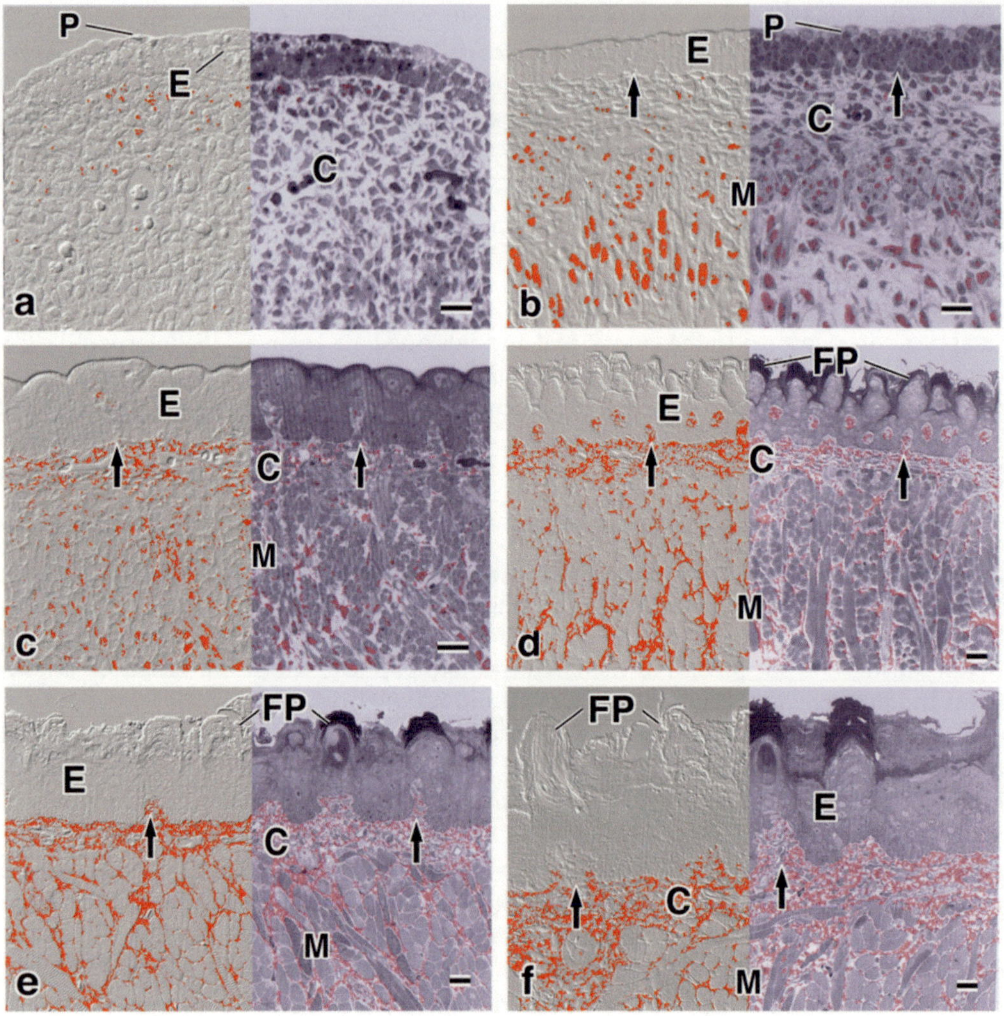

Fig. 6. Combinations of laser-scanning micrographs that show the localization of immunoreactive type II collagen (Alexa Fluor 488; red) and DIC images (left half of each micrograph) or images obtained by LSM in the transmission mode (right half of each micrograph) that show the histology and cellular morphology of semi-ultrathin sections of the lingual body.
(a) Frontal section from a fetus on E15;
(b) sagittal section from a fetus on E17;
(c) sagittal section from a fetus on E19;
(d) sagittal section from a juvenile on P0;
(e) sagittal section from a juvenile on P7; and
(f) frontal section from a juvenile on P14.
E, Dorsal lingual epithelium; P, periderm; C, connective tissue; M, muscle; FP, filiform papillae; and arrow, connective tissue papillae. Bars = 10 μm.

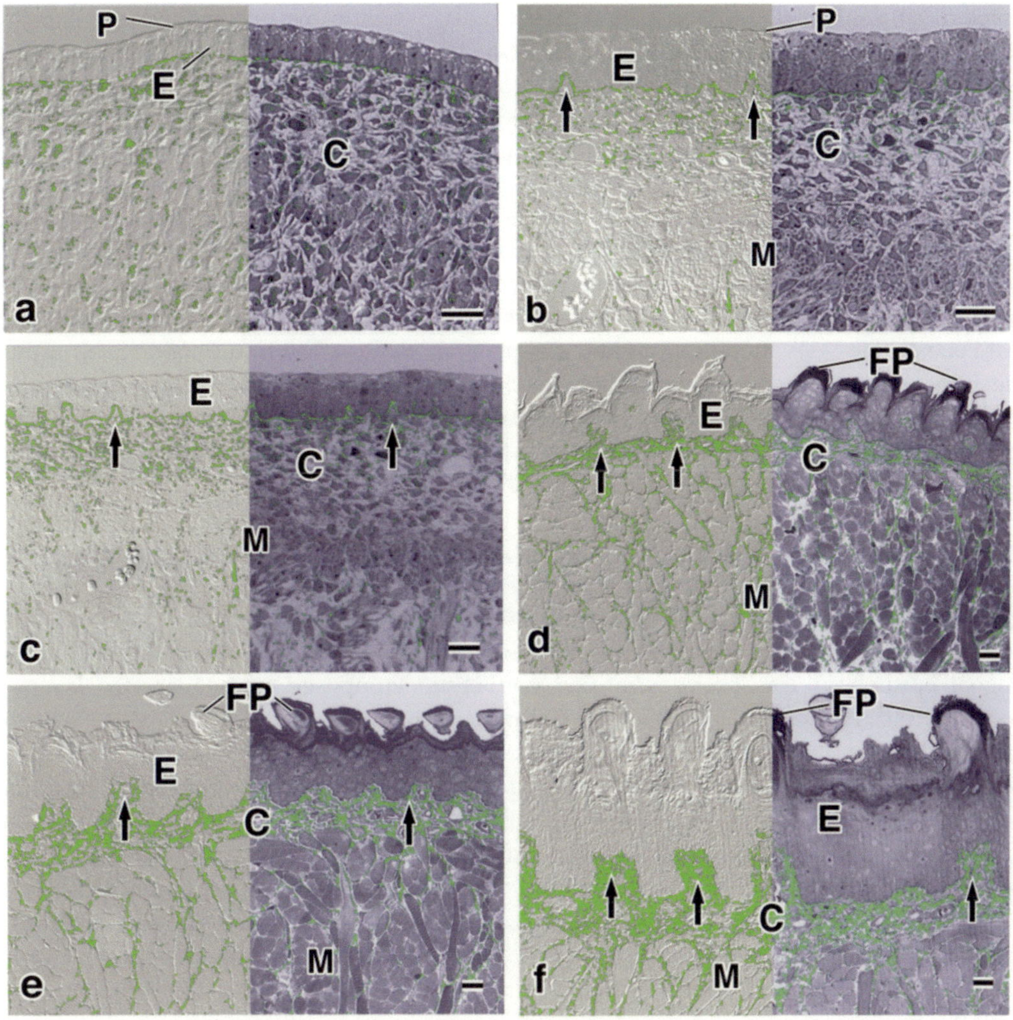

Fig. 7. Combinations of laser-scanning micrographs that show the localization of immunoreactive type II collagen (Alexa Fluor 633; green) and DIC images (left half of each micrograph) or images obtained by LSM in the transmission mode (right half of each micrograph) that show the histology and cellular morphology of semi-ultrathin sections of the lingual body.
(a) Frontal section from a fetus on E15;
(b) sagittal section from a fetus on E17;
(c) sagittal section from a fetus on E19;
(d) sagittal section from a juvenile on P0;
(e) sagittal section from a juvenile on P7; and
(f) frontal section from a juvenile on P14.
E, Dorsal lingual epithelium; P, periderm; C, connective tissue; M, muscle; FP, filiform papillae; and arrow, connective tissue papillae. Bars = 10 μm.

Immunoreactivity specific for CIII was distributed widely in the mesenchymal connective tissue in fetuses on E15, at which time the lingual epithelium was composed of one or two layers of cuboidal cells and the lingual muscle was barely recognizable. Immunoreactivity specific for CIII was clearly detected on the lamina propria in fetuses on E17 and on E19, and it was relatively distinct just beneath the lingual epithelium. Immunoreactivity specific for CIII was sparsely distributed on the connective tissue around the developing lingual muscle. In fetuses on E19, the epithelium became clearly stratified and squamous. At postnatal stages from birth to P14, keratinization of the lingual epithelium advanced gradually with the development of filiform papillae. In newborns on P0, myogenesis of the tongue was almost complete. The intensity of the fluorescence due to immunoreactivity specific for CIII at postnatal stages was almost same as that on E19. However, fluorescence just beneath the lingual epithelium had disappeared on P14. The immunoreactivity around the fully mature muscle was relatively distinct from P0 to P14. Thus, CIII appeared in conjunction with the morphogenesis of filiform papillae and the keratinization of the lingual epithelium, as well as in the connective tissue that surrounded the lingual muscle during myogenesis of the rat tongue (Iwasaki et al., 2008).

3.4 Localization of immunofluorescence specific for CII and CIII in the lingual mucosa during morphogenesis of gustatory papillae

Immunoreactivity specific for CII was recognized first in the mesenchymal connective tissue just beneath the circumvallate papilla placode in fetuses on E13. At this stage, most of the lingual epithelium was pseudostratified epithelium composed of one or two layers of cuboidal cells. However, the epithelium of the circumvallate papilla placode was composed of several layers of cuboidal cells. Immunoreactivity specific for CII was detected mainly on the lamina propria just beneath the lingual epithelium of the rudiment of the circumvallate papilla in fetuses on E15 and on E17, and slight immunostaining was detected on the lamina propria around the rudiment. In fetuses on E19, immunoreactivity specific for CII was widely and densely distributed on the connective tissue around the developing circumvallate papillae and on the connective tissue that surrounded the lingual muscle. Immunoreactivity specific for CII was sparsely distributed on the lamina propria of the central bulge. After birth, morphogenesis of the circumvallate papillae advanced gradually with the increase in size of the tongue. Immunoreactivity specific for CII was distinctively distributed in the lamina propria around circumvallate papilla, in the central bulge, and in the connective tissue that surrounded the lingual muscle. The examination of specimens by LSM, in transmission mode, after staining with toluidine blue, revealed details of the histology and cell morphology of the dorsal mucosa more effectively than examination of similar DIC images. The circumvallate papillary placode could be seen on the dorsal surface of the lingual root of fetuses on E13. The rudiment of the circumvallate papilla developed gradually in fetuses, and morphogenesis of the circumvallate papilla progressed significantly at postnatal stages. Except at early stages, CII appeared not only in the connective tissue of the lamina propria but also in the connective tissue papillae during the morphogenesis of the rat tongue. In addition, CII appeared in the connective tissue that surrounded the lingual muscle and its presence seemed to be related to the development of the endomysium and perimysium after completion of the myogenesis of the tongue (Figs. 8 and 9).

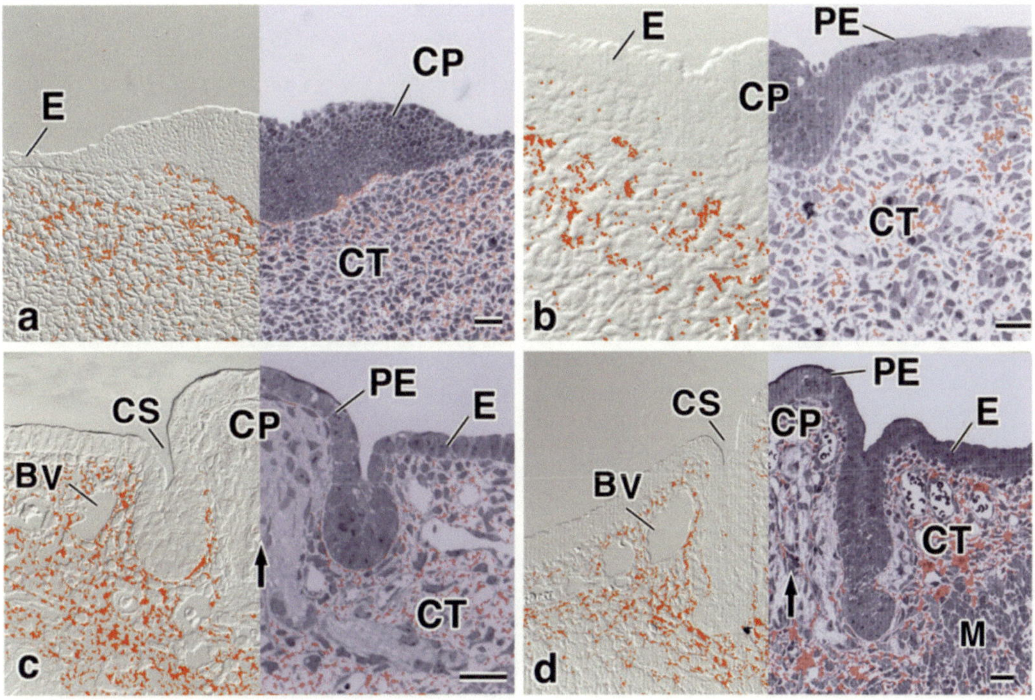

Fig. 8. Combinations of laser-scanning micrographs that show the localization (Alexa Fluor 488; red) of immunoreactive type II collagen and DIC images (left half of each micrograph) or images obtained by LSM in the transmission mode (right half of each micrograph) that show the histology and morphology of cells in semi-ultrathin sections of the developing circumvallate papillary area in fetal rats.
(a) Frontal section from a fetus on E13;
(b) frontal section from a fetus on E15;
(c) frontal section from a fetus on E17; and
(d) frontal section from a fetus on E19.
E, Dorsal lingual epithelium; PE, periderm; CT, connective tissue; M, muscle; CP, circumvallate papillary placode or the rudiment of the circumvallate papilla; CS, circular sulcus; BV, blood vessel; and arrow, connective-tissue papilla of the central papilla. Bars = 20 μm.

In an effort to identify a possible role for CIII in the morphogenesis of circumvallate papillae on the surface of the rat tongue, we examined its appearance by fluorescent immunostaining, in conjunction with DIC images and images obtained, after staining with toluidine blue, by LSM in the transmission mode. We analyzed semi-ultrathin sections of epoxy resin-embedded samples of the lingual mucosa of embryonic and juvenile rats, from E13 to P21. Immunoreactivity specific for CIII was recognized first in the mesenchymal connective tissue just beneath the circumvallate papillary placode on E13. At this stage, most of the lingual epithelium with the exception of the circumvallate papilla placode was pseudostratified epithelium that was composed of one or two layers of cuboidal cells.

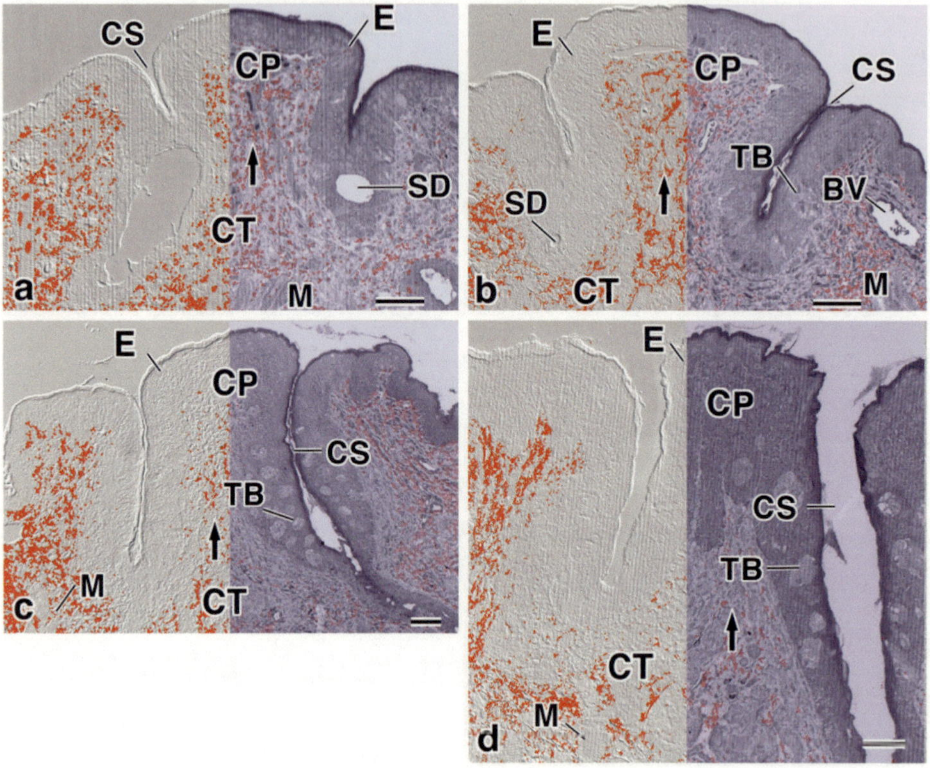

Fig. 9. Combinations of laser-scanning micrographs that show the localization (Alexa Fluor 488; red) of immunoreactive type II collagen and DIC images (left half of each micrograph) or images obtained by LSM in the transmission mode (right half of each micrograph) that show the histology and morphology of cells in semi-ultrathin sections of the developing circumvallate papillary area in juvenile rats.
(a) Frontal section from a fetus on E13;
(b) frontal section from a fetus on E15;
(c) frontal section from a fetus on E17; and
(d) frontal section from a fetus on E19.
E, Dorsal lingual epithelium; CT, connective tissue; M, muscle; CP, circumvallate papilla; CS, circular sulcus; SD, secretory duct of the Ebner's gland; TB, taste buds; and arrow, connective-tissue papilla of the central papilla. Bars = 50 μm.

However, the epithelium of the circumvallate papillary placode was composed of several layers of cuboidal cells. Immunoreactivity specific for CIII was detected mainly on the lamina propria just beneath the lingual epithelium of the rudiment of the circumvallate papilla and the developing circumvallate papilla in fetuses on E15 and on E17, and slight immunostaining was detected on the lamina propria around the rudiment. In fetuses on E19, immunoreactivity specific for CIII was widely and densely distributed on the connective tissue around the developing circumvallate papillae and, also, on the connective tissue that surrounded the lingual muscle. However, the immunoreactivity specific for CIII

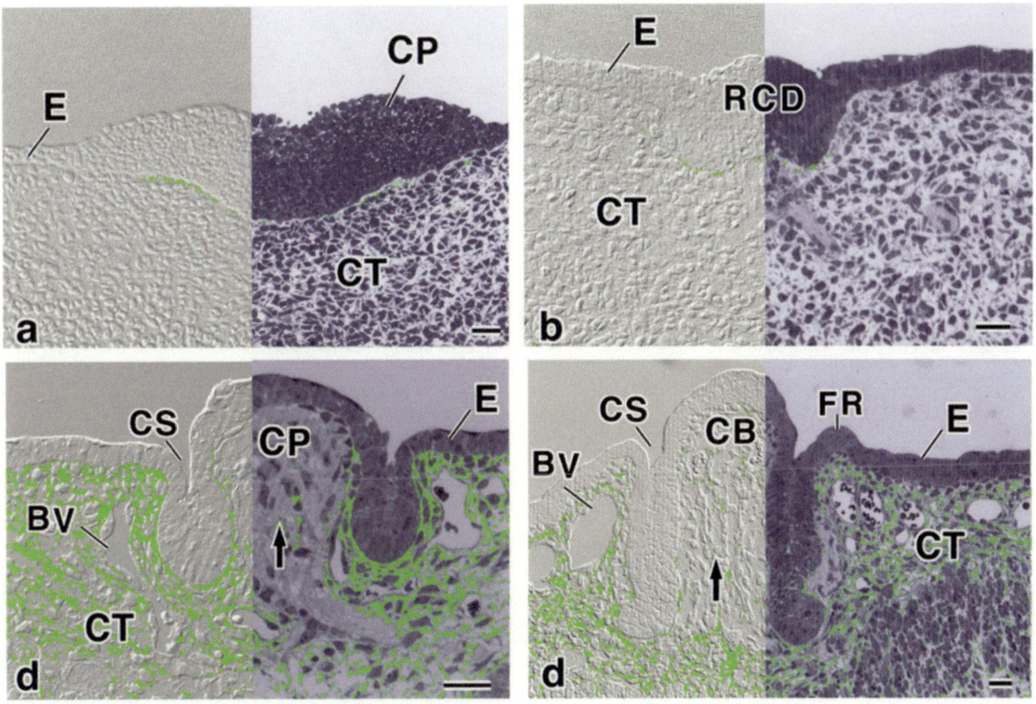

Fig. 10. Combinations of laser-scanning micrographs that show the localization (Alexa Fluor 633; green) of immunoreactive type III collagen and DIC images (left half of each micrograph) or images obtained by LSM in the transmission mode (right half of each micrograph) that show the histology and morphology of cells in semi-ultrathin sections of the developing circumvallate papillary area in fetal rats.
(a) Frontal section from a fetus on E13;
(b) frontal section from a fetus on E15;
(c) frontal section from a fetus on E17; and
(d) frontal section from a fetus on E19.
E, Dorsal lingual epithelium; PE, periderm; CT, connective tissue; M, muscle; CP, circumvallate papillary placode or the rudiment of the circumvallate papilla; CS, circular sulcus; BV, blood vessel; and arrow, connective-tissue papilla of the central papilla. Bars = 10 μm.

was sparsely distributed on the lamina propria of each central papillar structure. After birth, from P0 to P14, morphogenesis of the circumvallate papillae advanced gradually as the total volume of the tongue increased. At these postnatal stages, the intensity of the fluorescence due to immunoreactivity specific for CIII was distinctively distributed on the lamina propria around each circumvallate papilla, on each central bulge, and on the connective tissue that surrounded the lingual muscle. However, immunofluorescence was less distinct on the connective tissue that surrounded the lingual muscle. Thus, CIII appeared in conjunction with the morphogenesis of the circumvallate papillae, as well as in the connective tissue that surrounded the lingual muscle during myogenesis of the rat tongue (Figs. 10 and 11).

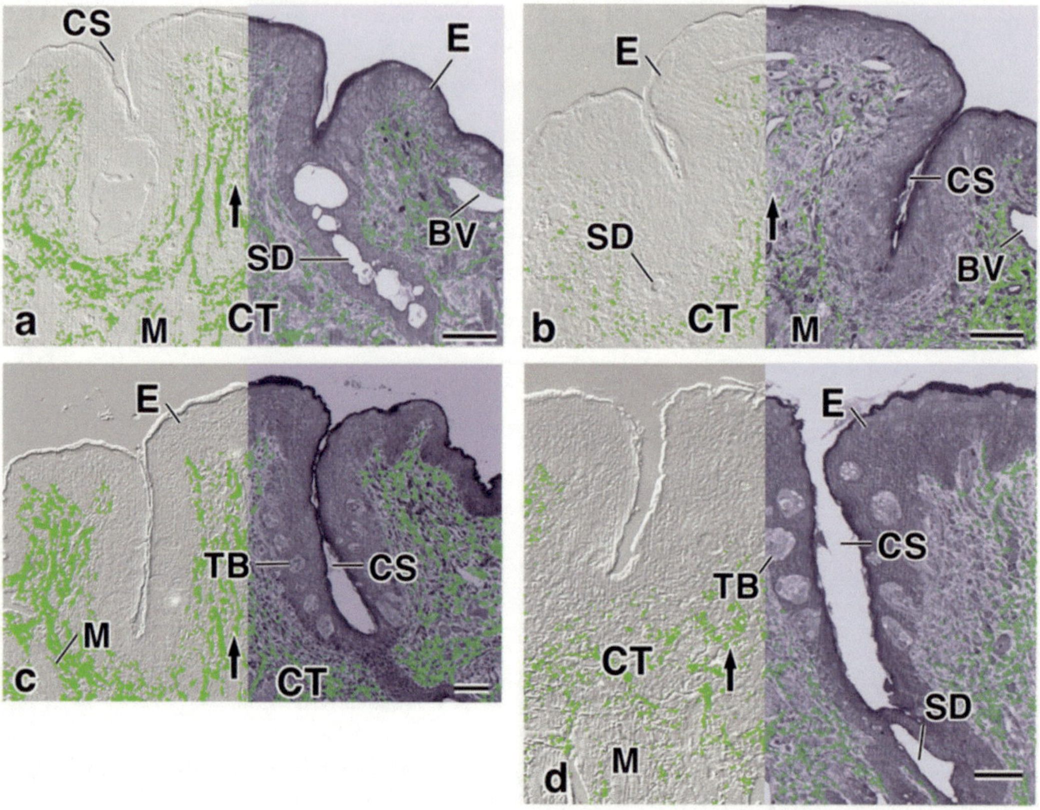

Fig. 11. Combinations of laser-scanning micrographs that show the localization (Alexa Fluor 633; green) of immunoreactive type III collagen and DIC images (left half of each micrograph) or images obtained by LSM in the transmission mode (right half of each micrograph) that show the histology and morphology of cells in semi-ultrathin sections of the developing circumvallate papillary area in juvenile rats.
(a) Frontal section from a fetus on E13;
(b) frontal section from a fetus on E15;
(c) frontal section from a fetus on E17; and
(d) frontal section from a fetus on E19.
E, Dorsal lingual epithelium; CT, connective tissue; M, muscle; CP, circumvallate papilla; CS, circular sulcus; SD, secretory duct of the Ebner's gland; TB, taste buds; and arrow, connective-tissue papilla of the central papilla. Bars = 50 μm.

3.5 Analysis of methodology

For the examination of sections by LSM, we found that microwave heating of specimens just before immunofluorescence staining was useful for the retrieval of antigens in epoxy resin-embedded specimens, even though this treatment had originally been developed for the retrieval of antigens in formalin-fixed, paraffin-embedded specimens. In our previous and present attempts to clarify the localization of immunoreactivity in tissues and cells, we used only a combination of immunofluorescence staining of semi-ultrathin sections and corresponding DIC images obtained by LSM. Using this technique, we were easily able to detect and localize immunofluorescence in the tongues of rat fetuses and juveniles at embryonic and postnatal stages (Iwasaki et al., 2003, 2006a, b, 2007a, b). In some of our previous studies, by contrast, we used LSM in the transmission mode to examine the same specimens as those in which we had monitored fluorescence in an effort to reveal histological and cell-morphological features more distinctly than those revealed in DIC images. By combining immunofluorescence images and the corresponding images obtained by LSM in the transmission mode, we were able to define the histological localization of various antigens more clearly than when we combined immunofluorescence images and the corresponding DIC images. Our method, using both light microscopy and confocal laser-scanning microscopy, should be applicable to various kinds of tissue and cell of which only very small amounts are available.

4. Conclusion

We have developed a technique, using a combination of immunofluorescence staining of semi-ultrathin sections of epoxy resin-embedded samples and the DIC images and images in transmission mode obtained by LSM, that provides detailed information about the immunolocalization of antigens and histological and cellular structures. To demonstrate the effectiveness of our method, we examined the immunofluorescence of immunostained K13 and K14 and that of immunostained CII and CIII and the corresponding DIC and transmission images during the morphogenesis of filiform papillae on the lingual epithelium of rat fetuses and juveniles. We demonstrated that our newly developed technique for localization of pairs of antigens should be useful for investigations of very small specimens, such as fetal tissues and organs.

5. References

Aoyagi, H., Asami, T., Yoshizawa, H., Wanichanon, C. & Iwasaki, S. (2008) Newly developed technique for dual localization of keratins 13 and 14 by fluorescence immunohistochemistry. *Acta Histochemica* 110(No. 4): 324-332.

Asami, T., Aoyagi, H, Yoshizawa, H., Wanichanon, C. & Iwasaki, S. (2008) Immunohistochemical expression of type II collagen in the lingual mucosa of rats during the morphogenesis of the tongue. *Archives of Oral Biology* 53(No. 7): 622-628.

Haraguchi, C.M. & Yokota, S. (2002) Immunofluorescence technique for 100-nm-thick semithin sections of epon-embedded tissues. *Histochemistry and Cell Biology* 117(No. 1): 81-85.

Iwasaki, S., Yoshizawa, H. & Kawahara, I. (1997) Study by scanning electron microscopy of the morphogenesis of three types of lingual papilla in the rat. *The Anatomical Record* 247(No. 4): 528-541.

Iwasaki, S., Yoshizawa, H. & Kawahara, I. (1999) Ultrastructural study of the relationship between the morphogenesis of filiform papillae and keratinization of the lingual epithelium in the rat. *Journal of Anatomy* 195(No. 1): 27-38.

Iwasaki, S., Aoyagi, H. & Yoshizawa, H. (2003) Immunohistochemical detection of the expression of keratin 14 in the lingual epithelium of rats during the morphogenesis of filiform papillae. *Archives of Oral Biology* 48(No. 8): 605-613.

Iwasaki, S., Yoshizawa, H. & Aoyagi, H. (2006a) Immunohistochemical expression of keratins 13 and 14 in the lingual epithelium of rats during the morphogenesis of filiform papillae. *Archives of Oral Biology* 51(No. 5): 416-426.

Iwasaki, S., Aoyagi, H. & Asami, T. (2006b) Expression of keratin 18 in the periderm cells of the lingual epithelium of fetal rats: visualization by fluorescence immunohistochemistry and differential interference contrast microscopy. *Odontology* 94(No. 1): 64-68.

Iwasaki, S. & Aoyagi, H. (2007a) Expression of keratin 14 in the basal cells of the lingual epithelium of mice during the morphogenesis of filiform papillae: visualization by fluorescent immunostaining and confocal laser-scanning microscopy in the transmission mode. *Odontology* 95(No. 1): 61-65.

Iwasaki, S., Aoyagi, H. & Yoshizawa, H. (2007b) Immunohistochemical detection of epidermal growth factor and epidermal growth factor receptor in the lingual mucosa of rats during the morphogenesis of filiform papillae. *Acta Histochemica* 109(No. 1): 37-44.

Iwasaki, S., Asami, T., Wanichanon, C., Yoshizawa, H. & Aoyagi, H. (2008) Immunohistochemical analysis of type III collagen expression in the lingual mucosa of rats during organogenesis of the tongue. *Odontology* 96(No. 1): 12-20.

Iwasaki, S., Aoyagi, H. & Yoshizawa, H. (2011a) Localization of keratins 13 and 14 in the lingual mucosa of rats during the morphogenesis of circumvallate papillae. *Acta Histochemica* 113: in press.

Iwasaki, S., Aoyagi, H. & Yoshizawa, H. (2011b) Localization of type II collagen in the lingual mucosa of rats during the morphogenesis of circumvallate papillae. *Acta Zoologica* 92 (No. 1). 67-74.

Litwin, J.A., Yokota, S., Hashimoto, T. & Fahimi, H.D. (1984) Light microscopic immunocytochemical demonstration of peroxisomal enzymes in epon sections. *Histochemistry* 81(No. 1): 15-22.

Mayor, H.D., Hampton, J.C. & Rosario, B. (1961) A simple method for removing the resin from epoxy-embedded tissue. *Journal of Cell Biology* 9(No. 4): 909-910.

Rahkonen, O., Savontaus, M. (2003) Expression pattern of cartilage collagens and Sox9 during mouse heart development. *Histochemistry and Cell Biology* 120(No. 2): 103-110.

Shi S.-R., Key M.E. & Karla K.L. (1991) Antigen retrieval in formalin-fixed, paraffin-embedded tissues: an enhanced method for immunohistochemical staining based on microwave heating of tissue sections. *Journal of Histochemistry and Cytochemistry* 39(No. 6): 741-748.

12

Determination of Subcellular Localization of Flavonol in Cultured Cells by Laser Scanning

Rie Mukai[1], Junji Terao[1], Yasuhito Shirai[2],
Naoaki Saito[2] and Hitoshi Ashida[2]
[1]*Tokushima University*
[2]*Kobe University*
Japan

1. Introduction

Flavonoids are widely distributed in the plant kingdom including in edible plants such as vegetables and fruits. They are polyphenolic compounds comprising of fifteen carbons, with two aromatic rings connected by a three-carbon bridge (C_6-C_3-C_6). Flavonoids are divided into seven subclasses; the flavones, flavonols, flavanones, catechins, anthocyans, isoflavones and the chalcones, according to their basic skeletal structure (Fig. 1). They are known to have various beneficial pharmacological effects such as anti-cancer, anti-obesity, anti-inflammatory and anti-oxidative activities (Ahn, et al. 2008; Boots, et al. 2008; Chu, et al. 2007; Hsu and Yen 2006; Kuzuhara, et al. 2008; Murakami, et al. 2008). To investigate the beneficial effects of trace food components, it is important to understand their metabolism, absorption, tissue distribution, and subcellular localization. Recently, the metabolism, including absorption and tissue distribution, of trace food components has been investigated using cultured cells and experimental animals (de Boer, et al. 2005; Urpi-Sarda, et al. 2008; Wang, et al. 2003). To determine the physiological concentrations and chemical structures of flavonoid metabolites, high-performance liquid chromatography (HPLC) or HPLC combined with mass spectrometry (MS) is conventionally used (de Boer, et al. 2005; Mullen, et al. 2006; Urpi-Sarda, et al. 2008; Wang, et al. 2003; Yanez, et al. 2008). The subcellular localization of the flavonoids is currently not fully understood, although the preparation of subcellular fractions using centrifugation, followed by HPLC analysis to determine the cellular localization of trace food components, including flavonoids, has been described (Gagne, et al. 2006). Radioactive isotope labeling has also been used to estimate tissue localization of trace food components in animals (Hirosawa and Yamada 1981), and this method could be applied to estimating subcellular localization in cultured cells. However, cross contamination between fractions during the preparation of subcellular fractions can limit the accuracy of such techniques. In this chapter, we demonstrate the value of confocal laser scanning fluorescence microscopy in the subcellular localization of flavanoids, by the detection of autofluorescence in intact culture cells. During this study, we focused on the flavonol subclass, as this subclass features stronger autofluorescence properties than other flavonoid subclasses.

(1)

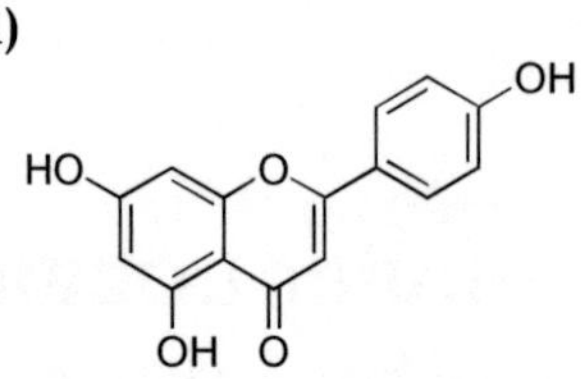

(2)

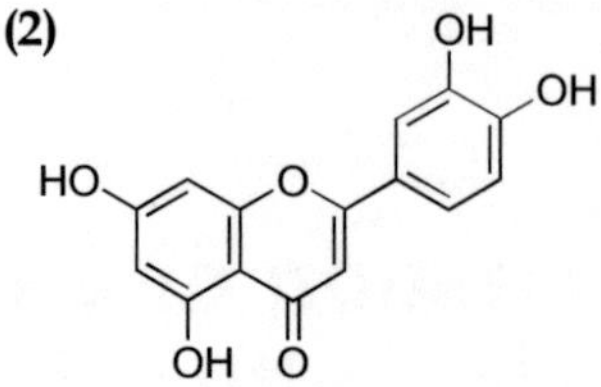

(3)

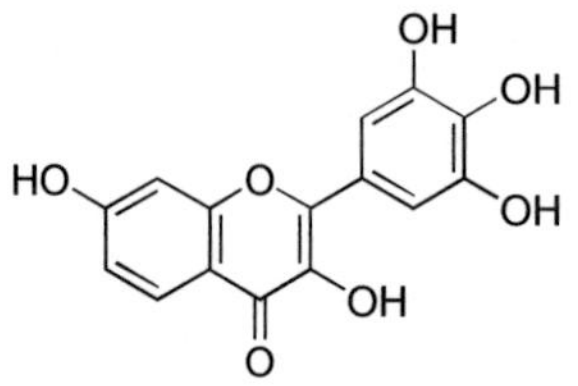

(4)

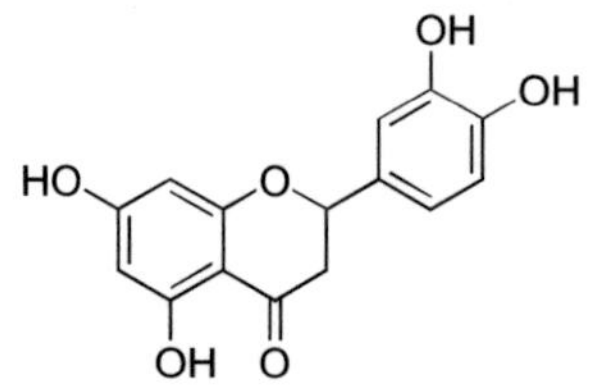

(5)

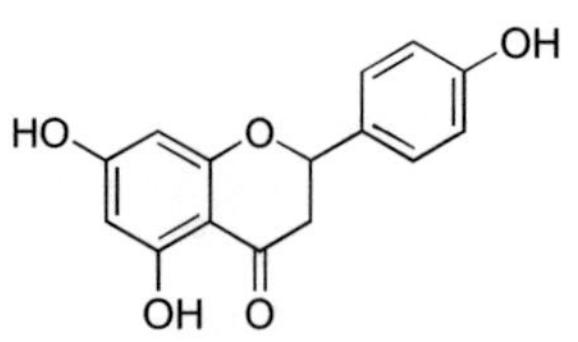

(6)

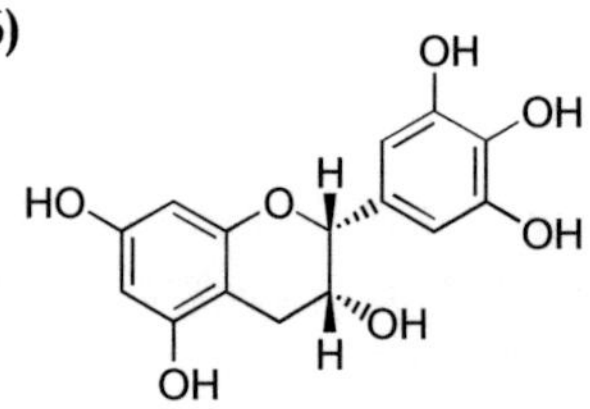

(7)

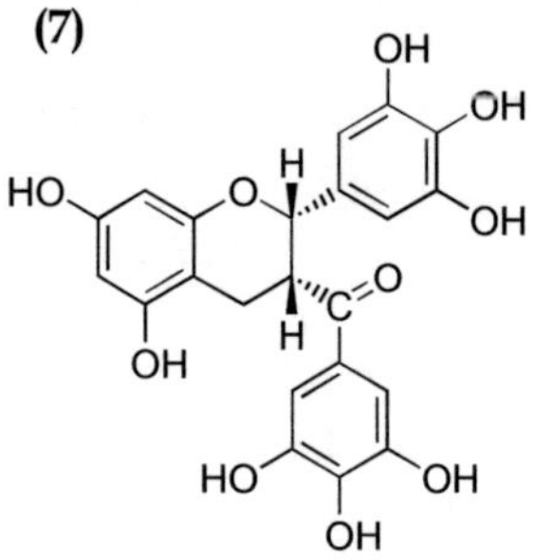

(8)

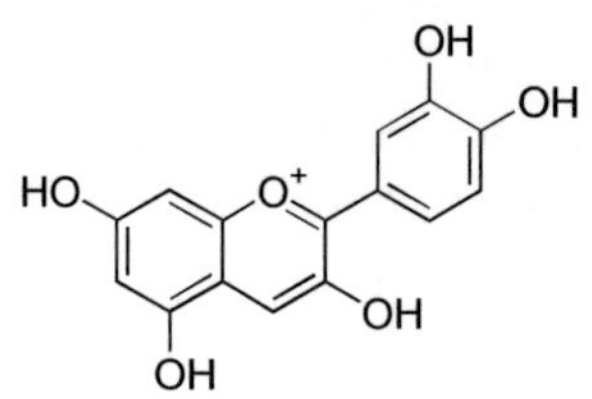

(9)

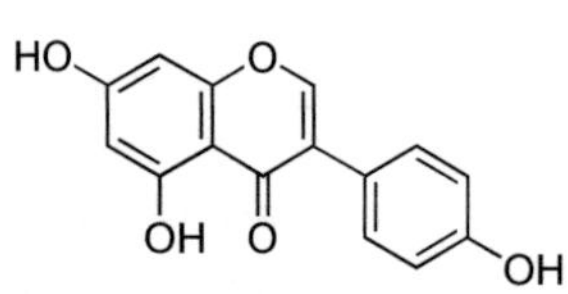

(10)

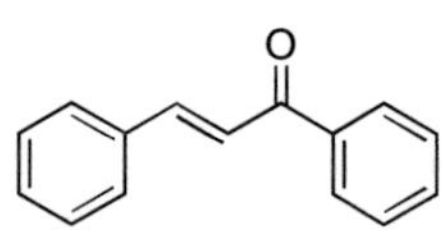

Fig. 1. Representative compounds from each flavonoid subclass
Apigenin (1) and luteolin (2) belong to the flavone subclass and robinetin (3) to the flavonol subclass. Eriodictiol (4) and naringenin (5) belong to the flavanone subclass while (-)-epigallocatechin (6) and (-)-epigallocatechin gallate (7) belong to the catechin subclass. Cyanidin (8), genistein (9) and chalcone (10) belong to the anthocyan, isoflavone and chalcone subclasses, respectively.

The flavonol subclass feature a hydroxyl group in the 3-position of the flavonoid structure (Figure 2). Flavonols are found in onions (*Allium cepa* L.), broccoli (*Barassica oleracea* var.

botrytis L.), and apples (*Malus pumlia* Mill.) (Sakakibara, et al. 2003). It has been reported that flavonols are found in the liver, kidney, muscle, heart, lung, brain, testes, spleen, thymus, bone, brown fat, and white fat of rats, and in the liver, lung, white fat, muscle, brain, kidney, heart, and spleen of pigs after consumption of flavonol containing food stuffs (Bieger, et al. 2008; de Boer, et al. 2005).

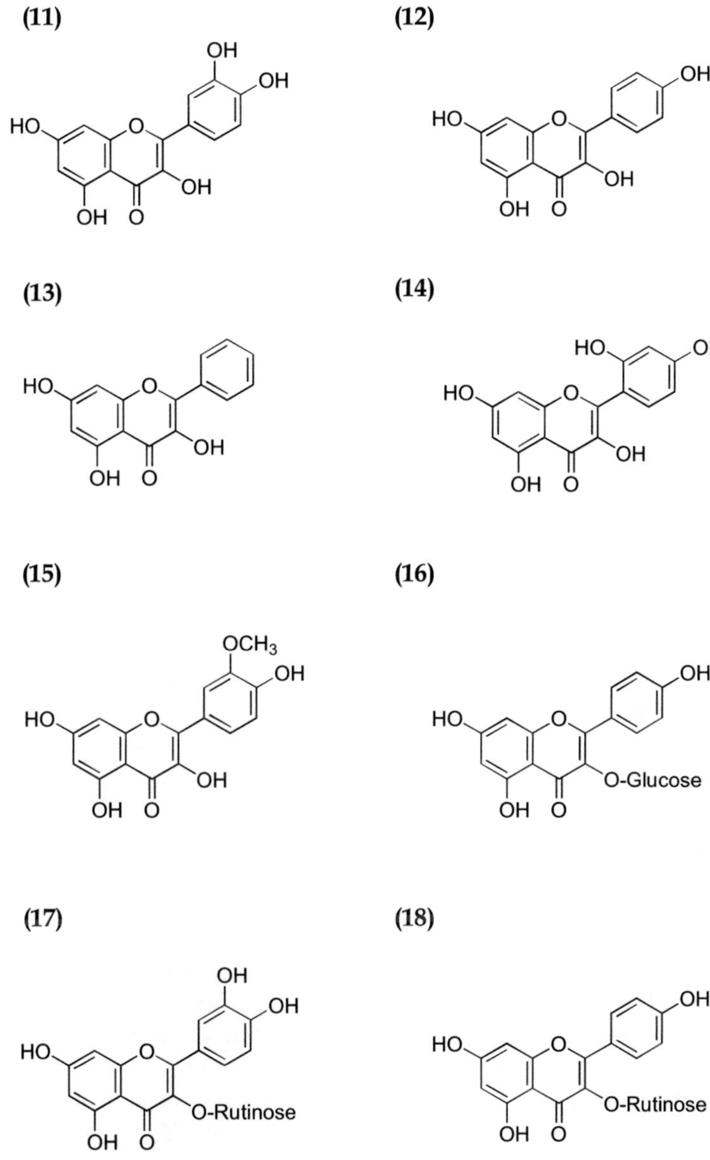

Fig. 2. Examples of the flavonol subclass. Structures of quercetin (11), kaempferol (12), galangin (13), morin (14), isorhamnetin (15), kaempferol-3-glucoside (16), quercetin-3-rutinoside (17), and kaempferol-3-rutinoside (18).

Thus, after dietary intake, flavonol becomes widely distributed in various tissues and is known to exert beneficial effects in many tissue types. In mouse liver, flavonol has been reported to reduce the toxicological effects of dioxin by suppressing the dioxin-induced activation of the aryl hydrocarbon receptor (Mukai, et al. 2009a). Flavonol has been shown to reduce lipid peroxidation in the aorta tissue of rabbits fed a high cholesterol diet (Kamada, et al. 2005). In the research that demonstrates the effect of flavonol on the brain, the nerve cell is generally used. In PC12, the nerve–like cells, flavonol glucuronides suppressed production of reactive oxygen species (Shirai, et al. 2006). The cellular absorption and metabolism of flavonoids in the intestines, human colonic adenocarcinoma Caco-2 cells have been used as a model. It was reported that flavonoids are incorporated into Caco-2 cells by passive diffusion (Walgren, et al. 1998) and/or the active sodium-dependent glucose transporter 1 (Walgren, et al. 2000b), and are excreted through an efflux transporter such as multidrug resistance-associated protein-2 (Walgren, et al. 2000a). Previously, it has also been reported that flavonol emits fluorescence in Caco-2 cells (Walgren, et al. 2000b) and platelet cells from healthy volunteers (Wright, et al. 2010) detectable by fluorescence microscopy. However, these reports did not address the subcellular localization of the flavonoids. To date, the fluorescence properties of flavonoids have not been taken advantage of in studies of flavonoid subcellular localization. Here, we present data showing the subcellular localization of flavonol aglycones in intact cultured cells using confocal laser scanning fluorescence microscopy.

2. Detection of flavonoid in cultured cell by fluorescence microscopy

Since liver tissue is known to accumulate flavonols due to its role in their metabolism, the mouse hepatoma cell line, Hepa-1c1c7 was used as a model cell-line in these studies. Hepa-1c1c7 cells were grown and maintained at 37 °C in α-minimum essential medium (α-MEM) containing 10% fetal bovine serum, 4 mM L-glutamine, 100 U/ml penicillin, and 100 mg/ml of streptomycin under a humidified atmosphere containing 5% CO_2. Hepa-1c1c7 cells were seeded onto a glass-bottomed culture dish and were grown until 80% confluent. Cells were incubated with fresh α-MEM supplemented with 5% fetal bovine serum for 24 h prior to treatment. The cells were treated with each flavonoid by addition of the solubilized flavonoid (in α-MEM containing 0.1% dimethyl sulfoxide) to the medium, and the cells were incubated for 70 min. It is noteworthy that the culture medium itself regularly showed fluorescence, hindering cellular visualization. To avoid this problem, common physiological buffers such as Krebs-Ringer HEPES buffer (50 mM HEPES, pH 7.4, 137 mM NaCl, 4.8 mM KCl, 1.85 mM $CaCl_2$, and 1.3 mM $MgSO_4$) were also applied to the cells during the incubation period. The fluorescence images obtained were shown in Fig. 3. Washing the cells with ice-cold phosphate buffered saline (PBS, pH 7.4) was important to obtain clear images (Fig. 4). After washing, the cells were fixed with 4% paraformaldehyde and 0.2% picric acid in 100 mM sodium phosphate buffer (pH 7.2) overnight at 4 °C in the absence of light. Then, autofluorescence from the flavonoid was observed under a confocal laser scanning fluorescence microscope (LSM 510 invert, Carl Zeiss, Jena, Germany) with excitation by an argon laser at 488 nm and a 515–535 nm band pass filter. A low energy laser was used during observation, but a full energy laser was used for generating images.

The uptake of flavonoids (listed in Table 1) into Hepa-1c1c7 cells was examined under a confocal microscope. When the cells were treated with five flavonol aglycones, namely kaempferol, galangin, isorhamnetin, morin and quercetin, green autofluorescence of the cells was observed (Fig. 5). In contrast, cells treated with either flavonol glycosides (kaemferol-3-glucoside, kaempferol-3-rutinoside and quercetin-3-rutinoside), or flavonoids belonging to other subclasses (flavone, apigenin, luteolin, naringenin, eriodictyol, (-)-epigallocatechin and (-)-epigallocatechin gallate), did not emit any autofluorescence (data not shown). The uptake of flavonol aglycones in Hepa-1c1c7 cells was observed with a confocal microscope. However, flavonol glycosides and compounds belonging to other subclasses could not be visualized.

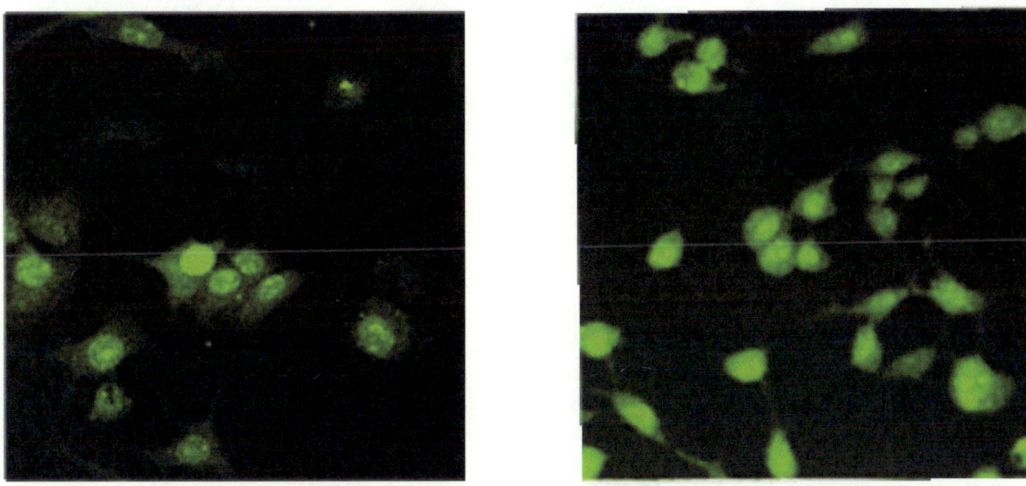

Fig. 3. Both medium and physiological buffer could be used during incubation in uptake experiments. Here, cells treated with the flavonol kaempferol at 50 μM in either α-MEM culture medium (left) or Krebs-Ringer HEPES buffer (right) for 70 min are shown.

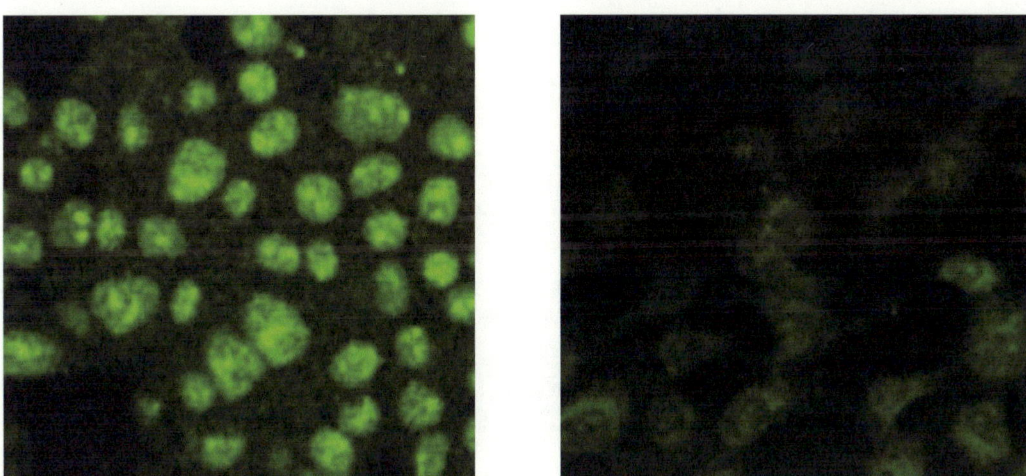

Fig. 4. Images to indicate the importance of cell washing for successful imaging .
Here, cells treated with the flavonol kaempferol at 50 μM, in α-MEM culture medium for 70 min, and either washed with PBS prior to imaging (left) or directly imaged (right) are shown.

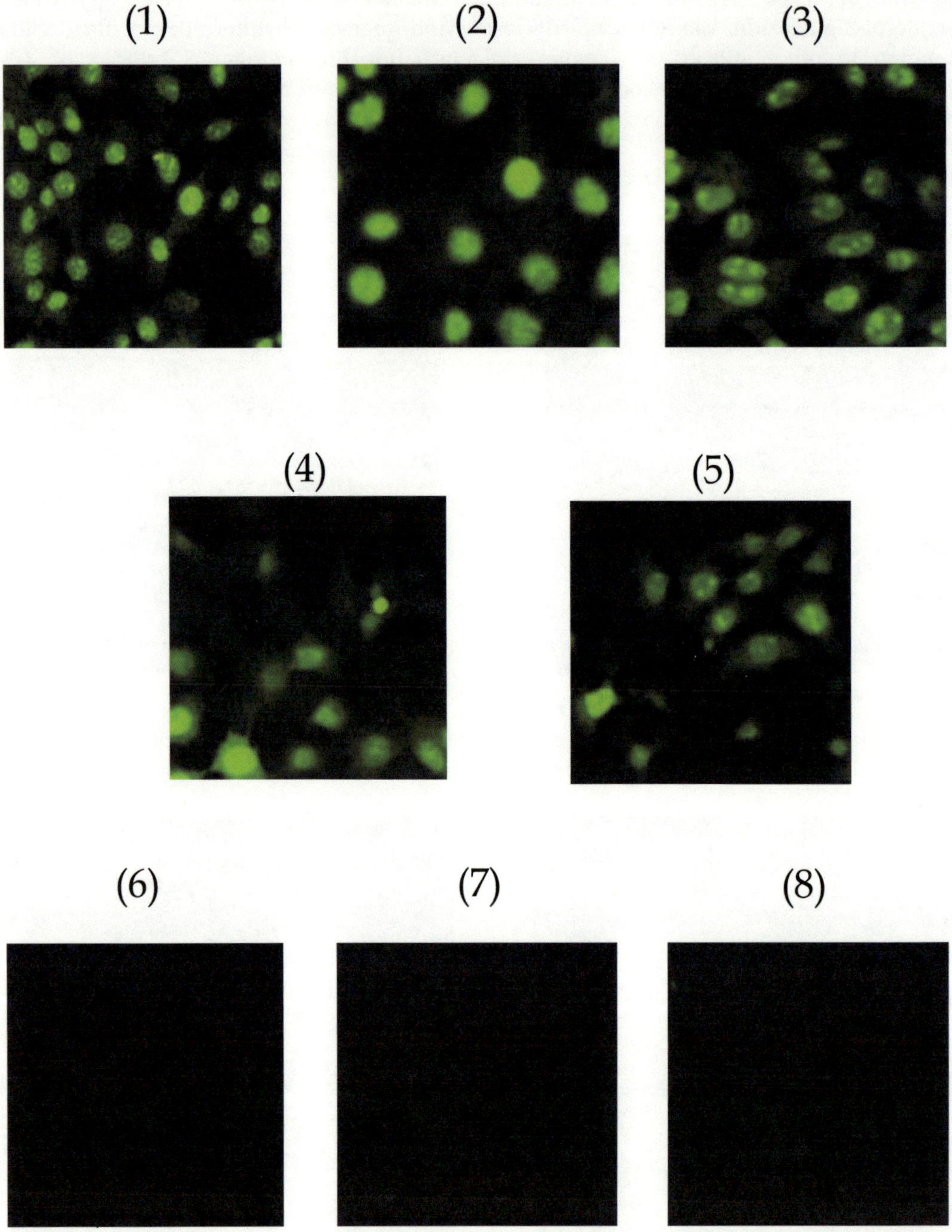

Fig. 5. Cellular uptake of various flavonoids as visualized by fluorescence microscopy. Hepa-1c1c7 cells were seeded onto glass bottom dishes and treated with (1) kaempferol, (2) galangin, (3) isorhamnetin, (4) morin, (5) quercetin, (6) kaempferol-3-glucoside, (7) kaempferol-3-rutioside and (8) quercetin-3-rutinoside at 50 μM for 70 min. Green fluorescence was measured under a confocal microscope. (Mukai, et al. 2009b)

Subclass	Name of Flavonoid
Flavonol	kaempferogalangin
	isorhamnetin
	morin
	quercetin
	kaempefrol-3-glucoside
	kaempferol-3-rhamnoside
	quercetin-3-hamnoside
Flavone	flavone
	apigenin
	luteolin
Flavanone	naringenin
	eriodictyol
Catechin	(-)-epigallocatechin
	(-)-epigallocatechin gallate

Table 1. Flavonoids examined in this study and their subclass categorization

The excitation spectra of selected flavonoids were measured using a fluorescence spectrophotometer (F2500, Hitachi, Tokyo, Japan) with an excitation wavelength range between 460 and 490 nm and fluorescence emission detected at a wavelength of 520 nm.

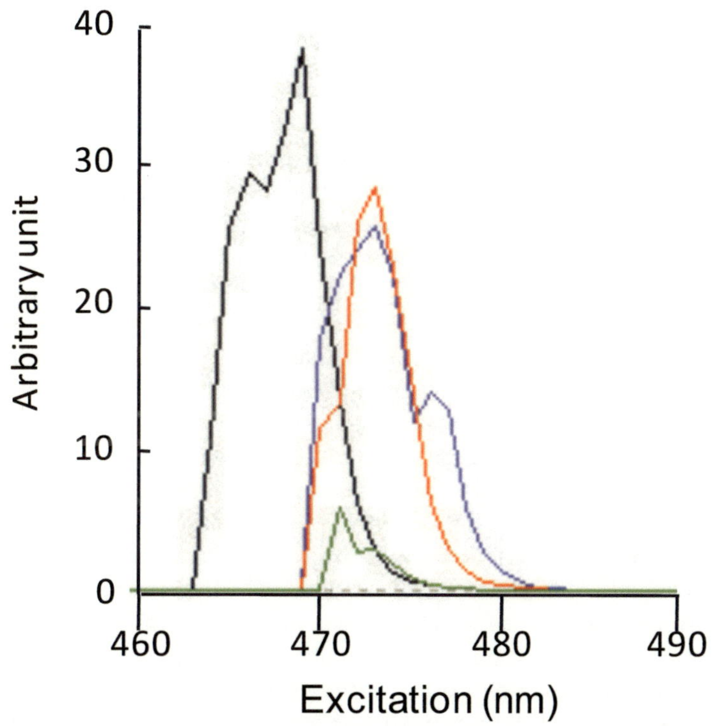

Fig. 6. Excitation spectrum of flavonoids. Excitation spectrum (ex. 460-490 nm) of kaempferol (black), quercetin (blue), quercetin-3- rutinoside (green), and naringenin (red) as measured using a fluorescence spectrophotometer. The fluorescence emission was detected at 520 nm.

These spectra indicate that kaempferol, quercetin, quercetin-3-rutinoside, and naringenin emits fluorescence with an excitation wavelength of 470-480 nm (Fig. 6), whereas apigenin, luteolin, and (-)-epigallocatechin do not (data not shown). These results indicate that only flavonols emit fluorescence and other flavonoid subclasses do not (with the exception of naringenin). Interestingly, not only do the flavonol aglycones, but also the glycosides, emit fluorescence.

3. Application of fluorescence microscopy to the estimation of flavonol cellular uptake

The viability of fluorescence microscopy as a means of estimating of the cellular uptake of flavonols, was tested. The detection limit and dose-dependency were determined and compared with a biochemical method using a radioactive compound. Fluorescent microscopic analysis was carried out using Hepa-1c1c7 cells treated with varying concentrations of kaempferol as a model compound (0, 10, 50 and 100 µM for 70 min).

The cells were treated with [^{3}H]-kaempferol at 1, 10 and 20 µM for 70 min. The cell lysate was transferred to vials with a scintillation cocktail, and the radioactivity incorporated into the cells was measured using a liquid scintillation counter. Data are represented as the mean ± SE (n=3). (Mukai, et al. 2009b)

Conditions for the fluorescent microscopic analysis are the same as those described in section 1. The uptake of radioactive flavonol by hepatocytes was tested as follows. Hepa-1c1c7 cells, seeded onto a 24-well culture plate at a density of 2 x 10^5 cells/well and incubated for 24 h, were treated with 1, 10 and 20 µM [^{3}H]-kaempferol (Moravek Biochemicals Inc.) in Krebs-Ringer HEPES buffer for 70 min. The cells were washed three times with ice-cold PBS and lysed by addition of 250 µL of 0.05 M NaOH and incubation for 4h. The lysate was transferred to a vial with a scintillation cocktail, and the radioactivity incorporated into the cells was measured using a liquid scintillation counter.

Monitoring the autofluorescence of the Hepa-1c1c7 cells after treatment with various concentrations of kaempferol using the confocal scanning microscope indicated that autofluorescence was too weak for imaging in cells treated with 10 µM kaempferol (Fig. 7) but green autofluorescence could be detected in cells treated with 50 µM or more. On the other hand, measurement of the incorporated radioactivity, reflecting kaempferol levels in the cells, increased in a dose-dependent manner and could be readily observed with cell treatment with only 10 µM kaempferol (Fig. 8). Under our experimental conditions, incorporation levels (ratio of the amount of uptake to the amount of treatment) were almost the same; about 7.1 to 7.6% of kaempferol was incorporated into Hepa-1c1c7 cells.

From these results, the sensitivity of the confocal microscopy method is lower than that of the biochemical method using radioactive compound. This may be because the former method only allows detection of incorporated flavonol aglycone, where the latter method allows detection of the aglycone as well as its metabolites. This indicates that the sensitivity of the confocal microscope method is likely to be dependent on the cell type analyzed, because the metabolism of flavonoids differs between different cell types (see section 3).

The time-dependence of kaempferol uptake was tested by monitoring the autofluorescence of cells treated with 50 µM kaempferol for incubation periods of 0, 1, 5, 10, 15, 30, 45, and 70 min, and the intensity of fluorescence was observed to increase in a time dependent manner (Fig. 9). After kaempferol treatment for 15 min, fluorescence was observed in the cells.

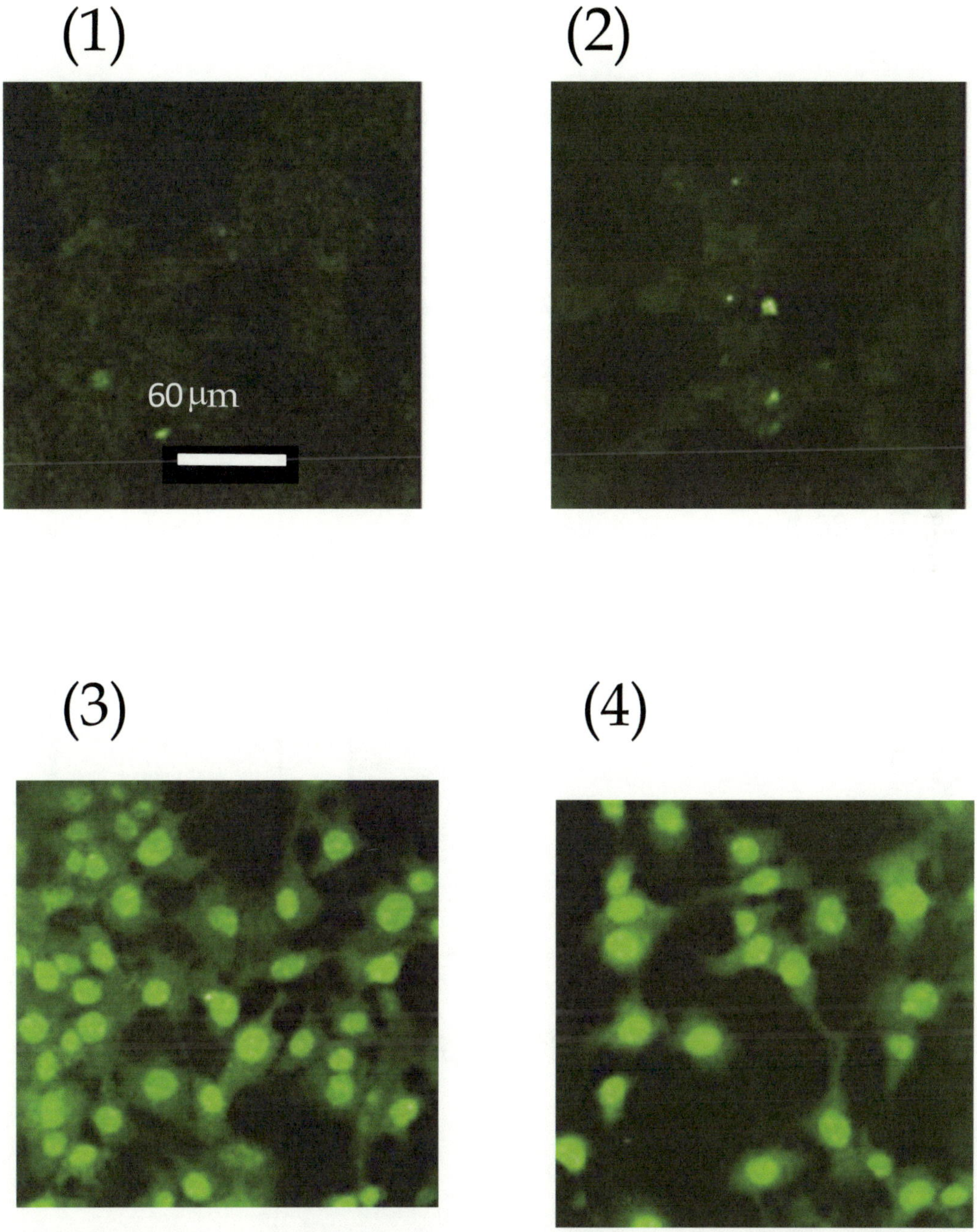

Fig. 7. Cellular uptake of kaempferol was visualized by fluorescence microscopy in Hepa-1c1c7 cells. The cells were seeded onto glass bottom dishes and treated with kaempferol at (1) 0 μM, (2) 10 μM, (3) 50 μM and (4) 100 μM for 70 min. Green fluorescence was observed under a confocal microscope (Ex. 488 nm- Em. 515-535 nm). (Mukai, et al. 2009b)

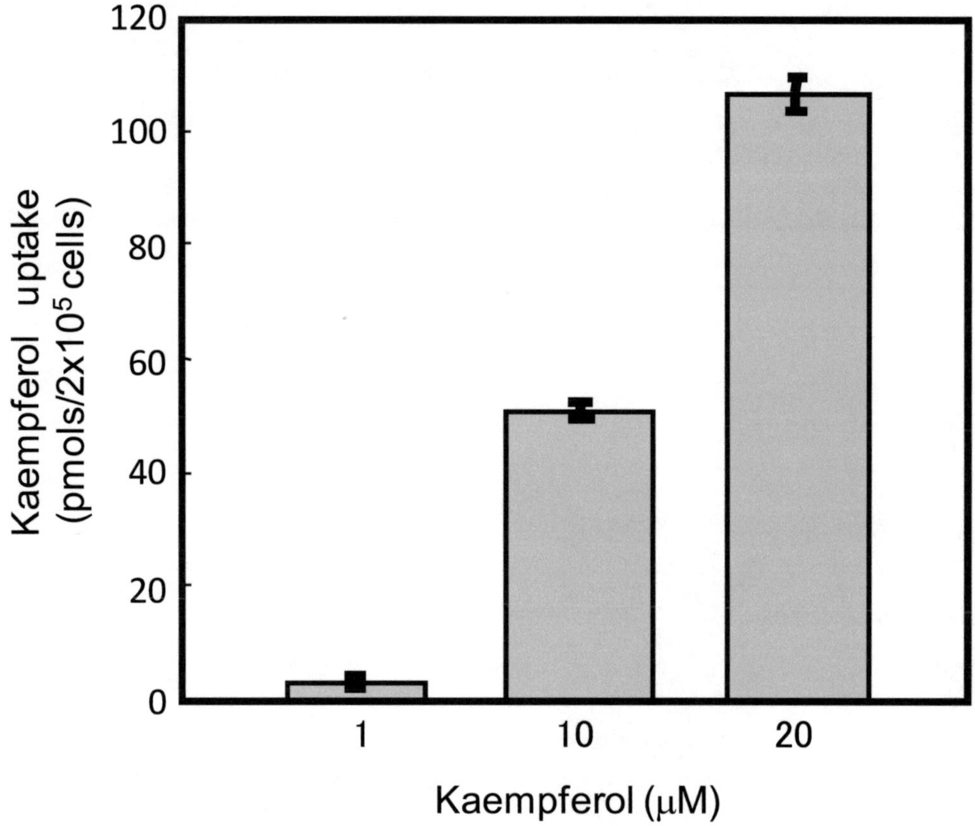

Fig. 8. The cells were treated with [^{3}H]-kaempferol at 1, 10 and 20 μM for 70 min. The cell lysate was transferred to vials with a scintillation cocktail, and the radioactivity incorporated into the cells was measured using a liquid scintillation counter. Data are represented as the mean ± SE (n=3). (Mukai, et al. 2009b)

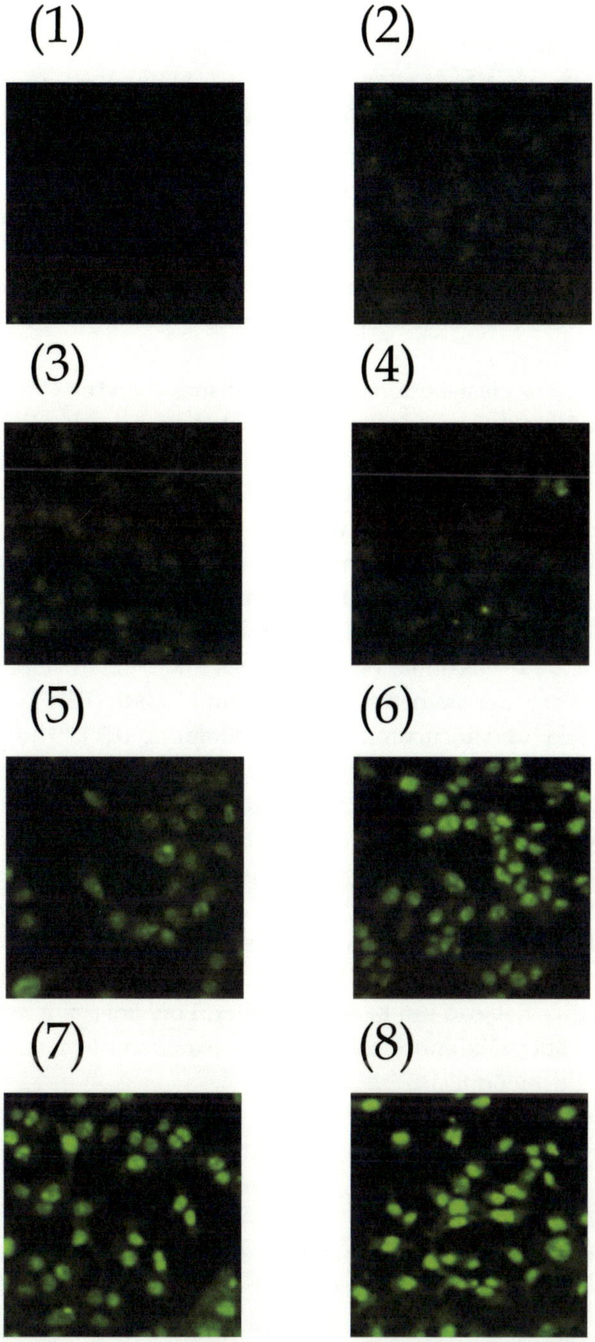

Fig. 9. Time-dependent cellular uptake of kaempferol in Hepa-1c1c7 cells.
The cells on glass bottom dishes were treated with kaempferol at 50 μM for (1) 0, (2) 1, (3) 5, (4) 10, (5) 15, (6) 30, (7) 45, and (8) 70 min. Green fluorescence was observed under a confocal microscope. (Mukai, et al. 2009b)

4. Determination of subcellular localization of flavonol in cultured cells

Culture conditions of Hepa-1c1c7 cells are described in sections 1 and 2. Human umbilical vein endothelial cells (HUVEC) were grown and maintained in a medium specially designed for endothelial cells obtained from Cell Applications Inc. (San Diego, CA). Neuro2A neuroblastoma cells were grown in Dulbecco's Modified Eagle Medium (DMEM) containing 10% BSA and penicillin-streptomycin at 100 U/ml and 100 µg/ml respectively, and after incubation for 24h, the cells were maintained in serum free DMEM containing penicillin-streptomycin. Caco-2 human colonic adenocarcinoma, were maintained in DMEM containing 15% BSA, 0.1% non-essential amino acids, 100 U/ml and 100 µg/ml streptomycin. Cells were maintained at 37 °C under a humidified atmosphere containing 5% CO_2.

All cells were seeded onto glass-bottomed culture dishes (MatTek Corp., Ashland, MA) and were grown until confluent. The cells were incubated with fresh culture medium for 24 h prior to treatment, and then treated with the test flavonols for 70 min. Kaempferol, galangin and quercetin were used as the model flavonols and were added as a solution at 100 µM in culture medium containing 0.1% dimethyl sulfoxide. The cells were washed with ice-cold PBS (pH 7.4) and then fixed with 4% paraformaldehyde and 0.2% picric acid in 100 mM sodium phosphate buffer (pH 7.2) overnight, at 4 °C and in the absence of light. To determine the subcellular localization of flavonols in intact mouse hepatoma Hepa-1c1c7 cells, HUVEC, mouse neuroblastoma Neuro2A cells and human colonic carcinoma Caco-2 cells, the nucleus was counter-stained with propidium iodide (PI, Fig. 10 and 11). The cells were washed with PBS, and incubated in PBS containing 0.3% tritonX-100 for 20 min at room temperature. The cells were incubated in 2X SSC (0.3 M NaCl, 0.03 M sodium citrate, pH 7.0) containing 100 µg/ml RNase for 30 min at 37 °C. The cells were rinsed 3 times in 2X SSC for 1 minute each time. The cells were incubated with 500 nM PI in 2X SSC for 30 min at room temperature and rinsed again as above. These cells were stored at 4 °C and used for measurement of autofluorescence within 24 h. Autofluorescence from the flavonol was observed using confocal laser scanning fluorescence microscopy with excitation by an argon laser at 488 nm and a 515–535 nm band pass filter. PI fluorescence was monitored at 543 nm (HeNe excitation) with a 560-615 nm band pass filter. Low energy laser excitation was used during observation, but maximum energy laser excitation was used for imaging the autofluorescence of the flavonols.

In hepa-1c1c7 cells and HUVEC (Figs. 10 and 11), the red fluorescence (PI) overlapped with the green fluorescence from the flavonols and yellow fluorescence was observed in merged images. (Mukai, et al. 2009b) Therefore, flavonols incorporated into Hepa-1c1c7 cells or HUVEC accumulated in the nucleus. In neuro2A (Fig. 11), the red fluorescence (PI) was separate from the green fluorescence of flavonol. The merged image therefore showed both green and red fluorescence, indicating that the flavonols incorporated into neuron 2A cells accumulated in the cytoplasm. In the case of Caco-2 cells (Fig. 11), the green fluorescence from flavonol completely covered the cells. In the merged image, orange fluorescence was observed, and flavonol fluorescence did not overlap with the PI. Thus, most of the flavonol was associated with the cellular membrane of Caco-2 cells, and did not penetrate into the cells.

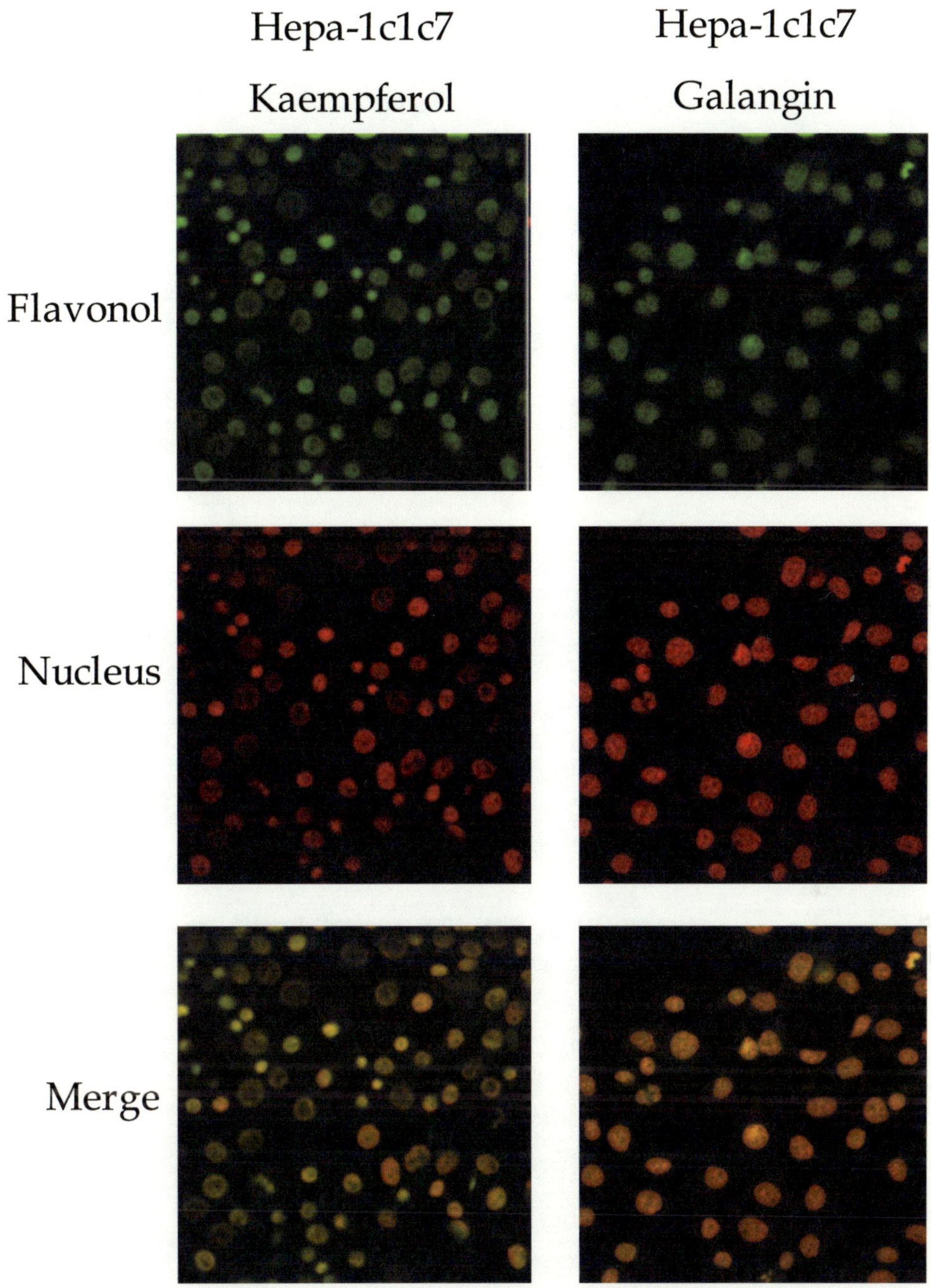

Fig. 10. Subcellular localization of kaempferol and galangin in Hepa-1c1c7 cells.
The cells were seeded onto glass bottom dishes, treated with kaempferol (left) or galangin (right) at 50 μM for 70min, and stained with PI. Green fluorescence (Ex. 488 nm- Em. 515-535 nm) from flavonol (top) and red fluorescence (Ex. 543 nm-Em. 560-615) from PI (middle) were monitored under a confocal microscope, and the images were merged (bottom). (Mukai, et al. 2009b)

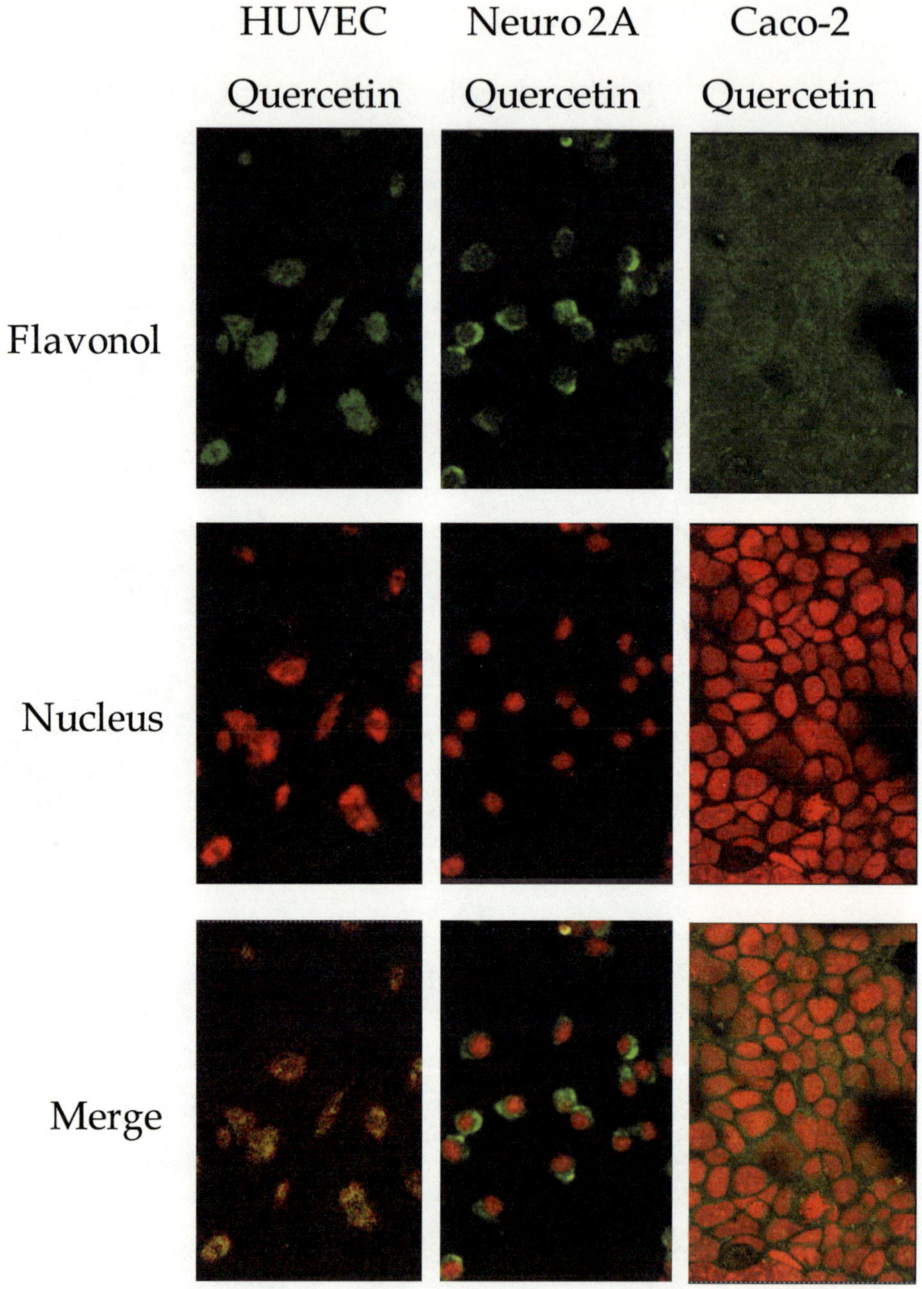

Fig. 11. Subcellular localization of quercetin in HUVEC, neuron 2A, and Caco-2 cells. HUVEC (left), neuron 2A (center), and Caco-2 cells (left) were seeded onto glass bottom dishes, treated with quercetin at 50 μM for 70min and stained with PI. Green fluorescence (Ex. 488 nm- Em. 515-535 nm) from flavonol (top) and red fluorescence (Ex. 543 nm-Em. 560-615) from PI (middle) were monitored under a confocal microscope, and the images were merged (bottom).

5. Discussion and conclusion

We describe here the use of laser scanning microscopy as an effective tool for the estimation of subcellular localization of flavonols in intact cultured cells. The results demonstrate that the subcellular localization of flavonol aglycones is different in different cell types: flavonol aglycones accumulate into the nucleus of Hepa-1c1c7 cells (Mukai, et al. 2009b) and HUVEC, into the cytoplasm of neuro 2A, and are associated with the cellular membrane of Caco-2 cells (Figs. 10 and 11). This method for subcellular localization is superior to HPLC and LC/MS methods, as it does not require pretreatments such as fractionation of the cells and extraction of the compound from the cells. Moreover, this method allows analysis of subcellular localization more rapidly than pretreatment methods. This method therefore allows estimation of flavonol aglycone localization in intact cells with non-fluorescent medium, however, it does require relatively high concentrations of the target compound to be used, and does not allow cellular concentrations to be quantified. When fluorescence-labeled compounds are used, it is necessary to consider whether the fluorescent label influences subcellular localization of the target compound. A further advantage of this method is therefore that changes in localization as a result of fluorescent labeling do not need to be considered, as the flavonol is used in its native form. As with all flavonol, observation should be carried out as quickly as possible and the sample should be kept in darkness, to avoid quenching of the autofluorescence, because flavonol itself easily undergoes decomposition.

A correlation between chemical structure and autofluorescence is observed, with the hydroxyl group at the C3-position in the flavonol skeleton being important to fluorescence, and no fluorescence observed in cells treated with other flavonoids. Even flavonols did not show autofluorescence after treatment of Hepa-1c1c7 cells if they were in the form of glycosides at the C3-position (Fig 4), although rutin (quercetin-3-rutinoside) showed fluorescence between ex. 470 - 480 nm without the presence of cultured cells (Fig.6). This difference may be due to a lack of uptake of the intact flavonol glycoside by hepatic cells. It has been reported that during the absorption process into the body, quercetin glycosides are hydrolyzed to the aglycone form by epithelial hydrolytic enzymes or enterobacteria in the intestinal tract (Murota and Terao 2003; Terao 2009). During absorption and metabolism, the flavonol aglycone is converted to a glucuronide or sulfate conjugate (Murota and Terao 2003). Therefore, it has been suggested that flavonol glycosides are unable to cross cellular and nuclear membranes without hydrolytic cleavage or glucuronic acid/sulfuric acid conjugation at the C3-position. The flavanone naringenin emitted fluorescence in the absence of cultured cells (Fig. 6), but not in the presence of cells (data not shown). It has been reported that the cellular uptake of flavanone naringenin is lower than that of flavonol (Tourniaire, et al. 2005), suggesting that the cellular uptake of naringenin is insufficient for observation by the fluorescent microscopy. To elucidate these points, further study is needed to clarify the structure-fluorescence relationship and absorption mechanism in cultured cells.

Cellular efflux is also an important event in flavonol metabolism, and flavonols are known to be excreted from cells through a cell membrane efflux transporter (Brand, et al. 2008; Ofer, et al. 2005). Treatment of cells with an inhibitor of the efflux pump such as P-glycoprotein increases the cellular concentration of flavonol and enhances its beneficial effects in cultured cells and experimental animals (Mukai, et al. 2009a; Wang, et al. 2005). It is also known that flavonols can have pharmacological effects in both the cytoplasm and nucleus of cultured cells. For example, flavonol affects the function of aryl hydrocarbon receptors in cytoplasm (Mukai, et al. 2010). In the nuclei of cancer cells, flavonols induce the oxidative cleavage of cellular DNA in the presence of copper ions and may therefore slow the progression of cancer growth (Hadi, et al. 2007). In this chapter, we have shown that subcellular localization of flavonols varies between cell lines (Fig. 10 and 11) although what factors mediate these differences in flavonol localization are not known. Further study is therefore needed to clarify these results and to investigate the relationship between the function and subcellular localization of flavonols in various cell lines. Flavonols have been detected in the liver, lung, white fat, muscle, brain, kidney, heart, and spleen of experimental animals by HPLC and LC/MS (de Boer, et al. 2005; Urpi-Sarda, et al. 2008; Wang, et al. 2003). It would therefore be of interest to investigate the subcellular localization of flavonols using fluorescence microscopy in other cell types in addition to those investigated in this chapter. It could be powerful tool for confirming the detailed subcellular localization by stains for cellular organelles and cytoskeletons such as cellular membrane, mitochondria, tubulins, and actins. In addition, cells have polarity for example enterocyte, which has apical side and basolateral one. The polarity of cell is involved in the intercellular and/or intracellular transport. To determine the relation ships between cell-polarity and flavonol-transport, three-dimensional observation using confocal fluorescent microscopy would be an excellent method for detecting localization.

This method is expected to be applicable for the study of subcellular localization for other fluorescent trace food components, drugs, and Chinese medicine components in various types of cells. It is hoped that this method will contribute to clarifying the relationship between subcellular localization of target chemicals and their beneficial and/or toxicological functions in various cell types in future.

6. References

Ahn, J.; Lee, H.; Kim, S.; Park, J.; & Ha, T. (2008). The anti-obesity effect of quercetin is mediated by the AMPK and MAPK signaling pathways. *Biochem. Biophys. Res. Commun.*, 373, 4, 545-549

Bieger, J.; Cermak, R.; Blank, R.; de Boer, V. C.; Hollman, P. C.; Kamphues, J. & Wolffram, S. (2008) Tissue distribution of quercetin in pigs after long-term dietary supplementation. *J. Nutr.*, 138, 8, 1417-1420

Boots, A. W.; Wilms, L. C.; Swennen, E. L.; Kleinjans, J. C.; Bast, A. & Haenen, G. R. (2008) In vitro and ex vivo anti-inflammatory activity of quercetin in healthy volunteers. *Nutrition*, 24, 7, 703-710

Brand, W.; van der Wel, P. A.; Rein, M. J.; Barron, D.; Williamson, G.; van Bladeren, P. J.; & Rietjens, I. M. (2008) Metabolism and transport of the citrus flavonoid hesperetin in Caco-2 cell monolayers. *Drug Metab. Dispos.*, 36, 9, 1794-1802

Chu, K. O.; Wang, C. C.; Chu, C. Y.; Choy, K. W.; Pang, C. P. & Rogers, M. S. (2007) Uptake and distribution of catechins in fetal organs following *in utero* exposure in rats. *Hum. Reprod.*, 22, 1, 280-287

de Boer, V. C.; Dihal, A. A.; van der Woude, H.; Arts, I. C.; Wolffram, S.; Alink, G. M.; Rietjens, I. M.; Keijer, J. & Hollman, P. C. (2005) Tissue distribution of quercetin in rats and pigs. *J. Nutr.*, 135, 7, 1718-1725

Gagne, S.; Saucier, C. & Geny, L. (2006) Composition and cellular localization of tannins in cabernet sauvignon skins during growth. *J. Agric. Food Chem.*, 54, 25, 9465-9471

Hadi, S. M.; Bhat, S. H.; Azmi, A. S.; Hanif, S.; Shamim, U. & Ullah, M. F. (2007) Oxidative breakage of cellular DNA by plant polyphenols: a putative mechanism for anticancer properties. *Semin. Cancer Biol.*, 17, 5, 370-376

Hirosawa, K. & Yamada, E. (1981) Localization of tritiated vitamin A in lymph nodes of the mouse: an autoradiographic study of vitamin A-storing cells. *Am. J. Anat.*, 161, 3, 299-310

Hsu, C. L. & Yen, G. C. (2006) Induction of cell apoptosis in 3T3-L1 pre-adipocytes by flavonoids is associated with their antioxidant activity. *Mol. Nutr. Food Res.*, 50, 11, 1072-1079

Kamada, C.; da Silva, E. L.; Ohnishi-Kameyama, M.; Moon, J. H. & Terao, J. (2005) Attenuation of lipid peroxidation and hyperlipidemia by quercetin glucoside in the aorta of high cholesterol-fed rabbit. *Free Radic. Res.*, 39, 2, 185-194

Kuzuhara, T.; Suganuma, M. & Fujiki, H. (2008) Green tea catechin as a chemical chaperone in cancer prevention. *Cancer Lett.*, 261, 1, 12-20

Mukai, R.; Satsu, H.; Shimizu, M. & Ashida, H. (2009a) Inhibition of P-glycoprotein enhances the suppressive effect of kaempferol on transformation of the aryl hydrocarbon receptor. *Biosci. Biotechnol. Biochem.*, 73. 7. 1635-1639

Mukai, R.; Shirai, Y.; Saito, N.; Yoshida, K. & Ashida, H. (2009b) Subcellular localization of flavonol aglycone in hepatocytes visualized by confocal laser scanning fluorescence microscope. *Cytotechnology*, 59, 3, 177-182

Mukai, R.; Shirai, Y.; Saito, N.; Fukuda, I.; Nishiumi, S.; Yoshida, K. & Ashida, H. (2010) Suppression mechanisms of flavonoids on aryl hydrocarbon receptor-mediated signal transduction. *Arch. Biochem. Biophys.*, 501, 1, 134-141

Mullen, W.; Edwards, C. A. & Crozier, A. (2006) Absorption, excretion and metabolite profiling of methyl-, glucuronyl-, glucosyl- and sulpho-conjugates of quercetin in human plasma and urine after ingestion of onions. *Br. J. Nutr.*, 96, 1, 107-116

Murakami, A.; Ashida, H. & Terao, J. (2008) Multitargeted cancer prevention by quercetin. *Cancer Lett.*, 269, 2, 315-325

Murota, K. & Terao, J. (2003) Antioxidative flavonoid quercetin: implication of its intestinal absorption and metabolism. *Arch. Biochem. Biophys.*, 417, 1, 12-17

Ofer, M.; Wolffram, S.; Koggel, A.; Spahn-Langguth, H. & Langguth, P. (2005) Modulation of drug transport by selected flavonoids: Involvement of P-gp and OCT? *Eur. J. Pharm. Sci.*, 25, 2-3, 263-271

Sakakibara, H.; Honda, Y.; Nakagawa, S.; Ashida, H. & Kanazawa, K. (2003) Simultaneous determination of all polyphenols in vegetables, fruits, and teas. *J. Agric. Food Chem.*, 51, 3, 571-581

Shirai, M.; Kawai, Y.; Yamanishi, R.; Kinoshita, T.; Chuman, H. & Terao, J. (2006) Effect of a conjugated quercetin metabolite, quercetin 3-glucuronide, on lipid hydroperoxide-dependent formation of reactive oxygen species in differentiated PC-12 cells. *Free Radic. Res.*, 40, 10, 1047-1053

Terao, J. (2009) Plant Phenolics and Human Health, In: Biochemistry, Nutrition, and Pharmacology, Fraga, C. G.,Ed., 185-196, John Wiley & Sons, 978-0-470-28721-7, Hoboken

Tourniaire ,F.; Hassan, M.; Andre, M.; Ghiringhelli, O.; Alquier, C. & Amiot, M. J. (2005) Molecular mechanisms of the naringin low uptake by intestinal Caco-2 cells. *Mol. Nutr. Food Res.*, 49.,10, 957-962

Urpi-Sarda, M.; Morand, C.; Besson, C.; Kraft, G.; Viala, D.; Scalbert, A.; Besle, J. M. & Manach, C. (2008) Tissue distribution of isoflavones in ewes after consumption of red clover silage. *Arch. Biochem. Biophys.*, 476, 2, 205-210

Walgren, R. A.; Karnaky, K. J. Jr; Lindenmayer, G. E. & Walle, T. (2000a) Efflux of dietary flavonoid quercetin 4'-β-glucoside across human intestinal Caco-2 cell monolayers by apical multidrug resistance-associated protein-2. *J. Pharmacol. Exp. Ther.*, 294, 3, 830-836

Walgren, R. A.; Lin, J. T.; Kinne, R. K. & Walle, T. (2000b) Cellular uptake of dietary flavonoid quercetin 4'-β-glucoside by sodium-dependent glucose transporter SGLT1. *J. Pharmacol. Exp. Ther.*, 294, 3, 837-843

Walgren, R. A.; Walle, U. K. & Walle, T. (1998) Transport of quercetin and its glucosides across human intestinal epithelial Caco-2 cells. *Biochem. Pharmacol.*, 55, 10, 1721-1727

Wang, F. M.; Yao, T. W. & Zeng, S. (2003) Determination of quercetin and kaempferol in human urine after orally administrated tablet of ginkgo biloba extract by HPLC. *J. Pharm. Biomed. Anal.*, 33, 2, 317-321

Wang, Y.; Cao, J. & Zeng, S. (2005) Involvement of P-glycoprotein in regulating cellular levels of Ginkgo flavonols: quercetin, kaempferol, and isorhamnetin. *J. Pharm. Pharmacol.*, 57, 6, 751-758

Wright, B.; Gibson, T.; Spencer, J.; Lovegrove, J. A. & Gibbins, J. M. (2010) Platelet-mediated metabolism of the common dietary flavonoid, quercetin. *PLoS One*, 5, 3, e9673

Yanez, J. A.; Remsberg, C. M.; Miranda, N. D.; Vega-Villa, K. R.; Andrews, P. K. & Davies, N. M. (2008) Pharmacokinetics of selected chiral flavonoids: hesperetin, naringenin and eriodictyol in rats and their content in fruit juices. *Biopharm. Drug Dispos.*, 29, 2, 63-82

13

Second Harmonic Generation Signal from Biological Materials Using Multi-Functional Two-Photon Laser Scanning Microscopy

Ali Hussain Reshak

[1]Institute of Physical Biology-South Bohemia University, Nove Hrady 37333

[2]School of Microelectronic Engineering, University Malaysia Perlis (UniMAP), Block A, Kompleks Pusat Pengajian, 02600 Arau Jejawi, Perlis

[1]Czech Republic

[2]Malaysia

1. Introduction

The nonlinear optical effect, such as two-photon (Denk et al., 1990) and three-photon (Maiti et al., 1997; Schrader et al., 1997; Wokosin et al., 1996) fluorescence significantly improved the depth resolution and reduced the background noise. The nonlinear optical techniques have been used to develop a new generation of optical microscopes with novel capabilities. These new capabilities include the ability of using near-infrared light to induce absorption, and hence fluorescence from fluorophores that are absorbed in the ultraviolet region. Other capabilities of nonlinear microscopes include: improved spatial and temporal resolution without the use of pinholes or slits for spatial filtering, improved signal strength, deeper penetration into thick, highly scattering tissue and confinement of photo-bleaching to the focal volume (Denk et al. 1990). The invention of nonlinear laser microscopy opened new opportunity to noninvasively examine the structure and function of living cells or tissues (Denk et al., 1990). Among different multiphoton implementations (Zipfel et al., 2003; Zumbusch et al., 1999), second harmonic generation imaging (Roth and Freund 1980; Freund et al., 1986; Campagnola et al., 2001; Yeh et al., 2002; Campagnola and Lowe, 2003; Cox et al., 2003) is particularly suitable for investigating non-centrosymmetric structures. Second harmonic generation (SHG) is a nonlinear optical process that occurs only at the focal point of a laser beam (Shen, 1989). SHG is not much younger than laser, but the application of such a weak process to the imaging of cellular structures and functions is quite new and notable (Campagnola and Loew., 2003). Advancement in mode-locked laser (instead of a continuous wave, mode-locked laser emits short pulses in the range of ns to fs) makes SHG imaging of cells possible since one do not need such a high intensity which in fact cooks the cells right a way. Using chiral chromophores, chiral SHG imaging can be applied to otherwise impossible symmetric structures (Yan et al., 2006). Second harmonic imaging microscopy (SHIM) is based on nonlinear optical effect known as second harmonic generation. A laser scanning microscope using second harmonic generation as a probe is shown to produce high-resolution images of transparent biological specimens. SHIM has been established as a viable microscope imaging contrast mechanism for visualization of cell and tissue structure and function. A second harmonic microscope obtains contrasts from variations in a specimen's ability to generate second harmonic light from the incident light, whereas a conventional optical microscope obtains its contrast by detecting variations in optical density, path length, or refractive index of the specimen. SHG requires intense laser light passing through a material with a non-centrosymmetric molecular structure. Second harmonic light emerging from an SHG material is exactly half the wavelength (frequency doubled) of the light entering the material. The two-photon-excited fluorescence (TPEF) is also a two photon process. The TPEF loses some energy during the relaxation of the excited state, while SHG is energy conserving.

In the two photon absorption (Figure 1a) a fluorophore is simultaneously excited from the ground state through a virtual state to either the first or second excited state. In case if the fluorophore is excited to the second excited state a rapid nonradiative decay to the first excited state generally results in the same emission independent of excitation mode. The excitation of the fluorescent molecules occurs because of the nonlinear response of the molecules to the electric filed. It is common to assume that the effective excitation profile is simply given by the square of the (Gaussian-Lorentzian) laser beam focused by the objective. Two-photo excitation fluorescence is a four level resonant process involving effectively simultaneous absorption of two low energy photons via a virtual state to cause electronic excitation to a real excited molecular vibronic state followed by fluorescent emission. This emission will have an angular distribution determined by the molecular configuration and a fluorescence decay time determined by the electronic upper state lifetime (Figure 1a). The emitted photon has a wavelength slightly greater than the half wavelength of the excited source. For centrosymmetric or randomly-oriented fluorophores, the emission is isotropic. The fluorescence decay time is in the order of picoseconds or longer. Hence, fluorescence is incoherent.

Let us start the discussions with an overview of the nonlinear optical process from the point of view of physics, in general the nonlinear polarization can be written as;

$$P = \chi^{(1)}E^1 + \chi^{(2)}E^2 + \chi^{(3)}E^3 + \cdots \tag{1}$$

where P is the induced polarization. $\chi^{(n)}$ is the nth order nonlinear susceptibility, and E is the electric field vector. The first term describes the normal absorption and reflection of the light, the second term is the second harmonic generation (SHG) and the third one is third harmonic generation (THG). The SHG is of primarily electronic origin and is a non-resonant process. In the SHG two photons of the same wavelength coalesce to a virtual state within the specimen to form a single photon of energy exactly twice that of the incident photons (Figure 1b). The SH photon is generated almost instantaneously (within a few fs) so that the SH signal is coherent.

It is well known that the SHG does not arise from an absorptive process; the intense laser field induces a nonlinear polarization in a molecule(s) resulting in the production of a coherent wave which is twice times the incident frequency. The magnitude of the SHG wave can be enhanced when the energy of the SH signal overlaps with an electronic absorption band (Heinz et al., 1982). Let us go back to eq. 1, we will find that the SHG can result from both non-centrosymmetric environment and electric quadrupole interaction from samples with a large change in optical dielectric constant between the interfacial regions. This interaction will give rise to the SHG signal from centrosymmetric regions, but this

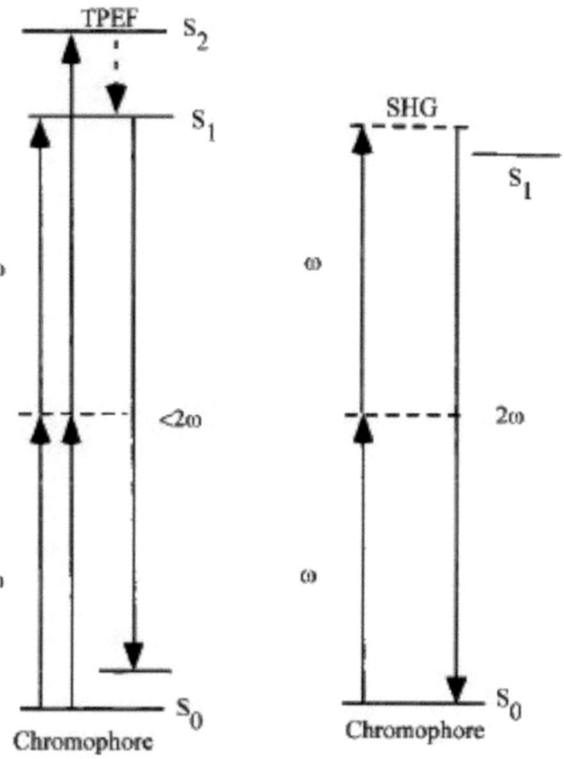

(a) (b)

Fig. 1. **(a)** The photo-physical pathways for two-photon excitation fluorescence. **(b)** Resonance enhanced SHG.

contribution is very weak in comparison with the interfacial components. So the SHG will form the basis of a high resolution nonlinear optical imaging that possesses all of the benefits of multiphoton excited fluorescence microscopy. Resulting in reduction of the photo-bleaching and photo-toxicity, moreover using mode-locked laser infrared wavelength also will reduces the photo-bleaching and photo-toxicity.

The multi-functional technique enables us to see the true architecture of the specimen which doesn't produce auto-fluorescence. In order to acquire fluorescence imaging for these samples one needs to stain those with fluorescence dyes in order to acquire images representing the true architecture of the specimen. Sometimes the staining harm the living tissue (some of these dyes are toxic) so for such sensitive tissues one prefers to use the SH imaging. Moreover, it is well known that during an investigation of new samples and acquiring fluorescence images, looking in very details at the true architecture of the specimen is required to exposure it to the laser beam more than the well-known sample. The sample will undergo photo-bleaching and photo-damage, because there is a certain limit for the laser power and the time of exposure the sample to the laser beam. While the SH imaging process does not undergo photo-bleaching and photo-damage because of being a coherent process, which occurs within a few femtoseconds, and is known to leave no energy deposition to the interacted matters due to their virtual energy conservation characteristic, it is an advantage of the SH imaging when one needs to work with sensitive samples. Thereby one can investigate the true architecture of the sensitive samples and the new samples. Then one can move only the filter cube (Beamspliter cubes, see Fig. 2a) in order to acquire fluorescence images after well investigation of the true architecture of the sample. The other advantage of SHG is the ability of imaging highly ordered structural proteins without any exogenous labels (Campagnola et al. 2001). The SHG is an intrinsic process, imaging of intrinsic compounds avoids the complications of slicing and labeling, and samples can be investigated under physiological conditions.

Advances in the developments of second harmonic generation microscopy have provided researchers with novel means by which non-invasive visualization of non-biological and biological specimens can be achieved with high penetration and high spatial resolution. The emitted SHG photon energy is the same as the total absorbed excitation photon energy which provides the optical non-invasive nature desirable for microscopy applications (Gao et al. 2006). The inhomogeneity inherent to most biological specimen, and in particular, the internal structure of various cells, leads to generate high quality SHG images without any preconditioning such as labeling or staining that might induce undesirable effects in the living cell (Reshak, 2009). Historically, resolution in fluorescence optical microscopy has been limited by the Rayleigh criterion. The Rayleigh criterion states that two images are just resolved when the principal maximum (of the Fraunhofer diffraction pattern) coincides with the first minimum of the other (Born and Wolf, 1980). Techniques with better resolution than the Rayleigh criterion have recently been established, among which is harmonic excitation light microscopy (Frohn et al., 2000).

Second harmonic generation microscopy is a newly emerging microscopic method. Both backward and forward generated second harmonic imaging can reveal non-centrosymmetric and inhomogeneous structure of the object. In SHG microscopy, second harmonic light is generated at the focal point of a tightly focused short pulse laser beam. If the medium at the focal point is centro-symmetric and homogenous, the second harmonic waves generated before and after the focal point interfere destructively, resulting in zero net SHG (Boyd, 1992). However, when there are inhomogeneities near the focal point, such as an interface between two media, or if the material is non-centrosymmetrical, the symmetry along the optical axis breaks and a measurable amount of second harmonic is generated (Reshak et al, 2007; Lukins et al., 2003)). Due to its nonlinear nature, the second harmonic light is generated only in close proximity to the focal point (Yelin and Silberberg, 1999). This, in turn, leads to high lateral resolution and inherent optical sectioning of the sample facilitating the construction of three dimensional images (Mason, 1999; Helmchen and Denk, 2005).

The use of infrared light to excite fluorophores in light-scattering tissue has added benefits (Helmchen and Denk, 2005). Longer wavelengths are scattered to a lesser degree than shorter ones, which helps to increase penetration depth and image quality. In addition, the non-linear nature of the process is less likely to cause damage outside of the focal volume.

In laser scanning two photons excited fluorescence microscopy, molecular excitation of fluorophores is achieved by the simultaneous absorption of two photons (within ~ 5 fs) (Helmchen and Denk, 2005), which also provide intrinsic optical sectioning. The excitation of fluorophores usually requiring single-photon absorption in the ultraviolet is now achieved with a stream of strongly focused subpicosecond pulses of red laser light. The fluorescence emission increases quadratically with the excitation intensity so that fluorescence and photo-bleaching are confined to the vicinity of the focal plane as expected for cooperative two-photon excitation (Mason, 1999). This technique also provides unprecedented capabilities for three-dimensional, spatially resolved photochemistry, particularly photolytic release of caged effector molecules.

Similar to TPEF, the SHG is produced in only a small focal volume, permitting high resolution 3D optical sectioning of thick tissues. In contrast to TPEF, the SHG is an intrinsic and a coherent process. Coherent constructive or destructive interference of SHG provides extra hints to the ultrastructure of the sample details and their organization. SHG is very efficiently generated in chloroplasts (Chu et al., 2001). Chloroplasts in celery showed a signal in the SHG image which did not co-localize with the autofluorescence of the chlorophyll. Crystalline starch in starch grains is typically organized with the crystallites in a radial fashion, yielding a characteristic cross image in polarized light (Clowes and Juniper, 1962). This in turn means that SHG image will be orientation dependent (Cox et al., 2004). The significant SHG seen in biological materials arises from low local symmetry and the large nonlinear coefficient typical for biological molecules and structures (Lukins et al., 2003; Mason, 1999; Helmchen and Denk, 2005).

The SHG signal unlike fluorescence is highly asymmetric due to the phase matching condition (Cheng et al., 2002; Moreaux et al., 2000; Williams et al., 2005; Mertz and Moreaux, 2001). The forward and backward SH signals depend on the axial size of the object. If the axial size of the object is of the order of the SH wavelength or greater then the object exhibits forward directed SHG, whereas if the object axial size is less than one tenth of the wavelength, it is estimated to produce nearly equal backward and forward signals (Moreaux et al., 2000; Williams et al., 2005; Mertz and Moreaux, 2001; Reshak, 2009).

SHG strongly depends on the state of the polarization of the laser light and the orientation of the dipole moment in the molecules that interact with the light. It is therefore advantageous to control the laser's state of polarization, to maximize SHG (Lukins et al., 2003; Reshak, 2009).

Although the SHG in biological tissue was first demonstrated two decades ago (Freund et al., 1986; Fine and Hansen 1971; Roth and Freund 1980), SHG has only recently been used for biological imaging applications (Georgiou et al., 2000; Campagnola et al., 1999; Gauderon et al., 1998; Moreaux et al., 2000) because the SHG and TPEF involve different contrast mechanisms, they can be used in tandem to provide complementary information regarding tissue structure and function (Zoumi et al., 2002). The SHG signals depend on the orientation, polarization, and local symmetry properties of chiral molecules, whereas the TPEF results from the nonlinear excitation of molecules fluorescence (Zoumi et al., 2002).

Researchers have been interested in optical methods for driving micro machines for a few years, with options often focusing on light-induced rotation of absorbing microscopic particles using elliptically polarized laser beams or beams with helical phase structures. These experiments, however, involved trapping absorbing particles and thus limited power to avoid heating effects. This power restriction, in turn, limited rotation rates to a few hertz.

Chloroplasts are the photosynthetic organelles of plant cells. They are commonly believed, and virtually all textbooks show them, to be ellipsoids of revolution flattened along the axis (Rezende-Pinto, 1972), or shaped like a dinner plate, flattened with curried edge (Sarafis, 1998). Most of the studies on their structure and shape have been made with light and electron microscopy (Wildman et al., 1980). Such studies, however, are largely limited to observation of two dimensional cross-sections because of the limited depth of field of the optical microscope and the need to use thin sections in electron microscopy. While studies by light and electron microscopy have been extensive, the preparation methods for these investigations involve treatment with fixatives and staining agents. Such treatment may alter the shape and the gross morphology of the organelles. Furthermore, a problem with assessing chloroplast shape is that the angle of presentation is usually a matter of chance. Tactile maneuvering of chloroplasts has not widely attempted. To overcome these problems the Multi-functional Two-Photon Laser Scanning Microscopy is used. The MF-2PLSM recently established (Reshak, 2009) by combing three platforms of laser scanning microscopy; the fluorescence microscopy, harmonic generation microscopy and polarizing microscopy. With MF-2PLSM one not need to stain the sample by utilizing the second harmonic imaging (SHI). Utilizing the non-invasive nature of the second harmonic imaging to record the rotation of the chloroplast with increasing/ decreasing the laser's power and follow the movements and the orientation of the chloroplast while illuminated by linearly polarized laser and increase the laser's power to see the react of the chloroplasts. The reaction of the chloroplasts comes as movement and rotation, trying to protect the photosynthetic system from being damaged while it is exposures to high intensity laser. The chloroplasts reorient themselves to make the orientation of the dipole moment in the molecules perpendicular to the laser's polarization direction. This rotation leads to very less amount of light will penetrate inside the chloroplasts. Then we try to increase the laser's power to monitor the behaviors of the chloroplast and as a consequence the chloroplasts move and rotate faster, which makes it possible to manufacture micromotors for micromechanical systems. The goal is to extend the concept of the optical torque on in vivo birefringent biological particle to a new optical method for driving micromachines.

2. Laser sources and imaging system

The experimental arrangement is shown in Fig. 2. We used an inverted i-mic 2 microscope (Till-Photonics, Grafelfing, Germany), (http://www.till-photonics.com/Products/imic.php) equipped with Ti:sapphire femtosecond laser with a tuning range of 690 nm to 820 nm. The laser is a Tsunami 3941-M3B pumped by a Millennia-V, 5W solid-state pump laser (Spectra-Physics). This laser was used to generate linearly polarized pulses at 810 nm, 20 mW, and 80 fs pulse width for fluorescence excitation and SHG. A beam expander was used to fill the back aperture of the objective. The excitation light was directed onto a pair of galvanometer XY scanners (Yanus-Till-Photonics) (http://www.till-photonics.com/Products/imic.php). The scanned excitation light was focused onto the specimen through the microscope objective (Olympus uplanApo/IR 60×/1.20 water immersion) in order to scan the specimen in the x-y direction at the focal plane. The stage of the microscope is driven by a computer controlled motor to take the specimen to different z positions following each x-y scan. The scanning mirrors are metal coated (silver) with a good thermal resistance (Diaspro, 2001). Further components from the setup in Fig. 2 are the dichroic mirrors (mirror 1: Omega 475DCLP for SHG (Fig. 3a), interference blue emission filter 1 (Fig. 3b), mirror 2: Q565LP for TPEF (Fig. 3c), emission filter 2: red glass 665 nm, PMTs: Hamamatsu R6357, objective 1: Olympus uplanApo/IR 60×/1.20 water immersion, objective 2: Zeiss 40×/1.2W korr).

The collecting efficiency of SHG signals is highly dependent on the numerical aperture (N.A.) of the objective (Cox et al., 2004; Han et al., 2005; Pantic-Tanner and Eden, 1999; Faludi-Daniel et al., 1978; Sarafis, 1998). Additional infrared beam block filters (HQ700SP-2p 58398) were placed in front of each PMT to ensure that illumination light was effectively suppressed and only TPEF and SHG signals were recorded. For SHG imaging, optical filtering is achieved with an interference filter centered on the expected SHG frequency. The configurations of the two PMTs were identical for both TPEF and SHG imaging. This setup

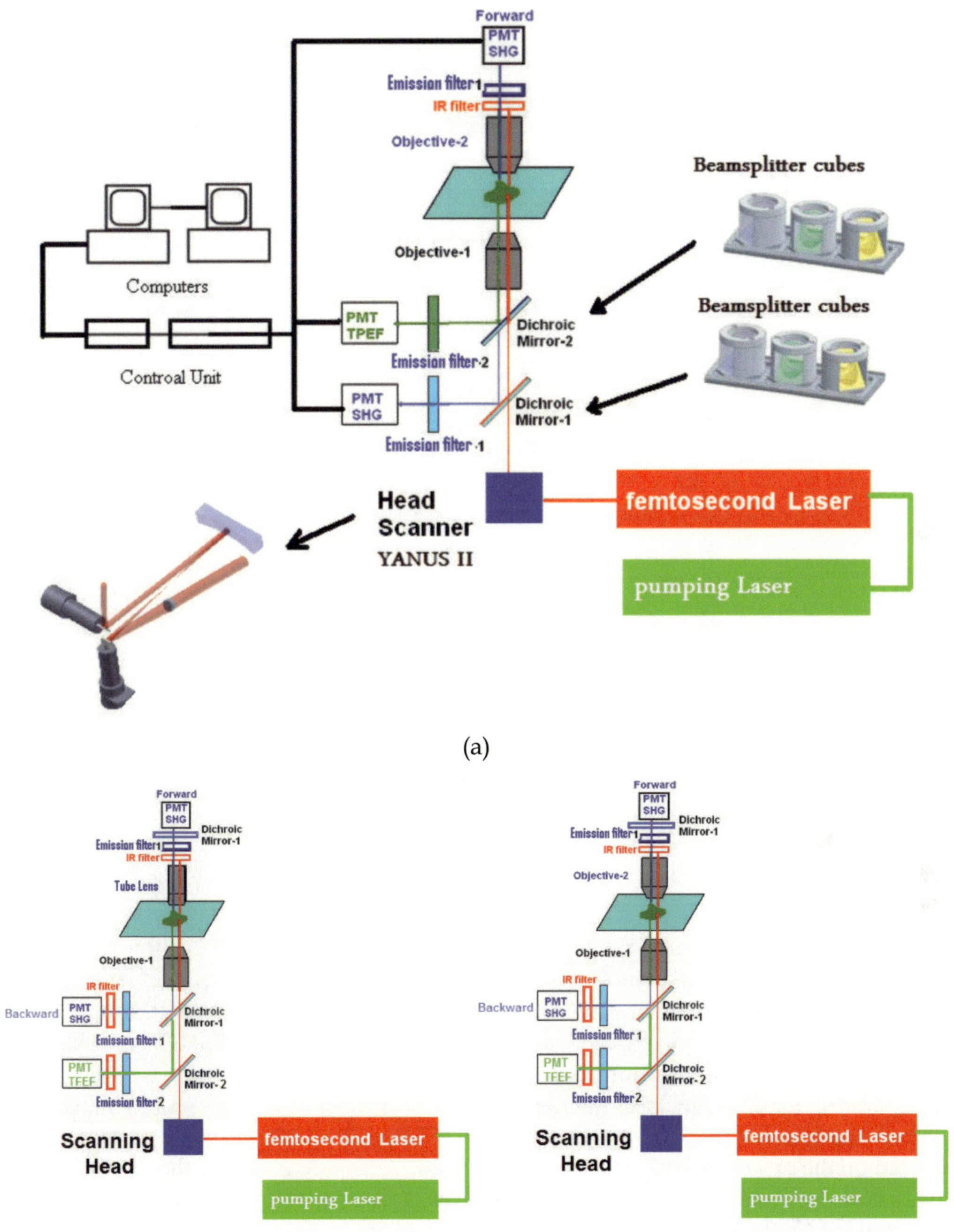

Fig. 2. (a) The experimental setup using two objectives for collecting the forward and the backward SHG signals. (b) Tube lens for collecting forward SH signals and objective for collecting the backward SH signals. (c). Objective for collecting forward SH signals and another objective for collecting the backward SH signals.

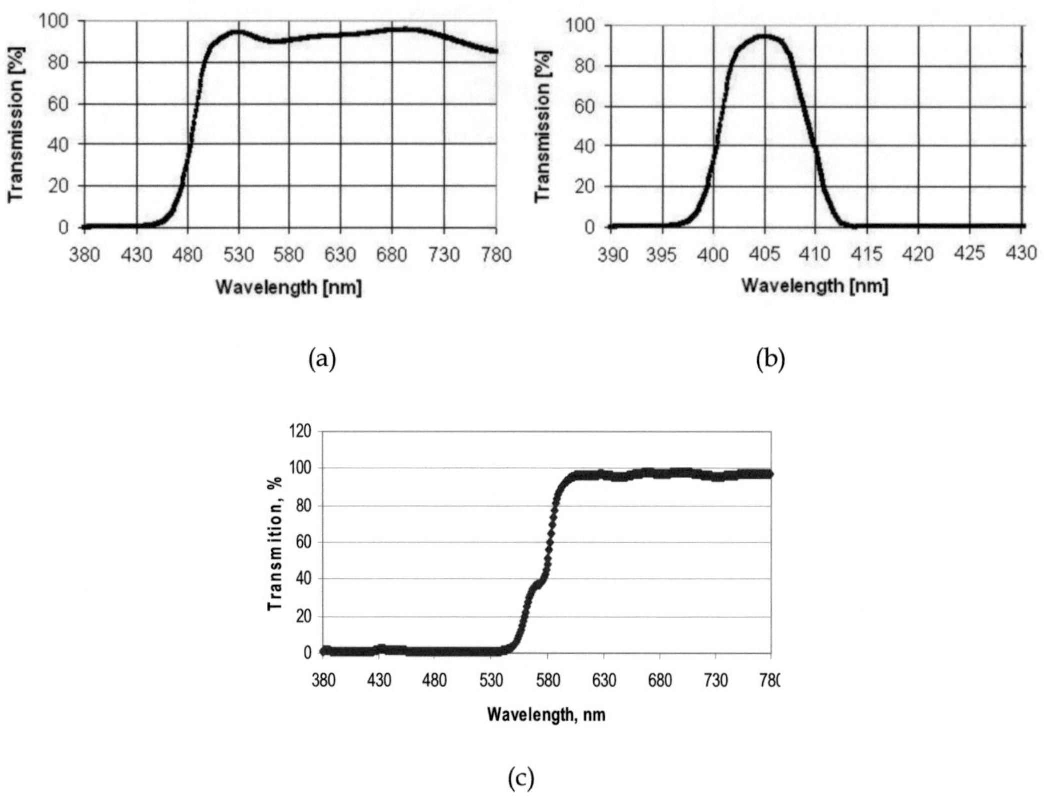

Fig. 3. The characterizations of the filters used for collecting the SHG and fluorescence; (a) Dichroic mirror Omega 475CLP for SHG. (b) Emission filter 1, for SHG. (c) Dichroic mirror Q565LP.

will enable the simultaneous measurement of SHG (Han et al., 2005) in the forward and backward directions. The signals from the photomultipliers are reconstructed by a computer into images. Images were obtained in stacks stepping along the z-axis with 0.5 μm steps. The microscopy is controlled via a standard high-end Pentium 4 PC and linked to the electronic control system via an ultrafast interface.

For separating the SH signals in the forward and backward directions and to suppress the fluorescence the dichroic mirrors Omega 475DCLP and interference blue emission filter centered on the expected SHG frequency were used. To ensure that only the SH signals were recorded and effectively suppressed any other signals an additional infrared beam block filters (HQ700SP-2p 58398) were placed in front of each PMT. In order to enhance the forward SH signals different objectives with different numerical aperture and immersion medium were used: Tube lens (f=16 mm, d=12 mm, PCX Lens fromEdmund), (http://www.edmundoptics.com/onlinecatalog/displayproduct.cfm?productID=2595), Olympus uplanFLN 10×/0.3, Zeiss 40×/1.2 korr, water immersion, and Olympus uplanApo/IR 60×/1.20 water immersion.

3. Material

In this work the moss *Plagiomnium affine* plant was used (see Fig. 4). This moss plant has a leaf of single cell layered. A leaf was detached from plants of the moss *Plagiomnium affine* by a watchmakers forceps from a darkened part of the moss plant. The leaf was mounted between two coverslips in water and the edges of the coverslip of smaller dimensions was sealed to the lower larger coverslip by means of nail varnish and allowing it to dry. The paired coverslips were placed on the stage of a Till-Photonics microscope and illuminated with a Titanium sapphire laser at 810 nm (linearly polarized laser). The objectives were aligned relative to one another and focused on the leaf. A set of images was captured over a time interval of half an hour where the chloroplasts changed their SHG due to their re-orientation.

Because of some uncertainty in the chloroplast imaging as shown by Chu et al. 2001, who looked at chloroplasts containing starch grains which are strong sources of second harmonic light we used a moss *Plagiomnium affine* with large chloroplasts devoid of starch by keeping the plants in the dark for approximately three weeks. The typical size of these chloroplasts vary between 5-10 μm. Short pulses should be used and average laser power should be kept low to prevent heating of the sample as well as unwanted one-photon absorption and to reduce the risk of highly nonlinear photodamage (Denk et al 1995). $\lambda/2$ plate was used to control or maximize the status of the laser's polarization at the sample.

Fig. 4. The moss *Plagiomnium affine* plant.

4. Results

Fig. 5, show time-series images using the experimental setup shown in Fig. 2. The backward SHG signal was collected using the objective Olympus uplanApo/IR 60×/1.2 water immersion. In this figure we labeled some chloroplasts as a reference for the discussion of the rotation and the movements of the chloroplast. The SH signal showed no signs of bleaching during acquisition of repeated images from a given area showing no damage to the structure. This is to be expected since SHG is a coherent process, unlike fluorescence, and no energy is lost (Mark et al., 2003). There is a signal from membrane-filled regions devoid of starch grains. The SHG intensity altered dramatically with time as seen in the time-series of images shown in Fig. 5 which is analyzed in Fig. 6.

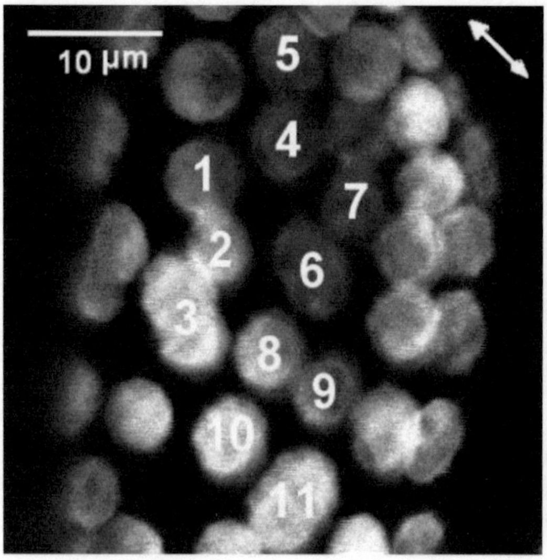

Fig.5a-Backward SHG 60x 1.2w

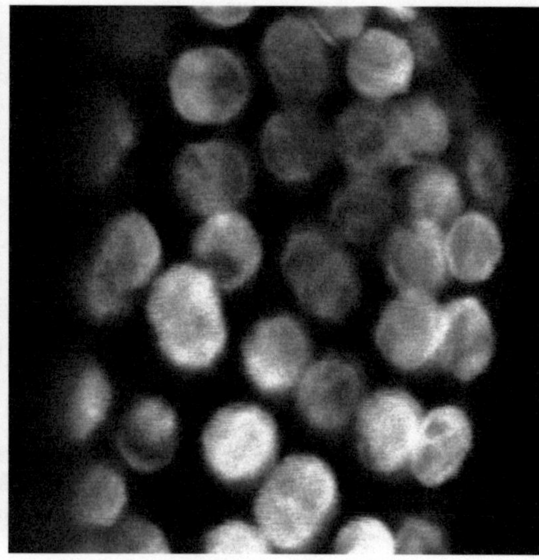

Fig5b-Backward SHG 60x 1.2w

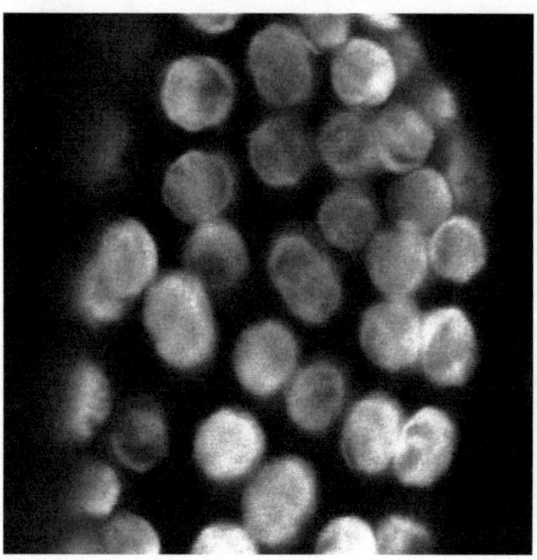

Fig.5c-Backward SHG 60x 1.2w

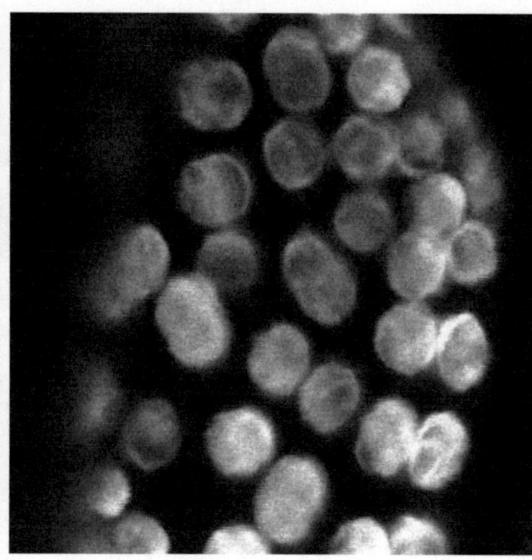

Fig.5d-Backward SHG 60x 1.2w

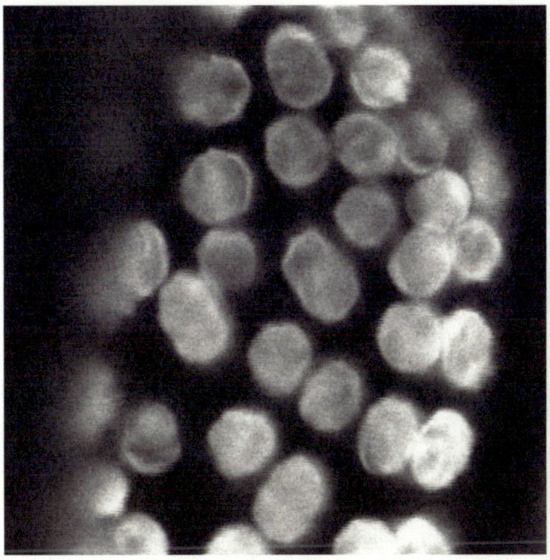

Fig. 5e. Backward SHG 60x 1.2w

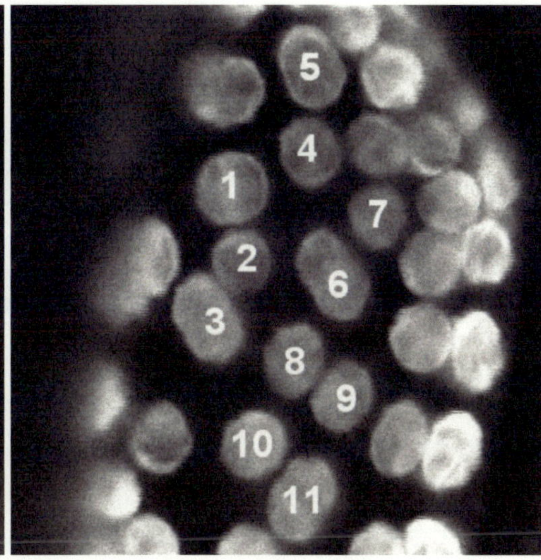

Fig. 5f. Backward SHG 60x 1.2w

Fig. 5. Time-series SHG images of *Plagiomnium affine* leaf showing the movement and the rotation of the chloroplast (a,b,c,d,e,f) Backward SHG signal using the objective Olympus uplanApo/IR60x/1.2 water immersion. The double arrow shows the direction of polarization of the 810 nm laser beam. Some chloroplast labeled by numbers 1,2,3,... as a reference for the discussion the rotation and the movements of the chloroplast in the time-series images.

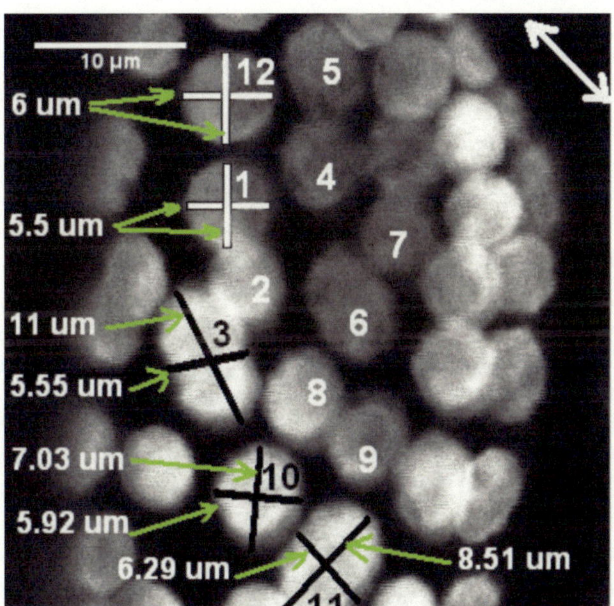

Fig. 5aa. (the first image of the time series SHG images-Fig.5a) shows the size of some selected chloroplast in the first stage of being illuminated by linearly polarized laser.

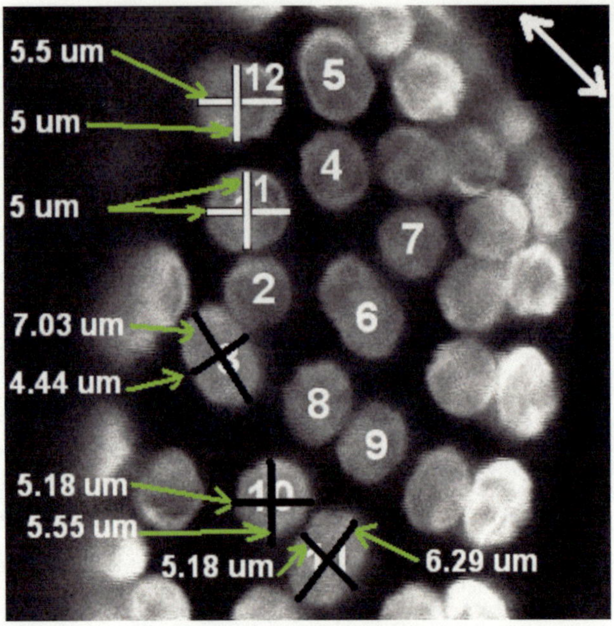

Fig. 5ff. (the last image of time series SHG images-Fig.5f) Shows the size of some selected chloroplasts in the final stage when it is illuminated by linearly polarized laser. It is clear that the sizes of these chloroplasts have changed due to the rotation of the chloroplasts when they were being illuminated by linearly polarized laser.

Fig. 6 shows the time-dependence of the average backward SHG signal in figure 5 of the inboxes each containing a chloroplast (see corresponding numbers in Fig. 5a). Using successive cross-correlations image drift was compensated for. We notice significant changes in the SHG intensity over time. The chloroplasts rotated resulting in changing the axial aspect of the chloroplast (the changing in the intensity signal). This rotation causes a reduction/ increase in the brightness (intensity) of the signals depending on the aspect presented to the laser.

In Figs. 5aa and 5ff, we measured the size of some selected chloroplasts at the first stage of being illuminated by linearly polarized laser and at the final stage when they rotate. It is clear that the apparent sizes of these chloroplasts have been changed due to their rotation when they illuminated by linearly polarized laser. Chloroplasts rotate as a result of the cells which contain them being illuminated by blue light (Haupt and Scheuerlein, 1990).

As axial thickness increased the SHG signal will decrease in intensity. The SHG stems from chloroplasts devoid of starch, thus clearly demonstrating the generation of SH from the non-centrosymmetric membranes opposed to each other in the grana. It is known that chloroplasts exhibit strong birefringence, with high local values, mostly originating from grana and inclued starch grains (Garab et al., 2005; Goedheer, 1955). The size and sign of the birefringence are such that the resulting anisotropic interaction with a linearly polarized laser beam significantly contributes to the torque orienting the chloroplast. Optical re-orientation by linearly polarized light occurs when $n_e > n_0$, and thus optical field tends to orient the director parallel to its electric vector (n_e and n_0 denote the extraordinary and ordinary refractive indices, respectively). Taking into account their asymmetric shape

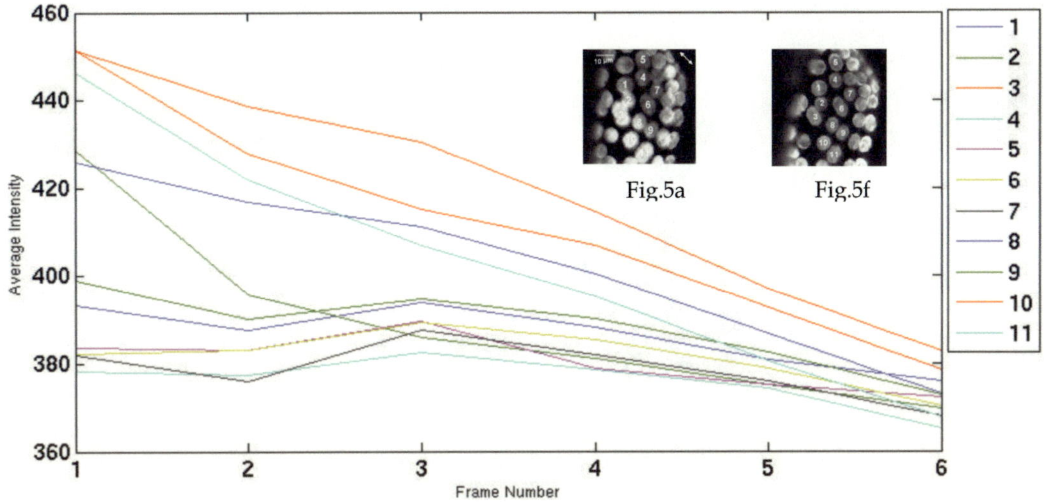

Fig. 6. Time dependence (corresponding to the backward generated SH shown in figure 5a-f) of the chloroplast-averaged backward SHG signal as numbered in Fig. 5a.

and the birefringence, biological entities may align in accordance with a preferential axis at the focus of the laser beam.

Figs.7 to 10 show the forward and the backward SHG structure of *Plagiomnium affine* leaf. The backward SHG signal was collected using the objective Olympus uplanApo/IR 60×/1.2 water immersion and the forward SHG signal was collected with a Zeiss 40×/1.2 water immersion objective, which show that the collecting efficiency of SHG signals is highly dependent on the numerical aperture (N.A.) of the objective (Cox et al., 2004; Han et al., 2005). In the right panel of each image the multicolor overlays of the respective forward (green) and backward (red) images were shown.

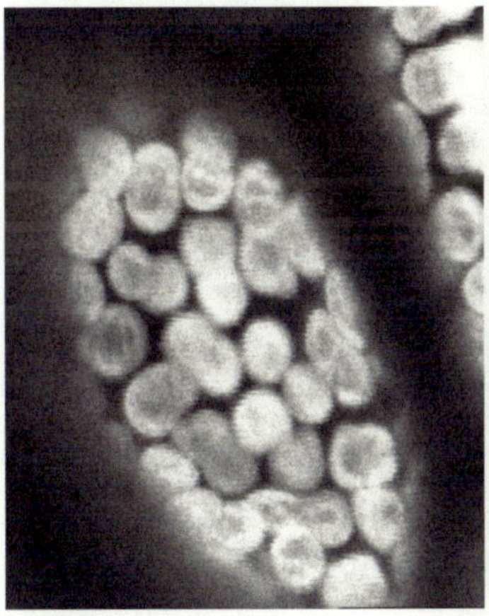

Fig. 7a.
Backward SHG 60x 1.2w

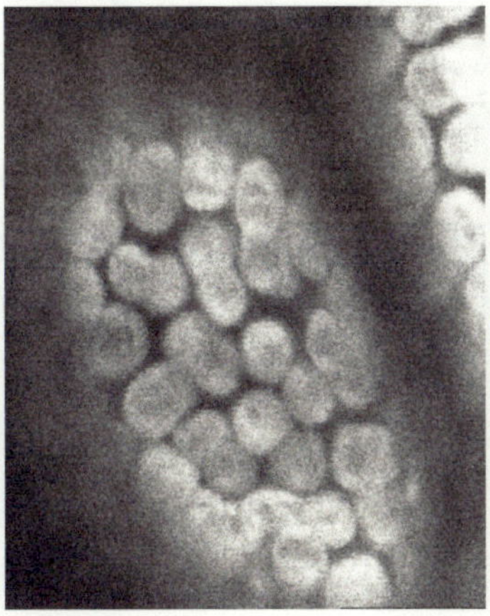

Fig. 7b.
Forward SHG 40x 1.2w

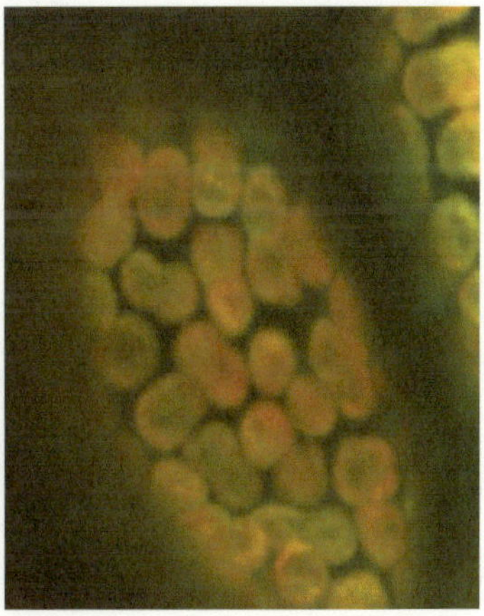

Fig. 7c.
Forward/Backward SHG

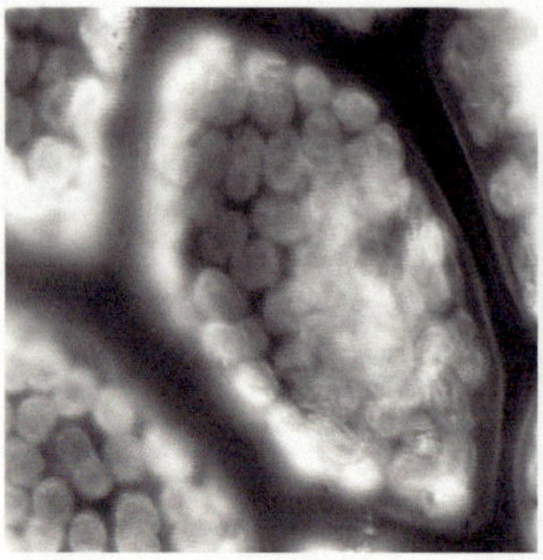

Fig. 8a.
Backward SHG 60x 1.2w

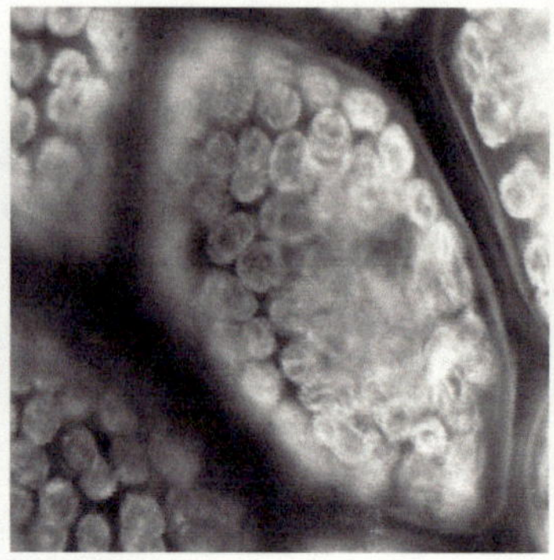

Fig. 8b.
Forward SHG 40x 1.2w

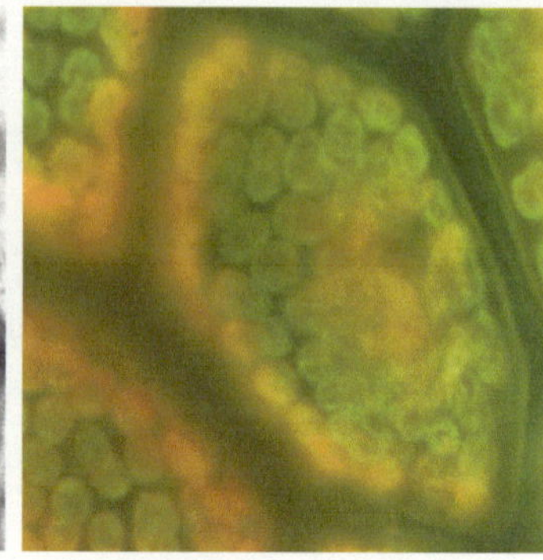

Fig. 8c.
Forward/Backward SHG

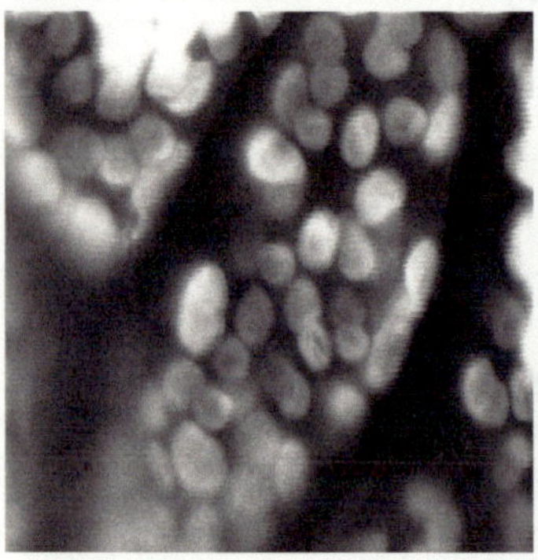

Fig. 9a.
Backward SHG 60x 1.2w

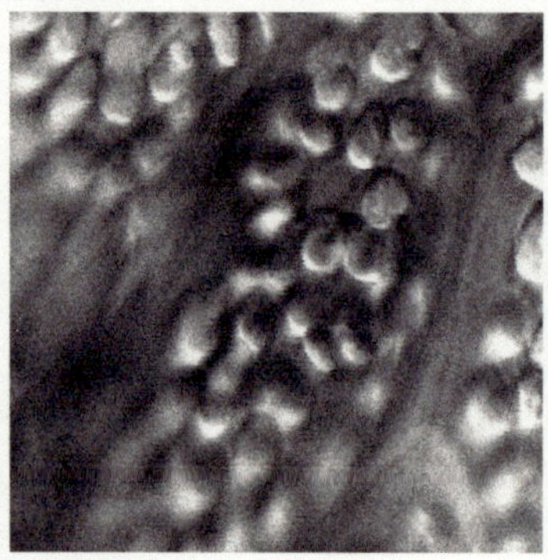

Fig. 9b.
Forward SHG 40x 1.2w

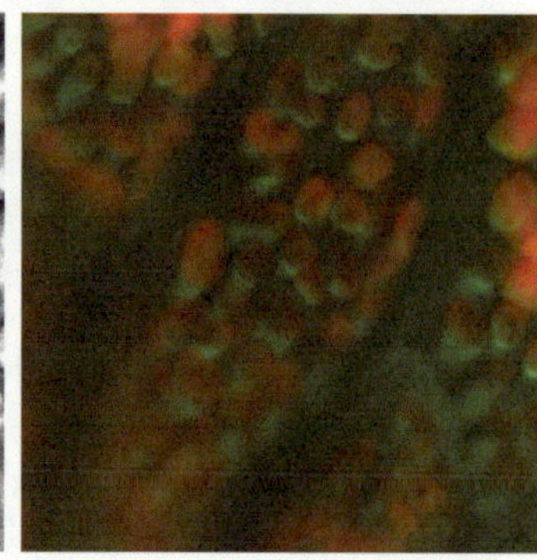

Fig. 9c.
Forward/Backward SHG

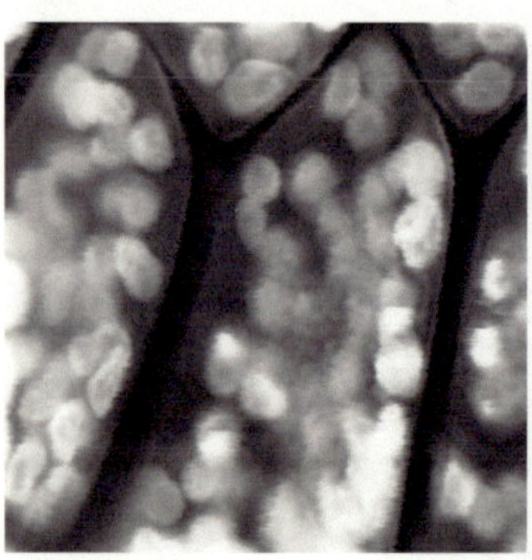

Fig. 10a.
Backward SHG 60x 1.2w

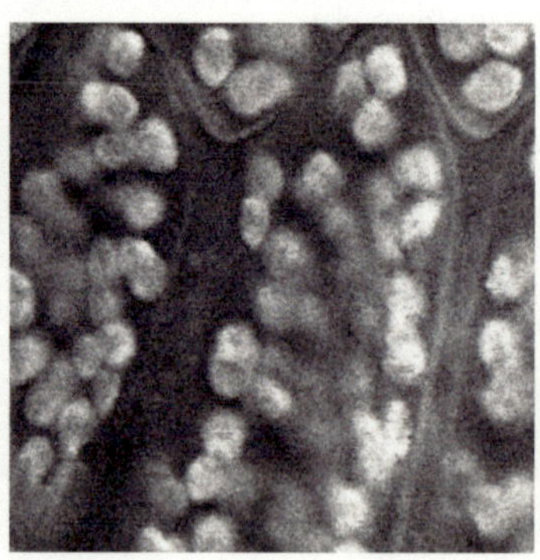

Fig. 10b.
Forward SHG 40x 1.2w

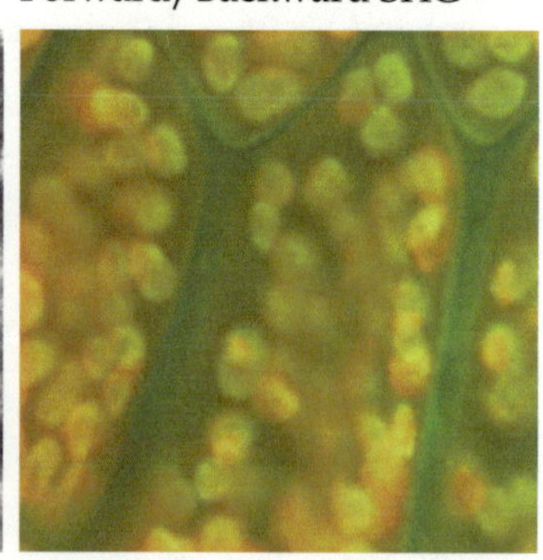

Fig. 10c.
Forward/Backward SHG

Figs. 7 to 10. shows SHG structure of *Plagiomnium affine* leaf. (a) Backward SHG signal using the objective Olympus uplanApo/IR 60×/1.2 water immersion. (b) Forward SHG signal was collected by a ZEIS 40×/1.2 water immersion objective. (c) Multicolor overlays of the respective forward (green) and backward (red) images.

Fig. 11 shows the fluorescence which comes from the chloroplast, i.e the same area which produces the SHG signals. It is well know that the SHG is very efficiently generated in chloroplasts (Chu et al., 2001). Chloroplast in celery showed a signal in the SHG image which did not co-localize with the autofluorescence of the chlorophyll and the signals were mostly due to starch. We have used the dichroic mirror Q565LP (Fig. 3c) which reflects the wavelength from 540 nm and above. We have to mention that here we did not use an emission filter for this filter set, so the PMT collected all the fluorescence starting from 540 nm and above.

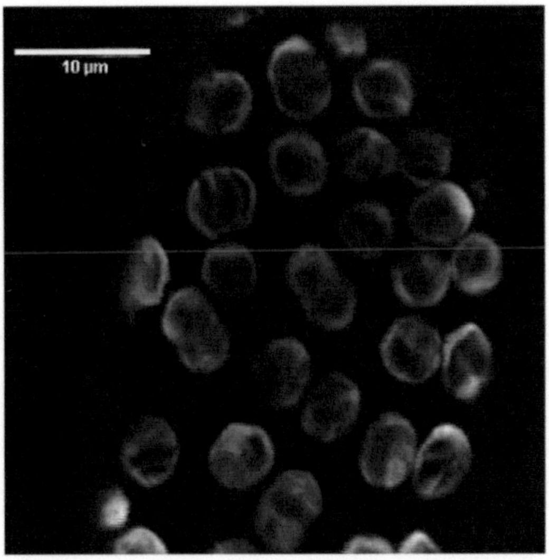

Fig. 11. Shows the fluorescence coming from the chloroplast, i.e the same area which produces the SHG signals.

5. Discussion

Chloroplasts in *Plagiomnium affine* leaves which were imaged came from a single cell layer lamina. The chloroplasts were disposed parallel to the upper and lower walls in all the cells of the lamina with the chloroplasts main axis oriented within the image plane. SHG signals were obtained in both directions with the forward signal predominating and these increased considerably when the chloroplasts rotated upon being illuminated by linearly polarized pulses of paired photons which are effectively equivalent to blue light at half the wavelength of the irradiating laser (Moreno et al., 2004). This chromatically equivalent blue range caused chloroplasts to rotate in a light causing orientation parallel to the side walls of the leaf cells. Starch grains are not involved thus eliminating a possible source of confusion. SHG indicates the non-centrosymmetric nature of the chloroplast membranes especially when aggregated into grana.

The chloroplast grana and intergranal membranes are distinguished by their birefringence (Pantic-Tanner and Eden, 1999). A low intrinsic birefringence is expected in intergranal membranes and a high birefringence where the membranes are stacked. The grana were expected to give a stronger SH signal than the intergranal regions with unstacked membranes. During the rotation of the chloroplast caused by linearly polarized high light intensity designed to present a smaller interface to the incoming light, the SHG increased considerably compared with the intergranal regions. Faludi-Daniel et al. 1978, refer to the structural distinctness of the chloroplast regions. Sarafis, 1998, reviewed and showed the range of chloroplast structure in higher plants in both deep shade plant leaves and dry resurrection plant leaves indicating a large variation in granal size which is expected to be shown by future second harmonic studies of these chloroplasts. We notice that the light absorption and the amount of fluorescence observed depend on the orientation of the chloroplast with respect to the polarization of the exciting photoactive light (Lukins et al., 2003).

6. Conclusions

We have demonstrated the simultaneously measured forward and backward SHG for the chloroplast of the *Plagiomnium affine* leaves. The SH signals are generated by a tightly focused short pulse laser beam. The SH signals are collected by two objectives to form forward and backward second harmonic digital images. These signals are dependent on the axial aspect of the chloroplast and that the back-scattered image was brighter than the forward scattered image. The image intensity also depended on the orientation of the chloroplast in relation to the illuminating polarization direction. A light-induced re-orientation in dependence of the illumination intensity could be observed. We found that the chloroplasts are moving and rotating when we expose the leaf to laser light for long time.

7. Enhancing the resolution of the forward Second Harmonic imaging of the multi-functional two-photon laser scanning microscope

Figs. 12a-e show the forward and backward SH imaging of *Plagiomnium affine* leaf using different objective and immersion imaging medium as shown in Figs. 2, 13 and 14. The backward SHG signal was collected using the objective Olympus uplanApo/IR 60×/1.2 water immersion and the forward SHG signal was collected by using different objectives; Tube lens(f=16 mm, d=12 mm, PCX Lens from Edmund), Olympus uplanFLN 10×/0.3, Zeiss 40×/1.2 korr, water immersion, and Olympus uplanApo/IR 60×/1.20 water immersion. The resolution of the forward SHG signals increased dramatically with raising the numerical aperture when the objective is designed to operate with an immersion medium such as water between the front lens and the specimen coverslip, which shows that the collecting efficiency of SHG signals is highly dependent on the numerical aperture of the objective as seen in the series of images shown in Figs. 12a-e. The SHG stems from chloroplast devoid of starch, such clearly demonstrating the generation of SH from the non-centrosymmetric membrane apposed to each other in the grana.

From Figs. 12a-e, we note that by using low N.A. tube lens with air gap, to collect the forward SH signal we gain an image with gray background (see Fig. 12b). This gray background results from the interface glass-air, also the rays, that come from the specimen will bend away from the front lens of the upper objective due to the big difference in the refractive indices of the glass (1.5) and air (1.0) (see Figs. 13a, 14a, and 15). A step toward, in order to modify the forward SH images the tube lens was replaced with the objective Olympus uplanFLN 10×/0.3 (low N.A. with air gap), a significant increase in the intensity of the image was noticed (see Fig. 12c) indicating that more rays (diffraction orders) were collected by the Olympus uplanFLN 10×/0.3 objective. But the gray backgrounds are still there because of glass-air interface. Dramatic enhancement was noticed when the Olympus uplanFLN 10×/0.3 objective was replaced by Zeiss 40×/1.2W korr objective (see Fig. 12d) indicating that by using high N.A. water immersion objective higher diffraction orders were collected (see Figs. 13b, 14b and 15). This enhancement attributed to the fact that the refractive index of the water (1.33) is closer to the refractive index of the glass (1.5) then the oblique rays will bend toward the front lens of the objective resulting to represent the true architecture of the specimen. In order to gain better quality images for the forward SH images the objective Zeiss 40×/1.2W korr was replaced by Olympus uplanApo/IR 60×/1.20 water immersion objective (see Figs. 12e, 14c and 15).

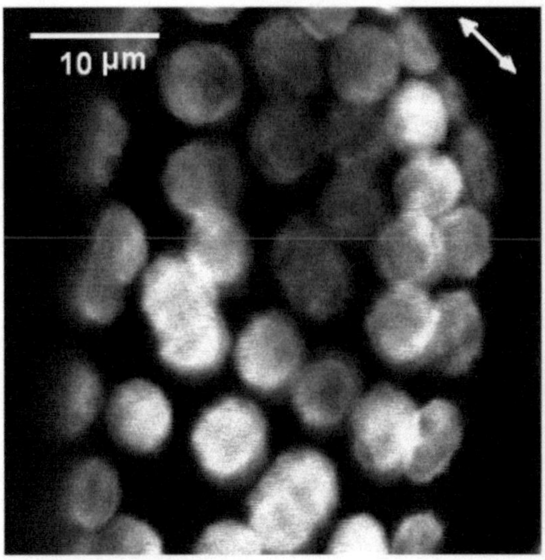

Fig. 12a.
Backward SHG with Olympus 60x 1.2w

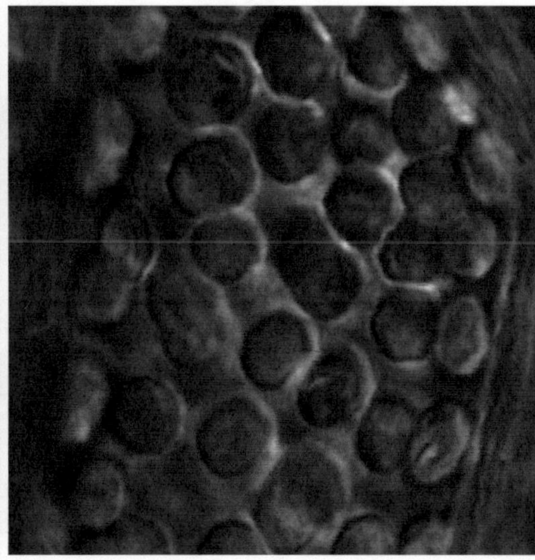

Fig. 12b.
Forward SHG with Tube lens

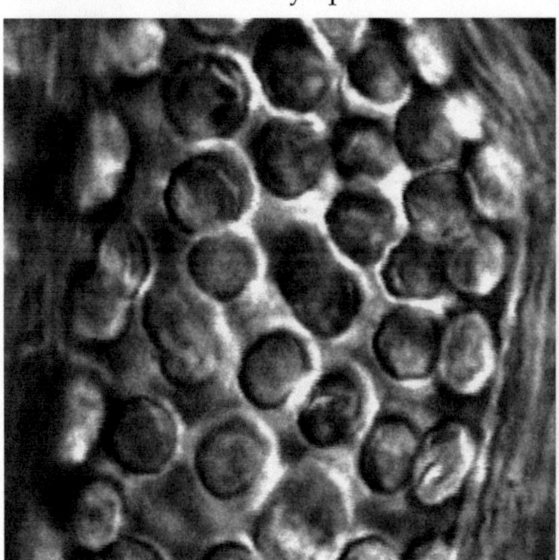

Fig. 12c.
Forward SHG with Olympus 10x

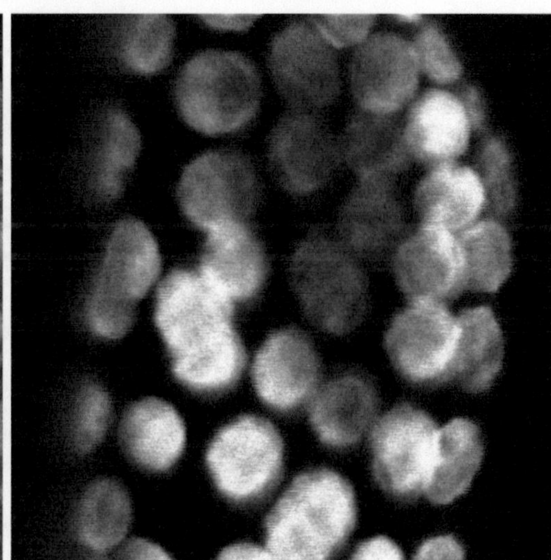

Fig. 12d.
Forward SHG with Zeiss40x 1.2w

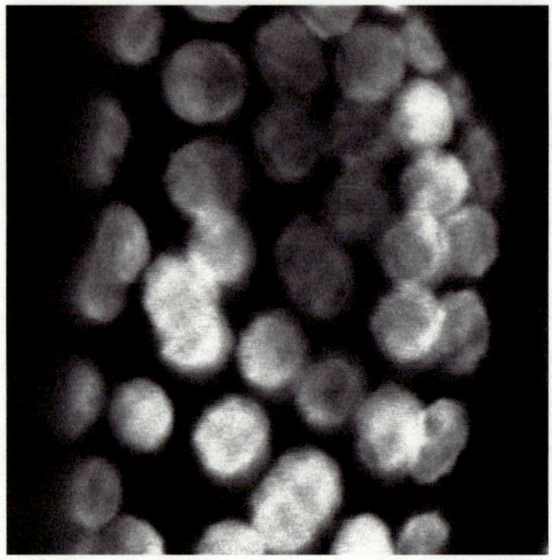

Fig. 12e. Forward SHG with Olympus 60x 1.2w

Fig. 12. SHG images of *Plagiomnium affine* leaf show the chloroplast (a) Backward SHG signal collected by using the objective Olympus uplanApo/IR 60×/1.2 water immersion. The double arrow shows the direction of polarization of the 810 nm laser beam. (b) Forward SHG signal using Tube lens. (c) Forward SHG signal using Olympus uplanFLN 10×/0.3 (d) Forward SHG signal using Zeiss 40×/1.2W korr objective (e). Forward SHG signal using the objective Olympus uplanApo/IR 60×/1.2 water immersion.

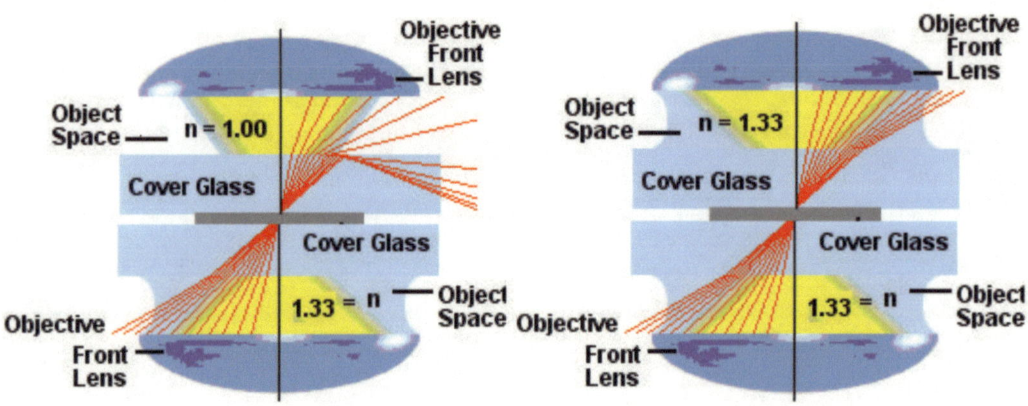

(a) (b)

Fig. 13. The forward and backward objective with different numerical aperture and immersion medium: (a). air for the upper objective (which is used to collect the forward SH signals) and water for the lower objective (which is used to collect the backward SH signals). (b). water for both upper and lower objectives.

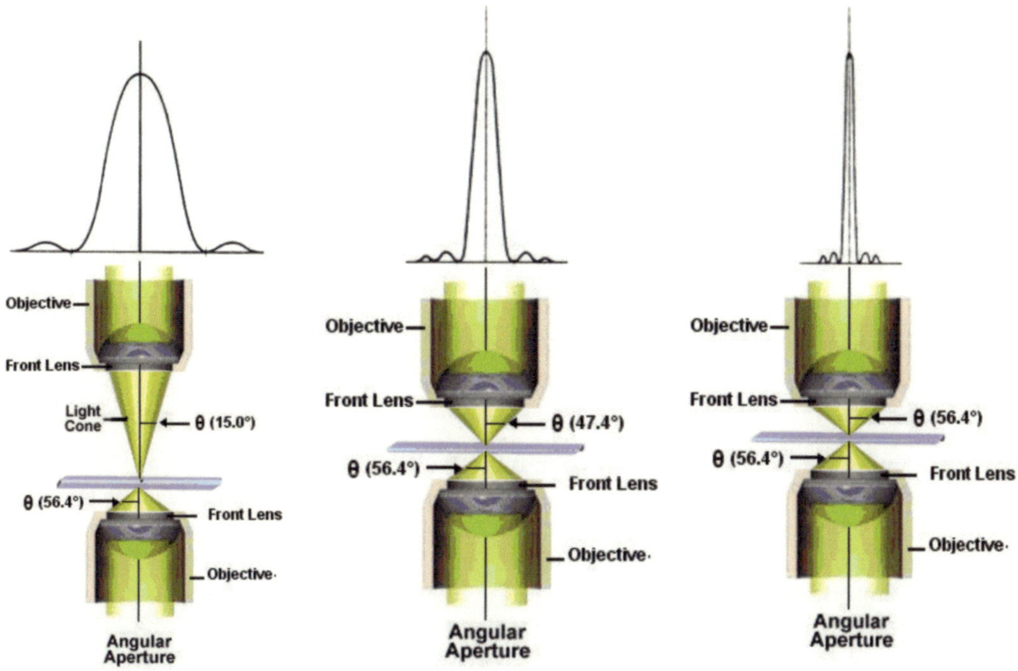

(a)	(b)	(c)
Forward with Olympus 10x	Forward with Zeiss 40x 1.2w	Forward with Olympus 60x 1.2w
Backward with Olympus 60x 1.2w	Backward with Olympus 60x 1.2w	Backward with Olympus 60x 1.2w

Fig. 14. The forward and backward objectives with different numerical aperture and immersion medium: (a). Olympus uplanFLN 10×/0.3 with air gap for the upper objective (which is used to collect the forward SH signals) along with the diffraction orders which can be collocated by the upper objective, and Olympus uplanApo/IR 60×/1.2 water immersion for the lower objective (which use to collect the backward SH signals). (b). Zeiss 40×/1.2W korr objective for the upper objective along with the diffraction orders which can be collocated by the upper objective, and Olympus uplanApo/IR 60×/1.2 water immersion for the lower objective. (c). Olympus uplanApo/IR 60×/1.2 water immersion for the upper objective along with the diffraction orders which can be collocated by the upper objective, and Olympus uplanApo/IR 60×/1.2 water immersion for the lower objective.

When the light hits the object it diffract, a single beam of light will split into several different diffraction orders bents at increasing angles from the original incident light. The easiest way to understand this property of light is to consider what happens when a beam of light shines through a pinhole into a dark background: The obtained images look like a negative target with a large central disk of light surrounded by a series of thin circles of light of decreasing brightness as it goes further away from the center.

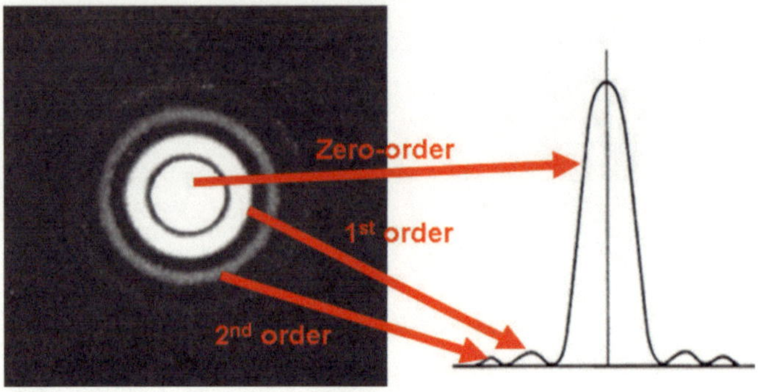

Fig. 15. The diffraction orders which are suppose to be collected by the objectives.

8. Discussion

The light-gathering ability of a microscope objective is quantitatively expressed in terms of the numerical aperture, which is a measure of the number of highly diffracted image-forming light rays captured by the objective. Higher values of numerical aperture allow increasingly oblique rays to enter the objective front lens, producing a more highly resolved image. An important concept to understand the image formation is the nature of diffracted light rays intercepted by the objective. Only in cases where the higher (1st, 2nd, 3rd, etc.) orders of diffracted rays are captured, can interference work to recreate the image in the intermediate image plane of the objective (see Figs. 14 and 15). When only the zeroth order rays are captured, it is virtually impossible to reconstitute a recognizable image of the specimen. When first order light rays are added to the zeroth order rays, the image becomes more coherent, but it is still lacking the sufficient detail. It is only when higher order rays are recombined, that the image will represent the true architecture of the specimen. This is the basis for the necessity of large numerical apertures (and subsequent smaller Airy disks, see Fig. 15) to achieve high-resolution images with an optical microscope.
The brightness of an image formed by an objective at a fixed magnification increases with the diameter of the angular aperture (the angle of the cone of light collected by the objective) see Fig. 14. Light rays emanating from the specimen proceed through air (or an immersion medium) that lies between the coverslip and the objective front lens. The angular aperture is expressed as the angle between the microscope optical axis and the direction of the most oblique light rays captured by the objective. The numerical aperture is expressed as:

$$\text{Numerical Aperture (N.A.)} = n \cdot \sin\theta \qquad (2)$$

where n is the refractive index of the medium in the object space (between the coverslip and the objective front lens) and θ is one-half the angular aperture (see Fig. 14). The value of n varies between 1.0 for air and 1.33 for a majority of immersion water utilized in optical microscopy. The angular aperture, which varies with the objective focal length, is the maximum angle of image-forming light rays emanating from the specimen that the objective front lens can capture when the specimen is focused. As the objective focal length decreases, the maximum angle between the specimen and the outside diameter of the objective front lens increases, causing a proportionate increase in the angular aperture. From equation (2), it is obvious that numerical aperture increases with both angular aperture and the refractive index of the imaging medium. Theoretically, the highest angular aperture obtainable with a

standard microscope objective would be 180 degrees, resulting in a value of 90 degrees for the half-angle utilized in the numerical aperture equation. The sine of 90 degrees is one, which indicates that numerical aperture is limited not only by the angular aperture, but also by the imaging medium refractive index. A majority of microscope objectives are designed to operate with air (which has a refractive index of 1.0) as the imaging medium between the coverslip and the objective front lens. This yields a theoretical maximum numerical aperture of 1.0, but in actual practice, the highest numerical aperture for a dry objective is about 0.95 (the angular aperture half-angle equals approximately 72 degrees).
The visual field brightness (B) of the microscope is determined by;

$$B \,\alpha\, (N.A.)^2/(M_{\text{objective}})^2 \tag{3}$$

where O_{bjective} is the objective lens magnification. Following equation (3) one can conclude that the brightness will increase as the numerical aperture increases and / or as the objective magnification decreases.
The resolving power of an objective lens is measured by its ability to differentiate two lines or points in an object. As the resolving power increases, the smaller the distance between the two lines or points to be distinguished can be. The more light that is collected by the objective lens the better resolution is and therefore as N.A. increases, so does the resolving power. The resolving power can be determined by:

$$\text{Resolving Power} = 0.61 \,.\, \lambda\, /N.A. \tag{4}$$

When the light rays come out of an axial object point enters a lens, the light rays with a larger N.A. are subjected to stronger refraction power and cross the optical axis in positions with larger differences from the ideal image formation position. As a result spherical aberrations are proportional to the $N.A.^3$.
The SH signals were obtained in both directions. These signals increased considerably when the chloroplasts rotated upon being illuminated by linearly polarized femtosecond pulses laser of paired photons which are effectively equivalent to blue light at half the wavelength of the irradiating laser (Cox et al., 2004; Reshak et al., 2009; Reshak, 2009). This chromatically equivalent blue range caused chloroplasts to rotate under the linearly polarized pulses laser causing orientation parallel to the side walls of the leaf cells. Starch grains are not involved thus eliminating a possible source of confusion. SHG indicates the non-centrosymmetric nature of the chloroplast membranes especially when aggregated into grana. The chloroplast grana and intergranal membranes are exhibit strong birefringence with large variations (Pantic-Tanner and Eden, 1999; Garab et al., 2005). A low intrinsic birefringence is expected in intergranal membranes and a high birefringence where the membranes are stacked. The grana were expected to give a stronger SH signal than the intergranal regions with unstacked membranes. During rotation of the chloroplast caused by linearly polarized high light intensity designed to present a smaller interface to the incoming light, the SHG increased considerably compared with the intergranal regions.

9. Conclusions

The enhancement of the forward direction of SH imaging was demonstrated by using set of different numerical aperture objectives with different immersion medium. The experiment shows that the numerical aperture is limited not only by the angular aperture, but also by the refractive index of the immersion imaging medium. The SHG signals are dependent on the numerical aperture and the refractive index of the immersion imaging medium. The brightness of the image depends on the axial aspect of the chloroplast. The image intensity also depends on the orientation of the chloroplast in relation to the illuminating polarization direction. A light-induced re-orientation in dependence of the illumination intensity could be observed (Reshak et al., 2009; Reshak, 2009).

10. Rotation of birefringent biological microparticles by linearly polarized laser using multifunctional two-photon laser scanning microscope

As it has been mentioned above, the chloroplast produced very strong birefringence which is originated from grana and the stacked regions of the thylakoid membranes. The chloroplast grana and intergranal membranes are distinguished by their birefringence (Pantic-Tanner and Eden, 1999). A low intrinsic birefringence is expected in intergranal membranes and a high birefringence where the membranes are stacked. When the incident beam of the linearly polarized light interacts with the chloroplast a considerable amount of torque will generates resulting in movements of the chloroplasts. Since the chloroplasts are non-centrosymmetric particles, measurable amount of second harmonic will generates without photo-damage or photo-belching by utilizing the fact that the SHG is energy conservation process. In this case one can expose the chloroplast to the laser radiations without risk. The intensity distribution of the SHG signals which comes from the chloroplast while illuminated by linearly polarized laser is shown in Fig. 16. Following these images, one can see the intensity distribution of the SH signal reduces as a consequence of the rotation during the exposure time, as axial thickness increases (when the chloroplast rotate around the short axis - the axis in the plane of the paper) the SHG signal will decrease in intensity, thus one can conclude that the chloroplast attempt to rotate and move away from the laser light in order to protect the photosynthetic system from being exposed to high intensity.

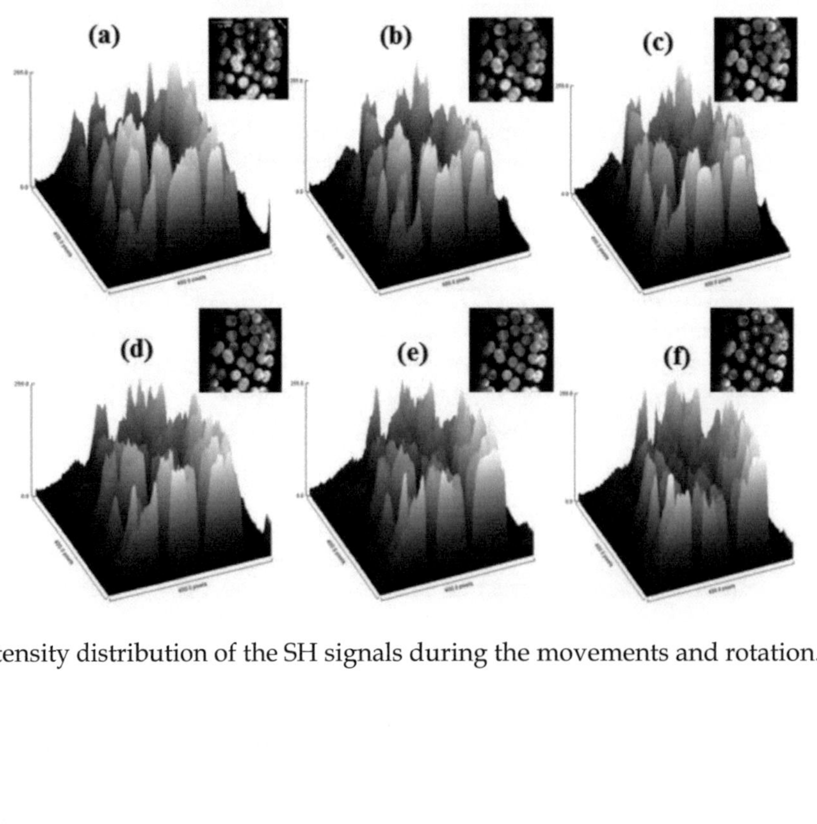

Fig. 16. Intensity distribution of the SH signals during the movements and rotation.

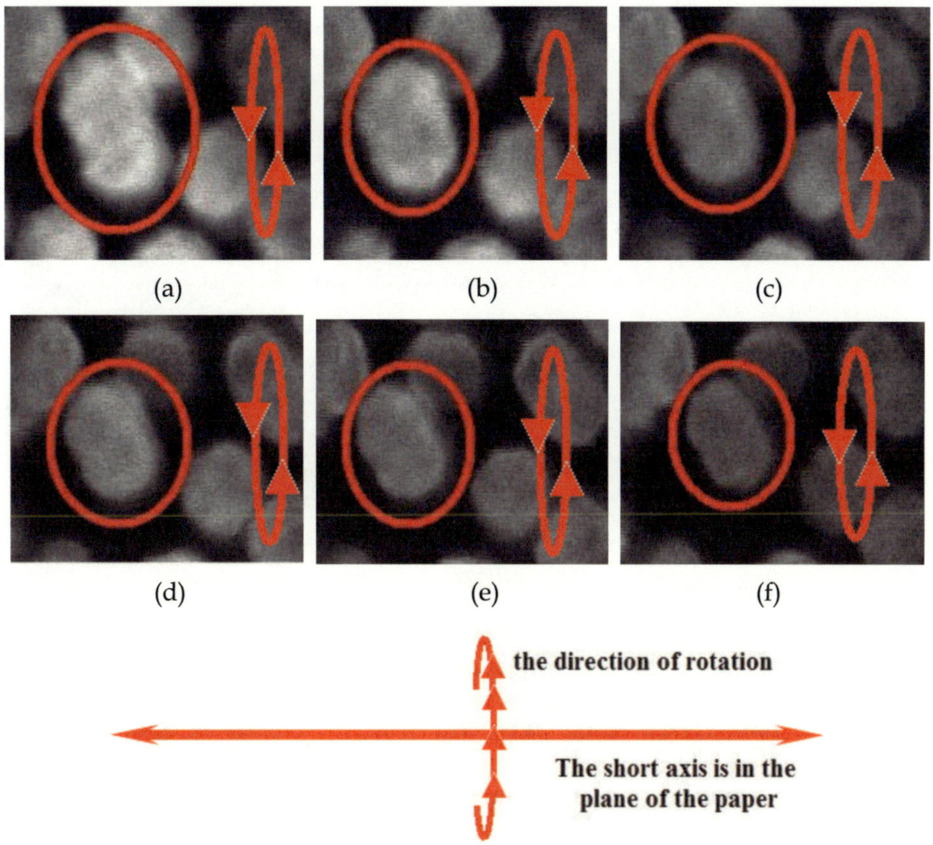

Fig. 17. Selected chloroplast illustrates the rotation around certain axis (axis parallel to the plane of the paper).

Fig. 17, shows the rotation of the chloroplast around certain axis (axis parallel to the plane of the paper), by this rotation of the chloroplast the dipole momentum in side the molecules oriented to be perpendicular to the polarization's direction of the laser beam. The SHG signal of the chloroplasts was collected using the objective Olympus uplanApo/IR 60×/1.2 water immersion. The SH signal showed no signs of bleaching during acquisition of repeated images from a given area, and thus showing no damage to the structure. This is to be expected since SHG is a coherent process, unlike fluorescence, and no energy is lost (Gorrell, 2003).

11. Discussion

Previous studies (Bayoudh et al., 2000) demonstrate that displacement of chloroplast inside the cell is extremely difficult, presumably due to chloroplast adhesion to the cytoskeleton and connections between organelles. In the present study we demonstrate that displacement and rotation of the chloroplast inside the cell were done using linearly polarized laser without photo-damage or photo-belching. For this experiment the setup in Fig. 18 was used. A linearly polarized laser beam was used to induce second harmonic generation signals from the chloroplast in *Plagiomnium affine* leaves. Since the light can carry angular

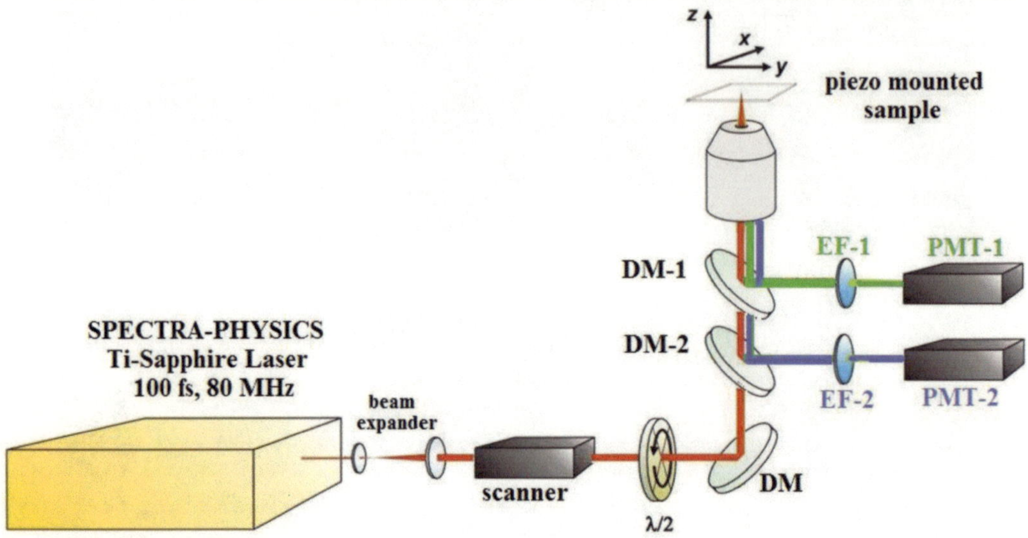

Fig. 18. The experiment the setup

momentum as well as linear momentum, the incident beam of linearly polarized laser becomes elliptically polarized and gains angular momentum after encroaching on a birefringent particle whose optical axis is not parallel to that of polarization of the light (Higurashi et al. 1999; Garab et al., 2005). The resulting change in angular momentum generates a reaction optical torque. Since the chloroplast exhibit a strong birefringence, the resulting anisotropic interaction with the linearly polarized laser beam strongly contributes to the optical torque orienting the chloroplast makes it possible to manufacture micromotors for micromechanical systems. The optical torque can be controlled by changing the laser's power following the formula (Higurashi et al. 1999);

$$\tau_{linear} = -\frac{P}{\omega}\sin\left(2\pi\frac{R}{\lambda}\right)\sin(2\theta) \qquad (5)$$

where τ is the optical torque generated by the linearly polarized laser acting on birefringent particle, P is the power of the linearly polarized laser at the particles, ω and λ are the frequency and the wavelength of the linearly polarized laser, respectively, θ is the angle between the polarization direction and the slow birefringence axis of the particle (the axis along which the index of refraction is largest), to obtain considerable amount of the optical torque the angle θ should be greater than zero, moreover the optical torque should be strong enough to overcome the Brownian motion, the latter is randomly oriented, and R is the retardation of the particle, defined as $R = \Delta n \times d$, where Δn is the birefringence defined as $\Delta n = n_e - n_0$, and d is the thickness of the particle. From this formula it is clear that laser power and the retardation term play major role in reducing / increasing the optical torque at a certain frequency and wavelength. The contribution of retardation part is come from the property of the particle so one has to select thick particle which possesses large birefringence in order to gain considerable amount for R. The laser's power P is one property of the setup meaning that it is under control, and one can increase / reduce the power to a certain limit without causing photo-damage or photo-belching to the sample by utilizing the fact that the SHG is known to leave no energy deposition to the interacted matters due to their virtual energy conservation characteristic. Since both P and R are proportional to τ, a particle with large R helps to use less laser's power, however by increasing the laser's power one can gain considerable amount of τ, which eventually overcomes the Brownian motion and the adhesion power of the chloroplast to the cytoskeleton and connections between organelles resulting in increasing the speed of the rotation. When a chloroplast inside a cell was rotate and displaced from its original position the nearby chloroplasts were rotating and displaced as well and followed the nearest neighbor. That is attributed to the fact that the chloroplasts were connected to each other by fine cytoplasmic strands or microfilaments. Thus, in general to maximize the signal short pulses should be used and average laser power should be kept in limit to prevent heating of the sample. We should emphasize that the asymmetric shape will contribute to the optical torque. Optically induced torque is always a result of transfer of angular momentum from light to a particle with conservation of momentum as an underlying principle. Consequently, rotation can be induced by a beam of light that carries angular momentum or by a beam that carries no angular momentum but where angular momentum is induced in the beam by the particle.

It is well known from Maxwell's theory that electromagnetic radiation carries both energy and momentum. The momentum may have both linear and angular contributions; angular momentum has a spin part associated with polarization (Allen et al., 1992) and orbital part associated with spatial distribution (Bijersbergen et al., 1992). Any interaction between radiation and matter is inevitably accompanied by an exchange of momentum. This often has mechanical consequences some of which are related to radiation pressure.

Light can carry angular momentum in two distinct forms; spin angular momentum, associated with the polarization state of the beam and orbital angular momentum, associated with the spatial distribution of the beam (Padgett and Allen, 2000). Optical torque is produced if either of these two forms of angular momentum is altered during the scattering of light.

The optical torque acting on the particle is equal and opposite to the rate of change of the angular momentum of the laser beam as it passes through the particle. Since the chloroplasts are birefringent particles, the angular momentum is transferred to the polarization state of the transmitted beam. Before acting on birefringent particles the linearly polarized light composed of equal quantities of left and right circular polarization, so it contains no net angular momentum. Generating a torque on chloroplasts (birefringent particles) causes an imbalance in the power of the left and right circular components of the transmitted beam, the spin torque acting on the chloroplasts is;

$$\tau = \left(P_{Right} - P_{Left}\right) / \omega_0 \tag{6}$$

where P_{Right} and P_{Left} are the right and left circularly polarized components, respectively; ω_0 is the optical angular frequency. If there is no particles then $P_{Right} = P_{Left}$.

This work describes experiments using linearly polarized laser to probe chloroplast. The linearly polarized laser was used to induce motion and tumbling of the chloroplast in cells of living leaf tissue.

12. Conclusions

We have demonstrated the movements and the rotation of the chloroplast of the *Plagiomnium affine* leaves while being illuminated by linearly polarized laser. Since the chloroplast produced very strong birefringence, the resulting anisotropic interaction with the linearly polarized laser beam strongly contributed to the optical torque orienting the chloroplast which makes it possible to manufacture micromotors for micromechanical systems. The torque can be controlled by changing the laser's power. The laser power and the retardation part play a major rule in reducing /increasing the optical torque at a certain frequency and wavelength. The contribution of retardation part comes from the property of the particle so one has to select thick particle which possess large birefringence in order to gain considerable amount for R. The laser's power P is one property of the setup that means it is under control, and one can increase/reduce the power to certain limit without cause photo-damage or photo-belching to the sample by utilizing the fact that the SHG is known to leave no energy deposition to the interacted matters due to their virtual energy conservation characteristic. Since both of P and R, proportional to τ, a particle with large R it helps to use less power, however, by increasing the laser's power one can gain considerable amount of τ which is overcomes to the adhesion power of the chloroplast to the cytoskeleton and connections between organelles resulting in increasing the speed of the rotation.

13. Acknowledgements

This work was supported from the institutional research concept of the Institute of Physical Biology, UFB (No.MSM6007665808), the program RDI of the Czech Republic, the project CENAKVA (No. CZ.1.05/2.1.00/01.0024), the grant No. 152/2010/Z of the Grant Agency of the University of South Bohemia. The School of Microelectronic Engineering, University Malaysia Perlis (UniMAP), Block A, Kompleks Pusat Pengajian, 02600 Arau Jejawi, Perlis, Malaysia.

14. References

Allen L., Beijersbergen M. W., Spreeuw R.J.C. and Woerdman J. P., Phys. Rev. A 45 (1992) 8185.

Born, M., Wolf, E., 1980. Principle of Optics. Cambrige University Press, Cambrige

Boyd, R., 1992. Nonlinear Optics. Academic, New York.

Bayoudh S., Mehta M., Rubinsztein-Dunlop H., Heckenberg N. R. and Critchley C, Journal of Microscopy, Vol. 203, 214-222 (2000).

Beijersbergen, M. W., Allen, L., van der Veen, H. E. L. O., and Woerdman, J. P., 1993, Opt. Commun., 96, 123

Campagnola P. J., Wei M. D., Lewis A., and Loew L. M., High-resolution nonlinear optical imaging of live cells by second harmonic generation Biophys. J. 77, 3341–3349 (1999).

Campagnola P. J., Clark H.A., Mohler W.A., Lewis A. and Loew L.M.,"Second-harmonic Imaging Microscopy of Living Cells," J. Biomed. Opt. 6, 277-286 (2001)

Campagnola PJ, Loew LM. Second-harmonic imaging microscopy for visualizing biomolecular arrays in cells, tissues and organisms., Nat Biotechnol. 2003 Nov;21(11):1356-60.

Cheng J., Volkmer A., and Xie X.S., "Theoretical and experimental characterization of coherent anti-Stokes Raman scattering microscopy" J. Opt. Soc. Am B 19, 1363-1375 (2002).

Cox G., Kable E., Jones A., Fraser I., Manconi F. , and Gorrell M. D., "3Dimensional Imaging of Collagen Using Second Harmonic Generation," J. Struct. Bio., 141, 53-62 (2003)

Cox G., Moreno N., and Feijó J., Second harmonic imaging of plant polysaccharides. Journal of Biomedical Optics 10, 024013_1-6 (2004).

Clowes F.A.L. and Juniper B.E, Plant Cells. Blackwell Scientific Publication, Oxford and Edinburgh. 546pp, 1968.

Chu S., Chen I., Liu T., Chen P. C., Sun C. K., and Lin B. L., Multimodal nonlinear spectral microscopy based on a femtosecond Cr:forsterite laser, Optics Letters, Vol. 26, Issue 23, pp. 1909-1911 (2001).

Cheng J., Volkmer A., and Xie X.S., "Theoretical and experimental characterization of coherent anti-Stokes Raman scattering microscopy" J. Opt. Soc. Am B 19, 1363-1375 (2002).

Denk W., Piston D. W. and Webb W. W., Two-photon molecular excitation in laser scanning microscopy, Handbook of Biological Confocal Microscopy, ed J B Pawley (New York; Plenum) pp 445-58 (1995).

Denk W., Strickler J.H., and Webb W.W., "Two-Photon Laser Scanning Fluoresence Microscopy," Science 248, 73-76 (1990).

Diaspro A., Building a two-photonmicroscope using a laser scanning confocal architecture, in Methods in Cellular imaging, A. Periasamy, Ed., pp. 162-179, Oxford University Press, New York (2001).

Faludi-Daniel A., Bialek G.E., Horvath G., Sz.-Rozsa Z., and Gregory R. P. F., Differential light scattering of granal and agranal chloroplasts and their fragments, Biochem. J. 174, 647-651 (1978).

Frohn J. T., Knapp H. F., and Stemmer A., True optical resolution beyond the Rayleigh limit achieved by standing wave illumination, Proc. Natl Acad. Sci. USA 97, 7232-7236 (2000).

Freund I., Deutsch M., and Sprecher A., "Connective Tissue Polarity, Optical second-harmonic microscopy, crossed-beam summation, and small-angle scattering in rat-tail tendon," Biophy. J. 50, 693-712 (1986).

Fine S., and Hansen W. P., Optical Second Harmonic Generation in Biological Systems Appl. Opt. 10, 2350–2353 (1971).

Gao L., Jin L. X. P., Xu J. W. Y., Ma Hui, and Chen D., Reconstruction of complementary images in second harmonic generation microscopy, Optics Express 14, 4727-4735 (2006).

Garab G., Galajda P., Pomozi I., Finzi L., Praznovszky T., Ormos P., Van Amerongen H., Alignment of biological microparticles by a polarized laser beam. Eur Biophys J 34, 335-343 (2005).

Georgiou E., Theodossiou T., Hovhannisyan V., Politopoulos K., Rapti G. S., and Yova D., Second and third optical harmonic generation in type I collagen, by nanosecond laser irradiation over a broad spectral region, Opt. Commun. 176, 253–260 (2000).

Gauderon R., Lukins P. B., and Sheppard C. J. R., Three-dimensional second-harmonic generation imaging with femtosecond laser pulses, Opt. Lett. 23, 1209–1211 (1998).

Goedheer JC. Orientation of the pigment molecules in the chloroplast. Biochim Biophys Acta 16:471–476 (1955).

Gorrell, M.D., Wang, X.M., Levy, M.T., Kable, E., Marinos, G., Cox, G., McCaughan, G.W., 2003. Interahepatic expression of collagen and fibroblast activation protein (FAP) in hepatitis C virus infection. Adv. Exp. Med. Biol. 524, 235–243.

Haupt W. And Scheuerlein R., Chloroplast movement. Plant Cell Environm 13, 595-614 (1990).

Han M., Giese G., and Bille J. F., Second harmonic geration imaging of collagen fibrils in cornea and sclera, Optics Express 13, No.15, 5791-5796 (2005).

Heinz, T.F., Chen, C.K., Ricard, D., Shen, Y.R., 1982. Phys. Rev. Lett. 48, 478–481.

Helmchen F. and Denk W., Deep tissue two-photon microscopy, Nature Methods 2, No.12, 932-940 (2005).

Higurashi E., Sawada R., and Ito T., Optically induced angular alignment of trapped birefringent micro-objects by linearly polarized light. Phys. Rev. E 59:3676-3681 (1999)
(http://www.till-photonics.com/Products/imic.php)

Lukins P. B., Xu P., Rehman S. and Sheppard C.J.R., Comparsion of two photon optical microscopies: second harmonic generation and fluorescence imaging, Science, Technology and Eduction of Microscopy: an overview, Volume 2, A Mendez-Vilas, editor, Kluwer/Formatex, Badajoz, ISBN 84-607-6699-3, PP. 712-719 (2003)

Maiti S., Shear J. B., Williams R. M., Zipfel W. R. and Webb W. W., "Measuring serotonin distribution in live cells with three-photon excitation," Science 275, 530-532 (1997).

Mark D. Gorrell, Xin M. Wang, Miriam T. Levy, Eleanor Kable, G. Marinos, Guy Cox, and G. W. McCaughan., Interahepatic expression of collagen and fibroblast activation protein (FAP) in hepatitis C virus infection, Advances in Experimental Medicine and Biology 524, 235-243 (2003)

Mertz J., and Moreaux L., "Second harmonic generation by focused excitation of inhomogeneously distributed scatterers", Opt. Commun. 196, 325-330 (2001).

Mason W. T. , Fluorescent and Luminescent Probes for Biological Activity, second Edition, Academic Press, London (1999).

Moreaux L., Sandre O., and Mertz J., "Membrane imaging by second harmonic genration microscopy" J. Opt. Soc. Am B 17, 1685-1694 (2000)

Moreno N., Feijo J., and Cox G., Implementation and evaluation of a detector for forward propagated second harmonic signals, Micron, 35, 721-724 (2004).

Pantic-Tanner Z. , and Eden Don, "calculation of protein form Birefringence using the finite element method" Biophysical Journal 76, 2943-2950 (1999).

Padgett M, Allen L (2000). Light witha twist in its tail, Contemp. Phys. 41:5, 275-285.

Reshak A. H., Sarafis V., Heintzmann R., Second harmonic imaging of chloroplasts using the two-photon laser scanning microscope, Micron 40, 378–385 (2009).

Reshak A. H., Second harmonic generation from thick leaves using the two-photon laser scanning microscope, Micron 40, 455–462 (2009).

Reshak A. H., High second harmonic generation signal from muscles and fasciapig's muscles using the two-photon laser scanning microscope, Journal of Microscopy, Vol. 234, 280–286 (2009).

Reshak A. H., Enhancing the resolution of the forward Second harmonic imaging using the Two-Photon Laser Scanning Microscope, Micron 40, 750-755 (2009).

Reshak A. H. and Auluck S., The linear and nonlinear optical properties of WS_xSe_{2-x} (x=0.5, 1.5, and 2.0) Physica B 393, 88-93 (2007).

Rezende-Pinto, M. C., On the birefringency of chloroplast in higher plants: convergence of results found by the use of the polarizing, light and electron microscopes. Port. Biol. Series A. 12, 267-270 (1972).

Roth S. and Freund I.,"Coherent Optical Harmonic Generation in Rat-tail," Opt. Commun. 33, 292-296 (1980).

Roth S., and Freund I., Imaging cells and extracellular matrix in vivo by using second-harmonic generation and two-photon excited fluorescence Opt. Commun. 33, 292–296 (1980).

Sarafis V., Chloroplasts: A structural approach. J Plant Physiol, 152(2-3), 248-264 (1998).

Schrader M., Bahlmann K. and Hell S. W., "Three-photon-excitation microscopy: theory, experiment and applications," Optik 104, 116-124 (1997).

Shen, Y. R. Surface properties probed by second-harmonic and sum-frequency generation, Nature 337, 519 (1989).

Wildman, S.G., Charlene, A.J. & Atchison, B. A., Light microscopic analysis of the three-dimensional structure of higher plant chloroplasts, position of starch grains and probable spiral arrangement of stroma lamellae and grana. Bot. Gaz. 141. 24-36 (1980).

Williams R.M., Zipfel W.R. and Webb W.W., " Interpreting second-harmonic generation images of collagen I fibrils" Biophy. J. 88, 1377-1386 (2005)

Wokosin D. L., Centonze V. E., Crittenden S. and White J., "Three-photon excitation fluorescence imaging of biological specimens using an all-solid-state laser," Bioimaging 4, 208-214 (1996).

Yan, P.; Millard, A. C.; Wei, M.-D.; Loew, L. M. "Unique Contrast Patterns from Resonance-Enhanced Chiral SHG of Cell Membranes" J. Am. Chem. Soc. 128, 11030 (2006).

Yelin D., and Silberberg Y., Laser scanning third harmonic generation microscopy in biology, Optics Express 5, No.8, 169 (1999).

Yeh A.T., Nassif N., Zoumi A. and Tromberg B.J., "Selective corneal imaging using combined second-harmonic generation and two-photon excited f luorescence," Opt. Lett. 27, 2082-2084 (2002)

Zipfel W.R., Williams R.M., and Webb W.W., "Nonlinear magic: multiphoton microscopy in the biosciences,"Nature Biotech. 21, 1369-1377 (2003).

Zumbusch A., Holtom G.R., and Xie X.S., "Three-Dimensional Vibrational Imaging by Coherent Anti-Stokes Raman Scattering," Phys. Rev. Lett. 82, 4142-4145 (1999).

Zoumi A., Yeh A., and Tromberg B. J., Imaging cells and extracellular matrix in vivo by using second harmonic generation and two photon excited fluorescence, PNAS 99, 11014-11019 (2002).

14

On the Airborne Lidar Contribution in Archaeology: from Site Identification to Landscape Investigation

Nicola Masini[1], Rosa Coluzzi[2] and Rosa Lasaponara[2]
[1]*CNR-IBAM (Institute for Archaeological and Architectural Heritage)*
[2]*CNR-IMAA (Institute of Methodologies for Environmental Analysis)*
Italy

1. Introduction

Historically, aerial photography has been the first remote sensing technology extensively used for surveying surface archaeological remains as well as for detecting underground archaeological structures through the reconnaissance of the so-called "soil" and "crop marks" (Crawford, 1929). Soil-marks are changes of colour or texture due to the presence of surface and shallow remains. Crop-marks frequently appear as differences of height or colour in crops which are under stress due to lack of water or deficiencies in other nutrients caused by the presence of masonry structures in the subsoil. Crop-marks can also be formed above damp and nutritious soil of buried pits and ditches. Such marks are well visible from aerial photos, especially during the spring season.

Nowadays two new technologies have strongly improved the performance of remote sensing in archaeology: (i) the Very High Resolution (VHR) satellite images and (ii) the airborne laser scanning.

The launch in 1999 of IKONOS, the first satellite sensor which acquires VHR imagery, opened new perspectives in the field of archaeo-geophysics.

The main advantages of VHR satellite imagery compared to aerial photos, are the synoptic view, the multispectral properties of the data and the possibility to extract geo-referenced information.

The use of data processing algorithms, from classifications methods to geo-statistics, from Principal Component Analysis to convolution filtering, enable us i) the extraction of land patterns useful for palaeo-geographic and palaeo-environmental investigations (Masini & Lasaponara, 2006); ii) the discrimination of surface archaeological remains from the surroundings (De Laet et al., 2007).

From 1999 up to now, the spatial resolution of satellite data has strongly increased, thus providing also valuable support to site discovery by means of soil/crop marks detection. The multispectral bands, available at a resolution four times lower than panchromatic channel, could be pan-sharpened by using image fusion algorithms available in several image processing software routines, thus allowing us to emphasize moisture and vegetation changes linked to the presence of buried archaeological deposits (Lasaponara & Masini, 2007).

The "great run" of satellite technology for reaching the resolution of aerial images seems have arrived at the end with GeoEye1 (launched in September 2008) which provides 41 cm panchromatic and 1.65 m multispectral imagery.

However, for archaeological applications, VHR satellite as well as aerial images (including hyperspectral data) still present limitations in detecting all the possible features of cultural interest.

We refer to archaeological remains covered by dense vegetation (forest, wood etc.) and, in many cases, to microrelief in bare-ground sites linked to the presence of anthropogenic earthworks and shallow remains.

In the first case, satellite images are capable to only detect big structures covered by forest. In such regard, we cite the identification of a Maya settlement in the jungle of northeast Guatemala (Garrison et al., 2008).

As concerns the second limitation, the visibility of micro-relief depends on many factors, such as off-nadir viewing angle of the collected imagery, time of image acquisition, view geometry, sun angle (micro-topographic relief variations are more visible in early morning or late evening) and surface characteristics (the presence of surface chaotic building material could make the detection of geometrical microrelief pattern very difficult; see, for example Lasaponara & Masini, 2005).

The above-said restrictions of optical imagery could be overcome by airborne laser scanner (ALS), also referred to as Light Detection And Ranging (LiDAR). It provides direct range measurements mapped into 3D point clouds between a laser scanner and the earth's topography.

ALS sensors can penetrate vegetation canopies allowing the underlying terrain elevation to be accurately modelled. Therefore, it is a powerful tool for recognizing and investigating archaeological heritage in wooded areas, usually well preserved due to the vegetation cover which protects the sites from erosion and from possible damage of mechanical ploughing.

Currently, a LiDAR survey could be carried out by two different types of ALS sensor systems (fig. 1) : (i) conventional scanners or discrete echo scanners and (ii) full-waveform (FW) scanners. The first, generally, delivers only the first and last echo, thus losing many other reflections. The second is able to detect the entire echo waveform for each emitted laser beam, thus offering improved capabilities especially in areas with complex morphology and/or dense vegetation cover.

Nowadays the majority of published studies are based on data collected by conventional ALS, for the management of archaeological monuments (Barnes, 2003), landscape studies (Shell & Roughley, 2004; Challis, 2006) and archaeological investigations to depict microtopographic earthworks in bare ground sites (Corns & Shaw, 2008) and in forested areas (Sittler 2004; Devereux et al., 2005; Crutchley, 2008; Gallagher & Josephs, 2008).

The potential of FW LiDAR for archaeological purposes has been assessed in a few studies, among which, for sake of brevity, we cite the study of an Iron Age hill fort covered by dense vegetation (Doneus et al 2008) and the investigations performed on two medieval settlements, located on bare ground hilly places (Lasaponara & Masini, 2009; Lasaponara et al., 2010).

This chapter is organized as follows: in Section 2 the available laser scanner technology is described; in Section 3, we focus on methodological issues, from data filtering to post processing; in Section 4 we deal with the state of the art of ALS in Archaeology; in Section 5 we show the investigation results obtained from two tests sites; finally, conclusions follow in section 6.

2. Conventional and Full-waveform ALS

ALS is an active remote sensing technique that provides direct measurements of the earth's topography, mapped into 3D point clouds.

The laser scanner, mounted to an aeroplane or helicopter, emits near infrared pulses, at a frequency rate of 30.000 to 100.000 pulses per second, into different directions along the flight path towards the terrain surface.

Each pulse could be reflected one or more times from objects (ground surface, vegetation, buildings, etc.), whose position is determined by computing the time delay between emission and each received echo, the angle of the emitted laser beam, and the position of the scanner (determined using differential global positioning system and an inertial measurement unit).

There are two different types of ALS (fig. 1): i) conventional scanners based on discrete echo and ii) FW scanners.

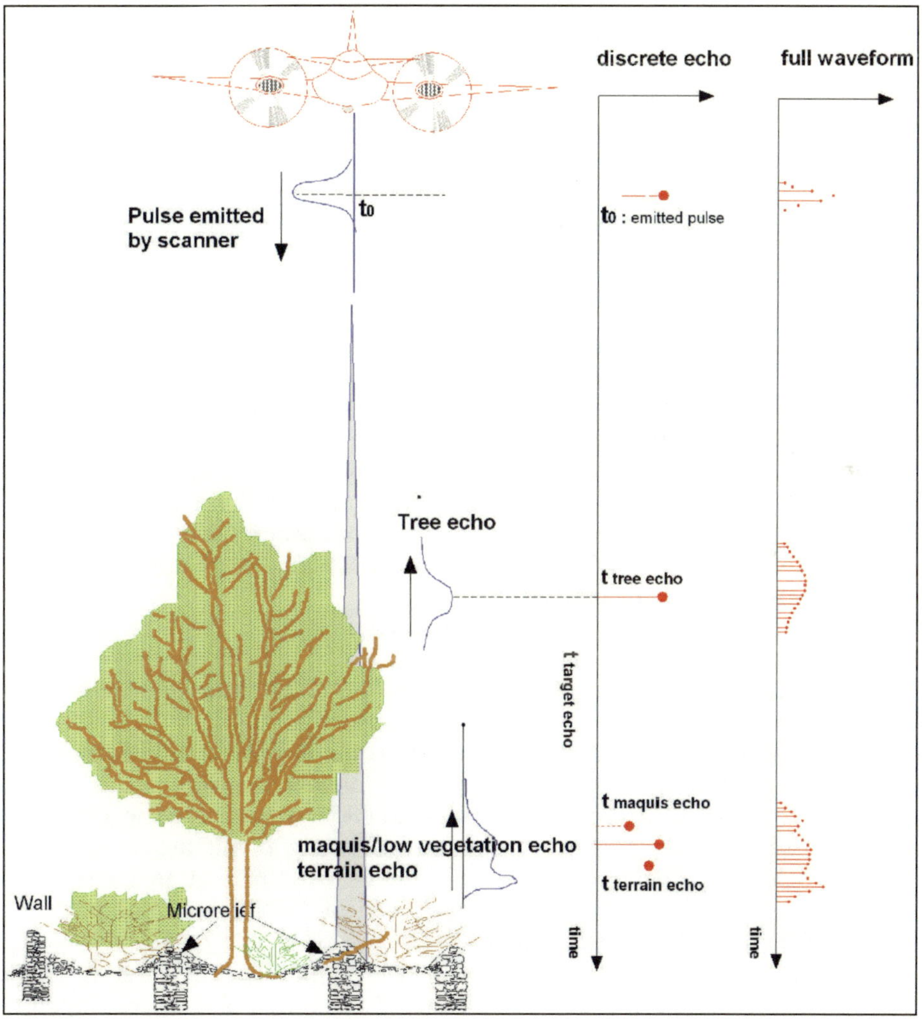

Fig. 1. Conventional and full-waveform ALS.

The first detects a representative trigger signal for each laser beam (see fig. 1).
The second digitizes the complete waveform of each backscattered pulse; thus allowing us to improve the classification of terrain and off terrain objects, such as low vegetation, buildings, and other man-made structures lying on the terrain surface (Doneus et al, 2008). This enables us to obtain DTMs with accuracy less than 0.1m and therefore to detect archaeological structures and earthworks even under dense vegetation cover.

3. Theoretical consideration on ALS data processing

3.1 Theoretical consideration on the extraction of DTM from ALS

In order to obtain a Digital Terrain Model from airborne laser scanning raw-data processing is essential. This process is generally called filtering., namely the discrimination of point clouds into terrain and off-terrain points and the elimination of erroneous points, such as, low points and aerial points.
The following step is the classification (also joined with a segmentation process) of raw LiDAR data, which allocates off-terrain points into specific classes, defined "a priori" (i.e. before applying the classification).
Many different algorithms have been published for ALS data filtering. A list of the most commonly used filters is reported below: (i) Morphological filtering, (ii) Progressive densification, (iii) Surface based filtering, (iv) Segment based filtering, (v) Spline interpolation filtering.

i. Morphological filtering: this group is based on the concept of mathematical morphology, a set of theoretical method of image analysis which provides a quantitative description of geometrical structures based on a set of operators. Morphological filtering was published by Lindenberger (1993), Vosselmann (2000), Sithole (2001), Roggero (2001), Lohmann et al. (2000).
   Filtering methods of Lindenberger (1993), Vosselmann (2000), Sithole (2001) and Roggero (2001) are applied to point clouds; whereas Kilian et al. (1996) and Lohmann et al. (2000) methods are applied to raster data.
ii. Progressive densification: this group is based on the classification of the whole data set starting with a small given point cloud and increasing them iteratively. The most popular progressive densification method is the progressive triangular irregular network (TIN) densification devised by Axelsson (2000). Another similar method was proposed by Sohn and Dowman (2002).
iii. Surface based filtering: in this case the whole point cloud belong to terrain surface and then, iteratively, points are removed according to a step-by-step refinement of the surface description. Surface based filters were proposed by Kraus and Pfeifer (1998) and Elmqvist et al. (2001).
iv. Segment based filtering: this group is based on the concept that classification is not based on single points but on segments, a set of neighbouring points with similar properties. In general, the point cloud segmentation is performed in object space or features space. In a step to step process, neighbouring points are merged to form a segment as long as their properties are similar with respect to some thresholds. Segment based filtering methods were proposed by Sithole and Vosselmann (2005), Sithole (2005), Tovari and Pfeifer (2005).
v. Spline interpolation filtering: this method was proposed by Brovelli et al. (2001; 2003) and implemented in the GIS GRASS software. To classify point clouds, first a bilinear spline interpolation and later a bicubic spline interpolation are performed.

A comparison and performance evaluation of these different filtering algorithms were provided by Sithole and Vosselmann (2004). They suggested that filtering methods can be categorized into the following four groups, on the basis of the structure of bare earth points in a local neighbourhood: (i) slope-based, (ii) block-minimum, (iii) surface-based and (iv) clustering/segmentation.

i. Slope based algorithms measure the difference in slope (or height) between two points, and assume that the highest point belongs to an object if the slope is higher than a given threshold value.
ii. Block-minimum algorithms assume as discriminant function a horizontal plane with a corresponding buffer zone above it, which defines a region in 3D space where bare earth points are expected to reside
iii. Surface-based filtering methods assume as discriminant function a parametric surface with a corresponding buffer which defines a region in 3D space where ground points are expected to reside.
iv. Finally, clustering/segmentation filtering approach assumes that any clustered points belong to an object if their cluster is above its neighbourhood.

Among the above-said filtering groups, the surface-based category appears to provide the best results in separating points on a ground surface from other points (Sithole and Vosselman, 2004). Examples of surface-based algorithms are: Axelsson (2000), Briese and Pfeifer (2001), Elmqvist (2001), Sohn (Sohn and Dohman, 2002), Wack and Wimmer (2002).

- Axelsson's (1999) algorithm is based on a progressive Triangulation Irregular Network (TIN) densification. Starting from a coarse TIN surface (obtained from reference points which are neighbourhood minima), new points are added, though an iterative way, if they meet criteria based on distances to TIN facets and angles to the vertices of the triangle.
- Briese and Pfeifer (2001) is a hierarchic based method. Starting from an approximate surface it performs interpolations in each hierarchy level by assuming weight values based on vertical distance of the points to the same approximate surface, thus allowing us to carry out the classification.
- Elmqvist algorithm(2001) is based on the concept of membrane floating upwards from beneath the point cloud, which defines the form of the bare-Earth.
- Sohn algorithm (Sohn and Dowman, 2002) uses a two-step (downward and upward) progressive densification of a TIN. The first operates a triangulation of four points closest to the corners of the rectangular bounds of the point cloud. The lowest point within each triangle is added to the triangulation. This process is repeated until no triangle has a point beneath it. The second step is performed in order to extract other bare-Earth points not caught by the downward.
- In the Wack and Wimmer (2002) algorithm a raster DEM is generated from a raw point cloud in a hierarchical approach.

The circumstances under which the filtering methods could meet difficulties and limits (Sithole and Vosselman, 2004) are generally the following: i) outliers; ii) spatial and morphological object complexity; iii) attached objects; iv) low vegetation; v) and geomorphological discontinuities.

i. Outliers could be low (caused by multi-path errors and errors in the laser rangefinder) and high (birds, low-flying aircraft, or errors in the laser range-finder).

ii. The spatial and morphological object complexity is a circumstance which typically characterizes a urban environment. In particular the filtering algorithms are likely to fail in presence of very large objects, very small objects (elongated objects, low point count, such as vehicles), very low objects (walls, cars), and complex shape.
iii. The attached objects are objects spanning the gaps between bare-earth surfaces such as buildings on slopes, bridges, natural/artificial ramps.
iv. As concerns the vegetation, the classification problems are mainly related to vegetation on slopes and low vegetation.
v. Finally, the most typical geomorphologic discontinuities are due to steep slopes and sharp ridges.

An overview of the different algorithms for the filtering of airborne laser scanning data is provided also in Pfiefer (2003), Sithole (2005).

3.2 Theoretical consideration on general details on Digital Elevation Model

Digital Elevation Model (DEM) is defined as any digital representation of the continuous variation of the relief over space (Burrough & McDonnel, 1998). More information about DEM are in Morre et al. (1991); Weible and Heller (1991); Wilson and Gallant (2000).

A Digital Terrain Model (DTM) is a model of the bare earth surface in digital form, a Digital Surface Model (DSM) is a model in the bare earth and objects are attached to it, such as buildings and vegetation.

In general DEM can be assembled, depending on the source and/or preferred method of analysis, into three different data structures: a) a Grid-DEM which is a regular square matrix where each pixel is an elevation (see fig.2a) b) The Triangulated Irregular Networks (TIN) model is a surface as a set of contiguous, non-overlapping triangles. Within each triangle the surface is represented by a plane. Each vertex is a known elevation value. An example of TIN is shown in fig. 2 b; c) The Contour structure is based on the concept that landscape can be divided into small, irregular shaped polygons based on contours lines and their orthogonals. An example of Contours is shown in fig. 2c

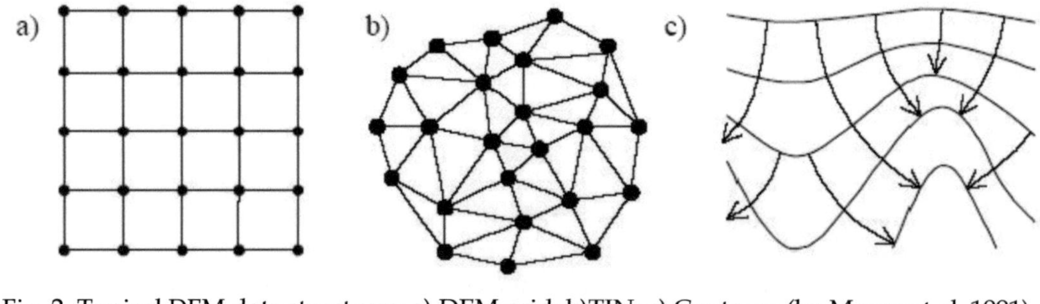

Fig. 2. Typical DEM data structures: a) DEM grid; b)TIN; c) Contours (by Moore et al.,1991).

Recently, a new classification was proposed by Hengl and Evans (2009). They divide DEM in two big groups: vector-based and raster-based. TIN-DEMs and Contour-DEMs are part of the vector-based group; Grid-DEMs are raster-based group.

Grid-Dem was the most widely used data structure in the past due to its simplicity, but it has the following disadvantages: i) it is not able to represent abrupt changes in elevation easily; ii) important details of the land surface in flat areas are missed; iii) it increases the

difficulty to calculate specific catchment areas accurately; iv) the computed upslope flow paths is tending to zig zag across the landscape.
Moreover, Grid-DEMs appear as a continuous-surface but in reality they are not continuous. For that when the DEM pixel dimension is selected it will not be possible to understand if it is representative of an abrupt change of the land surface.
The grid heights are typically determined by surface interpolation and approximation methods like inverse distance weighting, moving last squares, linear prediction, or kriging.
A method, often used for altitude data, is inverse distance weighting (IDW: for this and other interpolation techniques). In this method, the estimated value of a grid cell depends on its distance to neighbouring data points. In general, the greater the distance, the smaller the data point's influence on this value. This relation depends on an exponent that is defined by the user. In addition, the user determines the radius within which data points are used to calculate a grid cell value. The radius and exponent are determined on the basis of expert judgment. The disadvantages of this method are: (i) its tendency to smooth out small-scale relief, and (ii) the clustering effect around data points.
Another widely used interpolation method is a Triangulated Irregular Network (TIN).
This method produces a network of triangles that connect all data points. The values of the grid cells are calculated using the slope and shape of the triangles. The user determines (based on expert judgment) the maximum length of the triangle sides and an exponent.
TIN-DEM is able to incorporate discontinuities and is efficient to represent roughness terrain because the density of the triangle can be varied easily. For that it can represent easily abrupt changes on the land surface. On the contrary it is not as good for graduated changes of the land surface because TIN is not continuous. This causes that abrupt changes appear as unnatural effects related. The most popular interpolation to create a TIN-DEM is the Delaunay triangulation, based on Voronoi diagrams.
The final interpolation method herein listed is Kriging, considered to be based on the most solid theoretical principles. This method assumes that, a variable's value can be estimated through the data's spatial characteristics. The spatial characteristics are modelled in a variogram, where the squared difference of pairs of data points is plotted as a function of their spacing. Based on the position of these points in the variogram, a mathematical function is generated and then used for the interpolation. The way the researcher interprets the spatial relationship is explicitly expressed via the type of function chosen and the method to obtain the best fit. Thus, the choice of the interpolation parameters is (to a certain degree) objective, since it is based on the variogram. The disadvantages of kriging are the complexity of the method and the difficulties of filtering out the natural trends in larger study areas.
For the representation of DSM and DTM the same data structures and interpolation methods are in use.
A comprehensive discussion on the generation of 3D terrain models can be found in Pfiefer (2005).

4. ALS in Archaeology: overview of applications

In this section, we will provide a brief overview on the application of aerial LiDAR in archaeology in different countries. In the last decade, several national agencies acquired LiDAR data for different monitoring purposes mainly linked to environmental issues. The availability of these data has strongly encouraged investigations in the field of archaeology.

For these reasons, the majority of studies are mainly focused on the assessment of LIDAR capability in archaeology and generally exploited the data processing already done for other purposes. Therefore, this overview does not report the data processing or the methodological approaches because they are generally skipped in the available in literature, but it summarizes the significant experiences and results achieved by archaeologists in different environments.

4.1 Germany

One of first papers on the use of ALS in archaeology was published by Sittler (2004). The authors exploited the data acquired (2000-2004) for obtaining an accurate DEM for the entire Germany , in the framework of the project "Land and Survey bureau of the Baden-Wurttemberg State". This project aimed at providing comprehensive altimetry data-set with a resolution at around 1 meter and an accuracy at around 50 cm in height.
Sittler (2004) used part of this data set to analyze woodlands near Rastatt (30 km south of Karlsruhe, in South West Germany, including a sandy and dry flat terrace near the River Rhine. The visual analyses enabled the detection of patterns of the earlier medieval landscapes, including earth mounds and ridge and furrow structures. Moreover, using the 3D analyst extension of ArcView 3.2, quantitatively analyses were carried out to extract sizes of the detected patterns, such as, surface area, volume, length and width of the ridge and furrow as well as surface roughness and undulation.
On the basis of these successful investigations, subsequently (May 2009), the State Office for Cultural Heritage Management of Baden-Wurttemberg, launched a three-year new project aimed at obtaining an archaeological mapping of Baden-Wurttemberg using high resolution ALS data, covering an area of 35.751 km2.
Within this project, Hesse (2010) developed and implemented a new tool for archaeological prospection: the Local Relief Model (LRM). It is based on the removal of large-scale landscape forms from the data, thus allowing us to obtain local and small-scale elevation differences. In particular, LRM measures correctly and directly heights and volumes of small-scale features and extracts details of local topography without using numerous combinations of illumination azimuth and elevation.

4.2 Netherlands

In the Netherlands, the Dutch Ministry of Public Works initiated the setup of the so-called Actueel Hoogtebestand Nederland (AHN), namely `Up to date height data base of The Netherlands'. LiDAR data were collected from 1996 to 2004. The database consists of interpolated airborne laser altimetry data covering the whole country and contains at least one point per 4 m2 outside forests and one point per 16 m2 inside forests.
Humme et al. (2006) used a part of this dataset to study a Bronze Age village and 2500-year-old Celtic field system, near Doorwerth (East of the Netherlands). They proposed a method to filter the large scale topography component out by using a kriging interpolation method. Using this method the authors enhanced road beds, foot-paths and the earth walls surrounding the Celtic field system.
Van Zijverden and Laan (2003) used Lidar data for predictive modelling in the Holocene parts of the Netherlands (site of Eigenblok, in the municipality of Geldermalsen) in addition to the conventional data source such as soil, geomorphologic, geologic and palaeogeographic maps.

4.3 United Kingdom

In 1999, the Environment Agency of UK commissioned a LiDAR survey (fig. 2) for monitoring river corridors and coastal areas of England and Wales. (Brown, 2008). In general flights were organized to collect 5 - 10 points m2 and, after an appropriate processing, the LiDAR data were typically supplied in 2 km square tiles with a 2 m grid resolution in ESRI ASCII grid format. These data comprised a single return without the possibility to access and separate First Pulse, Last Pulse or intensity data.

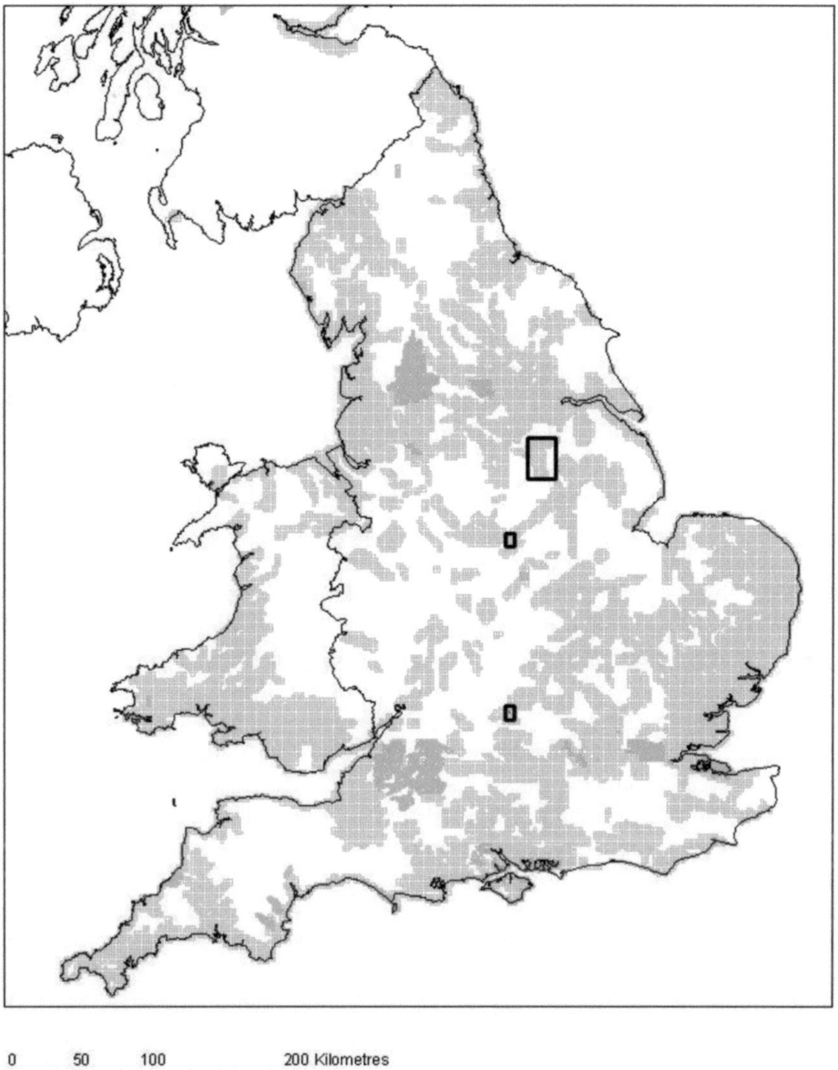

Fig. 3. Map showing the extent of the 2005 LiDAR coverage in grey (courtesy of Environment Agency)

Since 2002, such huge Lidar datasets has been exploited for archaeological prospections (Holden et al., 2002; Challis, 2006; Challis & Howard, 2006; Challis et al., 2008).

By exploiting different way of visualization of DTMs (shading procedure and vertical exaggerations), Holden et al. (2002) was able to identify and record the slight earthwork traces of a Roman Fort at Newton Kyme (in West Yorkshire) which was under the plough for decades. This site was characterized by earthworks, less than 1 m in height, and, for that, it had been missed by previous traditional aerial surveys.
This study led the English Heritage to commission a LiDAR survey specifically for archaeological investigation of the famous Stonehenge World Heritage Site. The project took place in the period March-July 2001 and was focused on the assessment of how many known sites not completely levelled by ploughing could be detected in the Stonehenge using LiDAR survey. To this aim, DTM and DSM were derived from the last and first pulse LiDAR point cloud, respectively, with a ground resolution of 1 m. Data interpretation was facilitated using color coding or continuous grey scale to enhance topographic details; whereas, "a digital sun", i.e. changing the direction of the digital illumination on the DSM was applied to visualize features.
The results of this investigation exceeded any expectations, as described in Bewley et al. (2005). In detail, LiDAR strongly contributed to the study of the Stonehenge landscape and to the detection of unknown sites. This was done by means of: i) the elimination of vegetation which hid features of archaeological interest, thus allowing to analyze their "inter-visibility" respect to the monumental area and to explore their spatial relationships; ii) the determination of the "Stonehenge view" of the other monuments using the Viewshed analysis; iii) interpretation of Relief shaded images to enhance the topographic location of unknown and known Neolithic sites; iv) Integration of LiDAR with images from CASI (Compact Airborne Spectrographic Imager) to provide information of vegetation, geomorphology and other features useful for the study of the landscape.
LiDAR data provided by the Environment Agency were used by Challis (2006) to map alluvial geomorphology. The author presents three different way of producing a DTM suitable for geoarchaeological purposes: (1) a 3x3 "grid cell variance filter" to remove areas of the DSM above a threshold limit (set at a 66.6° slope gradient), and replace them with new elevation values obtained by interpolating the gaps, (2) a high-pass filter to remove landscape clutter; (3) the generation of a simulated first and last-pulse from a single return. Challis suggests that all these three methods produce DTM with artifact but, for most geoarchaeological applications, the access to the first and last pulse data is desirable; whereas, the removal of landscape clutter is not a critical point to be addressed and, therefore, it can be skipped. Challis showed that LiDAR products, in particular the DSMs, are particularly effective for mapping features in floodplains dominated by lateral channel movement and desiccated peat. Whereas, it is less effective in upper river banks, which are dominated by rapid erosion with poor survival of palaeo-landscape features as well as in lower river banks where accretion is the dominant process.
Later, Challis et al. (2008) assessed the potential of LiDAR to enhance existing records of the historic environment of the River Dove valley. These data were compared with the existing inventory of sites and with a selected sample of vertical aerial photographs. Authors grouped earthwork features of archaeological interest in different categories including: agricultural remains (such as ridge and furrow), settlement remains, quarries and other features considered as evidence of past human activity. Each feature was listed and digitized as a polygon within a GIS environment, and compared with pre-existing records of the historic environment held by the local authorities. First, the number of sites found by LiDAR was compared with the number of sites in the existing Historic Environment

Records; then a selected number of aerial photographs was used to assess their extension. Around the 84.4 % of the archaeological features captured by the LiDAR survey were previously unknown. Anyway, it should be considered that some known features already recorded were not recognized by LiDAR. The majority of these were cropmarks or artifacts and buildings

Crutchley (2006) used LiDAR data derived from UK Environmental Agency to analyze four different case studies (Southrey, Barlings cemetery, Stixwould, Bardney environs). He employed a vertical scale exaggeration of elevation to analyze the LiDAR data for microrelief identification. Crutchley showed that Lidar clearly has potential for recording certain site types and especially in highlighting relationships between sites in the broader landscape, nevertheless, he highlighted the importance of not using LiDAR data alone, but as a part of all readily available sources.

Barnes (2003) studied the Salisbury Plain Area, located in the heart of Wiltshire (Southern England),. In this area, extensive remains dating back from the Neolithic (ca. 4000-2400 BC) to the late Roman period (5th century AD) are still visible as earthworks. Around 2300 individual sites and burial mounds were recorded. Two different LiDAR data set, acquired in January and February 2001, were used along with CASI imagery, which were very useful to map bare ground, to identify archaeological earthworks and different types of scrub.

Devereux et al. (2005) conducted a study in a prehistoric hill-fort at Welshbury Hill, in the Forest of Dean, Gloucestershire. The prehistoric earthwork was scanned using a conventional ALS Optech ALTM 3033 system of the Unit for Landscape Modelling of the Cambridge University. Two separate surveys of the site were conducted with a spatial detail at (i) 4 points per square meters and (ii) 1 point density per square meters respectively. Point clouds were converted to 0.25 m and 1 m grid respectively. To ensure the maximum laser penetration, the surveys were conducted in winter (on February 2004), when the leaves are falling and understory is at minimum. The DTMs revealed a possible Bronze Age field system varying the direction of the illumination source to easily detect feature on hill shaded DEMs.

4.4 Greece

In Greece, Rowland and Sarris (2007) used LiDAR combined with multi-sensor airborne remote sensing data from CASI and Airborne Thematic Mapper in order to locate the presence of exposed and known buried archaeological remains in Itanos (Eastern Crete). The LiDAR data were acquired using an Optech ALTM 3033 high-resolution airborne laser scanner with a point density of 1 per square meter. In this research DEM derived from LiDAR was used to identify the presence of new archaeological features such as, abandoned terraces and a circular depression.

4.5 Ireland

In Ireland, in the framework of the Discovery Programme aerial LiDAR were acquired in 2007 using the system FLI-MAP 400. Some studies were carried out on the basis of this dataset to map and identify archaeological features, among them, we cite Carns and Shaw (2009) Data were acquired for two different sites: (i) abandoned medieval settlements in Newtown Jerpoint (Kilkenny) and (ii) a prehistoric hillfort in Dún Ailinne (Kildare). The two surveys were carried out at two different resolutions (Newtown Jerpoint 50 pts/m2, Dún

Ailinne 15–30 pts/m2) in order to successfully record subtle topographic features in the two different investigated environs. The authors suggest that LiDAR can well record subtle features of archaeological interest at high spatial resolution, with a great level of definition, in short time and with a cost-effective way.

4.6 Belgium

In Belgium, Werbrouck et al. (2009) performed an interdisciplinary landscape study concerning the history of the settlement and environments in the north of Ghent (Flanders). This study is manifold and aimed to create a 'clean' topographical surface to be (i) investigated by archaeologists and (ii) used by soils scientist in their modelling procedures. The study area covers around 1400 km^2 which corresponds to the northern part of the Pleistocene valley of the Scheldt River.

Stal et al., (2010) focused on investigating remains of trenches of the First World War around Mount Kemmel, in Flanders. According to the authors, even if some of these trenches still exist, they have today subtle height differences with the surrounding surface, and, therefore, can not be detected by conventional techniques, such as fieldwalking and/or aerial photography, but LiDAR can overcome these drawbacks. An aerial survey was carried out in 2008 with an average point density of around 5 points per square meter. They obtained (i) a DSM from a random division of points interpolated with a grid at fixed resolution of 50 cm; and (ii) a DTM after filtering out non-ground points at the same resolution as DSM (50 cm). Then, the DTM was manipulated using filter also based on convolution.

On the basis of the obtained results, the authors pointed out that LiDAR DTM was a very useful tool for detecting subtle remains of trenches of the First World War, but, the data processing and filtering techniques are a critical step. This requires a careful evaluation and selection of the procedures useful to emphasize and detect microrelief. For example, for this study, Laplace and high-pass filter were not satisfactory, whereas Sobel filter and the pseudo-hillshade offered good performance.

4.7 Austria

The limits of conventional ALS in discriminating the low vegetation and underlying terrain have been dealt with by Pfeifer et al. (2004).

Doneus et al.(2008) investigated the potential of full-waveform airborne laser scanning (RIEGL Airborne Laser Scanner LMS-Q560), to investigate an Iron Age hillfort located in a forested area called Purbach, in Austria. Authors discuss in detail the LiDAR data processing and filtering. According to the authors the full-waveform scanner allowed more off-terrain points to be removed from the raw data than a conventional ALS and this creates a better terrain model. A good filtering between terrain and off-terrain points is very important to improve the archaeological detection of subtle micro-topographic features such as barrows obscured by forest. The resulting DTM reveals the entire hillfort with even subtle structures, as for example small shallow depressions on top of round barrows, which result from looting. A comparison with a detailed topographic mapping of the visible archaeological traces from the 1960's demonstrated that even very low earthwork features. As for example round barrows with a vertical extension of 20 cm (or even less), were identified in the DTM even tough been missed by the original trained surveyors in the field.

4.8 Italy

In Italy Coren et al. (2005) used LiDAR data along with hyperspectral images to improve information on the archaeological area of Aquileia (UD, North-East of Italy). Hyperspectral data allowed the identification of specific humidity, vegetation and thermal conditions, whilst accurate geometric information were provided by LiDAR. Feature detection was carried out using different filters, such as, High filter, Low filter, Laplacian Filter, etc.
Danese et al. 2008 focused on the processing of DTM obtained from LiDAR using the Viewshed Analysis to obtain information about the extension of the area under the "visual control" of a Mediaeval castle clinging to the top of a hil. Moreover, the authors also attempted a reconstruction of the medieval landscape based on the estimation of the location of the cultivated areas within given time slices. The LiDAR- DTM was also processed using site catchment analysis based on the distance that could be travelled out from the focus during the course of a day's journey. Results from this study allowed the identification of potential land uses obtained for one/two hour(s) Site Catchment.
Lasaponara et al. (2010a) focused on the potentiality of the latest generation of airborne ALS in the detection and spatial characterization of microtopographic relief linked to archaeological features. The investigations were carried out for Monteserico, an archaeological area in the Basilicata Region (Southern Italy) characterized by complex topographical and morphological features, which make air/space prospection very difficult. The LiDAR survey allowed the detailed identification of small surface relief and differences in height produced by surface and shallow archaeological remains (the so-called shadow marks), which were not visible from ground or from optical data set (aerial photo and satellite images). Using the high resolution DTM obtained from LiDAR the authors reconstructed the urban shape of the medieval village in great detail. The authors pointed out that the DTM-LiDAR data is a powerful instrument for detecting surface discontinuities relevant for investigating cultural features.
Moreover, the same author groups evaluated the capability of LiDAR data (Lasaponara et al. 2010b) to detect and discriminate micro-topographic relief linked to archaeological remains from natural geomorphological features. Results from the analyses performed in processing the DTM-LiDAR using geostatistical methods pointed out that the LiDAR is a powerful instrument for detecting macro and micro elevation changes, which are generally very critical to evaluate. The DTM obtained from LiDAR provided a sound basis for geomorphological interpretation, useful to detect surface discontinuities (e.g. breaklines, lineaments) and forms as well as to identify surface features relevant for geomorphological processes of the study area.
Finally, the authors suggested that despite the great potential of LiDAR in archaeology, the use of ALS data encounters serious challenges and still requires specific research, addressed to both (i) pre-processing (filtering and classification) to obtain detailed DTM and also to (ii) the post-processing to extract information (pattern extraction, classification). This challenge has been partially addressed by Coluzzi et al (2010) who defined the data processing chain along with the threshold-based algorithm for the classification of ground and non-ground points and for the detection of archaeological remains. The classification of laser data was performed using a strategy based on a set of "filtrations of the filtrate" (for more detail see section 4). Appropriate criteria for the classification and filtering were set to gradually refine the intermediate results in order to obtain the vegetation heights and to discriminate between canopy, understory and micro-topographic relief linked to terrain or earthwork. To test the algorithm performance, some sample areas on hill environments with different morphological

features and cover types, were processed and analyzed. Results from these investigations pointed out that the devised data processing enables the detection of micro-topographic relief in sparsely as well as in densely vegetated areas. The most important facts to cope with different environmental situations are mainly linked with (i) the resolution of the acquired data set and (ii) the data processing chain specifically devised for archaeological purposes.

4.9 USA

In the USA LiDAR in Archaeology represents a very small percentage of the applications of LiDAR. For sake of brevity we cite the follwing experiences.

Harmon et al. (2006) assessed the utility of 1 m resolution LiDAR for studying historic landscapes in two eighteenth-century plantation sites located near the Chesapeake Bay, in the state of Maryland. DSM LiDAR used in this study was obtained from the first return, whilst DTM from the last return. Relief detection was carried out by a visual analysis and also using enhancement of DEM based on hillshade surface models and contours maps.

Gallagher and Josephs (2008) used LiDAR to detect pre and post-European sites in the dense woodland of Isle Royale National Park (Michigan, USA). LiDAR data were collected with a conventional sensor and filtered by using TerraScan software. Grid DTM was derived from the last return (ground level) processed with a spatial resolution of 2 meters.

LiDAR bare-Earth models were used to 'see through' the vegetation in an effort to: (i) identify cultural features prior to the implementation of a pedestrian reconnaissance survey; (ii) aid in the development of a more informed survey strategy; and (iii) produce more efficient and cost-effective research design.

The identification of potential archaeological features from the LiDAR- DEM was based on four visual criteria and the degree to which the features appeared anthropogenic versus non-anthropogenic:

i. Shape: large or small; linear, sinuous, rectilinear, circular, conical, or cubic; mounded or depressed.
ii. Pattern: isolated, clustered, aligned, scattered.
iii. Texture: degree of smoothness or coarseness, based on the frequency of tonal changes on the image.
iv. Shadow: it provides an impression of the feature's shape in profile and can be a primary aid in feature recognition. Finally, shadows can also act to obscure irrelevant features.

Thirty-two potential archaeological features were interpreted from the imagery; 18 were previously recorded. A field survey enabled the localization of the larger number of features (previously recorded or newly discovered). Romain and Burks (2008) used LiDAR data for studying a 2000-year-old road that was mapped several times in the 1800's and was subsequently destroyed by urbanization or cultivation in Ohio. A segment of road was preserved in a wooded area with 30 cm embankments rising above either side of the path. Several profiles taken along the road showed that its morphology matched another ancient road segment in a different part of the state. In order to locate the parallel-walled Road leading from the Newark Octagon toward Ramp Creek, the authors used conventional LiDAR survey, which send out between 2000 and 5000 pulses per second. DEM was obtained at 2 meters. Results form Romain and Burks investigation showed that LiDAR provide not only surface maps, but, also new useful information extracted from the profile tools.

Chase et al. (2010) applied LiDAR-derived images in a tropical region, the jungle in Caracol, Belize, to study a very important ancient Maya site. The survey carried out by Optech GEMINI Airborne Laser Terrain Mapper (ALTM) covered a total area of around 200 sq km. The LiDAR derived product were a 1-m (DEM) for bare earth, and a 1-m Canopy Surface Model (CSM) for canopy top points. LiDAR data helped to reconstruct the topography of the landscapes, but, also structures, causeways, and agricultural terraces - even those with relatively low relief of 5 to 30 centimeters. Moreover, they were useful to demonstrate the ability of the ancient Maya to radically modify the landscape in order to create a sustainable urban environment.

5. A LiDAR approach for archaeological purpose

In the following section a methodological approach based on the use FW LiDAR is shown for two study cases in Southern Italy (fig. 4). One is Monte Serico, a bare-ground site located in Basilicata dating back to Medieval Age. The DTM has been used to identify microrelief related to the urban fabric of the medieval settlement.
The other study case is the Wood of Incoronata (in Apulia) which covers an interesting palaeohydrographical pattern.

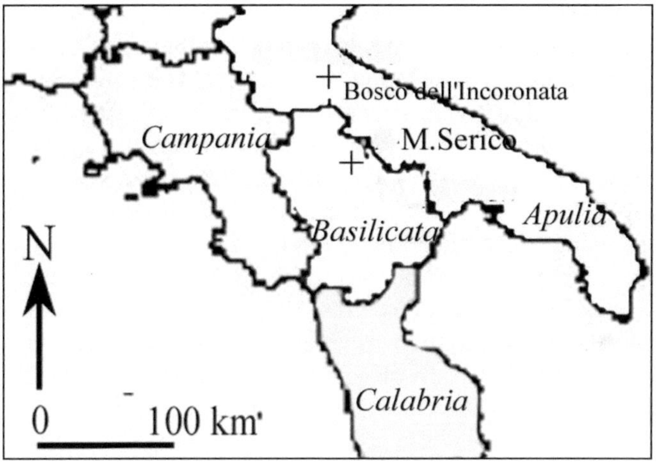

Fig. 4. Location of the study cases.

5.1 From data filtering to classification

The identification of archaeological features (form earthworks to surface structures) for both bare and densely vegetated areas, requires a very accurate DTM. To this aim, it is crucial to carry out the classification of terrain and off terrain objects by applying adequate filtering methods. In the examined study cases, we adopted the progressive Triangulation Irregular Network (TIN) densification method by Axelsson (2000). The algorithm starts from a coarse TIN surface obtained from reference points which are neighbourhood minima. Then new points are added, in an iterative way, if they meet criteria based on distances to TIN facets and angles to the vertices of the triangle. This algorithm has been embedded in Terrasolid's Terrascan commercial software (http://www.terrasolid.fi/en/products/terrascan). For its implementation, some parameters included maximum building size, terrain angle, iteration angle, iteration distance, and maximum edge length have been assigned.

The initial setup involved importing all the necessary raw data into the processing software, applying coordinate transformations and calibration, which is based on the comparison of the laser data produced by different flight passes which overlap each other.

Later, both DSMs and DTMs have been obtained from the classification, which was herein performed using a strategy based on a set of "filtrations of the filtrate". The workflow can be summarized as follows: i) Low point Classification; ii) Isolated points Classification; iii) Air points; iv) Ground Classification; v) Classification of points below surface; vi) Classification of points by class; vii) Classification of points by height from ground for different heights.

The data classification process started by including all the point cloud into a single class, called the default class. Then, the elimination of outliers points has been performed through classification of : (i) "low points", (ii) "isolated points", and (iii) air points. The first has found single points or groups of points with a height lower than 0.5 m compared to the other points within a ray of 5 m. The second routine has identified isolated points such as points present in the air (for example birds, etc.). The third one has detected points present in the air not classified as isolated points.

The following processing step has been based on the Axelsson TIN model (Axelsson, 2000) in an attempt to define a "ground" surface. To accept or reject points as being representative of the "ground" it has been necessary to define some geometric threshold values, which prescribe possible deviations from the average topographic surface.

A triangle of the primary mesh is progressively densified by adding a new vertex to a point inside it. The "Classification of points below surface" allowed us the identification of points under the surface level, such as wells or similar. Such classification was performed setting the standard deviation value at 8 with 0.01 m tolerance value.

The latest two classifications (vi and vii) identified and classified points according to a given class or height, respectively. All points left into the default class have been considered as vegetation. Finally, using "Classification of points by height from ground for different heights" three classes have been considered low (< 0.25), medium (0.25 to 2 m) and high (> 2 m). Further classification enabled the discriminations of cars, walls, buildings, vegetation types, etc. Finally, the DTM was created using Terra Modeller on the basis of the classification of terrain and off terrain objects performed using the whole processing chain from (i) to (vii) step.

5.2 Post processing: the shading procedure approach.

In order to emphasize archaeological features with particular reference to micro-relief shading procedures have been used. Several routines, embedded in commercial software allow different solutions, such as the visualization of the elevations by using color graduations and the slope of the terrain, in order to identify the portions of the terrain that are relatively flat vs those that are relatively steep.

For the visualization of elevations it is useful to enable Hill Shading option to view elevation data as shaded relief. With this option shadows are generated using the loaded elevation. To do it, it is necessary to light the DTM by an hypothetical light source. The selection of the direction parameters (zenith angle z and azimuth angle) depends on the difference in height and orientation of the micro-relief of potential archaeological interest. Single shading is not the most effective method to visualize and detect micro-relief. If features and/or objects are parallel to the azimuth angle, they will not create a shadow. As a result, it would not be possible to distinguish them.

The problem could be solved by observing and comparing DTM scenes shaded by using different angles of lighting, as done for the two study cases and presented in the following sections.
In addition the different shaded DTM scenes have been processed by using the Principal Components Analysis (PCA) (Stal et al. 2010), which is a linear transformation which decorrelates multivariate data by translating and/ or rotating the axes of the original feature space, so that the data can be represented without correlation in a new component space. For our application, the PCA transformed the input shaded DTMs in new components in order to make the identification of distinct features and surface types easier. The major portion of the variance is associated with homogeneous areas, whereas localized surface anomalies will be enhanced in later components, which contain less of the total dataset variance. This is the reason why they may represent information variance for a small area or essentially noise and, in this case, it must be disregarded.
Finally, convolution filtering techniques (Laplacian, directional, Gaussian High Pass) have been performed

5.3 Monte Serico case study

5.3.1 Study area and previous investigation

Monte Serico site is found on a hill located with an elevation of around 590 m a.s.l. which faces over a wide territory characterized by hills and plain crossed by the Basentello river in the Northeast side of the Basilicata Region (Southern Italy, see fig. 4).
From a geological point of view, the stratigraphic sequence is composed of Subappennine Clays, Monte Marano sands and Irsina conglomerates. Sporadic herbaceous plants grow over the investigated area.
Historical sources state that around the 11th century, a castle was built on the hill; whereas a village is attested to the 13th century and gradually abandoned between the end of the 14th and the first half of 15th century. Today the only buildings remaining are the castle and a church (see A and B in fig. 4). On the southern side of the hill, the presence of earthenware, pottery and crumbling building materials, indicates the existence of a buried settlement. The latter has been discovered in 1995, by means of aerial photos (Masini 1995). The use of QuickBird images allowed us to improve the spatial characterization of the urban shape (Lasaponara & Masini 2005).
A more detailed reconstruction of the urban fabric has been obtained by LiDAR survey carried out by GEOCART on 20th September 2008 using a full-waveform scanner, RIEGL LMS-Q560 on board a helicopter.
The data filtering and classification has been performed as described in section 4.1. Table 1 shows the threshold values assigned to classify ground and non ground points. The classification allowed us to obtain a DTM which puts in evidence the urban fabric characterized by a radio-centric pattern on the southern slope of the hill which develops according to the level curves (see fig. 5). Such features are also visible from the 1995 aerial image (Masini 1995) and the satellite data (Lasaponara & Masini 2005).
The DTM (fig. 5) shows other micro-relief which exhibit an alignment in the E-W direction, at southwest of the hill.
The geomorphological study based on the analysis of DTM allowed to better discriminate, respect to the available optical dataset, the features of archaeological interest from those linked to geomorphological phenomena (erosion and creep) (Lasaponara et al. 2010).
In the study of Lasaponara et al. (2010), the only post processing has been performed by using spatial autocorrelation statistics. It was mainly aimed at enhancing geomorphological

features. It allowed us to better survey landslide niches on the south-western foot of the hill and linear erosion phenomena on the Southern slope, and discriminate the morphological step and the lithological boundary between the Irsina conglomerates and the Monte Marano

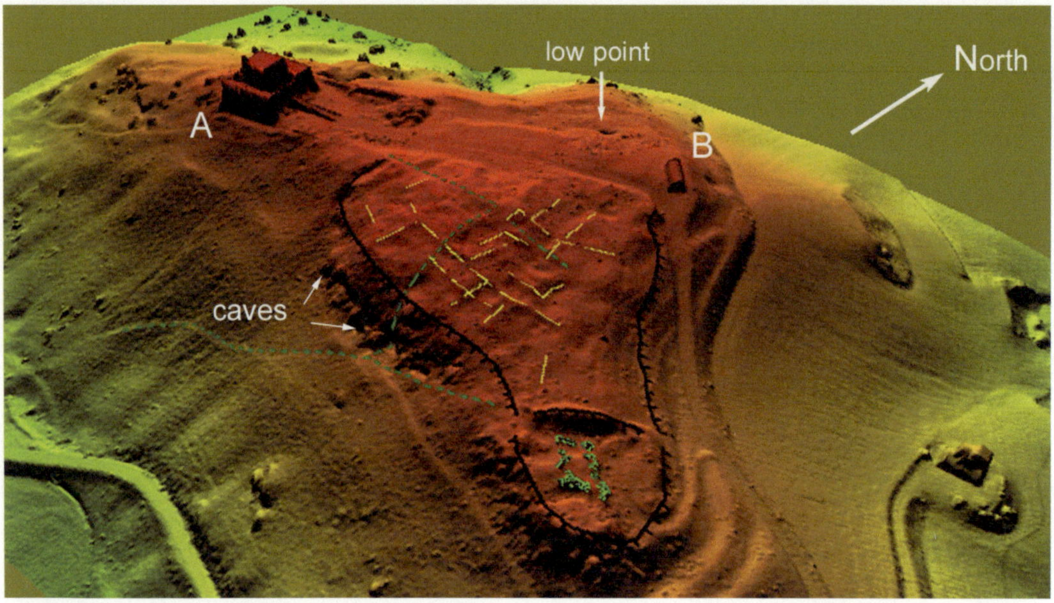

Fig. 5. Monte Serico study case: 3d DTM with archaeological interpretation of micro-relief. The picture shows a radio-centric urban shape, the castle (A), the church (B), some caves along the southern morphological step and the location of a point below surface, related to the presence of an under-ground cave.

Parameter	Maximum building size (m)	Terrain angle (°)	Iteration angle (°)	Iteration distance (m)	Max edge (m)
value	60	88	20	0,80	5

Table 1. Threshold values assigned to classify ground and non ground points

5.3.2 Post processing of DTM to improve the knowledge of Monte Serico: aims and results

The same DTM already analyzed in Lasaponara et al. (2010) is herein object of further post processing using different shaded DTM scenes. For Monte Serico study case, the comparative analysis of different shaded DTM seems to be particularly suited, due to its complex morphology which makes the accurate identification of archaeological micro-relief very difficult.

The visual analysis of shaded DTM, obtained from different light sources, highlights additional archaeological features. In particular the shaded DTM at azimuth equal to 90° (fig. 6a) shows rectangular micro-relief at the centre of the southern slope. Whereas in the shaded DTM at azimuth equal to 360°, several other micro-relief could be observed at south eastern angle of the scene (fig. 6d).

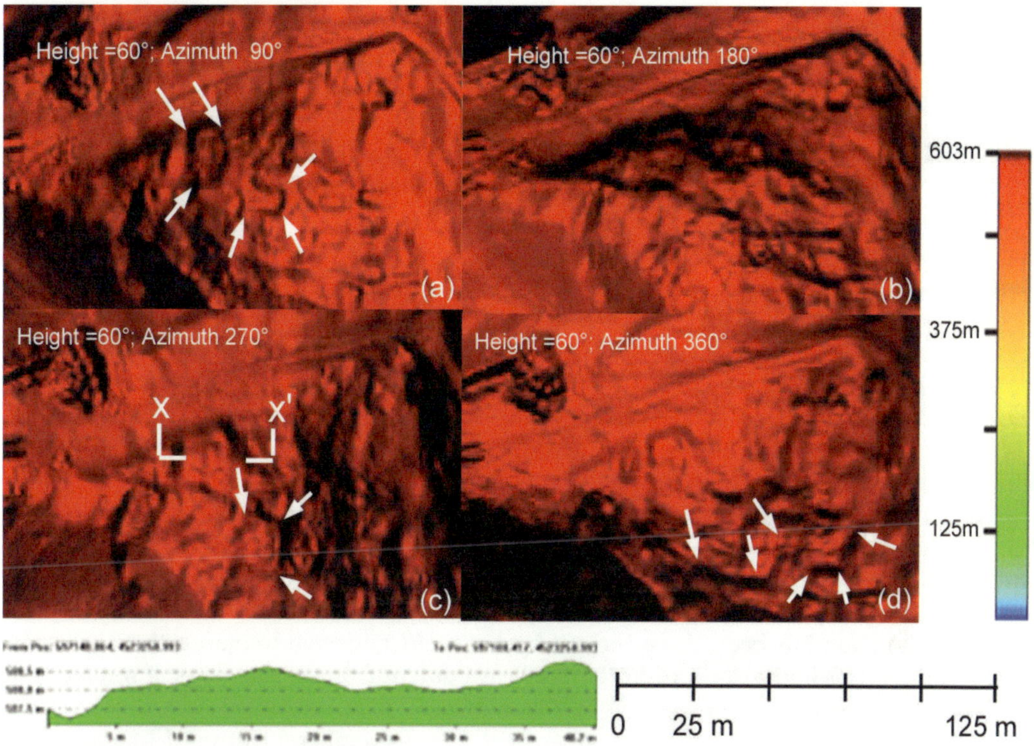

Fig. 6. Shaded DTMs at zenith angle of 60° and azimuth angles, from a to d, respectively equal to 90°, 180°, 270° and 360°.

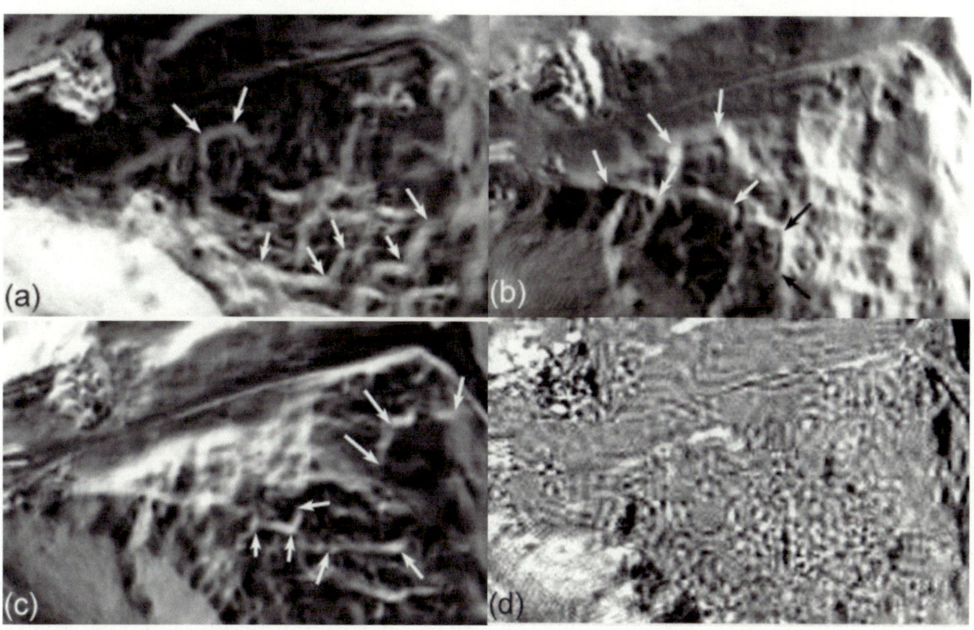

Fig. 7. The first four principal components.

These shaded DTMs have been further processed using Principal Component Analysis. It provided additional information on the urban fabric and emphasized archaeological features, already visible from shaded DTMs.
In particular, the result of the first component (PC1) is the average of all input shaded DTMs and looks like an edge thinning of the pattern of micro-relief (see fig. 7a).
The second component, PC2 emphasizes some micro-relief at southeast, mostly aligned in E-W direction.
The third component, PC3, improves the visibility of features at South and North of the shaded DTM.
Finally, the fourth component, PC4 is disregarded because it contains substantially noise.

5.4 Wood of Incoronata

The second study area, herein considered, is the natural park of Bosco dell' Incoronata that has an extension of around 1060 ha, with 162 ha of woodland (*Quercus pubescens*) and 115 ha of prairies.
The study area is located 12 km away from Foggia within the Tavoliere delle Puglie in the Apulia Regione (see fig. 4).
The investigated area is an important site from the naturalistic, historical and archaeological point of view. Bosco dell'Incoronata is an ancient lowland forest that was still present in the medieval time, and has been characterized by long and intensive human activity probably from Neolithic to Middle Ages (Mazzei, 2003) as evident from archaeological remains and historical documentation. As regards to the medieval time, historical record attests that Frederick II of Hohenstaufen (26 December 1194 – 13 December 1250) used to spend long periods in Foggia, which was a strategic position to reign over a vast territory extending from German to Sicily.
During Frederick's kingdom two royal residences, "*Palacium Pantano*" in S. Lorenzo and the "*Palacium dell'Incoronata*", were specifically built or restored for the imperator. Both of them were located very close to Foggia. Over the years "*Palacium Pantano*" has been widely investigated and partially restored, whereas the location of "*Palacium dell'Incoronata*" is still unknown today. It is thought that such location is very close to the Bosco dell'Incoronata and probably within the medieval forest area which is still present today and also known as the Frederick's woodland.
Our investigations have been mainly focused on the identification of traces of ancient landscapes and paleao-environmental features in order to improve knowledge about the transformation of the territory. Knowledge about palaeo-landscape features still fossilized in the modern landscapes is a crucial point and an invaluable data source for performing detailed archaeological investigations, for the identification of the environmental changes and of the underlying processes.
Unfortunately, in this area traces of past human activities are quite subtle and scarcely visible from aerial photographs and optical satellite images, because of the intensive agricultural activity of the whole area. Arable lands appear everywhere from images shown in figure 8 a s a result of the major post-war land reforms. This long and intense agricultural activity, along with the use of agricultural equipment and machinery for production, has generally destroyed traces of past landscapes.
Nevertheless, subtle microtopographic relief may be still preserved and visible from a detailed DTM.

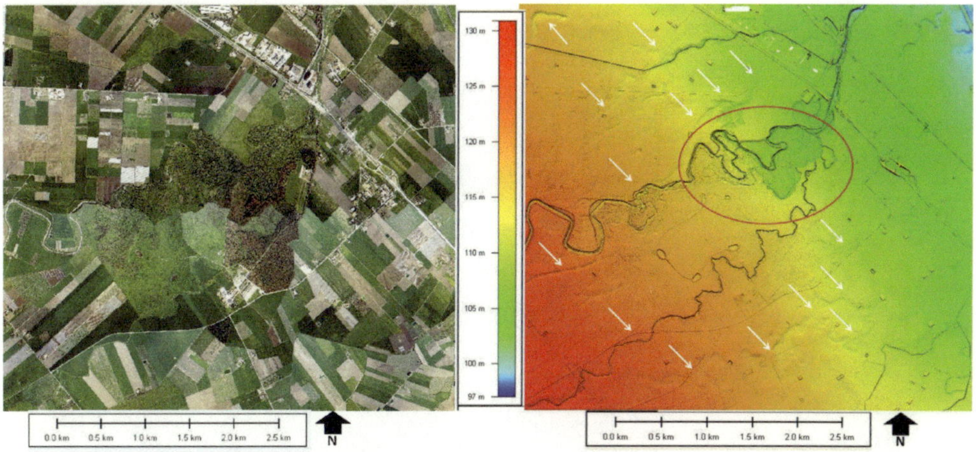

Fig. 8. Bosco dell' Incoronata within the Tavoliere delle Puglie (Southern Italy). (left): Orthophoto at a spatial resolution of 0.15 m. (right): DTM . White arrows denote the palaeodrainage system. The red circle show the complex system hidden by the woodland.

In order to throw light on the landscape changes an investigation based on the use of LiDAR data has been made (Lasaponara et al. 2010). The survey has been carried out by using a full-waveform scanner, RIEGL LMS-Q560 on board a helicopter. The data filtering and classification has been performed as described in section 5.1. Table 2 shows the threshold values assigned to classify ground and non ground points.

In our previous study (Lasaponara et al. 2010) we showed how spatial autocorrelation statistics applied to the DTM-LiDAR can help to detect surface discontinuities and microtopographic relief linked to palaeoenvironmental features.

In this study case, we want test how LiDAR data can contribute to the knowledge of the historical landscape, exploiting and processing different shaded DTMs.

Figure 8-left shows the orthophoto of the area, acquired simultaneously with the LiDAR data. It has a spatial resolution of 0,15 m. In the image we can recognize the woodland, which is very close to the Cervaro river.

Parameter	Maximum building size (m)	Terrain angle (°)	Iteration angle (°)	Iteration distance (m)	Max edge (m)
value	60	88	4	1,40	5

Table 2. Threshold values assigned to classify ground and non ground points for Bosco dell'Incoronata.

The DTM extracted from the LiDAR data (see fig. 8, right) visualizes the general topography of the valley and reveals extensive geomorphological details of the plain, not visible from the orthophoto (fig. 8, left), such as a wide and complex drainage system (see circle red in fig. 8, right) covered by dense tree canopy.

Other palaeo-riverbeds along with modern rivers and channels, on agricultural areas, are also more evident in the DTM (see white arrows, in fig. 8, right) than in orthophoto.

To better identify the palaeo-hydrographical pattern, different shading views of DTMs have been observed. Then the PCA is applied on the shaded DTMs.

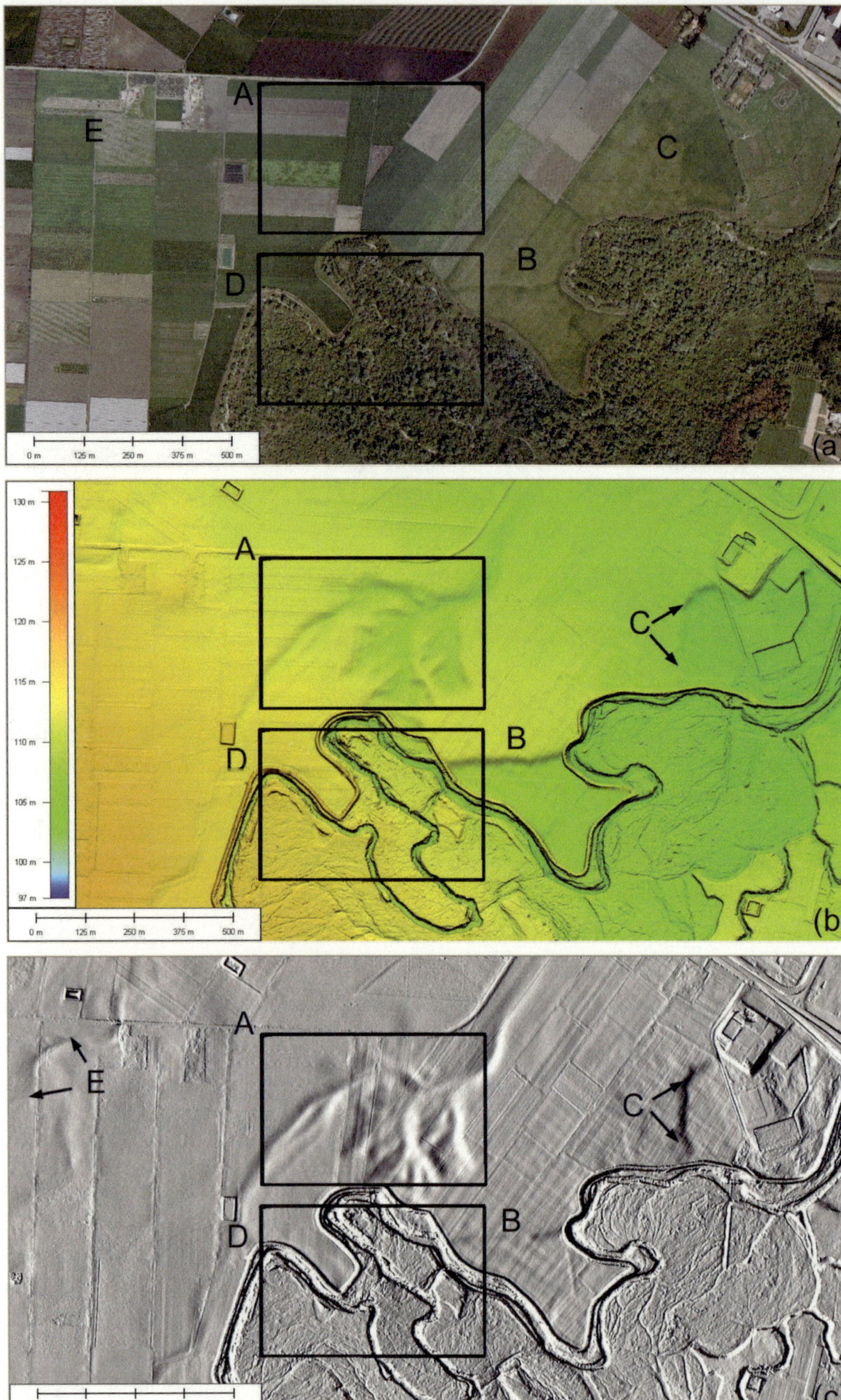

Fig. 9. Detail of the study area. Orthophoto (a); shaded DTM at azimuth angle equal to 90° (b); PC1 result (c).

Figures 9a-c, shows the orthophoto, a shaded DTM view and the first component (PC1) of a a part of the study case.
The comparison of these pictures puts in evidence the improvement of
i) the shaded DTM scene respect to the orthophoto; ii) and PC1 compared to the DTM, respectively.
In particular the DTM (fig. 9b), highlights the palaeoriverbeds A and the riverbed D which are not and less visible in the orthophoto, respectively (fig. 9a) whereas the edge of the terraces B and C are visible from both the orthophoto and DTM.
Finally, PC1 (fig. 9c) puts in evidence the above mentioned features (emphasizing any of them, such as A), and other features not visible in the DTM, such as the palaoeriverbed E and some land divisions which overlap on the palaoriverbed C.

7. Conclusions

In this chapter we offered on overlook of Airborne Laser Scanning for archaeological purposes. In particular in the first part (sections 2-4), we have dealt with the potential and limitation of the available laser scanner technology, the rational basis of LiDAR data processing (from data filtering to classification) and the State of Art of ALS in Archaeology.
In the second part (section 5) we showed the potential of using and processing point clouds surveyed by an Aerial full-waveform laser scanner on two sites of archaeological and natural interest. Their characteristics did not allow to investigate the two sites with the same effectiveness by means of remotely sensed optical data.
The first one is non vegetated hilly plateau, with several microrelief evidence linked to the existence of a buried settlement dating back to Middle Ages.
The second one is a wood which covers a palaeodrainage basin whose study is important for the reconstruction of the palaeoenvironmental setting.
The employed methodology has been based on two main steps. 1) the classification of terrain and off terrain objects performed using a strategy based on a set of "filtrations of the filtrate"; 2) the post processing based on comparing DTM scenes shaded by using different angles of lighting and following processing by PCA.
Such approach allowed us to improve, respect to the available optical data, the identification and interpretation of : i) microrelief for reconstructing the urban shape of the medieval site; ii) and the palaeoenvironmental features of the wooded site.

8. Reference

Axelsson, P. (2000). DEM generation from laser scanner data using adaptive TIN models. In: *IAPRS*, Vol. XXXIII, B4, Amsterdam, Netherlands, pp. 111–118.

Barnes, I. (2003). Aerial remote-sensing techniques used in the management of archaeological monuments on the British Army's Salisbury Plain Training Area, Wiltshire, UK. *Archaeological Prospection*, Vol. 10, pp. 83-91.

Bewley, R.H.; Crutchley, S.P.& Shell, C. (2005). New light on an ancient landscape: lidar survey in the Stonehenge World Heritage Site'. *Antiquity*, Vol. 79, pp. 636-647

Briese, C. & Pfeifer, N. (2001). Airborne laser scanning and derivation of digital terrain models. In: *Optical 3D Measurement Techniques*, Gruen, A., Kahmen, H. (Eds.), pp. 80– 87., Technical University, Vienna, Austria.

Briese, C.; Pfeifer, N. & Dorninger, P. (2002). Applications of the robust interpolation for DTM determination. In: *Photogrammetric Computer Vision. International Archives of Photogrammetry and Remote Sensing*, Kalliany, R.; Leberl, F. & Fraundorfer, F. (Eds), Vol. XXXIV, 3A, Graz, Austria, pp. 55–61.

Brovelli, M.A.; Reguzzoni, M.; Sansò, F. & Venuti G. (2001). Modelli matematici del terreno permezzo di interpolatori a spline. *Bollettino SIFET*, Supplemento Speciale, Vol. 2, pp 55-80.

Brown A.G. (2008). Geoarchaeology, the four dimensional (4D) fluvial matrix and climatic causality. *Geomorphology* , Vol. 101, pp. 278–297

Burrough, P. A. & McDonnell, R.A. (1998). *Principles of Geographical Information Systems.* Oxford University Press, Oxford, pp. 333.

Challis, K. (2006). Airborne Laser Altimetry in Alluviated Landscapes. *Archaeological Prospection*, Vol. 13, No. 2, pp. 103-127.

Challis, K., & Howard, A.J. (2006). A review of trends within archaeological remote sensing in alluvial environments. *Archaeological Prospection*, Vol. 13, No 4, pp. 231-240.

Challis, K.; Kokalj, Z.; Kincey, M.; Moscrop, D. & Howard, A.J. (2008). Airborne lidar and historic environment records. *Antiquity*, Vol. 82, No 318, pp. 1055-1064.

Chase, A.F.; Chase, D. Z.; Weishampel, J. F.; Drake, J. B.; Shrestha, R. L.; Slatton, K. C.; Awe, J. J. & Carter W. E. (2010). Airborne LiDAR, archaeology, and the ancient Maya landscape at Caracol, Belize. *Journal of Archaeological Science*. doi:10.1016/j.jas.2010.09.018

Coluzzi, R.; Masini, N. & Lasaponara, R. (2010). Flights into the past: Full-Waveform airborne laser scanning data for archaeological investigation. *Journal of Archaeological Science*, doi: 10.1016/j.jas.2010.10.003.

Coren, F.; Visintini, D.; Fales, F. M.; Sterzai P.; Prearo G. & Rubinich, M. (2005). Integrazione di dati laser scanning ed iperspettrali per applicazioni archeologiche, Atti 9a Conferenza nazionale ASITA, Catania 15-18 novembre.

Corns, A. & Shaw, R. (2008). High resolution LiDAR for the recording of archaeological monuments & landscapes. In: *Advances in Remote Sensing for Archaeology and Cultural Heritage Management*, Lasaponara R. and Masini N (Eds), Aracne, Roma, pp. 99-102.

Crawford, O.G.S. (1929). Air photography for archaeologists. *Ordnance Survey Professional Papers*, New Series, Vol. 12, HMSO, Southampton.

Crutchley, S. (2006). Light detection and ranging (lidar) in the Witham Valley, Lincolnshire: an assessment of new remote sensing techniques. *Archaeological Prospection*, Vol. 13, No 4, pp. 251-257.

Crutchley, S. (2008). Ancient and modern: Combining different remote sensing techniques to interpret historic landscapes. In: *Advances in Remote Sensing for Archaeology and Cultural Heritage Management*, Lasaponara, R. & Masini, N. (Eds), Aracne, Roma, pp. 103-106.

Danese, M.; Biscione, M.; Coluzzi, R.; Lasaponara, R.; Murgante, B. & Masini, N. (2009). An Integrated Methodology for Medieval Landscape Reconstruction: The Case Study of Monte Serico. In: *Computational Science and Its Applications*, Gervasi O. et al. (Eds.), Springer-Verlag Berlin Heidelberg, part. I, LNCS 5592, pp. 328–340.

De Laet, V.; Paulissen, E. & Waelkens, M. (2007). Methods for the extraction of archaeological features from very high-resolution Ikonos-2 remote sensing imagery, Hisar (southwest Turkey). *Journal of Archaeological Science*, Vol. 34, pp. 830-841.

Devereux, B.J.; Amable, G.S., Crow, P. & Cliff, A.D. (2005). The potential of airborne lidar for detection of archaeological features under woodland canopies. *Antiquity*, Vol. 79, pp. 648–660.

Doneus, M.; Briese, C.; Fera, M. & Janner, M. (2008). Archaeological prospection of forested areas using full-waveform airborne laser scanning. *Journal of Archaeological Science*, Vol. 35, No 4, pp. 882-893.

Elmqvist, M. (2001). Ground estimation of laser radar data using active shape models. In: *Proc. OEEPE workshop on Airborne Laserscanning and Interferometric SAR for Detailed Digital Elevation Models*, 1 - 3 March, OEEPE Publication no. 40, 8 pp. (on CD-ROM).

Gallagher, J.M. & Josephs, R.L. (2008). Using LiDAR to Detect Cultural Resources in a Forested Environment: an Example from Isle Royale National Park, Michigan, USA. *Archaeological Prospection*, Vol. 15, pp. 187–206.

Garrison, T.G.; Houston, S. D.; Golden, C.; Inomata, T.; Nelson, Z. & Munson, J. (2008). Evaluating the use of IKONOS satellite imagery in lowland Maya settlement archaeology. *Journal of Archaeological Science*, Vol. 35, pp. 2770–2777

Harmon, J. M.; Leone, M. P.; Prince, S. D. & Snyder, M. (2006). Lidar for archaeological landscape analysis: a case study of two eighteenth-century Maryland plantation sites. *American Antiquity*, Vol. 71, No 4, pp. 649-670.

Hengl, T. & Evans, I.S. (2009). Mathematical and Digital Models of the Land Surface. In: *Geomorphometry. Concepts, Software, Applications*, Hengl, T. & Reuter, H. I. (Eds), pp. 31-63

Hesse, R. (2010). LiDAR-derived Local Relief Models (LRM) – a new tool for archaeological prospection. *Archaeological Prospection*, Vol. 17, pp. 67-72, doi: 10.1002/arp.374.

Holden, N.; Horne, P. & Bewley, R. (2002). High-resolution digital airborne mapping and archaeology. In: *Aerial Archaeology: developing future practice*, Bewley, R. H. & Raczkowski, W. (Eds), NATO Science Series, sub-series I: Life and Behavioural Sciences, pp. 173-180. Nieuwe Hemweg 6b, 1013 BG Amsterdam, Netherlands: IOS Press. Available from
http://ads.ahds.ac.uk/catalogue/archive/lidar_eh_2008/downloads.cfm.

Humme, A.; Lindenbergh, R. & Sueur, C. (2006). Revealing Celtic fields from lidar data using kriging based filtering. In: *Proceedings of the ISPRS Commission V Symposium*, Dresden, 25–27 September, Vol. XXXVI, part 5.

Kraus, K. & Pfeifer, N. (1998). Determination of terrain models in wooded areas with airborne laser scanner data. *ISPRS JPRS*, Vol. 53, pp. 193–203.

Lasaponara, R. & Masini, N. (2005). QuickBird-based analysis for the spatial characterization of archaeological sites: case study of the Monte Serico Medioeval village. Geophysical Research Letter, 32(12), L12313.

Lasaponara, R. & Masini, N. (2007). Detection of archaeological crop marks by using satellite QuickBird. *Journal of Archaeological Science*, Vol. 34, pp. 214–221.

Lasaponara, R. & Masini, N. (2009). Full-waveform Airborne Laser Scanning for the detection of medieval archaeological microtopographic relief. *Journal of Cultural Heritage*, Vol. 10S, pp. e78–e82

Lasaponara, R.; Coluzzi, R.; Gizzi, F.T. & Masini, N. (2010). On the LiDAR contribution for the archaeological and geomorphological study of a deserted medieval village in Southern Italy. *Journal Geophysics Engineering*, Vol. 7, pp. 155-163.

Lasaponara, R.; Coluzzi, R. & Lanorte, A. (2010). On the LiDAR contribution for landscape archaeology and palaeoenvironmental studies: the case study of Bosco dell'Incoronata (Southern Italy). *Advances in Geoscience*, Vol. 24, pp. 125–132.

Lindenberger, J. (1993). *Laser-Profilmenssungen zur topographischen Gelandeaufnahme*, Ph. D. thesis.

Lohmann, P. (2000). Segmentation and filtering of laser scanner digital surface models. *IAPRS*, vol. 34 (2, WG II/2, August 22-23, Xi'an, China), pp. 311–315.

Masini, N. (1995). Note storico-topografiche e fotointerpretazione aerea per la ricostruzione della '"forma urbis" del sito medievale di Monte Serico. *Tarsia*, Vol. 16-17, pp. 45-64.

Masini, N. & Lasaponara, R. (2006). Satellite-based recognition of landscape archaeological features related to ancient human transformation, *Journal of Geophysics and Engineering*, Vol. 3, pp. 230-235.

Mazzei, M. (2003). Levate aeree per la conoscenza e la gestione del territorio. In : Lo sguardo di Icaro : le collezioni dell'Aerofototeca nazionale per la conoscenza del territorio, Guaitoli, M. (Ed), p. 115, 8888168125.

Moore I. D.; Grayson, R. B. & Ladson, A. R. (1991). Digital Terrain Modelling: A Review of Hydrological, Geomorphological and Biological Applications. *Hydrological Processes*, Vol. 5, pp. 3-30.

Pfeifer, N. (2005). A subdivision algorithm for smooth 3D terrain models. *ISPRS Journal of Photogrammetry and Remote Sensing*, Vol. 54, pp. 95-104.

Pfeifer, N.; Gorte, B. & Elberink, S.O. 2004. Influences of vegetation on laser altimetry e analysis and correction approaches. In: Proceedings of Natscan, Laser-Scanners for Forest and Landscape Assessment e Instruments. Processing Methods and Applications, Thies, M., Koch, B., Spiecker, H. & Weinacker, H. (Eds.), International Archives of Photogrammetry and Remote Sensing, Volume XXXVI, pp. 283e287. Part 8/W2.

Roggero, M. (2001). Airborne laser scanning: clustering in raw data. *International Archives of Photogrammetry and Remote Sensing*, Vol. XXXIV, B3/W4, Annapolis, pp 227-232.

Romain, W. F. & Burks, J. (2008) LiDAR Imaging of the Great Hopewell Road. available from
http://www.ohioarchaeology.org/joomla/index.php?option=com_content&task=view&id=231&Itemid=32

Shell, C. & Roughley, C. (2004). Exploring the Loughcrew landscape: a new airborne approach. *Archaeology Ireland*, Vol. 18, No 2(68), pp. 22-25.

Sithole, G. (2001). Filtering of laser altimetry data using a slope adaptive filter. *International Archives of the Photogrammetry, Remote Sensing and Spatial Information Sciences*, Vol. XXXIV(Pt. 3/W4), pp. 203- 210.

Sithole, G. (2005). *Segmentation and Classification of ALS Data.* PhD thesis, TU Delft.

Sithole, G. & Vosselman, G. (2004). Experimental comparison of filtering algorithms for bare-earth extraction from airborne laser scanning point clouds. *ISPRS Journal of Photogrammetry and Remote Sensing*, Vol. 59, No(1-2, pp. 85–101.

Sittler, B. (2004). Revealing Historical Landscapes by Using Airborne Laser Scanning. A 3-D Modell of Ridge and Furrow in Forests near Rastatt (Germany). In: Proceedings of Natscan, Laser-Scanners for Forest and Landscape Assessment - Instruments, Processing Methods and Applications., Thies, M.; Koch, B.; Spiecker H. & Weinacker, H. (Eds), International Archives of Photogrammetry and Remote Sensing, Volume XXXVI, Part 8/W2, pp. 258-261.

Sohn, G. & Dowman, I. (2002). Terrain surface reconstruction by the use of tetrahedron model with the MDL Criterion. *International Archives of the Photogrammetry, Remote Sensing and Spatial Information Sciences*, Vol. XXXIV (Pt. 3A), pp. 336- 344.

Stal, C.; Bourgeois, J.; De Maeyer, Ph.; De Mulder, G.; De Wulf, A.; Goossens, R.; Nuttens, T. & Stichelbaut, B. (2010). Kemmelberg (Belgium) case study: comparison of DTM analysis methods for the detection of relicts from the First World War. In: Proc. 30th EARSeL Symposium: Remote Sensing for Science, Education and Culture.

Tovari, D. & Pfeifer, N. (2005). Segmentation based robust interpolation - a new approach to laser data filtering. *ISPRS Workshop Laser Scanning 2005*.

Van Zijverden, W. K. & Laan, W. N. H. (2003).. Landscape reconstructions and predictive modeling in archaeological research, using a LIDAR based DEM and digital boring databases. In: *Archeologie und computer*, workshop 7. Vienna, Austria 2003. Available from: http://www.archeologie.leidenuniv.nl/content_docs/research/vanzijverden_laan_2005_landscape__reconstructions.pdf [Accessed 10 September 2007].

Vosselman, G. (2000). Slope based filtering of laser altimetry data. *International Archives of the Photogrammetry, Remote Sensing and Spatial Information Sciences*, Vol. XXXIII (Pt. B3), pp. 935- 942.

Wack, R. & Wimmer, A. (2002). Digital terrain models from airborne laser scanner data—a grid based approach. *International Archives of the Photogrammetry, Remote Sensing and Spatial Information Sciences*, Vol. XXXIV (Pt. 3B), pp. 293–296.

Weibel, R. & Heller, M. (1991). Digital Terrain Modelling. In: *Geographical Information Systems, principles and applications*, Maguire, D. J., Goodchild, M. F. & Rhind., D. W. New York: John Wiley And Sons, pp. 69-297.

Wilson , J. P.& Gallant, J. C. (2000). Digital Terrain Analysis. In: *Terrain Analysis: Principles And Applications*, Wilson, J. P. & Gallant., J. C. (Eds), (New York: John Wiley & Sons), pp 1-27.

15

Characterization of Ancient Ceramic Matrices with High Resolution Microscopy Methods

S. Mohammadamin Emami[1,2], Johannes Volkmar[1] and Reinhard Trettin[1]

[1]Institut für Bau- und Werkstoffchemie, Universität Siegen, Paul-Bonatz-Str. 9-11, 57068 Siegen

[2]Art Unversity of Isfahan, Departement of Conservation, Isfahan, Hakim Nezami St. P.O. Box 1744

[1]Deutschland

[2]Iran

1. Introduction

Potteries, ceramics and their production methods in the ancient world are one of the most important items for researching about the cultural heritage materials (Riederer, 1994). The processing of material being used for making ceramics and mineralogical investigations for understanding the production process, mainly firing, are among the most important goals of the present investigations.

Scientific studies on archaeological ceramics, based on laboratory methods carried out over decades and are known as ceramography (All-Saad, 2002). Studying the old objects give not only substantial information about the manufacturing in the past but also provides information about the processing of structures in such primitive materials (Letsch & Noll, 1983). Recognition of the evolution of ceramics is among the most important questions that has attracted the focus of archaeological researches. Apart from them, these ceramics being one of the oldest industries and arts, has become nowadays as one of the most modern research sciences for human being. In archaeological researches is also important to know about the procedure of manufacturing (in this case ceramics) in the daily life of nations (Fazeli et al, 2001). Beside archaeological investigations, cramography methods are useful and suitable ways for classification ancient ceramics and could help archaeologists to classify them with high accuracy. This investigation resulted in more accurate classification of some ceramics found in Persepolis World Heritage Site and showed high quality manufacturing process which used to make these ceramics (Maggetti, 1982).

The great consideration on these kinds of ceramics is due to their variation of raw materials which have been used for manufacturing, and additionally ceramics from Achaemenian period are recognized as transfer materials between two considerable dynasties in Iran, namely Elamite to Achaemenian. This transfer is marked as transfer through out of evidences and prominently also technical evidences.

To characterization ceramic matrices, some samples of ancient ceramic pieces from Persepolis (Takht-e Djamshid) world heritage site were selected. These samples were examined with Atomic Force microscopy (AFM-PFM) (Emami et al, 2008 & Maggetti, 1982).

The experiments get information about the topography of ceramic's surface and generating of the surface. Also grain boundaries were recognized and AFM pictures gave sufficient information on sintering around a grain and the spreading of this process from the grain boundary into the matrix of the ceramic. The sintering has occurred complimentary well in these samples. To characterization of the raw materials that used for making of these ceramics, QXRD method is recognized. The calculated and measured peaks are refined by Ritveld refining methods to realize about the crystalline structures. According to XRD and polarization microscopy studies the research materials can be divided in one main group considering their matrix structure. This group proves siliceous matrix which mainly consist of SiO_2 and Pyroxene component which show us the stability of the pyroxene solid solution in the area of 750 - 870°C. Calcareous matrices which mainly consist of CaO, $CaCO_3$ and a minor content of SiO_2 are also observed as non complimentary sintered ceramics.
In fact by using of chemical- mineralogical methods, one seeks is to find the using technique of crude material in the past by means of laboratory methods and not merely through comparative comparison of the potteries of a particular investigated area. Investigations based on material characterization also let to creating a technical chronology with respect to the technique that practised in each area.

2. Experimental procedure

For chemical- mineralogical investigation and characterization of the ceramic matrices, there are different disciplines which used in the following chapters. These methods and the aim of each method describe schematically in the Figure 5.
The main purpose of this investigation's discipline is to pay attention to the textural formation of the material as well as finally sinter structure of traditional ceramics based on mineralogical- chemical properties in such system.

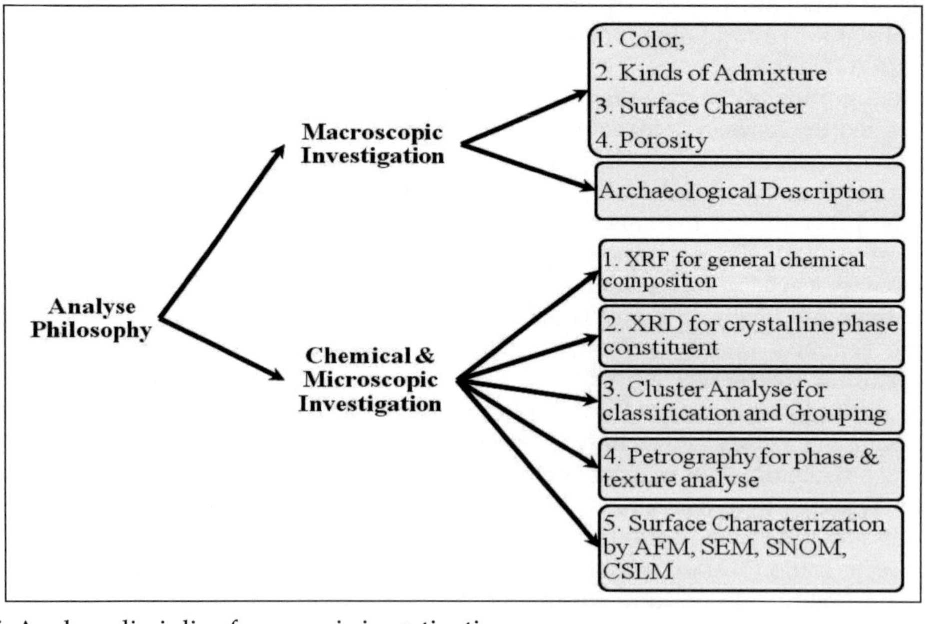

Fig. 5. Analyze discipline for ceramic investigation

3. Sample descriptions

3.1 Macroscopic investigation

The shape, typology and the color of the objects are drastically variable and can be grouped as follows:

1. Potteries without paintings, having a very single form, the color of which being red, green or buff.
2. Yellow, bright brown potteries having very fine designs in the form of parallel lines.
3. Potteries with simple forms and paintings produced by exerting pressure on the body.

In general potteries in the region are mostly produced by clay, soil and earthen material. The color of these potteries is classified as bright, having very low porosity, and for the same reason, have a rather high toughness. The color of some samples varies from the corn to the surface because of uncontrolling firing temperature (Fig. 1). In many samples typical admixtures are visible on the texture of the ceramics (Fig. 2). They could be the remains of

Fig. 1. Ceramic with different colors in the matrix as a function of uncontrolling firing.

Fig. 2. A part of a Vase structure varies from the surface to the core and also different admixtures

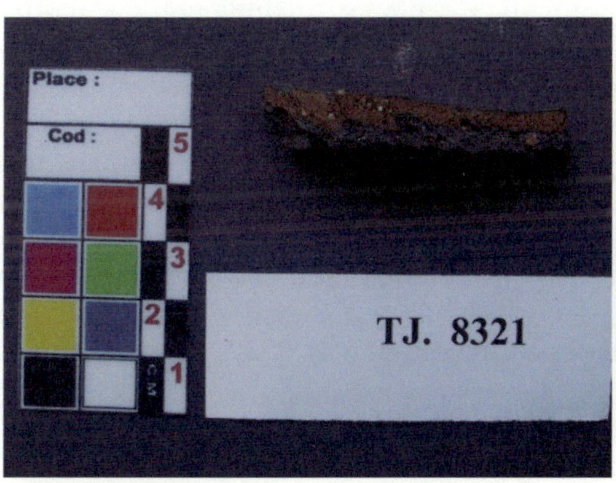

Fig. 3. Sandwich structure with different matrix texture in ceramics.

Fig. 4. Salt recrystallization on the surface through capillary effect of high porosity ceramics.

organic materials such as hairs, straws. In some cases the sandwich- and glassy structure are also noticeable (Fig. 3). These kinds of ceramics consist of three layers, the outer layers are reddish and the inner part of the ceramic structure is dark. These ceramics have glassy echo by knocking on their surface. The corrosion effects and the salt recrystallization on the surface of the ceramics are mostly observed. These could occur as a function of high porosity of ceramic's body which is very important for traditional ceramics (Fig. 4).
Their typology mostly includes long mouths and in some cases spherical shapes. In addition to these, there exist some bowls with a flat bottom and bevelled rims.

3.2 Chemical- mineralogical investigations

3.2.1 Chemical composition of potteries

Identification of the chemical composition of samples has been carried out by X-Ray Fluorescence Analyses (XRF). The aim is to find the correct chemical formula and combination of the samples which could also obtained an statistical distribution of elements that consist the pottery's matrix in system $CaO+MgO - SiO_2 - Al_2O_3$ (Noll, 1991). In later phases, some complementary methods will be used for grouping the materials. Among them one can mention by the "Scanning Electron Microscope" methods.
The chemical composition of all the samples is given in Fig. 6. From the whole range of samples, there are only 6 sub-samples that will be discussed in this paper. From the chemical standpoint, the samples are characterized by high SiO_2 and approximately amount of CaO contents. Another important composition in these samples is Al_2O_3. The ceramic's matrix is classified in general into two groups; the samples of group 1 are defined as sandy matrix, and the second group is defined as a calcareous matrix. The origins of these components depend on the raw materials which were used for manufacturing. Sand is the most common raw materials that make up the main clay mineral deposits in this area. The application of quartz had two considerable advantages with respect to its physical and chemical properties;
Firstly, Quartz is a hard mineral (H=7 on the Mohs scale), secondly, Quartz shows a high physico-chemical resistance against acidic conditions (acidic solution or acidic atmosphere). Indeed, here is another point that made quartz unfavourable for such procedures. Quartz changes its structure three times (decomposition) while reaching its melting point, and after this point it will cross over to the glass form. This could to be a hypothesis, why have the oldest ceramics still good texture as well as body for research purposes, and why optically they seem to be glassy (with uncontrolled temperature in the open furnace) (Emami et al, 2009). Quartz rich wares create a glassy rattle by the knocking on their surface. Perhaps this might be a way to distinguish quantitatively the age difference between some ceramic wares. The existence of carbonate could have three reasons. Firstly, the carbonate content seems to originate from calcareous stone, such as calcite and dolomite that contain the main geological formation on the field; secondly, the calc produced as a secondary phase through the firing process as a result of contamination of ceramic fragments (Ca^{2+}, Fe^{2+}, Mg^{2+}) and third, calc is used in the ancient time as flux.
XRF results proved that these kinds of ceramics are SiO_2 rich ceramics (Fig. 6) and there exist large distinction between them and the ceramics from south-west Iran (Emami et al, 2008). Their matrices are glassy and consist of many crystal fragments that have a magmatic origin. In some samples there exist also sandwich structures in the ceramics as layering structure with different colors. The inner layer is darker than the other and it comes through high SiO_2 content (Emami et al, 2009).

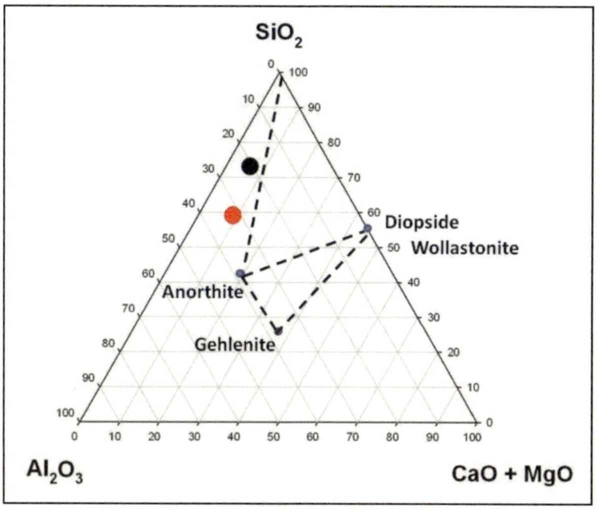

Fig. 6. Classification of two ceramic groups from Persepolis in system Al_2O_3 - SiO_2 - CaO + MgO. According to this classification the ceramics fit in to the high SiO_2 - Al_2O_3 category.

3.2.2 Identification of crystalline phases in pottery's texture with Rietveld refining method

XRD analysis carried out for identifying Mineralogical phases in the ceramic matrices. To gain a high-quality interpretation from X-Ray Diffraction and the refining of the results according to the Rietveld Method, the chemical and mineralogical investigations and phase analyses of ceramic samples carried out qualitatively and either quantitatively. The Rietveld method allows the option of classification of phase identification and compensates for line broadening induced by particle size effects. This method of research using Powder Diffraction was recorded by the International Union of Crystallography Commission (Mccusker et al, 1999). The topics from these kinds of refinements are as follows; (i) classification of data, (ii) peak shape description, (iii) background determination, (iv) Fourier analysis and (v) refinement of profile parameter (Mccusker et al, 1999). In this application, the results will be presented and discussed based on the final calculated diffractogram. For these samples, the X-ray generator operated with a copper target at 45kV and 40mA. The powder pattern of ceramics was measured between $2\theta = 3°$ and $2\theta = 61.80°$, using a scanning rate $2\theta = 0.01°$ per minute. This large scanning area was required because the important clay minerals, such as illite, kaolinite and layer silicates such as biotite and muscovite, show the major peak intensity over these ranges. On the other hand, some phases, such as quartz and calcite, that, indeed, play a large role in our research, show the typical peaks for matching their identity between $2\theta = 51°$ to $2\theta = 58°$. The samples showed the minimum intensity of 5 - 58 counts, and a maximum intensity between 1029 – 2828 counts. These differences occurred as the origin of materials and the methods for pottery production were different (Fig. 7).

Quartz, calcite and some chain silicates, such as pyroxenes, are the major minerals which are contained in the ceramic bodies. Feldspars, plagioclases and some modification of melilite groups are the minor phases in the ceramics. Micas and clays minerals in different modifications, and some other minerals, such as nepheline, are present as trace phases, which are calculated in these materials. Quartz appeared mainly as trigonal/rhombohedral in form of α- quartz. This structure of quartz is stable at low temperatures under 573°C, and the best crystallographic preferred orientation appeared at d [Å] = 3,34 (101), 4,26 (100) and 1,54 (211). The last peak is the important strong line of quartz because it does not overlap with another mineral phases, such as calcite, analcim or orthoclase. Over these ranges and with this strong line, it was possible to identify the quartz exactly. In many cases, Qz + Mus or either Qz + Ana showed a sharp overlapping over the main peaks at $2\theta = 26.64$. For using the Rietveld method in order to obtain a better quality of fit in the calculation such, it is better to refine other phases such as muscovite or analcim, with quartz. Calcite appeared with the best orientation at d [Å] = 3,04 (100), 2,28 (111) and 1,92 (011) and 2θ = 29,2° &39,4° & 47,0° (Emami et al, 2008). The next important group within these materials are the chain silicates, such as pyroxenes. Miscibility gaps of pyroxenes in general were apparent between calcic-clinopyroxenes and pigeonite, and therefore an intermediate composition may be possible at high temperature (Emami et al, 2009).

Extensive solid solution is possible among members of the calcic, calcic-sodic and sodic pyroxenes. The best preferred orientation is over 2θ = 29-31° and 34,5-36,5° (Emami et al, 2008). The next important calculated phases are the framework silicates, the so called feldspars & plagioclases. These phases appeared in all miscibility formations. The main minerals in these groups are alkali feldspars, such as Orthoclase, sanidine & microcline. The best-preferred orientation of plagioclases occurred between Qz and Cc. The greatest intensity of these groups of minerals occurs between 2θ = 27-30° and 2θ = 35-40°. In our case study, gehlenite ($Ca_2[Al_2SiO_7]$ as a common minerals from melilite miscibility appeared, and this structure required substitution at some compositions as $MgSi_2 \leftrightarrow Al$, AlSi to $MgSi_2 \leftrightarrow Si$, AlAl.

According to XRD studies, and with respect to available phases, the potteries from Persepolis can be divided in two groups as follow:

Group I, Mostly consist from quartz and granitic fragments; among them one can specify as Muscovite and Microcline. This kinds of samples includes % 54,16 of the samples. Other important phases in this texture are magnetite, hematite, enstatite, diopside, anorthite, orthoclase and nepheline. Hematite and magnetite are the high temperature phases in the traditional ceramic technology (Maggetti & Galetti, 1982). Enstatite and diopside are the secondary phases that occurred during firing

Group II., This group in general includes Calcite and Quartz. The chemical composition of this group of potteries varies drastically depending on pyroxene, muscovite and plagioclase. Stone fragments and calcareous fragments also exist in this group which is an evidence for different techniques in pottery making. This group includes some % 37,5 of the whole. In general it should be concentrate that Pyroxene is more abundant in potteries which are an evidence for the use of different admixtures in the soil of that region due to the variation in available quarries of the field. The variation of the crystalline phases in these ceramic textures based on XRD-Patterns demonstrate on Fig. 8.

3.2.3 Petrological and petrographical investigation through polarized light microscopy

The aim of "thin width samples" studies is to confirm the existence of identified phases through XRD method as well as relation between these phases and surrounded matrix from mineralogical point of view. Of greatest importance for these studies, is to investigate the binding process on the grain boundaries that accomplished the structure of the pottery at least as well as determine the kind of used temper and pasta. These kinds of phases and structures within them they reveal themselves in the background structure are twining, color, corona at the boundary of grains, double refraction of light, crystal network and

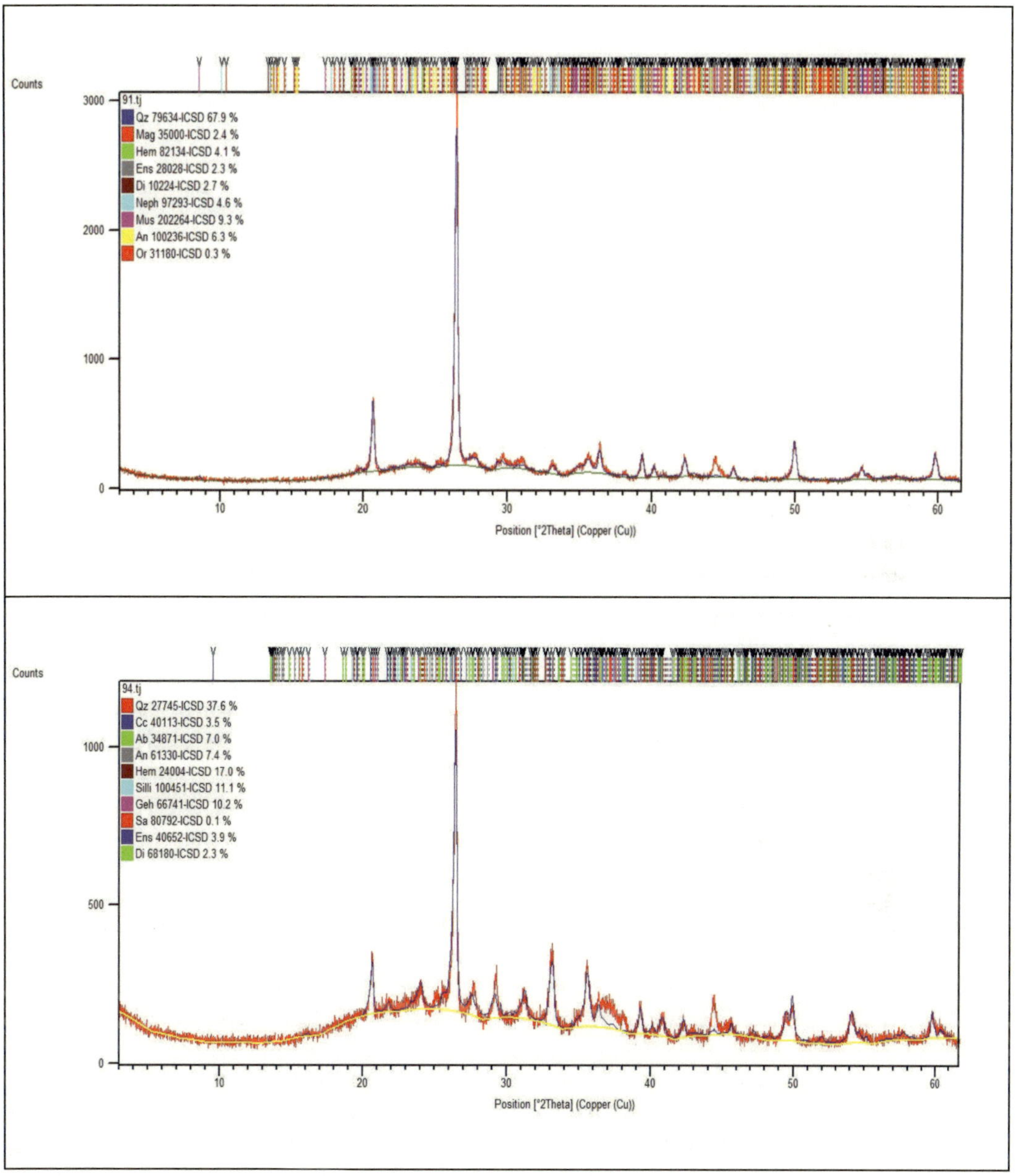

Fig. 7. XRD Patterns from two different samples from Persepolis after selective Dissolutions of Phases.

Qz = Quartz, Cc = Calcite, Di = Diopside, En = Enstatite, Ab = Albite, An = Anorthite, Or = Orthoclase, Sa = Sanidine, Mag = Magnetite, Hem = Hematite, Neph = Nepheline, Mus = Muscovite, Sili = Silimanite & Geh = Gehlenite

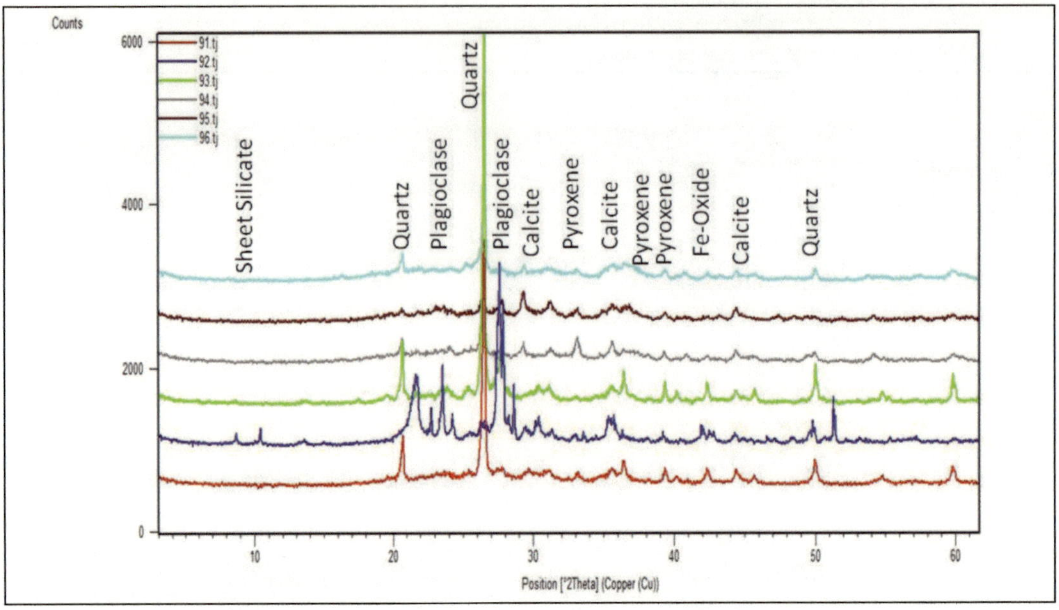

Fig. 8. The variation of the ceramic texture based on XRD-Patterns. It demonstrates the different structure based on quartz, plagioclase and pyroxene contents.

1. *Quartz;* which in most cases represent itself in the form of spherical as well as edge aggregates with different sizes. Structure of quartz with their specific "dark angel" is observed as undulatory extinction (Fig. 9). Quartz is seen in many cases together with Ferric ores. Potteries consist of abundant amount of quartz in their structure proved dark color than the others. The color becomes more inclined towards to reddish according to heating range and also kiln atmosphere.
2. *Calcite;* is observed in all samples and demonstrates different specification from optical aspect. Investigations have shown that calcite reveals itself in spherical or broken form. Collectively with calcite there is aspect of ferric ore in form of hematite layers (Fig. 10). This layer seems to be rebuilt from MgO, FeO or other secondary oxidation products. Calcite is among phase which is observed in the pottery's matrix and determined also the region's soil conditions as well as alteration.
3. *Feldspar;* appear in different modification as a function of unlike conditions; and for the same reason, reveal their phase in different structures. Feldspars appear mainly in form of albit as well as serisite-albit, or as plagioclase with microlcline modification. One can identify these structures through the specific twining structures or through detecting light direction in them. Plagioclase has seen less frequently in the pottery's structure; while serisite has been seen much more frequently as alteration products of albite (Fig. 11, 12). In addition to mentioned ores, some other phases observed also in pottery's textures; among them biotite and pyroxene (Fig. 13) are the typical phases in high temperature range. According to the microscopic observation these phases don't categorised as primary phases in pottery's structure. The structure of consist phases and sub-minerals get information about the authenticity of the soil used for pottery making. According to the microscopic investigations the soil that used for pottering has the same origin.

4. The group of clay minerals are among the most important constituents in the ceramic texture. The clay minerals provide their structure twice, firstly by addition of water to clay and secondly come through firing as secondary phases in the texture. Apart from them, one should consider that the size of the clay minerals is not coarse enough for discovering their structures with light microscopy. These minerals studied mainly through scanning electron microscopy as well as atomic force microscopy.
5. Admixtures that added to the clay source for pottering mainly consist of organic shale as well as straw and agricultural remains from the surrounding environments. In many cases fossils are the fingerprint for characterization of the soils (Fig. 14). The qualities of material as well as the unique similarities between these soil reservoirs nearby Persepolis and those which were used for pottering suggest them to be one of the sources which used for pottering in Persepolis complex.

3.2.4 Application of SEM investigation on interpretation of sinter structure

SEM investigation and observation lead to find some interesting results about the mechanical failures. In traditional ceramics, sintering occurs around 1100 - 1200°C, depends on the raw materials as well as the firing atmosphere. Sinter structure is mainly a compact structure with high porosity effect. The constituent minerals change their textures during sinter and followed by don't looked like the origin shape. One of the most evidently fractures, are the damaging effect on the surface of investigates materials.

Effectively, the origin of the fracture could be traced to the defect in the structure of the ceramics through the pottery manufacturing process. The cracks pass from their origin toward to the other side of the samples surface. The cracks could be classified as a capillary cracks, which are very fine and can be defined as microcracks, or shrinkage cracks, which happened by changing the physical- or thermodynamically coefficients of the surroundings environments.

According to Fig. 15, the structure of Aluminosilicates observed that reacted with the environment. Such disruption of aluminosilicate suggests that these kinds of minerals tend to change its structure easily. In this example the Matrix is mainly consist of lime or clay-rich raw material (Maggetti & Schwab, 1982). Alkaifeldspars with typical crossed cleavage are also observed as the high temperature phases in the quartz-rich matrix (Fig. 16) (Mathé, 2007). The bright structures suggest probably salt recrystallization on the surface of crystal. The conversion of quartz-α (low-quartz) to quartz-β (high-quartz) will take place with 5% volume expansion namely; quartz splitting. For the reason that this operation occur in a very short period, such cracks increased in the mainly part of the ceramics (Fig. 17).

The cracks appeared in a material with high density. Such cracks on the surface of collected grains could happen only after changing the ambient decomposition. In the middle of the samples, there are tiny cracks which proceed and propagate themselves from diagonally from the bottom to the top left hand on the surface. These cracks could be identified as secondary cracks, which appeared around the grain after weathering or diffusion (Fig. 18).

As a matter of fact, during the firing and burning process the quartz rich matrix and the calcareous matrix show different damage models. In the calcareous matrix, deformations normally increase over a large area on the samples and the cracks are wide enough to have secondary reactions among them. According to figure 18 in argillaceous and quartz rich matrices, the cracks normally are found on the quartz grain and demonstrate internal extension during the decomposition of α-quartz to β-quartz. In many cases, the crack's opening accomplish if another aggregate dropped out from the matrix.

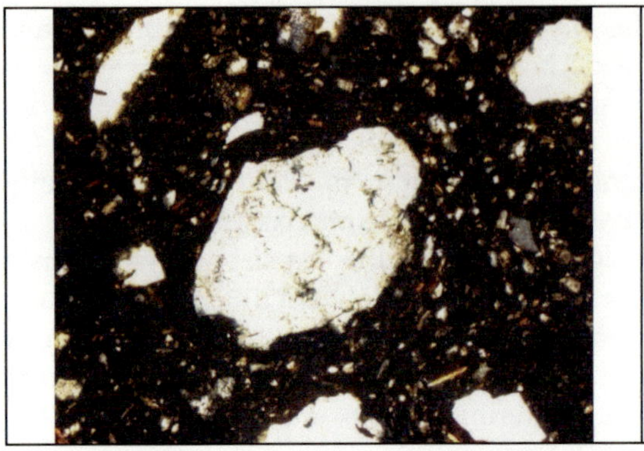

Fig. 9. Quartz with different size and typical undulatory extinction. In the matrixconsist of stone fragment. x 10. Pol + Ana.

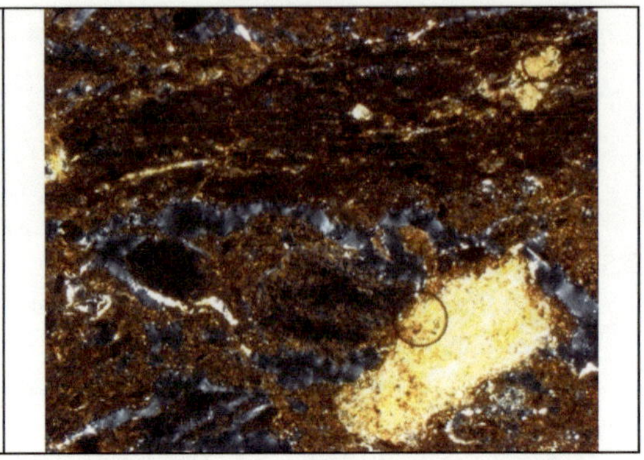

Fig. 10. Calcareous matrix consist of spherical calcitic aggregate sarounded with Ferro- material. x 5. Pol + Ana.

Fig. 11. Albit with typical zoneing structure as myrmekite texture. This structure appeared through rebuild of radiactive elements in the interatomic places of crystal structure. x 10. Pol +Ana.

Fig. 12. Sericitization in Albit as a fine lamel structure. In a matrix fron stone fragment quartz is seen too. x 20. Pol +Ana.

Fig. 13. Biotite as mineral aggregate whit characteristic red color. x 20. Pol +Ana.

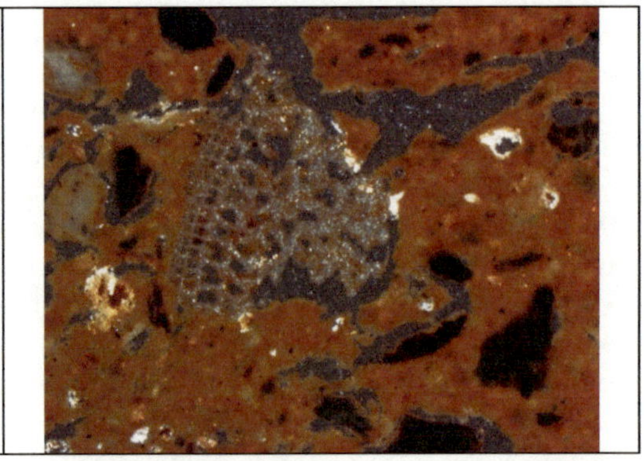

Fig. 14. Characteristic high Ca-matrix with organic admixture as fossils shale that comes through the soils. x 20. Pol +Ana.

Fig. 15. Decomposition of aluminosilicate in a quartz rich matrix through high sintering

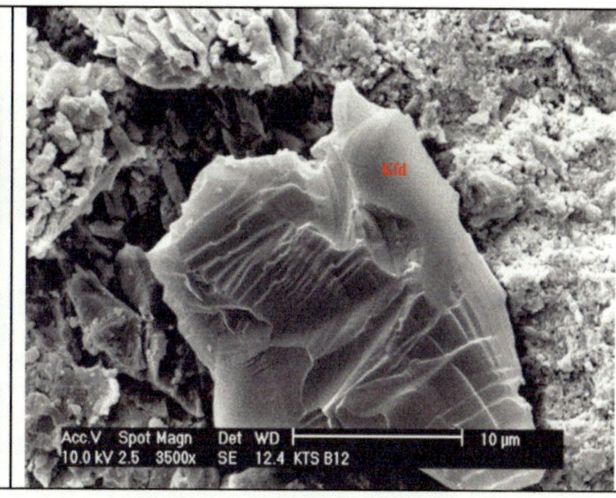

Fig. 16. Alkalifeldspar with typical twinning structure with calc as well as salt recrystallization.

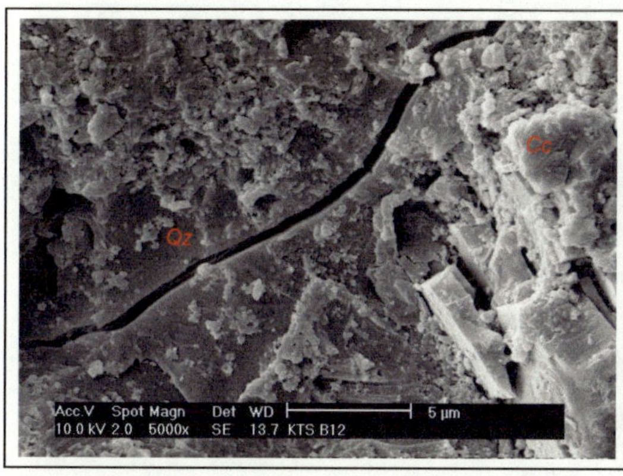

Fig. 17. Quartz splitting by 550°C and the primary cracks that happen during firing process.

Fig. 18. Secondary cracks on the surface of the ceramic's matrix as a function of firing process.

3.2.5 Confocal and atomic force microscopy

Confocal as well as AFM microscopy methods divided us to find out about the characterization and to obtain some physical and topographical properties of the surface. The homogeneous or inhomogeneous of the material's surface, and the frequent distribution of such elements such as Fe, Mg, Ca, can give us a result about the discolouring of the ceramic bodies and, on the other hand, the roughness and therefore potential diffusions under the microscope (Emami et al, 2008).

Ceramic texture been used to characterize many aspects of atomic and molecular arrangements. Confocal microscopy allows one to show the properties of the surface and textures of the materials, which includes a three-dimensional reconstruction of the specimen's surface and fractures over large time scales. The ceramic surface consists of a solid state with many phases in equilibrium. The structure and hardness of the surface and the character of mineral boundaries depend strongly on the type of the particles and their physical properties. Cracks, which happened through shearing forces or internal shrinkage or external forces, could be a reason for deformation and damage. Such microscopic observation helps one to connect the surface dynamics to the macroscopic damage behaviour. By use of the AFM method, it is be possible to measure the interaction between a solid surface and extremely fine pike (Emami et al, 2009). In general, this involves scanning an area in non-connected mode, which is carried out by a kind of sensor called a cantilever. In this case, a laser beam is focused on the cantilever and reflected to the detector. The detector measures the deflection the cantilever and, in the following, of the surface. To observe the cracks and deformation areas, it is important to have a fresh break on the surface and, for this reason, it is better to have a piece or sample instead of the thin sections. Physical and mechanical interferences on the surface have been interpreted by AFM method through topographic and phase images (Fig. 19).

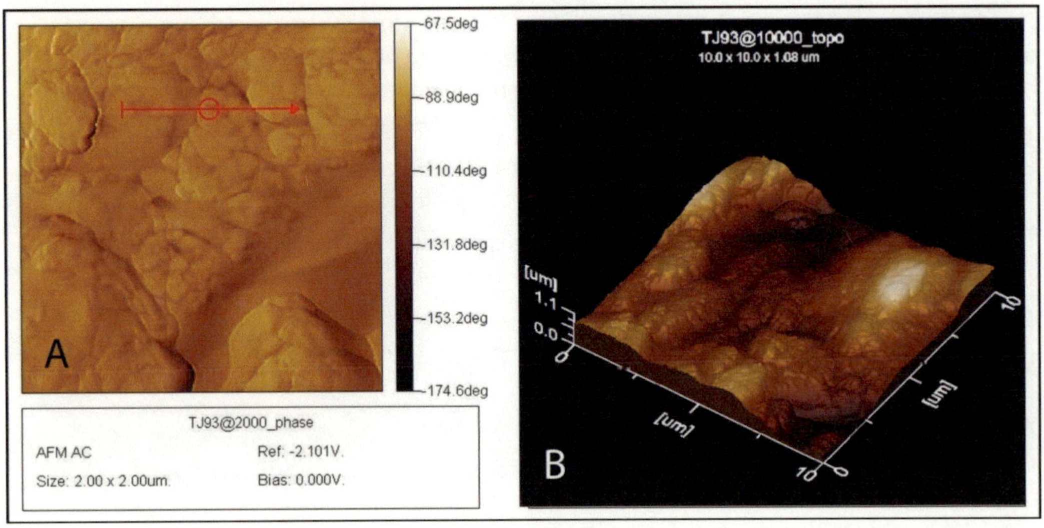

Fig. 19. AFM image of the topography on the surface in Non Contact Mode. A: is the phase image that illustrate a homogeneous matrix depends on candilevel flections factor. B: Topography image and the born of cracks on the surface of grains.

4. Results and discussion

The chemical analysis of investigated materials concludes that the materials respond as high quartz (argilleous), materials. The physicochemical properties of their materials depend on the components included in them. The chemical components also control the kinds of weathering as well as damaging process. There are many important factors during the firing process of ceramics that have influences on the reaction conditions; increasing and decreasing the temperature range and the chemical composition of the gases in the kiln. The twin-oxidation reaction in manufacturing of potteries ($C/CO/CO_2$ & Fe_2O_3/Fe_3O_4) will proceed under high oxygen fugacity and a wide range of temperature. According to these reactions, it is necessary to consider that firstly, water delivered from the system and, based on the free lime constituents the CO_2 concentration will increase in the atmosphere of kiln. These two reactions suggest to the dehydration of clay materials or dissociation of carbonates.

Apart from them, the ceramics that have a high amount of quartz, demonstrated a higher damaging process rather than the materials which contain a calcium or magnesium rich matrix. The ceramics with a high content of Mg proved refractory properties in their structure.
The effect of different carbonate decompositions tend to understanding about the pottery production process. The best wares and products were those wares which were fired and burned rapidly, and cooled very slowly. These procedures took place in closed kilns with exactly controlled temperature during the manufacturing process.

5. Conclusion

Analytical chemistry and petrological-mineralogical investigation as well as microscopy methods, such as polarization, scanning electron microscopy (SEM), confocal laser scanning microscopy (CLSM) and atomic force microscopy (AFM) have a great role on the study about ancient materials as interdisciplinary subjects.
Apart from them prominent damage features have been reliably identified and characterized, showing that mineralogical- chemical investigations can involve other techniques in archaeology in form of experimental archaeology or archaeometry. In general, phase analytical methods (XRD, and the refinement with the Rietveld methods as complimentary methods) showed meaningful results to explain as well as illustrate ancient technologies. Microscopic methods are complementary studies to reinforce the interpretations and observations.
The varieties of experiments and the reproducibility of the results make these analyses as good fundamental choice for distinguishing and explaining questions in archaeological problems as well as material science discipline.
Research methods regarding to the archaeological feature based on the technical and scientifically interpretations are interdisciplinary way to connect gaps between archaeology and material science.

6. References

All-Saad, Z., (2002), '*Chemical Composition and Manufacturing Technology of a Collection of Various Types of Islamic Glazes Excavated from Jordan*', Journal of Archaeological Science, 29, pp: 803-810.

Emami, S. M. A., Volkmar, J., Trettin, R., (2008), *Quantitative Characterization of Damage Mechanisms in Ancient Ceramics by Quantitative X-ray Powder Diffraction, Polarization Microscopy, Confocal Laser Scanning Microscopy and Non-contact Mode Atomic Force Microscopy,* Surface Engineering, vol. 24, issue 2, pp: 129-137

Emami, S. M. A., Kowald, T., & Trettin, R., (2009), *Mineralogical and Chemical Investigation on the Recrystallization Process during Sintering in Phase-Interface Areas in Ancient Ceramic Matrices,* Materials and Manufacturing Processes, issue 24, pp. 934-941.

Fazeli, H., Coningham, R., Pollard, A., (2001), *Chemical characterisation of late neolithic and chalcolithic pottery from the Tehran plane, Iran,* Iran vol. 1, pp: 55-71.

Letsch, J., Noll, W., (1983), *Mineralogie und Technik der frühen Keramiken Thessaliens,* N. Jb. Miner. Abh. 147, pp: 109-146.

Maggetti, M., (1982), '*Archaeological ceramics*', Washington, DC, Smithsonian Institute Press, pp: 121–134.

Maggetti, M., Galetti, G., (1982), '*Die Referenzgruppe(n) Lousana - mineralogische und chemische Untersuchungen der keramischen Produktion der Töpferwerkstätte Berna*', Jahresbericht der Schweiterischen Gesellschaft für Ur- und Frühgeschichte, vol. 65, S. 109-132.

Maggetti, M., Schwab, H., (1982), '*Iron age fine pottery from Chatillon-s-Glane and the Heuneburg*', Archaeometry, vol. 24, S. 21-36.

Mathé, V., Meunier, A., Lévêque, F., (2007), '*Anthropic acceleration of a natural clay mineral reaction in marshland soils (Atlantic Coast, France)*', Clay Minerals, vol. 42, No. 1, S. 1-12.

Mccusker, L. B., Von Dreele, R. B., Cox, D. E., Louer, D., & Scardi P., (1999), "*Rietveld Refinement quidlines*". Journal of Appliede Crystalography, Issue 32, 36-50

Noll, W., (1991), *Alte Keramiken und ihre Pigmente,* E.Schweizerbrat´sche Verlagsbuchhandlung Stuttgart.

Riederer, J., (1994), *Echt und falsch,* Springer Verlag Berlin Heidelberg, ISBN: 3540578935.

16

Application of Non-Destructive Techniques to the Recording and Modelling of Palaeolithic Rock Art

Diego Gonzalez-Aguilera, Pablo Rodriguez-Gonzalvez, Juan Mancera-Taboada, Angel Muñoz-Nieto, Jesus Herrero-Pascual, Javier Gomez-Lahoz and Inmaculada Picon-Cabrera

High Polytechnic School of Ávila, University of Salamanca
Hornos Caleros, 50, 05003, Ávila
Spain

1. Introduction

1.1 Motivation

According to Spanish legislation and international organisations such as UNESCO, the Council of Europe and the European Union, historical and artistic heritages should be preserved and placed in the service of society for either cultural, scientific or educational purposes. Thus, it is necessary to document and invest in their preservation, restoration, rehabilitation and/or archival for subsequent scientific studies (Elwazani, 2003), in addition to documenting their dissemination and social value. Archaeological remains are part of peoples' historical artistic heritage and represent a testimony of their past. The attitude and sensibility towards this cultural inheritance say much about our future as a society.

The documentation of general heritage and particularly archaeological heritage properties is indispensable before performing any type of measure or intervention. Such documentation implies recording, storing, cataloguing and measuring the elements that compose the heritage property. With regards to archaeological heritage, these tasks are developed both with elements that comprise movable heritage (for example, bone and lithic industries) and with elements that comprise non-movable heritage (for example, buildings, fields and caves). The geometric component of archaeological documentation begins with measurement and has the objective of a graphical representation, using multiple strategies and support. The act of measuring implies a quantification of the spatial characteristics of an object, especially its shape, dimensions, orientation and location, both in its immediate environment and in relation to the global geographic context.

In the geographic documentation of heritage, the measurement and graphical representation are indivisible tasks. Particularly, the traditional procedures of documentation and graphical representation are based on employing equipment that has a low cost and is easy to use. Despite the effectiveness of these procedures, they exhibit a series of inconveniences, such as the large amount of time required to acquire data, the limit of work within a two-dimensional scope, the loss of information during data transfer and the need to physically access the structure. Primarily, it is difficult to obtain reliable geometric data on irregular structures, as is the case for archaeological remains, which means that the precision obtained from this application is relatively low (Fossati et al., 1990). In recent years, due to the spectacular development of geomatic techniques and the evolution of the content and structure demanded in the documentation of heritage (Cannataci et al., 2003), a considerable increase has been noted in the use of techniques such as close-range photogrammetry and laser scanning, which provide reliable graphical and metric information on objects. These techniques allow digital archives to be obtained in various representation formats: 3D models in CAD format, 3D visualisations, conventional planes in 2D, orthoimages, reconstructions and virtual animations. However, despite the advances, the generation and visualisation of 3D models in archaeology are not as in demand as would be expected, due to various factors including (a) the high cost of 3D, (b) the difficulties for non-experts to easily obtain quality 3D models, (c) the belief that 3D is mainly an additional aesthetic factor and (d) the difficulty of integrating 3D "worlds" with other classic 2D archaeological data.

The metric documentation of heritage is an excellent opportunity that allows us to integrate technology, history, development and innovation so that various experts in this field have the possibility of applying precise, non-destructive tools for the documentation, modelling and analysis of traces of Palaeolithic art. In addition, and just as important, the possibility of virtual recreation combined with the variable of time will allow for preservation of places with massive visits, present hypothesis and recreation of the site's evolution, without forgetting the possibility that handicapped will be able to enjoy and appreciate this type of places without difficulty in access.

The **objectives** of this study revolve around three general lines of investigation:

1. Three-dimensional digitisation of Palaeolithic art corresponding to two caves in the eastern part of Asturias by means of advanced data collection systems that are non-destructive, non-invasive and automated.
2. The development of information processing strategies that will provide three-dimensional reconstruction and the acquisition of cartographic products that correspond to the Palaeolithic art of two subterranean caverns in the eastern part of Asturias.
3. The implementation of a Spatial Information System that permits the management and diffusion of the Palaeolithic art found in caves.

1.2 State of art

In this subsection a brief state of art about non-destructive techniques applied to rock art recording and modelling is presented. Furthermore, this subsection has been developed following a twofold level: the sensor level, in which a description of the main sensors applied to rock art recording is provided; and the processing level in which the main algorithms and strategies for rock art modelling are described.

- *Methods and Surveying Equipment*. Classic surveying methods based on measuring angles, distances and variations in height have been used in various studies on caves and subterranean cavities (Lopetegui et al., 2004). The equipment used for these purposes ranges from expeditious teams consisting of runners with tape measures, who measure short distances directly without much precision, followed by the classic theodolites, which permit the measuring of points based on indirect methods such as simple or multiple intersections (Ghilani and Wolf, 2006) supported by stadimetric systems; there are also modern complete stations to directly and simultaneously measure angles and distances without the need for reflective elements, considerably reducing field work. Currently, the use of popular GPS positioning systems has been added to these surveying methods to include the positioning and global georeferencing of the caves in the mapping of the area.

- *Methods and Photogrammetric Equipment.* From close-range photogrammetric methods, it is possible to derive measures of the images and to generate scale three-dimensional reconstructions. In addition, the images previously corrected for perspective and distortion can serve as an information source, providing realism to the reconstructed 3D models. In this sense, close-range photogrammetry entails an adequate system for measurement when dealing with the reconstruction of simple objects with a high degree of precision and a low cost (Chandler and Fryer, 2005). Thus, professionals in photogrammetry have developed different approaches to metrically document prehistoric art, from the more classic photogrammetric techniques based on the use of stereoscopic vision (Rivett, 1979), to three-dimensional modelling based on multiple convergent images (Ogleby, 1999). However, even though photogrammetric methods represent a universal, low-cost alternative, their application in subterranean scenarios such as caves and subterranean caverns is less common, as the conditions of accessibility and visibility are very restrictive and the geometries represent large complexities with alternating concave and convex shapes. Certainly, the successful application of photogrammetric systems in caves and subterranean caverns will demand time, skill and advanced knowledge on the part of archaeologists and pre-historians.
- *Methods and Laser Equipment.* Currently, terrestrial laser scanners constitute the latest breakthrough in the field of measurement, allowing massive amounts of information to be captured on the shape of objects by means of the measurement of angles and distances. They do not require operator intervention for visual determination and obtain thousands of points per second. There are variants of these instruments according to their measuring principle and range: long-reach laser scanners have ranges that vary from two to three metres, as a minimum distance, to a maximum distance of one kilometre, providing clouds of points with a precision similar to that of a total surveying station, thereby decreasing the measurements and increasing the distance between the object and the laser scanner; medium-range laser scanners vary from one to two metres, as a minimum distance, to a maximum distance of 350 metres, which is ideal for applications related to both the exterior and interior of immovable heritage sites (Robson-Brown et al., 2001; Cavers et al., 2008); short-range laser scanners are used for recording movable elements (Donelan, 2002; Trinks et al., 2005) and characteristically obtain precisions of up to five microns with a maximum measurement distance of two metres.

The development of the instruments, methods and geomatic techniques reviewed and applied to Palaeolithic art in subterranean spaces indicates that the production of distinct cartographic products requires one to consider the complexity of the object.

The application of new advances in the field of geomatics to the documentation of Palaeolithic deposits demands an adaptation of the methods and techniques, which are aimed at obtaining graphic and infographic products. The morphometric complexity (presence of concavities and convexities) of these environments has the consequence of generating multiple occlusions for data capture. Therefore, the combination of sensors and geomatic techniques is the recommended solution (Beraldin et al., 2006).

1.3 Summary

On this basis, this chapter on "Laser Scanning, Theory and Applications" presents an approach that has been developed and applied to two emblematic Palaeolithic caves. The structure of the chapter goes as follows: After this Introduction, in the second part, we will tackle with the specific methodology developed using non-destructive techniques and articulated in four steps: planning and data acquisition; ii) data processing; iii) modelling and cartographic products generation; iv) spatial information system development. In the third part, we will outline and discuss the results obtained. A final part will devoted to the main conclusions and the expected future developments.

2. Methodology

This section includes a detailed description of the approach developed: from the design and execution of the field campaign to the processing schemes that allow us to deliver three-dimensional models, cartographic products, as well as the spatial information system which represents the best tool to manage and diffuse the Palaeolithic Rock Art.

Given the geometric and radiometric characteristics of the caves under study, laser scanning appears to stand out as the most ideal geomatic technology to undertake the documentation and three-dimensional reconstruction, as this technique provides quality results for complex surfaces without requiring direct contact with the object or the availability of lighting within the cave. However, laser scanner technology suffers from various inconveniences that should be taken into consideration: a high cost, an unorganised and complicated nature of capturing information and the impossibility of providing data that incorporates true colours at a high resolution. Therefore, the methodology presented here proposes an integration with other non-destructive geomatic techniques, such as close-range photogrammetry, to provide an integral documentation and three-dimensional reconstruction while serving as the basis for the development of a spatial information system that permits the management of this archaeological heritage and its value in society.

The methodology developed here presents two clearly differentiable work phases:

- *Field work.* During this phase, the different sensors, a digital camera, a panoramic camera and a terrestrial laser scanner are configured with their basic parameters, and data acquisition is carried out following the protocols and basic rules that are particular to underground sites.
- *Laboratory work.* During this phase, the acquired information is individually processed and subsequently integrated to obtain cartographic products along with a spatial information system that adjusts to the real needs demanded by archaeologists, pre-historians and heritage managers.

Figure 1 summarises the methodology applied to the non-destructive techniques developed and practiced in two caves with Palaeolithic art in the eastern part of Asturias (Spain).

2.1 Equipment used

Below are the most relevant technical characteristics of the geomatic sensors employed during data collection in the fieldwork phase:

- A Nikon D80 digital reflex camera with a Nikkor AF DX fisheye lens is used, providing a field-of-view of close to 180° and reducing the number of shots necessary to generate panoramic images. To guarantee the immobility of the point of view and the angular regularity in the direction of the shot axis, this camera is set on a panoramic head that offers five degrees of freedom, including three translations (X, Y and Z) and two rotations (horizontal and vertical).

- A Canon 500D digital reflex camera is combined with the previous camera to capture detailed material. The obtained images are stored in RAW format for the digital development stage.

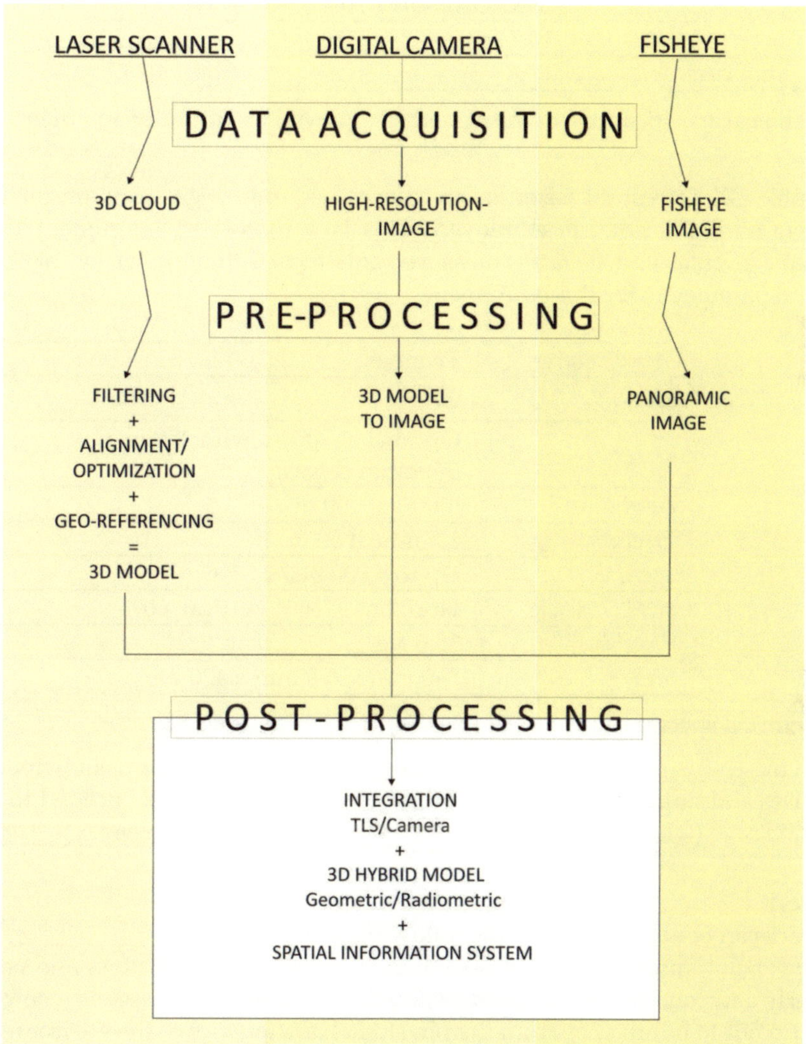

Fig. 1. Methodology for the non-destructive techniques developed and practiced in caves with Palaeolithic art in the eastern part of Asturias (Spain).

	NIKON D80	CANON 500D
Type of sensor	CCD (DX format)	APS-C CMOS
Resolution of sensor	10 MP	15 MP
Image size	3,872 x 2,592 pixels	4,752 x 3,168 pixels
Sensor size	23.6 x 15.8 mm	22.3 x 14.9 mm
	NIKKOR AF DX lens	CANON EF-S lens
Focus	18-70 mm	18-55 mm
Angle of view	76° - 22.5°	74.3° - 27.8°
	NIKKOR Fisheye lens	
Focus	10.5 mm	
Angle of view	175°	

Table 1. Technical specifications of the cameras employed in the study and their respective lenses.

- A Trimble GX Terrestrial Laser Scanner, of a medium range based on the operational principle of flight time, is employed. This is a motorised instrument that permits automatic angular and distance measurements in real time. Refer to Table 2 for more technical parameters for this equipment.

Manufacturer	Trimble
Model	GX
Range	Optimal at 200 m, with 350 m of overscan capacity
Resolution	3 mm at 100 m
Precision	1.5 mm at 50 m
Speed	Up to 5,000 pts/s
Field of vision	Horizontal: 360° Vertical: 60°
Weight	13.6 Kg
Size	340 mm x 270 mm x 420 mm

Table 2. Technical specifications of Trimble GX, Terrestrial Laser Scanner.

Along with these sensors, different accessories for each piece of equipment (tripods, rotating tripods, lenses and targeting plates) as well as the auxiliary materials needed to adequately illuminate the cave (wiring, generators and cold light sources) were used.

3.2 Fieldwork

The fieldwork can be summarised in the following phases:

a. *Planning*: Before undergoing any documentation work, the objectives and needs should be clearly and concisely established, especially in subterranean spaces where conditions may turn out to be adverse, from the physical characteristics of the environment such as the lighting, humidity and accessibility, to the technical characteristics that should be resolved with the correct methodology and materials.

   The process of taking photographs is significantly affected by the complicated work conditions in a cave (the lack of lighting, heightened humidity, presence of dust and so forth), resulting in difficulties in obtaining adequate photographs. Therefore, it is necessary to consider a few basic principles when determining the photographic material and equipment. The recommended equipment includes a photographic lens that is as bright as possible and has a short focal length to maximise angular coverage in narrow spaces. The reflex camera should be protected due to the possibility of airborne dust and/or water in the interior of the cave. With respect to the management of the camera, the lighting conditions require manual focus along with the lowest possible sensitivity (ISO) to reduce the possibility of noise in dark areas of the images. In addition, it is necessary to employ a tripod for long exposure times and a remote control to avoid unexpected vibrations when taking the photos. Acquiring images in RAW format allows for better image quality (no artefacts), an adjustment of white balance and a reduction of noise in the image.

With respect to lighting, it is necessary to place lamps at the site (according to its dimensions) to provide indirect light. It is also necessary to use a free bulb for dynamic illumination, that is, to add light to the darkest zones of the site during the exposure time of the camera.

b. *Laser shots*: Laser data are acquired according to some rules and basic principles established specifically for interiors and subterranean locations.

Planning for obtaining the data with the laser scanner is an essential task because it determines the subsequent information processing. In the case of subterranean caverns, there is a set of relevant factors that influence the planning for the job of data capture, and these are listed as follows.

The first factor to consider is the complexity of the geometry. The detailed heterogeneity of the concavities and convexities present in the interior of a cave, along with the occlusions (shadows) generated by stalactites, stalagmites and columns, limit the reach and effectiveness of the laser scans. Consequently, a greater number of stations are required, and the equipment performance is lower.

Secondly, the need to station the scanner in successive positions throughout the cave requires the visualisation of reference points common to several shots to facilitate the fit between them. Given the lighting conditions of the cave and the geometric conditions, this problem requires the use of artificial aiming devices (targets and areas) that are placed in the overlapping zones of contiguous laser stationing points, such that the fitting among images can be performed to guarantee the geometric quality of the set.

Another factor to consider when planning laser shots is the height of the ceilings in the halls and corridors of the cave. The physical constitution of the laser equipment limits the scan to a deflection angle of 40° above the horizontal, thus establishing another unfavourable condition for the execution of scanning work.

A final factor in the planning of laser shots is the arrangement, shape and geometric characteristics of the relevant elements for data acquisition. In general, the position of the scanner must be accommodated in a manner that optimises the representation of the areas of greatest archaeological and/or artistic interest.

Work in underground tunnels, however, presents some advantages over work performed above ground. One advantage is that the environmental conditions (lack of light, humidity and temperature) are very stable, which benefits the homogeneity of the conditions of the shots from both geometric and radiometric points of view. Additionally, there is no interference from objects foreign to the site, such as moving vehicles or people, or from static elements such as signalling or urban furniture, which introduce noise and disturbances to the process.

c. *Taking high-resolution images*: High-resolution images are acquired by means of a conventional digital camera following the principles and basic rules of close-range photography. These rules can be summarised in the following series of basic instructions:
1. Recognise the object by means of a simple sketch representing the layout or structure, taking into account the limitations of this process; e.g., limited visibility, hidden locations and obstacles. This recognition will allow for the definition of the areas of the object that will be modelled by means of multiple convergent images and of other areas that will be modelled by stereoscopic procedures. In general, the latter procedure is employed when the surface of the cave is irregular and continuous, and the former procedure is employed when the surface is simple enough to be able to be represented by a series of discrete points identified in the images (for example, flat surfaces with paintings).
2. The distance of the shot should correspond to two basic contradictory rules. On the one hand, the camera should be situated as close as possible to the object so that the scale will be as large as possible (precision criterion). On the other hand, greater distances will yield wider photographic coverage (efficiency criterion). This paradox is resolved according to the technical requirements of the project.
3. For modelling based on multiple convergent images, each area of the object should be preserved in at least three images taken from three different points of view, while maintaining the same distance to the object (if possible) and assuring that there is enough overlap among different areas of the object. A possible way to perform this task is to establish a path that is parallel to the borders of the object by shooting three images for each given interval of time: one of these with a shot angle that is more or less perpendicular to the wall and the other two with slopes of 45º to the left and right of the initial direction.
4. To achieve stereoscopic modelling, each area of the object should be preserved within the stereoscopic coverage zone of a pair of frames. The geometric paradigm to materialise this configuration is the so-called normal case in which all of the elements under consideration are orthogonal among themselves. In essence, the base (the vector between the two shots) is horizontal and parallel to the object (to the median plane of the object), and the axes of the shot are perpendicular to the object.
5. For definition of the Cartesian plane associated with the object, it is essential to determine both the distance between the points that appear in the images and the vertical direction. This yields the scale along with the basic orientation of the model.
In addition to these images, which give way to geometric and radiometric modelling of the cavern, a second series of high-resolution images should be taken with the goal of providing texture to the point cloud or triangle mesh (if applicable) obtained by the laser scan. These images should consequently cover the surfaces captured with the laser. This process is highly recommended to facilitate the subsequent process of the projection of photographic information over the modelled surface so that the point of the photographic shots approximately coincides with the point of the laser position.

d. *Taking panoramic images*: A panoramic image is an image that catches the entire viewing spectrum. It may be cubic, cylindrical or spherical, but always has the distinctive characteristic that it is viewed from the inside so as to create the sensation of being immersed in a given scene. Such representations are not metric but transmit a feeling of great believability and consequently have great communicative power. They can also be linked with other panoramic images through hyperlinks to create complex virtual visits that allow for the documentation of remote and inaccessible places in a way that is authentic, easy and visually attractive.

When shooting panoramic images, there should be no parallax. That is, it is necessary to guarantee that all of the images are taken from a single point of view.

The geometric foundation of these images is based on the projection of the images onto a mathematical surface (cube, cylinder, sphere) so that a single "image" is created from several initial images. These images can be stored in standard image formats (JPEG, GIF, TIFF and BMP) and be viewed by means of the appropriate software so that the user can choose the viewing directions by moving the mouse.

The following matters must be considered regarding the protocol for shooting panoramic images in subterranean caverns:

1. An objective with a large angular field (fisheye) should be used so that the number of shots in the field will be reduced to a minimum; in our case, there were six horizontal images (one each 60º) and one more image in the direction of the zenith.

2. To ensure perfect immobility of the point of view and also to control the regularity of the distribution of shot directions, a swivel attached to a tripod should be employed to facilitate the above-mentioned five degrees of freedom.

3. In addition to these geometric considerations, some precautions regarding lighting should be taken into account. As in the case of high-resolution images, it is necessary to provide lamps throughout the scene to provide indirect light, avoiding overexposed areas. Furthermore, to ensure the most homogenous illumination possible for the entire scene, a movable bulb that provides light to the least lighted areas (recesses and so forth) should be used. In addition, the availability of lighting in the directions that are farthest from the shot axis will prevent blurry photos.

Although lighting is an important factor, it is not a critical one: it is not necessary for all of the individual shots that make up a spherical panoramic image (seven for our particular configuration) to have exactly the same exposure, as the subsequent process of image mosaicking can obtain a smooth radiometric gradient.

e. *Establishment of control points*: Various singular points materialised by artificial markers (cards and spheres) are distributed throughout the entire cave and raised with the terrestrial laser scan with a double purpose: to register all of the sensors (cameras and laser scanner) under the same reference system and to georeference both sets of data (point clouds and images) in the already existent archaeological framework provided by the archaeologists. Furthermore, the definition of the reference system permits the establishment of a clear and unambiguous altimetric system (point of zero dimension) that is fundamental in the monitoring of the excavation. Finally, a global geo-referencing of each cave is provided based on external control points acquired with GPS. To this end, the last stage of the alignment consists on georeferencing the model according to the official Datum, ETRS89, using the three closest GNSS stations of the work area and the official EGM08 geoid model.

2.3 Laboratory work

Three sequential phases are applied: the first leads the independent pre-processing of each acquired data set (high-resolution images, panoramic images and laser data); the second phase establishes a record of the sensors (digital camera-laser scanner, panoramic camera-laser scanner and digital camera-panoramic camera); the third and final phase leads to a modelled set with all of the hybrid data and generates a system of spatial information that allows for the management and diffusion of the caves and their art.

Independent information processing

a. Pre-processing of high-resolution images.

0. The first step is the digital development of RAW images. As the RAW format allows for the storage of 10 to 16 bits per channel in comparison to 8 bits for the JPEG format, the image has enough degrees of freedom to modify the exposure, white balance, sharpness and tones.

1. Then, the internal geometry of the camera (or cameras if more than one is used in the shooting process) is determined. That is, the position of the point of view (centre of the object) is determined with respect to the CCD sensor, as well as the radial and tangential distortion of the object. These parameters are calculated in the process of geometric calibration in which the values are established to allow for mathematic modelling. The calibration process is performed according to a specific protocol that consists of taking a series of photographs of a mesh to form the geometric grounds that define a pattern of known dimensions. The extraction and correspondence of the points of interest of each image and their subsequent calculation by means of a bundle adjustment process allows the desired parameters to be obtained.

2. Subsequently, the external geometry of the cameras is determined. That is, the positions of the cameras are calculated with respect to the object at the moment of the shot. This step also requires the calculation of the orientation of the shot axis. It is a process of inverse intersection (spatial resection), whereby once various control points are known for the object, as identified by two or more photographs, the spatial and angular positions of the cameras are determined.

3. Once the anterior steps have been resolved, the identification of two or more homologous images of elements (points, lines) allows for reconstruction in the space of the lines that correspond to the rays of light from the object. Therefore, it also allows the inverse process (direct intersection) to be performed and permits the reconstruction of the object based on the images. Thus, this process requires at least two images of the same element of the object to be managed (Figure 2).

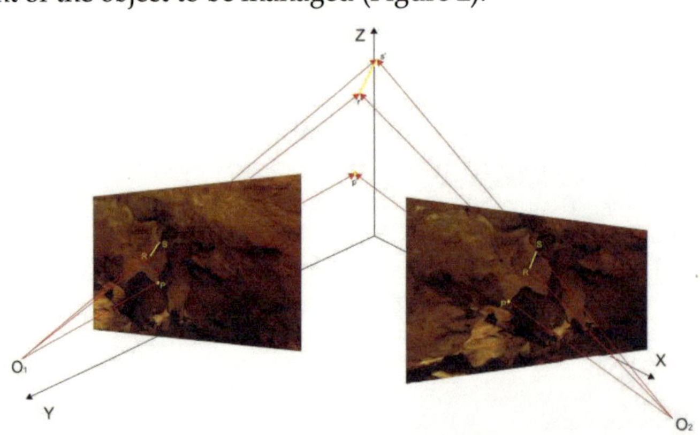

Fig. 2. The object can be reconstructed from the reconstruction of homologous perspective rays (direct intersection).

In the case of a stereoscopic image, this process takes advantage of the ability of human vision to fuse two images (planes), taken from points sufficiently close together, to generate a three-dimensional model.

1. Obtain pairs of images that correspond to the left eye and the right eye.
2. Allow the left image to only be viewed by the left eye and the right image to only be viewed by the right eye (with anaglyphs, polarising filters or stereoscopes). If this is achieved, a stereoscopic view is formed in the brain of the observer; this will be better established if the photographic base (the distance between the two points of view) is greater than the physiological distance between the two eyes.
3. Introduce a separate measuring device for each image to provide the observer with a three-dimensional browser ("floating mark") that allows the 3D dimensions of the object to be derived (Figure 3). To achieve this, the relative positions that the cameras occupy at the moment of the shooting should be rigorously recreated. The geometric model used for this is identical to that used in the anterior case.

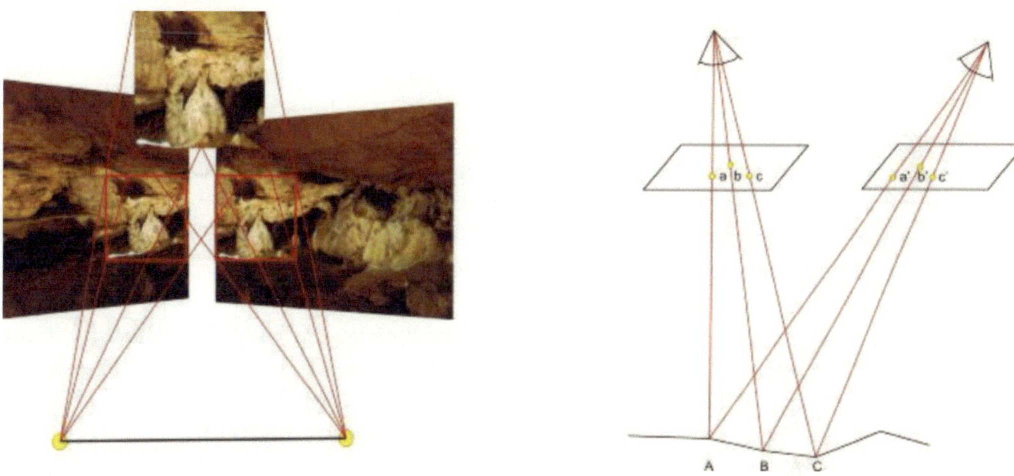

Fig. 3. Left: Principle of the stereoscopic pair. The visualisation of the left image by the left eye alone and of the right image by the right eye alone leads to the plastic display of the photographed object. Right: Principle of the "floating mark": the left and right marks are interposed in the optical path of the left and right eye, leading to the perception of a single mark that runs the depth of the object, that is, the mark "floats".

b. Panoramic image processing.

The processing of this type of image consists of the following basic steps:

1. Link the adjacent images, for which it is essential that the original images contain a determined overlap or covering between them. This step can be performed by identifying homologous points, that is, common points between the images.
2. Once a geometric model is established common to all of the images, the pixels of each of them are then projected onto a single geometric surface, a sphere, cylinder or cube. This is also the time to correct the effects of geometric distortions of the lens, which are considerable when using a fisheye lens.
3. To store these panoramic images in the normal rectangular format, we should take into account that a cylinder has a surface that can be projected onto a plane, while a sphere does not. In other words, in the case of a sphere we need to employ a

mathematical model that allows a sphere to be projected onto a plane, most appropriately the equirectangular projection.
4. The final step consists of the completion of the panoramic image. With spherical images, their lower part will show the panoramic head and the tripod if the image has been taken in the nadir direction, or there will be an empty area if the nadir image has been omitted. The normal image processing technique provides possibilities to complete these zones so that the sphere does not present any empty areas.
Note that these panoramic images exhibit considerable distortions such that there is no sense in direct visualisation. To explore panoramic images, additional algorithms should be implemented that have the function of selecting a sector of the panoramic image and presenting it, free of distortions, on a display window. It is essential that this sector can sweep, under user control, the "latitude and longitude" of the entire sphere.

c. Basic pre-processing of laser data: purification and filtration. Pre-processing of the laser data consists of a long and laborious task, where the organisation, depuration, fusion, georeferencing and filtration of the point cloud are essential to obtain a precise metric product, in 3D, of reality.
The large quantity of information generated by the laser scanner makes it necessary to purify and filter the point cloud. This task consists of eliminating those points that are not of interest for the documentation of the object. The presence of noise or unnecessary data is one reason for using filtering and segmentation tools to purify the data. The purification should be carefully performed to avoid the elimination of relevant information. Segmentation allows for the isolation and extraction of information that is important for the inventorying and cataloguing of objects such as prints, drawings and paintings.

d. Advanced pre-processing of laser data. This consists of aligning the shots and generating the meshes. The alignment or recording of the different shots or scans allows the positioning of the object to document a single reference system. A local or global reference system is established through the definition of an origin and the directions of the X, Y and Z axes, with the data laser in a single common system. For this, it is necessary that the shots with the data laser have overlaps greater than 25% to achieve sufficient precision in the alignment (fusion of clouds) and to avoid problems with or the impossibility of aligning the shots.
The fusion of distinct point clouds can be done in two ways: (i) automatically, through the use of artificial signals (cards, spheres, etc.) that are automatically recognised by the laser scanner; (ii) manually, though the identification by the user of at least three homologous points corresponding to the object. In both cases, the resolution of the alignment will consist of the calculation of a solid-rigid three-dimensional transformation consisting of three translations and three rotations.
In contrast, the aligned laser point cloud constitutes a mass "report" of points that is discontinuous and lacks a geometric structure to facilitate manipulation. Therefore, it is necessary to create a graphic model in the shape of the surface, allowing the point cloud to be managed and exploited. Particularly, a mesh structure is created from the raw point cloud based on the technique of 2.5D Delaunay triangulation and the incremental method (Bourke, 1989), incorporating improvements in geometric and topological constraints such as the maximum length sides of triangles and the topology of adjacent triangles.

Once the model of surfaces is constructed, it can be explored and processed with different goals such as acquiring the curvature, mapping the texture or calculating the volume.

e. Integration of data from the laser scanner-digital camera sensors.

The joint processing of information corresponding to complex scenarios, such as caves and subterranean caverns, consists of a multidisciplinary field of investigation with various challenges and difficulties, especially if one aspires to fuse the information from high-resolution laser scanner and digital camera sensors. Thus, the non-destructive techniques of close-range photogrammetry and laser scanning can be jointly exploited, with the aim of achieving comprehensive three-dimensional reconstruction in these complex scenarios.

The problem of integrating (recording) the high-resolution images and the laser model is intimately related to the problem of positioning the camera, which provides a projection model between the 3D laser data (object space) and the 2D image (image space). This projection is characterised by a rigid transformation (rotation and translation) together with the model of the camera (calibration), also known as the determination of the external and internal parameters, respectively. Rigid transformation will consider the correspondence between the 3D points of the object space and the 2D points of the image space as input data, while the model of the camera establishes the form in which the points are projected onto the image plane.

As a result of the registry of both sensors, a re-projection of the high-resolution images can be established over the three-dimensional laser models (Figure 4) to obtain photo-realistic models of the cave and the panels.

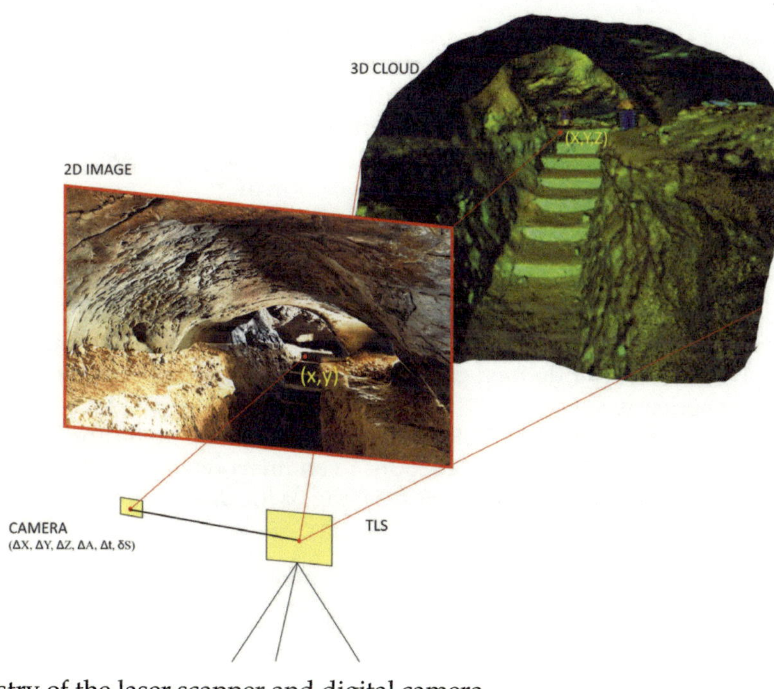

Fig. 4. Registry of the laser scanner and digital camera

Hybrid modelling of the information

f. Acquiring the derived hybrid products: orthophotos.
   An **orthophoto** is currently considered an efficient, low-cost product that allows the radiometric and geometric quality of the object or scene to be united into the same format. The capacity of the laser scanner to acquire dense three-dimensional models in only a few minutes, along with the development of information integration strategies, has allowed the process of orthoprojection to be applied to complex scenarios such as caves and subterranean caverns.
   Therefore, along the lines of the previously established integration of the laser scanner and digital camera sensors with the idea of continuing to broaden the products derived from the hybrid modelling of information, below are the steps followed in the generation of orthophotos that correspond to prehistoric panels. For this purpose, the previously registered high-resolution image and the improved and completed laser model are considered as input data.
   Using the method of anchor points (Kraus, 1993), it is possible to project onto a photographic texture. This method consists of applying an affine transformation to each one of the planes formed by the optimised triangular mesh, which was obtained from the point cloud determined by the laser. Through the condition of collinearity, the pixel coordinates of the vertices of the mesh are calculated, and the mathematical model of the affine transformation directly relates the pixel coordinates of the registered image and of the orthophoto (Figure 5).
g. Generation of a spatial information system
   The rigorous graphical representation of the subterranean caverns as a tool for management, monitoring, intervention, interpretation, visualisation or disclosure may be very important, but it can also be a basic tool for the structuring of information relative to the object.
   As in the case of geographic information systems, graphical information may be used to provide a structure to another type of information of a literal character or of a graphical character distinct from any other characteristic (sound, video). The graphical file is thus converted into a database that can be efficiently employed by an expert or by a researcher to perform consultations or determine analysis processes.

The main characteristics of the spatial information system are:

a. Completeness: the entire cave is documented, including areas that were being analysed at the time.
b. Interactivity: panoramic images that cover the entire scenario are generated, allowing any element of the cave to be examined.
c. Complementary documentation: the panoramic images have links to detailed photographs, videos, laser derived products and literal information, complementing the navigation experience.
d. Spreading: it is possible to distribute the information and spread the cultural heritage by remote access or by physical installations in an interpretation centre. The system can also be modified to make it a system for teaching or for disseminating knowledge to the public.

3. Case studies

In this section two relevant Palaeolithic caves located in the north of Spain (Asturias), La Loja and Buxu, have been recorded and modelled through the approach described. A brief

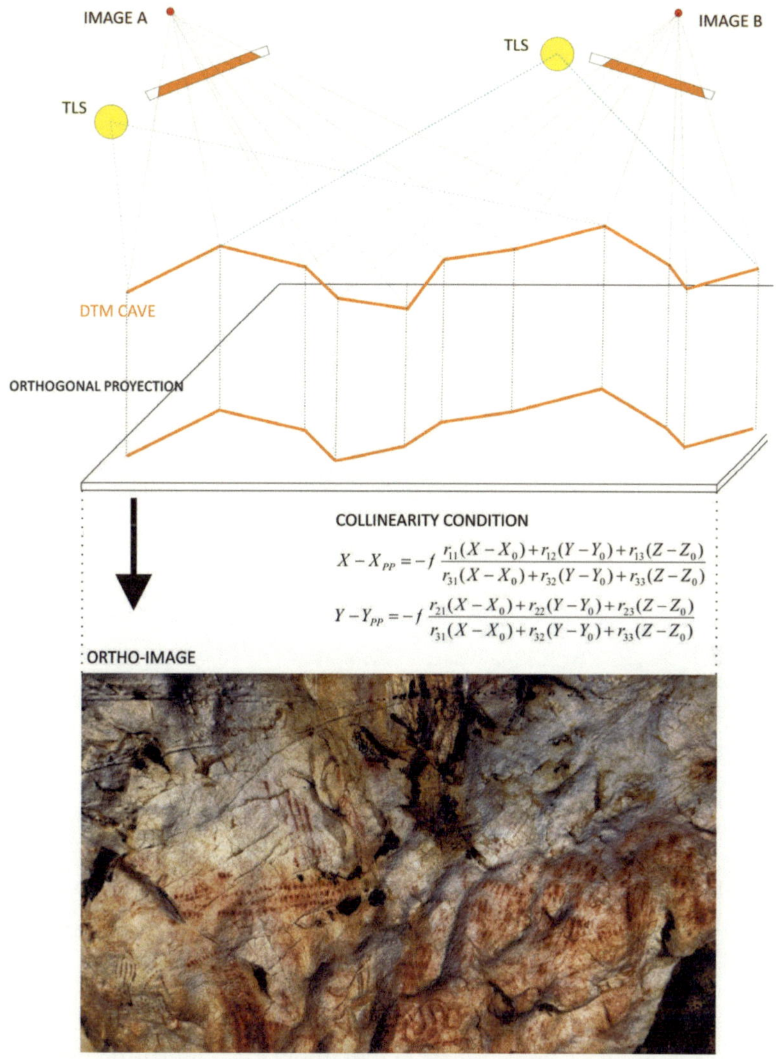

Fig. 5. Orthoprojection process and orthophoto generated on a prehistoric panel

description of the importance of the each cave is provided remarking the main Palaeolithic art of each cave. Furthermore, from a technical point of view, a brief comment about the constraints of each case of study and the way they have been solved is discussed.

3.1 Buxu cave

The Buxu cave is located 2 km from Cangas de Onis, Asturias, Spain. It was declared a historic site on June 29th, 1985. It was discovered in 1916. Archaeological excavations were performed in the vestibule of the cave, bringing to light the remains of the Upper Palaeolithic. Among the manufactured artistic objects, there was a sculpture of a bird carved on the tusk of a cave bear. The deepest parts of the cave contain the rock art that has brought renown to the cave throughout entire world. The art consists of abstract signs and animal figures carved or painted onto the rocks of the walls. Among the signs are enigmatic tectiforms, a type of gird or net. They were carved by the Solutrean occupants of the archaeological deposit of the vestibule. Among the animal figures are some goats carved or painted in black, and the figures from the deepest shrines of the cave include two beautiful horses carved in great detail and a deer shown during the rutting period (Figure 6).

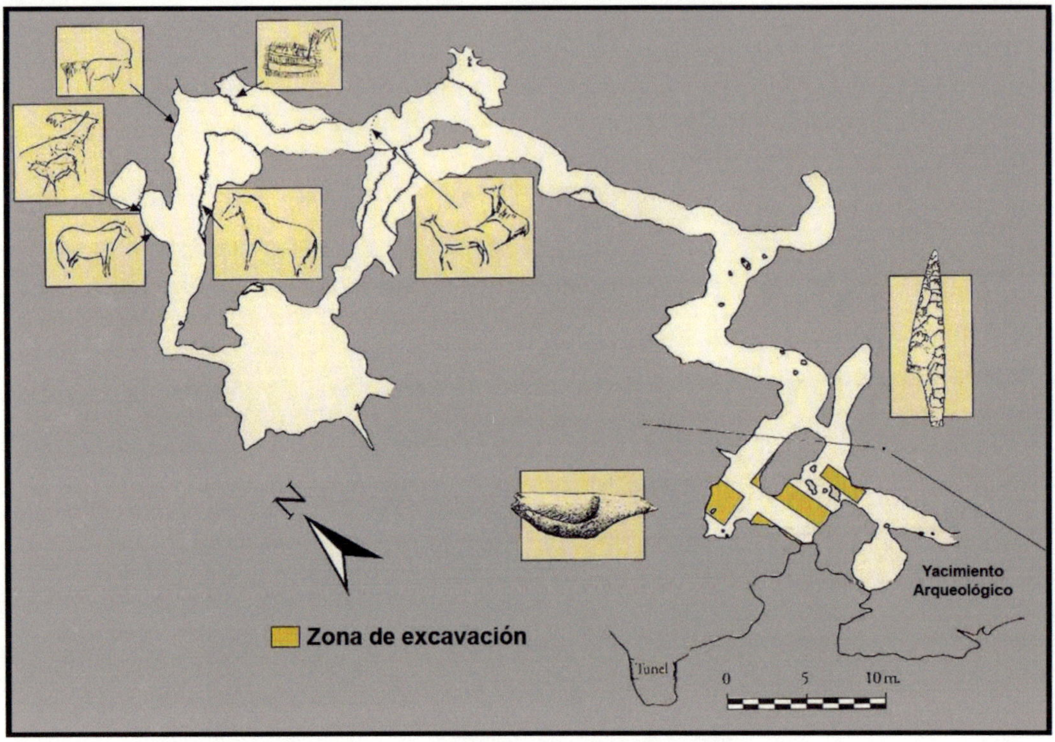

Fig. 6. Layout of the Buxu cave.

During the acquisition of laser data, we used a Trimble GX flight-time laser scanner. As is observed in Table 3, the resolution of the mesh varies from 10 and 20 mm to 10 m and 5 mm to 10 m in the panels and areas of artistic interest. The data were obtained from a local reference system, materialising the vertical axis, and had an overlap of around 30%. The number of raw points captured was 35 million. After filtration and optimisation of the points, the complete model had 33 million points. The number of hours used with the laser scanner was 50. During the initial planning, 25 scanner positions were set. The difficulty of mobility in the interior of the cave and the small spaces, combined with the complexity of the geometry of the cave, led us to perform a larger number of positions than that originally established, resulting in 34 positions. In addition, restrictions of the apparatus had to be considered, that is, the vertical angle and the minimum distance from the object to the scanner (2 m), increasing the data acquisition time and thus requiring more positions. The alignment of the point clouds was performed by means of manual alignment with homologous points and three-dimensional solid-rigid resolution.

Nº laser stations	34
Step mesh	5-10-20 mm to 10 m
Reference system	Local (level)
Overlap	30%
Nº points captured	$35*10^6$
Nº points optimised in model	$33*10^6$
Nº laser hours	50 h

Table 3. Summary of laser data of the Buxu cave

For the spherical images, we used a Nikon D80 camera with a fisheye lens and a photographic swivel. Cold light was used to illuminate the scenes. A total of 126 spherical images were shot, forming the spatial information system of the cave (18 stations) and allowing a virtual visit to the interior of the cave.

The terrestrial laser scan is a "non-destructive" technique that allowed a series of derived products to be obtained from a mathematical model, such as volumes, areas, transversal sections and so forth. The most relevant is the orthoimage. In Buxu, we obtained orthoimages from the rock paintings and the carved tectiforms, as can be observed in Figures 7 and 8.

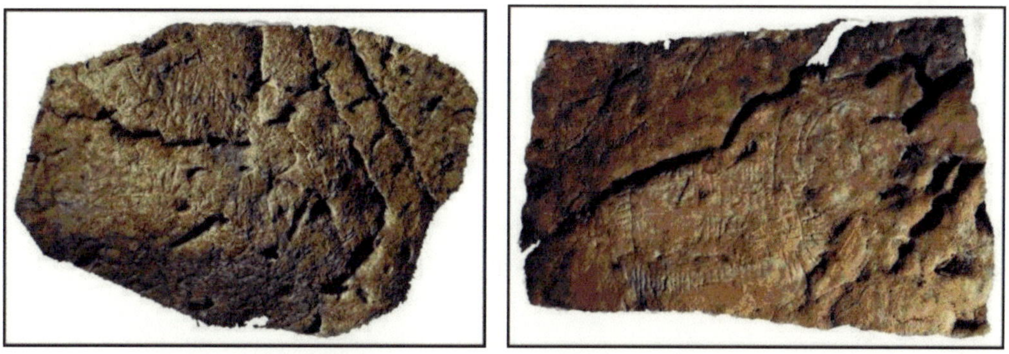

Fig. 7. Large tectiform (left) and carved tectiforms (right). Buxu cave.

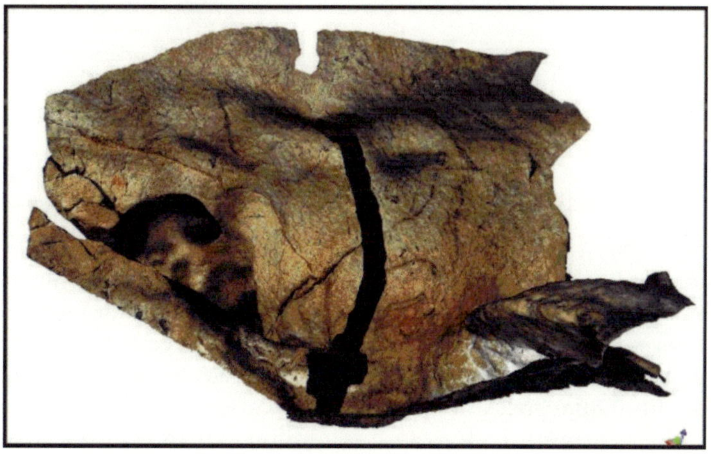

Fig. 8. Horse, goat and large deer. Buxu cave.

In addition, we can obtain other products such as the layout of the cave from the laser model (Fig. 9).

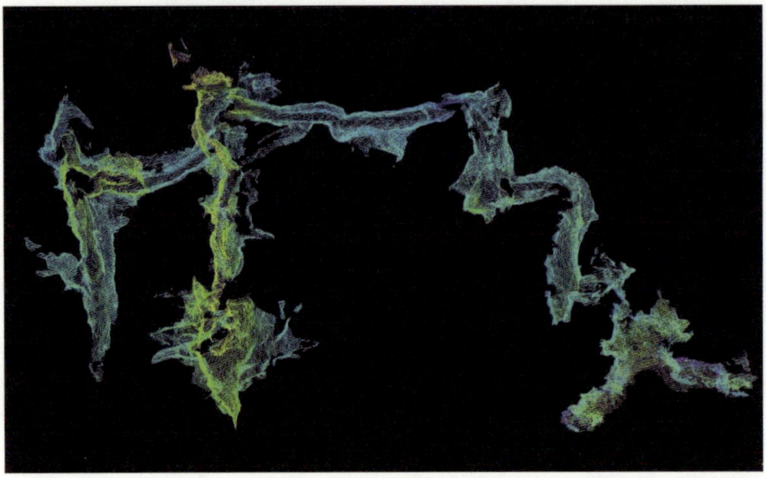

Fig. 9. Layout obtained from the laser data. Buxu cave.

3.2 La Loja cave

This cave is one of the most important known Palaeolithic enclaves of the Peñamellera Baja. Its prehistoric importance was established for the first time in 1908 by prehistorians H. Brehuil, M. Mengaud and H. Alcalde del Río. It is a small cave, just 102 m in total length, and its mouth opens over a limestone wall about 4 m from the ground. The first painting is found about 2 m from the entrance. It is a red symbol. At around 50 m from the entrance, about 4.5 m from the ground, there is a large panel of carvers measuring 1.80 x 0.70 m. The figures represented on this panel are animals (Figure 10).

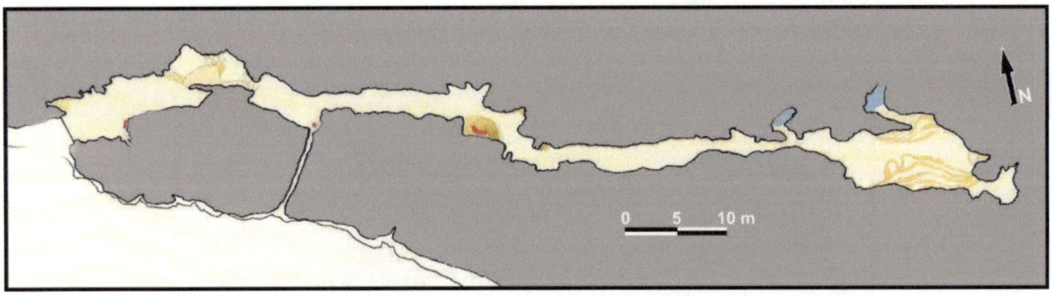

Fig. 10. Layout of the La Loja cave.

In this work, we used two Trimble GX laser scanner equipments, with the advantage of reducing the amount of time in the field. The major disadvantage of working in the cave of La Loja is the presence of water, both in the floor of the cave and in filtrations from the ceiling. Nothing can be done regarding this problem except to work with some appropriate safety measures and to take special care with each scanner position. As we can observe in Table 4, there were 29 positions with a mesh resolution that varies from 10 and 20 mm to 10 m and 5 mm to 10 m in the panels and areas of interest of the rock art. As in El Buxu, we worked with a local reference system, materialising the vertical axis, and had an overlap of around 45%. The number of raw points captured was 28 million. With the filtered and optimised model, we have a complete model of 22 million points. The number of hours employed with the laser scan was 35 hours. In this case in particular, the mobility within the cave is greater than that in El Buxu, but it also has higher ceilings that require the use of a professional swivel to incline the equipment and register those areas. The restrictions of the apparatus in the vertical angle and the minimum distance to the object to be scanned provide the same problems, resulting in an increase in the time to capture data. The alignment of the point clouds was performed through manual alignment with homologous points and Helmert 3D resolution. For the virtual visualisation, 112 spherical images were obtained for a total of 16 panoramas.

N° laser stations	29
Step mesh	5-10-20 mm to 10 m
Reference system	Local (level)
Overlap	45%
N° points captured	$28*10^6$
N° points in optimised model	$22*10^6$
N° laser hours	35 h

Table 4. Summary of laser data in the La Loja cave.

The Canon 500D camera was employed with its corresponding lens (see Table 1) following the recommendations for photography in the interior of caves for the radiometric record (see section 2.2). Below are the orthoimages obtained from the triangulated model and the radiometry (Figs. 11 and 12).

Fig. 11. Red sign. La Loja cave.

Fig. 12. Large panel. Orthoimage. La Loja cave.

4. Concluding remarks and future perspectives

This section will remark the essence of the research together with the main advantages and disadvantages of the approach.

The **benefits** derived from this study can be summarised as follows:

- The cataloguing of Palaeolithic art information is an indispensable element for the acts of conservation, maintenance and restoration.
- The integration of Palaeolithic art information into a spatial information system is an essential tool for its management.
- The diffusion and enhancement of this information through web services can multiply the outreach of the information.
- An enhancement of digital accessibility and an elimination of the barriers for people with disabilities can be achieved by means of virtual reality and increased technology.
- The model for action can be extended to other environments as a result of the methodological and instrumental developments that have been established.

In turn, the main disadvantages of this methodology are the following:

- There are technological limitations related to laser instruments, including the difficulty of finding a reasonable compromise between the capture speed, capture distance range and precision. Given the novelty of this technology, future improvements are expected in the performance of the equipment that will allow optimisation of the fieldwork.
- The capacity of the information processing is conditioned by the volume of information that must be handled. The design of more efficient processing algorithms and the

development of informational equipment will allow the laboratory work time to be reduced.

- The proposed workflow has a certain complexity in the processing and data integration phase. Therefore, this phase is susceptible to simplification through the automatisation of certain tasks (the elimination of noise, the optimisation of the triangular mesh and so forth).
- It is difficult to incorporate TIN models in environments of 3D visualisation that are flexible and efficient. When these difficulties are overcome, a greater popularisation in the use of these products will be achieved, and there will consequently be an increase in the demand for them.

5. References

Beraldin, J., Blais, F., Cournoyer, L., Picard, M., Gamache, D., Valzano, V., Bandiera, A., Gorgoglione, M., 2006. Multi-Resolution Digital 3D Imaging System Applied to the Recording of Grotto Sites: the Case of the Grotta dei Cervi, In: 7th International Symposium on Virtual Reality, Archaeology and Cultural Heritage VAST, Cyprus.

Bourke, P., 1989. An Algorithm for Interpolating Irregularly-Spaced Data with Applications in Terrain Modelling, In: Pan Pacific Computer Conference, Beijing, China.

Chandler, J., Fryer, J., 2005. Recording aboriginal rock art using cheap digital cameras and digital photogrammetry, In: XX CIPA Internatinal Symposium, Torino, Italy.

Cannataci, J., Rivenc, R., Borg, C., Guidi, G., Beraldin, J., 2003. E-heritage: the future for integrated applications in cultural heritage, In: XIX CIPA Symposium, Antalya, Turkey.

Cavers, G., Hepher, J., Hale, A., 2008. A comparison of laser scanning and photogrammetric techniques used to record rock art at Ormaig, Argyll, Scotland, In: 6th World Archaeology Congress (WAC6), Dublin, Ireland.

Donelan, J., 2002. Making prehistory, Computer Graphics World, March, 32-33.

Elwazani, S., 2003. Procedure for evaluating performance of measured survey methods, In: XIX CIPA Symposium, Antalya, Turkey, pp. 644-649.

Fossati, A.; Jaffe, L.; Abreu, M., 1990. Rupestrian Archaeology. Techniques and Terminology. A Methodological Approach: Petroglyphs, Cerveno. In: Ricerche Archeologiche, 1(I). Edizioni della Cooperativa Archeologica Le Orme dell Uomo: Val Camonica, Brescia, Italia.

Ghilani, C.D., Wolf., P.R., 2006. Adjustment computations; spatial data analysis (4th ed.), John Wiley & Sons, New Jersey, 611 p.

Kraus, K. 1993. Advanced Methods and Applications. Vol.2. Fundamentals and Estándar Processes. Vol.1. Institute for Photogrammetry Vienna University of Technology. Ferd. Dummler Verlag. Bonn.

Lopetegui-Galárraga, A., Valle-Melón, J.M., Rodríguez-Miranda, Á., 2004. Modelo de representación y difusión de cavidades. Cueva de Praile Aitz I (Deba, Guipúzcoa), In: VIII Congreso Nacional de Topografía y Cartografía (TOPCART), Madrid.

Ogleby, C.L., 1999. From Rubble to Virtual Reality: Photogrammetry and the Virtual World of Ancient Ayutthaya, Thailand, The Photogrammetric Record 16, 651-670.

Rivett, L.J., 1979. The application of photogrammetry to the recording of rock art and archaeological sites in the Kakadu National Park, Univ. of Melbourne, Dept. of Surveying, Melbourne, 64 p.

Robson Brown, K.A., Chalmers, A., Saigol, T., Green, C., D'Errico, F., 2001. An Automated Laser Scan Survey of the Upper Palaeolithic Rock Shelter of Cap Blanc, Journal of Archaeological Science 28, 283-289.

Trinks, I., Díaz-Andreu, M., Hobbs, R., Sharpe, K., 2005. Digital rock art recording: visualising petroglyphs using 3D laser scanner data, Rock Art Research 22, 131-139.

17

Laser Scanning Confocal Microscopy Characterization of Conservation Products Distribution in Building Stone Porous Network

Zoghlami, K.[1] and Gomez Gras, D.[2]
[1]*Université 7 Novembre à Carthage, Faculté des Science de Bizerte*
[2]*Universidad Autónoma de Barcelona, Departament de Geologia*
[1]*Tunisia*
[2]*Spain*

1. Introduction

Stone has been used as a building material for thousands of years. It has long been recognised as a material of great durability and superior artistic quality, the foremost choice for buildings associated with status, power and religion. The pyramids in Giza, Burial Chambers in the UK and The Prehistoric Temples Of Malta were all built from stone over 4000 years ago and are still standing. The use of stone in construction has declined over the last hundred years, but it remains an aristocrat of building materials.
Building stone derives from one of three naturally occurring rock types:
Igneous rocks: Rrocks are buried deep within the Earth, they melt because of the high pressure and temperature; the molten rock (magma) can then flow upward or even be erupted from a volcano onto the Earth's surface. When magma cools slowly, usually at depths of thousands of feet, crystals grow from the molten liquid, and a coarse-grained rock forms. The best example is Granite.
When magma cools rapidly, usually at or near the Earth's surface, the crystals are extremely small, and a fine-grained rock results. The fame example is Basalts. A wide variety of rocks are formed by different cooling rates and different chemical compositions of the original magma. Both varieties correspond to a hard and non-porous rock.
Sedimentary rocks: are formed at the surface of the Earth, either in water or on land. They are layered accumulations of sediments-fragments of rocks, minerals, or animal or plant material. Most sedimentary rocks become cemented together by minerals and chemicals or are held together by electrical attraction; some, however, remain loose and unconsolidated. Consolidate Sedimentary rocks are Soft and fairly porous rock widely used as building materials, the best examples are sandstone and limestone.
Metamorphic rocks: Hard and non-porous rock formed from pre-existing rock that has been transformed by intense heat or pressure within the Earth's crust. The process of metamorphism does not melt the rocks, but instead transforms them into denser, more compact rocks. New minerals are created either by rearrangement of mineral components or by reactions with fluids that enter the rocks. Some kinds of metamorphic such as gneiss and biotite schist are two strongly banded or foliated. Pressure or temperature can even change

previously metamorphosed rocks into new types. The most metamorphic rocks used in construction are marble and slate.

There are huge variations within each of these rock types, caused by specific mineralogy and geology conditions, and while any stone can be used for building, they each have constraints that make them more or less suitable for different purposes. Granite, sandstone and limestone can all be used for building walls, but slate is only suitable for roofs and floors. Some types of granite can contain mineral salts that cause spalling, where the outer face of stone falls off; slate can contain harmful minerals that break down on exposure to the atmosphere causing stone damage; and sandstone can be too porous and fragile for load-bearing structures. An understanding of how the rock material was formed will reveal how it can be used in a building, what its limitations are, how it will weather over time and how it can be treated for its conservation.

The term rock refers to the bulk volume of the material, including the grains or crystals as well as the contained void space. The volumetric portion of bulk rock that is not occupied by grains, crystals, or natural cementing material is termed porosity. That is to say, porosity is the ratio of void volume to the bulk volume. This void space consists of pore space between grains or crystals, in addition to crack space. There are several types of porosity. In endogenous rocks, porosity is mainly fissural type. However, in sedimentary rocks, the amount of pore space depends on the degree of compaction of the sediment, and on the formation history of the rock. We can found one or more porosity types in the same rock. This may lead to the presence of one or several families of pore and therefore one or more pore networks. This pore networks can be interconnected or not.

As well as many engineering materials pore structure influence the physical and mechanical properties of building stone, including elastic module, compressibility, thermal conductivity, poroelastic parameters, strength, and failure behavior.

The transport properties in particular depend critically on the geometry of pores (Adler, 1992; Dullien, 1992; Sahimi, 1995). In fact, size, shape, connectivity, and the tortuosity of the pore space affect the permeability of the porous medium more strongly than the total void fraction available to transmit that flow (Petford et al., 1999).

Secondly, the study of porosity is determinative in to assess and predict rock weathering phenomenon (Zoghlami et al. 2004). In fact, rock decay depends not only on the rate of porosity which can be determined by various means but also on the porous network characterizations. Critical parameters such as pore connectivity, size distribution and tuortuosity play a major role in governing decay in porous rocks. As an example, consider a quartzarenite with a bulk porosity of 20-25 % which 70% of pores are micropores and 30% are macropores, and another quartzarenite with the same porosity but 30 % are micropores and 70% are macropores. The first one is strongly more alterable than the second one.

So the study of porosity is decisive in the field of building materials, first to predict its durability and then to assess the effectiveness of applied conservation products.

2. Determination of pore system configuration

2.1 Previous techniques

Many techniques have been used to estimate the pore space volume such as the point-counting in thin section according to Gazzi–Dickinson's method (Ingersoll et al., 1984). The technique consists of saturating the pore space of the rock sample with colored resin in order to make it more visible and to facilitate the point-counting by the petrographic miroscope.

Connected porosity can be also measured by introducing gas or liquid into the rock. Thanks to the volume of the introduced fluid, porosity can be calculated according standards such as UNE-EN 1936.

Mercury porosimetry is probably the most commonly indirect technique used to characterize the pore space allowing calculating the total connected porosity and its pore size distribution. But mercury porosimetry only measures pore access and real pore size was usually misestimated. This numerical model cannot describe faithfully the porous network configuration as the porous network is incorrectly simulated as a collection of cylindrical non-interconnected tubes that is very far from true pore shapes and network configuration. Therefore, the geometric complexity of pore space based on pore throats and pore bodies leads to ambiguities in the physical interpretation of mercury data and of other indirect methods. For example, as shown in Figure 1, although the average radius of both channels is similar, a fluid flow in channels B will show different properties from one moving through channel A. Hence, rocks having pores type A are more resistant to the freezing-thaw weathering than rock with a pore type B.

Images obtained by conventional techniques such as scanning electron microscopy are commonly used by several authors to characterize the pore network configuration but since SEM can only analyse the surface of simple (planar section or sample fragment), information about pore shape, pore throat and pores interconnection are not available.

Several attempt to obtain information about pore shape and pore interconnection have been processed (Krohn, 1988; Wong et al., 1989, Fredrich et al., 1993; Berryman and Blair, 1986; Thompson, 1991) allowing a tow-dementional (2D) representation which insufficient to describe a tree-dimentional (3D) objects such as the pore media.

Several researchers aware of this limitation developed innovative techniques to overcome this handicap (Pittman and Duschatko, 1970; Bourbie and Zinszner, 1985; Myer et al., 1992; Zinszner et al., 1997). Some techniques are effective like proposed by Lin et al. (1989) but are time consumer and tedious as critiqued by Fredrich (1999). X-ray computed tomography to generate 3D images has been used to image porosity distribution (Wellington and Vinegar, 1987), however, quantitative measurement are complicated (Brown et al., 1993; Fredrich et al 1994). More over a typical resolution is about 0.1-1mm and is not sufficient to detail characterization of pores in the majority of rocks. Baldwin et al. (1996) and Doughty and Tomusta (1997) used the magnetic resonance imaging (MRI) for the same purpose, but

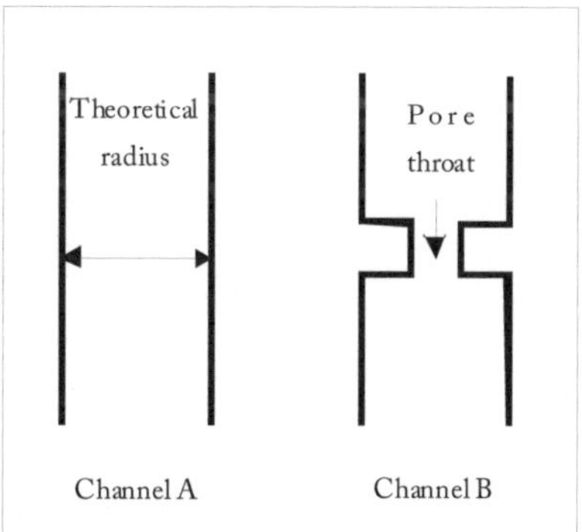

Fig. 1. Two channel geometries (A and B), with the same mean radius but with B containing a pore throat

although the higher resolution than the X-ray CT (25µm) is not sufficient to characterize the pores rock media since the average pore radio of the majority of natural rock is about a few microns and can reach the nanometer.

2.2 Contribution of LSCM

LSCM can afford high-resolution optical images with depth selectivity. Thanks to its ability to acquire in-focus images from selected depths, images are acquired point-by-point and reconstructed with a computer, allowing three-dimensional reconstructions of pore network typology. Its high resolution allows also a better precision in 2D studies than with others conventional microscopy techniques. This high performance is needed in this kind of study to provide details and precision about connectivity, pore distribution and pore shape, critical parameters, to control fluid transfer and hence rock durablity. Several authors (Petford and Miller, 1990, 1993; Montoto et al., 1995; Pironon et al., 1998; Fredrich, 1999; Petford et al., 1999; Menéndez et al., 1999 and Zoghlami et al., 2004) obtained a very good results in improving the 3D porous network reconstruction leading to better understand mercury porosimetry data, usually misinterpreted due to a lack of the 3D pore network configuration.
The technique consists of saturating the pore space of the rock sample with a very low viscosity epoxy (~60 cps) such as Epofix resin from Struers (Zoghlami and Gomez, 2004) doped with a fluorochrome such as araldite AY105 doped with Pylam Blue, with a butanone hardener (Petford et al., 1998). The wetting characteristics of the simple are generally improved if the sample is first flushed with a solvent such as acetone Fredrich (1999). Impregnation should be realized under vaccum to force the epoxy into the void space as deeper as possible.
Fredrich in 1999, decribed the methodology as following:

- Thorough mixing of the epoxy and the fluochrome added as a powder may take up to an hour, air is typically incorporated during the process
- The solution (fluochrome + epoxy) must be left a sufficient time under vacuum to allow a complete degassing before sample impregnation.
- Samples are impregnated under vacuum and left overnight to allow a complete penetration into the void space.
- Curing during X time at 60°C.
- Samples are slabbed and mounted on a glass slide.
- Finally, sample must be polished using standard abrasive polishing techniques, typically to a final finish with 0.05 µm alumina.

Fredrich in 1999 has demonstrated by direct observation and measures that this procedure allows a successful epoxy penetration into craks about 200 nm of diameter. The choice of the fluorochrome is significant (Wilson, 1990). In fact, The absorption peak should be well matched to the wavelength as confocal system are communly equipped with either an argon ion laser with lines at 488 nm and 568, or a Krypton-argon mixed gas laser with lines at 488 nm, 568 and 647 nm. Fluorochromes must be resistant to photobleaching and must be excited effectively with one of the before cited wavelengths. Fluorochromes as Rodamine B used at a concentration of 1:200 dissolved well en epoxy and displayed a very good results. Epodye yellow dye from Struers used at a concentration of 5 gr/L displayed very good results (Zoghlami et al., 2004). Beside the influence of the used epoxy and fluorochrome on the images quality, there are many other facture that affect mainly gray level and the brightness of images. Among this factures, we can cite the opening of the confocal aperture, the laser intensity and the gain of the photomultiplier. The user must balance each of this simultaneously to arrive at an optimum result.

3. Conservation product distribution

In rock conservation field, some chemical products such as consolidants and water-repellents are used as a protection and conservations means of deteriorated building rocks, especially in historical constructions and archeological objects.

The utility of a consolidant lies in re-establishing the cohesion of the particles in a deteriorated stone (Dukes, 1972; Torraca, 1975; Alessandrini et al., 1975). In addition, a good consolidant should meet performance requirements concerning durability, depth of penetration (Young et al., 1999), effect on stone porosity, effect on moisture transfer (Borselliet al., 1990; Dell'Agli et al., 2000) compatibility with stone, and effect on appearance (Biscontin et al., 1975). The porosity and pore size distribution of a stone may have a major effect on its durability. For example, the resistance of a given type of stone to frost and salt damage decreases as the proportion of fine pores increases (Hudec, 1978; Camaiti and Amoroso, 1997).

Water repellents are used to prevent or reduce water penetration into stonework, thus reducing its rate of decay. Parameters concerning treated rock durability, depth of water repellent penetration (Young et al., 1999), treated rock porosity, moisture transfer (Borselli et al., 1990; Dell'Agli et al., 2000) and rock appearance (Biscontin et al., 1975) are used to evaluate the quality of the water repellent. In turn, most of these characteristics mainly depend on the water repellent distribution in the pore network of the rock.

The application of consolidants and water repellents often causes changes in the pore-size distribution of the treated rocks (Esbert, 1993; Villegas et al., 1995; Alvarez De Buergo et al., 2004; Zoghlami and Gómez-Gras, 2004). A decrease or increase of micropores, are usually detected by mercury intrusion porosimetry, both effects are usually interpreted simplistically in the same way. When an increase in microporosity is detected, it is interpreted as a total sealing of the missing pores. When a decrease in microporosity occurs, it is interpreted as a partial sealing of originally larger pores (Esbert and Díaz-Pache, 1993). This general interpretation, applied to different kinds of rocks, is due to the lack of information about the real 3D configuration of the porous network in each type of rock. Some changes are accepted, while others are not. For exemple, a stone consolidant that reduces pore size without plugging may therefore be harmful in special conditions mainly if the weathering factor is freezing or salts. So, changes in treated rock must be controlled, mainly those affecting pore size distribution and pore network configuration as they govern the others properties. Moreover, without knowing the initial configuration of the porous network or the spacial distribution of the consolidant and the water-reppelent, it is very difficult to correctly interpret the effect of these products.

Mercury porosimetry was always used to characterise the porous network configuration of both untreated and treated rocks with conservation products (Sasse et al., 1993). But as shown previously, information provided by mercury porosimetry is insufficient to precisely understand the distribution of the conservation polymer within the rock's porous network and should be complemented with other techniques.

Scanning electron microscopy (SEM) has been the most commonly applied direct technique used in determining the distribution of conservation products in porous rocks (Esbert et al., 1990; Piacenti et al., 1993; Paterno and Charola, 2000; Alvarez De Buergo and Fort, 2001). Nevertheless, SEM allows analysis of sample surfaces only (rock fragments or thin sections).

Hence, in these works, only 2D images of the surface of a treated sample were obtained, the effect of the consolidant and water-repellent in the porous network have to be deduced from mercury porosimetry data. In addition, the use of 2D images introduces further problems related to the interconnection of the porous network, as these often mask the true 3D topology (Petford et al., 1999).
In order to correctly interpret the effect of these products, it is essentially to know the initial configuration of the porous network and the spacial distribution of conservation products. SEM-EDX and X-ray micro-computed tomography, combined with specific 3D software, had been employed to determine the impregnation depth and distribution of consolidants and water repellents in several building materials (Ruiz de Argandoña et al., 2003; Mees et al., 2003; Cnudde et al., 2004; De Vetter et al., 2006; Rodríguez-Rey et al., 2006). Nevertheless, this arrangement of techniques is not easily available and is presently limited to detect pores larger than 10µm. However, these techniques are not sufficient to characterize most of the natural rocks used as building materials if the average pore size of these materials can always reach the nanometer.
To improve the direct observations of conservation products distribution within the porous network of the building stone we proposed (Zoghlami and Gómez; 2004; Zoghlami et al., 2008) the LSCM as an innovative technique in the field of materials conservation. During the following case study, we will try to demonstrate how LSCM allows getting information, direct observation and measurement about porosity properties and conservation products distribution within the rock pore network.

4. Case study

4.1 Used materials

To realize the experiment we used sandstone widely used in the construction of the Tunisian monuments such as the aqueduct of Carthage and Uthina Romain site. It is a non-cemented quartz-arenite, lithified by compaction and composed of quartz grains (69-84 %), feldspars (mainly orthoclase; 0-1,1%), porosity (17-25%) and clay minerals as matrix (0-11%). Due to its homogenous mineralogical composition (Figure 2) and the simplicity of its pore network configuration formed by the void space let between the grains of quartz, constitute an ideal rock to show the contibution of LSCM in this field and how images obtained by LSCM can allow a better understood and evaluation of conservation product effectiveness.

Fig. 2. Microphotograph in thin section showing, in blow porosity and in white grains of quartz. Scale 400µm

To better understand the distribution of the consolidant, it is necessary to first to have an idea about the porous network configuration of the sandstone.

Total porosity of the rock, determined by mercury intrusion porosimetry, showed high values (17-25%). Macroporosity (pore diameter >15 µm,) according to Goñi et al. (1968) and Bousquié (1980) represents a percentage of between 83.38 and 89.18%, whereas microporosity ranges from between 8.65 and 14.78%, indicative of the sandstone's macroporous character (Figure. 3). The mode of the pore-access size varied from between 20 and 30 µm, representing 60% to 80 % of total porosity.

Mercury porosimety results show that most of the pores (>80%) have a pore access diameter of between 20 and 40 µm, depending on the rock's grain size. The rest of the pores (<20%) present a diameter pore access inferior to 15 µm (micropores). However, this method only allows obtaining quantitative porosity data; it does not provide information on the arrangement of porosity, nor the way that pores are interconnected.

Fluorescence images of thin sections show that this sandstone has a very simple porous structure constituted by pore throats and pore bodies (Figure 4A). In general, pores displayed channel-like shapes with diameters smaller than 40 µm (Figure 4B). Mega pores reaching up to 300 µm in diameter were also observed. As the images of thin sections obtained by fluorescence microscopy were 2D, it was not possible to infer the degree of interconnection between both pore types; the real configuration of the porous network could not therefore be determined.

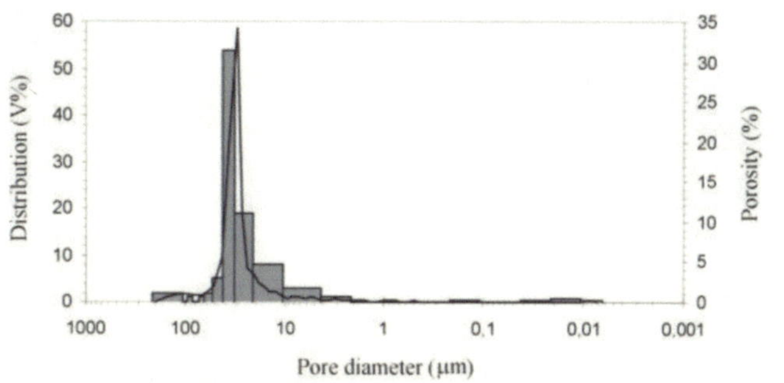

Fig. 3. Pore size distribution of the untreated sandstone simple

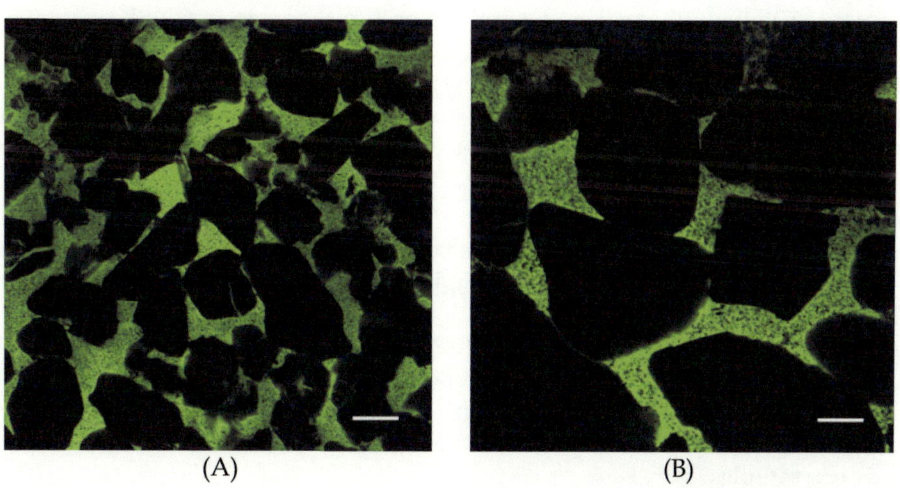

Fig. 4. LSCM images. (A) General aspect of the sandstone porosity. Bar length: 80 µm. (B) Detailed image of pores (communicated channels). Bar length: 40 µm. (C).

As mercury porosimetry only measures pore access, real pore size was measured using LSCM. The obtained results showed that in fine-grained sandstones, pore sizes varied from between 50-60 µm, and could reach up to 120 µm. In the medium-grained sandstones, the average main pore size was around 200 µm, reaching up to 600 µm, giving a macroporous character to the rock that allowed optimal interconnection between the pores that facilitated fluid circulation. The 3D reconstruction of the porous network (Figure 5) showed that it was constituted by a single pore system whose configuration only depends on grain arrangements and degree of compaction. The porous network was constituted by channels (<40 µm in diameter) that may occasionally expand, giving rise to megapores of up to 300 µm in diameter.

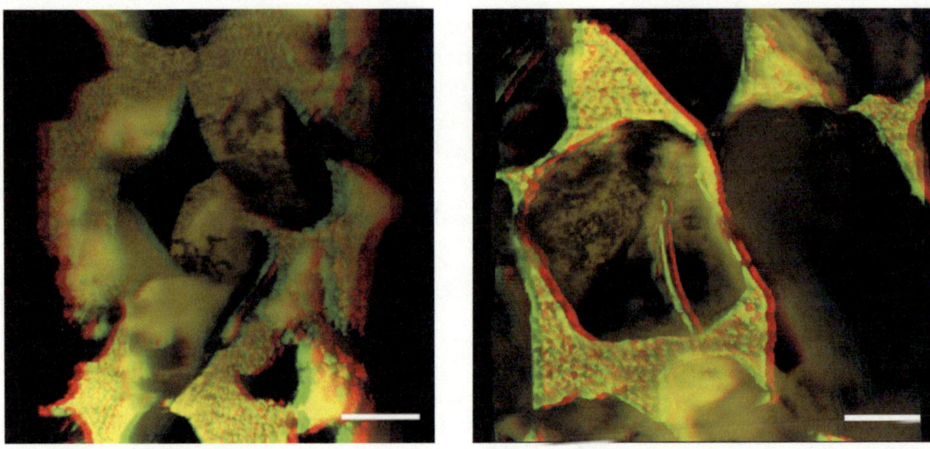

(A) (B)

Fig. 5. (A) 3D reconstruction of the porous network. Bar length: 40 µm. (B) 3D pore detail. Bar length: 40 µm.

Combined LSCM and mercury intrusion porosimetry data allowed recognition of the fact that porosity was present as large pores, intercommunicated by channels that constitute pore accesses. Although of smaller size with respect to the main pores, these pore accesses were still large (20-40 µm).

Hence, the 3-D reconstruction of the porous network in the sandstone facilitated not only an understanding of the effect of consolidants on pore-size distribution and on the network configuration of the rock, but also allowed correct interpretation of the porosimetry data.

4.2 Preparing samples

In order to determine the distribution of the consolidant within the porous network using LSCM (Zoghlami and Gómez, 2004), sample were prepared in accordance with the following procedure:

- Addition of powdered fluorochrome to the consolidant solution (5gr/1l of consolidant).
- The obtained solution was left for one hour to allow total dissolution of the Fluorochrome in the consolidant solution.
- Brush application of the doped consolidant and watter repellent solution to the sandstone samples (5x5x5 cm).
- Curing treated samples for one month with doped consolidant to permit complete polymerisation of the conservation products.

- Preparation and mounting of polished planar sections (thickness > 100 µm) on a glass slide.

4.3 Consolidant distribution into the pore network of the sandstone and LSCM contribution

In order to determine the percentage of porosity and its pore size distribution, IMP was used to detetect changes between untreated and treated sandstone simples with the consolidant.
Compared to untreated samples, samples treated with consolidants show a moderate decrease in total porosity. Consolidants affected pore-size distribution by producing a decrease in macroporosity and an increase in microporosity, especially for pores of less than 10 µm in diameter. Figure 6 shows that a decrease occurred in the amount of pores having sizes of between 20 and 40 µm, the most abundant range in this rock. Additionally, there was the appearance of pores having a smaller diameter (<10 µm). The presence of this new pore population might be explained by a partial sealing of pores of between 20 and 40 µm in diameter. The effect of partial sealing in different treated rocks has already been discussed by Esbert and Díaz-Pache (1993).

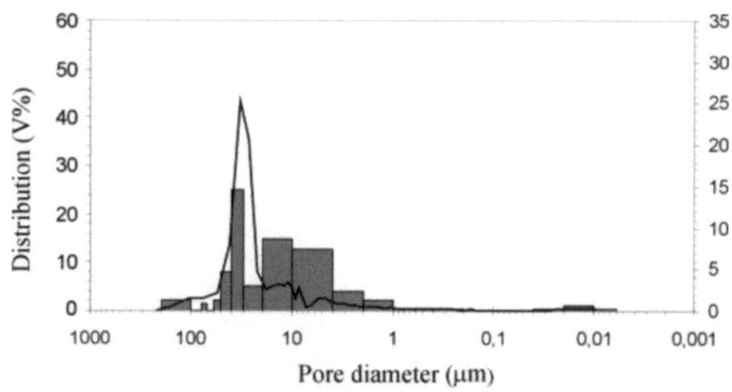

Fig. 6. Pore size distribution of the treated sandstone simple with consolidant (2)

In thin sections, LSCM observations showed that grain surfaces were covered by a discontinuous coating film. When grains were very close, consolidants usually formed meniscus plugging only in the small throats (diameter <40 µm) (Figures 7).
3D reconstruction (Figure 8A) showed that consolidants filled the pore-network throats (pore diameter of less than 40 µm), whereas megapores were covered by a very thin (1-2 µm) coating film of consolidant (Figure 7B). It is worth observing that the consolidant used in our experiment developed cracks reaching up to 10 µm in diameter. These cracks developed a network of small channels interconnected between both themselves and the megapores (Figure 8B).
As shown previously, mercury porosimetry data from samples treated with consolidants showed a microporosity increase with respect to untreated samples, a decrease in pores with diameters of between 20 and 40 µm and the development of new pores with diameters smaller than 10µm (Fig. 3). Nevertheless, LSCM images demonstrated that the main pores sizes (mega pores) were not affected by the application of consolidants, due to the development of a very thin coating film around the grains. 3D reconstruction showed that consolidant is concentrated within throats having a pore diameter smaller than 40 µm and that cracks developed in the polymer could reach up to 10µm in size. Therefore, the increase in microporosity can be attributed to the formation of microfissures probably due to the

type of catalyst used; this probably destabilizes the polymer used as consolidant, as reported by Brus and Kotlik (1996). These results also showed that a correct interpretation of the mercury porosimetry data needs to consider the formation of micro-cracks during the application of a consolidant.

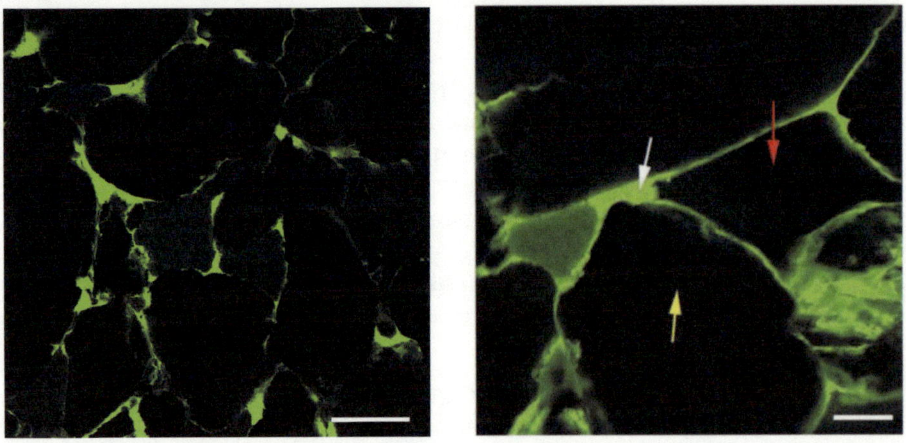

(A) (B)

Fig. 7. Fluorescence microscope images. Distribution of consolidant (A) and (B).
Red arrow: porosity; yellow arrow: grain; white arrow: consolidant. Bar length: 80 µm.

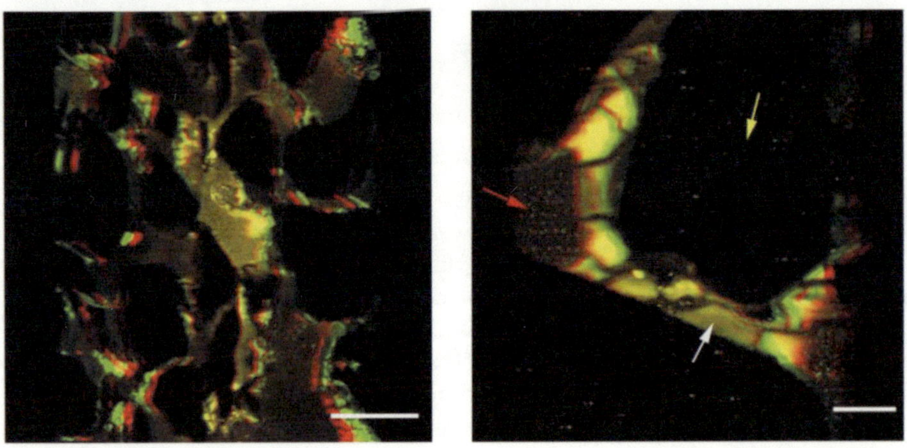

(A) (B)

Fig. 8. 3D reconstruction. (A) Image shows that consolidant is concentrated in throats. Bar length: 80 µm. (B) Image shows microfissures in consolidant. Bar length: 40 µm.

4.4 Watter repellent distribution into pore network of the sandstone and LSCM contribution

The studied treated and untreated samples used for this experiment were obtained from the same stone prism (10x5x5 cm) that was divided into 2 cubes (5x5x5 cm). One of them was treated with a doped water repellent solution, as explained below, and used for the LSCM

imaging and mercury intrusion porosimetry (Figure 9). The other one was treated with plain water repellent and employed in SEM observations, taking into account that the penetration depth of the water repellent applied on the stone cube was 1.5 cm, which was determined by microdrops, wetting and salt crystallisation tests (Zoghlami, 2003). The used water repellent is a Polymethyl-siloxan commercialized in water emulsion by several companies.
In order to distinguish the water repellent and determine its distribution within the pore network using LSCM, the conservation product had to be doped, so that part of the sandstone samples were prepared following the methodology previously described.

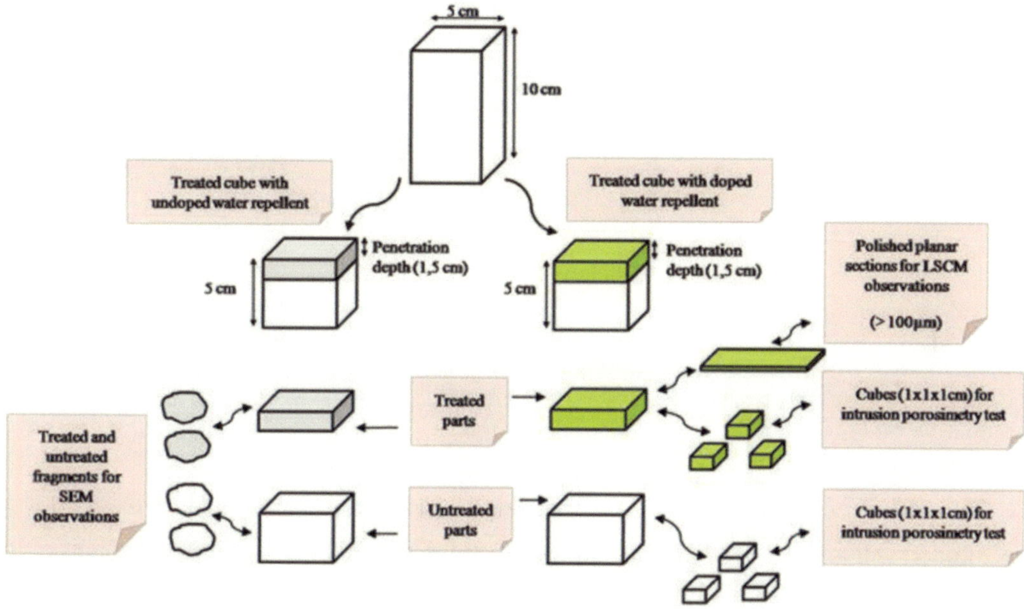

Fig. 9. Chart of sample distribution for the experimental procedures
60x35mm (300 x 300 DPI)

The porosity of the samples treated with water repellent was measured by mercury intrusion porosimetry. The data obtained did not change much the pore size distribution with respect to the untreated samples (Figure 10 A and B). However, a small reduction of total porosity occurred (2.29%) which affected both macro and microporosity percentages. The porosity reduction was more pronounced in macroporosity (1.66%) than in microporosity (0.63%). Nevertheless, this affected the porosity distribution, as the microporosity distribution value decreased, from 10.01 to 8.01%, and therefore the macroporosity distribution value increased from 89.99 to 91.99%. All pore sizes decreased in diameter as was observed in the shift towards the right hand-side of the pore size intervals in the porosity distribution histograms. Similarly, the average pore diameter decreased from 33.40 to 32.21 µm and the pore diameter mode adjusted varying from the 30-40 µm to the 20-30 µm intervals.
SEM observations of treated rock samples showed that the water repellent formed a continuous film that completely covered the sandstone grains (Figure 11) but did not occlude the macroporosity. The optical similitude under SEM between the conservation polymer, which formed a silica gel, and the quartz and feldspar grains of this sandstone disallowed the accurate thickness measurement of this film.

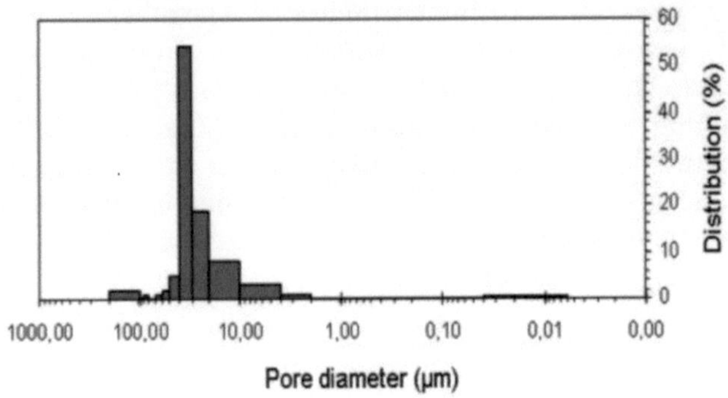

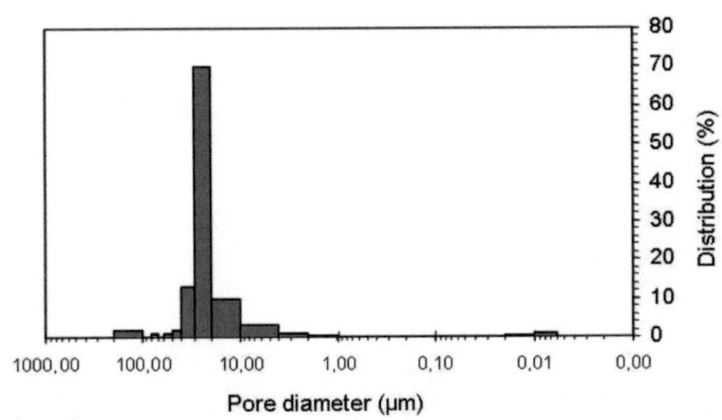

Fig. 10. Pore-size distribution diagrams obtained by mercury porosimetry test.
(A) Untreated sandstone (B) Treated sandstone with THE328 water repellent.

LSCM observations of water repellent in thin sections allowed the easy measurement of the coating film: it varied between 1.5 and 2 µm (Figure 12). Moreover, the water repellent polymer formed meniscuses where the sandstone grains were very close and plugged the throats whose diameters were less than 3-4 µm. The visualization and differentiation of the water repellent distribution through the pore network over a section 100 µm thick (Figure 12), which is possible using specific software, corroborated that the coating film had a homogeneous and continuous volumetric distribution in the three spatial directions.
The mercury intrusion porosimetry data reported, as expected, a small decrease (2.29%) in the total porosity of the treated samples. Although both macro and microporosity absolute values were reduced, the macroporosity distribution value increased by 2%. SEM imaging of the siliceous sample treated with THE 328 allowed the visualization of the water repellent polymer coating the grains and even the closure of some pores, but it was too difficult to measure the thickness of the water repellent film through the pore network over the penetration depth profile. It is therefore complicated to qualitatively explain, in this case, the porosity change in the treated samples with the SEM observations.
Alternatively, water repellent stained with fluorochrome was effortlessly observed under LSCM as green marks (Figure 12 and 13), thus achieving a successful differentiation from the siliceous components of the rock and its porosity. It was therefore simple to determine, with LSCM, the effective penetration depth of the product and to measure the thickness of the protective film formed over the grains as well as the plugged pore diameters. Fluorescence and 3D imaging of water repellent distribution and porosity by LSCM helps

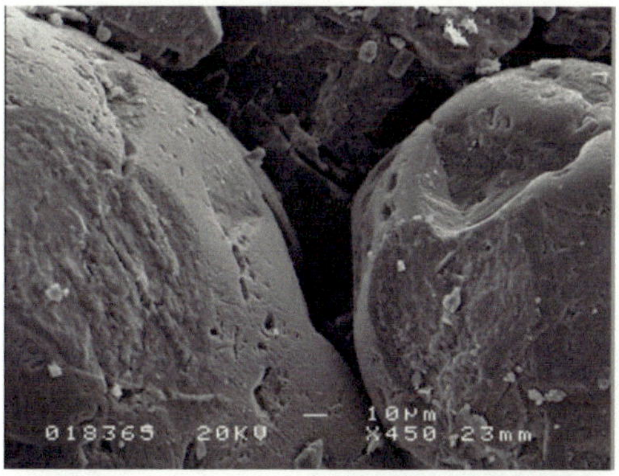

(A)

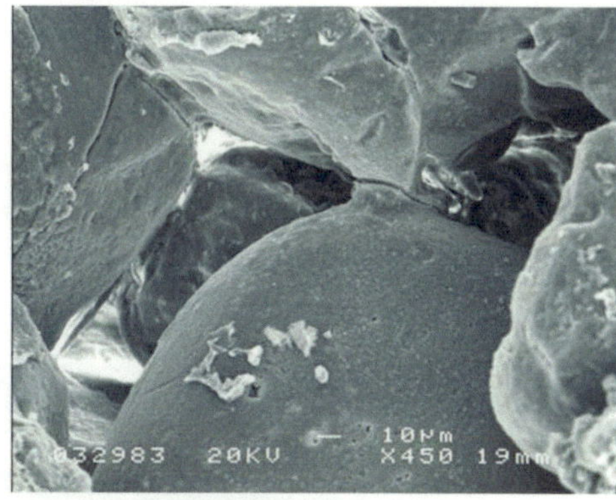

(B)

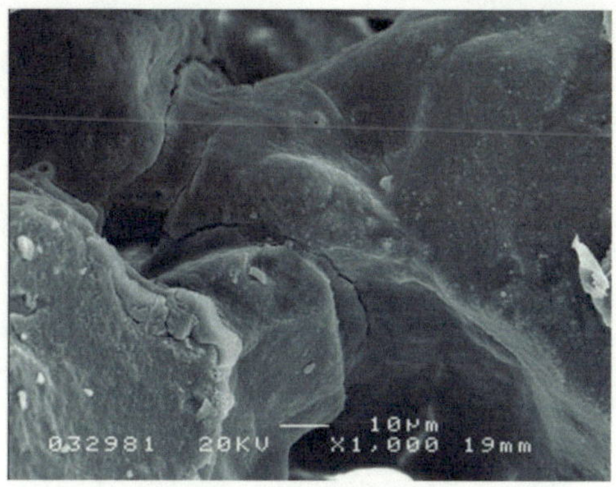

(C)

Fig. 11. SEM images of the sandstone. (A) Untreated sample. (B) Large view of the sample treated with THE328 water repellent. (C) Detailed photograph of the water repellent film between grains.

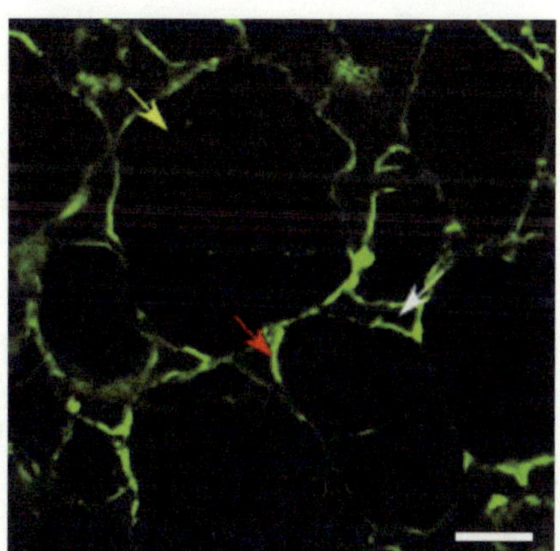

Fig. 12. Thin sections of a sample treated with THE328 water repellent observed with LSCM under fluorescence light. (A) Regular view where the red arrow points to the coating film (green), the yellow arrow a quartz grain, and the white arrow the porosity. Bar length: 40 µm.

with the interpretation of even the minor changes in porosity. This case study shows that the decrease of 3-4 µm in the diameter of all pores is the root of the shift towards the right hand side of the pore size intervals in the distribution histogram. As the displacement also affected the mode interval or average pore diameter, this shift had an impact in the new distribution between macro and microporosity in the samples after treatment because the modal pore size was close to the 15 µm value (the limit between macro and microporosity fields).

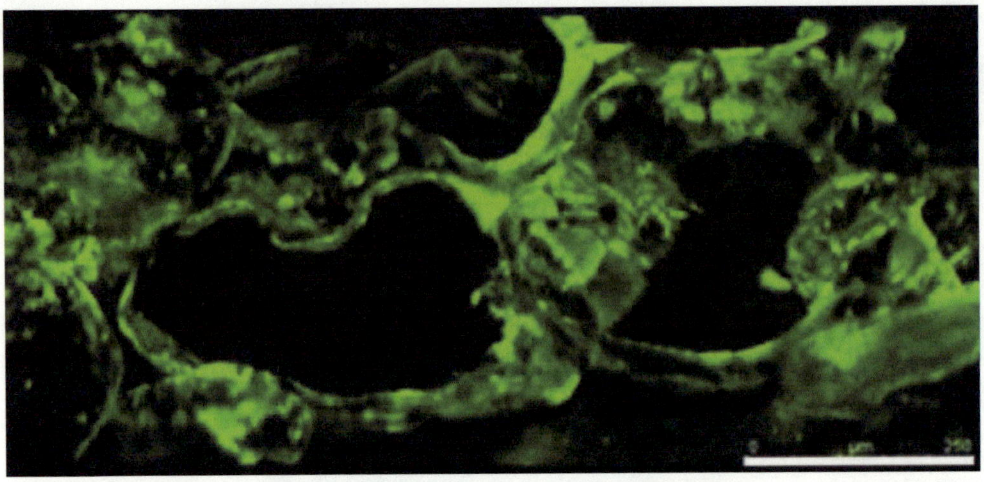

Fig. 13. 3D reconstruction obtained by LSCM of a section 100 5m thick of the water repellent distribution within the sandstone porous network. Bar length: 250 5m.

5. Conclusion

As demonstrated by the case study below, LSCM has proved to be a useful technique in achieving a tridimentional reconstruction of a rock's porous network, and in understanding quantitative porosity data obtained by mercury porosimetry. In this study, LSCM has been used for the first time to determine the 3D distribution of conservation products within the pore network of the sandstone. Thorough observations of consolidant and a watter repellent distribution within the rock's pores has obtained enabling an inderstanding of the consolidant's and water reppelent effect on the pore-network configuration and interconnection, both of these being important properties affecting rock durability, since they control fluids flow. So a correct assessment of the effectiveness of these products has been allowed and the usefulness of LSCM in building-stone preservation studies has therefore been demonstrated.

Data obtained confirms that LSCM is a very useful technique that complements and helps to integrate the quantitative porosity data as well as SEM observations.

The principal limitation of this technique is the reduced penetration depth. Until now, penetration depth about 250µm can be achieved before significant image degradation is apparent. But this is insufficient especially in rock materials to describe faithfully the pore media. In reality, deeper penetration can be obtained by greater laser intensity. Nevertheless, a high laser intensity increases the risk of photobleaching.

6. References

Adler, P.M. 1992. *Porous media: Geometry and transport*, 544 pp., Butterworth-Heinemann Series in chemical Enginneering, Stoneham, Mass.

Alessandrini G.; Del Fa, C.M.; Rossi-Doria, P.; Tabasso, M. & Vannucci, S. (1975). Treatment of Stone in Monuments. A Review of Principles and Processes. *Proceeding of the International Symposium of the Conservation of Stone I*, pp. 635-650, Bologna. Italy.

Alvarez De Buergo, M.; Fort, R. & Gómez-Heras, M. (2004). Contributions of Scanning Electron Microscopy to the assessment of the effectiveness of stone conservation treatments. *Scanning*, Vol. 26, 41-47.

Alvarez De Buergo, M. & Fort ,R. (2001). Basic methodology for the assessment and selection of water-repellent treatments applied on carbonatic materials. *Progress in Organic Coatings*, Vol.43, 258-266.

Álvarez De Buergo, M.; García Calleja, M.A.; Gónzalez Limon, T. & Soriano Carillo, J. (1994). Estudio de la eficacia y durabilidad de los tratamientos de conservación aplicados a los materiales pétreos del Claustro de San Juan de Duero en Soria. *Ingeniería Civil*, Vol. 9, 89-97.

Baldwin, C.A.; Sederman, A.J.; Matntelle, M.D.; Alexander, P. & Gladden, L.F. (1996). Determination and characterization of the structure of a pore space from 3D volume images. *J. Colloid Interface Sci.*, Vol. 181, 79-92.

Berryman, J.K. & Blair, S.C. 1986. Use of digital image analysis to estimate fluid permeability of porous materials: Application of tow-point correlation functions. *J. Appl. Phys.*, Vol.60, 1930-1938.

Biscontin, G.; Frascati, S. & Marchesini, L. (1975). Colour Variations in Old Bricks and Stones as a Result of Consolidation with Resin. *Proceeding of the International Symposium of the Conservation of Stone I.*, pp 741-747, Bologna. Italy.

Borselli, G. ; Camaiti, M. ; Pasetti, A., Maravelaky, P. & Matteoli, U. (1990). Protettivi impiegati nella conservazione dei materiali lapidei: storia, impieghi, sviluppi. *L'Edilizia*. 1/2: 67-78.

Bourbie, T. & Zinszner, B. (1985). Hydrolic and acoustic properties as a function of porosity in Fontainebleau sandstone. *J. Geophys. Res.*, Vol. 90, 11524-11532.

Brown, G.O.; Stone, M.L., & Gazin, J.E. (1993). Accuracy of gamma ray computerized topography in porous media. *Water Res.*, 29: 479-486.

Bousquié, P. ; Pellerin, F.M. ; Struillou, R. & Arnould, M. (1980). Contribution de la porosimetrie au mercure a l'étude de quelques propriétés de roches carbonatées. In: Wolters R, editor.Materials and engineering geology. *Bull Int Assoc Eng Geol.*, Vol. 22, 225-232.

Brus, J. & Kotlík, P. (1996). Consolidación de la piedra con mezclas de Alcoxisilanos y polímeros acrílicos. *Cuadernos sobre conservación*, Vol. 41, N°2.

Camaiti, M., & Amoroso, G.G. (1997). Scienza dei materiali e restauro. *La pietra: dalli mani degli artisti e delli scalpellini a quelle dei chimici macromolicolari* , 320 pp, Alinea ediytice srl- Firenze.

Cnudde, V.; Cnudde, J.P., Dupuis, C. & Jacobs, P.J.S. (2004). X-ray micro-CT used for the localization of water repellents and consolidants inside natural building stones. *Mater Charact*, Vol. 53, 259-271.

Dell'Agli, G.; Ferone, C., Mascolo, G.; Marino, O. & Vitale, A. (2000). Durability of tufaceous stones treated with protection and consolidation products. *9th international congress on deterioration and conservation of stone*. Venice, Italy. pp 379-386.

Doughty, D.A. & Tomutsa, L. (1997). Imaging pore structure and connectivity in high resolution NMR microscopy. *Int. J. Rock Mech. Min. Sci.* 30: 691-697.

Dullien, F.A.L.(1992). *Porous media: Fluid transport and pore structure*, 2nd ed., Academic Press, New York.

Esbert, R.M. & Díaz-Pache, F. (1993). Influence of petrographical characteristics in the penetration of consolidants in porous building stones. *Mater construcc.* Vol. 43, Nº230. 25-36.

Esbert, R.M.; Grossi, C.M.; Valdeon, J.; Ordaz, F.J.; Alonso, F.J. & Marcos, R.M. (1990). Laboratry study for stone conservation at the Cathedral of Murcia. *Mater construcc.* Vol. 40, 217. 5-15.

Esbert, R.M. (1993). The use of surface protectants: the current situation in Spain. *Mater Construcc.* Vol. 43, Nº 229. 5-16.

Fredrich, J.T.; Greaves, K.H. & Martin, J.W. (1993). Pore geometry and transport properties of Fantainebleau sandstone, Int. J. Rock Mech. Min. Sci., Vol. 30, 691-697.

Fredrich, J.T., Martin, J.W. and Owen, L.B., 1994. *Application of X-ray computerized tomography for non-destructive mesurment of physical properties of NTS tuff*, Deffens Nuclear Agency, DNA-TR-93-189, 121 pp., Alexandria, Virginia.

Fredrich, J.T. (1999). 3D Imaging of porous Media using Laser Scanning Confocal Microscopy with Application to Microscale Transport Process. *Phys. Chem. Earth*, Vol. 24, 551-561.

Gómez-Gras, D., & Zoghlami, K. (2003). Procedencia de las areniscas de la unidad superior de la Fm. Fortuna en el NE de Túnez. *Geotemas*, Vol. 5, 109-111.

Goñi, J. ; Ragot, J.P. & Sima, A. (1968). Méthode d'etude du champ microfissural des mineraux et des roches et possibilités d'application en géologie. *Bull. Bur. Rech. Geol. Min.* Section 2: Geologie des Gites Mineraux, Vol. 4, 51-86.

Hudec, P.P. (1978). Rock Weathering on the Molecular Level, In: Decay and Preservation of Stone, Engineering Geology Case Histories. *The Geological Society of America. Boulder*, Vol. 11, 47-51.

Ingersoll, R.V.; Bullard, T.F; Grimm, J.P.; Pickle, J.D. & Sores, S.W. (1984): The effect of grain size on deterial modes: a test of Gazzi-Dickinson point-counting model. *Jour. Sediment. Petrol.*, Vol. 54, 103-116.

Krohn, C.E. (1988). Sandstone fractal and Euclidean pore volume distribution. *J. geophys. Res.* Vol. 93, 3286-3296.

Menéndez, B.; David, C. & Darot M. (1999). A study of the Crack Network in Thermally and Mechanically Cracked Granite Samples using Confocal Scanning Laser Microscopy. *Phys. Chem. Earth*, Vol. 24, 627-632.

Montoto, M.; Martinez-Nistal, A.; Rodriguez-Rey, A.; Fernandez-Merato, N. & Soriano P. (1995). Microfractography of grantic rocks Ander confocal scanning laser microscopy. *J. microscopy*, Vol. 177, 138-149.

Myer, L.R.; Kemeny, J.M.; Zheng, Z.; Suarez, R.; Ewy, R.T. & Cook, N.G.W. (1992). Extensile cracking in porous rock under differential compressive stress, In: *Micromechanical Modelling of Quasi-Brittle Materials Behavior*, V.C.Li (ed), Appl. Mech.Rev., Vol. 45, 263-280.

Paterno, M.C. & Charola, A.E. (2000). Preliminary studies for the consolidation of Guadalupe tuff from the Philippines. *9th international congress on deterioration and conservation of stone*. Venice, Italy. pp 155-163.

Petford, N., Davidson, G., & Miller, J.A. (1999). Pore structure determination using Confocal Scanning Laser Microscopy. Phys. Chem. Earth., Vol. 24, 563-567.

Petford, N.; Miller, J.A. (1990). SLM confocal microscopy: an improved way of viewing fission tracks. *J. Geol. Soc. London,* Vol. 147, 217-218.

Petford, N. & Miller, J.A. (1993). The study of fission tracks and other crystalline defects using CSLM. *J. Microscopy,* Vol. 170, 201-212.

Piacenti, F., Camaiti, M., Mangalli del Fà C. & Scala, A. (1993). Fluorinated Aggregating materials for stone. *Conserv Stone and other Mater,* pp 740-747.

Pironon, J.; Canals, M., Dubessy, J., Walgenwitz, F. & Laplace-Builhe, C. (1998). Volumetric reconstruction of individual oil inclusions by CSLM. *Eur. J. Mineral* ,10: 1143-1150.

Pittman, E.D. & Duschatko, R.W. (1970). Use of pore casts and scanning electron microscopy to study pore geometry. *J.Sed. Petr.,* Vol. 40,1153-1157.

Rodríguez-Rey, A.; Ruiz de Argandoña, V.G; Calleja L, Suárez del Rio, L.M., Celorio, C. (2006). Consolidants Influence on the Sandstones Capillarity. X-ray Study. *Advances in X-ray Tomography for Geomaterials*. ISTE, Ltd., UK, pp. 381-387.

Ruiz de Argandoña, V.G.; Rodríguez-Rey, A., Celorio , C., Calleja, L. & Suárez del Río, L.M. (2003). Characterization by X-ray computed tomography of water absorption in a limestone used as building stone in the Oviedo Catedral (Spain), Mees F, Swennen R, Van Geet M and and Jacobs P. (eds): Applications of X-ray Computed Tomography in the Geosciences. *Geological Society, London, Special Publications,* Vol. 215, 127-134.

Sahimi, M. (1995). *Flow and transport in porous media and fractured rock,* 482 pp.,VCH, Weinhein, FRG.

Sasse, H.R.; Honsinger, D., Schwamborn, B. (1993). "PINS" - New technology in porous stone conservation. *Conservation of stone and other materials*. 705-715.

Thompson, A. H. (1991). Fractals in rock physics. Ann. Rev. *Earth Planet. Sci.* Vol. 19, 237-262.

Torraca, G. (1975). Treatment of Stone in Monuments. A Review of Principles and Processes *Proceeding of the International Symposium of the Conservation of Stone,* pp. 297- 315. Bologna, Italy.

Wellington, S.L. & Vinegar, J.J. (1987). X-ray computed tomography. *J. Pet. Tech.* Vol. 39, 885-898.

Villegas, R., Martín, L., Vale J.F. & Bello, M.A. (1995). Characterization and conservation of the stone used in the cathedral of Granada, Spain. *Mater Construcc.,* Vol. 45, 240. 17-35.

Villegas, R. & Vale Parapar, J.F. (1993). Evaluación de tratamientos de hidrofugación aplicados a piedras de catedrales andaluzas. III.- Ensayo de alteración acelerada en atmósfera contaminada. *Mater construcc.* Vol. 43, N°232, 25-38.

Wheeler, G. (1992). Comparative strengthening of several consolidants on Wallace sandstone and India limestone. *Proceedings of the 7th International Congress on Deterioration of Stone*. pp 1033-1042. Lisbon, Portugal.

Wilson, T. (1990). The role of the pinhole in confocal imaging systems. *In handbook of biological Confocal microscopy*, Pawley, J.B. (ed.), 113-126, Plenum Press, New York, NY.

Wong, T.F., Fredrich, J.T., and Gwanmesia, G.D., 1989. Crack aperture statistic and pore space fracture geometry of Westerly granite and Rutland quartzite: implication for an elasticcontact model of rock compressibility. *J. Geophys. Res.*, Vol. 94, N°10, 267-10, 278.

Young, M.E; Murray, M., Cordiner, P. (1999). Stone consolidants and chemical treatments in Scotland. *Report to historic Scotland*. 298 p.

Zinszner, B., Johnson, P.A., Rasolofosaon, P. N.J. (1997). Influence of change in physical state on elastic nonlinear response in rock: Significance of effective pressure and water saturation. *J. Geophys. Res. Vol.* 102, 8105-8180.

Zoghlami, K; Gómez-Gras D, Álvarez, A. & De Luxán, M.P. (2004). Intrinsic factors that condition the physical behaviour and the durability of the Miocene sandstones used in the construction of the Roman aqueduct of Zaghouan-Carthage. *Mater Construcc.Vol.* 54, N°276. 29-41.

Zoghlami, K., Gómez-Gras, D., Álvarez, A. & De Luxán MP. (2005). Evaluation of consolidating and water repellent treatments applied to the miocene sandstone used in Tunisian Heritage Monuments. *Mater Construcc.* Vol. 55, N°. 277, 25-39.

Zoghlami, K., Gómez-Gras, D. (2004). Determination of the distribution of consolidants and Interpretation of Mercury Porosimetry Data in a Sandstone Pore network Using LSCM. *Microsc Res Tech. Vol.* 65, 270-275.

18

Applications in Complex Systems

Thomas L. White[1], T. Bond Calloway[1], Robin L. Brigmon[1],
Kimberly E. Kurtis[2] and Amal R. Jayapalan[2]
[1]*Savannah River National Laboratory*
[2]*Georgia Institute of Technology*
United States

1. Introduction

1.1 Introduction

Laser scanning confocal microscopy (LSCM) is widely used in biological, semiconductor, geological and other material science fields. For most non-opaque applications, interior structures can be imaged. LSCM can perform optical sectioning by acquiring images point by point and then resolve the images using a computer to provide a 3-D profile of the sample (Corle, 1996). In the past ten to fifteen years, researchers have begun to develop LSCM techniques for applications with opaque samples. LSCM technology has allowed collection of scattered, reflected or fluorescence photons to provide depth profiling of opaque samples. More importantly, advances in software for LSCMs have allowed 3-D models to be developed that provide researchers an *in-situ* view of systems that until recently have only been understood macroscopically by measuring the bulk properties of the materials. These macroscopic measurements have only allowed a conceptual understanding of the microscopic solid-solid, gas-solid and gas-solid-liquid interactions. In this chapter, we will discuss the use of LSCMs for complex opaque systems. Specific applications with complex iron-aluminum and aqueous organic slurries and cement-based composites will be discussed.

LSCM is based on the principle of eliminating stray light from "out-of-focus" flare by means of confocal apertures. Images are acquired by scanning the sample with a fixed light source, choices of laser lines, detection filters, wavelength, and by recording the light reflected from the in-focus plane with a theoretical resolution of 0.2 μm (Carter, 1999). Tomography is accomplished by recording a series of consecutive layered images in both the x-y and x-z planes. LSCM allows for the study of both organic and/or inorganic samples with minimal preparation in real time. When used to observe the outermost sample surface/subsurface areas, LSCM requires no sample preparation beyond placement on the slide. The sample is thus free from artifacts induced by drying, sectioning, or similar adulterations that are required in other analytical procedures.

1.2 Radioactive waste slurries applications

While the concentrated radioactive Fe/Al slurries discussed in this section are unique to the defense related nuclear industry, fundamentally processing aqueous metal slurries is done in a variety of industries. Additionally, processing slurries containing complex aqueous organic molecules is relevant to a variety of industries. Many of the unit operations and much of the chemistry used to develop these waste processes was derived from industrial practices developed by the chemical and ceramics industries. LSCM has found applications in these industries as well. For example, LSCM has been used for examination of pore distribution (e.g. size and spatial arrangement) in ceramics for biomedical applications (Ren et al., 2005). Thus, the LSCM techniques described within this chapter are relevant to most any industry that processes opaque slurries.

The production of nuclear materials at Savannah River and Hanford Sites for the Department of Energy (DOE) resulted in the generation of radioactive waste that is currently stored in below ground storage tanks with a capacity of up to 1.3 million gallons. Each site has a goal of encapsulating the millions of gallons of radioactive waste slurry into borosilicate glass and/or grout. An involved infrastructure of facilities has been constructed or is under construction for the transportation, treatment, and immobilization of the radioactive waste slurry for long term storage.

The radioactive waste slurries are currently being immobilized in a borosilicate glass matrix using joule heated glass melters at the Savannah River Site (SRS) in the Defense Waste Processing Facility (DWPF) located near Aiken, South Carolina. A similar facility in West Valley, New York also immobilized waste generated from the reprocessing of spent nuclear fuel. Larger immobilization (vitrification) facilities are planned as part of the Hanford River Protection Project-Waste Treatment Plant (RPP-WTP) located near Richland, Washington. Highly radioactive waste is transferred to these facilities from underground tank farms. The insoluble solids content of the waste is limited by the design-basis rheological properties (e.g. the Bingham plastic yield stress and plastic viscosity) used to design the slurry transfer systems. Typically, these slurries are iron and/or aluminum hydroxides in a caustic solution.

These facilities have used or will use slurry-fed melters to safely immobilize the waste in a glass matrix. Glass forming chemicals or glass frit fabricated from glass formers is added to either the radioactive waste solutions or slurries. The resulting slurries (≈ 35 - 65 wt. % total solids) are sampled, analyzed, and then pumped to the melters. Although the DWPF process is currently operating successfully, production throughput is limited by certain bounding conditions used to design the plant (slurry rheology, glass property constraints, and in some cases, gas holdup). A simplified flow sheet of the Hanford waste treatment plant is shown in Figure 1.

The solids loading in the DWPF and future RPP-WTP melter feed slurries are limited by the rheological design bases of the mixing, sampling, and transport systems. It is desirable to increase the production rate and waste loading of the glass and therefore decrease the total quantity of waste glass produced from a total plant life cycle and cost perspective. Increasing the solids content of the melter feed would decrease the energy required to evaporate the water in the slurry, and would, therefore, increase the overall production (melt) rate of the immobilization process.

It is possible to modify the equipment used to mix, sample, and transport the waste slurry to the melter. The design and construction cost for any such modifications is very high due to the constraints (radiation and non-visible remote operation) imposed on the design and operation of radioactive waste immobilization processes. Therefore, adjustment of the rheological properties by trace chemical addition is being explored as one option to improve the overall production rate of radioactive waste vitrification processes.

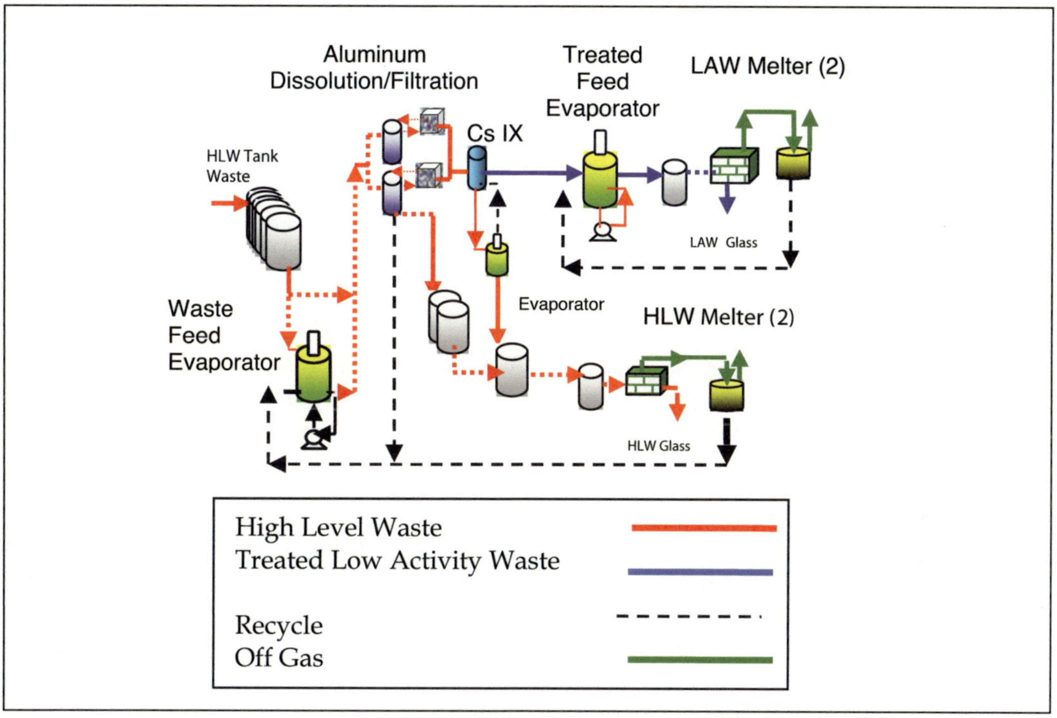

Fig. 1. Simplified flow sheet of the Hanford River Protection Project-Radioactive Waste Treatment Plant (RPP-WTP) with the dashed red/blue lines indicating unit operations involving slurries

The viscous nature of the radioactive waste slurries is also linked to operational and safety problems. Hydrogen gas produced by radiolysis can be held-up in the slurry. Understanding the hydrogen gas hold-up and release from the radioactive slurries remains a continuing research effort. Additionally, the viscous nature of these slurries causes air to be entrained in the slurry, which results in a foamy consistency, making the slurries difficult to pump. Similar problems are predicted to develop in the Hanford RPP-WTP (Kay et al., 2003, Zamecnik et al., 2003).

Nuclear waste is primarily composed of a sodium hydroxide salt supernate and transition metal sludge (See Figure 2). While the sludge is mostly inorganic oxides/hydroxides of aluminum, iron, nickel and manganese, the waste also contains sodium, carbonate, silica, oxalate, manganese, nitrite, phosphate, lead, zirconium, sulfate, potassium, nitrate as well as organic complexants and noble metals (Vijayaraghavan et al., 2006). Essentially most of the periodic table of elements is represented in nuclear waste derived from reprocessing nuclear fuel. Alumina rich sludge slurries tend to exhibit non-Newtonian rheological behavior due to floc formation increasing the viscosity and yield stress of the slurry (Ribeiro et al., 2004). At a high (≈ >45 - 50 wt. % total insoluble solids) enough solids content, the flow characteristics of the slurry can exceed the capability of the processing equipment.

Fig. 2. Non-radioactive chemical simulants of the nuclear waste containing sludge and salt supernates (Note the opaque nature of sludges that form concentrated (5 – 50%) Fe/Al slurries limiting traditional light microscopy methods)

During the DOE nuclear waste research and development effort of the 1980's, a process to remove radioactive cesium using sodium tetraphenylborate (NaTPB) to precipitate the cesium from salt supernate waste was developed. The process yielded a potassium and cesium tetraphenylborate aqueous slurry that proved to be difficult to process due to gas holdup, foaming and rheological issues (Calloway et al., 2001). Due to significant problems associated with the breakdown of tetraphenylborate to benzene, the process was eventually abandoned. Various flow sheets were explored in an effort to reduce or mitigate the effects of tetraphenylborate degradation (See Figure 3 and 4).

However, significant insight into the behavior of solid particles in slurries was developed as a part of an effort to reduce foaming in the process. These studies led to additional research into gas holdup, rheological and foaming control, and the need to better understand the behavior of particles, gases and liquids from a more fundamental microscopic level.

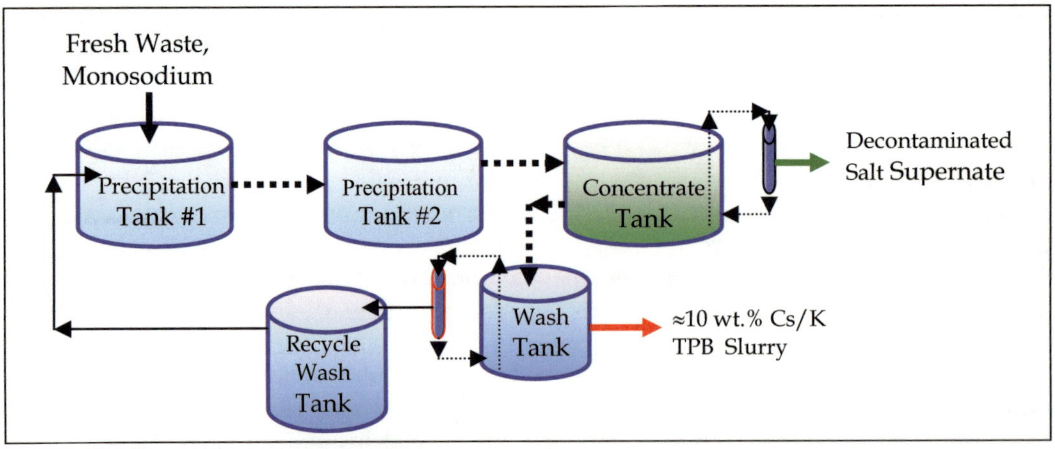

Fig. 3. Simplified flow sheet of NaTPB process to remove Cesium (Cs) from nuclear waste supernates (dashed lines indicate unit operations handling opaque slurries)

Prior to the development of LSCMs for use with opaque simulated radioactive slurries, only bulk slurry properties could be measured and the *in-situ* behavior of the particles, gases, liquids and more importantly the interface region between all three, was not well understood. These bulk measurements can also take considerable time and effort. Thus, practical development of engineering solutions to the various problems could only be accomplished empirically. Fundamental understanding of particle-gas-liquid interactions on a microscopic level has ultimately led to improved antifoam agents for nuclear waste processing facilities and should eventually help develop rheology modifiers that will allow increases in solids concentration in nuclear wastes which will increase plant throughput. Thus, further development of the LSCM slurry techniques described within this chapter should reduce the overall costs of many industrial slurry processes by furthering the understanding of the following process issues:

- Foaming/Antifoaming
- Rheology/Maximizing Solids Loadings
- Air Entrainment/Pump Cavitations

It is important for the reader to understand that LSCM is just one technique in an arsenal of analytical techniques that are available to understand complex opaque slurries. Traditional visual inspection, optical microscopy, and nuclear magnetic resonance, to more advanced neutron techniques available at national laboratories can be employed to further understand particle interactions within slurries.

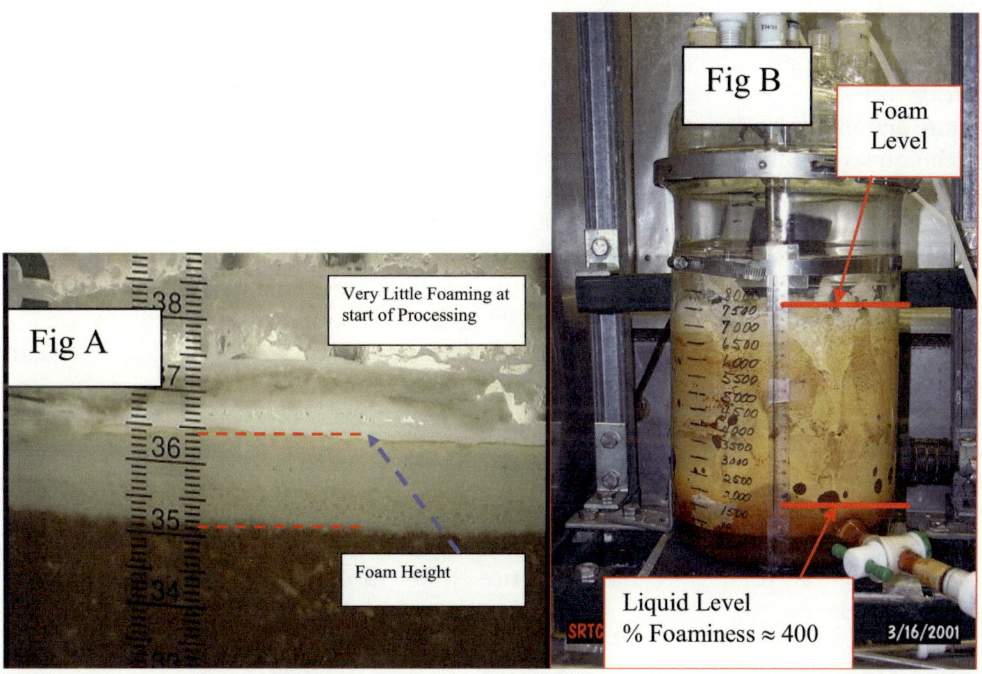

Fig. 4. Foaming and gas holdup in simulated nuclear Organic/Aqueous wastes - Fig. 4A shows foaming at the start of the process while Fig. 4B shows the dramatic change in foaming at the end of the process. The slurry is non-Newtonian and solid particles float when air is entrained in the slurry

1.3 Cement-based materials applications

This section will describe the application of LSCM to characterize the various interfaces that exist in cement-based materials, such as pastes, mortars, and concrete. While the use of LSCM for imaging construction materials is not as common as the use of stereomicroscopy (petrography) or scanning electron microscopy, advantages associated with LSCM over these more often applied techniques suggest that its use in this field could be increased. Using LSCM, samples can be imaged without much preparation, thus limiting the introduction of artifacts associated with drying, epoxy impregnation, or coating of samples for electron microscopy (Marusin, 1995). Because drying and coating are not required, *in situ* changes in microstructure during hydration or due to continuing deterioration can be studied using LSCM techniques (Collins et al., 2004a). Furthermore, the inherently porous surfaces, characteristic of cement-based composites as well as fracture surfaces, can be imaged by LSCM where surface asperities do not affect resolution or focus. With fracture surfaces, LSCM allows for quantification of roughness, which can be related to mechanical performance and fracture mechanics parameters (Kurtis et al., 2003; Mohr & Kurtis, 2006).

Thus, the development and use of LSCM as a characterization technique for cement-based and other construction materials has several advantages over conventional characterization methods. Here an innovative approach to the use of LSCM to examine aggregate/paste interfaces by imaging through aggregates is described (Collins et al., 2004a).

The microstructure formed at the interface between cement paste and aggregate in concrete is known to be generally weaker than both the bulk hydrated cement paste and aggregates typically used. As a result, this interface forms the "weak link" in concrete (Mehta & Monteiro, 2006). The properties of this Interfacial Transition Zone (ITZ) are thus very important and often limit the strength, stiffness and impermeability, among other properties, of concrete. The characterization of ITZ in concrete is therefore very important in understanding the physical structure and mechanical performance of cementitious composites.

A technique to study interfacial transition zones in concrete using LSCM has been developed. Here the technique is used to examine changes in concrete microstructure that occur due to a deleterious reaction called alkali-silica reaction (ASR) at the aggregate/paste interface. ASR occurs in concrete by the reaction of certain siliceous minerals found in some aggregates and alkalis which may be contributed by the cement. The reaction of such aggregates, in the presence of water, results in the formation of a potentially expansive alkali-silicate gel product. Expansion of the gel can cause cracking of concrete structures. Further details on the reaction can be found in Collins et al., 2004b.

Because the reaction essentially occurs between the aggregates and the alkaline pore solution, observations of the aggregate/paste interface are critical to understand the rate of reaction, the nature of the products formed, the mechanisms of damage, and the effects of various mitigation options. However, *in situ* imaging of the structure of the reactants present at the aggregate/paste interface, without introducing artifacts, has been challenging to achieve. This has been overcome by the use of the "through-aggregate" imaging method described herein.

2. Cement based materials – solid-solid system

2.1 Through-aggregate imaging of concrete

The through-aggregate imaging method was developed to characterize the internal aggregate/paste interfaces present in concrete. In this new technique, normal aggregate in mortar was replaced with transparent borosilicate glass beads. In addition to obtaining images at different depths on the surface of the sample as is usually carried out with the confocal microscope, the technique allows imaging into the concrete microstructure. Thus reactions occurring inside the microstructure, especially at the aggregate/paste interface can also be studied. Three-dimensional rotational images as shown in Fig. 5 were obtained by focusing on the interfaces at various z-depths.

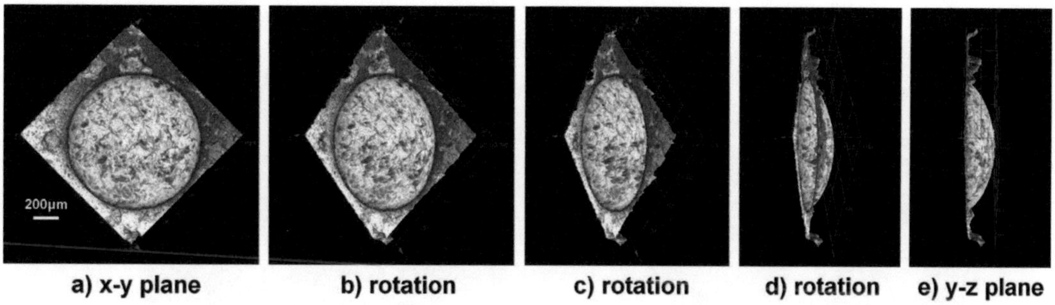

Fig. 5. Three dimensional LSCM rotational image of glass bead aggregate in mortar sample

For the study of ASR, the use of the reactive borosilicate glass beads as aggregate allows for the examination of the reaction, which is known to initiate at the aggregate/paste interface. Effective *in situ* examination of the progress of ASR was achieved using the new technique developed by imaging the same location of the sample at different stages of ASR deterioration. The samples were fixed on aluminum templates with three holes that were spaced so that the template could fit a custom-made sample stage in a unique position. Thus using the LSCM, the same locations were imaged and mapped in the x and y planes using a cross hair, indexible stage, and coordinate reader. Further details about the sample preparation and the imaging technique can be obtained from Collins et al., 2004a.

Analysis of the two-dimensional LSCM images, at various z-depths (or depths through the aggregate), shows changes in the reflected light intensity with depth and with the material being imaged (Fig. 6). At the surface of the sample, a higher light intensity was observed at the aggregate/air interface than at the cement paste/air interface, as cement absorbs more light (Fig. 6a). The intensity at the aggregate/air interface is the reflected light less the refracted light. As shown in Fig. 6b, the intensity at the aggregate/paste interface was observed to be lesser compared to the aggregate/air interface. The decrease in intensity at the aggregate/paste interface can be attributed to the multiple reflection and refraction that the light undergoes at the aggregate/air and aggregate/paste interface and also the absorption by the opaque paste at the aggregate/paste interface.

LSCM was also used to obtain two dimensional images (Fig. 7 & Fig. 8) of these mortars, similar to stereomicroscopy but with improved focus. These images can be used to track the progression of ASR damage in the sample with time. Typical indicators of damage seen in mortar or concrete undergoing ASR, such as cracking in the aggregate (Fig. 7b), increase in number of cracks and crack width (Fig. 7c), progression of cracks into the cement paste (Fig. 7c and 8b), and debonding at the aggregate/paste interface (Fig. 8a) was observed by LSCM. LSCM also allows observation of the changes occurring at the aggregate/air interface and the aggregate/paste interface with progressive ASR damage (Fig. 9). Fig. 9a shows the reaction product that was detected in the aggregate/air interface at the surface and which was observed to extend around the aggregate/paste interface to a depth of -0.19mm, as seen in Fig. 9b . Debonding due to ASR at the aggregate/paste interface was also observed in the LSCM images; it was characterized by a complete loss of intensity of the reflected light from the aggregate/paste interface.

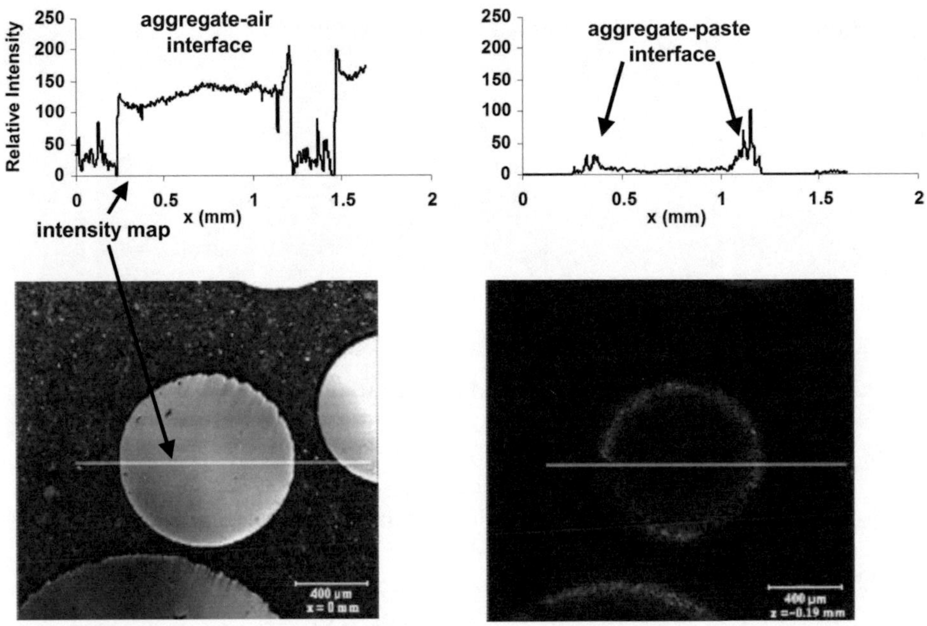

(a) (b)

Fig. 6. LSCM images and corresponding intensity maps of the reference sample at 2 days at depths of (a) 0 and (b) -0.19mm

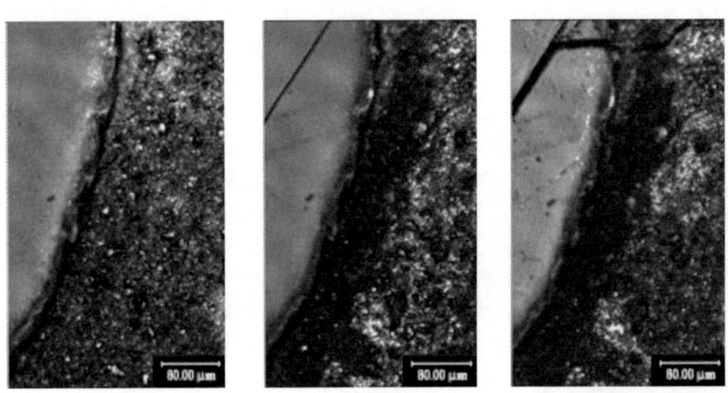

(a) (b) (c)

Fig. 7. LSCM images of crack formation and progression in the reference sample at (a) 2, (b) 7, and (c) 26 days

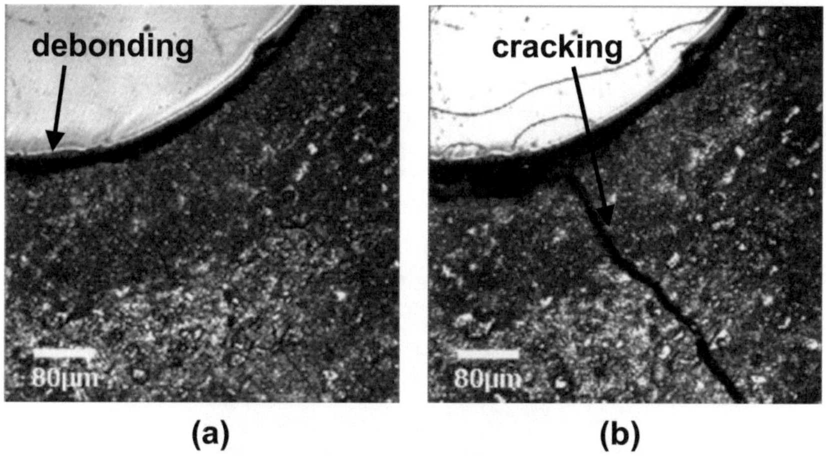

Fig. 8. LSCM images of debonding and crack progression in reference sample at (a) 7, and (b) 14 days

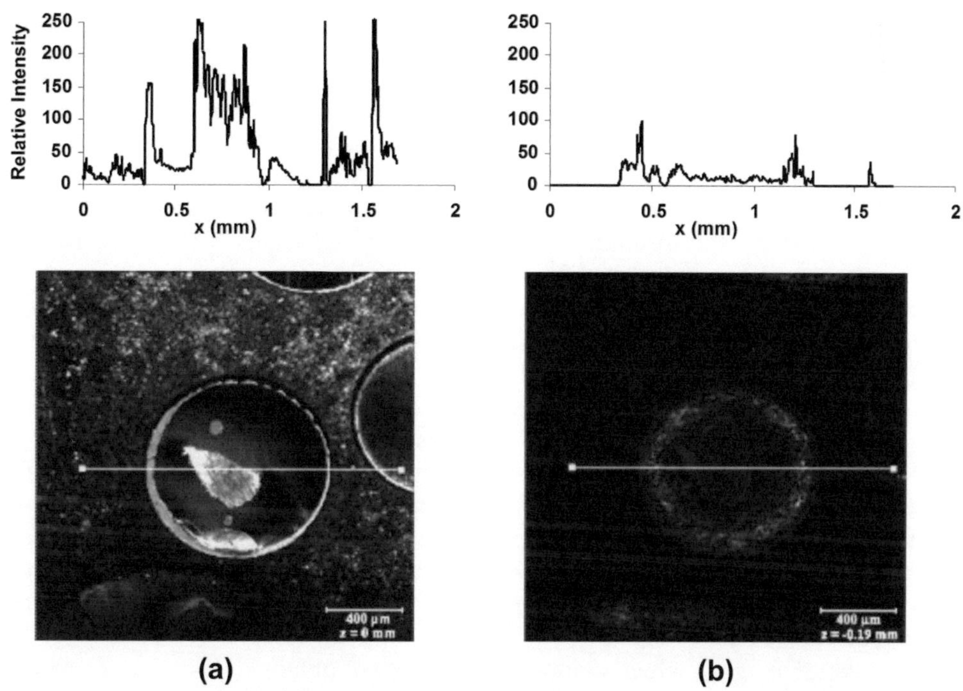

Fig. 9. LSCM images and intensity maps of the reference sample (corresponding to Fig. 6) at 67 days showing accumulation of product on the aggregate surface. The images were obtained at depths (a) 0 and (b) -0.19mm

2.2 Reaction products observed at aggregate/paste interface in concrete

The effect of addition of different lithium compounds (LiCl, $LiNO_3$ and LiOH) on ASR was also studied by LSCM. Lithium compounds have been shown to decrease expansion and damage associated with ASR (Kurtis & Monteiro, 2003; Ramachandran, 1998). Here, LSCM was used to examine the changes in microstructure, especially those occurring at the aggregate/paste interface, with the addition of individual lithium compounds at specific concentrations (measured by [Li] / [Na]) to the mix water.

Fig. 10. LSCM image of aggregate/paste interface in sample prepared with $LiNO_3$ (Li/Na = 0.50) at 3 days

LSCM imaging showed that the samples treated with lithium chemicals (where these chemicals were added to the mix water) did not demonstrate any physical manifestations of ASR damage nor did it result in cracking due to ASR. However reaction products with a different morphology (dendritic or crystalline) were observed at the aggregate/paste interface (Fig. 10) using the through-aggregate imaging technique. It is hypothesized that these dendritic reaction products could be less expansive compared to typical ASR gel products (Collins et al., 2004a). The light intensity at the aggregate/paste interface was also reduced tremendously due to the formation of the reaction product, because of the addition of another interface in the path of travel of light.

Thus, LSCM was shown to be a very powerful technique for characterizing interfaces in concrete and in the current study alkali-silica reaction was examined using a through-aggregate imaging technique. Some of the main advantages of the technique developed which could increase the use of LSCM to characterize various cement based materials are:

- Microscopy **into** the concrete microstructure was possible using the confocal microscope;
- Compared to scanning electron microscopic imaging, this method avoids introduction of artifacts as no special sample preparation techniques such as drying and epoxy impregnation are required; and
- LSCM plane images, when compared to images obtained using a conventional microscope, are better focused even when different locations of the sample are not at the same z-level.

3. Radioactive waste slurries – solid-liquid system

3.1 LSCM applications with radioactive waste slurries

Very few researchers have applied LSCM technology to opaque concentrated slurries. (Schmid et al., 2003) and (Thill et al., 1999) applied LSCM technology to activated waste water sludges to determine volumes, heterogeneity factors, compositions of bacterial population and aggregate size. (Kay et al., 2003) developed the first laser confocal images of concentrated (>10 wt. %, See Fig. 2) Fe/Al simulated radioactive sludges. The motivation for this effort was to begin to understand how slurry yield stress and viscosity could be improved and to understand the actual internal slurry structure.

The three dimensional representation shown in Figure 11 represents the true power of LSCM for opaque slurry applications (Kay et al., 2003). This technique allows the slurry to be analyzed in an *in-situ* condition. The microscope has the ability to make both two-dimensional pictures and three-dimensional representations of a sample. In the case of Figure 11, three-dimensional representations were made by scanning two-dimensional images at 1-micron increments. Image analysis software provided by Carl Zeiss, Inc. was used to stack the images together in a two dimensional image that provides a color gradient corresponding to the depth of the sample. These three-dimensional representations were used to understand the actual physical structure of the slurries. The slides with simulated waste (wet) samples were mounted using 2 drops of the material on a glass slide covered with a cover slip. A drop of oil was added to the top of the cover slip to view through oil immersion at 1300X. Slides were then examined and select images saved with a Laser Scanning Confocal Microscope (Model 310 Carl Zeiss, Inc., Thornwood, NY).

Figure 11 shows a three-dimensional representation of the RPP-WTP AZ102 slurry (Control, similar to opaque sludge shown in Fig. 2) using a LSCM. The scale in the upper left-hand corner shows the depth of the image. The scale in the lower portion of the image shows the horizontal scale. The red, green, and blue colors correspond to a depth of 0, 7.5, and 15 microns in the slurry sample. The AZ102 slurry particles shown in Figure 11 appear to be flocculated into larger (> 5 micron) size flocs. These flocs are suspended by smaller particles. Figure 11 appears to indicate the slurry is a flocculated, touching network of particles (Kay, 2003, Pugh, 1994). Figure 11 represents the first 2D three dimensional *in-situ* model of simulated radioactive high level waste slurry. Figure 12 is one of the two dimensional images that was used to develop the three dimensional model (one or the other 2D vs 3D) shown in Figure 11.

3.2 Dispersant use with slurries

A number of industries handle concentrated slurries such as ceramic (Chou & Senna, 1987), paint (Fujitani, 1996; Auschra et al., 2002), ink (Spinelli, 1998) and coal (Baxter & Habib, 1992). Transportation and processing of these slurries is often necessary highlighting the need to understand and control the solid-liquid mixture's response to flow. The rheological properties of the mixture are influenced by the solids content, particle size (Dabak & Yucel, 1987) and shape, particle interaction (Frith et al., 1987) and the solvent characteristics (Russel, 1987). In water, solids such as alumina can interact through interparticle forces to form flocs leading to non-Newtonian rheological behavior. These flocs immobilize some liquid essentially increasing the effective solid volume and thus the viscosity of the solid-liquid mixture increases (Garrido & Aglietti, 2001). Dispersants are added to improve the flow and handling characteristics of the slurry. Polymeric dispersants, such as polyacrylates, are designed with a hydrophilic segment and a hydrophobic segment. Solids are encapsulated by the dispersant with the hydrophobic portion absorbed onto the particles and the hydrophilic portion extending into the water. Deflocculation occurs with both steric

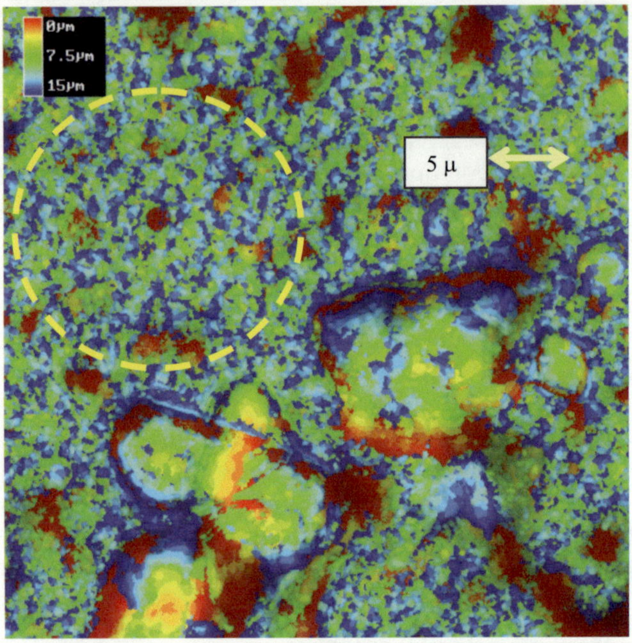

Fig. 11. Three dimensional representation of simulated AZ102 Hanford slurry with the area within the yellow dashed circle showing the most representative sample of the slurry

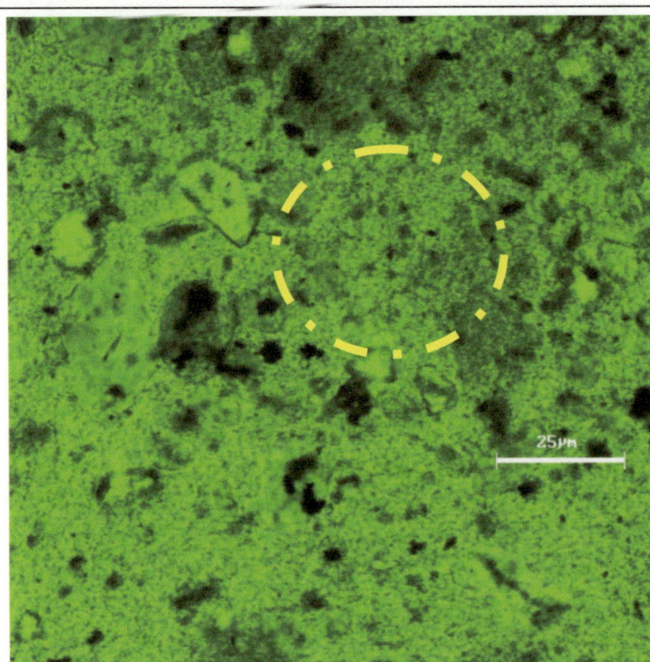

Fig. 12. Two-Dimensional view of simulated radioactive Hanford slurry AZ-102 with particles less than 25 micron (µm) in size shown in the dashed circle

and ionic stabilization of the particle. Treatment of solid-liquid mixtures with the appropriate dispersant can lead to near Newtonian rheological behavior and improved liquid flow properties especially at higher solids content (Ribeiro et al., 2004).

Other applications and references using LSCM technology to explore the effects of surfactants on various systems are discussed in the solid-liquid-gas systems section of this chapter.

3.3 The rheology of simulated radioactive waste with dispersant

The use of dispersants has been examined on simulated radioactive waste to improve flow properties of slurries (White et al., 2008, Kay et al., 2003). Plate-to-plate rheological measurements of the 20 wt% simulated waste slurries with and without polymeric dispersant showed at least 30% decrease in the yield stress for the dispersant containing samples as shown in Table 1. The baseline simulated waste slurry yielded a rheology curve that is typical for Bingham fluids (Figure 13) with a high initial yield stress (Dean et al., 2007). Equation 1 describes the two parameter rheology model that is used extensively in the drilling fluids industry to describe the flow characteristics of many different types of muds. These fluids require a certain amount of stress to initiate flow at the yield point (YP) and do not have a constant viscosity. The plastic viscosity (PV) is the slope of the shear stress (τ)/shear rate (γ) line above the yield point (YP). These results are consistent with what has been observed with Al-rich sludges in the ceramic industry as is the drop in yield stress when dispersant is added (Ribeiro et al., 2004). As more dispersant is added, the slurry becomes less pseudoplastic in character and trends toward Newtonian behaviour.

$$\tau = YP + PV(\gamma) \quad (1)$$

Where

τ = shear stress

YP = yield point

PV = plastic viscosity

γ = shear rate

	Yield Stress, Pa				
#	Description	Up Ave.	% Diff.	Down Ave.	% Diff.
1	Baseline	14.0	NA	13.2	NA
2	3000 ppm Dolapix CE64	7.79	44.2	5.97	54.8
3	3000 ppm Disperse-Ayd W28	8.83	36.7	7.65	42.1
4	3000 ppm Cyanamer P35	9.5	31.9	8.78	33.6

Table 1. Rheology data in duplicate for simulated radioactive waste slurries

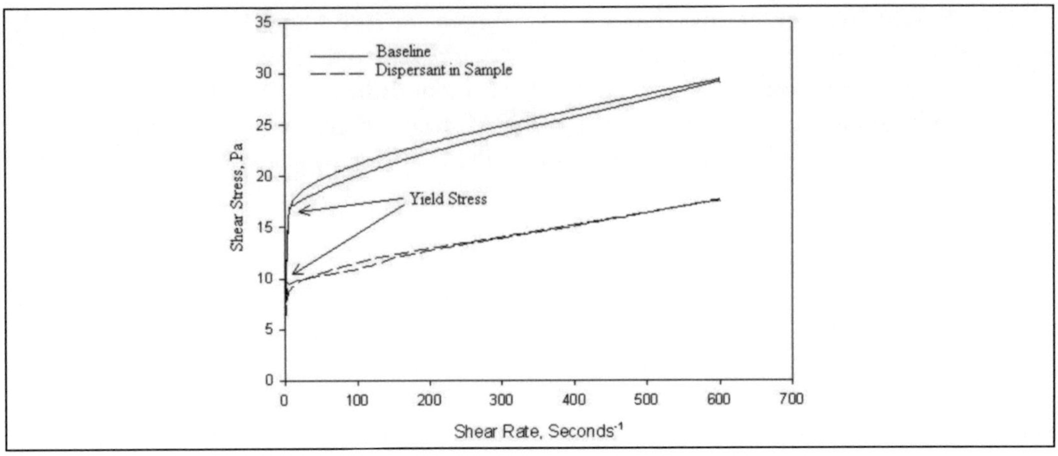

Fig. 13. Effect of dispersant on simulated radioactive waste slurries (Note the sample with dispersant has a lower yield stress moving to curve to more Newtonian behaviour)

3.4 Examination of simulated radioactive waste with dispersant

To better understand the effect of the dispersants, a direct method rather than a bulk method such as rheology was employed. LSCM offered a way of directly observing the solid-liquid mixtures with depth profiling on a micrometer scale. This capability allows the researcher to observe changes in the material due to dispersant addition and assess if particle stability is occurring. For the simulated radioactive waste work (White et al., 2008), samples were prepared on a Sedgewick Rafter slide that contains a chamber (50 mm l x 20 mm w x 1 mm d) suited for containing a drop of slurry. A glass cover was carefully placed over the liquid taking care to avoid bubble formation. This sample mounting method allows for viewing of the slurry as an undisturbed drop. The solid-liquid mixture (20%wt, 1-5 μm particles) was viewed under low (100x) magnification as shown in Figure 14 A & B. Large aggregates, many near 50 μm, can be observed in Figure 14 A that are not present in Fig. 14 B. The Z-contrast detailed the unique three-dimensional structural information as shown in Figure 14 C & D and was formed by repeatedly scanning at designated depths of 40 images at 10 μm intervals. The topography shows large aggregates present in image C without dispersant and not in image D where particles have been stabilized with dispersant. These images successfully portray the deflocculation of the simulated waste expected from the rheology results on a μm scale. It would be useful to continue developing methods such as specific dyes that highlight the dispersant for contrast and higher magnification to see below the μm scale to the particle edge.

4. Radioactive waste slurries – solid-liquid-gas systems

A majority of foams are comprised of two phases, typically a gas and liquid-either aqueous or organic-containing surfactants. However, foams containing three phases (i.e., gas, liquid and solid particles) are also encountered in industry; for example, in the processing of solid wastes, food, chemical, and agricultural products. Radioactive waste presents some of the most complex foam-forming systems (Vijayaraghavan et al., 2006; Calloway et al., 2001). Wasan et al. (2004) have identified at least two kinds of particles in such three phase gas-solid-liquid systems: hydrophilic (i.e., water-wet) and amphiphilic or Janus (i.e., biphilic, partially wetted by water). The Janus particles are formed due to the non-uniform

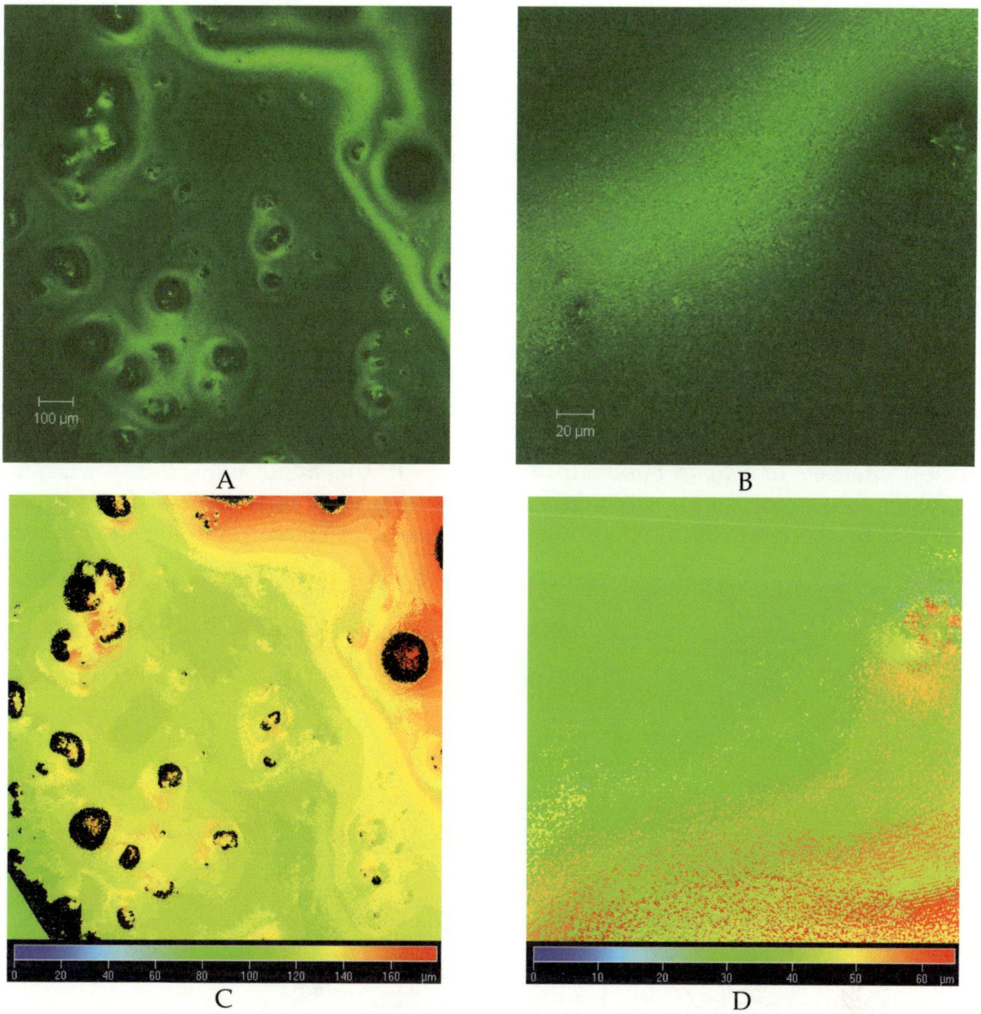

Fig. 14. LSCM images of baseline-simulated radioactive waste slurry (A & C) and simulated radioactive waste slurry with dispersant (B & D) (Note the loss in large aggregates)

surface energy of crystalline materials present in the wastes (Bindal et al., 2001, Binsk & Fletcher 2001). Solids which are not readily wetted with water (i.e., Janus), tend to migrate to a liquid-vapor/gas interface. At the interface, the solids essentially block the liquid in the foam from draining and therefore increase the stability and duration of the foam.

Various researchers have explored the use of LSCM to investigate the nature of the solid-liquid-gas interface in foams (Liu et al., 2009; Zhang et al., 2008; Fujii et al., 2006; Reed et al., 1997; Lau & Dickinson, 2005; Dong et al., 2010). Zhang and subsequently Liu investigated aqueous foam stabilized with Laponite particles modified by a surfactant. Laponite forms a nearly clear fluid in aqueous systems and makes an ideal model system for exploring various effects of surfactants and solids concentration on slurries. Typically, researchers investigating Laponite systems use Rhodamine B, which is negatively charged in a basic solution and has a maximum excitation wavelength at 543 nm. The modified Laponite particles were in effect "labeled" by the fluorescently active Rhodamine B. Excess

Rhodamine B was washed from the particle dispersions with deionized water. The LSCM was then used to image the particles "labeled" with Rhodamine B. In the Laponite application, the LSCM was used to image how the partially hydrophobic or "Janus" particles attached to the gas bubble surface. Fujii examined latex-stabilized aqueous foams by LSCM in which the laser was operated at wavelengths of 351 nm for excitation of the pyrene groups in the polystyrene matrix. Reed and co-workers developed an actual three-dimensional image of the foam lamella in which the 3D effect can be viewed with the appropriate red/green glasses. The liquid portion of the solution was ethanol and fluorescein allowing viewing of the liquid fraction of the foam, thus revealing the complex three dimensional structures. Lau used the LSCM to monitor a long sequence of foam images at closely spaced intervals to study the dynamic nature of foams over time.

Researchers at the Savannah River National Laboratory developed a LSCM method to capture a complete picture of the Janus particles at the liquid-vapor/gas interface. Figure 15 shows a two-dimensional image of potassium tetraphenylborate particles (e.g., Janus particles) attached to a foam bubble in the slurry. In this application, excess tetraphenyl borate ions in solution and possibly benzene and other trace organic breakdown products are excited by the laser. In this case, the "Janus" particles are surrounding the air bubble.

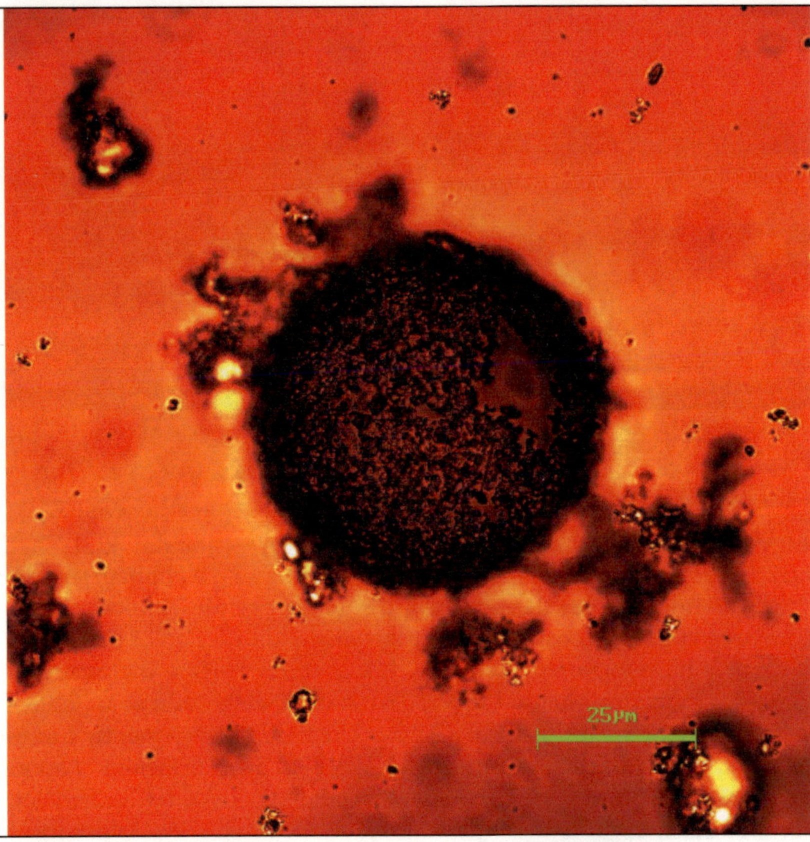

Fig. 15. LSCM image of simulated nuclear waste foams. Janus particles are clearly attached to the liquid-gas interface. Fundamental understanding of these systems has been used to develop new antifoams. Note size of bubble is greater than 25 um in diameter

The slides with simulated wet waste samples were mounted using 2 drops of the material on a glass slide covered with a cover slip. A drop of oil was added to the top of the cover slip to view through oil immersion at 1300X. Slides were then examined and select images saved with a Laser Scanning Confocal Microscope (Model 310 Carl Zeiss, Inc., Thornwood, NY). The slurry was not treated with antifoam agent (surfactant).
Further slides (See Figure 16) of the same NaTPB slurry sample reveal a complex network of foam within the slurry which also explains why the solid particles actually float to the surface of the solution.

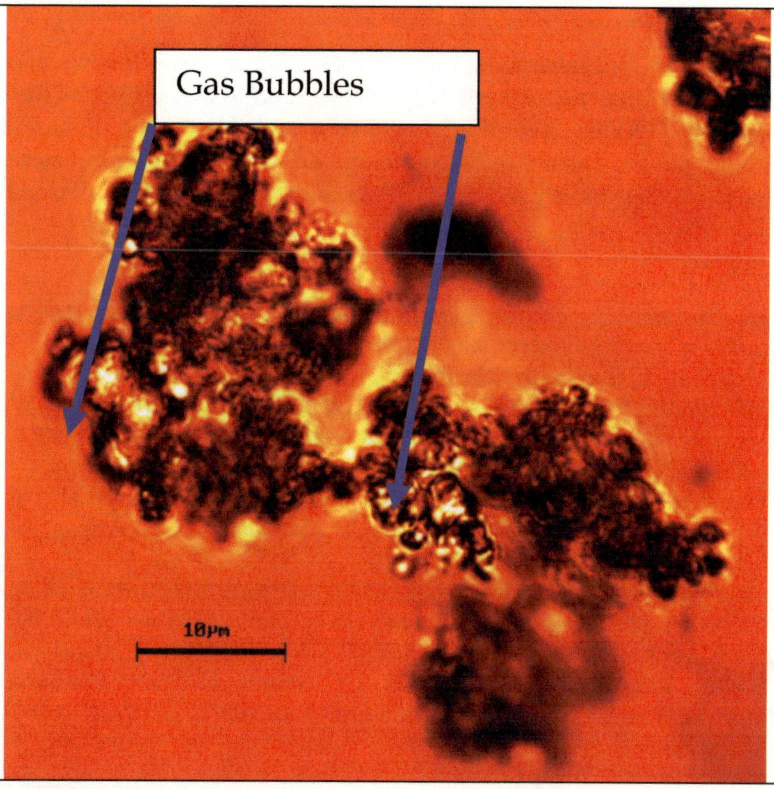

Fig. 16. View of NaTPB foam (Note the entrained bubbles within the solid particles- Entrained gas allows the solid particles to float to the surface)

5. Conclusion

This work demonstrates that LSCM is a useful tool for direct imaging of both complex solids and slurries. The integrity of cementitious materials may be examined by imaging the microstructure for cracking and debonding (Fig. 7 & Fig. 8) and various interfaces for the formation of reaction products (Fig. 9 & Fig. 10). Small scale changes such as gas bubble formation in slurries (Fig. 16), a critical factor in the process, may be highlighted. Particle stabilization with dispersant (Fig. 14) can be imaged and studied. Once a set of optimum parameters has been found for a given material, they can be used and quickly adjusted for rapid imaging. Considerable effort is needed to set up operating conditions such as the optimum laser, filter, and wavelength. In addition, sample preparation needs to be

considered. Several mounting formats were tested before the Sedgewick-Rafter slides were found to preserve the integrity of the slurry sample during imaging by eliminating drying or introducing other artifacts.
Future work should focus on improved image contrast using stains and flurochromes specific to the sample to highlight regions. Further, image improvement gains can be achieved by applying improved image processing software. Finally, novel methods of sample preparation, if required, prior to imaging are critical and should continue to be refined.

6. Acknowledgement

The authors Kurtis and Jayapalan would like to thank Courtney Collins for the contributions to this research project. They also acknowledge the support to this project by the U.S. National Science Foundation POWRE Award CMS-0074874 and Grant No. CMMI-0825373. Any opinions, findings, and conclusions or recommendations expressed in this material are those of the authors and do not necessarily reflect the views of the National Science Foundation.

7. References

Aguilera, J. & Stanley, D. (1999). Examining food microstructure, In: *Microstructural Principles of Food Processing and Engineering*, (2nd), (19 - 22), Aspen Publishers, 0-8342-1256-0, Gaithersburg, Maryland

Aushra, C.; Eshstein, E. ; Mühlebach, A.; Zink, M. & Rime, F. (2002). Design of new pigment dispersants by controlled radial polymerization. *Progress in Organic Coatings*, 45, 2-3, (October 2002) (83-93), 0300-9440

Baxter, L. & Habib, Z. (1992). The effect of surfactants on disaggregation of coal-water slurry particles during combustion. *Combustion and Flame*, 90, 2, (August 1992) (199-209), 0010-2180

Bearsley, S.; Forbes, A. & Haverkamp, R. (2004). Direct observation of the asphaltene structure in paving-grade bitumen using confocal laser-scanning microscopy, *Journal of Microscopy*, 215, 2 (August 2004), (149-155), 0022-2720

Bindal, S.; Nikolov, A.; Wasan, D.; Lambert, D. & Koopman, D. (2001). Foaming in simulated radioactive waste, *Environmental Science & Technology*, 35, 19, (October 2001) (3941-3947), 0013-936X

Binsk, B. & Fletcher, P. (2001). Particles adsorbed at oil-water interface: a theoretical comparison between spheres of uniform wettability and "Janus" particles, *Langmuir*, 17, 16, (August 2001) (4708 – 4710), 0743-7463

Calloway, T.; Baich, M. & Lambert, D. (2001). Fate of IITB52 Antifoam Across the Small Tank Tetraphenylborate Process, WSRC-TR-2001-00102, http://www.osti.gov/bridge/product.biblio.jsp?query_id=0&page=0&osti_id=783018

Carter, D. (1999). Practical considerations for collecting confocal images, In: *Methods in Molecular Biology, Confocal Microscopy Methods and Protocols*, Stephen W. Paddock, (122), (35-57), Humama Press, 0-89603-526-3, Totowa, New Jersey

Chou, C. & Senna, M. (1987). Correlation between rheological behavior of aqueous suspensions of Al_2O_3 and properties of cast bodies: effects of dispersants and ultrafine powders. *American Ceramic Society Bulletin*, 66, 7, (Jule 1987) (1129-1133), 0002-7812

Chou, K. (1989). Effect of dispersants on the rheological properties and slip casting of concentrated alumina slurry. *J. Am. Ceram. Soc.*, 72, 6, (September 1989) (1622-27), 0002-7820

Collins, C.; Ideker, J. & Kurtis, K. (2004a). Laser scanning confocal microscopy for *in situ* monitoring of alkali-silica reaction. *Journal of Microscopy-Oxford*, 213, (February 2004) (149-157), 0022-2720

Collins, C.; Ideker, J.; Willis, G. & Kurtis, K. (2004b). Examination of the effects of LiOH, LiCl, and $LiNO_3$ on alkali-silica reaction. *Cement and Concrete Research*, 34, 8, (August 2004) (1403-1415), 0008-8846

Corle, R., Kino, S. (1996). *Confocal Scanning Optical Microscopy and Related Imaging Systems.* Academic Press, 0-12-408750-7, San Diego, CA

Dabak, T. & Yucel, O. (1987). Modeling of the Concentration and Particle Size Distribution Effects on the Rheology of Highly Concentrated Suspensions. *Powder Technology*, 52, (October 1987) (193-206), 0032-5910

Dean, E.; Glowinski, R. & Guidoboni, G. (2007). On the Numerical Simulation of Bingham visco-plastic flow: Old and New Results. *J. Non-Newtonian Fluid Mech.*, 142, (March 2007) (36-62), 0377-0257

Dong, X.; Xu, J.; Cao, C.; Sun, D. & Jiang, X. (2010). Aqueous foam stabilized by hydrophobically modified silica particles and liquid paraffin droplets, *Colloids and Surfaces A: Physicochem. Eng. Aspects*, 353, 2-3 (January 2010) (181-188), 0927-7757

Frith, W.; Mewis, J. & Strivens, T. (1987). Rheology of concentrated suspensions: experimental investigations, *Powder Technology*, 51, (June 1987) (27-34), 0032-5910

Fujii, S.; Iddon, P.; Ryan, A. & Armes, S. (2006). Aqueous particulate foams stabilized solely with polymer latex particles, *Langmuir,* 22, 18, (August 2006) (7512-7520), 0743-7463

Fujitani, T. (1996). Stability of pigment and resin dispersions in waterborne paint. *Progress in Organic Coatings*, 29, 1-4, (September-December 1996) (97-105), 0300-9440

Garrido, B. & Aglietti, F. (2001). Effect of rheological properties of zircon-alumina suspensions on density of green casts *Materials Research* 4, 4, (September 2001) (279-284), 1516-1439

Ilievski, D.; Austin, P. & Whittington, B. (2003). Studies into the internal structure of gibbsite agglomerates, *Chem. Eng. Technol.* 26, 3, (March 2003) (363-368), 0930-7516

Karimi-Lotfabad, S. & Gray, M. (2000). Characterization of contaminated soils using confocal laser scanning microscopy and cryogenic-scanning electron microscopy, *Environmental Science & Technology*, 34, 16, (August 2000) (3408-3414), 0013-936X

Kay, E.; Calloway, T.; Koopman, D.; Brigmon, R. & Eibling, R. (2003). Rheology Modifiers in Radioactive Waste Slurries, *Proceedings of ASME Fluid Eng. Div. Summer Meeting* pp. 855-863, 0-7918-3696-7 , Honolulu, Hawaii, USA, July 2003, ASME, New York

Kurtis, K.; El-Ashkar, N.; Collins, C. & Nalk, N. (2003). Examining cement-based materials by laser scanning confocal microscopy, *Cement & Concrete Composites*, 25, 7, (October 2003) (695-701), 0958-9465

Kurtis, K. & Monteiro, P. (2003). Chemical additives to control expansion of alkali-silica reaction gel: proposed mechanisms of control. *Journal of Materials Science*, 38, 9, (May 2003) (2027-2036), 0022-2461

Lau, C. & Dickinson, E. (2005). Instability and structural change in an aerated system containing egg albumen and invert sugar, *Food Hydrocolloids*, 19, 1, (January 2005) (111-121), 0268-005X

Liu, Q., Zhang, S., Sun, D. & Xu, J. (2009), Aqueous foams stabilized by hexylamine-Laponite particles, *Colloids and Surfaces A: Physicochem. Eng. Aspects*, 338, 1-3 (April 2009) (40-46), 0927-7577

Marusin, S. (1995). Sample preparation - The key to SEM studies of failed concrete. *Cement & Concrete Composites*, 17, 4, (1995) (311-318), 0958-9465

Mehta, P. & Monteiro, P. (2006). *Concrete - Microstructure, properties and materials*, McGraw-Hill, 0-07-146289-9, New York

Mohr, J. & Kurtis, E. (2006). Fractography of fiber-cement composites via laser scanning confocal microscopy, *Proceedings of 16th European conference on fracture, measuring, monitoring, and modeling concrete properties: in honor of Surendra P. Shah*, 503-508, Alexandroupolis, Greece

Pugh, J. (1994). Dispersion and stability of ceramic powders in liquids, *Surface and Colloid Chemistry in Advanced Ceramics Processing*, Marcel Dekker, New York, pg. 127-188

Ramachandran, V. (1998). Alkali-aggregate expansion inhibiting admixtures, *Cement & Concrete Composites*, 20, 2-3, (April-June 1998) (149-161), 0958-9465

Reed, M.; Howard, C. & Shelton, C. (1997). Confocal imaging and second-order stereological analysis of a liquid foam, *Journal of Microscopy*, 185, 3, (March 1997) (313-320), 0022-2720

Ribeiro, M.; Tulyaganov, D.; Ferreira, J. & Labrincha, J. (2004). Production of Al-rich sludge-containing ceramic bodies by different shaping techniques, *Journal of Materials Processing Technology*, 148, (May 2004) (139-146), 0924-0136

Ren, F.; Smith, I.; Baumann, M. & Case, E. (2005). Three-dimensional microstructural characterization of porous hydroxyapatite using confocal laser scanning microscopy, *Int. J. Appl. Ceram. Technol.*, 2, 3, (2005) (200–211), 1546-542X

Russel, W. (1987). Theoretical approaches to the rheology of concentrated dispersions, *Powder Technology*, 51, (June 1987) (15-25), 0032-5910

Schmid, M.; Thill, A.; Purkhold, U.; Walcher, M.; Bottero, J.; Ginestet, P.; Nielsen, P., Wuertz, S. & Wagner, M. (2003). Characterization of activated sludge flocs by confocal laser scanning microscopy and image analysis, *Water Research*, 17, (May 2003) (2043-2052), 0043-1354

Selomulya, C.; Liao, J.; Bickert, G. & Amal, R. (2006). Micro-properties of coal aggregates: Implications on hyperbaric filtration performance for coal dewatering, *International Journal of Mineral Processing*, 80, 2-4, (September 2006) (189-197), 0301-7516

Spinelli, H. (1998). Polymeric Dispersants in Ink Jet Technology, *Adv. Mater.*, 10, 15, (October 1998) (1215-1218), 0935-9648

Tanaka, S.; Kato, Z.; Uchida, N. & Uematsu, K. (2003). Direct observation of aggregates and agglomerates in alumina granules, *Powder Technology*, 129, 1-3, (January 2003) (153-155), 0032-5910

Thill, A.; Wagner, M. & Bottero, J. (1999). Confocal scanning laser microscopy as a tool for the determination of 3D floc structure, *Journal of Colloid and Interface Science*, 220, 2, (December 1999) (465-467), 0021-9797

Vijayaraghavan, K.; Nikolov, A.; Wasan, D.; Calloway, B.; Crowder, M.; Stone, M. & Quershi, Z. (2006). Radioactive Waste Foams: Formation and Mitigation, *Journal of Environmental Engineering*, 132, 7, (July 2006) (716-724), 0733-9372

Wasan, T., Nikolov, D.; Shah, A., (2004), Foaming-antifoaming in boiling suspensions, *Ind. Eng. Chem. Res.*, 43, 14, (2004) (3812-3816), 0888-5885

White, T.; Stone, M.; Calloway, T.; Brigmon, R.; Eibling, R.; Nikolov, A.; Wasan, D. (2008). Understanding the effects of dispersant addition to slurry rheology using laser scanning confocal microscopy, *Separation Science and Technology*, 43, 9, (July 2008) (2859-2871), 0149-6395

Zamecnik, R., White, L.; Jones, M.; Hassan, M.; Eibling, E.; Duignan, R.; Crawford, L.; Calloway, B. (2003). Radioactive Waste Evaporation: Current Methodologies Employed for the Development, Design and Operation of Waste Evaporators at the Savannah River Site and Hanford Waste Treatment Plant, *Proceedings of ASME 2003 9th International Conference on Radioactive Waste Management and Environmental Remediation*, pp. 157-170, 0-7918-3732-7, Oxford, England, September 2003, ASME, New York

Zhang, S., Lan, Q., Liu, Q., Xu, J., Sun, D., (2008), Aqueous foams stabilized by Laponite and CTAB, *Colloids Surf.*, 317, 1-3, (March 2008) (406-413), 0927-7757

Visualization of Sorption Dynamics: Application of Confocal Laser Scanning Microscope Technique

Takashi Hasegawa[1,2], Kiyoshi Matsumoto[3] and Toshiro Matsui[2]
[1]*R&D Planning Division, Tobacco Business Headquarters, Japan Tobacco Inc., 2-2-1 Toranomon, Minato-ku, Tokyo 105-8422*
[2]*Bioscience and Biotechnology, Graduate School of Agriculture, Kyushu University, 6-10-1 Hakozaki, Higashi-ku, Fukuoka 812-8581*
[3]*Faculty of Biotechnology and Life Science, Department of Applied Microbial Technology, Sojo University, 4-22-1 Ikeda, Kumamoto 860-0082*
Japan

1. Introduction

The sorption of compounds such as liquid additives and vapor components into polymer materials is one of the most deteriorative phenomena for polymer products. In the field of material engineering, various additives are used in the process of manufacturing polymer products for the purpose of improving product functions. For example, in the manufacturing process of thermoplastics for plastic lapping and rubber materials, plasticizers are added to increase flexibility, toughness and/or transparency [1, 2]. In those cases, the diffusion of plasticizer in polymer materials is associated with the stability of product quality [3-5]. Food industries aim to avoid the serious deterioration due to the loss of flavors responsible for product quality, through sorption into packaging materials [6-8].

In past decades, numerous studies on sorption behavior of chemicals into polymer materials have been reported, mainly on the basis of their physicochemical properties [3-5, 9, 10]. One perspective is an investigation of diffusion and distribution of plasticizers themselves in polymer materials [3-5], and another is an estimation of the influence of co-existing plasticizers on the sorption behavior of penetrant compounds [9, 10]. In these studies, the sorption amount of penetrant chemicals has been determined by measuring a change in weight and/or by extracting absorbed penetrants with solvents in evaluating the diffusion kinetics of penetrant chemicals. However, no *in situ* examination of penetration or distribution of absorbed chemicals has been performed due to the lack of real-time and non-destructive assays. To overcome these problems, we have clarified the dynamic sorption behavior of flavors into polymer films [11-15]. The physical and chemical parameters of chemicals and/or polymers such as molar volume, free energy, free volume and solubility parameters have been clarified to be characterized as contributors for flavor sorption [11-13] by determining the sorption dynamics with our proposed aqueous and vaporous penetration methods [14, 15]. The proposed convenient penetration methods, however, could not provide any *in situ* information on the real-time sorption behavior of penetrants in polymers.

Although observational methodology using time-of-flight secondary ion mass spectrometry (TOF-SIMS) was applied to obtain the cross-sectional distribution of penetrants in polymer films or fibers [16], destruction treatment of samples was needed to determine the concentration of penetrants in each sectional region.
The lack of non-destructive observational methodology, thus, led us to apply the use of a confocal laser scanning microscope (CLSM) for real-time monitoring of penetrants in polymers.
CLSM has been widely used for observation and/or analysis of mass transfer phenomena in the field of physiology and pharmacology with fluorescent labeling technique or fluorescent reagent mixture [17-23]. In physiology, there has been reports on CLSM usage by researchers for monitoring the formation process of poly(ethylene glycol) micro- and nano-gels [17] or surface cross-linked structure of poly(vinyl alcohol) [18]. In pharmacology, CLSM has been applied to analyze the transport phenomenon of protein [19, 20] or the dynamics of glucose transporter in lipocyte [21].
Recently, the material engineering field has recorded reports on attempts to apply CLSM. For example, there was an observation by CLSM [22] on the absorption behavior of glycerin containing fluorescent dye into the fiber as an analytical method for evaluating the hydrophilicity of the surface of nylon fiber. In addition, to investigate the plasticizing effect of supercritical fluid treatment for polypropylene, the distribution of a fluorescent probe mixed in the film has been analyzed by CLSM [23].
In this chapter, we reviewed the application of CLSM technique to real-time and non-destructive observation of fluorescent reagent sorbed into polymer materials [24-26]. Subsequently, a novel observational methodology for sorption behavior was established by CLSM in section 2. In section 3, effects of various additives on sorption behavior of fluorescent reagent were evaluated. Lastly, effects of film types and film depth on diffusion coefficient of fluorescent reagent were examined as a further development of CLSM methodology in section 4. As a fluorescent reagent and a polymer material, perylene and cellulose acetate (CA) were used, respectively. CA is widely used in various industrial fields such as for cigarette filters, clothing, water purification and medical applications [27, 28], and there are a number of studies reported on the sorption behavior of water and alcohols into CA [29-31]. Additives for dissolving penetrant perylene were glycerol triacetate (GTA), triethylene glycol diacetate (TEGDA), 1,3-butylene glycol diacetate (BGDA), paraffin liquid and polyethylene glycol 200 (PEG), in which GTA, TEGDA and BGDA are known as the plasticizers for CA [32, 33], and paraffin liquid and PEG for additives of CA filters in cigarette industrial field [34, 35].

2. Establishment of CLSM methodology for visualizing sorption behavior

2.1 Fluorescent properties of samples

Closed-system CA film (CCA) for CLSM analysis was prepared according to the following procedure. CA flake with a degree of substitution of 2.5 (Mw = 1.0×10^5 g/mol, Figure 1) and kindly supplied from Daicel Chemical Industries Ltd. (Osaka, Japan) was dissolved in acetone to make a 2.0 w/v % solution without further purification. An aliquot (200 μl) of CA solution was then put into a film-making-unit with a penicillin cup and cover glass as shown in Figure 2. Excess acetone was vaporized at room temperature for 15 h, and the prepared CA film onto the surface of film-making-unit was then dried *in vacuo* overnight. The prepared films were 95 ± 5 μm thick.

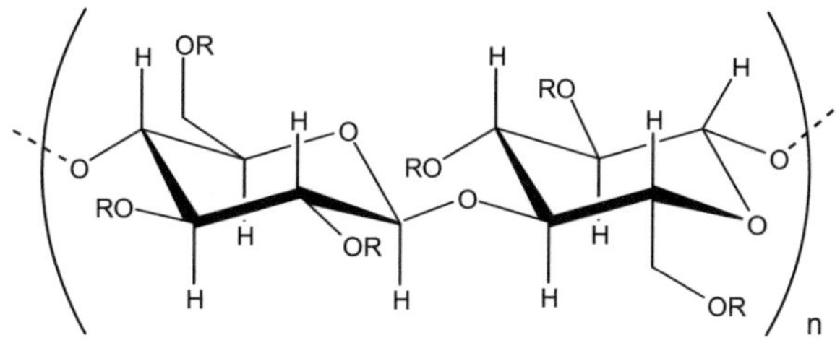

Fig. 1. Chemical structure of CA.

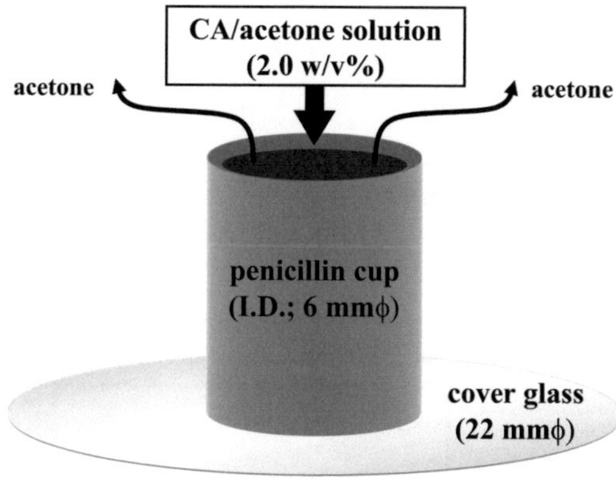

Fig. 2. Preparation of CCA sample on film-making-unit.

And then, perylene (Mw = 252.3 g/mol, λ_{ex} = 411 nm, λ_{em} = 470 nm, Wako Pure Chemical Industries Ltd., Osaka, Japan) and GTA (Mw = 218.2 g/mol, Sigma-Aldrich, Tokyo, Japan) were mixed to make perylene/GTA mixture at room temperature by dissolving perylene in GTA to reach a molar ratio of from 2.2 to 11 × 10^{-5}. The molar ratio of perylene, C_{per}, was defined as formula (1).

$$C_{per}\,[-] = \frac{Perylene\ amount\,[mol]}{Perylene\ amount\,[mol] + GTA\ amount\,[mol]} \qquad (1)$$

To evaluate fluorescent properties of CCA sample, GTA and perylene/GTA mixture, the fluorescent spectra were obtained by a CLSM system (CLSM system A1, Nikon Corporation, Tokyo, Japan) equipped with a dry objective lens (CFI Plan Apo VC 20×, Nikon Corporation, Tokyo, Japan). The spectrum of CCA sample was obtained by placing a film sample on a microscope directly, while the spectra of GTA and perylene/GTA mixture were obtained by pipetting 4 μl of the solution onto a cover glass placed on a microscope. A diode laser (408 nm, 290-330 μW) was used as an excitation laser, and the fluorescence excited from the around cover glass surface was detected. Figure 3 shows the fluorescent spectra of CCA sample, GTA and perylene/GTA mixture when excited by a diode laser at 408 nm. The spectra clearly demonstrate that only perylene/GTA mixture had significant perylene-induced fluorescent intensity, indicating that any interfering fluorescences may be excluded to investigate the distribution dynamics of penetrant, perylene.

Fluorescent intensity of perylene/GTA mixture was measured by the above-mentioned CLSM conditions at room temperature. The intensity of reflected fluorescence in the range

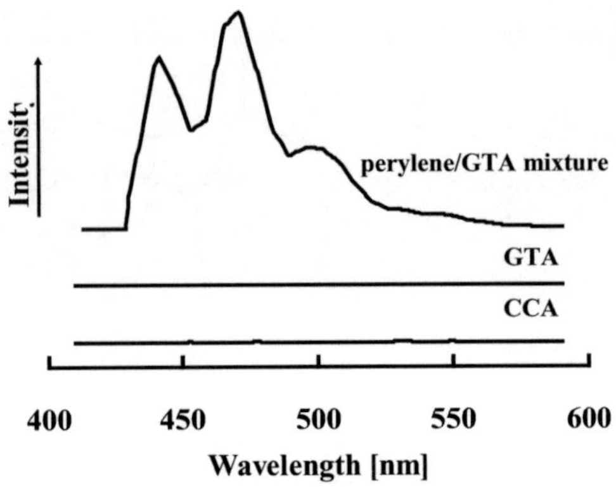

Fig. 3. Fluorescent spectra of CCA sample, GTA and perylene/GTA mixture.

of 450 ± 25 nm of wavelength was measured from confocal images around the cover glasses surface by using an optical filter. Figure 4 shows the relationship between the C_{per} and fluorescent intensity in each mixture. Three replicates of CLSM analysis of perylene (± standard deviation, SD) were performed for this study. As a result, the fluorescent intensity presented a good linearity (r = 0.998) with C_{per} ranging from 2.2 to 11 × 10^{-5}.

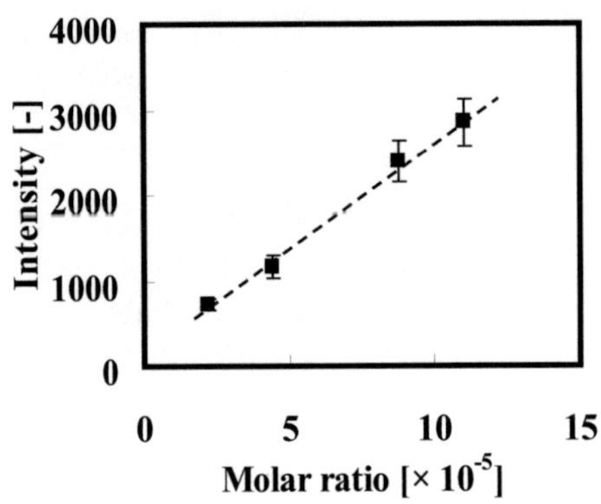

Fig. 4. Relationship between molar ratio of perylene and fluorescent intensity.

2.2 Effect of CLSM conditions on visualized sorption behavior

Perylene in CCA sample was observed by CLSM at several analytical conditions. That is, the fluorescent intensity of distributed perylene in CCA sample was measured as follows. Prior to perylene measurement by CLSM, the air-contact surface of the CCA sample was determined by transmission image using a halogen lump. Then, CLSM analysis was carried out at 60 min after placing 4 μl of perylene/GTA mixture onto CCA sample. Confocal images were obtained by scanning the diode laser in the cross-sectional region of the CCA sample with intervals along the Z-axis (Figure 5); the depth from the air-contact surface of the CCA sample were ± 10 or ± 30 μm with intervals at 1 or 3 μm, respectively. The confocal imaging area was approximately 640 × 640 μm on the X-Y surface, and both low- (scanning speed = 1 fps, flame size = 512 × 512 pixels) and high- (scanning speed = 30 fps, flame size = 512 × 512 pixels) speed-modes of scanning were selected for observation. Under each condition of analysis, C_{per} was 8.8 × 10^{-5}.

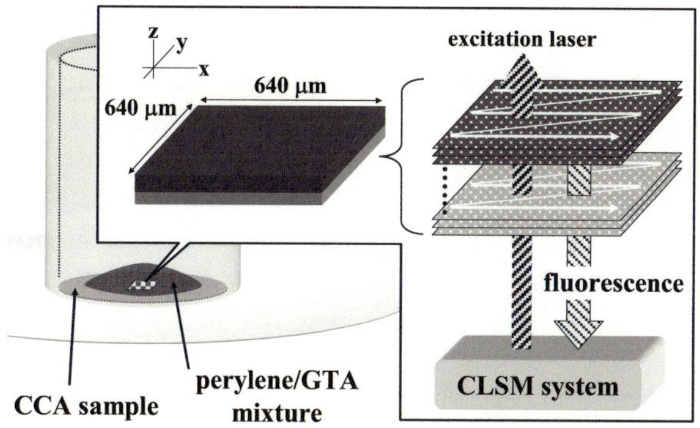

Fig. 5. Observation of perylene/GTA mixture in CCA sample by CLSM.

Figure 6 shows the chemical images of X-Y surface at 0, 2, 4, 6 and 8 μm-depth of CCA sample treated with perylene/GTA mixture under the CLSM conditions of low-speed-mode, ± 10 μm range and 1 μm intervals. Considering the influence of asperity on the air-contact surface of CCA sample, approximately 4.0×10^4 square μm of area of X-Y surface was selected from the observed areas for analysis. The blue color derived from the fluorescent intensity of perylene/GTA mixture was detected at 60 min at each depth of CCA sample. It was clear that the higher the depth of the film, the darker the blue color induced by sorbed perylene.

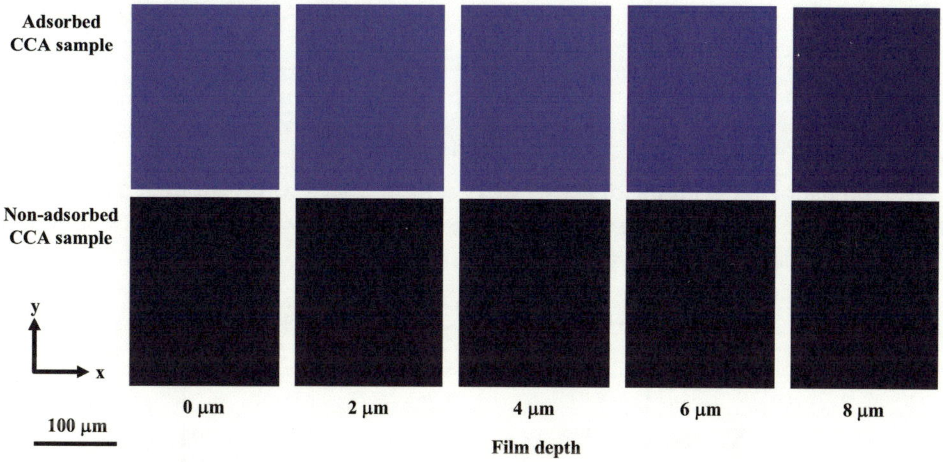

Fig. 6. Chemical images of X-Y surface at several depths of CCA sample at 60 min after the addition of perylene/GTA mixture obtained under conditions of low-speed-mode, ± 10 μm range and 1 μm intervals.

Figures 7-9 show changes in fluorescent intensities at each depth of the CCA sample. The intensity was calculated as an average of each pixel in a given analytical area. Three replicates of CLSM analysis of perylene (± SD) were performed for this study. In the horizontal axis, the 0 μm-depth represents the air-contact surface of the CCA sample. The results of sorbed perylene at the low-speed-mode, in the range of ± 10 μm and 1 μm intervals are shown in Figure 7, while the results of sorbed perylene at the low-speed-mode, in the range of ± 30 μm and 3 μm intervals are shown in Figure 8. Under both CLSM conditions, the perylene-induced intensity was detected within 9-μm depth of the CCA sample and became lower with film depth. The fluorescence phenomena revealed that after starting the diffusion of absorbed perylene on the air-contact surface of the CCA sample, perylene dynamically sorbed into inner CCA sample with time for 60 min. By considering that at a film depth of higher than 12 μm the perylene-induced intensity was not detected in Figure 8, the scan range of ± 10 μm along the Z-axis would be sufficient for monitoring the perylene sorption behavior in this study. In contrast, at a high-speed-mode of scanning in the range of ± 10 μm (Figure 9), the increase in intensity of sorbed perylene was detected within almost the same (9 μm) depth of the film as that in Figure 7. Thus, the scanning speed of high-speed-mode with a wide range along the Z-axis in a rapid-CLSM analysis in seconds could be useful for dynamic and rapid observation of penetrant distribution in CCA sample. To evaluate the effect of optical path length on fluorescent intensity induced by perylene, 12 μl of perylene/ GTA mixture was analyzed as mentioned above with the use of a film-

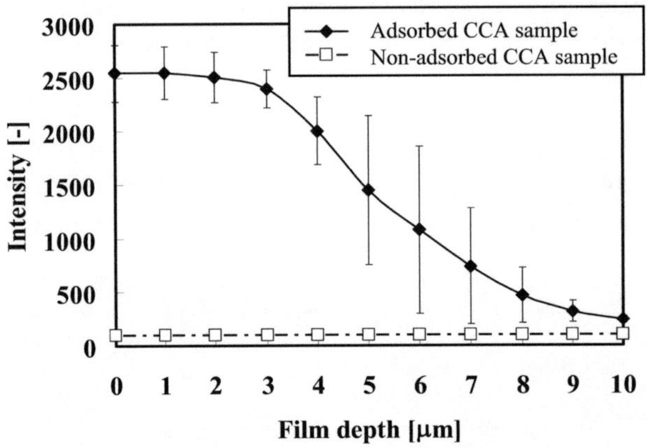

Fig. 7. Relationship between the depth of CCA sample and fluorescent intensity derived from perylene/GTA mixture obtained under conditions of low-speed-mode, ± 10 μm range and 1 μm intervals.

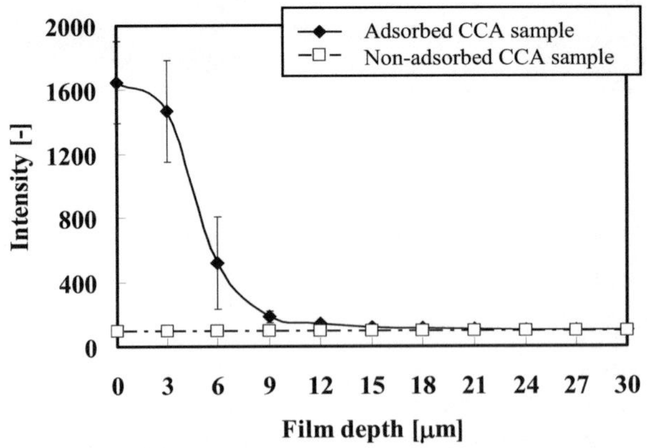

Fig. 8. Relationship between the depth of CCA sample and fluorescent intensity derived from perylene/GTA mixture obtained under conditions of low-speed-mode, ± 30 μm range and 3 μm intervals.

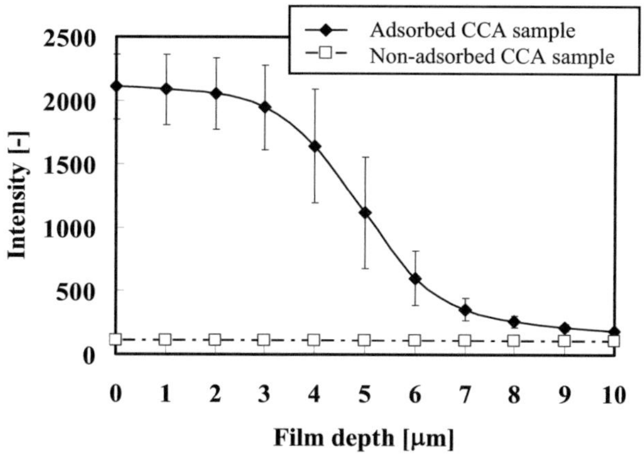

Fig. 9. Relationship between the depth of CCA sample and fluorescent intensity derived from perylene/GTA mixture obtained under conditions of high-speed-mode, ± 10 μm range and 1 μm intervals.

making-unit (without CA film). Figure 10 shows the relationship between the path length of CLSM and the fluorescent intensity of perylene. On the vertical axis, fluorescent intensity was calculated as an average of each pixel in the observation area at 1 min after pipetting the mixture onto the film-making-unit. On the horizontal axis, the value represents the distance from the top of the cover glass. As a result, the longer the distance, the lower the fluorescent intensity, suggesting that the power of the excitation laser or the intensity of the fluorescence would decay with the increase of optical path length. This result suggests that although the decay behavior of fluorescent intensity derived from perylene in the CCA sample was not clarified, the gradient of the amount of perylene must be sharper than gradients of the observed fluorescent intensities in Figures 7, 8 and 9.

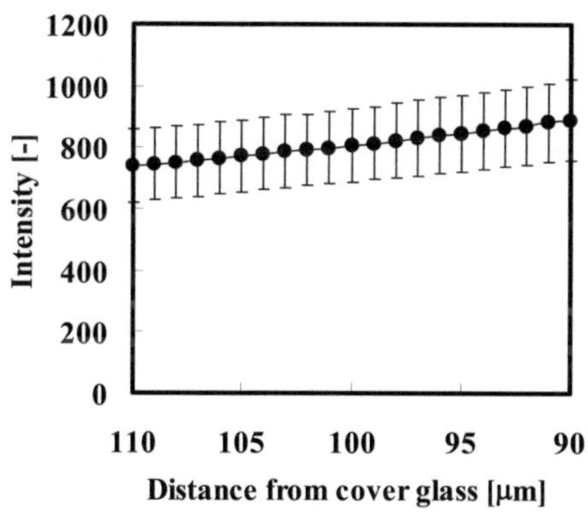

Fig. 10. Relationship between optical path length and fluorescent intensity derived from pelyrene/GTA mixture.

As to the sorption of compounds into polymer materials, physical and chemical parameters of compounds and/or polymers such as molar volume, solubility parameters, free energy and free volume determine the dynamics [11-13]. When taking particular note of the solubility parameter (SP), the affinity of a compound for polymer material is explained by two-dimensional distance (δ_c) defined as formula (2) [11],

$$\delta_c = \left[\left(\delta_{1np} - \delta_{2np} \right)^2 + \left(\delta_{1p} - \delta_{2p} \right)^2 \right]^{1/2} \quad (2)$$

where δ_t is the total SP value and δ_{np} and δ_p are nonpolar and polar components of δ_t, respectively, and the smaller the δ_c, the higher the affinity. Table 1 presents the SP values of CA, perylene and GTA, and the δ_c value of perylene and GTA from CA. The SP values of CA were as previously reported [36] and those of perylene and GTA were calculated using computational software, Molecular Modeling Pro (version 6.0.1, ChemSW, Inc., CA). The δ_c value of GTA was smaller than that of perylene and it was considered that the affinity of GTA for CA was higher than that of perylene. As a plasticizer for CA, GTA should accelerate the sorption of other compounds into polymer material, and it was thought that GTA had some effect on the sorption behavior of perylene into the CCA sample obtained in this study [10]. Therefore, in other words, this CLSM methodology can explain the sorption behavior of liquid additives such as GTA using the fluorescent reagent like perylene.

Compounds	SP [MPa$^{1/2}$]			δ_c* [MPa$^{1/2}$]
	δ_t	δ_{np}	δ_p	
CA	25.1	21.6	12.7	-
perylene	26.0	26.0	1.3	12.2
GTA	27.2	26.3	6.8	7.6

* Distance to CA

Table 1. Solubility parameter (SP) value and two-dimensional distance of compounds.

2.3 Calculation of diffusion coefficient of fluorescent reagent

Figure 11 shows change of fluorescent intensity of perylene in each depth of CCA sample measured by a CLSM system every 3 minutes to 30 minutes after dropping a 4 μl-perylene/additive mixture onto CCA sample. CLSM conditions were low-speed-mode and 1 μm intervals. On the horizontal axis, the 0 μm-depth indicates the air-contact surface of CCA sample. Fluorescent intensities in each depth were standardized by subtracting CCA-based intensity and subsequently dividing the obtained intensity by that of 0 μm-depth each time. As a result, the fluorescent intensity derived from perylene became higher in each depth over the course of time.

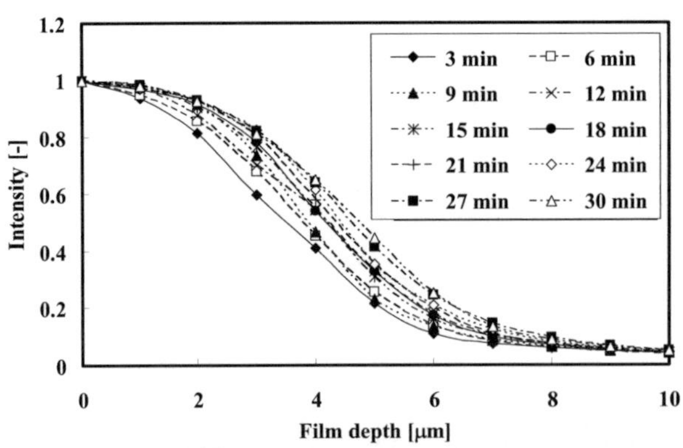

Fig. 11. Relationship between the depth of CCA sample and fluorescent intensity at several times after addition of perylene/GTA mixture.

From the results shown in Figure 11, the diffusion coefficient of perylene, D, was calculated based on Fick's second law given as formula (3),

$$\frac{\partial C}{\partial t} = D\frac{\partial}{\partial x}\left(\frac{\partial C}{\partial x}\right) \tag{3}$$

where C is a concentration of perylene, t is the time after the addition of mixtures and x is the film depth. The left side of the equation was obtained as the slope of standardized intensity, S_i, versus t in each depth of film because the intensity had a good linearity for the molar ratio of perylene (Figure 4). Then, the second term of the right side was calculated as the slope of a change of standardized intensity in 2 μm-depth range, $\partial S_i/\partial x$, versus x. Finally, D was calculated as the coefficient of both sides of the equation for each depth of film and was averaged. Considering a good agreement with Fick's second law, the calculation was performed in a particular film depth range where both S_i to t and $\partial S_i/\partial x$ to x relations had linearity. As shown in Figures 12 and 13, high linearity ($r > 0.969$) was obtained in each mixture and D was calculated in the range of 3 to 8 μm-depth of CCA sample and the value was 1.7×10^{-15} m^2/s when mixed with GTA and added on CCA sample.

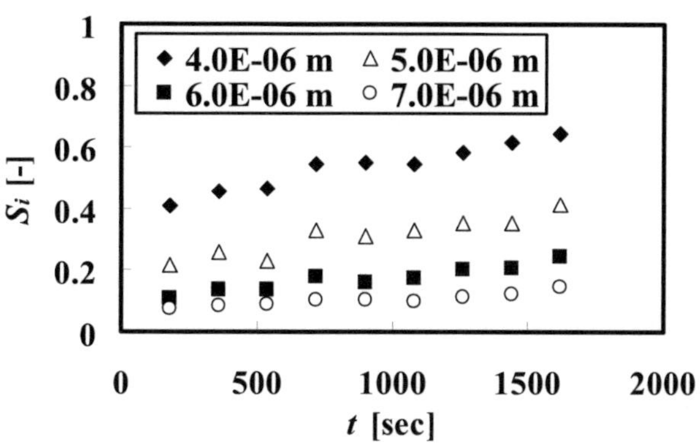

Fig. 12. Relationship between time after mixtures addition and standardized intensity.

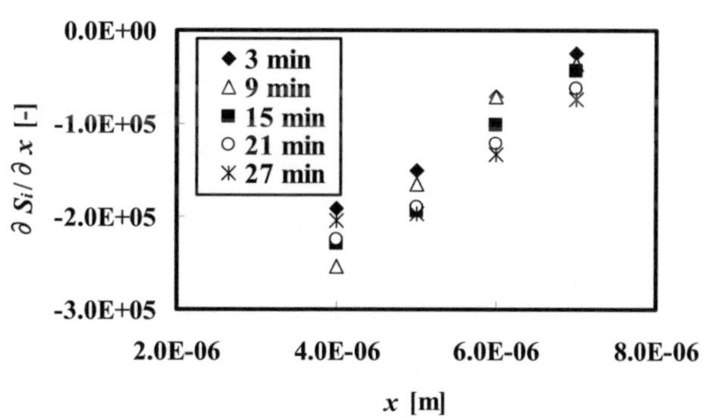

Fig. 13. Relationship between film depth and change of standardized intensity.

In this section, we visualized the dynamic distribution of perylene in CA film by applying a CLSM system for the first time. Analytical conditions of CLSM such as the scanning range along the Z-axis and the scanning speed did not affect the results and diffusion coefficient of perylene in CCA sample was obtained according to Fick's second law. This methodology has an enormous advantage for evaluating the sorption behavior of penetrant into polymer; the proposed non-destructive confocal optics system enables direct cross-sectional analysis to be performed without any destruction of sorbed film.

3. Effect of additives on sorption behavior of fluorescent reagent

3.1 Difference of diffusion coefficient

As additives mixed with perylene and added to CCA sample, TEGDA (Mw = 234.3 g/mol, Tokyo Chemical Industry Co., Ltd., Tokyo, Japan), BGDA (Mw = 174.2 g/mol, Daicel Chemical Industries Ltd.), paraffin liquid (C ~ 28, Mw ~ 395 g/mol, Nacalai Tesque, Inc., Kyoto, Japan) and PEG (C ~ 4, Mw ~ 200 g/mol, Wako Pure Chemical Industries Ltd.) were used further to GTA. Each perylene/additive mixture was prepared at room temperature by dissolving perylene with an additive to achieve a desired molar ratio in the range of 2.1 to 13 $\times 10^{-5}$ of perylene defined as formula (4).

$$Molar\ ratio[-] = \frac{Perylene\ amount\ [mol]}{Perylene\ amount\ [mol] + Additive\ amount\ [mol]} \tag{4}$$

Fluorescent spectra of perylene/additive mixtures, additives and CCA sample when excited by diode laser at 408 nm are shown in Figure 14. The spectra clearly demonstrated that perylene/additive mixtures showed significant perylene-induced fluorescent intensity at around 470 and 503 nm, while no fluorescent intensities were observed for additives and CCA sample. This proves that any interference in fluorescence from additives or CA film could be excluded for an *in situ* penetration monitoring of perylene in CCA sample. Figure 15 shows the relationships between molar ratio of perylene in perylene/additive mixtures and fluorescent intensities. As a result, a good linearity ($r > 0.978$) was observed in each mixture in the range of 2.1 to 13 $\times 10^{-5}$ molar ratio of perylene.

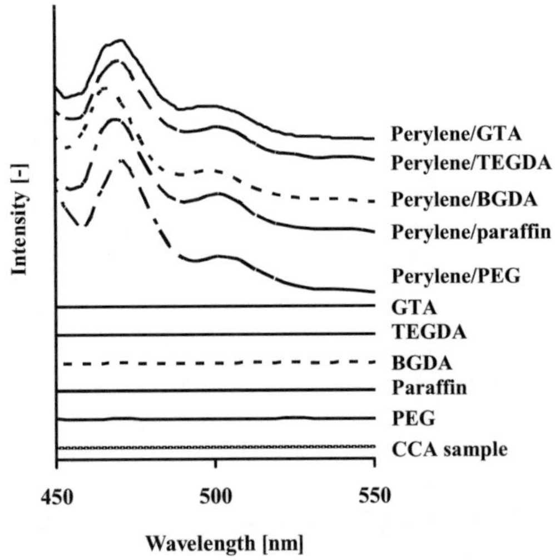

Fig. 14. Fluorescent spectra of perylene/additive mixtures, additives and CCA sample.

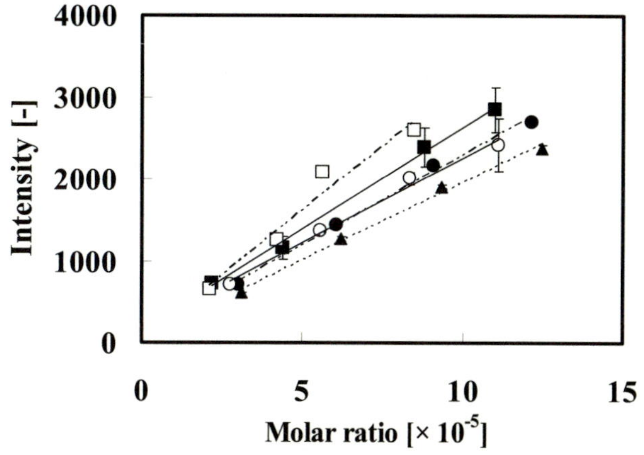

Fig. 15. Relationships between molar ratio and fluorescent intensity of perylene mixed with additives: GTA (■), TEGDA (●), BGDA (○), paraffin liquid (□) and PEG (▲).

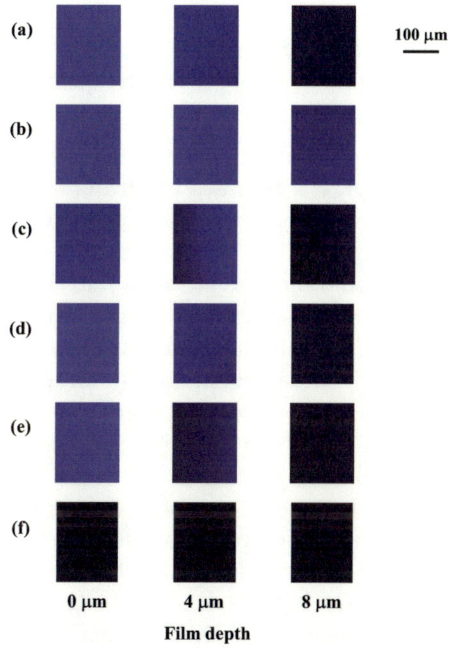

Fig. 16. Chemical images of CCA sample at several depths 9 minutes after addition of perylene/additive mixtures: (a) /GTA, (b) /TEGDA, (c) /BGDA, (d) /paraffin liquid, (e) /PEG and (f) control (without addition).

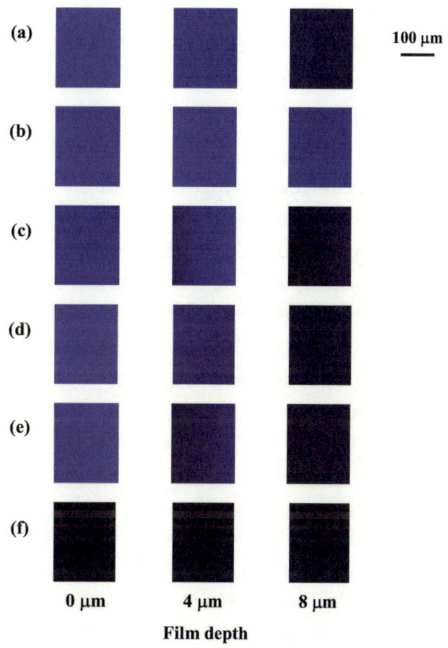

Fig. 17. Chemical images of CCA sample at several depths 21 minutes after addition of perylene/additive mixtures: (a) /GTA, (b) /TEGDA, (c) /BGDA, (d) /paraffin liquid, (e) /PEG and (f) control (without addition).

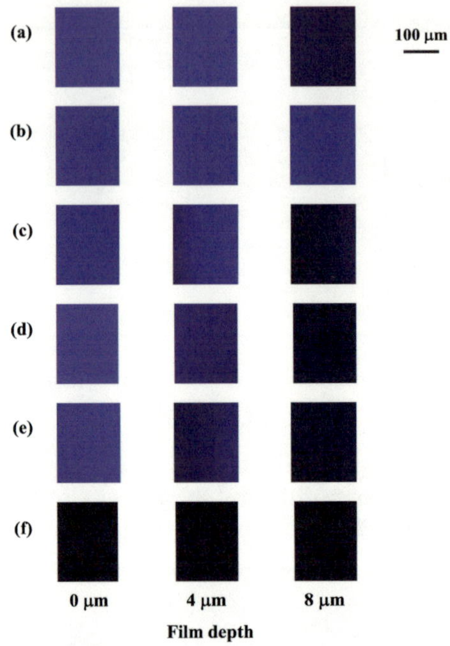

Fig. 18. Chemical images of CCA sample at several depths 30 minutes after addition of perylene/additive mixtures: (a) /GTA, (b) /TEGDA, (c) /BGDA, (d) /paraffin liquid, (e) /PEG and (f) control (without addition).

Change in fluorescent intensity of perylene in CCA sample during sorption experiments at room temperature was measured by a CLSM system every 3 minutes to 30 minutes after dropping a 4 μl-perylene/additive mixture (molar ratio of perylene was $8.8 \pm 0.5 \times 10^{-5}$) onto CCA sample. Chemical images at 0, 4 and 8 μm-depth of CCA sample treated with perylene/additive mixtures 3 minutes after the addition onto the film are shown in Figure 16 as well as the time-course images obtained at 21 and 30 minutes after the addition (in Figures 17 and 18, respectively). CLSM conditions were identical with the aforementioned section 2 except that speed-mode was low and the intervals were 1 μm. Considering the influence of asperity on air-contact surface of CCA sample, approximately 4.0×10^4 square μm of areas were selected from the observation areas for analysis. The blue color derived from the fluorescent intensity of perylene was detected in each image. It was clear that the higher the depth of the film, the darker the blue color induced by sorbed perylene. The results in Figures 16, 17 and 18 prove that the brightness of blue color became higher over time after the mixture addition, especially for GTA, TEGDA and BGDA. These results indicate that the sorption of perylene/additive mixture progressively proceeded from the top of CCA sample to inside of the film. The brightness and its change were the highest for perylene/TEGDA mixture.

To quantify these behaviors, fluorescent intensities were calculated as an average of each pixel in a given analytical area for perylene/additive mixture. Figure 19 shows changes of relationships between the film depth and the standardized intensities, in which the intensity was averaged through three replicates of CLSM analysis. As a result, the change in intensity over time was the largest in perylene/TEGDA mixture. In particular, the standardized intensity in perylene/TEGDA mixture increased by a factor of 0.37 at 8 μm-depth from 3 to 30 minutes after the addition of mixture. Perylene-induced intensities also increased clearly

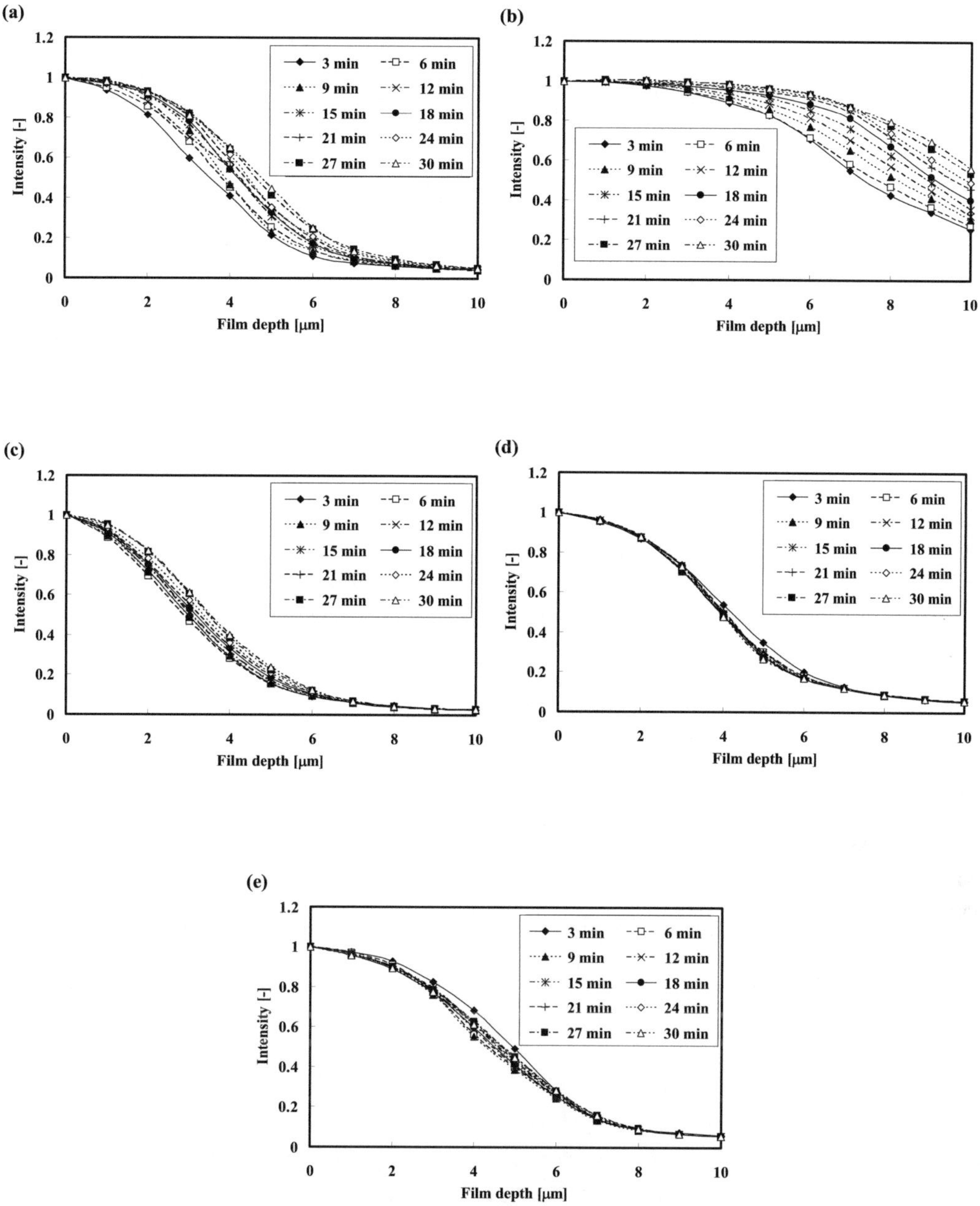

Fig. 19. Relationship between the depth of CCA sample and fluorescent intensity at several times after addition of perylene/additive mixtures: (a) /GTA, (b) /TEGDA, (c) /BGDA, (d) /paraffin liquid and (e) /PEG.

when mixed with GTA and BGDA, and their increment was a factor of 0.24 at 4 µm-depth and 0.12 at 3 µm-depth, respectively. For paraffin liquid and PEG, on the other hand, changes in intensities were as little as 0.01 at any film depth, though the intensities were detected within almost 9-µm depth of CCA sample.

From the results shown in Figure 19, the diffusion coefficient of perylene, *D*, was calculated based on Fick's second law as mentioned in section 2. Figure 20 shows the *D* values of perylene in CCA sample when mixed with additives. The value varied depending on the additives and became the highest in the mixture with TEGDA (8.9×10^{-15} m^2/s). The order was TEGDA > GTA (1.7×10^{-15} m^2/s) > BGDA (1.3×10^{-15} m^2/s) > PEG (0.54×10^{-15} m^2/s) > paraffin liquid (0.34×10^{-15} m^2/s) in descending order, indicating that the additives greatly affected the diffusion behavior of perylene in CCA sample.

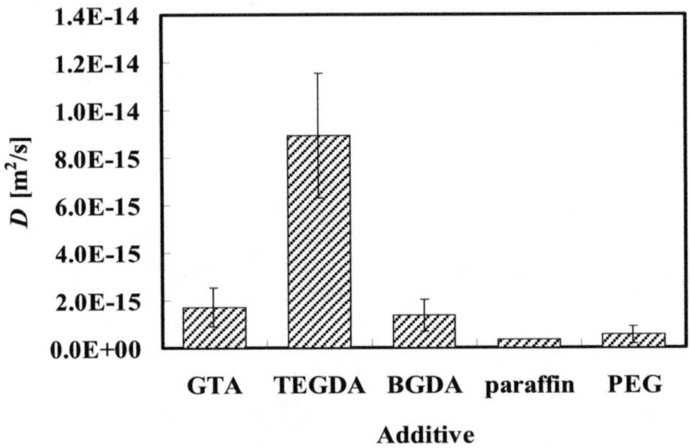

Fig. 20. Diffusion coefficient of perylene mixed with additives in CCA sample. Data are the mean ± SD (n = 3).

3.2 Effect of physicochemical properties of additives

The above *D* values suggest that the sorption behavior of perylene in CCA sample was largely affected by the physicochemical properties of additives. Table 2 displays SP values of perylene, additives and CA. SP values of TEGDA, BGDA, paraffin liquid and PEG were calculated using with Molecular Modeling Pro. δ_c values of additives from perylene and CA were calculated with the formula (2), and Figure 21 shows an index of chemical affinities, $1/\delta_c$. The chemical affinities of additives with perylene or CA would affect the perylene-additive compatibility or sorption of the additive itself into CCA sample.

Compound	SP [MPa$^{1/2}$]		
	δ_t	δ_{np}	δ_p
perylene	26.0	26.0	1.3
GTA	27.2	26.3	6.8
TEGDA	25.9	25.3	5.7
BGDA	18.9	18.4	4.1
paraffin	16.7	16.7	0
PEG	22.5	22.0	4.5
CA	25.1	21.6	12.7

Table 2. SP value of compounds.

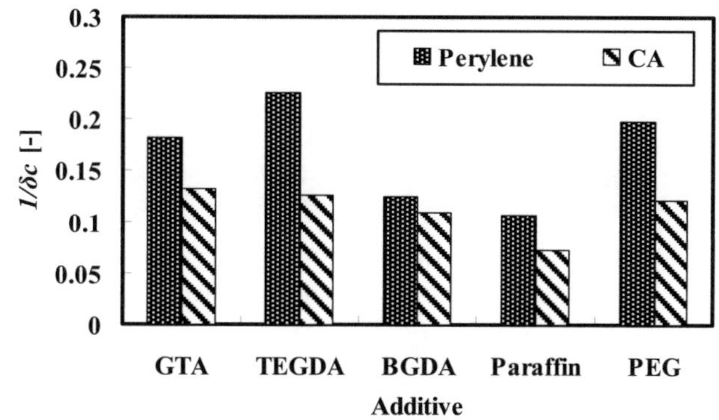

Fig. 21. Chemical affinity of additives with perylene and CA.

It was considered that the highest D value in perylene/TEGDA mixture was due to the high affinity of perylene with TEGDA. In contrast, the lowest D value in perylene/paraffin liquid mixture was caused by the low affinity of perylene with both paraffin liquid and CCA sample. For GTA, BGDA and PEG, difference of D could not be explained only by chemical affinities. Although the $1/\delta_c$ values from CA were almost the same, the $1/\delta_c$ values from perylene were PEG > GTA > BGDA in descending order. This order did not agree with that of D shown in Figure 20; namely, the diffusion coefficient of perylene in CCA sample added with PEG was lower than the expected value.

Physical parameters such as the molecular size and viscosity of chemicals were then taken into consideration as external affecting factors. When considering the sorption of additives into CA film, the diffusion coefficient of additive, D', may be influenced by its molecular diameter, d, and viscosity, η. The relationships of those parameters are given by Stokes-Einstein equation as follows:

$$D' = \frac{kT}{3\pi d\eta} \tag{5}$$

where k and T are Boltzmann's constant and absolute temperature, respectively. Table 3 presents the sphere-equivalent molecular diameter and viscosity of additives. Sphere-equivalent molecular diameters were calculated by Molecular Modeling Pro. And viscosities were measured at 25 °C by digital viscometer (DVL-BII, Toki Sangyo Co., Ltd., Tokyo, Japan). As a result, the diffusion coefficient of PEG in CA film was calculated to be lower than those of GTA and BGDA on account of the higher viscosity of PEG. Thus, the lower D of perylene in PEG (Figure 20) would be caused by the poor diffusion of PEG itself in CCA sample. For perylene/paraffin liquid mixture, it was thought that the lowest D of perylene was due not only to low chemical effects but also to the physical properties of paraffin liquid such as high molecular size and high viscosity.

In this section, the effects of various additives, GTA, TEGDA, BGDA, paraffin liquid and PEG, on the sorption behavior of perylene into CCA sample were compared. The sorption behavior of perylene was visualized dynamically by CLSM methodology while diffusion coefficients were calculated according to Fick's second law. Taking these findings together, it was confirmed that the diffusion coefficient of perylene was influenced by both chemical affinities for additives or CCA sample and the diffusivity of additive in CCA sample. The highest diffusion coefficient of perylene, 8.9×10^{-15} m^2/s, obtained in TEGDA mixture was due to the high chemical affinity of TEGDA with perylene ($1/\delta_c$ value was 0.23). Meanwhile, in the case of perylene/GTA, BGDA or PEG mixture, the difference in diffusion coefficient of perylene (1.7 × 10^{-15}, 1.3 × 10^{-15} and 0.54 × 10^{-15} m^2/s, respectively) could not be explained only by the chemical affinities between additives and perylene ($1/\delta_c$ values were 0.18, 0.12 and 0.20, respectively). Based on physical parameters, it was suggested that the lower diffusion coefficient of perylene with PEG into CCA sample would be attributed to the high viscosity of PEG. The lowest diffusion coefficient of perylene, 0.34 × 10^{-15} m^2/s, when mixed with paraffin liquid, would be caused by both the low chemical affinity and the low diffusivity of paraffin liquid in CCA sample.

Additive	Molecular diameter d [nm]	Viscosity η [cPs]
GTA	0.73	17.0
TEGDA	0.71	9.8
BGDA	0.64	3.2
paraffin	0.96	147.5
PEG	0.71	48.8

Table 3. Molecular diameter and viscosity of additives.

4. Effect of film types and film depth on sorption behavior of fluorescent reagent

4.1 Difference of diffusion coefficient caused by multiple stratification treatment of film

As a different type of CA film from CCA sample, open-system CA film (OCA) was prepared by putting an aliquot (4 ml) of CA solution (2.0 w/v % in acetone) into glass petri dish (27 mmϕ), vaporizing excess acetone for 7 days at room temperature. After drying *in vacuo* overnight, film was peeled off from petri dish and fixed between two neodymium magnets (Figure 22, 23 mmϕ of inner diameter and 1 mm thick). The prepared OCA sample was 60 ± 10 μm thick.

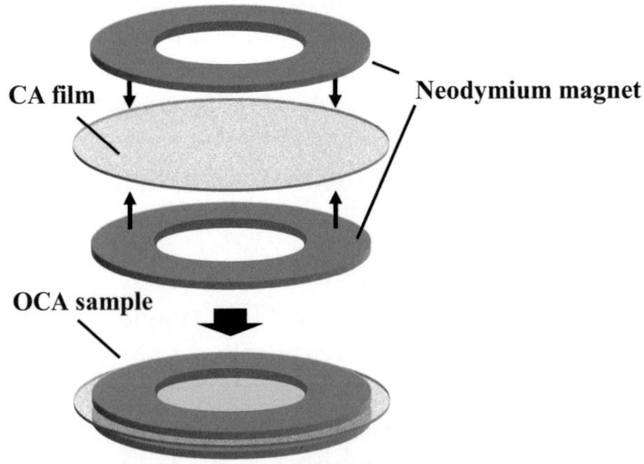

Fig. 22. OCA sample fixed between neodymium magnets.

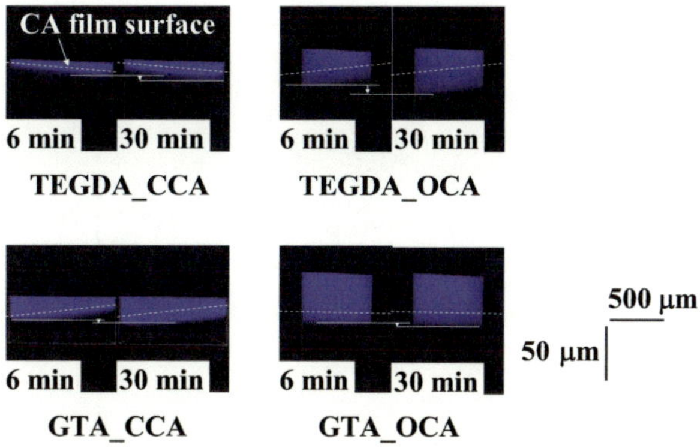

Fig. 23. Time-course cross-sectional chemical images (X-Z surface) of CCA and OCA samples.

Perylene/GTA or perylene/TEGDA mixtures were added on the CCA or OCA sample and the sorption behavior of perylene in films were observed by the CLSM methodology. CLSM conditions were identical with the aforementioned section 3 except that analytical depth from the air-contact surface of film samples was 20 μm with intervals of 1 μm. The molar ratio, defined by formula (3), in each perylene/GTA and perylene/TEGDA mixture was 9.1×10^{-5} and 8.8×10^{-5}, respectively.

Three-dimensional chemical images of CA film samples treated with perylene/additive mixtures were reconstructed by the chemical images of X-Y surface obtained at several depths of film samples. Figure 23 shows cross-sectional chemical images (X-Z surface) of CCA and OCA samples at 6 and 30 minutes after the addition of perylene/additive mixtures. Although there are some influences of asperity or slant on air-contact surface of film samples, the blue color derived from the fluorescent intensity of perylene was detected in each image. It was clear that the blue color expanded to deeper area over time after the addition of mixtures in each CA film sample. These results describe that the sorption of perylene/additive mixture proceeded progressively from the top of film samples to inside of the film. With respect to the effect of the additives on the perylene-sorption, the change was higher for perylene/TEGDA mixture in both CCA and OCA samples. Intriguingly, perylene-sorption behavior in the OCA sample became more intense than that in the CCA sample especially for perylene/TEGDA mixture.

To quantify this behavior, fluorescent intensities derived from perylene in CA film samples were calculated as an average of each pixel in X-Y surfaces. Approximately 4.0×10^4 square μm of areas were selected from the observation areas for analysis. Figures 24-27 illustrate variations of intensities in each depth of CA film samples, in which the intensity was averaged through three replicates. On the horizontal axis, the 0 μm-depth indicates the air-contact surface of film samples. Fluorescent intensities in each depth were standardized by subtracting CA-based intensity and subsequently dividing the obtained intensity by that of 0 μm-depth each time.

Consequently, the standardized intensity (I_S) of each depth increased progressively from 6 to 30 minutes after adding mixtures. Concerning the CCA samples (Figures 24 and 25), I_s increased by a factor of 0.34 at 8 μm-depth and 0.22 at 4 μm-depth in perylene/TEGDA and perylene/GTA mixtures, respectively. For the OCA samples (Figures 26 and 27), on the

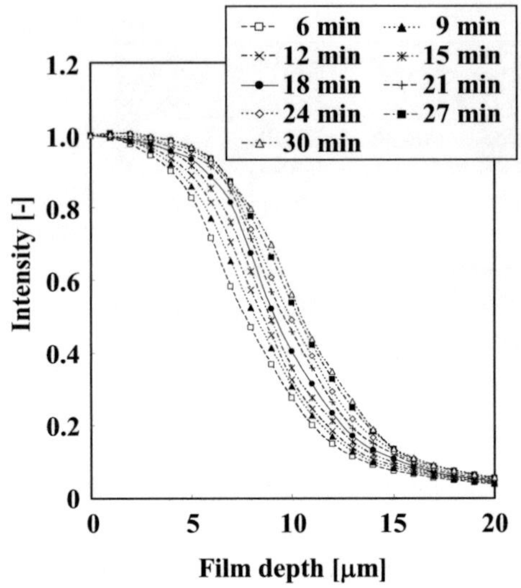

Fig. 24. Relationship between the depth of CCA sample and I_s at several times after addition of perylene/TEGDA mixture.

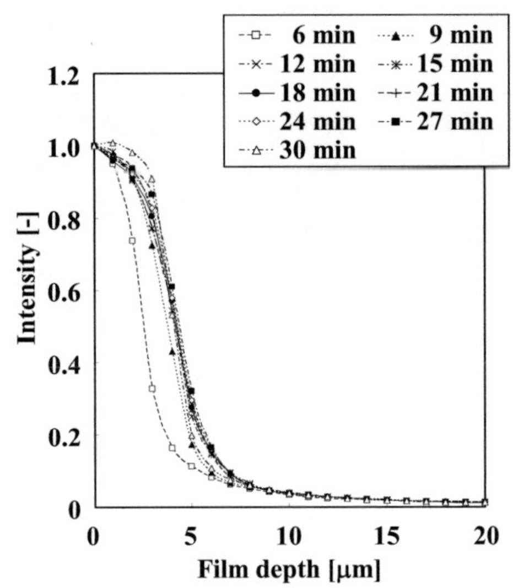

Fig. 25. Relationship between the depth of CCA sample and I_s at several times after addition of perylene/GTA mixture.

other hand, each increment was a factor of 0.26 at 14 μm-depth and 0.20 at 9 μm-depth in perylene/TEGDA and perylene/GTA mixtures, respectively. From the results illustrated in Figures 24-27, the diffusion coefficient of perylene, *D*, was calculated in the same way as described in section 2. Figure 28 displays the *D* values of perylene in CCA and OCA film samples when mixed with additives. Concerning the CCA samples, *D* values were calculated in 10 to 13 and 3 to 8 μm-depth range in perylene/TEGDA and perylene/GTA mixtures, respectively. For the OCA samples, *D* values were calculated in 14 to 17 and 9 to 11 μm-depth range in perylene/TEGDA and perylene/GTA mixtures, respectively. It was

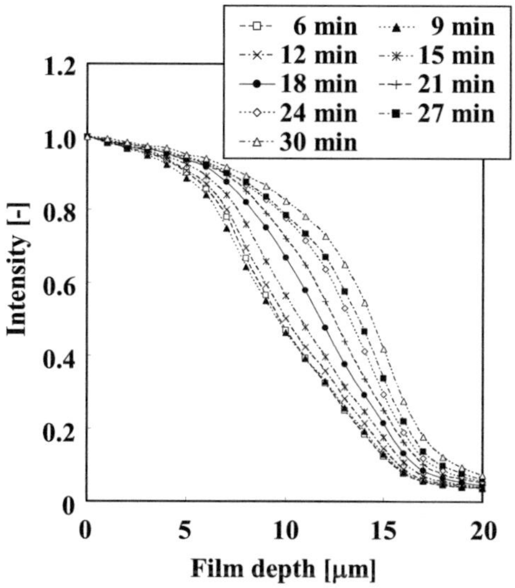

Fig. 26. Relationship between the depth of OCA sample and I_s at several times after addition of perylene/TEGDA mixture.

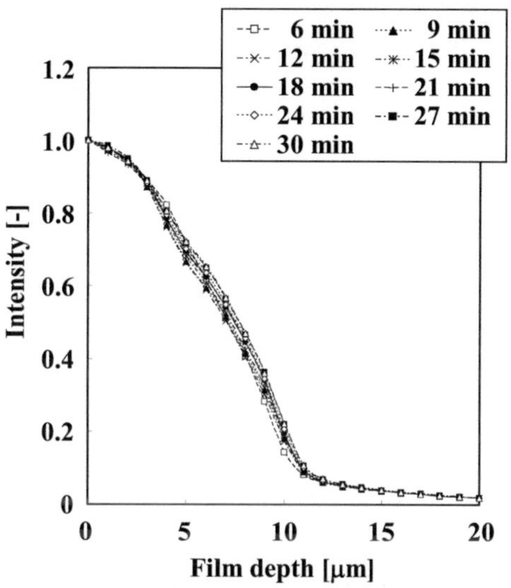

Fig. 27. Relationship between the depth of OCA sample and I_s at several times after addition of perylene/GTA mixture.

confirmed that D value varied depending on the kind of additives and the CA film type. In both cases of CCA and OCA samples, D values of perylene when mixed with TEGDA (CCA: $8.9 \pm 2.6 \times 10^{-15}$ m^2/s, OCA: $11 \pm 5.1 \times 10^{-15}$ m^2/s) were higher than those of GTA (CCA: $1.7 \pm 0.83 \times 10^{-15}$ m^2/s, OCA: $3.3 \pm 2.2 \times 10^{-15}$ m^2/s). And it was found that regardless of the kind of additives, D values of perylene in CA film became higher for the OCA samples (TEGDA: $11 \pm 5.1 \times 10^{-15}$ m^2/s, GTA: $3.3 \pm 2.2 \times 10^{-15}$ m^2/s) than for the CCA samples (TEGDA: $8.9 \pm 2.6 \times 10^{-15}$ m^2/s, GTA: $1.7 \pm 0.83 \times 10^{-15}$ m^2/s).

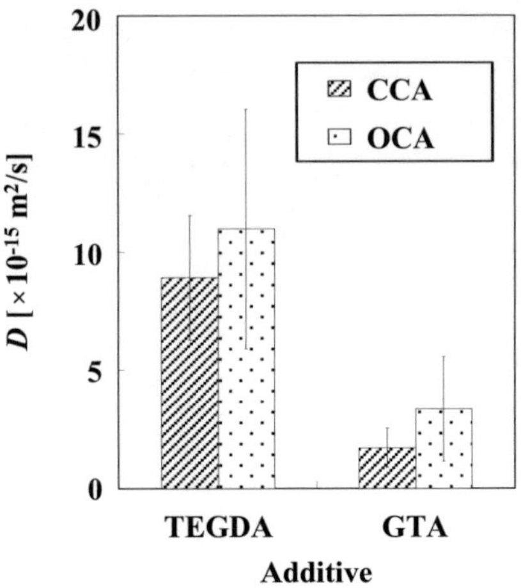

Fig. 28. Diffusion coefficients of perylene in CCA and OCA samples when mixed with TEGDA or GTA. Data are the mean ± SD (n = 3).

As described in section 3, diffusion coefficient of perylene in CCA sample was affected by physicochemical properties of additives, and regarding the higher *D* value of perylene mixed with TEGDA (8.9×10^{-15} m^2/s) than that of mixed with GTA (1.7×10^{-15} m^2/s), it was attributed to higher chemical affinity of perylene with TEGDA based on estimation with solubility parameters (SPs). Similarly in the case of OCA sample, as the above results in this study suggested, the order of the *D* value of perylene, TEGDA mixture (11×10^{-15} m^2/s) > GTA mixture (3.3×10^{-15} m^2/s), was attributed to the difference of chemical affinity of perylene with additives.

Regardless of the kind of additives, *D* values of perylene in OCA samples were higher than those of in CCA samples. Figure 29 represents shifts of I_s in each depth of CCA and OCA samples 1, 2 and 24 hours after the addition of perylene/TEGDA mixture. With the exception that cross-sectional region for scanning was 0-40 μm-depth from air-contact surface of film samples, experimental conditions and analytical protocols were same as mentioned above. As for the CCA sample, no investigation was carried out on a progressive sorption of perylene into the film but on a swing-over of perylene-sorption to some extent after 1 hour. I_s at 1, 2 and 24 hours after the addition of perylene/TEGDA mixture were 0.015, 0.014 and 0.012 at 20 μm-depth and 0.0034, 0.0035 and 0.0033 at 40 μm-depth, respectively. This result pointed out that perylene-sorption came to equilibrium around 1 hour after the addition of mixture, and the distribution ratio of perylene at 40 μm-depth was 0.3 % compared with the air-contact surface (0 μm-depth) of the film. For the OCA sample, on the other hand, it was identified that sorption of perylene into the film was progressing even at 24 hours after the addition of mixture. I_s at 1, 2 and 24 hours after the addition of perylene/TEGDA mixture were 0.30, 0.88 and 1.1 at 20 μm-depth and 0.017, 0.27 and 1.1 at 40 μm-depth, respectively. This result represented that perylene was homogeneously-distributed in OCA sample by 24 hours after the mixture addition as opposed to CCA sample.

Studies on gas permeation through polymer films revealed that the permeability of multilayer was not always correspondent to the multiplication of those of each consisting monolayer [37] or the permeability of multilayer differed depending on the side of coating layer (inside or outside) [38]. Considering the cover glass of the CCA sample as a coating layer of CA film, the above results clarified that diffusion coefficient and distribution of perylene in CA film largely vary with multiple stratification treatment.

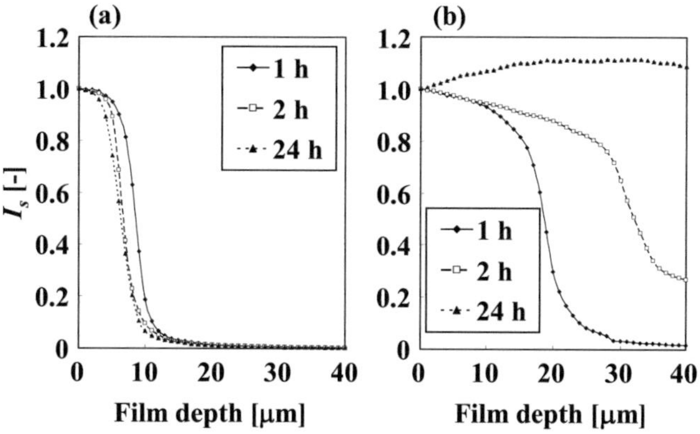

Fig. 29. Relationship between the depth of CA film and I_s at several times after addition of perylene/TEGDA mixture: (a) CCA and (b) OCA.

4.2 Film depth dependence of diffusion coefficient

To promote more understandings for sorption behavior of perylene into CA film, diffusion coefficients at each film depth were calculated pursuant to Fick's second law. Figure 30 describes relationships between film depth and diffusion coefficient of perylene in CCA and OCA samples. Greatly interesting to note, these results pointed out clearly that there were strong negative correlations between the two, *i.e.* the deeper the film depth, the lower the diffusivity of perylene regardless of type of additives and films. As is well known, Fick's second law is based on hypothesis that although concentration gradient penetrating to film depth transforms over the time of course, rate of its transformation is constant and D is obtained as a constant number. These results, however, contradicted the hypothesis. Comparing the slopes obtained approximate linearization, absolute values of OCA samples (TEGDA: 3.9×10^{-15}, GTA: 2.2×10^{-15}) were higher than those of CCA samples (TEGDA: 2.0×10^{-15}, GTA: 0.64×10^{-15}). It means that influence of film depth increase on decline of diffusivity was larger for OCA samples than for CCA samples.

As factors causing the film depth dependence of diffusion coefficient, two viewpoints were considered. The first factor was the effect of additives mixed with perylene. With respect to the diffusion of penetrant in polymer materials, free volume theory was proposed in 1960s [39] and recently, it has been generally thought that amorphous portion in polymer material is dominating diffusion pathway [40-46]. It was considered that with the diffusion of TEGDA or GTA, amorphous portion, *i.e.* diffusion pathway for perylene, gradually increased on account of CA solubility [33] of additives. As a result, D value of perylene in superficial part of CA film was higher than that of deep part. The second factor was the concentration dependence of diffusion coefficient. It has been reported that the higher the concentration of penetrant in experimental system, the higher the diffusion coefficient of penetrant in the polymer materials [47, 48]. In polymer materials, the concentration of penetrant should be different in accordance with the distance from liquid/solid or vapor/solid boundaries. It was considered that the aforementioned film depth dependence of D value directly indicates the concentration dependence of diffusion coefficient in the CA film. As for the CCA sample, perylene had a lower D value than that obtained above (3.8×10^{-15} m^2/s < 8.9×10^{-15} m^2/s) when mixed with TEGDA at lower molar ratio ($0.28 \times 10^{-5} < 9.1 \times 10^{-5}$). Moreover, this result proves that the diffusion coefficient of perylene had the concentration dependence.

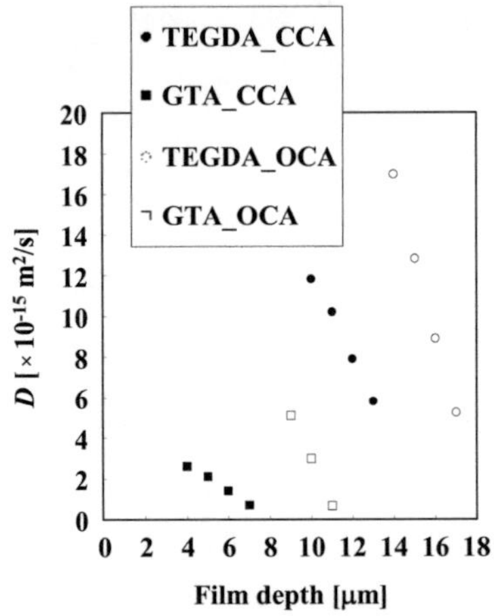

Fig. 30. Relationship between CA film depth and diffusion coefficient of perylene.

In this section, the effects of kinds of additives, TEGDA and GTA, and film types, CCA and OCA, on the sorption behavior of perylene into CA film were evaluated. Through dynamic monitoring of the perylene-penetration by CLSM methodology, overall diffusion coefficients of perylene in CA film as well as its film depth dependence were calculated according to Fick's second law to be assessed.

Higher chemical affinity of TEGDA with perylene than that of GTA commonly caused the higher diffusion coefficient of perylene both for CCA sample (TEGDA: 8.9×10^{-15} m^2/s > GTA: 1.7×10^{-15} m^2/s) and OCA sample (TEGDA: 11×10^{-15} m^2/s > GTA: 3.3×10^{-15} m^2/s). What is really interesting was that perylene had higher diffusivity in OCA sample than CCA sample beyond types of additives. In conclusion, it was simultaneously clarified that when mixed with TEGDA, perylene appeared homogeneously-distributed in OCA sample by 24 hours after the mixture addition (I_s = 1.1 at 40 μm-depth of CA film) as opposed to CCA sample (I_s = 0.0033 same as above). These results demonstrated that diffusion coefficient and distribution of perylene in CA film largely vary with multiple stratification treatment.

From the results on the film depth dependence of diffusion coefficient, greatly interesting to note, it was explained clearly that the deeper the film depth, the lower the diffusivity of perylene regardless of type of additives and films. Although this finding contradicted the hypothesis of Fick's second law, gradual increase in diffusion pathway for perylene caused by additive diffusion as well as concentration dependence of perylene's diffusion coefficient were considered as factors.

5. Conclusions

In this chapter, the application of CLSM technique was described to observe the fluorescent reagent sorption behavior into polymer material in real-time and non-destructively. In section 2, the dynamic distribution of perylene into CA film (CCA sample) was visualized for the first

time, and the analytical conditions of CLSM such as the scanning range along the Z-axis and the scanning speed did not affect the results. In section 3, the effects of various additives, GTA, TEGDA, BGDA, paraffin liquid and PEG, on the sorption kinetics of perylene into CCA sample were compared by calculating diffusion coefficients of perylene according to Fick's second law. As a result, it was revealed that the diffusion coefficient of perylene was influenced by not only chemical affinities for additives or CCA sample and the diffusivity of additive in CCA sample but also by physical parameters such as molecular size or viscosity of additives. In section 4, effects of film types and film depth on diffusion coefficient of fluorescent reagent were examined as a further development of CLSM methodology. It was directly clarified that diffusion coefficient and distribution of perylene in CA film were declined with multiple stratification treatment. Thus, the deeper the film depth, the lower the diffusivity of perylene regardless of type of additives and films caused by additives diffusion into the film as well as concentration dependence of perylene's diffusion coefficient.
The aforementioned results and findings could be directly achieved owing to the advantages of CLSM system, that is, real-time and non-destructive visualization method for sorption dynamics. It was proved that our proposed CLSM methodology should be beneficial to promote better understandings for sorption behavior in the field of material science and engineering industries.

6. Acknowledgment

This study was conducted with the technical support of Nikon Instech Co., Ltd., Tokyo, Japan. We are grateful to Mr. Toyofumi Kameoka and Ms. Takayo Furuya for their operational support of the CLSM system.

7. References

[1] Hofmann, G. H.; Lee, W. C. J. Vinyl Add. Tech., 2006, 12, 33.
[2] Deanin, R. D.; Shah, N. A. J. Vinyl Tech., 1983, 5, 167.
[3] Ware, R. A.; Tirtowidjojo, S; Cohen, C. J. Appl. Polym. Sci., 1981, 26, 2975.
[4] Bouajila, J.; Dole, P.; Joly, C.; Limare, A. J. Appl. Polym. Sci., 2006, 102, 1445.
[5] Sakata, I.; Senju, R. J. Appl. Polym. Sci., 1975, 19, 2799.
[6] Shimoda, M.; Nitanda, T.; Kadota, N.; Ohta, H.; Suetsuna, K.; Osajima, Y. J. Jpn. Soc. Food Sci. Technol., 1984, 31, 697.
[7] Leufven, A.; Hermansson, C. J. Sci. Food Agric., 1994, 64, 101.
[8] Sadler, G. D.; Braddock, R. J. J. Food Sci., 1991, 56, 35.
[9] Kalachandra, S.; Turner, D. T. J. Polym. Sci.: Part B: Polym. Phys., 1987, 25, 697.
[10] Kalaouzis, P. J.; Demertzis, P. G.; Kontominas, M. G. Packag. Technol. Sci., 1993, 6, 261.
[11] Matsui, T.; Nagashima, K.; Fukamachi, M.; Shimoda, M.; Osajima, Y. J. Agric. Food Chem., 1992, 40, 1902.
[12] Matsui, T.; Fukamachi, M.; Shimoda, M.; Osajima, Y. J. Agric. Food Chem., 1994, 42, 2889.
[13] Fukamachi, M.; Matsui, T.; Shimoda, M.; Osajima, Y. J. Agric. Food Chem., 1994, 42, 2893.
[14] Shimoda, M.; Matsui, T.; Osajima, Y. J. Jpn. Soc. Food Sci. Technol., 1987, 34, 402.
[15] Matsui, T.; Shimoda, M.; Osajima, Y. J. Jpn. Soc. Food Sci. Technol., 1989, 36, 52.

[16] Hellsing, M.; Fokine, M.; Claesson, Å.; Nilsson, L. -E.; Margulis, W. Appl. Surf. Sci., 2003, 203-204, 648.
[17] Van Thienen, T. G.; Demeester, J.; De Smedt, S. C. Int. J. Pharm., 2008, 351, 174.
[18] Wu, L.; Brazel, C. S. Int. J. Pharm., 2008, 349, 144.
[19] Hubbuch, J.; Linden, T.; Knieps, E.; Ljunglöf, A.; Thömmes, J.; Kula, M. R. J. Chromatogr. A, 2003, 1021, 93.
[20] Hubbuch, J.; Linden, T.; Knieps, E.; Thömmes, J.; Kula, M. R. J. Chromatogr. A, 2003, 1021, 105.
[21] Yamaguchi, S. J. Kyorin Med. Soc., 2003, 34, 139.
[22] Wang, C.; Zhu, L.; Qiu, Y. J. Appl. Polym. Sci., 2008, 107, 1471.
[23] Wang, Y.; Yang, C.; Tomasko, D. Ind. Eng. Chem. Res., 2002, 41, 1780.
[24] Hasegawa, T.; Matsui, T.; Matsumoto, K. J. Appl. Polym. Sci., 2010, 116, 1552.
[25] Hasegawa, T.; Matsui, T.; Matsumoto, K. J. Appl. Polym. Sci., 2010, 116, 1710.
[26] Hasegawa, T.; Matsumoto, K.; Matsui, T. J. Appl. Polym. Sci., *in press*.
[27] Kutowy, O.; Thayer, W. L.; Tigner, J.; Sourirajan, S.; Dhawan, G. K. Ind. Eng. Chem. Prod. Res. Dev., 1981, 20, 354.
[28] Lance, S. K.; Stephen, G. W.; Kamal, Z. I. Anal. Chem., 1989, 61, 303.
[29] Perrin, L.; Nguyen, Q. T.; Sacco, D.; Lochon, P. Polym. Int., 1997, 42, 9.
[30] Roussis, P. P. Polymer, 1981, 22, 768.
[31] Roussis, P. P. Polymer, 1981, 22, 1058.
[32] Ohnishi, A.; Maeda, K.; Endo, Y.; Akinaga, Y.; Uehara, M. Senbai Chuken-ho, 1972, 114, 105.
[33] Maeda, K.; Anzai, Y.; Sawakuri, T.; Noguchi, K. Senbai Chuken-ho, 1975, 117, 109.
[34] Kawamoto, J. Jpn. Pat. 5,068,527 (A), 1993.
[35] Baggett, M. S.; Morie, G. P. Beitr. Tabakforsch., 1975, 8, 150.
[36] Burke, J. Book and Paper Group Annual, 1984, 3, 13.
[37] Mrkić, S.; Galić, K.; Ivanković, M.; Hamin, S.; Ciković, N. J. Appl. Polym. Sci., 2006, 99, 1590. 38. Alger, M. M.; Stanley, T. J.; Day, J. Polym. Eng. Sci., 1989, 29, 639.
[39] Fujita, H. Fortschr. Hochpolym. –Forsh., 1961, 3, 1.
[40] Olkhov, A. A.; Vlasov, S. V.; Iordanskii, A. L.; Zaikov, G. E.; Lobo, V. M. M. J. Appl. Polym. Sci., 2003, 90, 1471.
[41] Yasuda, H.; Peterlin, A. J. Appl. Polym. Sci., 1974, 18, 531.
[42] Michaels, A. S.; Bixler, H. J. J. Polym. Sci., 1961, 50, 393.
[43] Villaluenga, J. P. G.; Seoane, B.; Compañ, V. J. Appl. Polym. Sci., 1998, 70, 23.
[44] Compañ, V.; Andrio, A.; López, M. L.; Alvarez, C.; Riande, E. Macromolecules, 1997, 30, 3317.
[45] Wang, L. H.; Porter, R. S. J. Polym. Sci., 1984, 22, 1645.
[46] Nikishin, E. L.; Chalykh, E. A.; Avgonov, A.; Kulenznev, V. N.; Neverov, A. N. Colloid J., 1979, 40, 660.
[47] Vrentas, J. S.; Duda, J. L.; Ni, Y. C. J. Polym. Sci., Polym. Physics Ed., 1977, 15, 2039.
[48] Sano, Y.; Yamamoto, S. J. Chem. Eng. Jpn., 1990, 23, 331.

20

An Inspection of Some Hydrogenated Carbon Fibers by Scanning Electron Microscopy and Confocal Laser Scanning Microscopy

Antonio Madroñero[1] and Jose Mª Amo[2]
[1]*Centro Nacional de Investigaciones Metalúrgicas*
[2]*Consejo Superior de Investigaciones Científicas*
Spain

1. Introduction

The huge and rapid development of hydrogen technology has led to great attention being paid to hydrogen absorbent carbonaceous materials (Dillon and Heben, 2001).

Because of this, all families of carbonaceous materials have been separately examined with the intention to clarify the physical mechanisms of the adsorption of hydrogen, looking for the possibility to optimize their capacity of hydrogen storage (Alcañiz-Monge and Roman-Martinez, 2008). Particularly, the carbonaceous nanostructures have been studied from the point of view of the location of adsorbed hydrogen atoms, the storage mechanisms, and the kinetic of adsorption/desorption (Nechaev, 2006). According to these studies, the carbonaceous nanostructures show to have a limited capacity of hydrogen storage only.

Interesting works have been conducted to establish the details that describe hydrogen insertion in carbonous absorbents. On the one hand, models consisting of the absorption and penetration of molecular hydrogen through the carbon network have been conducted (Seifert, 2004). On the other hand, other studies stress that the amount of absorbed hydrogen per weight depends on the micropore volume of the sample (Takagi et al., 2004). This indicates the importance of being able to examine the surface of the carbonous materials to determine the hydrogen housing locations.

For this study, we have selected vapor grown carbon fibers (VGCFs) as the carbonous material. The fibers were prepared via catalyzed decomposition of a gaseous hydrocarbon that produced a solid carbon deposit in the form of filaments. The germination and growth mechanism is explained as a vapor-liquid-solid (VLS) process (Gorbunov et al., 2002, Madroñero, 1995). Minute ferrocene crystals were utilized as germs to initiate fiber growth, as explained by (Serp et al, 1999). By using this growth method, the iron germ binds to the substrate and the fiber grows as a liquid drop on its tip; in this way, what can be expected is that the fiber will remain free of iron impurities.

Therefore, we have checked that the fibers utilized in this work do not contain iron in spite of them have been developed using ferric seeds (ferrocene). To this end, we have conducted their analysis via inductively coupled plasma mass spectrometry (ICP-MS) with Perkin Elmer® SCIEX ELAN 6000 equipment. Significant values for iron content were not obtained, with the determination limit of the technique, under the utilized working conditions, being 0.1 ppm.

The fibers thus obtained with cylindrical morphology comprise two carbonous phases. There is a central portion of greater crystalline perfection, and a peripheral portion of a more amorphous nature, as can be seen in figure 2.b. Therefore the structure of these fibers was labeled as sword in sheath (Madroñero et al., 1997).
The results of the above studies indicate that solid aromatic hydrocarbons, at the end of the solid hydrocarbon list, are capable of absorbing and storing hydrogen in a uniform manner (Denis, 2008). As the cortical layer of the fibers utilized in this work is formed partially by hydrocarbons of this type, the results of the aforementioned studies establishing that these types of fibers could store significant quantities of hydrogen are perfectly reasonable (Madroñero et al., 1995).

2. Experimental

2.1 Preparation of the fiber samples

The fibers were fabricated in a vertical oven formed by a 10 cm diameter quartz pipe where some strays or substrates were placed vertically, and where some minute grains of ferrocene were then placed. A mixture of 70% hydrogen and 30% methane flowed from the lower section of the oven to the outlet located in the upper section. Thus, the gas passed through the substrates, fostering growth of the carbon microfibers. The process temperature was 1,300 K. The details of the process are given by (Serp et al., 1999).
Subsequently, the density of these fibers was measured with a Micromeritics® make, Accupyc 1330 model gas pycnometer.
Figure 1. shows the cylindrical morphology of these fibers that, as a consequence of the VLS process that occurred during their growth, show continuous changes in the direction of the progression of the fiber tip. It is therefore a discontinuous growth, pursuant to a pull sequence.
In this work, a very simple inspection technique has been selected, consisting of subjecting the fibers to an oxidation process in air; this operation was conducted in the same oven as was utilized to produce the fibers. The oxidation temperature was 973 K with a 0.090 mol/min air flow and variable oxidation times ranging between 2 and 15 min.
With this superficial oxidation attacks were being made, preferably in the areas that were richest in hydrogen, given that we know that the oxidation of carbonous materials is greatly affected by the hydrogen content of the material (Zhou and McGinn, 2006). Moreover, and considering that oxidation is very active over time, successive layers of the material were progressively removed so that observation by SEM and CLSM microscopy allows appreciation of the locations of the hydrogen stored in the structure of the cortical layer in shallow layers, and not only on the surface. These techniques were utilized previously (Merino and Brandl, 2003).
The fabricated fibers have been divided up in order to conduct this study on two types of material. The first group is simply the grown VGCF fibers, without any type of superficial treatment; they showed a density of 1.419 g/cm^3.
The fibers of the second group were subjected to a careful annealing process to eliminate absorbed molecules that could interfere with the hydrogen (Park et al., 1999). After that, these fibers were subsequently subjected to annealing at 1,473 K and 0.5 10^{-3} bar with residual atmosphere of Ar. They had a density of 1,459 g/cm^3 and were labeled as "discharged fibers".

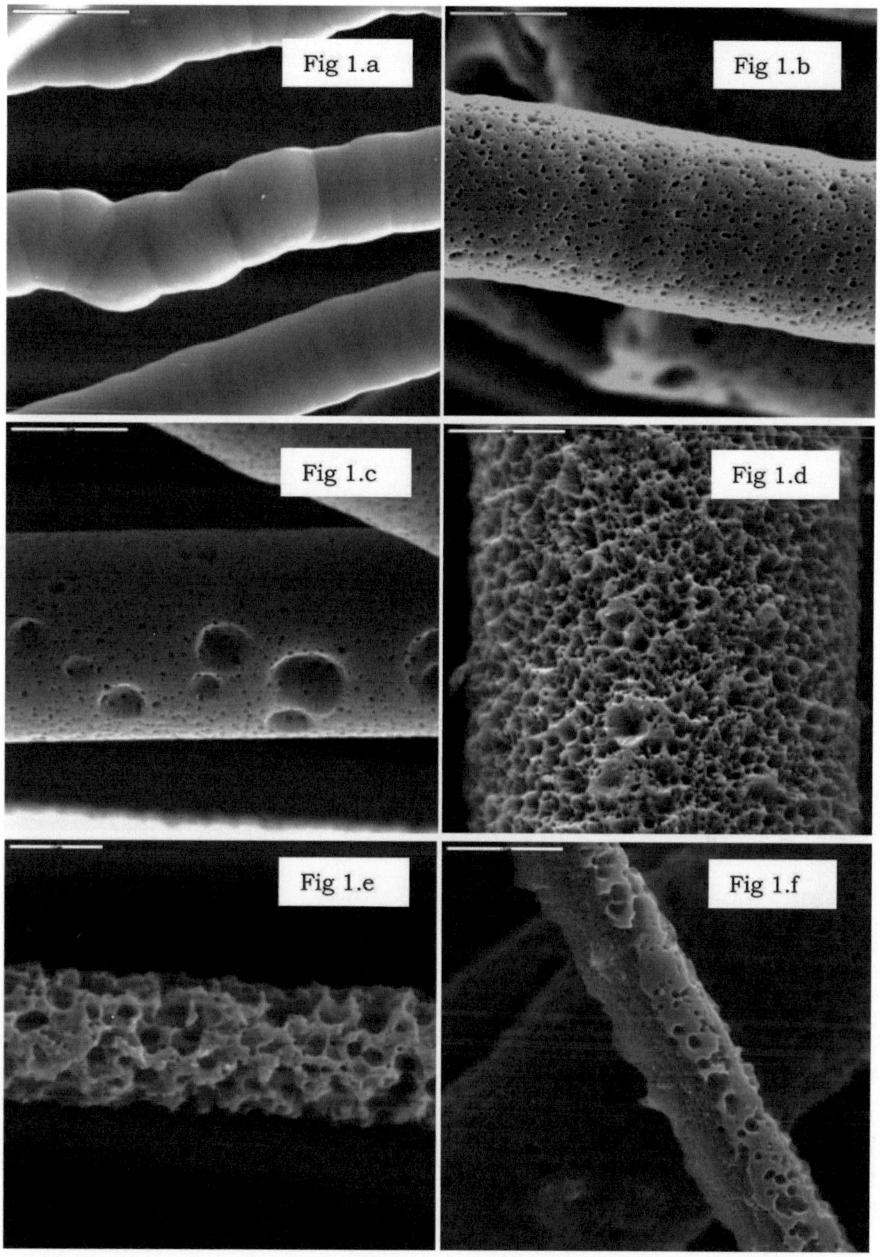

Fig. 1. Oxidation effects in as grown fibres observed by SEM. Scale bars: 1a, 1b, 1c, 1d, 1f = 5 μm; 1e = 2 μm

2.2 Characterization of the fibers by microscopy

2.2.1 Differences between an inspection of the fibers by SEM and by CLSM

SEM inspection is customary in these types of studies as it allows an image of the surface of fibers that show an increasing degree of erosion in such a way that with the entire array of all the images (see Fig. 1), the location of the accumulations of hydrogen on ever more interior layers can be identified. This is because annealing occurs simultaneously to the oxidation process and its effect is the formation of small pittings that mark the point where an accumulation of hydrogen, which had subsequently escaped leaving an empty space, had occurred.

The utilization of CLSM together with SEM has been successfully proven in cases of tracking of the alteration that occurs on some surfaces (Yamada et al., 2004). Results of the utilization of CLSM by reflection with laser light on inspection of glass fibers are known (Clarke and Davidson, 1993). The intensity of the reflected light depends on the angle of reflection and, moreover, thanks to the transparency of glass, the radiation penetrates into the cortical zone and a very interesting 3D image is obtained.

The thickness of the cortical layer [15] is obtained from:

$$d_z = \frac{0.45\lambda}{1-\cos\theta}$$

where λ is the wavelength of the laser radiation utilized, and θ is the angle of reflection.

In the case of VGCF, formed by carbon, what is to be expected is that the radiation can penetrate the thin carbonous layers (Veersamy et al., 2003) with a refraction index that depends on the proportion of sp^2 and sp^3 links, and therefore, what can be expected is that the CLSM will be capable of observing the cortical layer of the fibers, unlike the SEM which is only going to provide us with an image of the surface.

2.2.2 Inspection via SEM

The inspection was performed with Zeiss® model 960 equipment. The VGCF for this study were installed on sample trays subsequently plated with gold in an evaporator.

2.2.2.1 SEM inspection of VGCF fibers labeled as "grown".

Fig. 1.a shows the aspect of said fibers in initial state, from which they will be subjected to a 973 K oxidation process with different treatment times.

After an oxidation time of 2 minutes (see Fig. 1.b.) the outer picture of the fiber shows regular and uniform distribution of multiple pittings that appear to be sized from 0.04 to 0.16 μm.

The aspect of the fibers for which the oxidation time is extended to 4 minutes, the aspect of the fibers is shown in Fig. 1.c.; these fibers in addition to the pittings shown in the previous figure, have pittings that are slightly increased in size and large pittings 2~3 μm. And so we can assume that they have another formation mechanism.

When 8 minutes of oxidation is reached, the surface, as shown in Fig. 1.d., shows that all the pittings grew until the entire surface had been covered. We can still observe the pittings that were visible after 2 minutes of oxidation as well as those corresponding to the large pittings that appear in Fig. 1.c.

The carbonaceous material elimination process reaches its final stage in the state generated by 10-minute oxidation, as indicated in Fig. 1.e.

Fig. 1.f. shows that the matter removal process finishes when the cortical layer of the fiber has been eliminated. For more extended oxidation times, the nucleus will present a slower and more uniform attack without the large pittings that act as a cortical layer destruction mechanism, as we have seen in the previous pictures.

2.2.2.2 SEM inspection of the discharged fibers

Fig. 2.a. shows the initial state of the fibers we have labeled as "discharged".

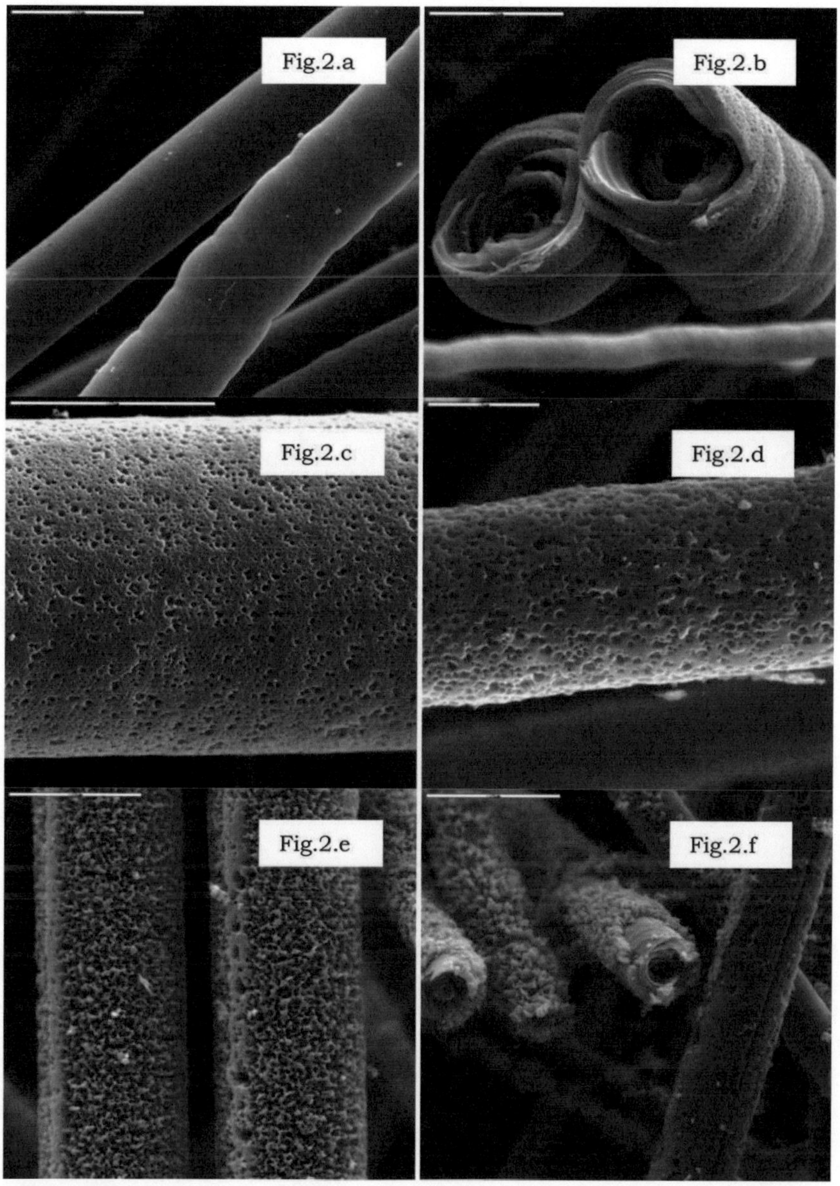

Fig. 2. Oxidation effects in annealed fibres observed by SEM. Scale bars: 2a, 2b, 2e, 2f, = 10 µm; 2c = 20 µm; 2d = 5 µm

Figure 2.b. shows the aspect of the fibers in a situation equivalent to those that had been shown in figure 1.b. The same appearance of small uniformly distributed pittings can be seen.
When 4 minutes have passed, the oxidation process shows differences to the "grown" fibers as can be seen if we compare Fig. 2.c. to Fig. 1.c. The large pittings to be seen in Fig. 1.c. do not form on the "discharged" fibers.
If we examine the effects of the oxidation process after 8 and 10 minutes, we see practically the same in the discharged fibers (Fig. 2.d. and 2.e) as we see in the grown fibers (Fig. 1.d. and 1.e.).
The final of the oxidation process seems similar if we compare Fig. 2.f. to Fig. 1.f., with both corresponding to an oxidation period of 15 min.

2.2.3 Inspection using confocal microscopy by reflection

Figures 3.a. and 3.b. show the images obtained with confocal microscopy of the fibers studied in this work. They refer to the same samples as in figures 1.a. and 2.a., and we can see that there is a certain degree of irregularity on the discharged fiber's surface that was not to be seen on the grown fibers.
The information provided by confocal microscopy observation of fibers subjected to oxidation during 4 minutes is very interesting. If we compare Fig. 3.c. to Fig. 1.c. we see that the large pittings are a result of the complete elimination of a carbon grain while, if we compare Fig. 3.d. to Fig. 2.c. we can see that the pittings are located on boundaries between grains. This information is not deducible from the images provided by SEM microscopy.
After 8 min of oxidation, the width of the furrows between grains has been increased (Fig. 3.e. and 3.f).
This is because, with the use of confocal microscopy the same evolution of the oxidation process as with SEM microscopy was seen, but with optic microscopy we obtain information indicating that oxidation is preferential in areas located between grains.

3. Discussion of the results

From all the above, it appears that in the fabrication of VGCFs, two types of defects appear: some minute and homogeneously distributed, and others that are more voluminous, scarcer and more irregularly distributed.
This scheme coincides with that already known via research in nuclear materials, where it was proven that hydrogen locates itself inside graphite in two types of housing (Atsumi, 2003), some trappings inside some grains with an activation energy for the escape of hydrogen from these 2.6~4.4 eV accumulations, and other trappings located on the boundary between grains with 1.3 eV energy. Therefore, the storage of hydrogen in these accumulations must be more intense.
This model of inhomogeneity in storage is more congruent with the storage model in a sp^3 type carbon matrix, the phase constituted mostly by semi-amorphous carbons. In this matrix grainier sp^2 type islets appear (Dasgupta et al., 1991).
The double structure of the network constituted by carbon atoms is well known. sp^2-bonded carbon is localized in graphitic clusters which consist of four or more fused sixfold rings (Robertson, 1986). These clusters are connected by sp^3-bonded carbon.

Large voids in curved form where hydrogen is stored (Gupta and Srivastava, 2001) are also described in other graphite materials conformed in fibrous form.
We have also been able to check that when a graphite material is subjected to irradiation with neutrons (Atsumi et al., 1996), as the size of the grains diminishes, the hydrogen storage capacity increases as the number of boundaries increases. The content in hydrogen can be up to 20 times higher.
Therefore, both types of accumulations can be identified with the cavities shown in Fig. 1.c. To this end, a not too prolonged oxidation time freed the hydrogen from the accumulations on the boundaries, but not the hydrogen stored in the more voluminous accumulations located in the center of a grain.
The images obtained with CLSM are excellent proof of these models. For example, in Fig. 3.c. related to grown fibers, two types of accumulations can be clearly distinguished. With the same oxidation time, Fig. 3.d. shows us that in the case of discharged fibers the boundaries between grains are wider. This is due to the fact that there was take out of hydrogen in the discharged fibers samples were during the annealing.
This granular image of the skin of the VGCF coincides with that which was already known as regards its microstructure via previous studies conducted with transmission electron microscopy TEM. Therefore, it is known (Madroñero et al., 1996) that the fine structure consists of some 50 nm plates that compact to form curved plates of some 250 nm thickness and few micrometers wide (Madroñero et al., 1998).
These macrostructures now enter within the limit of optic microscopy, as we have seen through the images shown in Fig. 3.
Nonetheless, the possibilities of CLSM do not cease here as the use of ultraviolet radiation while functionalizing the surface of fibers with ammonium (Lacerda et al., 2007), would allow the obtaining of images with fluorescence that could enable the provision of additional information in the future.
In any case, the possibilities of a technique as simple as the CLSM are remarkable. In the present study is seen how the structure of fibers grown from a hydrocarbon gas is clearly different of the grain observed in commercial carbon fibers. As an example, in Fig. 4 is shown the microstructure of an ex-PAN carbon fiber, while Fig. 5 shows the features of an ex-Pitch fibre. Both types of fibers are very representative of the carbon fibers that are used currently in the industry to manufacture composite parts.

4. Conclusions

1. Regarding the aspect of oxidized fibers, it can be said that the oxidation of these materials occurs at the points of dye corrosion where the hydrogen is concentrated. Therefore, the corrosion figures can be used for drawing the location map of the hydrogen accumulations.
2. In the problem of hydrogen location in the fibers studied in this work, it is interesting to note the existence of two types of accumulations revealed by the studies performed to understand the behavior of carbons in nuclear reactors.
3. The utility of confocal laser optical microscopy as a powerful tool to allow tracking of the increase in width of the boundaries while being abandoned by the resorbed hydrogen has been proven to be effective.

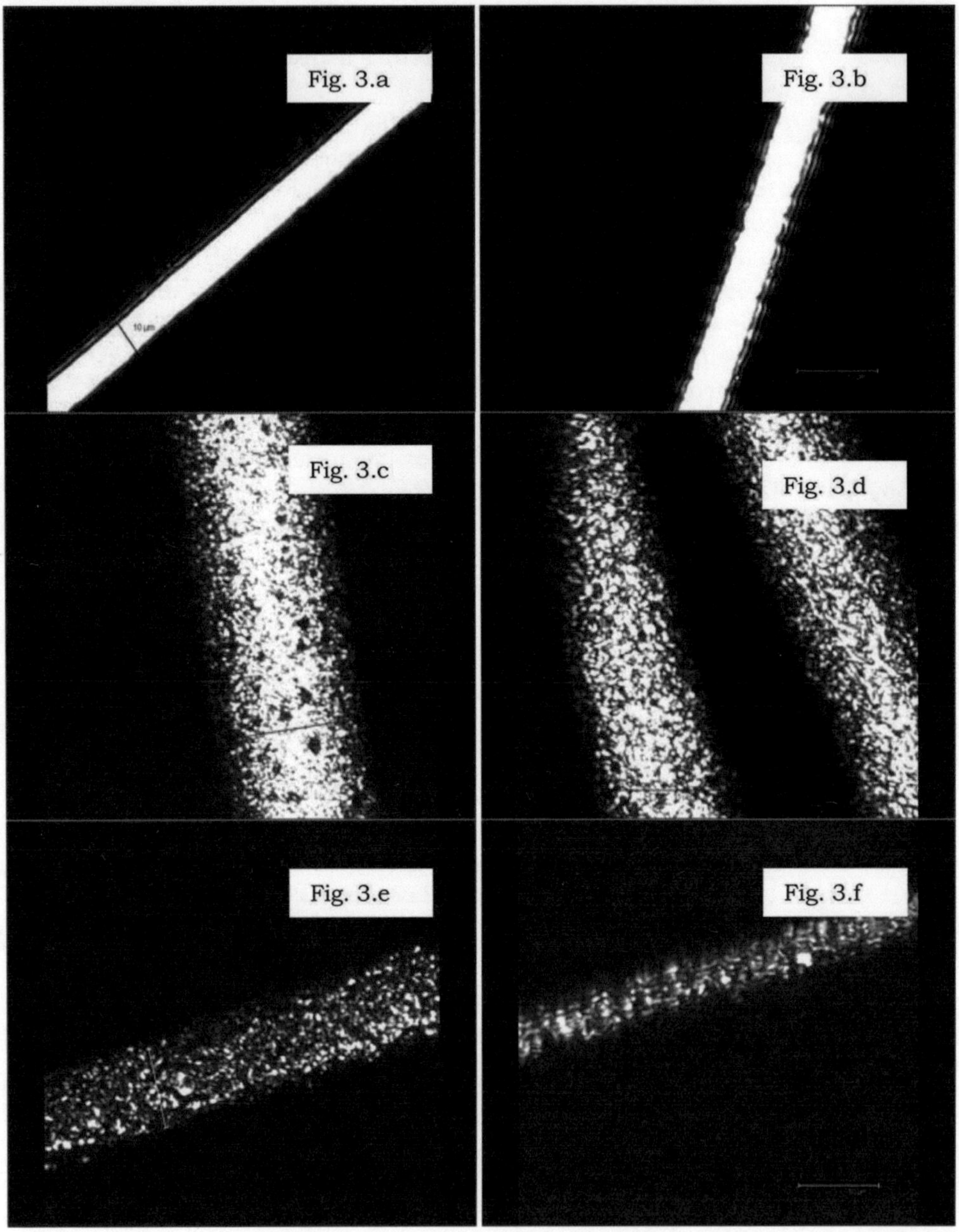

Fig. 3. Oxidation effects observed by confocal laser scanning microscopy. Scale bars: 3a, 3b, 3e, 3f, = 10 µm; 3c = 20 µm; 3d = 15 µm

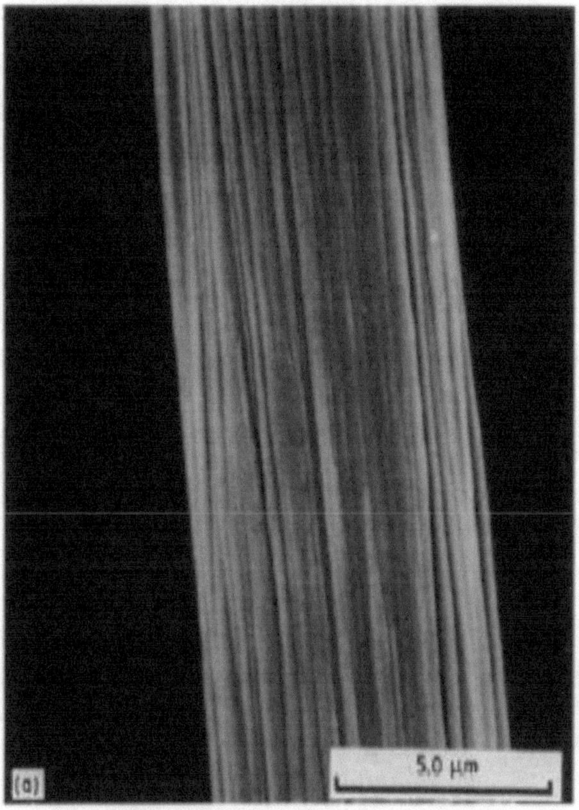

(a)

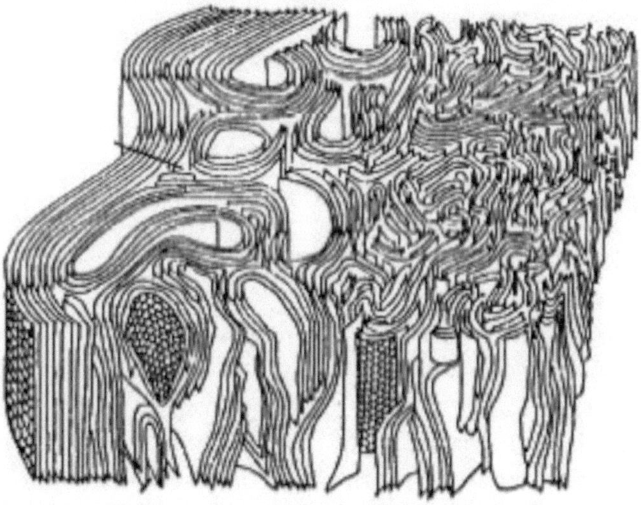

(b)

Fig. 4. Microstructure of a commercial carbon fibre type ex-PAN. a = SEM micrograph, b = inner architecture

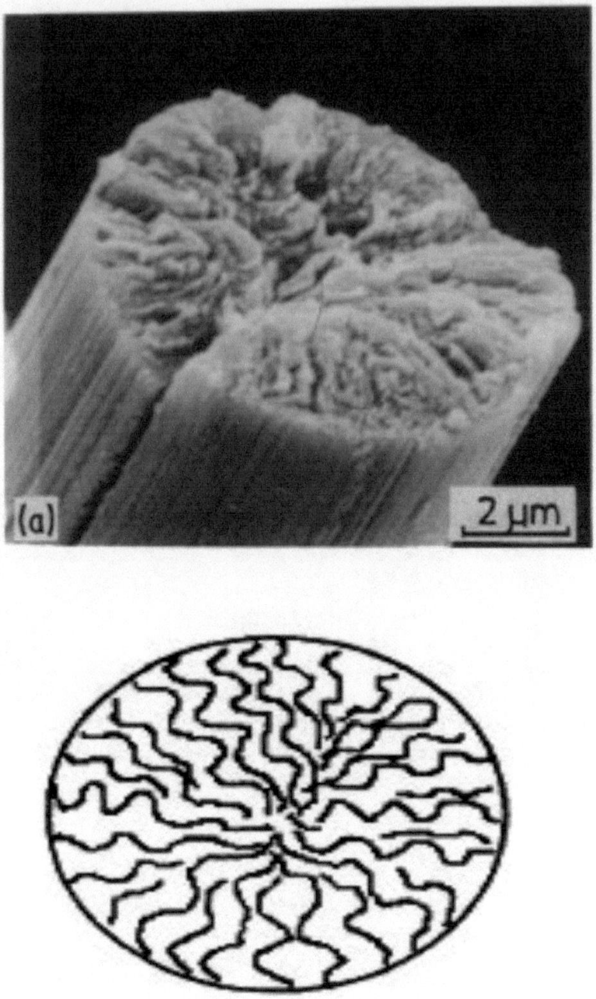

(b)

Fig. 5. Microstructure of a commercial carbon fibre type ex-pitch. a = SEM micrograph, b = inner architecture

4. The granular structure observed using confocal laser optical microscopy is in good agreement with the microstructure described in previous studies with TEM.
5. Unlike other carbonaceous materials that lose their hydrogen content slowly, the fibers studied in this work empty their deposits in a few minutes, provided they are annealed at the temperatures that have been utilized for thermal treatment processes in this study.

5. Acknowledgement

This work was performed with funds of the Project "Tailoring electronic and phononic properties of nanomaterials; towards improved thermoelectricity (nanoTHERM)", included in the Consolider-Ingenio Program of the Spanish Ministry of Science and Innovation.

6. References

Alcañiz-Monge J, Roma´n-Martinez MC. 2008. Upper limit of hydrogen adsorption on activated carbons at room temperature: A thermodynamic approach to understand the hydrogen adsorption on microporous carbons. Microporous Mesoporous Mater 112:510–520.

Atsumi H, Iseki M, Shikama T. 1996. Hydrogen behavior in carbon based materials and its neutron irradiation effect. J Nuclear Mater 233–237:1128–1132.

Atsumi H. 2003. Hydrogen retention in graphite and carbon materials under a fusion reactor environment. J Nuclear Mater 313–316:543–547.

Clarke A, Davidson N. 1993. Measurements of fibre direction in reinforced polymer composites. J Microsc 171:69–79.

Dasgupta D, Demichelis F, Tagliaferro A. 1991. Electrical conductivity of amorphous carbon and amorphous hydrogenated carbon. Phil Mag B 63:1255–1266.

Denis PA. 2008. Investigation of H_2 physisorption on corannulene ($C_{20}H_{10}$), tetraindenocorannulene ($C_{44}H_{18}$), pentaindenocorannulene ($C_{50}H_{20}$). C_{60}, and their nitrogen derivates. J Phys Chem C 112:2791–2796.

Dillon AC, Heben MJ. 2001. Hydrogen storage using carbon adsorbents: Past, present and future. Appl Phys A-Mater Sci Process 72:133–142.

Gorbunov A, Jost O, Pompe W, Graff A. 2002. Solid-liquid-solid growth mechanism of single-wall carbon nanotubes. Carbon 40:113–118.

Gupta BK, Srivastava ON. 2001. Further studies on microstructural characterization and hydrogenation behaviour of graphitic nanofibres. Int J Hydrog Energy 26:857–862.

Kino GS, Corle TR. 1989. Confocal scanning optical microscopy. Phys Today 42:55–62.

Lacerda L, Pastorin G, Gathercole D, Budle J, Prato M, Bianco A, Kostarelos K. 2007. Intracellular traficking of carbon nanotubes by confocal laser scanning microscopy. Adv Mater 19:1480–1484.

Madroñero A. 1995. Possibilities for the vapour-liquid-solid model in the vapour-grown carbon fibre growth process. J Mater Sci 30: 2061–2066.

Madroñero A, Ariza E, Verdú M. 1995. Adsorption of hydrogen in carbon fibres made from polyacrylonitrile precursor, and vapour grown. Eur J Solid State Inorg Chem 32:1115–1127.

Madroñero A, Ariza E, Verdú M, Brandl W, Barba C. 1996. Some microstructural aspects of vapour-grown carbon fibres to disclose their failure mechanisms. J Mater Sci 31:6189–6193.

Madroñero A, Verdú M, Froten L, Domínguez M. 1997. A diffusion model for sword in sheath failure mode in vapour grown carbon fibers. Adv Perform Mater 4:305–315.

Madroñero A., Verdú M, Issi JP, Martin-Benito Romero J, Barba C. 1998. Study by a modified scanning electron microscope fractography of hydrogenation process in vapour grown carbon fibres. J Mater Sci 33:2079–2085.

Merino C, Brandl W. 2003. Oxidation behaviour and microstructure of vapour grown carbon fibres. Solid State Sci 5:663–668.

Nechaev YUS. 2006. Hydrogen sorption by carbon nanostructures: Its nature, kinetics, and limiting capacity. Physics-Uspekhi 49:563–572.

Park C, Anderson PE, Chambers A, Tan CD, Hidalgo R, Rodriguez NM. 1999. Further studies of the interaction of hydrogen with graphite nanofibers. J Phys Chem B 103:10572–10581.

Robertson J. 1986. Amorphous carbon. Adv Phys 35:317–374.

Seifert G. 2004. Hydrogen on and in carbon nanostructures. Solid State Ion 168:265–269.

Serp PH, Madroñero A, Figueiredo JL. 1999. Production of vapour grown carbon fibres: Influence of the catalyst precursor and operating conditions. Fuel 78:837–844.

Takagi H, Hatori H, Soneda Y, Yoshizawa N, Yamada Y. 2004. Adsorptive hydrogen storage in carbon and porous materials. Mater Sci Eng B-Solid State Mater Adv Technol 108:143–147.

Veerasamy VS, Luten HA, Petrmichl RH, Thomsen SV. 2003. Diamonlike amorphous carbon coatings for large areas of glass. Thin Solid Films 442:1–10.

Yamada MK, Uo M, Ohkawa S, Akasaka T, Watari F. 2004. Non-contact surface morphology analysis of CO2 laser-irradiated teeth by scanning electron microscope and confocal laser scanning microscope. Mater Trans 45:1033–1040.

Zhou C, McGinn PJ. 2006. The effect of oxygen on the processing of mesocarbon microbeads to high-density carbon. Carbon 44:1673– 1681.

21

Confocal Laser Scanning Microscope: A Very Useful Tool in Multilayer Capsules

Liqin Ge
State Key Laboratory of Bioelectronics, Biological Science and Medical Engineering Department, Southeast University, Nanjing 210096, P. R. China

1. Introduction

Liposomes are widely used as drug carriers to deliver functional compounds into the membrane. [1-3] However, the liposomes from phospholipids are not stable even in the living system. Much attention has been drawn to the mixed polyelectrolyte/lipid system and all of these methods have been commonly recognized to stabilize the mixed film and assigned lipids some new features. [4-6] In order to prepare stabilized vesicles with a defined size, an alternative strategy has been developed by employing layer-by-layer technique. PE multilayers were assembled on the colloidal particles and the colloidal particles were selectivley sacrificed by dissolution to obtain hollow multilayered PE capsules. Phospholipids are adsorbed on these capsules. By this way, giant liposomes can be obtained with a defined size, shape and such giant liposomes are very stable from months to years. [7-19]
Based on our past work on the successful fabricating giant liposomes, a deeper and systematic understanding of the multilayer versatility regarding their functions and properties is clearly necessary. It was possible to change the permeability by pH, salt concentration and temperature for a limited number of systems. [8-10] In this paper, multilayered PE capsules fabricated with Layer-By-Layer (LBL) method were employed as supports and then the lipid bilayers were assembled on them, thus giant liposomes were obtained and their sizes were determined by the supports' size. Their permeabilities will be tuned by the lipid coating and the surface enzyme catalysis reaction. The small fluorescence molecule 6-CF diffusion behavior across the capsule's wall was studied by Confocal Laser Scanning Microscopy (CLSM), which may give information on the permeability control of "artificial cell". All the experimental results were got at room temperature.

2. Materials and methods

2.1 Materials

The sources of chemicals were as follows: poly(styrenesulfonate, sodium salt) (PSS, Mw 70,000) and poly(allylamine hydrochloride) (PAH, Mw 70,000) were from Aldrich. L-a-Dimyristoylphosphatidic acid (DMPA), 6-carboxyfluorescein (6-CF) were purchased from Sigma. All these materials were used as received.

The water used in all experiments was prepared in a three-stage Millipore Milli-Q Plus 185 purification system and its resistivity is higher than 18.2 MΩ cm^{-1}.
Positively charged melamine formaldehyde particles (MF particles) with a diameter of around 2.85 ± 0.09 µm were obtained from Microparticles GmbH, Berlin.

2.2 Polyelectrolyte shells prepared through layer-by-layer adsorption

Multilayer assembly was accomplished by the adsorption of polyelectrolyte at a concentration of 1 mg/ml, 0.5 M NaCl aqueous solutions. Oppositely charged polyelectrolyte species were subsequently added to the suspension followed by repeated centrifugation cycles. After the expected number of layers was adsorbed, 0.1 M HCl was used to remove the core (MF particles) and the obtained hollow polyelectrolyte shells were obtained. [14] PSS is used to form the first layer and the outermost layer is PAH, which is positively charged in order to bind the negatively charged phospholipids in the next step.

2.3 Preparation of phospholipid solution

The phospholipid DMPA was dissolved in a mixed solvent of chloroform and methanol (V_{CHCl3}:V_{MeOH} =1:1) with a concentration of 0.5 mg/ml. After the solvent was evaporated in a rotavap at 30 °C, water was added to a final lipid concentration (C_{DMPA} =0.025 mg/ml), and sonicated for 5 min. Lipid solution was added into polyelectrolyte shells and allowed 5 min for adsorption. The mixed solution was washed for three times with water in order to remove the non-adsorbed materials by centrifugation.

2.4 Measurements of confocal laser scanning microscopy (CLSM)

The images of capsules were obtained by a Leica confocal scanning system. A 100* oil immersion objective with a numerical aperture of 1.4 was used. In the experiment, 6-Carboxyfluorescein (6-CF) was selected as hydrophilic fluorescence dyes. The Raman spectra before and after reaction were also measured with CLSM.

2.5 Measurements of fluorescence recovery after photobleaching (FRAP)

In order to study the penetration of the dyes across the wall, the dyes inside the capsules were photo-chemically bleached. To do this, an argon ion laser from the CLSM emitting at a wavelength of λ= 488 nm was used. The laser beam was focused onto the selected area inside the capsule. The time for bleaching was long enough to ensure that almost all dye molecules in the selected area were bleached. Imaging was typically performed at rather low laser intensity. The interval between each image scan varied with the duration of the recovery established at an initial pilot experiment. Recovery was considered completely when the intensity of the photo-bleached region was stable and the curve was flat. For quantitative analysis, the fluorescence intensity was integrated by tracing a fixed area in the interior (ROI analysis system provided by the CLSM software), giving an intensity value for each time point.[20]
It should be pointed out that there is always a small amount of fluorescence dyes lost during repeatedly scanning and recovering. However, this will not influence the experimental results.

2.6 Measurements of single particle light scattering (SPLS)

Measurements were conducted on a home-made photometer equipped with an argon laser, Innova 305 from Coherent, with power track. The equipment has been described earlier. [11]

The dispersion is pressed through a thin capillary with an orifice of 0.1 mm at the end. Hydrodynamic focusing is applied. Particle concentration is adjusted to minimize coincidences. The scattering volume is 1.7×10^{-9} cm^{-3}. Forward scattered light pulses recorded from particles flowing through the scattered volume are detected between 5° and 10°. Intensity distributions are obtained with a resolution of about 0.5%.
The Rayleigh-Debye-Grans theory was used for evaluating the scattering data. The refractive index of the melamine core was assumed as 1.53. This value was obtained by extrapolating concentration dependent measurements of the refractive index of melamine formaldehyde solutions by means of an Abe refractometer. The refractive index of the polyelectrolyte multilayer was taken as 1.47. [22] The scattering intensity distribution of the core particles $\varphi(I)$ is converted into a distribution of the particle radius $\varphi(r)$. To this aim the recorded intensity is calculated as the superposition from the scattered light from the interior of the shell wall.

3. Results and discussions

3.1 Diffusion study of multilayered PE capsules

Fig.1a shows CLSM images of the obtained multilayered $(PAH/PSS)_5$ capsules, the intact capsules are the closed cycles (white arrow). The small fluorescence molecule 6-CF solution was mixed with the capsules' solution in order to study the capsule's permeability. Fig. 1b gives out a further magnification CLSM image. From the picture, one can see that both the inside and the outside of the capsules are bright, i.e., filled with fluorescence molecules, which present that the capsules are permeable to small molecules 6-CF. There are unclosed cycles (black arrow) which present the capsules are imperfect. AFM measurements will give more details about intact and imperfect capsules, shown in Fig.2. The brightest part is because of the water evaporation which results in the wall collapse of the capsules. Comparing Fig.2a with 2b, one can find that the intact capsule is a closed ball while the imperfect capsule is not a closed ball. Since the permeability of intact capsules can only be controlled by environment, we will study them in our research work.

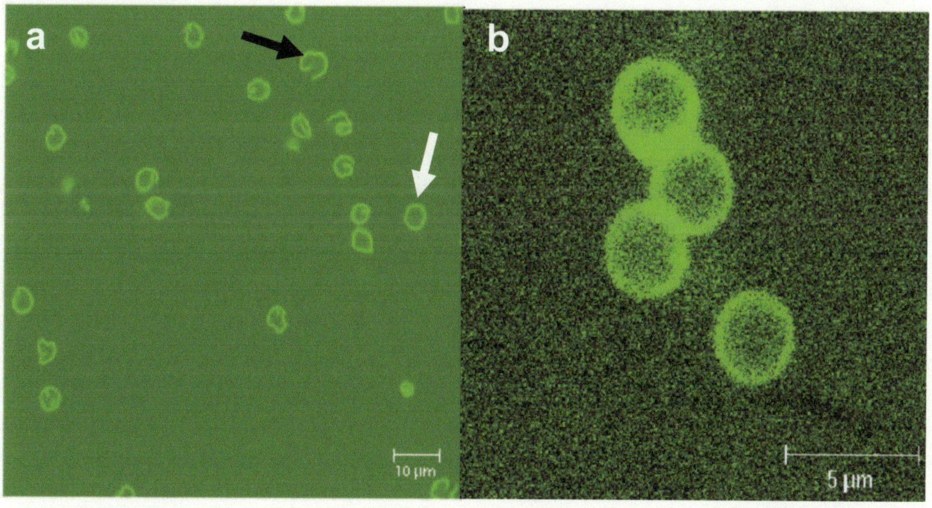

Fig. 1. CLSM images of multilayered PE capsules (a) many imperfact capsules (b) intact capsules

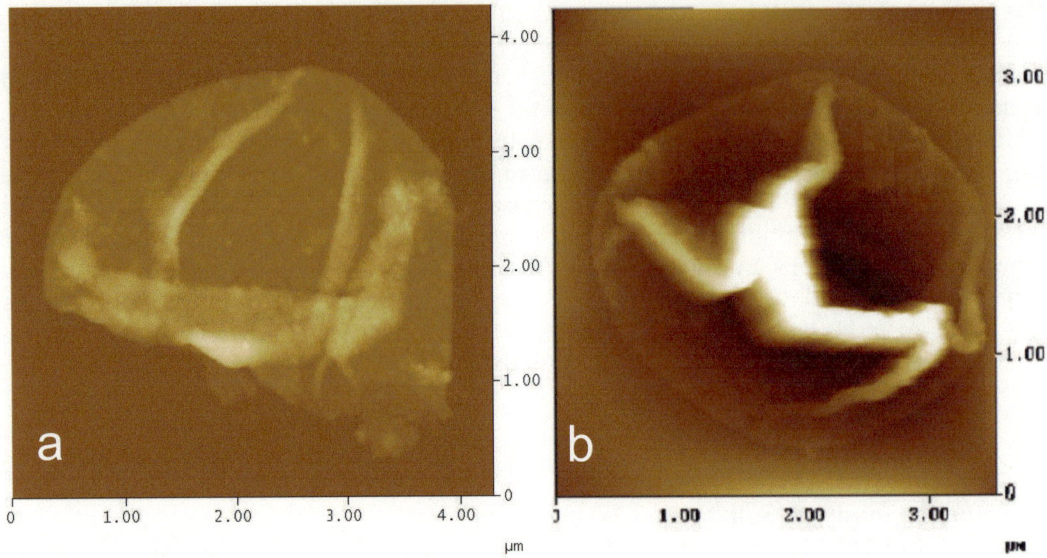

Fig. 2. AFM images of (a) un-intact capsules (b) intact capsules

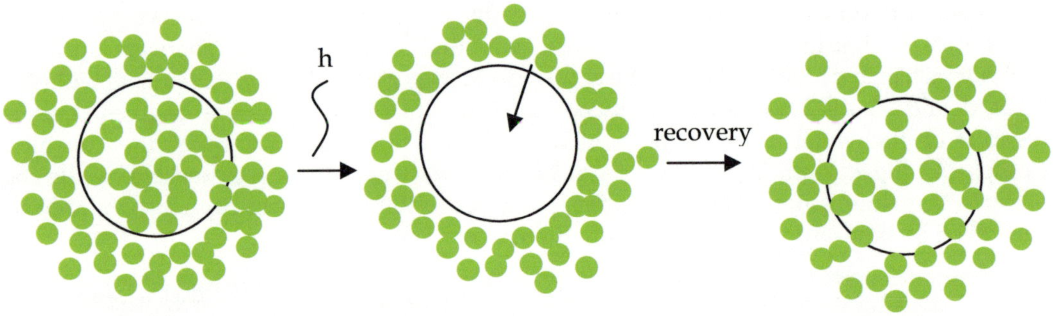

Fig. 3. Scheme for FRAP measurement

To quantify the penetration of the dye molecules, the dyes inside the capsule were bleached photochemically and the fluorescence intensity was recorded due to the penetration of the dye molecules from outside into the capsules with the time. [20] Fig. 3 gives the scheme of the FRAP (fluorescence recovery after photobleaching) experiment. Only permeable capsules can be carried such experiment. Firstly, a permeable capsule was grasped; secondly, the laser was focused onto the spot inside of the capsule, with a maximum bleaching power, most of the dye molecules lost their fluorescence. A dark center inside the capsule would be observed. The third step was to stop bleach the inside of the capsules; and the fourth step is that the outside fluorescence molecules will diffuse through the capsule's wall to reach the inside of the capsules; finally, after a certain time, the fluorescence molecules will arrive an equilibration state between inside and out side of the capsules. The recovery of the fluorescence intensity inside the capsules as a function with time will be recorded automatically by CLSM. Selected CLSM images were shown in Fig.4 during FRAP process and every image was correspond to one step, respectively.

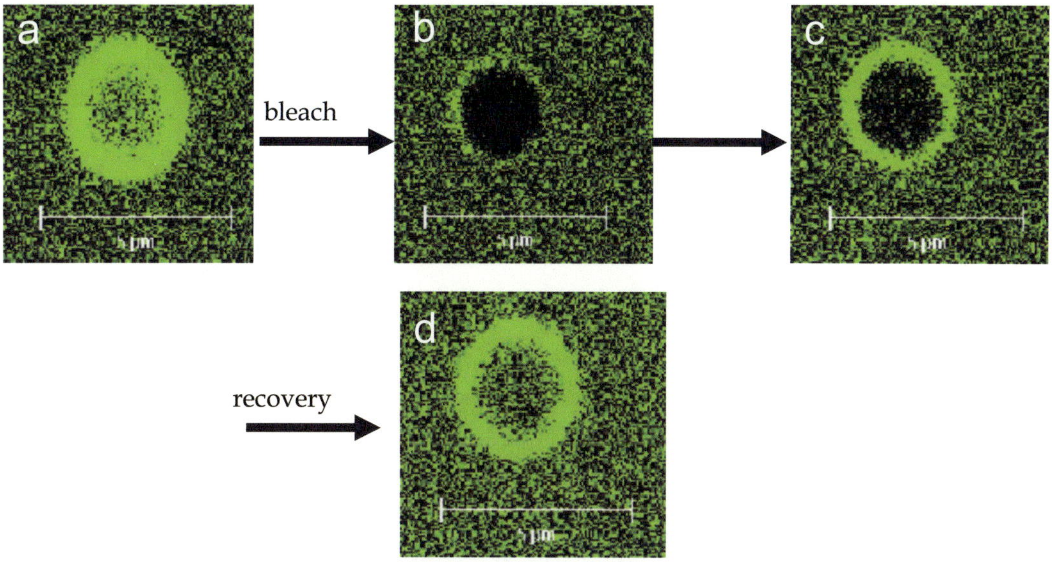

Fig. 4. CLSM images selected from FRAP measurements (a) original permeable capsule (b) bleached capsule (c) recovering capsule (d) recovered capsule

In order to quantity the diffusion coefficient for the small dye 6-CF passing through the multilayered wall, the obtained function between fluorescence intensity to the recovery time will be fitted as shown in Fig.5 (the fitted curve is marked as red line) and they can be described by Equation (1) [20-22]

$$I = I_0(1 - \exp(-t / T_D) \quad (1)$$

Where I_0, I presents the original fluorescence and final equilibrium fluorescence intensity, respectively. T_D is the curve's slope.
In solution, it will obey Fick law, and it can be written as the following Equation (2):

$$\frac{dc}{dt} = -A(C_0 - C) \quad (2)$$

Where C_0, C presents the dye concentration outside the capsules and inside the capsules, respectively. In solution, when small dye molecules pass through a wall with a certain thickness and diameter, A is equal to $1/T_D$. So, for the capsule's diffusion coefficient can be written as the Equation (3):

$$A = 3D / rh \quad (3)$$

Where r is the capsule's radius, h is the capsule's thickness, and D is the diffusion coefficient. Among them, the value for A can be obtained in the fitted function. r can be obtained from the template, h can be obtained by single particle laser scattering (SPLS) or roughly from Atomic Force Microscopy(AFM). For Fig.5, the A=2.05, r is about 2.85 ± 0.09μm, h is about 20 ± 5nm which is measured by SPLS shown in Fig.6 in this study. The diffusion coefficient can be obtained:it is 0.59×10^{-12} cm^2/S, which is higher than the one on the flat surface but is consistent with the reference. [13] The possible reason behind is that the structure of the capsule's wall (multilayered PE films) was damaged during the removal the core.

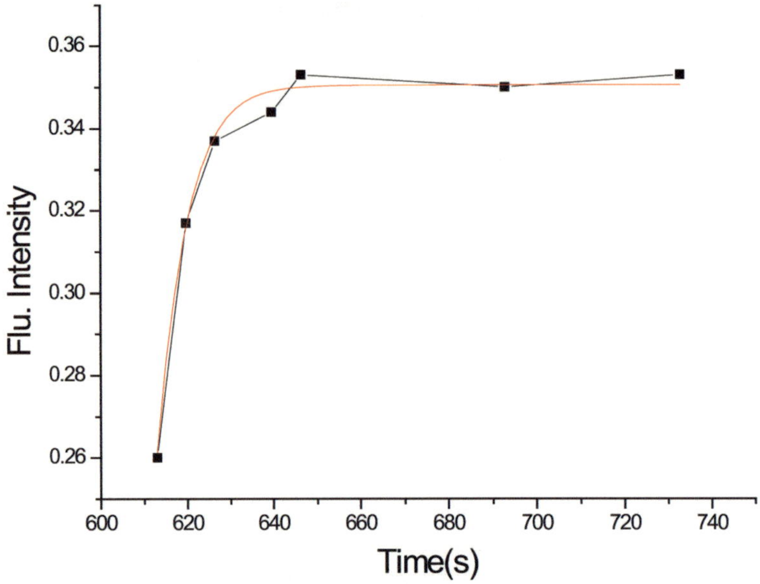

Fig. 5. Recovery profile for pure PE capsule measured by CLSM (black dot is recorded by CLSM, red line is the fitted curve)

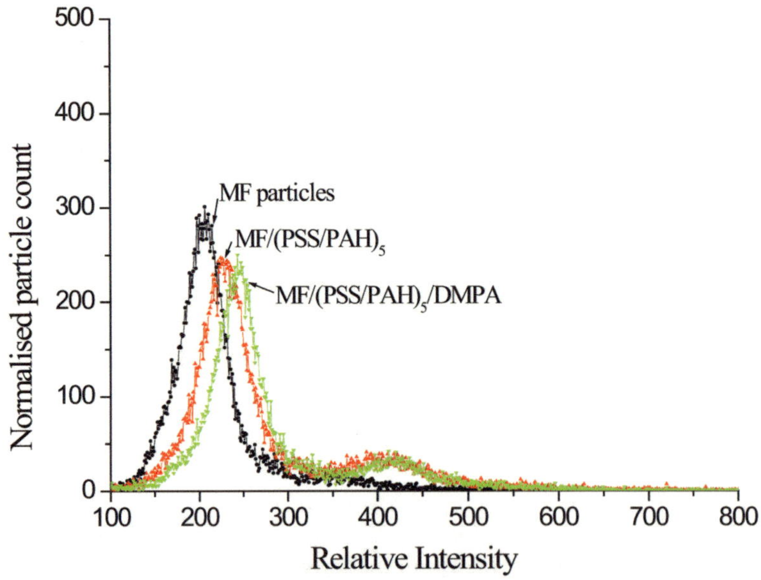

Fig. 6. Normalized light scattering intensity distributions of uncoated MF particles, MF coated with $(PSS/PAH)_5$ and $MF/(PSS/PAH)_5/DMPA$, respectively. The diameter of MF particles used in experiment is 1.08 ± 0.10 μm.

3.2 Assembly of phospholipid vesicles onto polyelectrolyte capsules

According to our previous research, [11-14] one can control the capsules' permeability to small molecules 6-CF by partly coated by phospholipids, which is still permeable to 6-CF but the capsules' surface will become more smooth than the pure multilayered PE capsules, i.e., the mean roughness of the pure multilayered capsules decreases from about 10 nm to 5-6 nm after the phospholipid partly coating on the capsules (the scheme shown in Fig. 7). [12] In this paper, we will study the diffusion behavior for small fluorescence molecules passing through the phospholipid partly coated capsules' wall by CLSM. The relationships between the recovery fluorescence intensity and the recovery time (black line) and the fitted function (red line) are shown in Fig.8. According to the Equation (3), one can obtained the diffusion coefficient is 0.49×10^{-13} cm^2/S. This value is much smaller than that of the pure PE capsules, therefore, the phopholipids on the surface prevent the dye passing through the wall quickly.

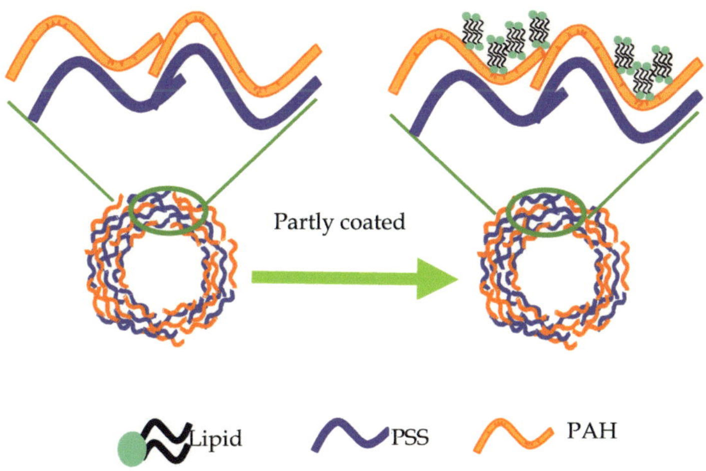

Fig. 7. Scheme for lipid partly coated PE capsules

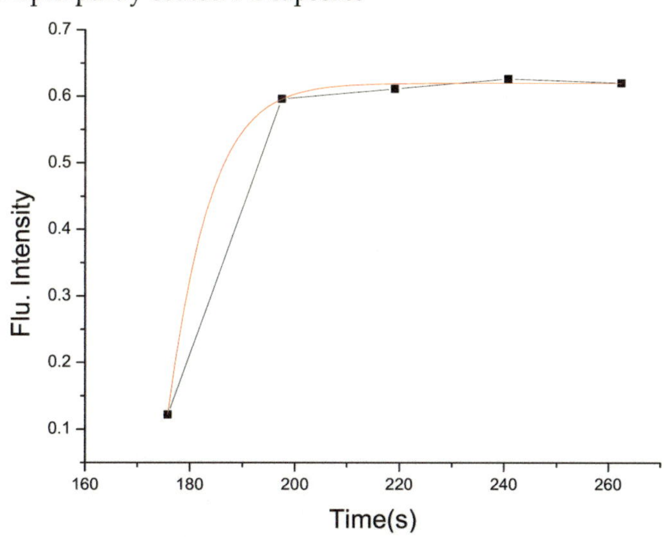

Fig. 8. Recovery profile for lipid partly coated PE capsules measured by CLSM (black dot is recorded by CLSM, red line is the fitted curve)

3.3 Lipid fully coated PE capsules in EtOH solution

According to our previous results [11-14], pure PE capsules are permeable to small molecules 6-CF, the fully coated with phospholipids capsules are impermeable to 6-CF. i.e., one can control the capsule's permeability to 6-CF by fully coated with phospholipids. The phospholipid fully coated capsules' surfaces become much smoother than the pure multilayered PE capsules', [12] the mean roughness decreased from about 10nm for the pure PE capsules to 3-4 nm for the lipid fully coated capsules measured by AFM. However, if such phospholipid fully coating multilayered capsules were treated with EtOH, the intact structure of the phopholipids on the surface will be destroyed and the capsules become permeable again to 6-CF. As a result one can realize the task to tune of the capsule's permeability. Small molecules 6-CF to such capsules' diffusion behavior was studied by CLSM and the recovery fluorescence intensity with a function of time was recorded in fig. 9 as black line and the red line is fitted function. According to Equation (3), the diffusion coefficient is 0.67×10^{-12} cm^2/S, it is similar to that of pure multilayered PE capsules. From this result, one also can see that after treated with EtOH, the intact structure of the lipids was destroyed completely which changes the capsule's permeability.

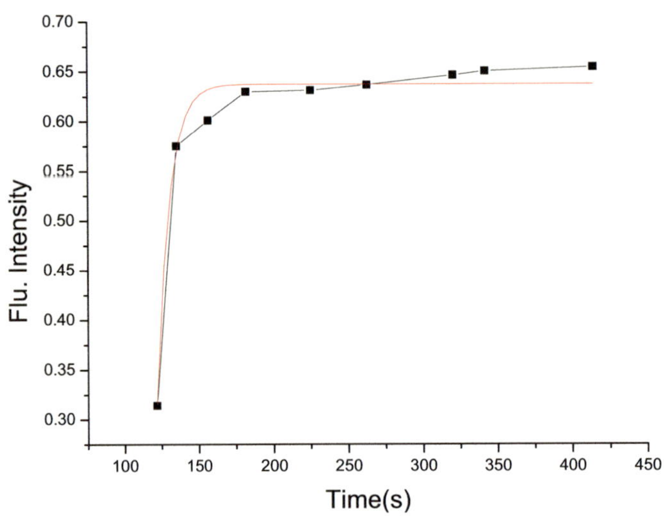

Fig. 9. Recovery profile for the product from EtOH treated lipid completely coated PE capsules (black dot is recorded by CLSM, red line is the fitted curve)

3.4 Lipid fully coated PE capsules catalyzed by PLA_2

PLA_2 is a small, stereo-selective, calcium-dependent enzyme that hydrolyzes the sn-2 ester linkage of phosphatidylcholine.[23-24] In the present work, we chose the mixed-lipid system, l-α-dimyristoylphosphatidic acid (DMPA), which is less hydrolyzed by PLA2 but is better suited for the attraction to the charged polyelectrolyte by electrostatic interaction, and l-α-dipalmitoylphosphatidylcholine (l-DPPC), which is easily cleaved by PLA2 (Fig.10), to cover the capsule's surface and obtained lipid coated multilayered capsules which are impermeable to 6-CF. By making use of the selectivity of the cleavage reaction, there are channels produced on the surface of hollow polyelectrolyte capsules to enable controlled the capsule's permeability, i.e., the product multilayered capsules are permeable to 6-CF. Confocal laser raman spectra (Fig.11) were employed to confirm that after enzyme catalyst reaction one of the products: fatty acid was removed after washing by centrifugation because no special -COOH stretch peak at 1750 cm^{-1} of fatty acid was observed. To study the 6-CF molecules passing through the product capsule's wall, FRAP measurement by CLSM was carried out and the result was shown in Fig.12 (black line is the recovery fluorescence intensity with a function of time, red line is fitted function for the black one). According to Equation (3), the diffusion coefficient is 0.55×10^{-13} cm^2/S. This value is about an order of

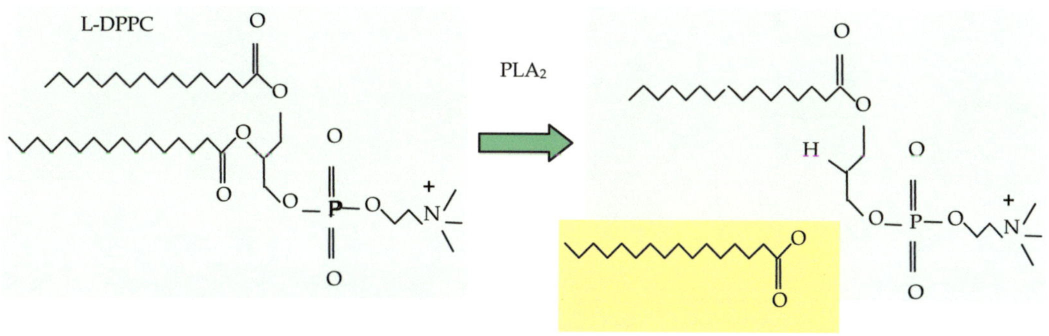

Fig. 10. Scheme for PLA_2 catalyzed l-DPPC reaction

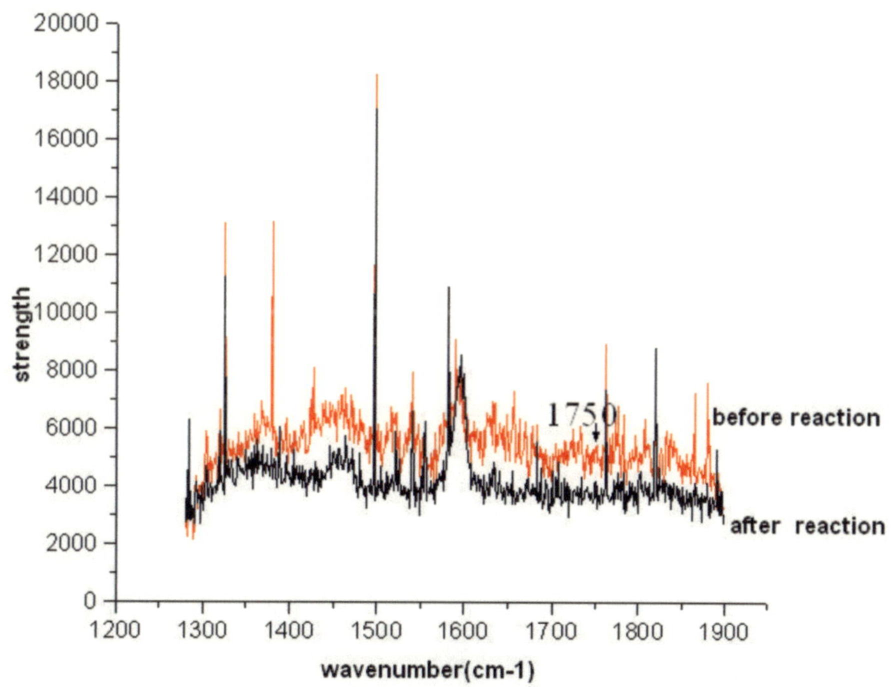

Fig. 11. Confocal laser raman spectra of capsules before and after reaction

magnitude lower than the value measured for the diffusion of dyes across polyelectrolyte multilayer capsules. It indicates that there may be some sealing of holes in the polyelectrolyte by the products of the reaction that have not been removed. [11]

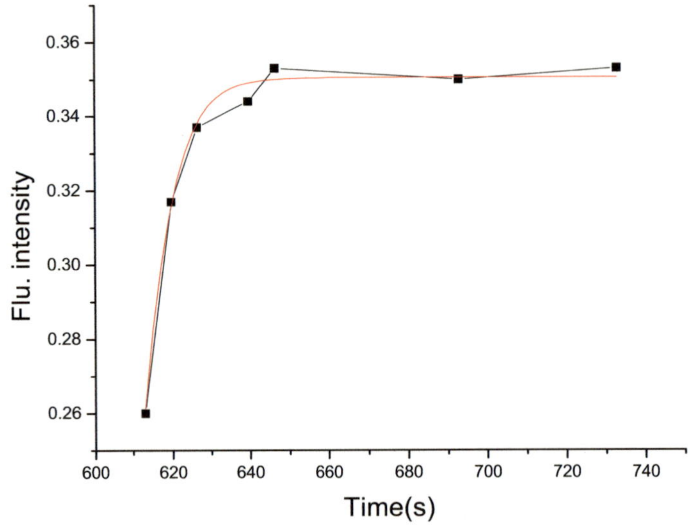

Fig. 12. Recovery profile for the product from PLA_2 catalyzed reaction on PE capsules (black dot is recorded by CLSM, red line is the fitted curve)

4. Conclusions

In this paper, Layer-By-Layer (LBL) self-assembly technique was employed to fabricate multilayered polyelectrolyte capsules, which are permeable for small fluorescence 6-CF molecules. In order to quantify dye's diffusion behavior across the multilayered polyelectrolyte films, FRAP measurement was carried out by CLSM and the diffusion coefficient was obtained as 0.59×10^{-12} cm^2/S, which is much larger than the one on the flat surface but is consistent with the reference. It might due to the damage of capsule wall structure during the removal the core. Lipid coating on the PE capsules makes capsule's permeability to 6-CF changing from permeable to impermeable. There are two ways to tune such lipid coated multilayered PE capsules' permeability, there are two ways: the first one is treated it with EtOH; the other one is to be catalyzed by phospholipids A_2. Both of them can reverse lipid coated multilayered PE capsules' permeability from impermeable to permeable to 6-CF. FRAP measurements were carried out to such two products and diffusion coefficients were obtained, 0.67×10^{-12} cm^2/S, 0.55×10^{-13} cm^2/S, respectively. Comparing it with the one of pure multilayered PE capsules and the one of lipid partly coated multilayered PE capsules, one can find that the diffusion coefficient of the product by EtOH treatment is similar to the one of the pure multilayered PE capsules. Thus, we can conclude that after treated with EtOH, the intact structure of the lipids was destroyed completely and there are almost no lipids on the surface. While the diffusion coefficient of the product by PLA_2 catalyzed reaction, it is similar to the one of lipid partly coated capsules, this indicates that there might be some sealing of the holes in the polyelectrolyte by the products of the reaction which have not been removed.

5. Acknowledgements

We acknowledge the financial support from the Excellent Young Teacher Supporting Program for Teaching and Researching from Southeast University, Nanjing, P.R. China.

6. References

[1] D. D. Lasic, Liposomes: From Physics to Application, Elsevier, Amsterdam, 1993.
[2] G. Steinberg-Yfrach, J. Rigaud, E.N. Durantini, A.L. Moore, D. Gust, T. A. Moore, *Nature* 1998, *392*, 479.
[3] Z. He, L.D. Kispert, R.M. Metzger, D. Gosztola, M.R. Wasielwski, *J. Phys. Chem. B.* 2000, *104*, 6302.
[4] R.V. Klitzing, H. Moehwald, *Macromolecules* 1996, *29*, 6901.
[5] C. Tedeschi, F. Caruso, H. Moehwald, S. Kirstein, *J. Am. Chem. Soc.* 2000, *122*, 5841.
[6] J. Hotz, W. Meier, *Langmuir* 1998, *14*, 1031.
[7] F. Caruso, R. A. Caruso, H. Moehwald, *Science* 1998, *281*, 1111.
[8] C. Gao, E. Donath, S. Moya, V. Dudnik, H. Moehwald, *Eur. Phys. J. E.* 2001, *5*, 21.
[9] G. B. Sukhorukov, L. Dane, J. Hartmann, E. Donath, H. Moehwald, *Adv. Mater.* 2000, *12*, 112.
[10] S. Moya, E. Donath, G.B. Sukhorukov, M. Auch, H. Lichtenfeld, H. Baeumler, H. Moehwald, *Macromolecules* 2000, *33*, 4538.
[11] L. Q. Ge, H. Möhwald, J. B. Li, *Chem-A. Eur.* 2003, *9*, 2589.
[12] L. Q. Ge, H. Möhwald, J. B. Li, *ChemPhysChem* 2003, *4*, 1351.
[13] L. Q. Ge, H. Möhwald, J. B. Li, Biochem. *Biophys. Res. Comm.* 2003, *303*, 653.
[14] L. Q. Ge, J. B. Li , H. Möhwald, *Colloids and Surfaces A* 2003, *221*, 49.
[15] K. Ariga, J. P. Hill, Q. M. Ji, *Phys. Chem. Chem. Phys.* 2007, *9*, 2319.
[16] B. G. De Geest, N. N. Sanders, G. B. Sukhorukov, J. Demeester, S. C. De Smedt , *Chem. Soc. Rev.* 2007, *36*, 636.
[17] J. B. Li, Y. Cui, *J. Nanosci. Nanotechnol.*, 2006, *6*, 1552.
[18] K. Ariga, T. Nakanishi, T. Michinobu, *J. Nanosci. Nanotechnol.* 2006, *6*, 2278.
[19] A. P. R. Johnston, C. Cortez, A. S. Angelatos, F. Caruso, *Current Opinion in Colloid & Interface Science* 2006, 11, 203.
[20] N. Klonis, M. Rug, I. Harper, M. Wickham, A. Cowman, L. Tilley, *Eur Biophys J.* 2002, *31*, 36.
[21] A. Alexei, G. B. Suckhorukov, Edwin Donath and Helmuth Moehwald, *J. Phys. Chem. B* 2001, *105*, 2281.
[22] R. Von Klitzing, H. Moehwald, *Macromolecules* 1996, *21*, 6901.
[23] D. W. Grainger, A. Reichert, H. Ringsdorf, C. Salesse, D. E. Davies, J. B. Lloyd, *Biochim. Biophys. Acta* 1990, *1022*, 146.

[24] G. Scherphof, B. V. Leeuwen, J. Wilschut, J. Damen, *Biochim. Biophys. Acta.* 1983, *732*, 595.

22

Integration of Laser Scanning and Imagery for Photorealistic 3D Architectural Documentation

José Luis Lerma[1], Santiago Navarro[1], Miriam Cabrelles[1], Ana Elena Seguí[1], Naif Haddad[2] and Talal Akasheh[3]

[1]*Universidad Politécnica de Valencia*
[2]*Hashemite University*
[3]*Cultech for Archaeology and Conservation*
[1]*Spain*
[2,3]*Jordan*

1. Introduction

Documentation of architectural and archaeological sites and monuments is an activity that requires the capture of information from different sources. Metric information is mandatory as the basis for documentation, information management, archiving, analysis, monitoring and dissemination activities, among others. In highly weathered environments, accurate 3D models are required for multi-temporal analysis over time in 4D, altogether with imagery. Experience has shown that it is possible to provide the necessary information with the required accuracy and completeness only by integration of multisource data (Georgopoulos & Ioannidis, 2006). Especially in large and complex monuments parallel use of geodetic and surveying measurements, photogrammetric data acquisition with imagery and terrestrial laser scans has proven to be the ideal combination (Haddad, 2007; Haddad & Ishakat, 2007). Furthermore, the multi-sensor integration of surveying data (not only terrestrial but also aerial) and architectural needs should be combined with a multi-resolution approach, spanning from few centimetres down to millimetres, if necessary, both in geometry and in texture (Guidi et al., 2008; Remondino et al., 2009). A general overview of the typical requirements and solutions for cultural heritage documentation based on purpose, product, methodology or eventual emphasis is presented in Patias (2006).
Modern technology has changed matters in documentation radically and promises to keep bringing rapid changes. Photographic and non-photographic (graphic) documentation tools are merging in one process, in which the digital photographic technology is the main base (Haddad, 2010; Haddad & Akasheh, 2005). Until the wide spread used of terrestrial laser scanning (TLS) for cultural heritage documentation, most of the multi-temporal information for documentation and conservation activities was based mainly on graphic documentation: 2D drawings of elevation plans, cross-section and last but not least, images (analogue or digital). However, photographs are easier to interpret and recognize than drawings; they contain information about surface detail and can provide information on the condition of a monument, before, during, and after restoration. Nowadays, the afore-mentioned documents are still requested by architects or building personnel but the analysis is starting

to be fully in 3D owing to: first, the simplified and comprehensive data acquisition of data sets in 3D mainly by terrestrial laser scanners and 3D cameras; second, availability of easy-to-use software that handles millions of features such as points, point splats, meshes and textures fully in 3D; third, availability of surveying equipment all over the world either image-based, range-based or satellite-based.

It is well-known that the input data (points with intensity and/or colour values) coming from TLS after registration has a limited impact on architectural and archaeological documentation in contrast to the widespread used of photographs (Fig. 1). In fact, the field of conservation has very slowly assimilated the advancements of technology (Eppich & Chabbi, 2006). The techniques range from simple hand-drawn sketches to sophisticated virtual reality representations. The point clouds are generally used for visualisation purposes in 3D, dissemination and extant surveys (Fig. 2). Out of the point clouds is possible to deliver, on the one hand, position and linear measurements, top and elevation plans, cross-sections and fly-through animations, on the other, efficient and accurate way to record and reconstruct digitally large-scale cultural heritage monuments and sites (Takase et al., 2003).

Despite the time consuming processing and required user's interaction, 3D models are preferred for automating the processing of delivering not only a large number of cross-sections but contours and volumetric deformations in the three dimensions. 3D models are also required to analyse thoroughly structural damage assessment (Olsen et al., 2010), deformation on surfaces (Schueremans & Van Genechten, 2009), alterations on materials and

Fig. 1. View of the Obelisk Tomb (upper storey) and the Bab As-Siq Triclinium (lower storey) in Petra/Jordan.

Fig. 2. 3D view of the four point clouds collected for the Obelisk Tomb and the Bab As-Siq Triclinium in Petra/Jordan.

decay mapping either standalone with topographic survey (Rüther et al., 2009) or in combination with image-based sensors such as visible non-metric cameras (Navarro et al., 2009) and thermal cameras (Cabrelles et al., 2009), mainly on large and inaccessible parts of the monuments. Fig. 3 displays four perspective views of one 3D model in which is possible to identify the effects of weathering on architectural features and elements such as cornice, arches, columns and obelisks.

The direction of the light helps to interpret the deteriorations on the surface of the stone monuments due to the concavities of the own high resolution 3D model (Fig. 3). Nevertheless, the incorporation of the texture onto the 3D model increases the ability of visualisation and understanding of the 3D models (cf. Fig. 3 vs Fig. 4). Texture can be acquired by the laser scanner itself (most of the present laser scanning systems support internal or attached digital cameras) or by the user with an external high resolution camera. High quality texture mapping frequently requires external draping of imagery onto the 3D model. With external photography, the user can select the most appropriate position and attitude to take the pictures, the appropriate exposures, and, last but not least, the model of the camera that fits best for the project on purpose. Definitely, the selected camera might not have the same features as the camera integrated in the TLS. Furthermore, there is neither restriction to take pictures with high-end cameras nor metric ones. The user might make use of compact off-the-shelf digital cameras, SLR cameras, high speed camera systems for scientific applications, etc. It is recommended to take pictures with a good camera body and lens to improve the quality of the photorealistic 3D modelling.

This chapter presents the integration of terrestrial laser scanning (TLS) and imagery to generate effectively high quality photorealistic 3D models. This integration is one of the most effective solutions existing nowadays to deliver metric products with high level of appearance (Georgopoulos & Ioannidis, 2006; Remondino et al., 2009; Lerma et al., 2010b). The TLS provides the 3D information which is used to compute eventually the triangulated 3D meshes of the monument, whereas the photogrammetric processing is used to texture

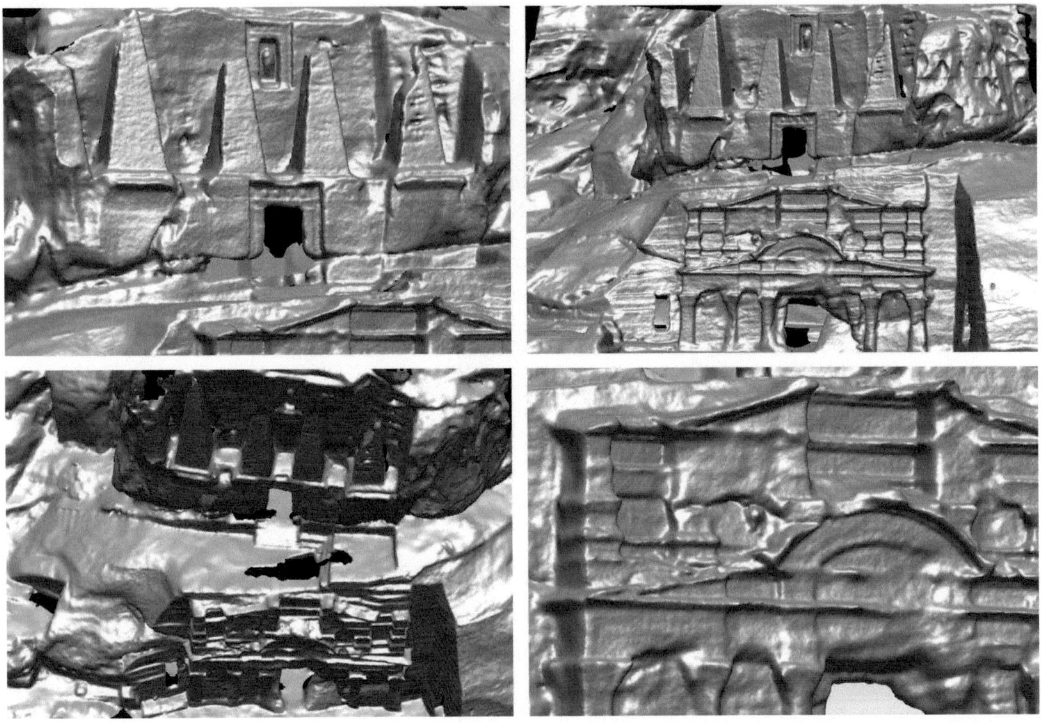

Fig. 3. Perspective views of the Obelisk Tomb and the Bab As-Siq Triclinium 3D model with the lights on: top left) left; top right) right; bottom left) top; and bottom right) Detail of the Triclinium.

accurately the content from the imagery onto the meshes. The approach is suitable to texture either complex architectural monuments such as towerlike tombs (Navarro et al., 2009), high altar and vaults (Biosca et al., 2007) or tiny archaeological engravings at high resolution with close up imagery (Lerma et al., 2010b). Texturing of the virtual monument is based on multi-images to select the best texture map, taking into account occlusions between the object and the set of imagery (Al-kheder et al., 2009).

Other reality-based solutions for 3D modelling of object and sites consider only image-based approaches (Remondino & El-Hakim, 2006; Cabrelles et al., 2010). The image-based photogrammetric solution can most of the times yield similar results as the integration of the range-based and the image-based approaches, but not always due to texture, light or geometrical conditions. The two kinds of data, image-based and range-based, can be integrated as proposed herein and used at different resolutions for complementary purposes and merged eventually together.

Our approach requires the extraction of point features in contrast to the alternative presented by Alshawabkeh & Haala (2004), owing to the high probability of success in weathered sandstone monuments. Instead of registering the images to the 3D model by minimising a photo-consistency cost function (Chetverikov et al., 2007), the well-known collinearity equations are used to relate all the images and the architectural monuments. The performance of the integrated photogrammetric approach presented herein is particularized for complex and highly weathered stone-carved architectural monuments in the Petra Archaeological Park in Jordan, the Obelisk Tomb and the Bab As-Siq Triclinium (Figs. 1-3).

The remainder of this chapter is structured as follows. Section 2 reviews the architectural documentation mission, specifying the requirements for the surveying, mapping and the modelling. Section 3 describes the context of the monument inside the Petra Archaeological Park, the state of conservation and the weathering effects regarding the Obelisk Tomb and the Bab As-Siq Triclinium. Section 4 is focus on the integration of TLS and photogrammetry, describing our approach for the production of high quality photorealistic 3D models. Section 5 presents the results achieved in the context of architectural documentation and a discussion of the results. Section 6 sums up the presented approach and what is still missing for optimal photorealistic 3D models of today.

2. Architectural documentation

Documentation is one of the principal ways available to give meaning, understanding, definition and recognition of the values of the cultural heritage. The importance of documentation may be undertaken as an aid to various treatment activities including: protection, restoration, conservation, preservation, identification, monitoring, interpretation, management of historic buildings and sites and cultural landscapes (Haddad & Akasheh, 2005; Haddad, 2007), in addition to creating a register of stolen movable objects. However, documentation is not only needed for proper conservation and preservation, but foremost to raise public awareness.

It is a fact that each documentation method has its own advantages and disadvantages. The method which should be applied in each case depends on various factors: cost and time, location facts, size extent of task (content as well as quantitative), accuracy class, kind of approach (preferable non-destructive), and style of results for presentation, interpretation and monitoring. In the photographic methods, the entire vicinity is quasi completely determined, point by point, and the filtration for the extraction of the essential points happens later, meanwhile a relevant difference arises in the acquisition process *in situ,* compared to a laboratory process. This concept is completely different from the graphic documentation process, where the *in situ* needs more time (Haddad, 2007; Haddad, 2010). However, a shadow always causes errors in the results, because certain picture information is lost.

A digital orthoimage allows users to point out and manage information about many elements on the documented architectural monument on different layers which is a very important issue in architectural conservation and preservation, for example architectural elements, shape relationships, construction techniques, texture of materials, historical phases, colour values, decorative elements and decay conditions (Agosto et al., 2005; Artesea et al., 2005). In architectural surveys of historical buildings using digital photogrammetry and TLS methodologies, the support of different specific skills is often required; hence it is crucial to choose the correct tools for a multidisciplinary analysis. Actually, architectural documentation is dependent on the instruments used (degree of reliability and precision), representation methods (flexibility degree and amount of information), research approaches (degree of exportability, interdisciplinary nature and transformation), and means of communication (degree of compatibility with other technologies and ability to diffuse) (Bornaz & Dequal, 2004; Artesea et al., 2005). In fact, when dealing with architectural documentation as in our case, it is important to represent materials, colours, decorations, physical and chemical decay and other phenomena.

3. The obelisk tomb and the Bab As-Siq Triclinium in Petra

Petra is in zone 36 North in the WGS 1984 UTM coordinate system. It forms a plateau (around 850 m above sea level) surrounded by higher mountains, the Sharah Mountain range, characterized by limestone deposits. At about 900-950 m elevation, Ordovician sandstone with its white colour and high porosity is found abundantly. The Precambrian sandstone can be found at a lower elevation and forming the largest geologic deposit in Petra. It is characterized by its beautiful multi-coloured mineralogical formations. After walking down the Petra Archaeological Park predominates a wide streambed called the Outer Siq. Most of the important monuments were carved from the Precambrian rock massif on both sides of the wadi (streambed or valley in Arabic). Ordovician monuments are mostly found in the Outer Siq.

It seems that Petra, the ancient Capital of the Nabataean Kingdom, flourished towards the end of the second century BC, and during the course of the first century BC, while most significant contact was with Ptolemaic Egypt during the first century BC, as reflected in Nabataean architecture. The city is remarkable both for its location and its construction. The Nabataeans were creative builders. From the cliffs they carved great altars, temples, tombs, theatres, and even an entire water collection and distribution system. These rock-cut monuments in the Petra Archaeological Park have substantial dimensions which are unique in antiquity. The architecture of Petra constitutes a main source of evidence, not only for the city, but also for the architecture of the late Hellenistic and Roman period. It stimulates a new speculation and evolution of the roughly studied Eastern Hellenistic architecture, which at the very least must be placed, in the much larger context of Hellenistic traditions in the Near East. In fact, during the Roman rule there were changes in patterns in architecture that allowed greater circulation and more freedom of interaction.

Several scholars have noted that the architectural façade formation images of the main monuments of Petra are shared with other Hellenistic and Roman architecture. In fact, the architectural treatments applied at the rock-cut Nabataean façade are maintaining the particularity of the late eastern Hellenistic stylistic architecture. These morphological stylistic elements made Petra a model of a metropolis during the Hellenistic and Roman periods. Actually, Nabataean Petra is the best model to understand the cultural interaction of late Hellenistic stylistic morphology reflected in the Tomb façade formation (Haddad, 1999). The architectural formation of the facades can roughly be divided into two main categories "oriental with Hellenistic doorways" and completely "Hellenistic". However, this means that there is reason to believe that the simpler façade formation, showing stronger oriental influence, would be older than the richly Hellenistic decorated ones such as the Al-Khazna. What visitors can see in most of the Nabataean rock-cut façade formation in Petra are some of the best preserved samples of the late Hellenistic morphology often in an appealing combination of oriental stylistic elements (Egyptian cavetto or the Assyrian crow steps and Western stylistic Pedimental doorways). Indeed, through the variety and richness of the decorative architectural elements of the façades (Zayadine, 1987), the Nabatean architects adopting the architectural conception of Hellenization, moved among different cultures create high artistic standards in architecture in cooperation with local tradition materials and landscape (Haddad, 1995; Haddad, 1999). This synthesis with these architectural elements is unique in the History of architecture, and is clear in the significant architectural treatment of the façade formation of the Obelisk Tomb and the Bab As-Siq Triclinium (1st century BC-1st century AD).

The architectural complex of the Obelisk Tomb and the Bab As-Siq Triclinium (Figs. 1-3) is the first major monument encountered when entering Wadi Musa on the way to the 1.2 km long Siq, the main gorge entrance to the ancient city of Petra, facing NW and dominating the left side of the road, a few meters down from a complex of three Djin blocks. In reality, the Obelisk Tomb is separated into two monuments, stacked on top of each other: the Obelisk Tomb (upper storey); and the Bab as-Siq Triclinium (lower storey), Fig 1. The Obelisk Tomb (also known as 'Nefesh' Tomb) is named after the four obelisks (approximately 7 m height) that decorate the top of the entrance of the tomb guarding the rock-hewn cave tomb entrance and was used for burials. The lower storey, the Bab as-Siq Triclinium is decorated in a more classical style and was apparently used for funeral banquets as many such chambers in Petra were used for memorial feasts in honor of the dead, a practice that was also common among the Greek and Romans. This rock-cut architectural connection between the Obelisk Tomb and the Bab As-Siq Triclinium is unique in tomb architecture of Petra.
The Obelisk Tomb houses the graves. The interior space consists of an approximately square chamber (5.80 m x 5.90 m, h. 4 m) with aboard recess in the back wall in a form of rectangular arcosolium (2.90 m x 1.70 m, h. 3.10 m), with two loculi (each one is approximately 2.50 m x 1.25 m, h. 2.30 m), carved in each side of the wall. The façade wall has a slightly raised band on both sides of the doorway and two splayed windows which emerge as slits on either side of the entrance doorway (Fig1). The doorway width is about 1.35 m and is approached by four steps; each one is 0.40 m in width.
The lower storey (Bab as-Siq triclinium) is approached by stairs on the left. Generally the façade is highly weathered. The interior space is approximately a square room (6.45 m x 7.40 m, h. 4.35), with benches on three sides. The left and the right sides-walls are plain; meanwhile the back wall has two loculi high up starting 2 m above the bench top. The triclinium façade (w. 15.60 m, h. 11.20 m), is more in the classical Nabataean style. It consists of an engaged order with six semi-columns and a broken pediment framing a segmental pediment and an upper dwarf order with a broken pediment as characteristic features of the so called Baroque style (McKenzie, 1990). The doorway pillars have anta-type capitals while the entablature has a Doric frieze.
This Nabataean architectural style mixed the Hellenistic visual conception with purely pharaonic features (Obelisks). Here the Nabataean architectural style is an interaction of two different architectural traditions; scenes in which Egyptian and Greek themes create a new dialogue and were juxtaposed in a duality or "double style" at the same façade, that consciously attempted to respect the uniqueness, integrity, and artistic heritage of each architectural tradition. Actualy, these two different historical architectural treatments at the same monument were not only decorative but they indicate how this family wanted to be seen for the eternity. The complex façades are conceived as an independent stylistic screen set in the front of the structure rather than organic and logical elements of the structure as a whole (Haddad, 1999).
Actually, the architecture of the rock-cut façades in Petra represents the final artistic model of the evolution of late Hellenized stylistic morphology (Haddad, 1999; Haddad & Ishakat, 2007). However, according to the architectural concept of the Eastern Hellenistic architecture, especially the luxury architecture of eastern late Hellenistic palaces reflected in the architecture of Nabataean Petra and the Roman wall painting of the so - called "second Pompeian style" must not only be considered and understood as an actual representation of the free standing models, but as a model of cultural interaction between the East and West and the architectural treatments of the Obelisk Tomb and the Bab As-Siq Triclinium consist of a model of this approach.

4. Geometric and appearance modelling by integration of laser scanning and photogrammetry

This section describes the methodology followed to deliver the high-resolution photorealistic 3D models on large and complex monuments. The general frame of the integration of the range-based and image-based data acquisition and processing is presented in Fig. 4 as part of the architectural documentation. The geometric and appearance modelling by integration of laser scanning and photogrammetry is focus on achieving both the required geometrical accuracy and maximum reliability in the texture mapping.

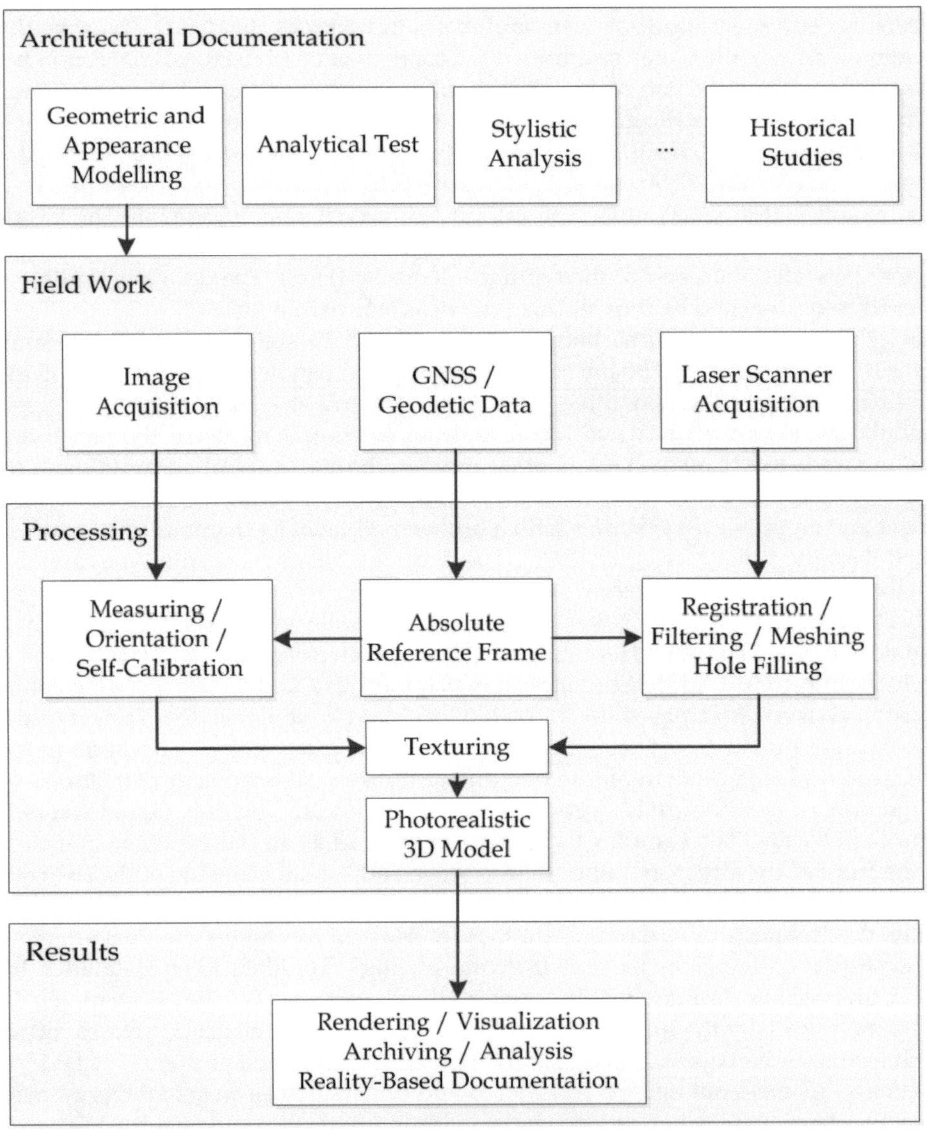

Fig. 4. General pipeline to deliver photorealistic 3D models from image sensors, laser scanners and geodetic instruments

The pipeline described in Fig. 4 focus the attention on two main activities: on the one hand, field work using different data acquisition systems or sensors, such as GNSS, laser scanner and cameras, opened to either terrestrial or aerial acquisition; on the other hand, parallel processing of the data coming from the different sensors. Each data acquisition system requires typical adjustments. The photogrammetric measurement of the image features previous to the orientation and calibration of the bundle block of imagery can start independently of the laser scanning processing and the geodetic computations, and vice versa. The laser scanning follows a typical workflow: noise reduction, registration, filtering, meshing, hole filling and smoothing, among others. Nevertheless, processing converges towards the end. In particular, after handling the geodetic data (mainly the GNSS network), the Cartesian coordinates of the control points are used to set up the local or official reference frame of the mapping, depending on the country and on the project. Moreover, texturing of the multiple images is carried out onto the 3D model to yield eventually the photorealistic 3D model. After the photorealistic 3D model, different activities and analysis can follow within the own architectural documentation mission such as metric reports and plots, stone surface monitoring, analysis and effects of damages. Out of the previous mission, there are applications that can be conducted best with the photorealistic 3D models, for instance, production of high definition movies, virtual reconstructions, interactive museums, or just print-outs for dissemination activities.

The photogrammetric approach presented herein combines the 3D models (developed from the laser scanning data) and the multiple high resolution external images to yield photorealistic 3D models with subpixel accuracy in image space (Lerma et al., 2010b). The approach consists of two steps which will be developed next:

1. Orientation and calibration of the bundle of images onto the laser scanner datasets.
2. ZI-Buffer for assigning the right texture mapping onto the DSM, after analysing occlusions and double-projections.

4.1 Orientation and calibration of the bundle of images onto the laser scanner datasets

The bunch of images used to texture the model has to be accurately positioned and oriented in the object space. This is a geometric requirement that needs to be satisfied to guarantee reality-based modelling from imagery. This step is conventionally achieved by means of the exterior orientation of the image, which relates the coordinates of an object point with the image coordinates of the image point through the perspective centre. The mathematical relationship between the image and object space reference system is set by the well-known collinearity equations (Slama, 1980; Kraus, 1993; Wolf & Dewitt, 2000; Mikhail et al. 2001).

When the camera is non-metric or uncalibrated, the interior orientation parameters of the camera have to be determined. A review of the different approaches to adjust the different set of additional parameters can be found in (Fraser, 1997; Gruen & Huang, 2001; Lerma, 2002). For close range applications with non-metric digital cameras, self-calibration bundle adjustment is the most satisfactory approach.

The strategy used to determine the exterior orientation parameters is presented in Lerma et al. (2010a), without adding baseline distance constraints. It can be summarize in the following steps (Fig. 5):

1. Manual, semi-automatic or automatic measurement of homologous image features.
2. Determination of the fundamental matrix of pairs of imagery

3. Relative orientation of image pairs and block initialization
4. Bundle block adjustment
5. Self-calibration bundle block adjustment

The determination of the blunders in the first and second step is considered crucial to compute quickly the different sets of orientation parameters. Although it is possible to take into account different sets of additional parameters for each image, it is recommended to fix the focus of the camera in order to simplify the self-calibration bundle block adjustment.

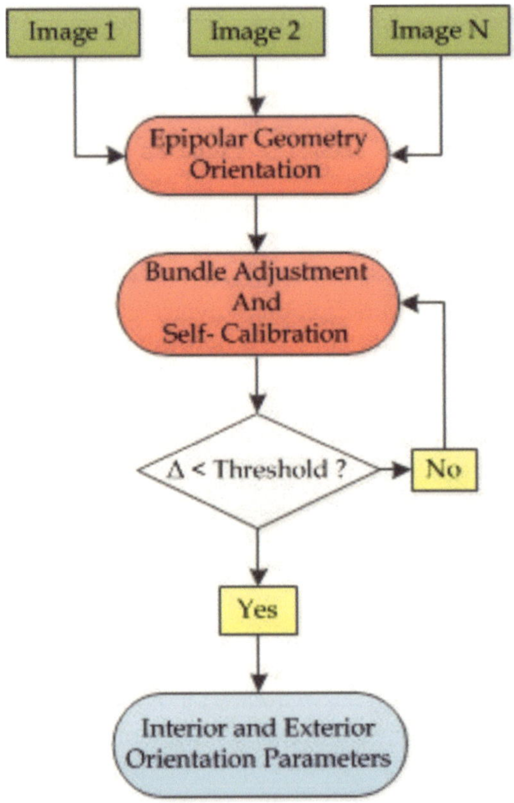

Fig. 5. Multi-image orientation strategy

The delta threshold (Δ) included in Fig. 5 means that the solution is iterative till all the requirements in the computation of the orientation parameters are satisfied. On the contrary, checking the precision of the different adjusted parameters is crucial to determine eventually high accuracy in the adjustment. Image residuals should be below or around one pixel to consider that the adjustment is correct, despite it not only depends on those values but in the geometry of the bundle of images, the geometry of the monument and the distribution of the image features to compute the afore-mentioned five steps.

The absolute reference frame established by the ground control points (GCPs) can be used at the beginning of the approach, in the middle or at the end. The classical way is to carry out the self-calibration bundle adjustment in the absolute reference frame at the beginning of the photogrammetric processing. It is recommended that the GCPs used in the self-calibration bundle adjustment be the same as those used for the transformation of the unified point cloud from the laser scanner onto the absolute reference system. As the number of GCPs usually is smaller than the number of required object points for the registration of the 2D images onto either the 3D point cloud or the 3D model, manual registration is unfortunately the common way required to perform a successful match between the range-based and the image-based datasets. This concept is also known in the literature as co-registration or fusion of 2D and 3D data sets. Nevertheless, it is possible to find in the research community some interesting approaches to automate the identification of 2D and 3D data (Forkuo & King, 2004, González-Aguilera et al., 2009). Unfortunately, the automatic transfer of homologous features is neither implemented in conventional software nor reliable enough although the future is promising. As stated in Remondino et al. (2009), this is the bottleneck of the texturing phase and it is still an interactive procedure.

4.2 Texturing from multiple images onto the 3D model

Texturing of single images onto the 3D model is commonly used in the photogrammetric and the computer vision communities. For the former, mainly to build up 2D products such as image rectifications (that correct the perspective), orthoimages and image mosaics; for the latter, to build up quick and fast texture mapping, most of the times requiring virtual-based appearance instead of reality-based appearance.

It is understood that a snapshot of a high quality photorealistic 3D model should be equivalent to a picture taken from the same camera position and attitude. Biosca et al. (2007) present a comparative study of texture mapping in 3D between conventional laser scanning software and the photogrammetric approach. The output textured 3D models coming from conventional software can not be considered of high quality compared to the presented photogrammetric approach. A summary of our implemented strategy is displayed in Fig. 6. A particularization of the texturing approach to build up high-resolution photorealistic 3D models in archaeological caves can be found in Lerma et al. (2010b).

The texturing of the multiple images is carried out onto one final 3D model obtained from the point clouds acquired by the TLS, image by image. The ZI-buffer strategy described by Amhar et al. (1998) is used to test the visible faces from the 3D mesh. From each image there is one output textured 3D model that will be eventually merged to deliver the final photorealistic 3D model. For that purpose, an additional best face perspective check is implemented considering different weights based on the normal angle and on the distance camera model.

5. Results and discussion

The results presented in this chapter were acquired basically with three kinds of instruments: first, a non-metric digital SLR camera Canon 1Ds Mark III which provides a maximum resolution of 5616x3744 pixels and fixed lens of 50 mm principal distance (Fig. 7 displays the four images used for texturing); second, a panoramic midrange time-of-flight terrestrial laser scanning Mensi GS 100 from which four point clouds were scanned at a resolution of 5 mm (one scan) and 10 mm (three scans), Fig. 2; third a Leica reflectorless total station from which 12 GCPs were measured following a pattern of three rows. All the measurements, the photogrammetric adjustments and the algorithms for texture mapping in 3D were performed with our in-house FotoGIFLE software developed by the first four authors at the Universidad Politécnica de Valencia. The final results are depicted in Fig. 8.

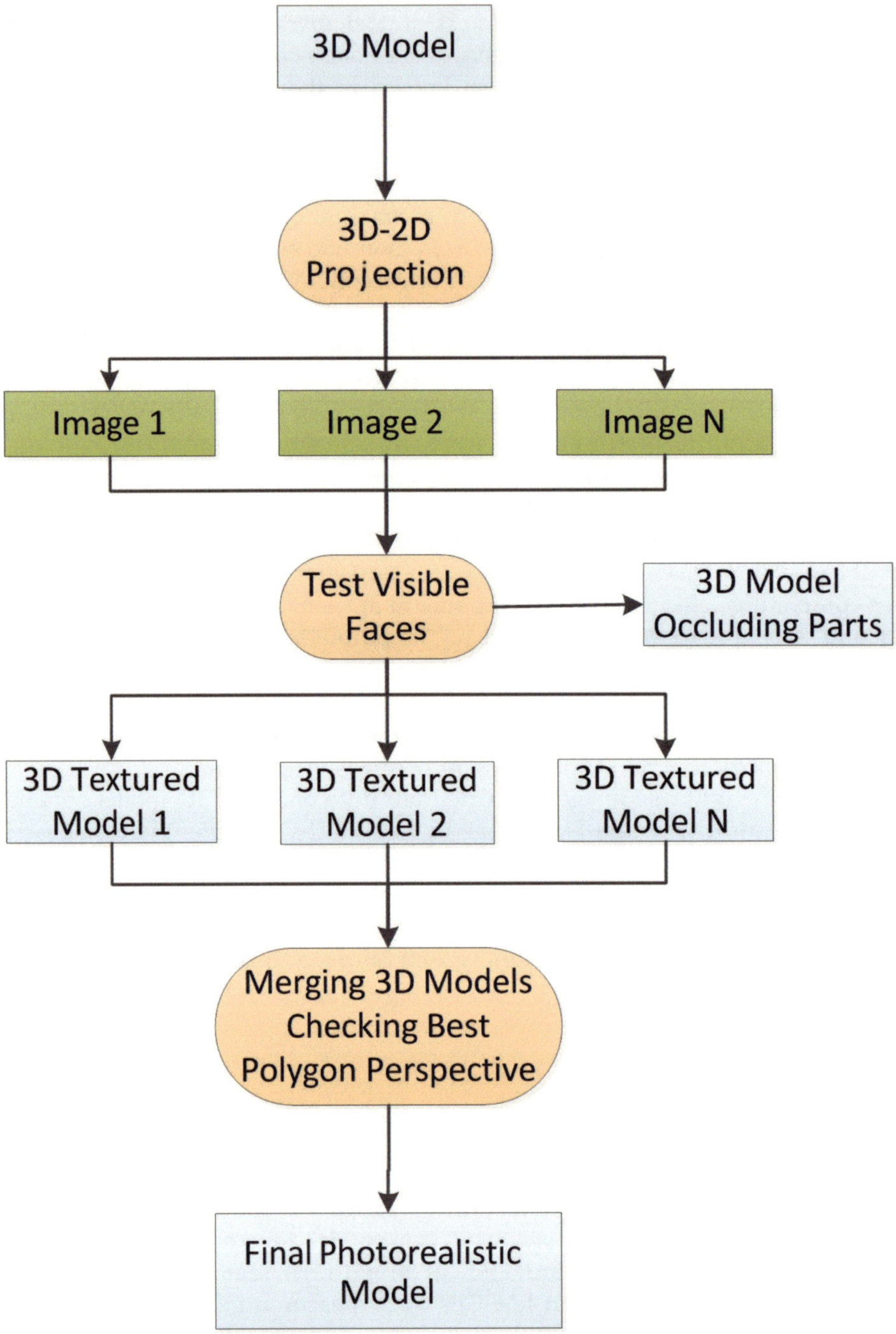

Fig. 6. Workflow used to build up photorealistic 3D models from multiple 2D images

Fig. 7. Set of images used for the photorealistic 3D model of the Obelisk Tomb and the Bab As-Siq Triclinium in Petra

The production of high quality photorealistic 3D models of large and complex monuments and sites is not free of issues such as data management, accessibility, emplacement and, last but not least, money. The results shown in Fig. 8 are mainly for the exterior façade without considering indoors and the upper side of the monument. For the top, additional equipment should have been considered such as the employment of unmanned aerial vehicles (UAVs), but due to economic restrictions that solution was not estimated.

Full data collection of these monuments with terrestrial laser scanning is not advisable. The best way to deliver fully covered indoor and outdoor high quality photorealistic 3D models, with different resolution levels (multi-resolution) depending on the requirements of the project, is integrating both image-based and range-based data, as stated in Forkuo & King (2004), Georgopoulos & Ioannidis (2006), Alshawabkeh, Y. & Haala, N., (2004), and Remondino et al. (2009). Nevertheless, our approach presented herein is valid whether the user extracts the metric information from the laser scanning or from the digital cameras.

Besides, the integration of different sensors and techniques requires that most of the steps be automated. It is preferable to have fully automatic solutions than semi-automatic, and better semi-automatic solutions than manual. The expertise of the user when matching homologous features either in 2D (for images standalone) or in 2D-3D (for images and point clouds or for images and meshes) can be ideally replaced in perfect scenarios; however, it is difficult to accomplish in real projects. Further research is advisable to succeed in everyday, everywhere projects with smaller, faster and lighter laser scanners and modern off-the-shelf digital cameras.

Fig. 8. Perspective views of the Obelisk Tomb and the Bab As-Siq Triclinium photorealistic 3D model seen from different points of view: top left) left; top right) right; bottom left) top; and bottom right) Detail of the Triclinium.

6. Conclusions and future research

This chapter has presented a successful approach for photorealistic 3D modelling of cultural heritage by integrating laser scanning technology and photogrammetry. The approach combines the 3D models (developed from the range-based data) and the multiple high resolution external images to yield photorealistic 3D models. The subpixel accuracy in the 2D-3D co-registration process between the images and the model is fundamental to match accurately the texture from the imagery onto the final 3D model without distortion.

The presented approach is not only suitable for yielding high quality perspective views and photorealistic 3D models but amazing reality-based movies (highly appreciated by managers, conservators and restorers of cultural heritage).

Today, the benefit of adding texture with low-cost digital cameras compared to the high-cost of terrestrial laser scanners is paramount. Future research should be focus on automating the matching between homologous features either in 2D for imagery, in

2D-3D for imagery and the model or even in 2D-4D and 3D-4D mainly for monitoring applications.

7. Acknowledgements

Authors would like to thank the support provided by the *Agencia Española de Cooperación Internacional para el Desarrollo* (AECID) to the project A/025999/09, and the Spanish Ministry of Science and Innovation project HAR2010-18620. Additional support to the Jordanian team from *Società Italiana per Condotte d'Acqua S.p.A.* is very much appreciated.

8. References

Agosto, E.; Ardissone, P. & Bornaz, L. (2005). The castle of Graines: different survey methodologies for the documentation of historical buildings. *The International Archives of the Photogrammetry, Remote Sensing and Spatial Information Sciences*, 36, 5/C34, (October 2005) 55-58, ISSN 1682-1750

Al-kheder, S.; Al-shawabkeh, Y. & Haala, N. (2009). Developing a documentation system for desert palaces in Jordan using 3D laser scanning and digital photogrammetry. *Journal of Archaeological Science*, 36, 2, (February 2009) 537-546, ISSN 0305-4403

Alshawabkeh, Y. & Haala, N. (2004). Integration of Digital Photogrammetry and Laser Scanning For Heritage Documentation. *The International Archives of the Photogrammetry, Remote Sensing and Spatial Information Sciences*, 35, B5, (July 2004) 424-429, ISSN 1682-1777

Amhar, F.; Jansa, J. & Ries, C. (1998). The generation of true orthophotos using a 3D building model in conjunction with a conventional MDT. *The International Archives of the Photogrammetry, Remote Sensing and Spatial Information Sciences*, 32, 4, (September 1998) 16–22, ISSN 1682-1750

Artesea, G.; Achillib, V.; Boattob, G.; Fabrisb, M.; Salemib, G. & Trecrocia, A. (2005). Peter Bernini in Calabria: The sculptures of the "Ss. Pietro E Paolo church" in Morano Calabro. *The International Archives of the Photogrammetry, Remote Sensing and Spatial Information Sciences*, 36, 5/C34, (October 2005), 91-94, ISSN 1682-1750

Biosca, J. M.; Navarro, S. & Lerma, J. L. (2007). Modelado tridimensional de una bóveda barroca mediante la combinación de láser escáner y fotogrametría, *Proceedings of the 7 Setmana Geomàtica*, pp. 1-9, Barcelona, Spain , February 2007

Bornaz, L. & Dequal, S. (2004). The solid image: an easy and complete way to describe 3D objects. *The International Archives of the Photogrammetry, Remote Sensing and Spatial Information Sciences*, 35, B5, (July 2004), 432-437, ISSN 1682-1777

Cabrelles, M.; Galcerá, S.; Navarro, S.; Lerma, J. L.; Akasheh, T. & Haddad, N. (2009). Integration of 3D laser scanning, photogrammetry and thermography to record architectural monuments. *The CIPA International Archives for Documentation of Cultural Heritage*, 22, (October 2009) 1-6, ISSN 2076-7730

Cabrelles, M.; Seguí, A. E.; Navarro, S.; Galcerá, S; Portalés, C. & Lerma, J. L. (2010). 3D Photorealistic Modelling of Stone Monuments by Dense Image Matching. *The International Archives of the Photogrammetry, Remote Sensing and Spatial Information Sciences*, 38, 5, (June 2010) 121-124, ISSN 1682-1777

Chetverikov, D.; Jankó, Z; Lomonosov, E. & Ekárt, A. (2007). Creating Photorealistic Models by Data Fusion with Genetic Algorithms, In: *Studies in Fuzziness and Soft Computing,* Mike Nachtegael, Dietrich Van der Weken, Etienne Kerre, Wilfried Philips (Ed.), 239–263, Springer, ISBN 10.1007/978-3-540-38233-1_9, Berlin

Eppich & Chabbi, 2006. Recording and Documenting Cultural Heritage – 3D Modeling for Conservation in Developing Regions, In: *Recording, Modeling and Visualization of Cultural Heritage,* Emmanuel Baltsavias, Armin Gruen, Luc Van Gool, Maria Pateraki (Ed.), 11-20, Taylor & Francis Group, ISBN 0 415 39208 X, London

Forkuo, E. K. & King, B. (2004). Automatic Fusion of Photogrammetric Imagery and Laser Scanner Point Clouds. *The International Archives of the Photogrammetry, Remote Sensing and Spatial Information Sciences*, 35, B4, (July 2004) 921-926, ISSN 1682-1777

Fraser, C. S. (1997). Digital Camera Self-Calibration. *ISPRS Journal of Photogrammetry and Remote Sensing*, 52, 4, (August 1997) 149-159, ISSN 0924-2716

Georgopoulos, A. & Ioannidis, C. (2006). 3D visualization by integration of multisource data for monument geometric recording, In: *Recording, Modeling and Visualization of Cultural Heritage,* Emmanuel Baltsavias, Armin Gruen, Luc Van Gool, Maria Pateraki (Ed.), 375-384, Taylor & Francis Group, ISBN 0 415 39208 X, London

González-Aguilera, D.; Rodríguez-Gonzálvez, P. & Gómez-Lahoz J. (2009). An automatic procedure for co-registration of terrestrial laser scanners and digital cameras. *ISPRS Journal of Photogrammetry and Remote Sensing*, 64, 3, (May 2009) 308-316, ISSN 0924-2716

Gruen, A. & Huang, T. S. (2001). *Calibration and Orientation of Cameras in Computer Vision,* Springer, ISBN 3540652833, Berlin

Guidi, G.; Remondino, F.; Russo, M.; Menna, F. & Rizzi, A. (2008). 3D Modeling of Large and Complex Site Using Multi-sensor Integration and Multi-resolution Data, *Proceedings of the 9th International Symposium on Virtual Reality, Archaeology and Cultural Heritage (VAST)*, pp. 85-92, ISBN 978-3-905674-14-9, Braga, December 2008, Eurographics Association, Germany

Haddad, N. (1995). *Doors and Windows in Hellenistic and Roman Greece, (Dissertation Theses in Greek)*, AristotileUniveristy of Thessaloniki, Thessaloniki

Haddad, N. (1999). Macedonia, Alexandria, Petra: Tomb Architecture, *Proceedings of the International Congress, Alexander the Great: from Macedonia to the Oikoumene*, pp. 161-171, ISBN 9609054811, Veria, May 1998, Greece

Haddad, N. & Akasheh, T. (2005). Documentation of Archaeological Sites and Monuments: Ancient Theatres in Jerash. *The International Archives of the Photogrammetry, Remote Sensing and Spatial Information Sciences*, 36, 5/C34, (October 2005), 350-355, ISSN 1682-1750

Haddad, N. & Ishakat, F. (2007). 3D Laser Scanner and Reflectorless Total Station: A Comparative Study of the Slots of El-Khazneh at Petra in Jordan. *The International Archives of the Photogrammetry, Remote Sensing and Spatial Information Sciences*, 36, 5/C53, (October 2007) 356- 361, ISSN 1682-1750

Haddad N. (2007). Towards Creating a Dialogue between the Specialized Technician and non Technician Users of the 3D Laser Scanner. *The International Archives of the*

Photogrammetry, Remote Sensing and Spatial Information Sciences, 36, 5/C53, (October 2007) 350-355, ISSN 1682-1750

Haddad, N. (2010). From Ground Surveying to 3D Laser Scanner: A Review of Techniques Used for Spatial Documentation of Historic Sites, Journal of King Saud University [Engineering Science], 2010, (in press)

Kraus, K. (1993). *Photogrammety. Fundamentals and Standard Processes, Vol. 1*, Dümmler, ISBN 3427786846, Bonn

Lerma, J. L. (2002). *Fotogrametría Moderna: Analítica y Digital*, Universidad Politécnica de Valencia, ISBN 8497052102, Valencia

Lerma, J. L.; Navarro, S.; Cabrelles, M. & Seguí, A. E. (2010a). Camera Calibration with Baseline Distance Constraints. *The Photogrammetric Record*, 25, 130, (June 2010) 140–158, ISSN 0031868X

Lerma, J. L.; Navarro, S.; Cabrelles, M. & Villaverde, V. (2010b). Terrestrial laser scanning and close range photogrammetry for 3D archaeological documentation: the Upper Palaeolithic Cave of Parpalló as a case study. *Journal of Archaeological Science*, 37, 3, (March 2010) 499-507, ISSN 0305-4403

McKenzie, J. (1990). *The Architecture of Petra*, Oxford University Press, ISBN 184217164X, United Kingdom

Mikhail, E. M.; Bethel, J. S. & McGlone, J. C. (2001). *Introduction to Modern Photogrammetry*, John Wiley & Sons, ISBN 0471309249, New York

Navarro, S.; Seguí, A. E.; Portalés, C.; Lerma, J. L.; Akasheh, T. & Haddad, N. (2009). Integration of TLS data and non-metric imagery to improve photo models and recording. A case study on Djin Block No. 9, Petra (Jordan), *Proceedings of the 15th International Conference on Virtual Systems and Multimedia*, pp. 58-63, ISBN 978-0-7695-3780-0, Vienna, September 2009, IEEE Computer Society, United Stated of America

Olsen, M. J.; Kuester, F.; Chang, B. J. & Hutchinson, T. C. (2010). Terrestrial Laser Scanning-Based Structural Damage Assessment. *Journal of Computing in Civil Engineering*, 24, 3, (May/June 2010) 264-272, ISSN 0887-3801

Patias, P. (2006). Cultural Heritage Documentation. *International Summer School "Digital Recording and 3D Modeling"*, Aghios Nikolaos, Crete, Greece, 24-29 April 2006

Remondino, F. & El-Hakim, S. (2006). Image-based 3D modelling: A review. *The Photogrammetric Record*, 21, 115, (September 2006) 269–291, ISSN 0031868X

Remondino, F.; Girardi, S.; Rizzi, A. & Gonzo, L. (2009). 3D Modeling of Complex and Detailed Cultural Heritage Using Multi-resolution Data. *Journal on Computing and Cultural Heritage*, 2, 1, (July 2009) 1-20, ISSN 1556-4673

Rüther, H.; Chazan, M.; Schroeder, R; Neeser, R; Held, C.; Walker, S. J.; Matmon, A. & Horwitz, L. K. (2009). Laser scanning for conservation and research of African cultural heritage sites: the case study of Wonderwerk Cave, South Africa. *Journal of Archaeological Science*, 36, 9, (September 2009) 1847-1856, ISSN 0305-4403

Schueremans, L. & Van Genechten, B. (2009). The use of 3D-laser scanning in assessing the safety of masonry vaults—A case study on the church of Saint-Jacobs. *Optics and Lasers in Engineering*, 47, 3-4, (March-April 2009), 329-335, ISSN 01438166

Slama, C. (1980). *Manual of Photogrammetry, Fourth Edition,* American Society of Photogrammetry, ISBN 0937294012, Virginia

Takase, Y.; Sasaki, Y.; Nakagawa, M.; Shimizu, M.; Yamada, O.; Izumi, T. & Shibasaki, R. (2003). Reconstruction with Laser Scanning and 3D Visualization of Roman Monuments and Remains in Tyre, Lebanon. *The International Archives of the Photogrammetry, Remote Sensing and Spatial Information Sciences,* 34, 5/W12, (July 2003) 325-329, ISSN 1682-1750

Wolf, P. R. & Dewitt, B. A. (2000). *Elements of Photogrammetry with Applications in GIS, Third edition,* McGraw-Hill, ISBN 0072924543, New York

Zayadine, F. (1987). Decorative Stucco at Petra and other Hellenistic Sites, In: *Studies in the History and Archaeology of Jordan III,* Adnan Hadidi (Ed.), 131-142, ISBN 0710213727, Routledge & Kegan Paul Books Ltd, United Kingdom

23

Application of Confocal Laser Scanning Microscopy to the In-situ and Ex-situ Study of Corrosion Processes

Rafael Leiva-García, José García-Antón and Mª José Muñoz-Portero
Univ. Politécnica de Valencia, Ingeniería Electroquímica y Corrosión (IEC), Departamento de Ingeniería Química y Nuclear
Spain

1. Introduction

1.1 Background of confocal microscopy

In 1955 Marvin Minsky developed the basic concept of confocal microscope when he was a Junior Fellow at Harvard University (Minsky, 1957). The principle of confocal microscopy is based on the rejection of the light from the planes out of focus. Minsky's design performs a point-by-point image construction by focusing a point of light sequentially across a specimen and then collecting some of the returning rays. In this way, Minsky avoided most of the undesirable scattered light that obscures an image if the entire specimen is illuminated at the same time. In addition, the light that the specimen returns passes through a second pinhole aperture that rejects rays that are out of the focal point. The remaining selected light rays are then collected by a photomultiplier and the image is gradually reconstructed using a long-persistence screen. A real image was not formed in Minsky's original microscope; the output from the photodetector was translated into an image of the region of interest. In Minsky's original design the image was built up on the screen of a military surplus long persistence oscilloscope with no facility for hard copy. Minsky said later that the image quality in his microscope was not very impressive due to the quality of the oscilloscope display and not due to lack of resolution achieved with the microscope itself (Minsky, 1988).

To obtain the image, Minsky scanned the specimen by moving the stage rather than the light rays. Using a 60 Hz solenoid to move the platform vertically and a lower-frequency solenoid to move it horizontally, Minsky obtained a frame rate around one image every 10 sec. **Fig. 1** shows a scheme of Minsky's invention.

Although Minsky built his microscope with two objective lenses he realized that the system could work using a single objective lens operated in a reflected light mode. This mode is, in fact, how all current commercial confocal laser scanning microscopes (CLSM) systems work. Minsky's invention remained largelly unnoticed, due most probably to the lack of intense light sources necessary for imaging and the computer horsepower required to handle large amounts of data. The great advances in computer, laser technology, and digital images acquisition have permitted to discover the application of the confocal microscope in a widely number of fields such as biology, geology or materials research. Laser light offers several advantages over other lights for confocal microscopy. Each kind of laser produces light of only one or a very few well defined wavelengths. The light is spatially and temporarily coherent. Spatially coherent light is light in which all the waves have the same frequency, direction, and polarization angle. Temporarily coherent light is light in which the waves have exactly the same phase and speed. This combination produces a beam of light that is bright and clearly focused. The high intensity of laser light is useful in getting a sufficiently bright spot of light onto the specimen, even when the specimen is dark.

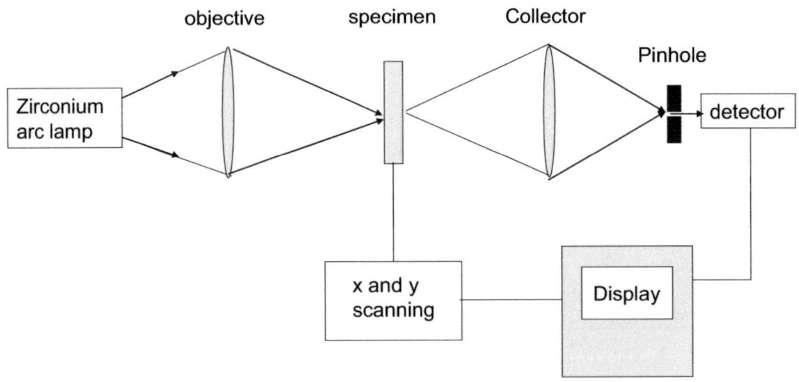

Fig. 1. Scheme of Minsky's invention, the first confocal microscope.

1.2 Advantages of confocal microscopy over to conventional optical microscopy

The main advantages of the confocal microscopy over the conventional optical microscopy can be summarised as follows:

a. **Higher resolution**. Resolution increases with the shorter wavelengths and is higher as the numerical aperture of the objective increases.
b. **Better contrast**. All undesirable light is neglected.
c. **Possibility to obtain optical slices**. By changing the pinhole aperture and the focal plane it is possible to obtain slices of different planes.
d. **Three dimensions reconstruction**. With the slices obtained in the different focal planes, it is possible to create a three dimensions image of the studied specimen.

1.3 Kinds of confocal microscopy systems

There are different kinds of confocal scanning techniques:

a. **Single beam scanning,** this system illuminates the specimen with only one light point and it is necessary to scan the specimen. There are two versions of this system: confocal stage scanning microscopy (CSSM) and confocal laser scanning microscopy (CLSM). The former the model designed by Minsky and uses a fixed laser beam, and the specimen is scanned using a motorized stage in the microscopy. The main characteristic of this system is the slow acquisition of images. Despite that, one of the advantages of this method is that the laser remains fixed and the aberrations due to the objective are lower, the optical response being constant during image acquisition. On the other hand, in confocal laser scanning microscopy the scan of the specimen is obtained by moving the laser beam. This movement is carried out using galvanometric mirrors that permit changing the incidence point on the specimen. Some of the advantages of this method are that the acquisition is faster than in the other system and the sample has not to be moved.

b. **Multiple-beam scanning**; in this system the microscope is equipped with a spinning disk containing an array of pinholes that allows scanning the specimen with multiple light beams at the same time. One of the advantages over single beam systems is the image acquisition speed. However, some disadvantages of this system are the low intensity of the images obtained and the thickness of the optical slice that is higher than in the CLSM systems.

1.4 CLSM operation principles

Fig. 2 illustrates the principal components and optical pathway in a confocal laser scanning microscope.

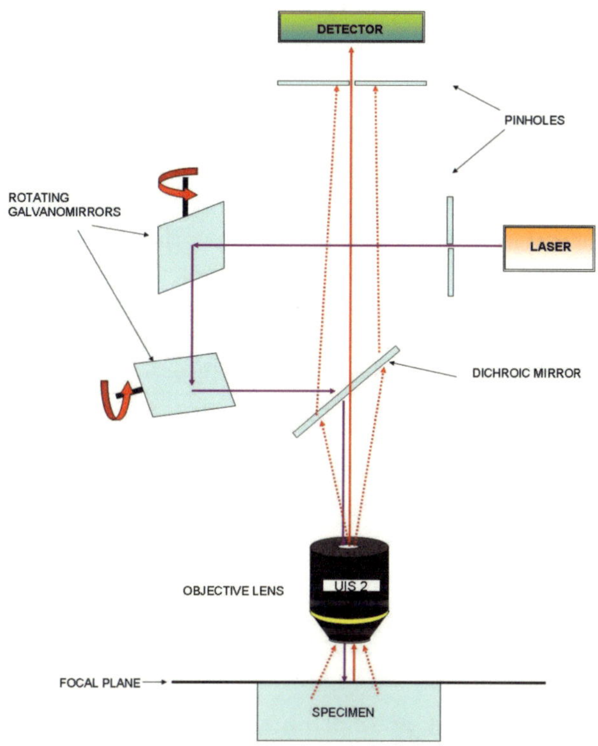

Fig. 2. Diagram of a confocal laser scanning microscope.

One or more lasers can be present in the CLSM systems. Some Kind of optical coupling mechanism such as optical fiber is used to allow the laser beam to enter the microscope system. Coherent light emitted by the laser system (excitation source) passes through a pinhole aperture that is situated in a conjugate plane (confocal) with a scanning point on the specimen and a second pinhole aperture positioned in front of the detector (a photomultiplier tube).Next, the laser beam is reflected by a set of scanning mirrors. These

mirrors are mounted on a mechanism that can move them very precisely and very rapidly. One mirror scans the beam in the X direction, the other in the Y direction. Together, these mirrors scan the beam in a raster fashion. The beam is then reflected to the back focal plane of the objective lens. The objective lens focuses the beam onto the specimen. The result of all this is that a spot of light is scanned over a small area of the specimen. The size of this area is equal to or smaller than the objective lens's field of view. By changing the raster size it is possible to obtain higher magnifications with the same lens.
If the specimen is fluorescent or reflective, some fraction of light is reflected back to the objective lens. This light travels backwards along the same path as the laser did. This light then must pass through a semitransparent mirror or a dichroic mirror that reflects it away from the laser and towards the detection system. The first part of the detection system is the pinhole aperture. The aperture serves to allow only a small central portion of the reflected light to reach the light detectors. This light will pass through a polarizer that will only allow laser light with a different polarization angle from the initial laser light to pass through. The light enters a photomultiplier tube type detector and then the signal is transmitted to the computer where the image of the focal plane is generated.

1.5 Main parameters of CLSM systems

There are different parameters that are important in CLSM systems.

a. **Laser wavelength and laser intensity**. The lower the laser wavelength is, the greater the CLSM resolution is. **Table 1** shows the different emission wavelengths of gas lasers use in CLSM. Higher intensities can be useful when the specimen is too dark and the light has to be reflected.

Laser	Wavelength (nm)			
	UV	Blue	Green	Red
Helium-Cadmium	325	442		
Helium-Cadmium (RGB)		442	534; 538	636
Low Power argon ion		488	514	
Water-cooled argon ion	351; 364	457; 488	514; 528	
Argon-Krypton mixed gas		488	568	647
Helium-neon (green)			543	
Helium-neon (red)				633

Table 1. Wavelengths of gas lasers used in CLSM.

b. **Pinhole diameter**. There is an optimal pinhole diameter for every lens. With this optimal pinhole the thickness of the slice is the lowest (maximum resolution in the Z axis). The thickness of the slices increases with the pinhole diameter.

c. **Gain of the photomultiplier**. With the gain, it is possible amplify the electrical signal generated by the photons detected in the photomultiplier. When the signal intensity is low, it can be improved by increasing the gain.

d. **Scan rate of the laser**. It is the number of lines per second that the laser scans. Resolution decreases at higher scan rates.

e. **Image size**. This parameter is defined as the number of pixels of the image. Resolution is improved with higher number of pixels.

2. The problem of corrosion

2.1 Importance of corrosion

Corrosion is an important problem of metallic materials used in the technological, mechanical and structural applications and can cause failures and accidents. The importance of corrosion studies is threefold. First of all, economic point, as material losses due to the progressive deterioration or sudden break of metallic components such as pipes, tanks, devices, hulls, etc, would be lower Another point is the preservation of natural resources, the world's resources of metallic materials are limited. Furthermore, the manofacture of metallic components involves the use of energy and water. Last but not least important than is the issue of environmental and industrial safety. An inadequate maintenance of metallic components against the corrosion can be cause of accidents or leaks of dangerous products to the environment.

2.2 Definition of corrosion

According to the American Society for Testing and Materials, corrosion can be defined as "The chemical or electrochemical reaction between a material, usually a metal, and its environment that produces a deterioration of the material and its properties" (ASTM G-15 international, 1994). More in the scope of this research, in electrochemical corrosion the metallic surface is the site for two electrode reactions, which happen in different places, the anodic reaction and the cathodic reaction. The reactions can be written as follows:

$$\text{Anodic Reaction} \qquad M \rightarrow M^{+n} + n\,e \qquad (1)$$

$$\text{Cathodic Reaction} \qquad Ox + n\,e^{-} \rightarrow Red \qquad (2)$$

Electrochemical corrosion takes place when the metallic materials are in contact with environments with electrolytic conductivity such as water, salt solutions or atmospheric humidity. Therefore, the presence of water molecules over the metallic surface is the necessary condition to produce electrochemical corrosion. The electrical contact between different metallic areas immersed in an electrolytic solution produces an electrical current due to the difference in potential between these areas. This electrical charge produces the movement of cathions and anions between cathodic and anodic areas. The metal surface that tends to solve (anodic area) is corroded in a process where the metallic atoms pass to the solution as positive ions leaving behind these electrons in the metal. On the other hand, the metal surface less prone to solve (cathodic area), remains immune to the corrosion attacks. This surface gains the electrons from the anodic area through the metallic mass, these electrons are supplied to an oxidant agent present in the electrolyte; this reaction is called reduction. The result of these electrochemical reactions is the deterioration of metals with the subsequent loss of the materials properties.

2.3 Corrosion tests

The measurement of the corrosion processes can be carried out by means of direct or indirect tests. The first one is a direct measure of the effect of the immersion of metals in an aggressive solution. One example of these tests is the weight-loss test, in this test the metal specimen is immersed in the solution under study under different conditions (temperature, time, concentration of aggressive anions, etc). Specimens are weighed before and after the test and the differences in weight are registered. The main disadvantage of this test is the long duration, sometimes months or years.

On the other hand, in the indirect measure tests, the electrical parameters (potential and current density) of the metallic specimens in the studied solutions are measured against a reference electrode. There are different indirect tests such as potentiodynamic curves, cyclic potentiodynamic curves, zero resistance ammeter test (ZRA), electrochemical impedance spectroscopy (EIS), etc. These tests are faster than the direct tests and provide very useful information about the corrosion processes.

2.4 Forms of corrosion

One of the main classifications of corrosion processes is related to attack morphology. As the location of the anodic and cathodic areas can vary, corrosion can be categorised as follows:

a. **Uniform corrosion**, the entire surface in contact with the electrolyte is attacked, and the loss of thickness and mechanical properties is similar in all the surface.
b. **Pitting corrosion**, occurs in localised surface areas, not greater than 1 or 2 mm^2, and goes growing inside the material forming a cavity. Despite the fact that the quantity of affected material can be low, the problems can be very important.
c. **Intergranular corrosion**, occurs in the metals after heat treatment that produces the depletion of the alloying elements in the grain boundaries. It is an attack on the grain boundaries of a metal or alloy. The loss of mechanical properties is really important in this corrosion process.
d. **Selective corrosion**, the attack is not distributed in a homogeneous way on the surface. In this attack, one of the components of an alloy undergoes damage or transformation.
e. **Galvanic corrosion**, this kind of corrosion is related with the destruction of a metal that is in contact with a more noble metal, when they are in solutions that allow the flow of current.
f. **Crevice corrosion** is produced at the region of contact of metals with metals or non metals in crevices or junctions that are insufficiently aerated and the renovation of the corrosive media is difficult.

Differences in the attack morphology can be established by observation of the surface during or after the corrosion processes. Therefore, it is very important to have optical tools that permit the analysis of those differences. At this point it is when the CLSM techniques show that they can be very useful in the study of the corrosion processes.

3. Previous corrosion studies with image acquisition systems

In previous works of this research group, different electro-optical devices were employed in order to obtain images of the specimen surface during the electrochemical tests. The first electro-optical devices used were the patented devices P-200002525 and P-200002526 (García Antón et al., 2000a; García Antón et al., 2000b; García Antón et al., 2001; García Antón et al., 2003). These devices allow observing the electrode surface in real-time as it undergoes electrochemical corrosion. The experimental device consists of two elements: the electrochemical unit and the image acquisition section. The electrochemical system is composed of the data acquisition equipment, which registers the electrical signal obtained from the corrosion processes that take place inside a horizontal electrochemical cell. The image acquisition system consists of a microscope-stereoscope and a colour video camera assembled to the optical device. **Fig. 3** shows a photograph of these devices.

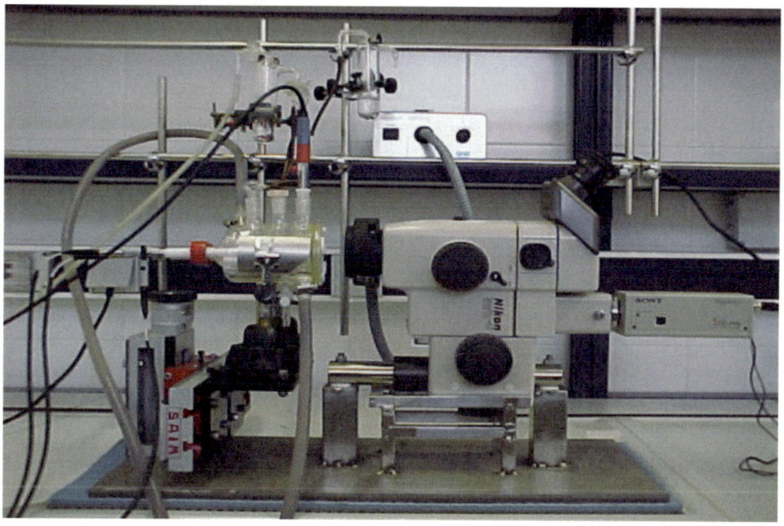

Fig. 3. Photograph of the patented electro-optical devices P-200002525 and P-200002526.

Some images obtained with these devices during potentiodynamic curves with copper (uniform corrosion) and duplex stainless steel (localised corrosion) are presented in **Fig. 4.** In these images, it is possible to distinguish between the generalised and localised processes. In the first one the entire surface of the copper is affected by the corrosion attack. In the case of stainless steel, corrosion begins at localised points.

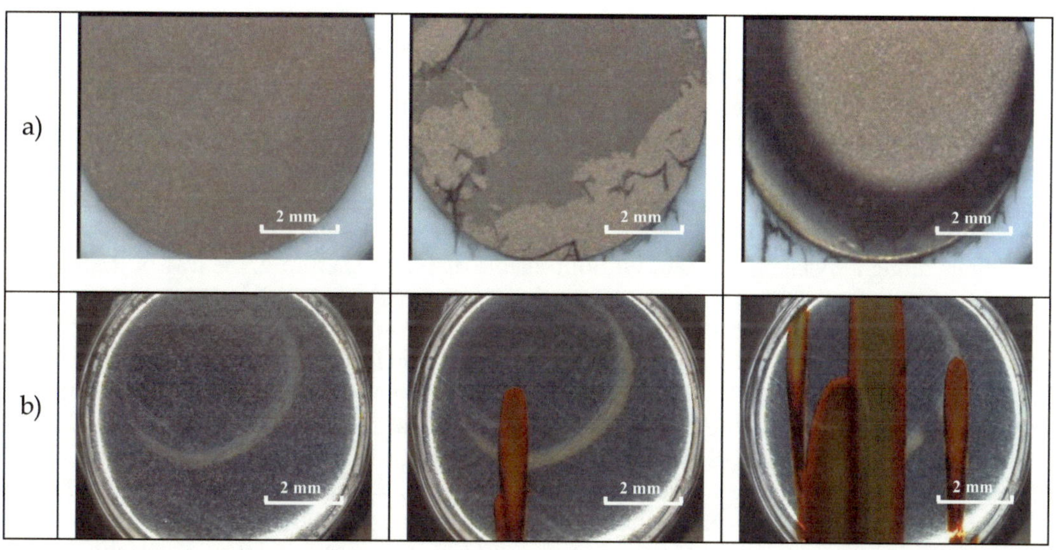

Fig. 4. Images of a) copper (generalised corrosion) and b) duplex stainless steel (localised corrosion) obtained with the electro-optical devices P-200002525 and P-200002526 during potentiodynamic curves in a 992 g/L LiBr solution at 25 °C.

An improvement of the previous electro-optical devices is the patented electrochemical cell P-200803389 (García-Antón et al., 2008). With this cell it is possible to visualise the specimens during the electrochemical tests and monitor the volume of gas generated in the

cathode and the anode. In this way, reactions like hydrogen evolution (a very important reaction in the research of fuel cells) can be studied.
Some limitations of these electro-optical devices are the impossibility to obtain three dimensional images, the highest magnification is 75x, and the impossibility to correct slanted surfaces. Higher magnifications in the observation of corrosion processes is a very important point because despite the fact that the effects of corrosion processes are visible at large scale, its mechanisms happen at a smaller scale. Some authors have used microscopic systems such as microcells, scanning electrochemical microscopy or atomic force microscopy in order to study in-situ corrosion processes (García et al., 2008a; García et al., 2008b; Martin et al, 2008). These experimental devices can be applied to different settings, such as heterogeneous materials (dual phase steels, welding, etc).

4. In-situ observation of corrosion processes

4.1 Description of the used CLSM

The confocal laser scanning microscope is an "Olympus LEXT 3100 OLS" with the following features:

a. The wavelength of the laser is 408 nm (Violet Laser Diode).
b. The horizontal resolution is 0.22 μm and the vertical resolution is 0.01 μm.
c. Three acquisition modes: optical light, non confocal laser, and confocal laser.
d. A motorized stage that permits moving quickly to the region of interest.
e. The change of the objective lens is executed electrically through PC control.
f. The raster of the laser scan can be reduced up to an additional magnification of 14x without need for changing the lens.

4.2 In-situ study of heat treatments

When stainless steels are heated improperly, they can undergo morphological changes that affect their corrosion resistance. During heat treatments new phases can appear or chromium carbides can precipitate at the grain boundaries. These processes deplete the alloying elements in certain areas that lose their corrosion resistance. This phenomenon is called **sensitisation to intergranular corrosion**. This kind of corrosion causes serious problems because it can propagate along a metal component without visible damage. Therefore, the loss of mechanical properties can occur without apparent signs, causing an accident.
Fig. 5 shows two images obtained by scanning electron microscopy (SEM) of a duplex stainless steel (DSS), Alloy 900, before and after heat treatment at 825 °C in an inert atmosphere for 1 hour (Leiva-García et al., 2009). In this image it can be observed how after heat treatment new phases appear and the percentage of one of the initial phases (ferrite) decreases. Duplex stainless steels are widely used in oil, chemical, petrochemical, nuclear, and marine industries. Duplex stainless steels have a two-phase microstructure, in which ferrite and austenite are present in relatively large separate quantities and in approximately equal volume fractions. Therefore, one important application of the CLSM to corrosion studies is the in-situ observation of metal evolution during heat treatment in order to associate the morphological changes due to heating with the further corrosion behaviour of the metal.
A "Limkam" heating stage can be accommodated in the CLSM stage in order to observe the metal evolution during heat treatment. The sample is placed inside the ceramic sample cup so that it is heated from underneath as well as from the sides; a ceramic heat shield is placed over the top to prevent heat from escaping this micro oven. The stage body and large

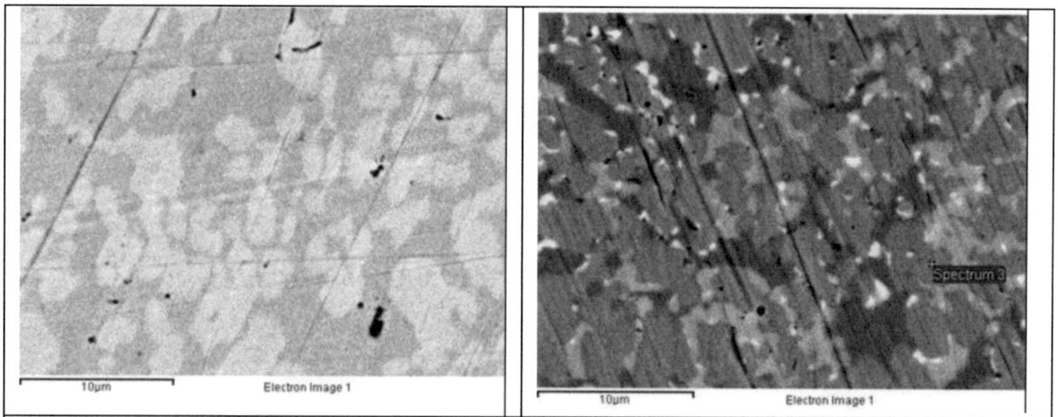

a) DSS in the as-received state	b) DSS after heat treatment at 825 °C

Fig. 5. Backscattered electron images obtained by SEM of a DSS in the unsensitised and sensitised state.

diameter quartz lid window are kept at a safe temperature by sealed circulating water. Furthermore, there is a gas-tight chamber for atmospheric control that allows carrying out heat treatments in an inert atmosphere. The maximum temperature that can be reached is 1500 °C, close to the melting point of different stainless steels.

Prior to heat treatment, the specimen was wet abraded from 220 Silicon Carbide (SiC) grit to a 4000 SiC grit finish, polished with 1 micron alumina and finally rinsed with distilled water. Then, the specimen was put in the crucible and the stage was closed in order to maintain an inert atmosphere. Prior to heating, an argon flow passed through the heating stage in order to purge the oxygen; during the test this argon atmosphere was maintained. After that a heating ramp was programmed from 25 °C to 825 °C and the temperature was held for 1 hour. Different images in 2 and 3 dimensions were obtained during this process. **Fig. 6** shows 2-D images of the DSS surface during one of these heat treatments obtained by means of the confocal laser scan. These images are the reconstruction of the different slices obtained during image acquisition. The evolution of the DSS surface can be analysed from the photographs of **Fig 6.**, during the heating ramp two phases are clearly visible (**Fig 6. a)**), the main difference being the expansion coefficient that produces different height evolution in the phases during heating. These differences in topography can be detected with the CLSM and this differentiation is higher as greater the resolution is. According to the literature there is a temperature interval (600 °C - 1050 °C) where many carbides, nitrides, intermetallic, and secondary phases can precipitate (Martins M. and Casteletti L.C., 2005; Pohl M. et al., 2007). **Fig. 6 b), c), and d)** show how these changes affect the specimen's surface. On the other hand, during heating at 825 °C (**Fig. 6 e) and f)**) the volume of the dark areas grows and this change is registered by the CLSM. These areas could be related with accumulations of the alloying elements. **Fig. 7** presents a 3-D image of the specimen at the end of the heat treatment.

After heat treatment, electrochemical tests can be carried out in order to relate the electrochemical behaviour with the morphological changes that happen as a consequence of heating.

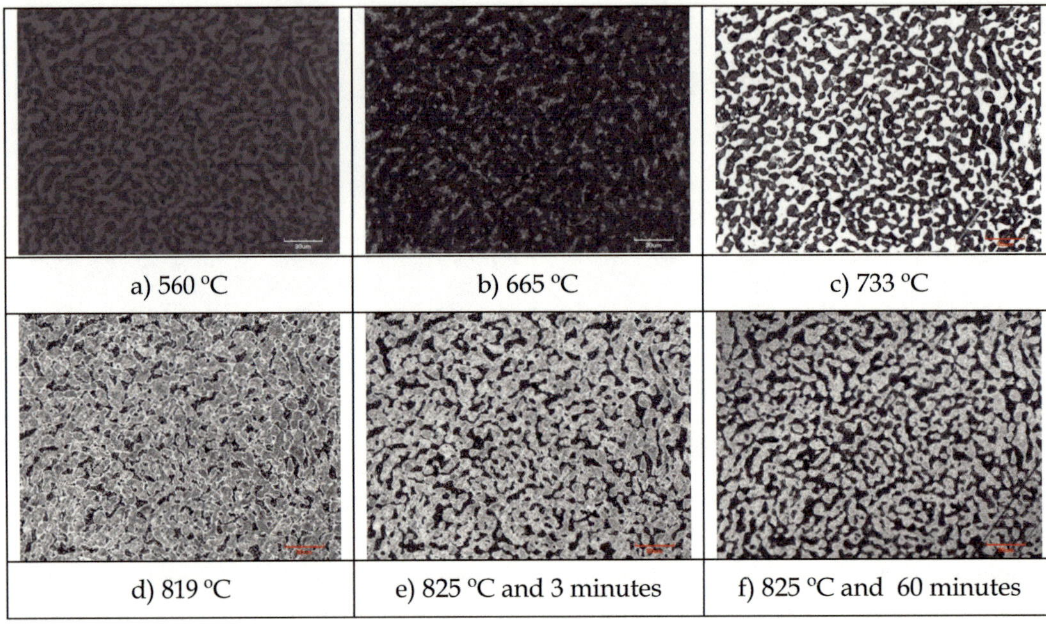

Fig. 6. 2-D images of a DSS specimen during heat treatment in an argon inert atmosphere at 825 °C for 1 hour.

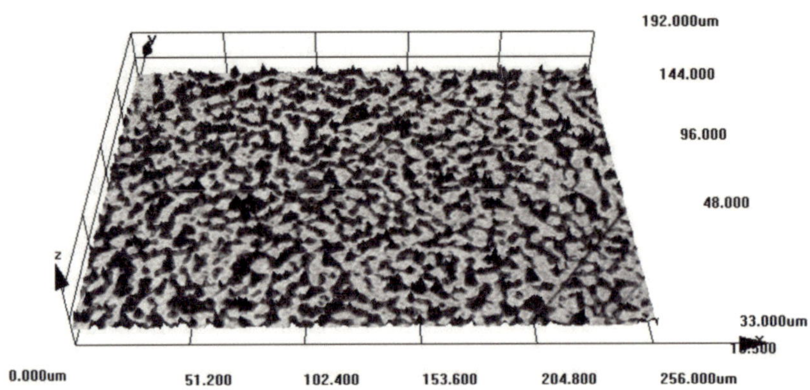

Fig. 7. A 3-D image of a DSS specimen during the last minute of a heat treatment in an argon inert atmosphere at 825 °C during 1 hour.

Another application of the heating stage could be the study of dry corrosion. Dry corrosion is defined as the corrosion that happens at high temperatures when the metal is in contact with vapour or gases. If the inert atmosphere is replaced by air or an oxygenated atmosphere, the growth of the oxide film with temperature and time could be studied by means of the CLSM.

4.3 In-situ study of corrosion attacks

A device called potentiostat was employed in the electrochemical tests. This electronic device provides a constant or variable potential with regard to a reference electrode. The counter electrode is used to close the electric circuit. This electrode is normally made of materials resistant to corrosion attacks in the medium studied (gold, platinum…). While the potential is imposed to the metallic specimen, the current density is registered. Another work option with the potentiostat is the galvanostatic mode. In this case a constant or variable current density is imposed and the potential variations are registered.

A minicell that can be placed in the CLSM stage was designed in order to observe corrosion attacks during the electrochemical tests (Leiva-García et al., 2010). The minicell permits using confocal laser scanning microscopy to observe the beginning of the microscopic corrosion process during the tests. **Fig. 8** shows a scheme of the minicell. The cell is made of glass and consists of two parts. The first part is a base to support the specimens. This base has four small supports to maintain the specimen in a horizontal position (**Fig. 8 c)**), and a frosted lateral surface to close the cell with the upper part. The upper part has the inlets and outlets of the cell. There are two inlets for the electrodes, the first of them is for the reference electrode, which is a silver-silver chloride with 3 M KCl reference mini-electrode. The second inlet serves to take out the electrical connection of the specimen. The counter electrode consists of two platinum filaments that pass through the glass and are connected outside the cell to the potentiostat. When the cell is closed, it is under a completely insulated atmosphere. Additionally, an inlet and an outlet serve to introduce the electrolyte into the cell from a glass container by means of a pump. The glass container has a thermostatic jacket that controls the temperature of the solution. It is also possible to bubble nitrogen in the solution to deaerate it. **Fig. 9** shows two photographs of the cell in the CLSM stage.

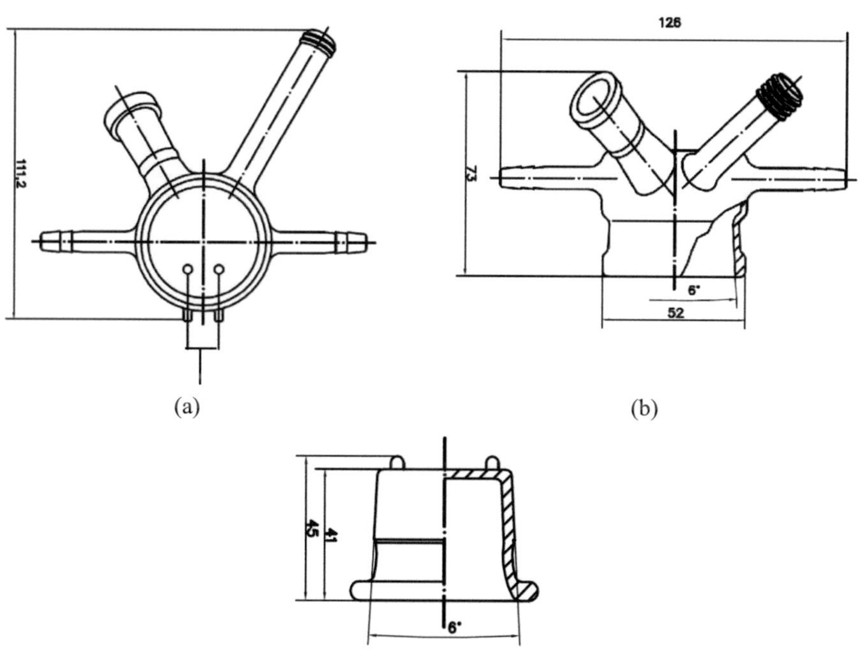

(a) (b) (c)

Fig. 8. Scheme of the parts of the minicell: (a) Main view of the top of the cell, (b) Cross view of the top of the cell, and (c) Cross view of the base of the cell. (Measurements have been made in millimetres).

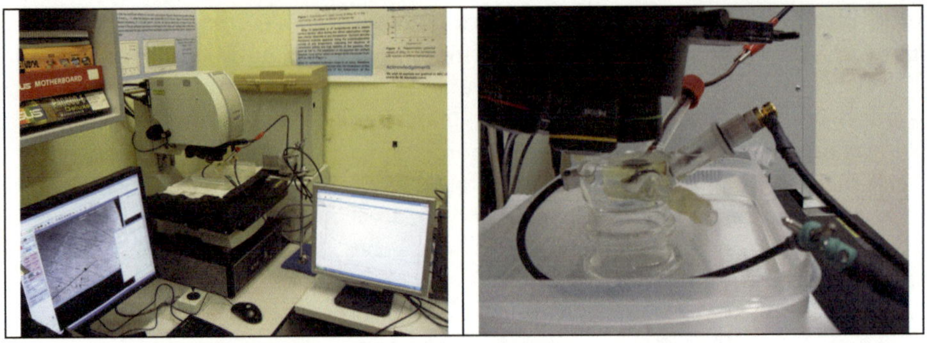

Fig. 9. Photographs of the minicell placed in the CLSM.

The material used in the minicell's tests was a duplex stainless steel, Alloy 900 (UNS 1.4462). The specimens were machined as shown in **Fig. 10**. These electrodes have a hole drilled on the base, where the electrical connection was made, and a cylindrically reduced rod 1.6 mm in diameter. The electrode was covered with an epoxy-resin; in this way only a circular area 1.6 mm in diameter was exposed to the corrosive solution. The electrical connection to the potentiostat was done by means of a conductor wire. Prior to the electrochemical tests, the specimens were wet abraded from 220 Silicon Carbide (SiC) grit to a 4000 SiC grit finish, and finally rinsed with distilled water.

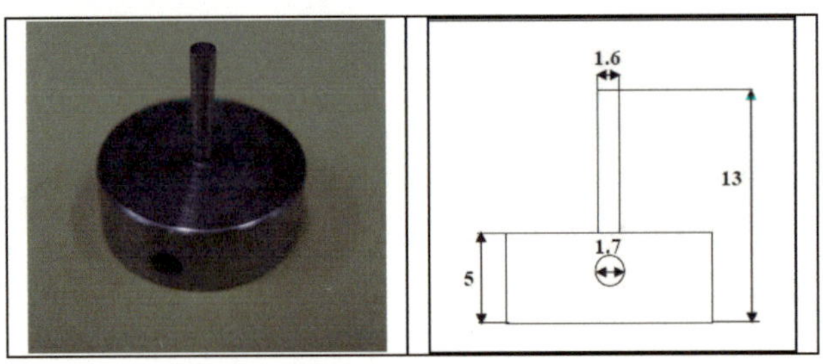

Fig. 10. Machined working electrode (measurements in millimetres).

Galvanodynamic curves were conducted in aqueous 992 g/L LiBr solutions at 25 °C in order to check the morphology of the attack with the CLSM. These tests allow controlling the current, and therefore, it is possible to control the corrosion rate when the damage begins. This issue is important because one of the objectives of the work is the study of the morphology of the first stages and propagation of corrosion; thus, it is very important to control the speed of the process. The H_2O-LiBr solution was selected because is the most commonly employed refrigerant/absorbent couple in absorption systems due to their favourable thermophysical properties. However, LiBr can cause serious corrosion problems on metallic components in refrigeration systems.

The 992 g/L LiBr aqueous solution was deaerated by bubbling nitrogen for 15 minutes, prior to immersion. Before each test, the specimen was immersed in the test solution for 1 hour at the open circuit potential (OCP). After the OCP test, the specimen's current density was reduced to 0 mA/cm^2 during 60 seconds so as to begin all the tests under the same conditions. Then, the galvanodynamic curves were recorded from 0 mA/cm^2 to positive current densities at a 10^{-5} mA/s sweep rate. Images of the surface of the working electrode were obtained during the tests with the CLSM. During the OCP an image of the tested specimen was obtained with the 200x objective. The field observed with the 200x objective is smaller than that of the specimen. Therefore, it is necessary to take different confocal images and later reconstruct the entire area of the specimen. **Fig. 11** shows an example of a specimen surface during OCP.

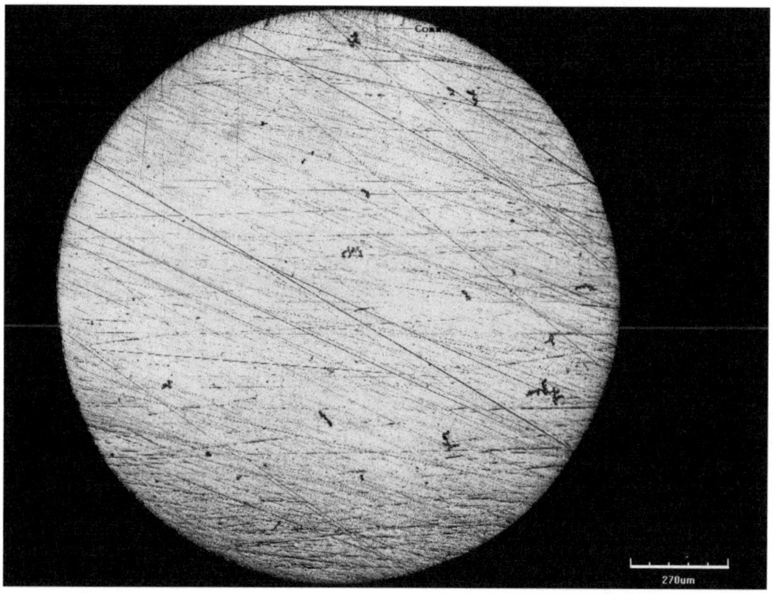

Fig. 11. Laser confocal image of the Alloy 900 specimen during OCP test in the LiBr solution. The image is a tiling reconstruction of different fields obtained with the 200x objective.

Fig. 12 shows different steps of the corrosion evolution during a galvanodynamic test. In these images, it can be observed how the corrosion process spreads inside the cloud of corrosion product and affects first one of the phases and then the other phase. When the cloud of corrosion product grows, it catalyses the corrosion in the new occupied area. According to some authors (Dornhege M. et al., 2007; Wu B. et al., 1997), the release of aggressive anions from pits and the subsequent diffusion of corrosion product over the electrode surface causes weakening of the protective passive film and each active pit is more likely to generate further pits in its surrounding.

The area of the corrosion expansion was calculated by image analysis as the projected area of the image. It was observed that the larger the damaged area, the lower the effective current density. The effective current density is related to the damage area, this area is the anodic area (where the dissolution of the metal happens) and the entire current of the test has to pass through this area. **Fig. 13** shows the evolution of the damaged area and the effective current density during a galvanodynamic test. At the beginning of pit formation the effective current density is very high, and it is at this point when the passive film is broken. This is in accordance with the literature (Szklarska-Smialowska, 2002), because the breakdown of the passive film should be accompanied by a very high local current density.

Therefore, the beginning of the corrosion processes is very fast and the damaged area grows quickly during the first stages of the attack. In addition, the damaged area (anodic area) /unaffected area (cathodic area) ratio is very unfavourable for the pit. The greater the damaged area, the more slowly the corrosion product spreads over a new unaffected area.

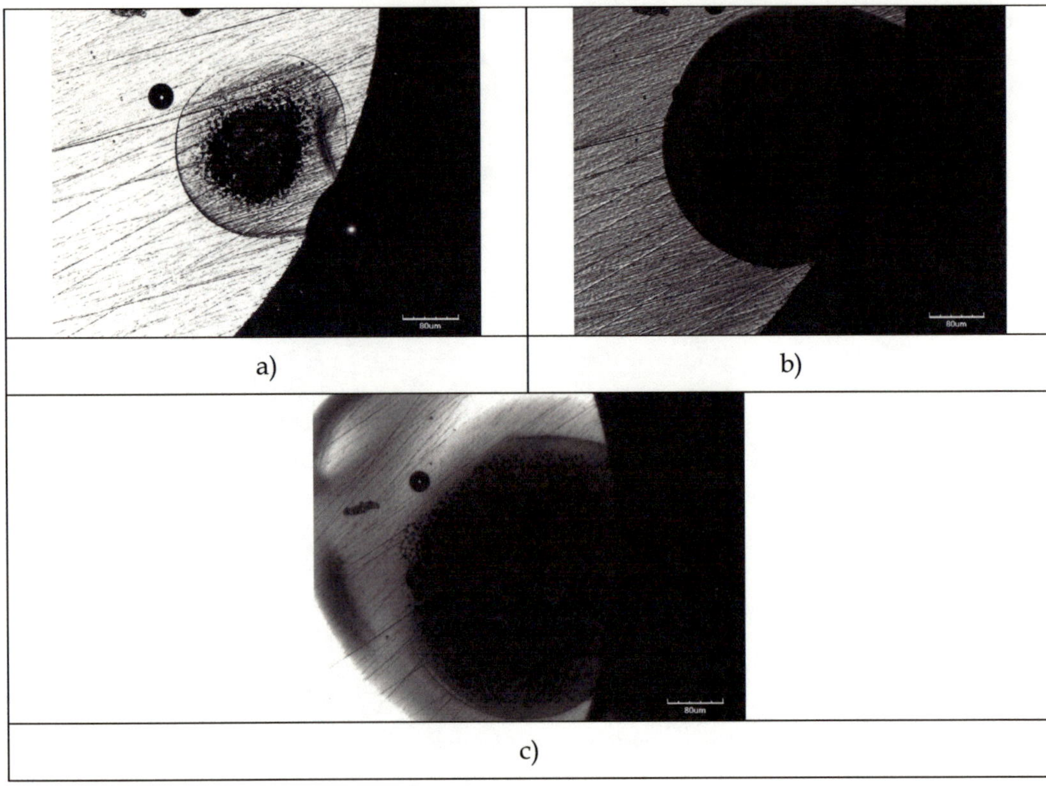

Fig. 12. Laser non-confocal images of the Alloy 900 specimen at different points of a galvanodynamic test in the LiBr solution at 25 °C.

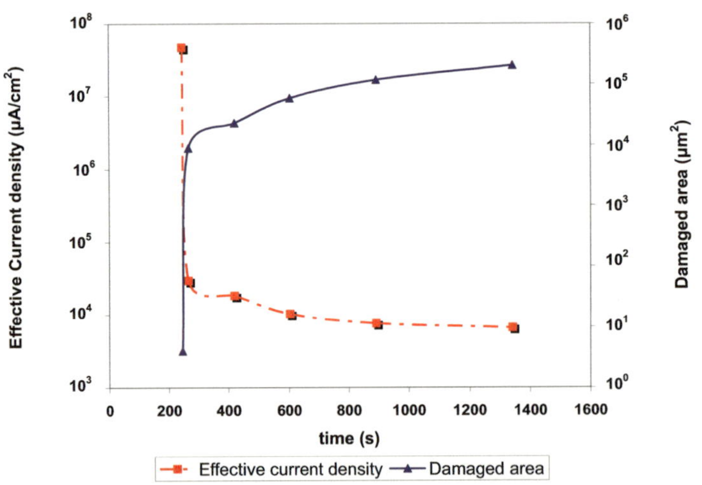

Fig. 13. Effective current density and evolution of the damage area during a galvanodynamic test in the 992 g/L LiBr solution at 25 °C.

The reason to consider the projected area as the effective area is the difficulty to measure on-line the real area during the tests, because there is a cloud of corrosion product over the affected pit. This point could be a drawback of the technique. However, this technique offers the possibility to detect the cloud of corrosion product with the CLSM because its different optical properties can offer additional information about corrosion processes. The volume of these clouds of corrosion product can be measured and related with the electrochemical signal. **Fig. 14** shows three different images of the growth of the cloud of corrosion product during a galvanodynamic test in the 992 g/L LiBr solution at 25°C.

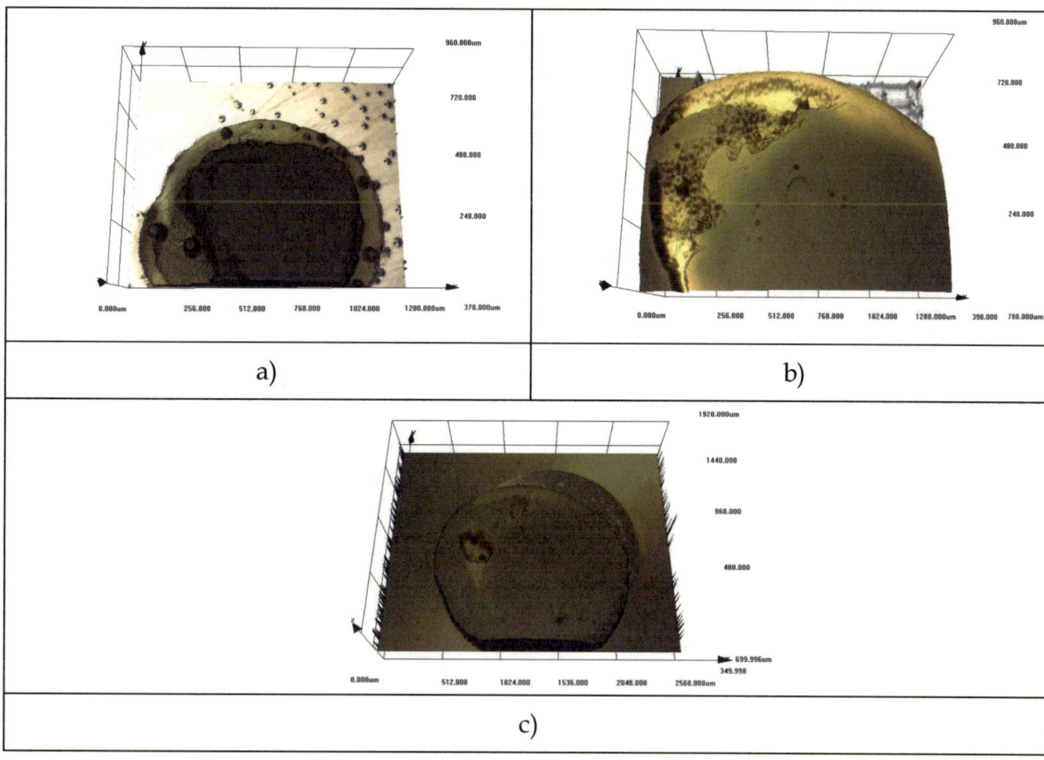

Fig. 14. 3-D colour images of the corrosion product over Alloy 900 during the galvanodynamic curve in 992 g/L LiBr solution at 25 °C.

5. Ex-situ visualization of corrosion effects

The evaluation of the specimens after the tests is an important part of the corrosion studies in order to analyse the morphology of the attack and to obtain metrological data of the damage. Confocal microscopy offers some advantages over conventional optical microscopy: correction of slanted surfaces, higher resolution, 3-D image generation, etc. Two applications of CLSM to ex-situ corrosion studies are presented below.

5.1 Ex-situ study of sensitised and unsensitised specimens

Fig. 15 shows two images of a highly alloyed austenitic stainless steel (Alloy 926) after a galvanodynamic curve in the 992 g/L LiBr solution at 25 °C. The first one corresponds to the alloy in the as-received state and the second one is the same alloy after heat treatment at 825 °C for 1 hour in an inert atmosphere.

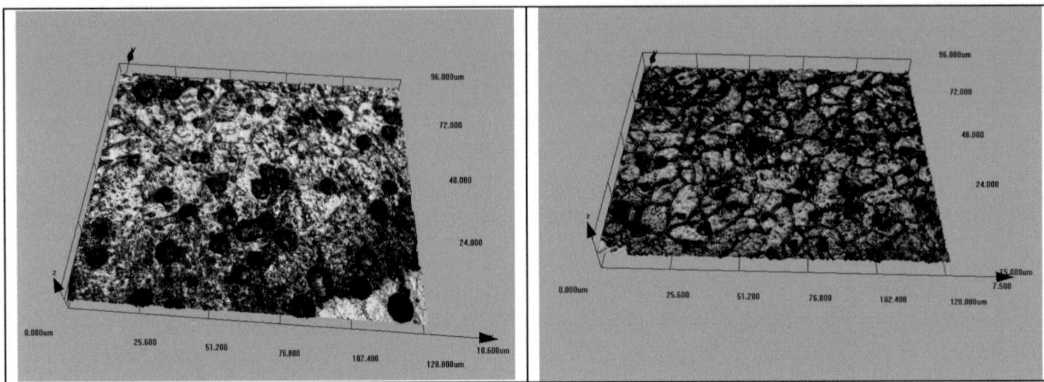

a) Alloy 926 in the as-received state	b) Alloy 926 in the sensitised state

Fig. 15. 3-D images in laser intensity of a highly alloyed austenitic stainless steel (Alloy 926) after a galvanodynamic curve in the 992 g/L LiBr solution at 25 °C.

The less corrosion resistant areas are those first attacked during the galvanodynamic tests. In the case of Alloy 926 in the as-received state, the formation of pits during corrosion spread is a random process because the probability of attack is the same in the entire surface. On the other hand in the sensitised Alloy 926, the attack grows along the grain boundaries. This kind of attack is due to the chromium carbide precipitation during heat treatment. These carbides deplete the alloying elements in the surrounding area, which are less corrosion resistant than the rest of the surface.

5.2 Ex-situ study of specimens under cavitation tests

In a recent work of this research group (Fernández-Domene et al., 2010), the regeneration of the passive film (protective oxide layer formed over stainless steels) after mechanical damage was studied. The passive film breakdown was produced by cavitation and the damage generated on the alloy surface was quantified by means of image analysis and CLSM. **Fig. 16** shows an image of a pit in Alloy 31 (highly alloyed stainless steel) under cavitation conditions in a 992 g/L LiBr solution at 25 °C. From the analysis of the images obtained with CLSM we can conclude that:

a. The attack morphology was different depending on the conditions. Under static conditions, the morphology of the damage generated on the electrode surface was uniform (although the initial mechanism was pitting corrosion). However, under cavitation conditions, the attack was localized and visible pits appeared on the surface.
b. Damage quantification on the Alloy 31 surface in 992 g/L LiBr solution during the potentiostatic tests, by cavitation– corrosion using Confocal Laser Scanning Microscopy and image analysis techniques showed that the volume, surface area and depth of these damages increased with increasing potentials. Moreover, a linear relationship between the damaged area and the applied potential was obtained.

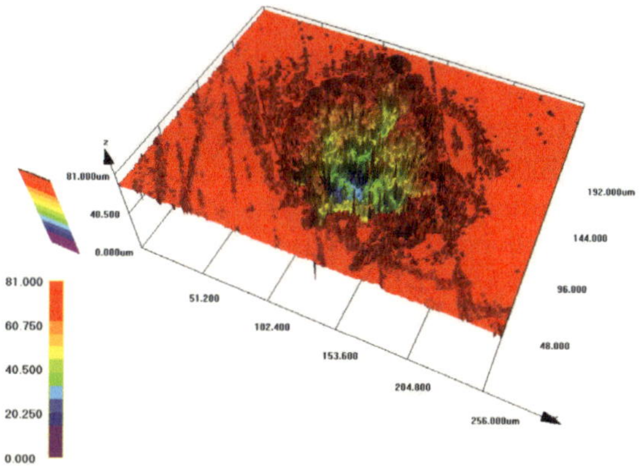

Fig. 16. 3-D Image of an Alloy 31 pit in 992 g/L LiBr solution, under cavitation conditions.

6. Conclusions

An important issue in corrosion studies is the analysis of attack morphology. CLSM presents different advantages over optical microscopy (higher horizontal and vertical resolution, 3-D reconstruction, correction of slanted surfaces…). Different applications of CLSM to corrosion research have been presented in this chapter. CLSM is a valuable technique for a better understanding of corrosion processes. The analysis of the in-situ processes has allowed us to observe morphological changes during heat treatments; these changes can be associated with further SEM analysis or electrochemical behaviour during corrosion tests. On the other hand, the electrochemical tests carried out using the minicell and CLSM help to observe in-situ the beginning and evolution of corrosion during electrochemical tests at high magnifications. Different metrological parameters like damage area or volume of the cloud of corrosion product can be obtained. Furthermore, the examination of the specimens after corrosion tests provides very useful and precise information about the morphology of the attack and its dimensions. Therefore, CLSM has demonstrated to be a powerful tool in the corrosion research.

7. Acknowledgements

We wish to express our gratitude to MEC (CTQ2009-07518), to FEDER, to the Generalitat Valenciana for its help in the CLSM acquisition (MY08/ISIRM/S/100) and to Dr. Asunción Jaime for her translation assistance

8. References

ASTM international (1994). ASTM G15-94 Standard Terminology Relating to Corrosion and Corrosion Testing. American Society for Testing Materials (1994), USA.

Dornhege M., Punckt C., Hudson J.L., & Rotermund HH. (2007). Spreading of corrosion on stainless steel. Simultaneous Observation of metastable pits and oxide film. *Journal of the electrochemical society,* 154, (november, 2007) C24-C27, ISSN 0013-4651.

Fernández-Domene R.M., Blasco Tamarit M.E., Garcia-Garcia D.M., & Garcia-Anton J (2010). Repassivation of the damage generated by cavitation on UNS N08031 in a LiBr solution by means of electrochemical techniques and Confocal Laser Scanning Microscopy. *Corrosion science,* 52, 10, (October, 2010), 3453-3464, ISSN 0010-938X.

García Anton J., Igual Muñoz A., Guiñón J.L., & Pérez Herranz V. (2000a). Electro-Optical Method by On-line Visualization of Electrochemical Process and Experimental Process. Spain P-200002525 . 2000a.

García Anton J., Igual Muñoz A., Guiñón J.L. & Pérez Herranz V. (2000b). Horizontal Electrochemical Cell by the Electro-Optical Analysis of Electrochemical Process. Spain P-200002526. 2000b.

García Anton J., Igual Muñoz A., Guiñón J.L. & Pérez Herranz V. (2001). A new technique for online visualization of the electrode surface under electrochemical corrosion processes. *Journal of Applied electrochemistry,* 31, 11, (November, 2001). 1195-1202, ISSN 0021-891X.

García Anton J., Igual Muñoz A., Guiñón J.L., Pérez Herranz V., & Pertusa-Grau J. (2003). Online visualization of corrosion processes of zinc and a Cu/Zn galvanic pair in LiBr solutions. *Corrosion,* 59, 2, (February, 2003), 172-180, ISSN 0010-9312.

García, C., de Tiedra M.P., Blanco Y., Martin O., & Martin F. (2008a). Intergranular corrosion of welded joints of austenitic stainless steels studied by using an electrochemical minicell. *Corrosion Science,* 50, 8, (August, 2008) 2390-2397, ISSN 0010-938X .

García, C., Martin F., de Tiedra P., Blanco Y., & Lopez M. (2008b). Pitting corrosion of welded joints of austenitic stainless steels studied by using an electrochemical minicell. *Corrosion Science,* 50, 4, (April, 2008) 1184-1194, ISSN 0010-938X.

Garcia-Antón J., Blasco Tamarit M.E., Garcia-Garcia D.M., Guiñón-Pina V., Leiva-Garcia R., & Pérez-Herranz V. (2008). Celda electroquímica de generación de gases para el análisis de procesos electroquímicos. (Spain) P-2000803389 . 2008.

Leiva-García R., Muñoz-Portero M.J., & García-Antón J. (2009). Evaluation of Alloy 146, 279, 900 and 926 sensitization to intergranular corrosion by means of electrochemical methods and image analysis. *Corrosion Science,* 51, 9, (September, 2009) 2080-2091, ISSN 0010-938X.

Leiva-García, R., García-Antón J., & Muñoz-Portero, M.J. (2010). Contribution to the elucidation of corrosion initiation through confocal laser scanning microscopy (CLSM). *Corrosion Science* 52, 6, (June, 2010) 2133-2142, ISSN 0010-938X.

Martin, F. A., C. Bataillon & J. Cousty. (2008). In situ AFM detection of pit onset location on a 304L stainless steel. *Corrosion Science,* 50, 1, (January, 2008) 84-92, ISSN 0010-938X.

Martins M. & Casteletti L.C. (2005). Heat Treatment Temperature Influence on ASTM A890 GR 6A Super Duplex stainless Steel Microstructure. *Materials characterization* 55, 3, (September, 2005) 225-233, ISSN 1044-5803.

Minsky, M. U.S. Patent No. 3013467. 1957.

Minsky, M. (1988). Memoir on inventing the confocal microscope. Scanning 10 (November, 1988), 128-138.

Pohl M., Storz O., & Glogowski T. (2007) Effect of intermetallics precipitations on the properties of duplex stainless steel. *Materials characterization* 58, 1, (January, 2007) 65-71 ISSN 1044-5803.

Szklarska-Smialowska Z. (2002). Mechanism of pit nucleation by electrical breakdown of the passive film. *Corrosion Science,* 44, 5, (May, 2002) 1143-1149, ISSN 0010-938X.

Wu B., Scully J.R., Hudson J.L., & Mikhailov A.S. (1997) Cooperative Stochastic Behavior in Localized Corrosion. *Journal of the electrochemical society* 144, 5, (May, 1997) 1614 - 1620, ISSN 0013-4651.

24

Registration between Multiple Laser Scanner Data Sets

Fei Deng
Wuhan University
China

1. Introduction

Laser scanners provide a three-dimensional sampled representation of the surfaces of objects with generally a very large number of points. The spatial resolution of the data is much higher than that of conventional surveying methods. Since laser scanners have a limited field of view, it is necessary to collect data from several locations in order to obtain a complete representation of an object. These data must be transformed into a common coordinate system. This procedure is called the registration of point clouds.

In terms of input data, registration methods can be classified into two categories: one is the registration of two point clouds from different scanner locations, so-called pair-wise registration (Rusinkiewicz & Levoy, 2001), and the other is simultaneous registration of multiple point clouds (Pulli, 1999; Williams & Bennamoun, 2001). However, the global registration of multiple scans is more difficult because of the large nonlinear search space and the huge number of point clouds involved.

Commercial software typically uses separately scanned markers that can be automatically identified as corresponding points. Akca (2003) uses the special targets attached onto the object(s) as landmarks and their 3-D coordinates are measured with a theodolite in a ground coordinate system before the scanning process. Radiometric and geometric information (shape, size, and planarity) are used to automatically find these targets in point clouds by using cross-correlation, the dimension test and the planarity test.

According to the automatic registration problems, several efforts have been made to avoid the use of artificial markers. One of the most popular methods is the Iterative Closest Point (ICP) algorithm developed by Besl & McKay(1992) and Chen & Medioni (1992). ICP operates two point clouds and an estimate of the aligning rigid body transform. It then iteratively refines the transform by alternating the steps of choosing corresponding points across the point clouds, and finding the best rotation and translation that minimizes an error metric based on the distance between the corresponding points. One key to this method is to have a good priori alignment. That means, for partially unorganized and overlapping points, if there is lack of good initial alignment, many other ICP variant don't work well because it becomes very hard to find corresponding points between the point clouds. Also, although a considerable amount of work on registration of point clouds from laser scanners, it is difficult to understand convergence behavior related to different starting conditions, and error metrics. Many experiment showed that the rate of convergence of ICP heavily relies on the choice of the corresponding point-pairs, and the distance function.

There are kinds of variants of ICP method to enhance the convergence behavior according to different error metrics, and point selection strategies. There are two common distance metrics. First error metric (Besl, 1992) is the Euclidian distance between the corresponding points, but it is highly time consuming due to the exhaustive search for the nearest point. Another one is the point-to-plane distance that described in (Chen,1991, Rusinkiewicz & Levoy ,2001), which uses the distance between a point and a planar approximation of the surface at the corresponding point. They found the point-to-plane distance to perform better than other distance measures. If there are a good initial position estimation and relatively low noise, ICP method with the point-to-plane metric has faster convergence than the point-to-point one. However, when those conditions cannot be guaranteed, the point-to-plane ICP is prone to fail (Gelfand et al.,2003).

The iterative closest compatible point (ICCP) algorithm has been proposed in order to reduce the search space of the ICP algorithm (Godin et al., 2001). In the ICCP algorithm, the distance minimization is performed only between the pairs of points considered compatible on basis of their viewpoint invariant attributes (curvature, color, normal vector, etc.). Invariant features coming from points, lines and other shapes and objects, moment invariants can be applied widely from classification, identification and matching tasks. Sharp(2002) presented the feature-based ICP method that also called ICP using Invariant Feature(ICPIF), which chooses nearest neighbor correspondences by a distance metric that represented a scaled sum of the positional and feature distances. Compared with traditional ICP, ICPIE converges to the goal state in fewer iterations, and doesn't need for a user supplied initial estimate. In order to overcome different point densities, noise, and partial overlap, Trummer(2009), extending an idea known form 2-dimensional affine point pattern matching, presented a non-iterative method to optimally assign invariant features that are calculated from the 3-dimentional surface without local fitting or matching. Compared to Sharp, this method doesn't use any initial solution.

As for line-based and plane-based literature, Bauer(2004) proposed a method for the coarse alignment of 3D point clouds using extracted 3D planes that they both are visible in each scan, which leads to reduce the number of unknown transform parameters from six to three. Remain unknowns can be calculated by an orthogonal rectification process and a simple 2D image matching process. Stamos and Allen (2002) illustrated partial task for range-to-range registration where conjugate 3D line features for solving the transformation parameters between scans were manually corresponded. Stamos and Leordeanu (2003) developed an automated feature-based registration algorithm which searches line and plane pairs in 3D point cloud space instead of 2D intensity image space. The pair-wise registrations generate a graph, in which the nodes are the individual scans and the edges are the transformations between the scans. Then, the graph algorithm registers each single scan related to a central pivot scan. Habib et al. (2005) utilized straight-line segments for registering LIDAR data sets and photogrammetric data sets though. Gruen and Akca (2005) developed the least squares approach tackling surface and curve matching for automatic co-registration of point clouds. Hansen (2007) presented a plane-based approach that the point clouds are first split into a regular raster and made a gradual progress for automatic registration. Jian (2005) presented a point set registration using Gaussian mixture models as a natural and simple way to represent the given point sets. Rabbani et al. (2007) integrated modeling and global registration where start out extracting geometric information for automatic detection and fitting of simple objects like planes, spheres, cylinders, and so forth, to register the scans. In Jaw(2007),an approach for registering ground-based LiDAR point clouds using overlapping scans based on 3D line features, is presented, where includes three major parts: a 3D line feature extractor, a 3D line feature matching mechanism, and a mathematical model for simultaneously registering ground-based point clouds of multi-scans on a 3D line feature basis. Flory(2010)illustrated surface fitting and registration of point clouds using approximations of the unsigned distance function to B-spline surfaces and point clouds. Nowadays, most of the laser scanners can supply intensity information in addition to the Cartesian coordinates for each point, or an additional camera may be used to collect texture. Therefore, further extension can simultaneously match intensity information and geometry under a combined estimation model. Kang (2009) presented an approach to automatic Image-based Registration (IBR) that refers to search for corresponding points based on the projected panoramic reflectance imagery that converted from 3D point clouds. In Roth (1999), a registration method based on matching the 3-D triangles constructed by 2-D interest points that are extracted from intensity data of each range image, was presented.

2. Iterative Closest Point (ICP)

2.1 Basic Concepts of ICP

The purpose of registration is to bring multiple image of the surveyed object in to the same coordinate system so that information from different views or different sensors can be integrated. Let f denote the imaging process, I is the image and O is the surveyed object, we can get:

$$I = f(O) \tag{1}$$

Considering the currently used surveying method, such as photogrammetry and laser scanning, we can say that f is an injection. What we have known is image fragments of the real world , and the destination of registration is then fitting all those fragments together. Mathematically, it is to find the relations between those f s. Concretely, the aim of registration is to find the rigid transformation T that can bring any pair of corresponding points (p_i, q_j) from the surfaces of two shapes P and Q representing the same point of the surveyed object into coincidence. The rigid transformation T is determined by

$$E(P,Q) = \iint_{\Omega} d(Tp(u,v),\ q(f(u,v),g(u,v)))^2 dudv = 0 \tag{2}$$

Where d is a distance function used to calculate the distance between two points, assume d is the Euclidian distance function, then

$$E(P,Q) = \iint_{\Omega} \|Tp(u,v) - q(f(u,v),g(u,v))\|^2 dudv = 0 \tag{3}$$

This function is usually used as the cost function in registration algorithms. For registration of point sets with point-to-point distance, it can be represented as discrete form:

$$\forall p_i \in P, \exists q_j \in Q \mid e_i = \|Tp_i - q_j\| = 0 \tag{4}$$

From the discrete representation we can directly get the solution idea of registration problem, finding corresponding points, in fact this can be treated as an absolute orientation problem without scaling which has been well solved by Horn(1997). He used unit quaternion to represent the rotation which simplifies this problem to a linear one and reduces this problem to a eigenvector problem, the rotation is estimated from finding the largest eigenvector of a covariance matrix, then translation can be found by the difference of the masses of two point clouds. More generally we can extend this to other features such as line segments, planar patches and some even more complex basic geometric models like spheres, cylinders and toruses (Rabbani, 2007).

Registration is a chicken-and-egg problem, if correspondences are know an estimation can be obtained by minimizing the error between the correspondences, if an initial estimation is known we can generate matches by transforming one point set to the other. The first approach is usually used for coarse registration and the second is for refinement. The ICP method belongs to the second approach and it is an excellent algorithm for registration refinement.

Since the introduction of ICP method (Besl & McKay, 1992, Chen & Medioni, 1991), it has been the most popular method for alignment of various 3d shapes with different geometric representation. It is widely used for registration of point clouds, and there have been many kinds of variants of basic ICP algorithm. However the basic concepts or workflow of ICP method is the same. It starts with two meshes and an initial estimate of the aligning rigid-body transform, then it iteratively refines the transform by alternately choosing corresponding points in the meshes through the given initial estimation and finding the best translation and rotation that minimizes an error metric based on the distance between them. Intuitively the ICP method chooses the reference point p and the closest point q of surface Q as a point pair, that's why it is called iterative closest point algorithm. The basic workflow of ICP method is as below,

---

Input: model P and Q with overlapping region, initial estimation of the rigid transformation T_i which transform P to the coordinate system of Q ;

- Select reference points p_j on P ;
- For every reference point p_j ,transform the reference point to the coordinate system of Q with T, the new point will be $p'_j = Tp_j$;
- For every transformed reference point p'_j, select the closest point q_i on Q which is closest to p'_j ;
- Using the point pairs (p_j, q_j) to calculate a new estimation of the rigid transform T_{i+1}
- Iteratively repeat this procedure until the difference between the two transformation T_{i+1} and T_i is little enough.

Rusinkiewicz & Levoy(2001) gave a Taxonomy of ICP variants according to the methods used in the stages of the ICP method. Salvi et al.(2007) gave a survey of recent range image registration methods also include some ICP methods. Because the ICP method has to identify the closest point on a model to a reference point, and thus affect the quality of the point pairs and the error metric directly, various definitions of closest point such as point-to-point, point-to-(tangent) plane and point-to-projection were proposed for accuracy or efficiency]. As an optimization problem, the stability is very important; kinds of sampling

method, outlier detection method and robust estimate method were proposed for ICP method.
In this part, we will first discuss the usually used point-to-point, point-to-plane and point-to-projection error metric and introduce some method proposed for the improving of stability of the ICP method.

2.2 Closest point definition and finding

As the quality of alignment obtained by this algorithm depends heavily on choosing good pairs of corresponding points in the two datasets, one of the key problems of ICP method is how to define the closest point, it influences the convergence performance and speed and accuracy of the algorithm directly. In order to improve the ICP method various variants of ICP method have been developed with different definition of closest point. Rusinkiewicz and Levoy's survey about variants of ICP introduced a taxonomy of some of these methods. According to him, these methods can be classified into several groups based on their error metrics, including direct point-to-point method, methods using normal of the points, such as point-to-(tangent) plane methods. Additional information such as range camera information such as point-to-projection method and intensity or colors can be used to reduce the search effort of correspondent point pairs or to eliminate the ambiguities due to inadequate geometric information on the object surface . All of these methods can be accelerated with KD-Tree searching (Simon, 1996), Z-buffer (Benjemaa & Schmitt, 1999) or closest-point caching (Simon,1996, Nishino & Ikeuchi, 2002).
Besl's original ICP method aims at the registration of 3d shapes, so he proposed kinds of definition of closest point for a given point on various geometric representations. We focus on the point set representation. Let Q be a point set with Nq points, d is an Euclidean distance calculator, the distance of closest point of Q to $\vec{p}$ of point set P equals to the distance between the closest point $\vec{q}_i$ of Q to the point $\vec{p}$,

$$d(\vec{p},Q) = \min_{i\in\{1,2,\cdots,N_q\}} d(\vec{p},\vec{q}_i) \tag{5}$$

And the residual for each pair of points $(\vec{p}_i,\vec{q}_i)$ is

$$e_i = T\vec{p}_i - \vec{q}_i \tag{6}$$

Besl's original work used naive point-to-point error metric, adopting the Euclidian distance between corresponding points as the error metric. It belongs to explicit method which has to detect corresponding points on the other surface. And the problem is the low convergence speed of the algorithm for certain types of data sets and initial estimation.
Laser scanning sensors often combined with image sensors, it is easy to obtain both the range and color information of the surveyed object. Weik (1997) used intensity information to detect corresponding points between two range images. Godin et al.(1994) first proposed iterative closest compatible point (ICCP) method with additional intensity information, then adopted this method with invariant computation of curvatures (Godin et al., 1995), and introduced a method for the registration of attributed range images (Godin et al., 2001). He gave the distance between two attributed points in a (3+m) dimensional space with m additional attributes like color (m=3).

$$d(\vec{p},\vec{q}) = (\vec{p}_x - \vec{q}_x) + \sum_{i=1}^{m} \lambda_i (\vec{p}_a^{\,i} - \vec{q}_a^{\,i}) \tag{7}$$

Where $\vec{p}_x$ and $\vec{q}_x$ are positions of the two points, $\vec{p}_a^{\,i}$ and $\vec{q}_a^{\,i}$ are the ith components of the additional attributes. Johnson & Kang(1997) took both the distance and the intensity or color difference into account. He found the closest point in a space with six freedoms, three for coordinates in the Euclidian space, and three for the color space. The distance between two corresponding point p with position (x_1^p, x_2^p, x_3^p) and color (c_1^p, c_2^p, c_3^p) and q with position (x_1^q, x_2^q, x_3^q) and color (c_1^q, c_2^q, c_3^q) is

$$d(\vec{p},\vec{q}) = [(x_1^p - x_1^q)^2 + (x_2^p - x_2^q)^2 + (x_3^p - x_3^q)^2 + \alpha_1 (c_1^p - c_1^q)^2 + \alpha_2 (c_2^p - c_2^q)^2 + \alpha_3 (c_3^p - c_3^q)^2]^{\frac{1}{2}} \tag{8}$$

where $\alpha = (\alpha_1, \alpha_2, \alpha_3)$ are scale factors that weigh the importance of color against the importance of shape. In order to eliminate the effect of shading which influence the intensity of the color. Johnson and Kang employ the YIQ color model which separate intensity (Y) from hue (I) and saturation(Q) and make the scale of the Y channel one tenth the scale of the I and Q channels. Let $(c_1, c_2, c_3) = (y, i, q)$, then $\alpha = (\alpha_1, \alpha_2, \alpha_3) = (0.1, 1, 1)$

However the point-to-point error metric doesn't take the surface information into account, this point-point distance based error metric suffers from the inability to "slide" overlapping range images (Nishino & Ikeuchi,2002), and may converges to a local minimum with a noise data set. Pottman & Hofer(2002) did a research on squared distance function to curves and surfaces, according to their work the point-to-point distance is only good when the two surface are aligned with a long distance after the initial transformation. As we have mentioned, a good initial estimation is essential to avoid running into a local minimum position for ICP registration method, that means the distance between the two surfaces should be close. Pottman and Hofer indicated that the point-to-tangent plane distance is exactly the second order Taylor approximant of the distance between the two surfaces.

In order to get a better registration result, point-to-plane metric is widely used (Chen & Medioni 1991; Dorai et al.,1997). Chen and Medioni employ a line-surface intersection method which can be subsumed in point-to-(tangent) plane methods. They treat the intersection point of the line l passing through the point $\vec{p}$ directing to the normal of the surface P at $\vec{p}$ as the closest point.

$$l = \{\vec{a} \mid \vec{n}_p \times (\vec{p} - \vec{a})\} \tag{9}$$

The green points indicate the real position of the correspondence, the red one is the identical point obtained from corresponding method. For point-to-point metric, it is the closest point to p. The calculating for point-to-plane metric is much more complex, we have to get the point q which is the intersection of the normal vector at p of surface P and the surface Q, then the distance from the point p to the tangent plane of Q at point q as the metric. The point-to-projection metric can be get easily from the projection of point p onto surface Q from the scanning station of Q.

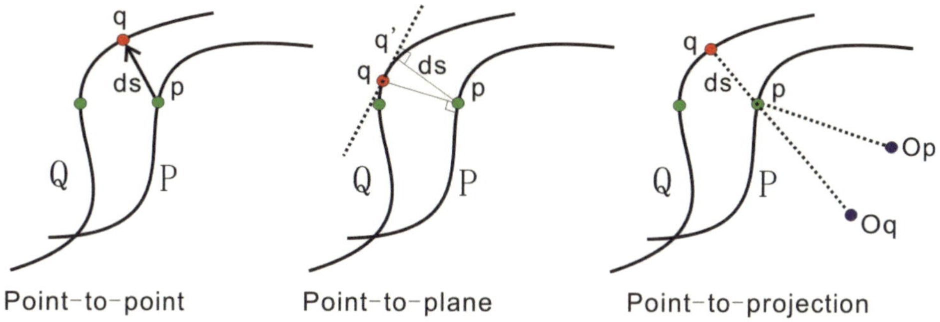

Fig. 1. From left to right are point-to-point point-to-plane and point-to-projection metric.

And the residual for each pair of points $(\vec{p}_i, \vec{q}_i)$ is

$$e_i = (T\vec{p}_i - \vec{q}_i) \cdot n_{\vec{p}_i} \tag{10}$$

By embedding the surface information into the error metric in this way, point-plane distance metric based methods tend to be robust against local minima and converge quickly with an accurate estimation. However the computation of point-plane distance is too expensive, given that we have gotten the initial estimation, we can take the advantage of the information to accelerate the closest point finding process, then point-to-projection methods using viewing direction to find the correspondence are proposed for efficiency (Benjemaa & Schmitt, 1999; Neugebauer, 1997; Blais & Levine, 1995).

The principal idea of Blais and Levine's points-to-projection method is to firstly project the point $\vec{p}$ backward to a 2D point $\vec{p}_Q$ on the predefined range image plane of the destination surface scanned at station O_Q, and then $\vec{p}_Q$ is forward projected to the destination surface Q from the station O_Q to get $\vec{q}$. The obvious drawback of this method is that the closest point between the two surfaces obtained can't represent the closeness of the two surfaces, consequently resulting in bad accuracy.

Park & Subbarao (2003) introduced the contractive projection point algorithm which take the advantage of both the point-to-plane metric's accuracy and point-to-projection metric's efficiency. He projects the source point to the destination surface then re-project the projection point to the normal vector of the source point, by iteratively applying the normal projection the point will converge to the intersection point of point-to-plane method.

As the Fig. 2 showed, first project p_0 onto a image plane viewed from the station O_Q, then find the intersection point q_{p_0} of the line of sight from O_Q to p_0 and the surface Q, and project this point onto the normal of the point p_0 of surface P to get the point p_1. Iteratively do forward projection and normal projection to p_i we can get the approximate correspondence of p_0.

Assume M_Q is the perspective projection matrix of view Q to the image plane I_Q, T_Q is the transformation matrix from the camera coordinate system of view Q to the world coordinate system, I_Q is a 2d image plane with respect to the data set Q which can be treated as a range image viewed from station O_Q. First back project $\vec{p}_0$ onto the image plane,

$$P_q = M_Q T_Q^{-1} \vec{p}_0 \tag{11}$$

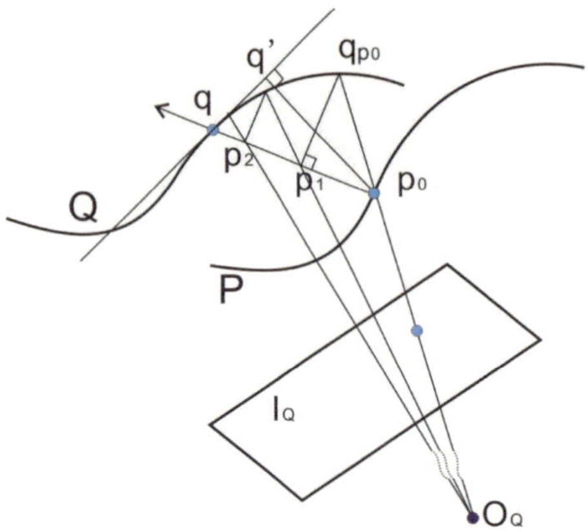

Fig. 2. Finding the closest point with the Park and Subbarao's method.

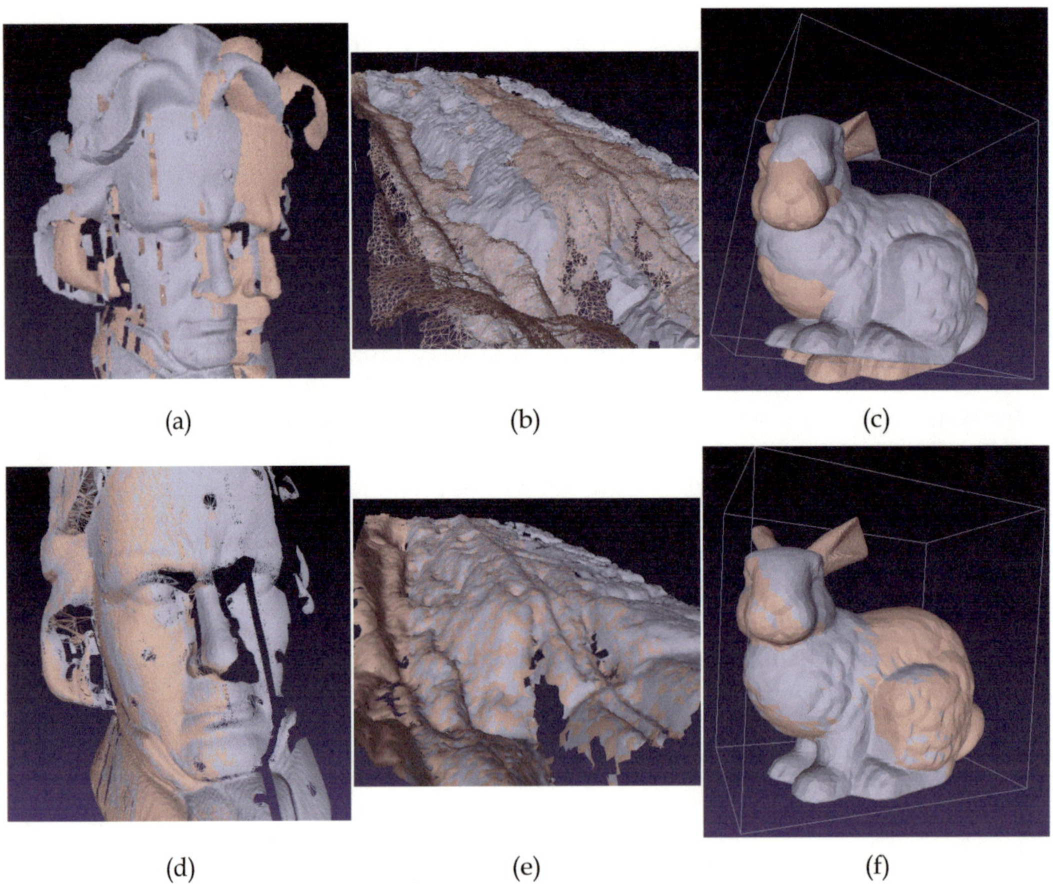

Fig. 3. The alignment with point-to-point ICP method. (a) ,(b)and (c) are initial transformed with the initial estimation of the transformed used as input in the ICP algorithm; (d) and (e) are aligned with the point-to-point ICP method, (f) is aligned with the point-to-plane ICP method.

Then forward-project the point P_q onto the surface Q, we get point q_{p_0}. In fact q_{p_0} is interpolated from the range image I_Q. Next is the so called normal projection which project the point q_{p_0} onto the normal vector $\hat{p}$ at $\vec{p}_0$ to get the point $\vec{p}_1$,

$$\vec{p}_1 = \vec{p}_0 + (\vec{q}_{p_0} - \vec{p}_0)\bullet\hat{p}\bullet\hat{p} \tag{12}$$

By iterating the similar projections to $\vec{p}_1$, $\vec{p}_i$ will converge to the intersection point $\vec{q}_s$ of point-to-plane method if i goes to infinity. Experimentally this method is fast for both pair-wise registration and multi-view registration problems.

2.3 Techniques for stability

Though we can find corresponding points through above methods, the quality of alignment obtained by those algorithms depend heavily on choosing good pairs of corresponding points. If the input datasets are noised or bad conditioned, the correspondence selection methods and the outlier detection methods play a very important role in the registration problem. In fact correspondence selection and outlier detection essentially mean to the same thing which keep good matches and eliminate bad matches.

First, we will give some details of Gelfand et al.'s method (Gelfand et al., 2003) to analyze geometric stability of point-to-pane ICP method based on 6x6 co-variance matrixes. By decompose the rigid transformation T into rotation R and translation $\vec{t}$, the residual of point-to-plane error metric can be written as

$$e_i = (T\vec{p}_i - \vec{q}_i)\cdot n_{\vec{p}_i} = (R\vec{p}_i + \vec{t} - \vec{q}_i)\cdot n_{\vec{p}_i} = (\vec{r}\times\vec{p}_i + \vec{t} - \vec{q}_i)\cdot n_{\vec{p}_i} \tag{13}$$

Where $\vec{r}$ is a (3×1) vector of rotations around the x, y, and z axes, and $\vec{t}$ is the translation vector. Replace the functor d in formulation (1) instead of (13) then we can get

$$\begin{aligned} E &= \sum_{i=1}^{k}((R\vec{p}_i + \vec{t} - \vec{q}_i)\cdot n_i)^2 \\ &= \sum_{i=1}^{k}((\vec{p}_i - \vec{q}_i)\cdot n_i + \vec{r}\cdot(\vec{p}_i\times\vec{n}_i) + \vec{t}\cdot\vec{n}_i)^2 \end{aligned} \tag{14}$$

Given an increment $(\Delta\vec{r}^T\ \Delta\vec{t}^T)$ to the transformation vector $(\vec{r}^T\ \vec{t}^T)$, the point-to-plane distance will correspondingly changed by

$$\Delta d_i = [\Delta\vec{r}^T\ \Delta\vec{t}^T]\begin{bmatrix} p_i\times n_i \\ n_i \end{bmatrix} \tag{15}$$

By linearizing the residual functions, we will get the covariance matrix

$$C = \begin{bmatrix} p_1 n_1 & \cdots & p_k n_k \\ n1 & \cdots & nk \end{bmatrix}\begin{bmatrix} (p_1 n_1)^T & (n_1)^T \\ \cdots & \cdots \\ (p_k n_k)^T & (nk)^T \end{bmatrix} \tag{16}$$

Formulation (15) tells us if $p_i\times n_i$ is perpendicular to $\vec{r}$ and n_i is perpendicular to $\vec{t}$, the distance to plane will always be zero, that means the error function (14) will not change.

This will cause problems when using the covariant matrix to calculate the transformation vector $(\vec{r}^T\ \vec{t}^T)$, because certain types of geometry may lead the matrix to be a singular one which means the answer will not be unique. So Gelfand et al. used the condition number of the covariance matrix as a measure of stability, and gave some simple shapes below which may cause problems.

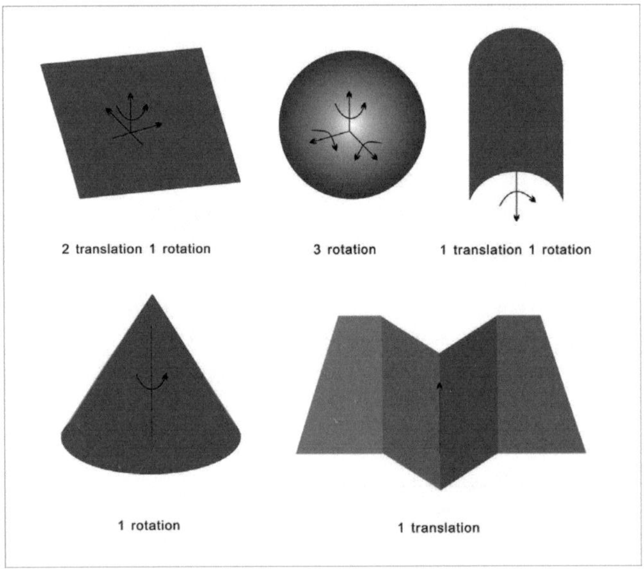

Fig. 4. Unstable shapes and the corresponding number and types of instability.

The original and early ICP methods (Besl & McKay, 1992, Chen & Medioni, 1991) usually chose all available points, some uniformly subsample the available points (Turk, 1994), others use random sampling method (Masuda et al.,1996). If too many points are chosen from featureless regions of the data, the algorithm converges slowly, finds the wrong pose, or even diverges, so color information is also used for point pairs selection (Weik, 1997). Normal-space sampling is another simple method of using surface features for selection (Johnson & Kang, 1997), it first buckets the points according to the normal vectors in angular space, then sample uniformly across the buckets. Rusinkiewicz & Levoy(2001) suggested another Normal-space sampling method which select points with large normals instead of uniformly sampling.

As sliding between datasets can be detected by analyzing the covariance matrix used for error minimization (Guehring, 2001), the adoption of weighting methods based on the contribution of point pairs to the covariance matrix is reasonable. Dorai et al(1997) improved Chen's method by weighting the residuals of those point pairs, extended it to an optimal weighted least-squares framework to weaken the contribution of the residual of outliers to the error function, as a point-to-plane method, the derivative of the standard deviation of point-to-plane distance is a good choice for the weighting factor. Simon(1997) iteratively adds and removes point-pairs to provide the best-conditioned covariance matrix, Gelfand et al. brought this method to the covariance-based sampling method. Assume L_i and x_i (i=1,…,6) are corresponding eigenvalues and eigenvectors of covariance matrix C, n_{p_i} is the normal vector at p_i, and $v_i = [p_i \times n_{p_i}, n_{p_i}]$. Then, the magnitude of $v_i \cdot x_k$ represent the error the pair of points will bring in. The covariance-based sampling aims to minimize the sum of this value to all of chosen pairs over the six eigenvector. That is when choosing the nth pair we have to guarantee

$$\min\{V \mid \max\{\sum_{i=1}^{n} v_i \cdot x_k, (k=1,\cdots,6)\}\} \tag{17}$$

Methods with thresholds are also proposed. The distance thresholder of Schutz et al.(1998) regards the point pairs whose distance is much greater than the distance s between the centers of masses of the two point clouds as outliers, that is if $\|Tp_i - q_j\|^2 > (c \cdot s)^2$, then the point pair (p_i, q_j) is an outlier, where c is an empirical threshold. Assuming the distances between corresponding point pairs are distributed as a Gaussian, Zhang(1994) introduced a statistical outlier classifier which examines the statistical distribution of unsigned point pair distances to dynamically estimate the threshold to distinguish which pairs are outliers. Dalley & Flynn (2002) did a comparison between original point-to-point ICP, point-to-point ICP with Schutz's and Zhang's classifier and point-to-plane ICP, he found that point-to-point ICP with Zhang's outlier detection method is sometimes even better than point-to-plane ICP method.

Generally the selection methods of point pairs, the outlier detection methods and threshold methods improve the stability of ICP method to certain conditions, this problem is far from well done, since no general method has been proposed to suit all of these bad conditions. The ultimate solution of these problems is to improve the quality of the datasets.

3. Feature based registration

The commonly used features for registration in 2D images are feature points, which are easy to find with various corner point detection methods, such as the very popular method Scale-invariant feature transform (SIFT)(Lowe, 2004). Contrarily point features of point cloud are hard to detect, instead line and plane features which can be easily extracted are widely used in feature based registration method. Line features are usually extracted from the intersection of plane features, plane features can be easily automatically extracted from point clouds with scan line or surface growing methods (Vosselman et al.,2004, Stamos, 2001). If we can found the matching between those features automatically the registration process can be automated without initial estimation or supply initial estimation for other algorithms, this is very important because fine registration methods such as ICP usually need initial estimation, if we can provide an initial transformation to the fine registration process, the whole procedure will be automated. Stamos's group used automatically extracted line features to do range-range image registration and range-2d image registration (Stamos & Leordeanu, 2003, Stamos, 2001, Chao & Stamos, 2005), they also proposed a circular feature based method (Chen & Stamos, 2006). He et al.(2005) introduced an interpretation tree based registration method with complete plane patches. Von Hansen(2007) grouped surface elements to large planes then adopted a generate-and-test strategy to find the transformation. Dold & Brenner(2006) and Pathak et al.(2010) brought geometric constraints into the matching of plane matches of the plane based registration method, and Makadia et al.(2006) proposed an extended Gaussian image based method to estimate the rotation automatically. Jaw & Chuang(2007) introduced a framework to integrate point, line and plane features together, and compared the integrated method with algorithms using such features separately, he found that the integrated method is much more stable than those separated ones. Rabbani et al.(2007) goes even far more, he integrated the modeling and registration together, simultaneously determined the shape and pose parameters of the objects as well as the registration parameters. In this chapter we will give the functional and stochastic models of point, line, and plane features based registration, and some of their properties.

3.1 Point features

Given a pair of point features $(\vec{p}_i, \vec{q}_i)$, Apply a rigid transformation T with a 3×3 rotation R and translation $\vec{t} = (t_x \; t_y \; t_z)^T$, we can get the mathematic model

$$v = R\vec{p}_i + \vec{t} - \vec{q}_i = 0 \tag{18}$$

In this model point pairs $(\vec{p}_i, \vec{q}_i)$ are measurements, R and $\vec{t}$ are parameters we need to estimate, so this equation can be treated as an adjustment of condition equations with unknown parameters. To solve this problem we need at least three point pairs, one for translation the other two for rotation. Given one point pair $(\vec{p}_1, \vec{q}_1)$, we can fix the point $\vec{p}_1$ with the point $\vec{q}_1$, then the point cloud P can rotate about $\vec{p}_1$, if another pair was known we $(\vec{p}_2, \vec{q}_2)$ can rotate $\vec{p}_2$ to $\vec{q}_2$, at this time the point cloud P can rotate about the line l cross $\vec{p}_1$ and $\vec{p}_2$, the third pair $(\vec{p}_3, \vec{q}_3)$ will align the point cloud P coincident with Q by rotating about the line l and fix $\vec{p}_3$ to $\vec{q}_3$.

There are various representation of rotation, such as Euler angle, $(\varphi, \omega, \kappa)$ system, roll-pitch-yaw system, rotation matrix, and quaternion system. Rotation matrix is the usually used representation for analysis, but it has nine elements, because there are only three freedoms for a rotation, there are six nonlinear conditions between the nine elements to form an orthogonal matrix which is troublesome in calculation. The first three methods represent the rotation with three angles; the difference between them is little in calculating process, so we will discuss them together as an angle system. The last quaternion method has been proven very useful in representing rotations due to several advantages above other representations; the most important one is that it varies continuously over the unit sphere without discontinuous jumps.

The point feature based method is the special case in absolute orientation without scaling, Horn [14] adopted quaternion representation of rotation into it and reduced it to a linear problem. Any rotation can be represented as a quaternion $q = [q_0 \; q_1 \; q_2 \; q_3]$, as rotation in R^3 has only three freedom, so the quaternion must be constrained by $\|q\| = \sqrt{q_0^2 + q_1^2 + q_2^2 + q_3^2} = 1$. The relation between the quaternion and the rotation matrix is

$$R = \begin{bmatrix} 2q_0^2 + 2q_1^2 - 1 & 2q_1q_2 - 2q_0q_3 & 2q_1q_3 + 2q_0q_2 \\ 2q_1q_2 + 2q_0q_3 & 2q_0^2 + 2q_2^2 - 1 & 2q_2q_3 - 2q_0q_1 \\ 2q_1q_3 - 2q_0q_2 & 2q_2q_3 + 2q_0q_1 & 2q_0^2 + 2q_3^2 - 1 \end{bmatrix} \tag{19}$$

Horn found that the matrix

$$M = \sum_{i=1}^{n} \vec{p}_i \vec{q}_i^T = \begin{bmatrix} S_{xx} & S_{xy} & S_{xz} \\ S_{yx} & S_{yy} & S_{yz} \\ S_{zx} & S_{zy} & S_{zz} \end{bmatrix} \tag{20}$$

Whose elements are sum of product of matched point pairs contains all the information required to solve the least square problem.

Let $N = \begin{bmatrix} S_{xx}+S_{yy}+S_{zz} & S_{yz}-S_{zy} & S_{zx}-S_{xz} \\ S_{yz}-S_{zy} & S_{xx}-S_{yy}-S_{zz} & S_{xy}+S_{yx} \\ S_{xy}-S_{yx} & S_{zx}+S_{xz} & -S_{xx}+S_{yy}-S_{zz} \end{bmatrix}$, the result unit quaternion will be the eigenvector corresponding to the most positive eigenvalue. Once the rotation is solved, the translation can be solved by calculating the direction from the mass of the rotated moving point set to the mass of the fixed point set.

The linear solution doesn't take the error of measurements into account, for better accuracy we will introduce the least square method below. No matter which representation is used, by taking the first-order partial derivatives of Eq. 18 with respect to the parameters, we can generally linearize Eq. 18 then adopt a least square method to get this problem solved. Because the coordinates of the feature points are also measurements which are not accurate, taking this into consideration we can even refine them by adding them to the parameters needed refined and take the first-order partial derivatives with respect to them. The initial value of the transformation can be obtained from the linear idea introduced above, then for every pair of points the general form of the linearized functional model is

$$A_{3\times 6} V_{6\times 1} + B_{3\times 6 or 3\times 7} \hat{x}_{6\times 1 or 7\times 1} + e_{3\times 1} = 0 \tag{21}$$

where V is the incremental parameter vector of the coordinates of the pair of points, $\hat{x}$ is the incremental parameter vector of the parameters of the transformation, e is the error vector of every pair of points.

Given n point pairs the functional model and statistical model will be

$$A_{3n\times 6} V_{6n\times 1} + B_{3n\times 6 or 3n\times 7} \hat{x}_{6\times 1 or 7\times 1} + e_{3\times 1} = 0 \tag{22}$$

$$D_{3n\times 3n} = \sigma_0^2 Q_{3n\times 3n} = \sigma_0^2 P_{3n\times 3n}^{-1} \tag{23}$$

where σ_0 is the mean error of unit weight along x, y, z directions for each pair of points, Q is the covariance matrix, P can be regarded as the weight matrix of each pair of points which indicate the matching degree along the three direction, if we assume the weight of all of the directions of a pair of points (p_i, q_i) is the same, e.g. w_i, the weight matrix for each correspondence will be $P_i = diag(w_i, w_i, w_i)$, the matrix P will be

$$P = diag(P_1, P_2, \cdots, P_n) \tag{24}$$

Apply condition adjustment with unknown parameters, we can get

$$\hat{x} = -N_{bb}^{-1} B^T N_{aa}^{-1} e \text{ where } N_{aa} = AQA^T, N_{bb} = B^T N_{aa}^{-1} B \tag{25}$$

3.2 Line features

Line segments are very common in survey especially man-made features, they are more obvious and the matches between line features are more intuitive. Because the limitation of the resolution of the scanner, we can't find accurate point-to-point matches in the datasets, the line feature based method gets out of the hard work of finding point-to-point match because of more accurate and robust line features. The general expression of line is the intersection of two plane, however with such an expression this problem can be extended to the plane based method, so we choose six parameters $(x_0\ y_0\ z_0\ n_x\ n_y\ n_z)^T$ to define the line, three $(n_x\ n_y\ n_z)^T$ for the unit vector of the direction which has one constraint, the other three $(x_0\ y_0\ z_0)^T$ for position of the line. Lines in R^3 can be represented as

$$\frac{x-x_0}{n_x}=\frac{y-y_0}{n_y}=\frac{z-z_0}{n_z},(n_x^2+n_y^2+n_z^2=1) \tag{26}$$

For each line correspondence, there are two constraints, the first one is the transformed unit direction must be the same, secondly the already known point $(x_0^P\ y_0^P\ z_0^P)^T$ on the line feature l_P detected from point set P should be on the corresponding line l_Q extracted from point set Q.

$$\vec{n}_Q = R\vec{n}_P \tag{27}$$

$$\frac{x'-x_0^Q}{n_x^Q}=\frac{y'-y_0^Q}{n_y^Q}=\frac{z'-z_0^Q}{n_z^Q},\ where\ \vec{x}'_Q=\begin{bmatrix}x'\\y'\\z'\end{bmatrix}=R\vec{x}_Q+\vec{t} \tag{28}$$

As the line feature correspondences are conjugate, we can get another function

$$\frac{x''-x_0^P}{n_x^P}=\frac{y''-y_0^P}{n_y^P}=\frac{z''-z_0^P}{n_z^P},\ where\ \vec{x}''_P=\begin{bmatrix}x'\\y'\\z'\end{bmatrix}=R^T(\vec{x}_P-\vec{t}) \tag{29}$$

If we regard $\vec{n}_P$ and $\vec{n}_Q$ as points, we can estimate the rotation with Horn's method directly just like the point based method, and then the translation can be also calculated directly from Eq. 28 and Eq.29.

For iterative least square method, the linearized functional model and statistic model are

$$A_{7n\times12}V_{12n\times1}+B_{7n\times6or7n\times7}\hat{x}_{6\times1or7\times1}+e_{7\times1}=0 \tag{30}$$

$$D_{7n\times7n}=\sigma_0^2Q_{7n\times7n}=\sigma_0^2P_{7n\times7n}^{-1} \tag{31}$$

For line feature based method, we need at least 2 pairs of nondegenerate line which are not parallel to get the transformation solved, given the first line correspondence, we can fix the line of the moving point cloud to the corresponding line of the fixed point cloud, the moving point cloud can still move along this line and rotate about this line, once the other nonparallel line correspondence is given, the moving point cloud is aligned to the fixed one.

3.3 Plane features

Plane is the most common feature in scene, and generally used in registration, for it is much more direct and accurate than point and line features. Just like what we have mentioned above, the line features based method also leads to the plane features based, since the line features are usually extracted from the intersection of detected planes, that is to say the general expression of line feature with the intersection of two plane is more accurate.
Each plane can be defined by three points or a plane equation with four parameters. usually there are numerous point measurements on each plane, estimating the function of the plane and applying them for the registration is much more reasonable.
The function of a plane can be represented as

$$n_x x + n_y y + n_z z + d = 0 \tag{32}$$

where (n_x, n_y, n_z) is the unit normal of the plane.

For each pair of plane match, $(n_x^P, n_y^P, n_z^P, d^P)$ from the moving point cloud P and $(n_x^Q, n_y^Q, n_z^Q, d^Q)$ from the fixed point cloud Q, we can get

$$n^Q - Rn^P = 0 \tag{33}$$

$$d^P - d^Q + (Rn^P) \cdot \vec{t} = 0 \tag{34}$$

Just like the line feature based method, the rotation can be also estimated directly, to get the translation, we can bring Eq. 33 to Eq.34, then

$$d^P - d^Q + n^Q \cdot \vec{t} = 0 \tag{35}$$

Which is a linear function which can be solved directly; we can use it to get an approximate estimation of the transformation.
The linearized functional model and statistic model of plane based method are

$$A_{6n\times8} V_{8n\times1} + B_{6n\times6or7n\times7} \hat{x}_{6\times1or7\times1} + e_{6\times1} = 0 \tag{36}$$

$$D_{6n\times6n} = \sigma_0^2 Q_{6n\times6n} = \sigma_0^2 P_{6n\times6n}^{-1} \tag{37}$$

We need at least two nonparallel planes to estimate the rotation, and from Eq. 34 we can see that we need at least three planes in nondegenerate mode to estimate the translation. Given a plane correspondence we can fix these two planes together, then the moving point cloud can rotate about the norm of the plane and translate on the plane, once the other correspondence is given the rotation is determined, but the moving point cloud can still translate along the intersection line of the first two planes, so we need the third plane correspondence to get this resolved.

	Measurements for each correspondence	Minimal number of correspondences	Num of functions for each correspondences
Point based	6(3 for each point)	3	3
Line based	12(6 for each line)	2	7
Plane based	8(4 for eanch plane)	3	6

Table 1. Summary and comparison of point, line and plane based registration

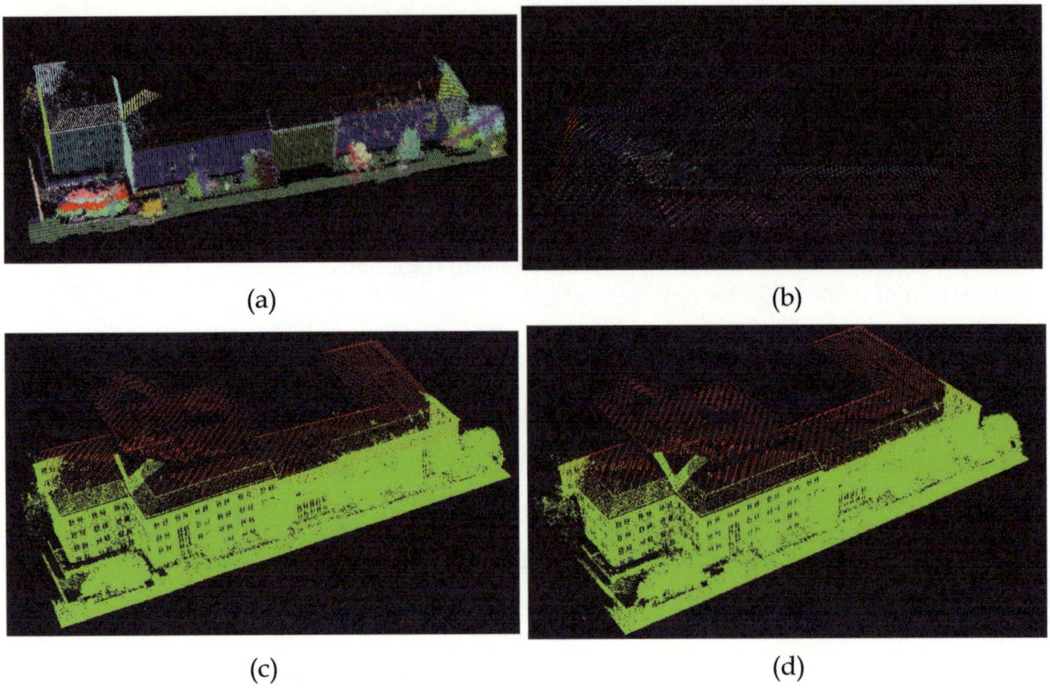

Fig. 5. The registration of ground laser based and airborne laser based point clouds of the civil building of Purdue University. (a) is the plane patches extracted from obtained by a ground laser; (b) is the plane patches extracted from obtained by a airborne laser. There is a great deal of difference of resolution between the two point cloud, the resolution of ground laser based is at least ten times higher than the airborne laser based one. (c) is the initial estimation from the approximate linear method; (d) is the refined alignment with point-to-plane ICP method.

3.4 Automatic matching of feature

We have introduced the basic concepts of feature based registration, however there is still a problem, which is how to automatically get corresponding features. Point features are the smallest elements in the 3D space, but they are so universal with little constraints and information that it is the very difficult to automatically find point matches. Line feature is also very difficult to find point matches, because the line are always extracted from the intersection of planes, thus the matching between line features can be deduced to plane matching problem. While planes have much information, including the plane equation and the points on this plane, so it can restrain the matching much better than point and line features.

Most of the automatic feature correspondence finding method use an exhaustively traversing method which taking all of the possible matches into consideration with some methods like geometry constraints(He et al.,2005), colors or intensity(Rabbani et al., 2007), statistic based method(Pathak et al.,2010) to cut off those obvious error matches and prune the searching space. We will generally introduce some of these methods below.

In He's work, he proposed a method with a interpretation tree and kinds of geometric constraints to generate plane correspondences. The interpretation tree includes all of the possible correspondences between two point sets. The ith layer of the interpretation tree are all possible features in point set P corresponding to feature q_i in point set Q.

The geometric constraints they introduced include area constraint, co-feature constraint and two matched feature constraint. The area constraint ensures the corresponding planes have almost the same area, and the co-feature constraint make the angle between a feature and its father is similar and ensure the difference of distance between the two features within a proper threshold. The two matched feature constraint assume the centroids of the corresponding planes are identical points, then two plane matches is enough for the registration; however it can't suit in such conditions where the planes are occluded and incomplete. But still we can generalize this constraint to three matched feature constraint according to the property described above, and stop the growing of the branch of the tree which has more than three matches.

Dold added boundary length constraint, bounding box constraint and Mean intensity value constraint to He's method. Boundary length constraint claims that the length of the boundary derived from the convex hull should be nearly the same, the ratio of the edges of the bounding box with one edge along the longest edge of the boundary is also proposed as an similarity between two corresponding planes, what's more, intensity information is also used in Dold's work.

Pathak introduced a statistic based method; the greatest difference between this method to the geometric based method above is the use of the covariance of the plane parameters and the covariance of the estimated transformation. The constraint or test used for pruning the searching space include size similarity test, given translation agreement test, odometry rotation agreement test, plane patch overlap test, cross angle test and parallel consistency test. The size similarity test is achieved by the use of the covariance matrix of the plane parameters which is proportional to the number of points in the plane. Given translation agreement test, odometry rotation agreement test need the initial value of the translation and rotation parameters which is not always available in practice, plane patch overlap test is to ensure that the estimated relative pose of the point sets have overlapped area, the cross angle test is similar with the geometric constraint which ensure the similarity of the angle between the correspondences, and the last parallel consistency test really can be classified to the cross angle test as a special situation where the cross angle should be zero. Size similarity test employs a threshold based method, each of the other tests generates a well-designed distribution, and uses an confidence interval to determine whether the correspondence is usable or not. Thorough introduction and derivation of this method can be found in Pathak's work(Pathak et al., 2010).

Makadia et al.(2006) proposed a reliable and efficient method which extended Gaussian images (EGI)(Horn, 1987). The rotation estimate is obtained by exhaustively traversing the space of rotations to find the one which maximizes the correlation between EGIs. it seems time consuming, but actually improves the efficiency by the use of the spherical harmonics of the Extended Gaussian Image and the rotational Fourier transform. Generally, the process is time-consuming, but by the use of the spherical harmonics of the Extended Gaussian Image and the rotational Fourier transform come up with by Makadia, the efficiency is improved remarkably. Once the rotation is estimated, the translation can also be estimated by maximizing the correlation between two point sets.

Another special method proposed by Aiger et al.(2008) adopted the invariability of certain ratios under affine transformation, and hence rigid transformation. This method extract all sets of coplanar 4-points from a point set that are approximately congruently related by rigid transforms to a given planar 4-points, and then test if the conjugate four-points satisfy the invariability of the such ratios.

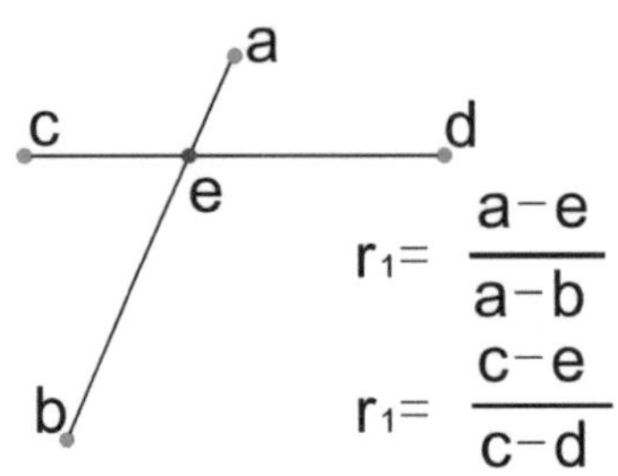

Fig. 6. e is the intermediate point e of the four-points {a,b,c,d}, the ratios r_1 and r_2 is invariant under rigid transformation.

4. Least square 3D surface matching

The Least Squares Matching(LSM) concept had been developed in parallel by Gruen (1984, 1985), Ackermann (1984) and Pertl (1984) for image matching which has been widely used in Photogrammetry due to its high level of flexibility, powerful mathematical model and its high accuracy. Gruen and Akca(2005) indicated that if 3D point clouds derived by any device or method represent an object surface, the problem should be defined as a surface matching problem instead of the 3D point cloud matching, and generalized the Least Squares (LS) image matching method to suit the surface matching mission. This method matches one or more 3D search surfaces to a 3D template surface through the minimizing of the sum of squares of the Euclidean distances between the surfaces. The unknown parameters are treated as stochastic quantities in this model with priori weights which can be used to eliminate gross erroneous and outliers, and this can improve the estimation very much. The unknown parameters of the transformation are estimated by the use of the Generalized Gauss–Markoff model of least squares. Because most of the point clouds in application just represent the surface of the scene, many registration missions can be treated as surface matching tasks. In this part we will give an introduction of the mathematical model of Least Square 3D surface matching which can be used for point cloud registration.

4.1 Mathematical model

Each surface can be represented of a function of three bases of the 3D world. For generalization the conjugate overlapping regions of the scene in the left and right surface can be represented as $f(x,y,z)$ and $g(x,y,z)$ no matter the representation of the surface is point cloud, triangle mesh, or any other methods. The mission is to make the two surfaces identical to each other, that is

$$f(x,y,z) = g(x,y,z) \tag{38}$$

Assume the errors caused by the sensor, the environment and the measure method are random errors, we can get the observation function as

$$f(x,y,z)-e(x,y,z)=g(x,y,z) \tag{39}$$

The sum of squares of the Euclidean distances between the template and the search surface elements can be used as the goal function of the least square minimization.

$$\sum\left\|\vec{d}\right\|^2=\min \tag{40}$$

Though the relation between two point clouds usually is a rigid transformation without scaling, Gruen and Akca use a seven-parameter similarity transformation to express the relationship of the conjugate surfaces. However the sole and core even the representation of this method will not change, the only difference is the parameter space. Adding a scaling parameter to the rigid transformation, the similarity transformation will be

$$\begin{bmatrix} x \\ y \\ z \end{bmatrix}=\vec{t}+sR\vec{x}_0=\begin{bmatrix} t_x \\ t_y \\ t_z \end{bmatrix}+s\begin{bmatrix} r_{11} & r_{12} & r_{13} \\ r_{21} & r_{22} & r_{23} \\ r_{31} & r_{32} & r_{33} \end{bmatrix}\begin{bmatrix} x_0 \\ y_0 \\ z_0 \end{bmatrix} \tag{41}$$

The formulation (39) can be linearized by Taylor expansion.

$$f(x,y,z)-e(x,y,z)=g^0(x,y,z)+\frac{\partial g^0(x,y,z)}{\partial x}dx+\frac{\partial g^0(x,y,z)}{\partial y}dy+\frac{\partial g^0(x,y,z)}{\partial z}dz \tag{42}$$

with

$$dx=\frac{\partial x}{\partial p}dp, dz=\frac{\partial x}{\partial p_i}dp, dz=\frac{\partial z}{\partial p}dp \tag{43}$$

Where $p=[t_x,t_y,t_z,s,\varphi,\omega,\kappa]^T$ is the parameter vector, get Eq. 41, 42 and 43 together we will obtain the linearized observation functions.

$$\begin{aligned} -e(x,y,z) &= g_x dt_x+g_y dt_y+g_y dt_y+(g_x a_{10}+g_y a_{20}+g_y a_{30})dm \\ &= (g_x a_{11}+g_y a_{21}+g_y a_{31})d\varphi+(g_x a_{12}+g_y a_{22}+g_y a_{32})d\omega \\ &= (g_x a_{13}+g_y a_{23}+g_y a_{33})d\kappa \end{aligned} \tag{44}$$

Then the general form of the observation model and the statistical model will be

$$-e=Ax-l \tag{45}$$

$$D=\sigma_0^2 Q=\sigma_0^2 P^{-1} \tag{46}$$

Until this step, we can say that it is almost the same with the point-to-point method, or point-to-point ICP because it also needs to find the correspondences with the initial estimation, and then iteratively improve the estimation of the transformation. To accelerate the searching of closest point, Akca and Gruen(2006) proposed an boxing structure, which partitions the search space into boxes, and correspondence is searched only in the box containing the point and its neighbors. What makes this method different from the ICP method is that the unknown transformation parameters are treated as stochastic quantities using proper a priori weights.

$$-e_b = Ix - l_b \tag{47}$$

$$D_b = \sigma_0^2 Q_b = \sigma_0^2 P_b^{-1} \tag{48}$$

where I is an identity matrix, l_b is a fictitious observation vector for the system parameters. The incremental vector and the standard deviation of the parameters can be got by

$$\hat{x} = (A^T PA + P_b)^{-1}(A^T Pl + P_b l_b) \tag{49}$$

$$\hat{\sigma}_0^2 = (v^T Pv + v_b^T P_b v_b) / r, \ \ where \ v = A\hat{x} - l, v_b = I\hat{x} - l_b \tag{50}$$

Generally the contribution of this method is the generalized mathematical model which can suit to any kind of representation of the surface and the method of treating the unknown parameters as stochastic quantities which established the control of the estimating of the parameters.

5. References

Ackermann, F., 1984. Digital image correlation: performance and potential application in photogrammetry. *Photogrammetric Record,* 11 (64), 429– 439.

Aiger, D., N. J. Mitra and D. Cohen-Or, 2008, 4-points Congruent Sets for Robust Surface Registration, *ACM Transactions on Graphics,* vol.27,3, pp.1-10

Akca, Devrim. 2003, FULL AUTOMATIC REGISTRATION OF LASER SCANNER POINT CLOUDS, ETH, Swiss Federal Institute of Technology Zurich, Institute of Geodesy and Photogrammetry, vol.1, pp.330-337.

Akca, D. and A. Gruen, 2006. Fast correspondence search for 3D surface matching. ETH, Eidgenössische Technische Hochschule Zürich, Institute of Geodesy and Photogrammetry.

Bauer, J., K. Karner, A. Klaus, R. Perko, 2004, Robust Range Image Registration using a Common Plane, *Proceedings of 2004 WSCG.*

Benjemaa, R., Schmitt, F., 1999. Fast global registration of 3D sampled surface using a multi-z-buffer technique. *Image and Vision Computing* 17, pp.113–123.

Bentley, J.L. Multidimensional binary search trees used for associative searching. *Comm. of the ACM* 1975, 18(9), 509–517.

Besl, P.J., McKay, N.D., 1992. A method for registration of 3-D shapes. *IEEE Transactions on Pattern Analysis and Machine Intelligence* 14 (2), 239–256.

Blais, G. and M. Levine, 1995, Registering Multiview Range Datato Create 3D Computer Objects, *Trans. PAMI,* Vol. 17, No. 8.

Chen, Y. and Medioni, G. 1991, Object Modeling by Registration of Multiple Range Images, *Proc.IEEE Conf. on Robotics and Automation.*

Chen, Y. and Medioni, G., 1992. Object Modeling by Registration of Multiple Range Images,*Image and Vision Computing,* 10(3), pp. 145-155.

Chen Y., Y.Shen, D. Liu. A Simplified Model of Three Dimensional-Datum Transformation Adapted to Big Rotation Angle, *Geomantic and Information Science of Wuhan Universti,* 2004,12,pp:1101-1105.

Dalley, G. and P. Flynn, 2002, Pair-wise range image registration: a study in outlier classification, *Comput. Vis. Image Und*, 87 (1–3), pp.104–115.

Dold, C. and C. Brenner, 2006, Registration of terrestrial laser scanning data using planar patches and image data, *International Archives of Photogrammetry and Remote Sensing*, pp.25-27.

Dorai, C., Weng, J. & Jain, A.K. 1997. Optimal registration of object views using range data. *IEEE Trans.on Pattern Analysis and Machine Intelligence* 19(10), pp.1131-1138.

Flory, S. and M. Hofer, 2010,Surface Fitting and Registration of Point Clouds using Approximations of the a Unsigned Distance Function, *Computer Aided Geometric Design*, Vol.27(1), pp:66-77

Gelfand, N.; Ikemoto, L.; Rusinkiewicz, S.; Levoy, M., 2003, Geometrically stable sampling for the ICP algorithm, 3-D Digital Imaging and Modeling(3DIM). *Proceedings. Fourth International Conference on 3D Digital Imaging and Modeling* , pp. 260- 267.

Chao, C. and Stamos, I., 2005, Semi-Automatic Range to Range Registration: A Feature-Based Method. In Proceedings of the Fifth international Conference on 3-D Digital Imaging and Modeling (June 13 - 16, 2005), *3DIM. IEEE Computer Society, Washington, DC*, pp.254-261.

Chen, C.C. and I. Stamos, 2006, Range Image Registration Based on Circular Features, *3DPVT*, pp.543-550.

Godin, G., D. Laurendeau, R. Bergevin, 2001. A Method for the Registration of Attributed Range Images. In: Proc. of Third Int. *Conference on 3D Digital Imaging and Modeling, Quebec*, pp. 179-186.

Godin,G., M. Rioux and R. Baribeau, 1994, Three-dimensional registration using range and intensity information, *Proc. SPIE Videonietrics III*, vol. 2350, pp. 279-290.

Godin, G. and P. Boulanger, 1995, Range image registration through invariant computation of curvature, *Proc. ISPRS Workshop From Pixels to Sequences, Zurich, Switzerland*, pp.170-175.

Gruen, A., 1984. Adaptive least squares correlation—concept and first results. Intermediate Research Project Report to Heleva Associates, *Inc. Ohio State University, Columbus, Ohio*, pp. 1- 13.

Gruen, A., 1985. Adaptive least squares correlation: a powerful image matching technique. South African Journal of Photogrammetry, *Remote Sensing and Cartography*, 14 (3), 175–187.

Gruen A. and D. Akca, 2005, Least squares 3D surface and curve matching, *ISPRS Journal of Photogrammetry & Remote Sensing* 59 (2005) 151- 174

Guehring J., 2001, Reliable 3D surface acquisition, registration and validation using statistical error models. *Proc. 3DIM.*

Habib, A., Mwafag, G., Michel, M., and Al-Ruzouq, R., 2005. Photogrammetric and LIDAR Data Registration Using Linear Features, *Photogrammetric Engineering & Remote Sensing*, Vol. 71, No. 6, pp. 699–707.

Hansen, V. W., 2007. Registration of Agia Sanmarina LIDAR Data Using Surface Elements, *ISPRS Journal ofPhotogrammetry and Remote Sensing.*

He, W., Ma, W. and Zha, H., 2005. Automatic registration of range images based on correspondence of complete plane patches, *Proceedings of the Fith International*

Conference on 3-D Digital Imaging and Modeling, Ottawa, Ontario, Canada, pp. 470–475.

Horn, B.K.P., 1987, Closed-form Solution of Absolute Orientation using Unit Quaternions, *Journal of the Optical Society A*, 4, pp.629–642.

Horn, B. K. P., 1987 Extended gaussian images, *IEEE*, pp.1671–1686.

Jaw, J.J., and T.Y. Chuang, 2007. Registration of LIDAR Point Clouds by means of 3D Line Features, *JCIE Journal of the Chinese Institute of Engineers.*

Jian B. and B. C. Vemuri, 2005, A Robust Algorithm for Point Set Registration Using Mixture of Gaussians, *Tenth IEEE International Conference on Computer Vision (ICCV'05)* Vol. 2, Beijing, China ,October 17-October 20

Johnson, A. and Hebert, M., 1997, Surface Registration by Matching Oriented Points, *Proc. 3DIM.*

Johnson, A. and S. Kang, 1997, Registration and Integration of Textured 3-D Data, *Proc. 3DIM.*

Kang, Z.; Zlatanova, S. Automatic registration of terrestrial scan data based on matching corresponding points from reflectivity images, *Proceeding of 2007 Urban Remote Sensing Joint Event, Paris, France, 11-13 April, 2007,*pp. 1-7.

Kang, Z.; Zlatanova, S.; Gorte, B. Automatic registration of terrestrial scanning data based on registered imagery, *Proceeding of 2007 FIG Working Wee; Hong Kong SAR, China, 13-17 May, 2007.*

Kang, Z.; Zlatanova, S. A new point matching algorithm for panoramic reflectance images. *Proc of the 5th International Symposium on Multispectral Image Processing and Pattern Recognition (MIPPR07). Wuhan, China, 15-17 November, 2007,* pp. 67882F-1-10.

Kang, Z., J. Li, L. Zhang, Q. Zhao, and S. Zlatanova, 2009,Automatic registration of terrestrial laser scanning point clouds using panoramic reflectance images, *Sensors, 2009,9,* pp:2621-2646, doi:10.3390/s90402621

Lowe, D. G., 2004, Distinctive Image Features from Scale-Invariant Keypoints, *International Journal of Computer Vision,* 60, 2, pp. 91-110.

Masuda, T., K. Sakaue, and N.Yokoya, 1996, Registration and Integration of Multiple Range Images for 3-D Model Construction, *Proc. CVPR 1996.*

Makadia, A., A. IV Patterson, and K. Daniilidis, 2006, Fully automatic registration of 3d point clouds. In CVPR '06: *Proceedings of the 2006 IEEE Computer Society Conference on Computer Vision and Pattern Recognition,* pp.1297–1304.

Neugebauer, P., 1997, Geometrical cloning of 3d objects via simultaneous registration of multiple range images, *In Proc. Int. Conf. on Shape Modeling and Application,* pp.130–139.

Nishino, K., K. Ikeuchi, 2002. Robust simultaneous registration of multiple range images, *The 5th Asian Conference on Computer Vision (ACCV2002),* pp. 23–25.

Park, S. and M.Subbarao, 2003. An accurate and fast point-to-plane registration technique. *Pattern Recogn.* Lett. 24, 16, pp.2967-2976.

Pathak,K., A. Birk, N. Vaskevicius and J. Poppinga. 2010, Fast registration based on noisy planes with unknown correspondences for 3-d mapping. *IEEE,* 26, pp.424–441.

Pertl, A., 1984. Digital image correlation with the analytical plotter Planicomp C-100. *International Archives of Photogrammetry and Remote Sensing,* 25 (3B), 874– 882.

Pottman, H. and M. Hofer., 2002, Geometry of the squared distance function to curves and surfaces. *Vienna Institute of Technology Technical Report No.90.*

Pulli, K., 1999. Multiview registration for large data sets. *Proceedings of 3-D Digital Imaging and Modelling (3DIM),* pp. 160–168.

Rabbani T., S. Dijkman, F. Heuvel, and G. Vosselman, 2007. An Integrated Approach for Modelling and Global Registration of Point Clouds, *ISPRS journal of Photogrammetry and Remote Sensing,* Vol. 61, pp.355-370.

Roth, G., 1999. Registration two Overlapping Range Images, *Proc. of Second Int. Conference on 3D Digital Imaging and Modeling, Ottawa,* pp. 191-200.

Rusinkiewicz, S., Levoy, M., 2001. Efficient variant of the ICP algorithm. *Proceedings of 3-D Digital Imaging and Modelling (3DIM),* pp. 145–152.

Salvi, J., C. Matabosch, D. Fofi, and Forest, 2007. A review of recent range image registration methods with accuracy evaluation. *Image Vision Comput,* 25, pp.578-596.

Schutz,C., T. Jost and H. Hugli, 1998, Semi-automatic 3d object digitizing system using range images, in: *Proceedings of the Third Asian Conference on Computer Vision,* vol. 1, pp. 490–497.

Sharp, G., S. Lee, and D. Wehe. ICP registration using invariant features. *IEEE Transactions on Pattern Analysis and Machine Intelligence,* 24(1):90–102, 2002.

Simon, D. A., 1996 , Fast and Accurate Shape-Based Registration, Ph.D. Dissertation, Carnegie Mellon University.

Stamos, I., 2001, Geometry and Texture Recovery of Scenes of Large Scale: Integration of Range and Intensity Sensing. Doctoral Thesis.a

Stamos, I. and M. Leordeanu, 2003. Automated Feature-based Range Registration of Urban Scenes of Large Scale, *Proceedings of IEEE Computer Society Conference on Computer Vision and Pattern Recognition,* Vol. 2, pp. 555-561.

Stamos, I. and P.K. Allen, 2002. Geometry and Texture Recovery of Scenes of Large Scale, *Computer Vision and Image Understanding,* Vol. 88, No. 2, pp. 94-118.

Trummer, M., H. Suesse, and J. Denzler, 2009 Coarse Registration of 3D Surface Triangulations Based on Moment Invariants with Applications to Object Alignment and Identification, *IEEE International Conerence on Computer Vision (ICCV) 2009,* pp: 1273-1279.

Tuhin K. Sinha, David M. Cash, Robert J. Weil, Robert L. Galloway, and Michael I. Miga, 2002,Cortical Surface Registration Using Texture Mapped Point Clouds and Mutual Information, *Lecture Notes in Computer Science: Medical Image Computing and Computer-Assisted Intervention - MICCAI 2002.*

Turk, G. and Levoy, M. 1994, Zippered Polygon Meshes from Range Images, *Proc. SIGGRAPH 1994.*

Vosselman,G.,Gorte,B.G.H., Sithole,G., Rabbani,T. , 2004, Recgnising structure in laser scanner point clouds. *The international Arhives of Photogrammetry, Remote Sensing and Spatial Information Siences,* vol.46,part8/W2, pp.33-38.

Weik, S. 1997. Registration of 3-D partial surface models using luminance and depth information. *Proc.of IEEE Int. Conf. on 3DIM:* 93-100, Ottawa, pp.12-15.

Williams, J., Bennamoun, M., 2001. Simultaneous registration of multiple corresponding point sets. *Computer Vision and Image Understanding* 81 (1), 117–142.

Zhang, Z., 1994, Iterative point matching for registration of free-form curves and surfaces, *Int. J. Comput Vision* 13 (2) ,pp.111–152.

25

Terrestrial Laser Scanning Data Integration in Surveying Engineering

Bahadır Ergün (PhD.)
Gebze Institute of Technology,
Geodesy & Photogrammetry Department
Turkey

1. Introduction

In terms of sensitivity and equipment, there are several differences between Close Range Photogrammetry (CRP) which is today often used for documentation of historical and cultural structures, restoration, relievo studies, and Terrestrial Laser Scanner (TLS) technology which is recently started to be used for 3D modelling of every kind of objects and geometric data acquisition.

Terrestrial laser scanning is a new technology that enables easy and fast data acquisition from objects with complex structures such as buildings, machines, etc. In recent years, some manufacturers designed and improved different systems for specific purposes [4]. The combination of terrestrial image photogrammetry and CRP method, which is also a new technology, provides us new opportunities in presentation of 3D photorealistic models, classification of real objects and creation of visual reality [9].

TLSs usage becomes more popular day by day since they present quick and effective solutions in the way of acquiring 3D geometric information on objects.

Some of the significant study areas that benefit from TLS are as follows:

- Archeology,
- Architectural restoration studies,
- Measurement of tunnels and roads,
- Urban modeling,
- Virtual factories, applications of virtual reality,
- Mining and infrastructure projects,
- Manufacturing controls,
- Crime scene investigations,
- Studies of industrial design, etc.

We can consider CRP and TLS as either independent methods or two complementary methods; therefore their area of application broadens day by day.

If TLS and CRP are compared with Laser Scanners;

- 3 dimensional points can be acquired directly
- Intense point cloud data can be obtained within the frequency of millimeters, in a short period
- Smooth data acquisition over difficult surfaces in terms of their form

- Opportunity of real image acquisition

If TLS and CRP are compared with Close Range Photogrammetry;

- High resolution
- Low costs
- Since data acquisition is possible via limning, little time

are the main elements and characteristics.

Terrestrial laser scanners allow the imaging of the object as a point cloud by scanning them as point series in horizontal and vertical directions, under a specific angle. Point cloud data contains polar coordinates (X, Y, Z) which are based on the scanner and a value of reflection density (tons) information for each point measured. These values (polar coordinate, etc.) can be printed in several formats, namely in txt format [10].

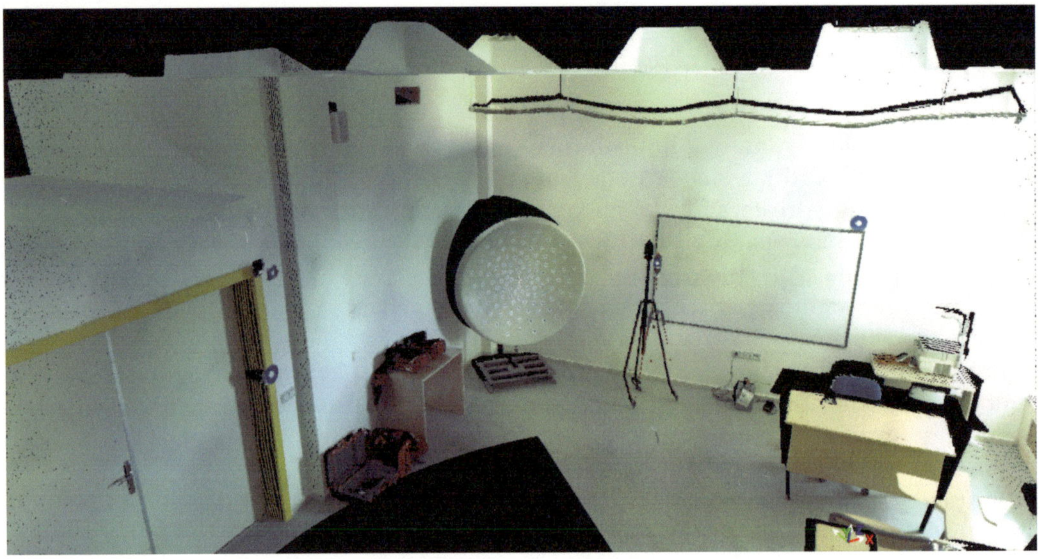

Fig. 1.1. Sample Point Cloud Data from TLS

Integration of scanning data from different station points in a single coordinate system is realized by the help of artificial targets. By coordinating the targets in question by the help of geodesic methods (Total Station etc.) object coordinates are obtained within the desired reference system.

TLS point cloud data is a very intense data set, and making a study with CAD (Microstation, Autocad, etc.) software necessitates computers with very potent hardware and very experienced operators. In order the TLS data to be used in terrestrial photogrammetry, points with known coordinates that will be used in image rectification should be picked up from the whole TLS point cloud data. The sample point cloud data from TLS was shown in Fig. 1.1.

The purpose of this chapter is to provide a filter creation in which surfaces in architectural photogrammetry can be automatically distinguished, so the operator's duty will be reduced under the density of point cloud taken from the surfaces via terrestrial laser scanner. Purpose of the algorithm is to obtain real surface of the scanned object and to reduce the volume of data. In this algorithm, vertical flat surfaces of the structure are auto-selected, irrelevant points are filtered. Depending on the object's surface, different methods should be

performed. These methods are Intersection of Plane Surfaces, Hough Transformation and a new technology based on surface projection etc.
With the improvement of application areas and the advantages mentioned, terrestrial laser technology is academically an interesting topic and continues to be studied.

2. Equipment and data structure of terrestrial laser scanners

Laser technology is an area of research since 1960's; however using laser technology for the sake of making measurements is relatively new concept. As laser scanning systems are innovated, they gain a broader area of application, too. By different companies, depending on their application specifications IQSun, Leica, Optech, Callidus, Trimbe, Riegl, etc. (Table 2.1) and depending on their distance lengths (1m - 1500m) some terrestrial laser scanners which are shown in Fig. 2.1, are manufactured.

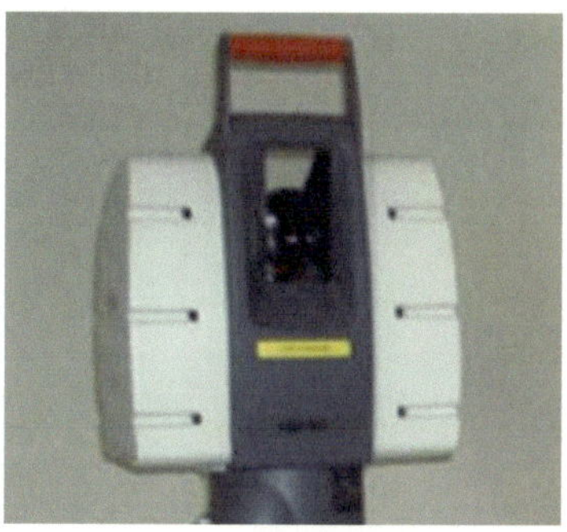

Leica -HDS 3000

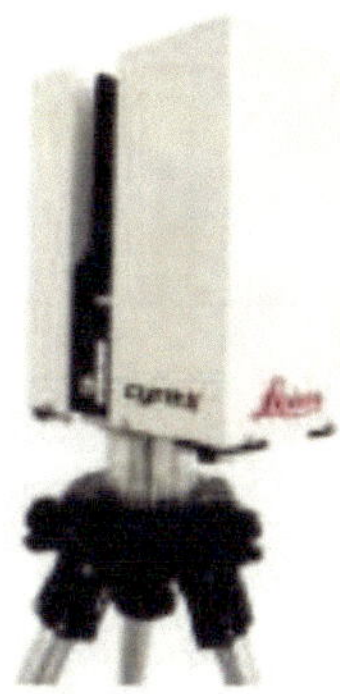

Leica -HDS 4500

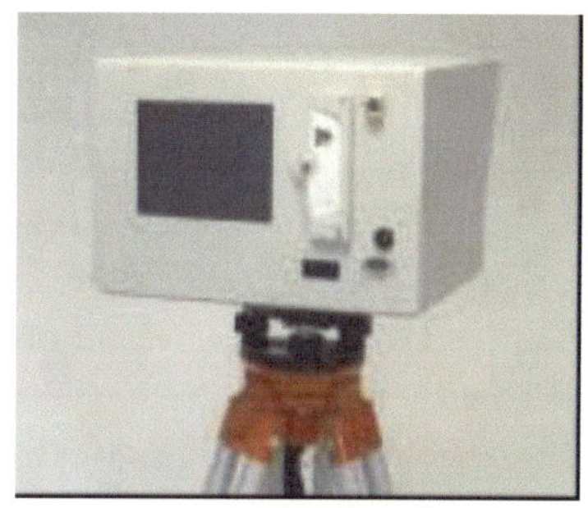

Optech -ILRIS 3D

Trimble- (Mensi) GS200

Riegl LMS-Z210i

Callidus -CP 3200

Fig. 2.1. Terrestrial Laser Scanners of Various Companies

Specifications of several laser scanners currently used are shown in Table 2.1 with respect to their measurement distances and scanning sensitivities. According to this table, TLS system's (Leica HDS 3000) point position accuracy can be acquired [19-23].

Firm	Type	Measurement Range(m)	Meas. Point	Scanning Angle	Distance Accuracy	Position Accuracy	Light Beam
Leica	Scan Station	300m (90% Reflection)	4000 point/sec.	$270^0(V)$ $360^0(H)$	4 mm @ 50 m	6 mm @ 50 m	4 mm @ 50 m
	HDS3000	300m (90% Reflection)	4000 point/sec.	$270^0(V)$ $360^0(H)$	4 mm @ 50 m	6 mm @ 50 m	4 mm @ 50 m
	HDS4500	53.5 m	500000 point/sec.	$310^0(V)$ $360^0(H)$	5mm+120pp m (%100yns. y.)	13.7mm @ 25 m	8.5 mm @ 25 m
	HDS6000	79 m (80% Reflection)	500000 point/sec.	$310^0(V)$ $360^0(H)$	5 mm @ 50 m	10 mm @ 50 m	3 mm + 0.22mrad
Optech	Ilris 3D	3m–1500m (80% Reflection)	2500 point/sec.	$310^0(V)$ $360^0(H)$	7 mm @ 100 m	8 mm @ 100 m	0.00974^0
Riegl	LMS-Z420	2m-1000m (80% Reflection)	12000 point/sec.	$0-80^0(V)$ $0-360^0(H)$	8 mm @ 50 m	10 mm @ 50 m	0.25 mrad
	LMS-Z390	1m-300m (80% Reflection)	11000 point/sec.	$0-80^0(V)$ $0-360^0(H)$	4 mm @ 50 m	6 mm @ 50 m	0.25 mrad
	LMS-Z210	4m-650m (80% Reflection)	12000 point/sec.	$0-80^0(V)$ $0-360^0(H)$	10 mm @ 50 m	15 mm @ 50 m	2.7 mrad
Z - F	Imager 5006	1m - 79 m	500000 point/sec.	$310^0(V)$ $360^0(H)$	mm @ 50 m	1 mm @ 50 m	0.22 mrad
Faro	LS 880	1 m - 80 m	12000 point/sec.	$320^0(V)$ $360^0(H)$	3 mm	5 mm	0.01^0

Table 2.1. Laser Scanners and Their Specifications

2.1 Equipment structure of terrestrial laser scanners

In this section, a Terrestrial Laser Scanner is used and as it is figured out in Fig. 2.2, the components of a Terrestrial Laser Scanner are:

- Scanner,
- Control unit,
- Power source,
- Stand

Fig. 2.2. Terrestrial Laser Scanner Used In This Application and Its Components

Scanning unit is the part that obtains 3D data in TLS system. They can be examined under three groups depending on their distance measuring method.

Measurement Model	Range (m)	Accuracy (mm)	Company
Flight Time	<100	<10	Optech, Leica, Mensi, Callidus, Riegl
	<1000	<20	Optech, Riegl
Phase Diff.	<100	<10	IQSun, Leica, Visimage, Z - F
Optical Trian.	<5	<1	Mensi, Minolta

Table 2.2. The Classification of TLSs According to Their Distance Measuring Method

As seen in Table 2.1 and 2.2, it is necessary to choose a scanner relevant to the scanning surface and the desired sensitivity. For instance, for a completed road project, in order to control the suitability of the manufactured product with respect to the project or in the volume measurement of huge excavation fields such as stone pits, sand, gravel, borrowed pits, TLSs like Optech, Riegl can be preferred, because they measure distance with Time of Flight principle (<1000) and a sensitivity of <20mm is more than satisfying in such

applications. However, in a surface with a reduced cross-sectional area, such as a tunnel, a machine of <100 m measurement range should be preferred [3].
The broadest distance measuring method of the scanners, which can be examined under three groups according to their distance measuring principle, is the time of flight method as shown in Table 2.2. The basis of TLS system shown in Fig 2.3, which uses time of flight distance measuring method, is the process of measuring distance by recording the time of flight. That is, it records the time of reflection of laser beam sent by pulse from the object's surface points, the moment of origin of laser beam that leaves the scanner and its time of return, namely the time of flight, so it measures the distance.

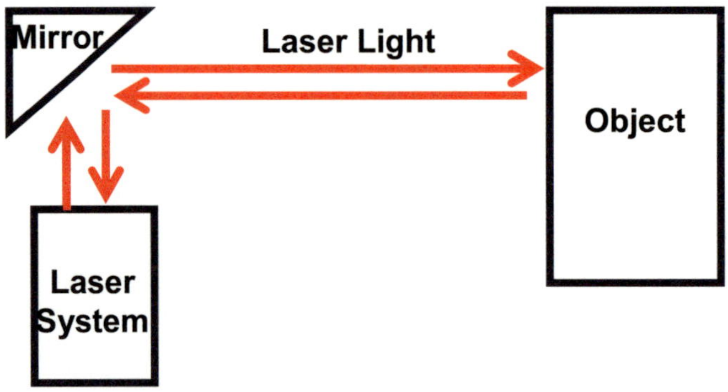

Fig. 2.3. Distance Measuring Method Based on Time of Flight

The second popular distance measuring principles used in TLSs are Phase Difference Method as shown in Fig. 2.4. Laser sent in this method is adjusted with a harmonic wave. The distance is calculated by the phase difference between transmitted and received waves. Most of TLSs, that effectively scans in short distances work under this principle [2].

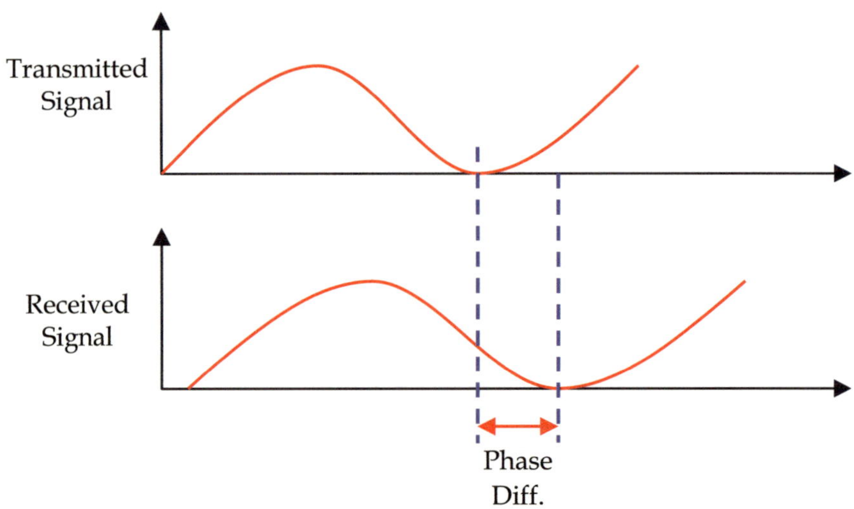

Fig. 2.4. Phase Difference System

Systems that are based on the third principle are systems which work with triangulation method (optical triangulation). Those systems, by performing scans of <1mm sensitivity and

within short distances, work under this principle. They are used in the area of industrial and sensitive applications and for small objects [2].
If scanning unit is inspected generally, the following items can be seen:

- Laser telemetry and
- Laser beam derivation unit.

Laser Telemetry Block Layout includes the following as shown in Fig. 2.5:

- Transistor laser or semi permeable laser diode (Transmitter),
- Detector (AGC) (Receiving Canal),
- Time differentiation, measurement unit (TDC),
- Transmitting and receiving optics

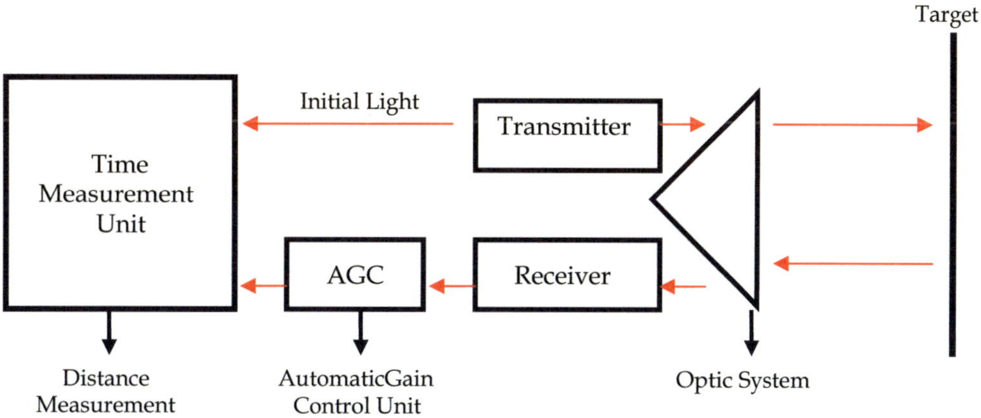

Fig. 2.5. Laser Telemetry Block Layout

Laser transmitter transmits an initial laser pulse which goes to the transmitter and serves to initialize the time measurement unit at the same time. Detector (AGC) is used in the detection of laser signal reflected from the surface and when the signal from measured object is detected by the detector, the time measurement unit initialized by the initial pulse also stops [1].
Distance of the target to the scanner can be calculated with the equation of;

$$R = c\frac{t}{2} \tag{2.1}$$

In equation (2.1), c represents the speed of light, t represents the time that passes during the contact of laser pulse to the measured object and its return.
Laser beam deviation unit is used to deflect the beam in vertical and if necessary horizontal direction and the following items are used in this process:

- Swinging flat mirrors,
- Swinging polygon mirrors,
- Galvanometric mirrors.

Scanning unit is generally high capacity, often portable computers, in which Terrestrial Laser Scanner software is installed and which is used for recording the whole scanning process and for recording and controlling the data obtained. Power source is composed of one or more batteries. Laser scanner gets its power from these batteries. TLSs are used with an adjustable stand, which can be adjusted depending on the station point used and its purpose of usage.

2.2 Data structure of terrestrial laser scanners

Laser scanners allow the imaging of the object as a point cloud by scanning them as point series in horizontal and vertical directions, under a specific angle. Scanner-centric polar coordinates are measured for each of the laser points. These are; slant distance to measured point P, the angle of measurement line with X-axis and horizontal plane α , and the slope angle of measurement line with horizontal plane φ. Terrestrial laser scanners accept their positioned point as their initial point as shown in Fig. 2.6. They perform a measurement completely based on their local coordinate systems [10].

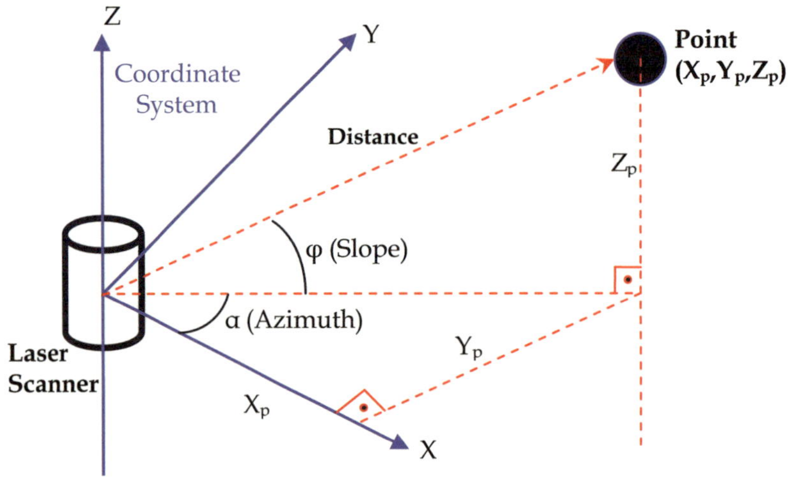

Fig. 2.6. TLS Local Coordinate System

Obtained point cloud data can be processed in various formats as coordinate, angle distance such as DXF for CAD models, ASCII for surface modelling, VRML format for visualization and txt or pts. Point cloud data can be obtained by the software, which varies depending on different TLS models used.

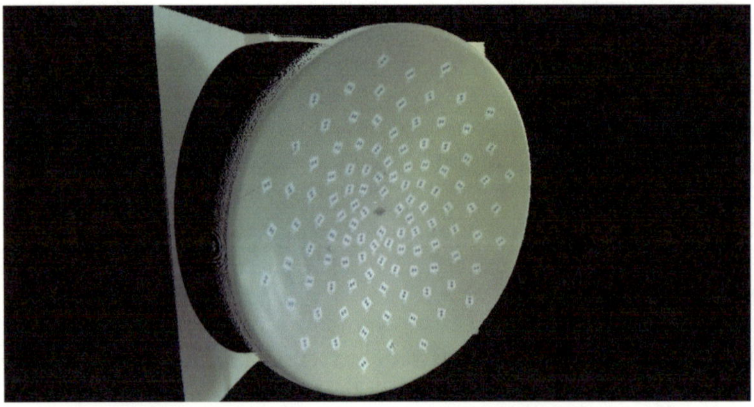

Fig. 2.7. Point Cloud Image from Cyclone 5.2

TLSs scan the surface of object by obtaining X,Y,Z Cartesian coordinates within second in the coordinate system centered on their established station point. Besides three dimensional coordinates, depending on the structure of scanned surface and measuring distance, the obtained data includes the density of returning signal in terms of RGB (Red, Green, Blue). By the help of recorded RGB density values, modeling of scanned object and environment becomes easier. This dense data acquired by scanning is called point cloud. The txt data formation for point cloud data is shown in Table 2.1.

X (m)	**Y (m)**	**Z (m)**	**Color Information**			
0.153229	0.521369	-0.004161	-76	91	115	113
0.270996	0.521319	-0.004880	-75	87	109	107
0.153229	0.467538	-0.009394	41	75	94	98
0.270874	0.467157	-0.006026	-167	81	100	97
0.216461	0.494419	-0.006889	-170	79	98	96

Table 2.1. txt Data Formation

The software exist in goal-oriented modules, in order to obtain raw data, to transform the data into a form that can be worked on, to perform texturing (when necessary) etc. In Fig. 2.7, point cloud data, which is visualized with RGB density values and scanned with Leica HDS - 3000 terrestrial laser scanner.

3. Data integration techniques for TLS in surveying

The use of TLSs in photogrammetric data production is a preferred method today. Points with known coordinates that will be used in image rectification should be picked up from the whole TLS data. Those points are defined as maximum and minimum points over object surfaces. However, because of different scanning angles within TLS data and dense point cloud data, reading and processing of these points over surfaces necessitate a difficult research and it is not time-effective for the operator that produces data. Especially in photogrammetric data creation, in the studies of architectural restoration and the protection of cultural heritage, the data are the images, converted vertically. Taking these images from TLS data, the point cloud set, and using them in photogrammetric image rectification can be done by means of an operator. In order to make this study independent from the operator, differentiating surfaces with different depth over point cloud data can be followed as an integration method [5-8].

As an integration method, in order to determine different surfaces of different depths automatically, a geometric filtering method research is made in this study. Thanks to this integration method, data required for image rectification which is a part of photogrammetric data production can be created automatically without any human factor over the point cloud data which forms TLS data. Data file that constitutes the basis of TLS data format coordinate information and the data that constitutes photogrammetric coordinate information are the same in terms of data coordinate information. However, in order to integrate photogrammetric data automatically, data of vertically converted photographic information should be selected accurately within this point cloud data. They should be classified with a filter algorithm used within this point cloud data. Cyclone 5.2 software is used in forming these TLS point cloud data. When coordinate information of the point cloud data obtained with this software is examined, an orderly processing and classification can not be seen in data contents. In this case, determination of point coordinate from point cloud data can be made by the help of operator, manually, on CAD interface of the software. 3D classification of the data is considered, by applying a filter with the help of a mathematical algorithm over raw data obtained for the classification of point position data of different depth-surfaces. We worked on a mathematical algorithm that would be able to make this classification and the point cloud data is automatically classified under MatLAB environment with an algorithm based on geometric location information. Thus, without any operational interference, it will be possible to obtain information photogrammetrically on point cloud data.

In measurement technology, besides the photogrammetric method, terrestrial laser scanning and modeling technology are intensely used in architectural documentation and cultural heritage studies. In the perception of engineering perspective, these two methods are independent of each other. However, these two methods can be combined with each other in the essential parts of architectural documentation and cultural heritage studies. While one is implementing these applications, the working procedure is so significant that if the amount of data increases, time consumption reaches higher levels. These two methods are compulsorily optimized for working in a collaborative way, especially in the large-scale studies. This necessity leads to the application of two types of integration. One of them is hardware integration, and the other one is data integration. Nevertheless, the laser scanner-measuring algorithm has some problems. The first problem stems from the huge amount of data. The point cloud of a historical building easily contains more than one million measured points. Such a huge amount of data causes problems on CAD documentation. In addition to this significant problem, the huge amount of data also causes problems in photogrammetrical documentation of surfaces, especially in the process of photogrammetric image rectification for building facades [11,12]. There is a variety in the methodology for such studies. Some studies combine laser scanner data and photogrammetry in cultural heritage recording [3,6]. Others apply different techniques for the treatment of laser data. Some of the techniques focus on the object segmentation of architectural facades [8,13,14,15]. In order to solve the above-mentioned confusion, this study presents a specific approach to obtain real surfaces from a scanned object on the one hand and a reduction of the data volume on the other. The algorithm is based on mathematical surface clustering techniques. Using this technique, it manages to extract planar surfaces, including parameters of depth. Additionally, surface points belonging to the designated surfaces are filtered. Once one has the planar surfaces of the buildings, several methods can be proposed to extract the feature bounds.

3.1 Mathematical models of data integration with filtering

In this section, a new data integration model is suggested for the operating processes in architectural documentation and cultural heritage studies, with the terrestrial laser scanning that uses new data production methods of photogrammetry and surveying. The mathematical model and algorithm in the presented integration model is programmed in MatLAB 7.0 and is tested with the original data set. The functional model is explained, benefiting from the object oriented programming languages in MS Windows system [8,13,18].

The point cloud data that is shown in Fig.3.1. is defined by distinct surfaces. Indoor areas are usually defined by plane surfaces such as walls. Figure 3.1 shows that point cloud data and different surface construction in three dimensional coordinate system [18].

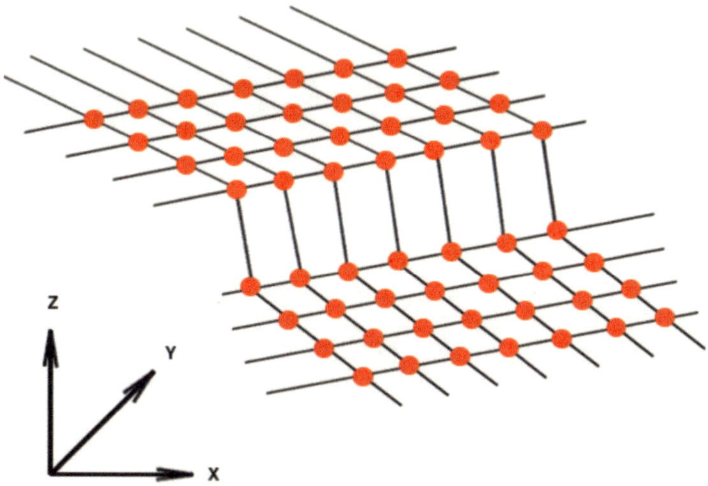

Fig. 3.1. Data Structure in 3D Coordinate System.

The algorithm used in this section was constructed by using the mathematical model of the surface. This surface function has four parameters:

$$\begin{aligned} ax_1 + by_1 + cz_1 + d_1 &= 0 \\ ax_2 + by_2 + cz_2 + d_2 &= 0 \\ &\dots \\ ax_n + by_n + cz_n + d_n &= 0 \end{aligned} \tag{3.1}$$

The first step of the algorithm was to define the reference surface. The reference surface was created by an operator by determining at least five points that belonged to same surface. This process was called adjustment computation for surface parameters in Eq. 3.1. After the process, Δz_i values were calculated separately for each point in the data set. The adjusted reference plane and the elevation coordinate system are explained in Fig. 3.2.

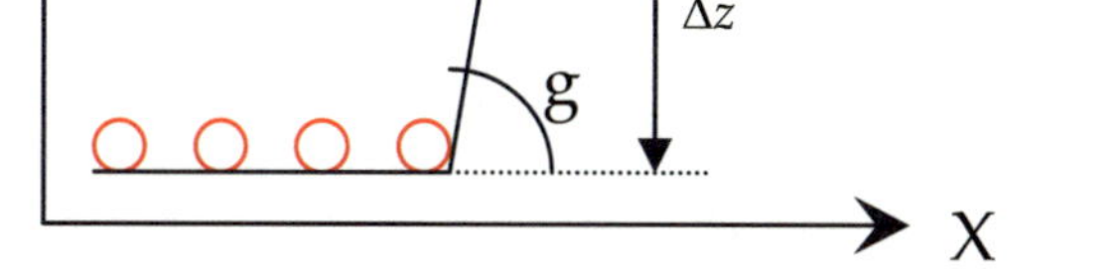

Fig. 3.2. Reference and Elevation Coordinate System for Point Cloud Data

These values were added to the data set for the column matrix. The second step of the algorithm was the determination process of the slopes of the points in reference to the

surface. These values were also embodied to the data set matrix as a column. In this case, the data set matrix was composed of five columns.
The suggested filtering function are shown as below [18].

$$f(p) = e^{\frac{\Delta z^2 . Cosg}{\sqrt{x^2+y^2}}} \tag{3.2}$$

x_i : x coordinate of a point.
y_i : y coordinate of a point.
g : Slope angle from the reference surface.
The mathematical model for the filtering function is based on the slope angle of the points, Δz_i by the reference plane and is parallel to the distance of the points from origin. In this filtering algorithm, the origin point of the coordinate system has to be on the reference plane. This algorithm has mathematical model such as in Eq.3.2. This obligation is basically set up by affine transformation [7, 8]. In this case, Δz_i points get closer to zero. Therefore, the result of the filtering function of the points, which appears on the reference plane, gets closer to 1. Thus, with this filter function one can select the points on the reference plane easily.

3.2 Application

In order to facilitate CRP and TLS integration, data of TLS is simplified by a filter in this dissertation study. For this reason, a 3D, object-based filter, based on point weights, which has a mathematical algorithm is developed and is applied in two different data sets under two stages. The dissertation includes the obtained results and used data sets and their accuracy analysis has been made. For the purpose of obtaining TLS point cloud data that will be used in the chapter, scanning's made with Leica HDS 3000 Laser Scanner in the classroom of GYTE Geodetic and Photogrammetric Engineering Department.
The developed filter has been coded in Matlab 7.0 software as the base programming language via the flowchart in Fig.3.3. A plane is made with the points determined by an operator in the first step and after balancing, plane equation parameters are obtained [18].
In the second step, depth differences Δz in all points are calculated with the obtained parameters. Surfaces are differentiated from each other by a filter containing the geometric surface model and Δz parameters which are determined according to the average surface parameters calculated. The filter matrix (T) which is manipulated by the filter created under the scope of this dissertation study helps us to compare the results, by the applying it to all of the points.
This algorithm was tested with an application in indoor planar surfaces that were placed in the classroom of GIT. The application is provided in the Fig.3.4. below.
At first, each data set was obtained with Leica HDS 3000 scanner identically because of the point density and plane area wideness [9,10]. A data set of two distinct surfaces with 232 points was chosen, which had an approximate difference of depth between the designated surfaces. This data set is shown in Fig.3.5.
The first surface was selected as a blackboard, and the second one was a wall surface. The difference of depth between the board and the wall was approximately 3 cm. The wall surface was determined as the reference surface and 5 significant points that were necessary for the calculation of the parameters of plane equation were chosen from the mentioned

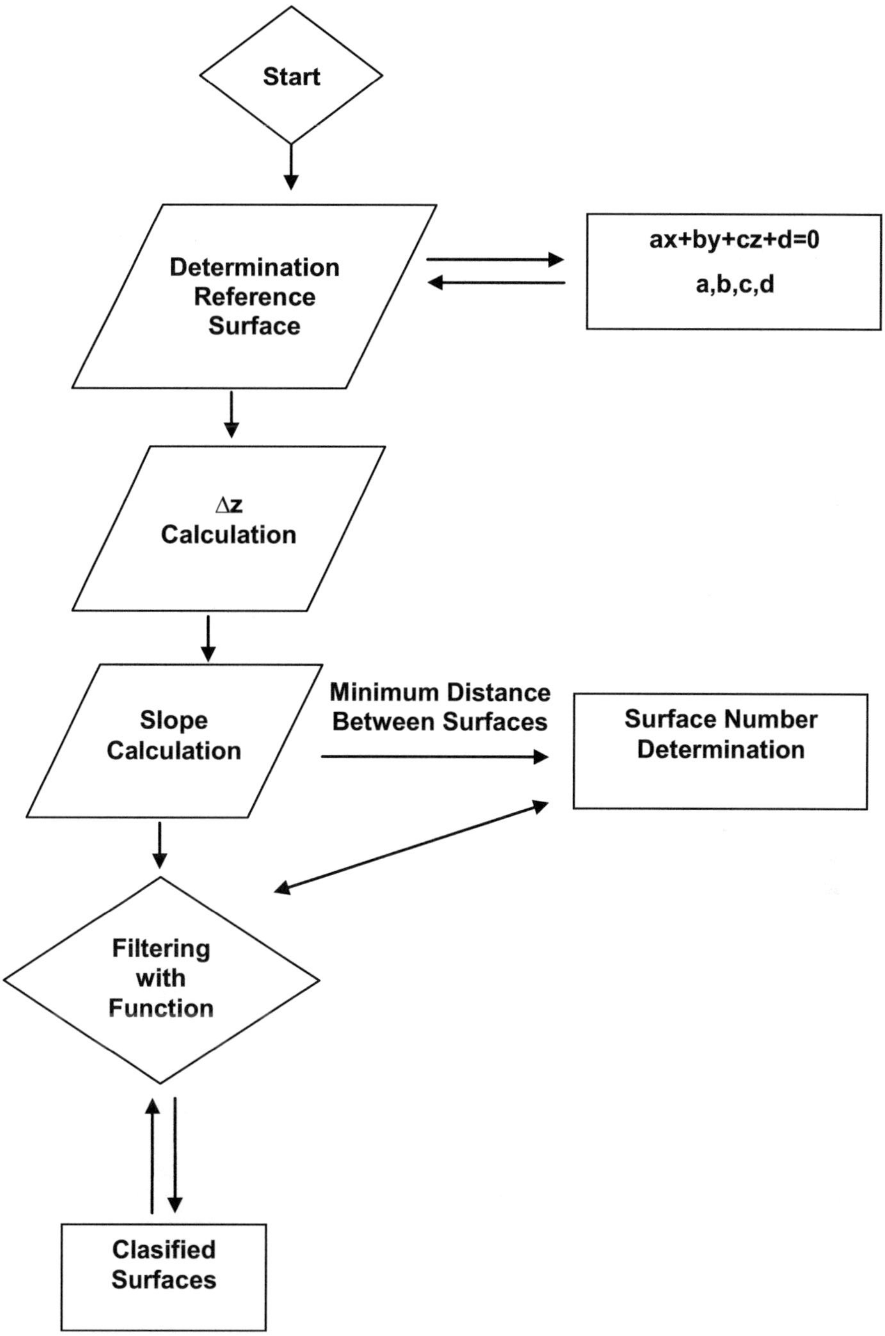

Fig. 3.3. Flowchart of The Filtering Code

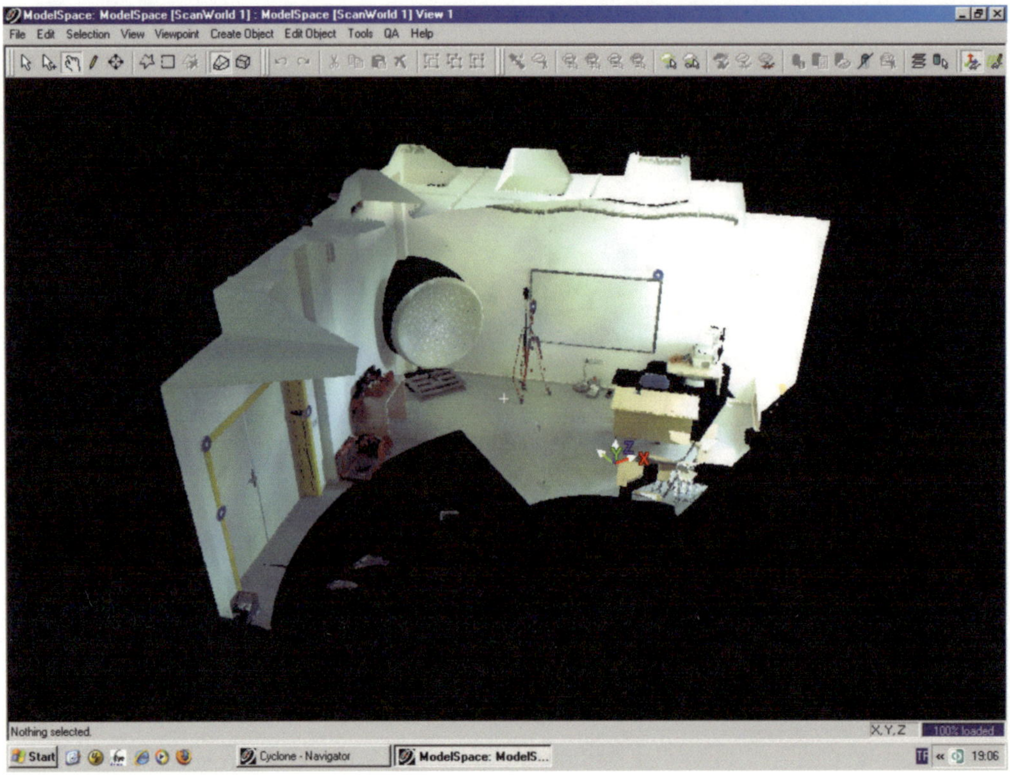

Fig. 3.4. Application Room within Point Cloud Data

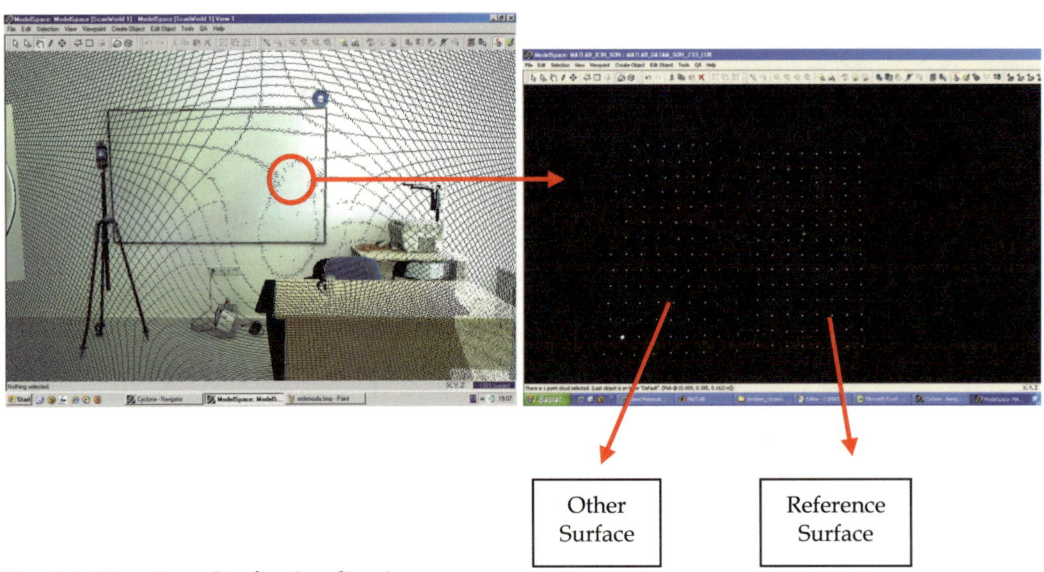

Fig. 3.5. First Data Set for Application.

surface. In line with those five points, the parameters of the plane and the height of reference surface were determined. The scanning frequency for the data set was 1 cm.
Once the group of points belonging to planar surfaces was obtained, different planes existing in our data were observed. With the filtering algorithm, separation of the planes from each other became possible. In this case, this algorithm was some kind of modification of the geometric method based on the plane theory, which enabled location of the data structure. Thus, it was not obligatory to specify the number of classes existing in the data. It was essential because the number of different plane orientations in a data set were unknown. Due to the algorithm, the points were obtained in separated groups that represented planes with a determined weight, by the Δz_i distance from the reference surface. Two points in parallel planes were still in the same group, but they were separated with the weightiness of the T Matrix. The reason for this process was to differentiate the points of different parallel planes and to ascertain the groups of points that were represented in T Matrix within the last column. The calculation of the weight of the points was done with formulation of Eq.3.2. The weights of the points were distributed between 0 and 1. This was the mathematical exponential function of distribution. The result for the classification of the first data set is shown in a Fig.3.6. below.

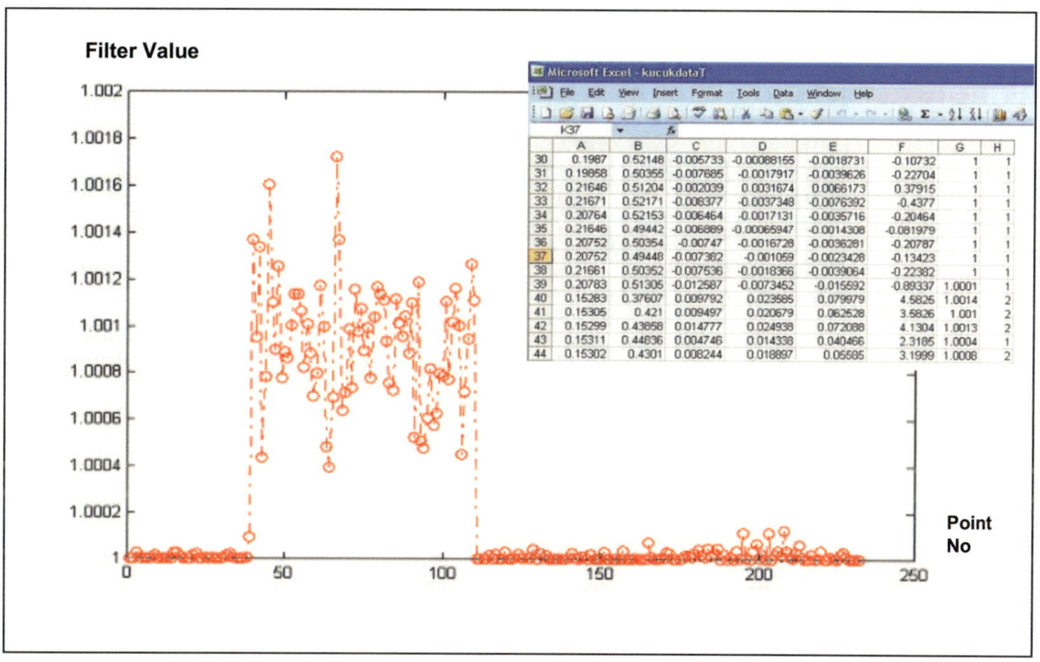

Fig. 3.6. The Classification for The First Data Set with Filtering Method

Secondly, an original data set containing four different surfaces and 3807 points was selected, and the difference of depth between the surfaces of the doors, borders, wall, and columns was changing from 7 cm. to 20 cm. Fig.3.7 shows that data set within four different surfaces.

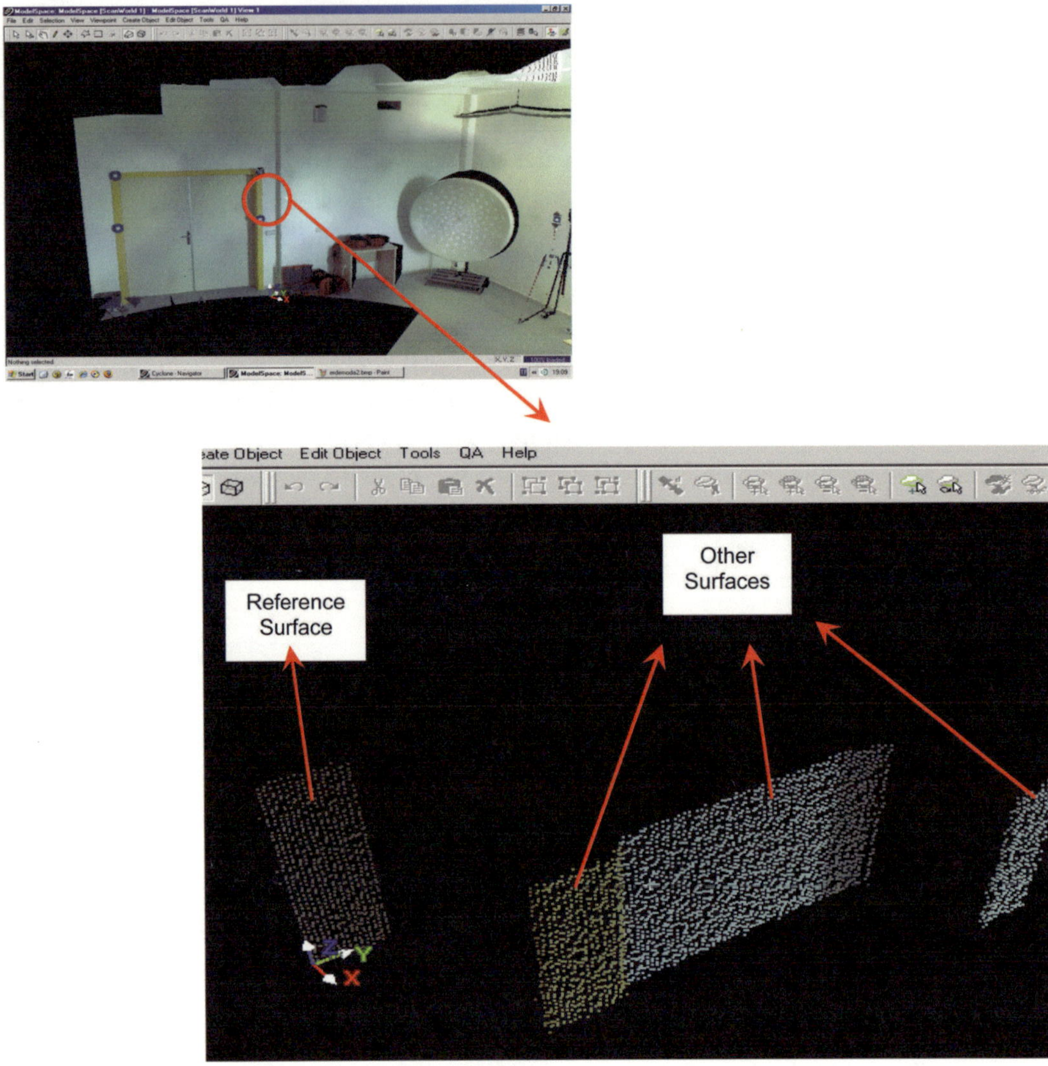

Fig. 3.7. Second Data Set From 4 Distinct Surfaces.

The result for the classification of the second data set is shown in Fig.3.8 above. Hence, the ASCII file was randomly composed with the coordinates of the surface. In this case, affine coordinate transformation was considered necessary. The Cyclone 5.2 software randomly collected the X, Y, Z data to the txt format. Therefore, the operator manually selected the points that were used in the rectification for all surfaces. This process caused extra time consumption and resulted in mismatching surface points when the deepness of the surfaces was in close proximity. With this algorithm, the operator was able to find the right coordinates automatically for image rectification from X, Y, Z ASCII file, whose points belonged to the surface for photogrammetric image rectification, without any additional coordinate transformation [14,17,18]. The data was collected from the hardware in accordance with the time, and the filtering algorithm could be applied directly [18].

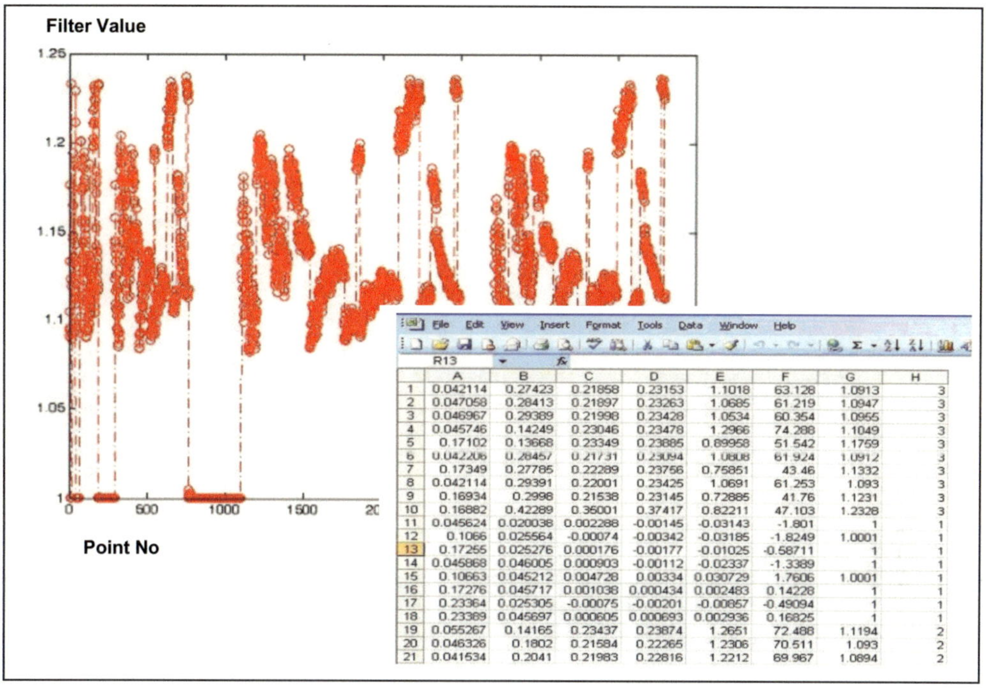

	A	B	C	D	E	F	G	H
1	0.042114	0.27423	0.21858	0.23153	1.1018	63.128	1.0913	3
2	0.047058	0.28413	0.21897	0.23263	1.0685	61.219	1.0947	3
3	0.046967	0.29389	0.21998	0.23428	1.0534	60.354	1.0955	3
4	0.045746	0.14249	0.23046	0.23478	1.2966	74.288	1.1049	3
5	0.17102	0.13668	0.23349	0.23885	0.89958	51.542	1.1759	3
6	0.042206	0.28457	0.21731	0.23094	1.0808	61.924	1.0912	3
7	0.17349	0.27785	0.22289	0.23756	0.75851	43.46	1.1332	3
8	0.042114	0.29391	0.22001	0.23425	1.0691	61.253	1.093	3
9	0.16934	0.2998	0.21538	0.23145	0.72885	41.76	1.1231	3
10	0.16882	0.42289	0.35001	0.37417	0.82211	47.103	1.2328	3
11	0.045624	0.020038	0.002288	-0.00145	-0.03143	-1.801	1	1
12	0.1066	0.025564	-0.00074	-0.00342	-0.03185	-1.8249	1.0001	1
13	0.17255	0.025276	0.000176	-0.00177	-0.01025	-0.58711	1	1
14	0.045868	0.046005	0.000903	-0.00112	-0.02337	-1.3389	1	1
15	0.10663	0.045212	0.004728	0.00334	0.030729	1.7606	1.0001	1
16	0.17276	0.045717	0.001038	0.000434	0.002483	0.14228	1	1
17	0.23364	0.025305	-0.00075	-0.00201	-0.00857	-0.49094	1	1
18	0.23389	0.045697	0.000605	0.000693	0.002936	0.16825	1	1
19	0.055267	0.14165	0.23437	0.23874	1.2651	72.488	1.1194	2
20	0.046326	0.1802	0.21584	0.22265	1.2306	70.511	1.093	2
21	0.041534	0.2041	0.21983	0.22816	1.2212	69.967	1.0894	2

Fig. 3.8. The Classification for The Second Data Set with Filtering Method

3.3 Accuracy analysis

The analysis of the filter program coded and accuracy given by the function necessitates a real effort and relatively long time. Since data sets contain data structure, a comparison between operational and visual differentiation is necessary for determining true accuracy. Because of that, accuracy ratios obtained in the application are obtained over two data sets and the resultant accuracies are determined. First data set and second data set point clouds are analyzed individually by the operator under CAD environment and for each point coordinate information, whether they belong to a surface or not are visually inspected and recorded. Those recordings and surface results obtained with the filter are compared. In these comparisons, for the two surfaces used and classified in the first data set, 2 of total 232 points are classified under different surfaces. In the classification of two surfaces with total 352 points, 17 points are seen to be on different surfaces. This accuracy represents approximately 90-98% accuracy for the first data set. The second data set is composed of total 3807 points and belongs to 4 surfaces. With the evaluation on this data set, 279 points are determined to be classified in different surfaces. This ratio is again approximately 92%. When these misclassified positions of the points are examined, the errors generally occur on the points which are most close to the surface limit in the surface transitions. This means the necessity of using the skewed parameters used in this filter model by the help of a better method. These skewed parameters should be optimized and their significance should be increased in the function [11-18].

Filtering Result					Raw Data for Surface 1			
X(m)	Y(m)	Z(m)	S.N.		X(m)	Y(m)	Z(m)	Control (m)
0,0785220	0,40471	-0,002459	1		0,0785217	0,404715	-0,024586	-0,0000003
0,0786130	0,37763	-0,002509	1		0,0786133	0,377631	-0,025091	0,0000003
0,0786440	0,38669	-0,00248	1		0,0786438	0,38669	-0,024799	-0,0000002
0,0788880	0,33255	-0,002592	1		0,0788879	0,350549	-0,025142	-0,0000001
0,0788880	0,35055	-0,002514	1		0,0788879	0,332547	-0,025916	-0,0000001
0,0984800	0,32326	-0,002011	1	*	0,078949	0,368543	-0,023656	-0,0195310
0,0789490	0,35953	-0,002412	1		0,078949	0,359534	-0,024119	0,0000000
0,0791320	0,34151	-0,002425	1		0,0791321	0,341505	-0,024246	0,0000001
0,0792850	0,39549	-0,0021	1		0,0792847	0,395494	-0,020997	-0,0000003
0,0875550	0,40479	-0,002581	1		0,0875549	0,404788	-0,025814	-0,0000001
0,2014800	0,32273	-0,001071	1		0,0877075	0,332633	-0,028482	-0,1137725
0,0877690	0,37766	-0,002582	1		0,0877686	0,377661	-0,025824	-0,0000004
0,0880740	0,36855	-0,002437	1		0,0880737	0,368551	-0,024369	-0,0000003
0,0984800	0,32326	-0,002011	1		0,0882568	0,395582	-0,02311	-0,0102232
0,0883790	0,35945	-0,002268	1		0,0883789	0,359452	-0,022675	-0,0000001
0,0888060	0,35038	-0,002105	1		0,0888062	0,350378	-0,021046	0,0000002
0,0972600	0,40465	-0,002362	1		0,0972595	0,404653	-0,023617	0,0000005
.	.	.	1		.	.	.	.
.	.	.	1		.	.	.	.

Table 3.1. Control Table for The First Data Set

Filtering Result					Raw Data for Surface 2			
X(m)	Y(m)	Z(m)	S.N.		X(m)	Y(m)	Z(m)	Control (m)
0,0328369	0,3457570	0,2140753	2		0,0328369	0,3457570	0,2140753	0,00000
0,0328674	0,3558072	0,2144537	2		0,0328674	0,3558072	0,2144537	0,00000
0,0328979	0,3607536	0,2148979	2		0,0328979	0,3607536	0,2148979	0,00000
0,0329285	0,3704611	0,2164029	2		0,0329285	0,3704611	0,2164029	0,00000
0,0329895	0,3659124	0,2146190	2		0,0329895	0,3905043	0,2174159	0,00000
0,0329895	0,3805984	0,2163755	2		0,0329895	0,3805984	0,2163755	0,00000
0,0329895	0,3905043	0,2174159	2		0,0329895	0,3659124	0,2146190	0,00000
0,0744934	0,1364970	0,2352521	2		0,0330200	0,3858122	0,2158834	-0,04147
0,0330811	0,3958577	0,2163480	2		0,0330811	0,3958577	0,2163480	0,00000
0,0330811	0,4009570	0,2163684	2		0,0330811	0,4009570	0,2163684	0,00000
0,0331726	0,4108378	0,2174723	2		0,0331726	0,4108378	0,2174723	0,00000
0,0332336	0,4069457	0,2131742	2		0,0332336	0,4166092	0,2150874	0,00000
0,0332336	0,4166092	0,2150874	2		0,0332336	0,4069457	0,2131742	0,00000
0,0364075	0,1546124	0,2197993	2		0,0364075	0,1546124	0,2197993	0,00000
0,0364990	0,1600594	0,2174183	2		0,0364990	0,1600594	0,2174183	0,00000
0,0408630	0,1320048	0,2336225	2		0,0365601	0,1745574	0,2190564	-0,00430
0,0365906	0,1699335	0,2174628	2		0,0365906	0,1699335	0,2174628	0,00000
0,0366211	0,1653550	0,2158285	2		0,0366211	0,1653550	0,2158285	0,00000
0,0746460	0,1325353	0,2306476	2		0,0366211	0,1795070	0,2190469	-0,03802
0,0366821	0,1897474	0,2175339	2		0,0366821	0,2040320	0,2203057	0,00000
0,0366821	0,2040320	0,2203057	2		0,0366821	0,1897474	0,2175339	0,00000
..	..	..	2		..	..	..	..

Table 3.2. Control Table for The Second Data Set

In Table 3.1 and 3.2, there are some samples from the tables created in Excel software, for the accuracy analysis of data sets that enters into the filter. In these tables, the first part is T analysis matrix outputs; the second part is the true values differentiated from the raw data, depending on the surface that they belong to. The column of difference represents differences of Coordinate X. When the column of difference gets the value of zero, the points are the same and in that line, the point is on the right surface [18].

4. Conclusion

The analysis of the developed software and determination of the surfaces has been made as a result of the analysis matrix obtained by the application made with the recommended filtering model and selected sample data.

In consideration of these data, feasibility and accuracy of the selected model is detected. With the statistical analysis, it has been determined that the model gives an accurate result for data under 9 mm scanning frequency with approximately 90% accuracy. However, since the applied model and filtering technology operates based on point data, the greater the point data, the higher the processing time and the lower the program speed. Additionally, local data and especially Lidar originated data should be inspected and the analysis of results should be performed in surveying engineering.

With this chapter, we observe that terrestrial laser scanners (TLS) and photogrammetric technologies need equipment (hardware) and data based integration as the first step of the process. Obviously, TLSs are independent equipments which depend on their own algorithm and optical (Laser) structures. Photogrammetric technologies are methods based on seen light and used with independent equipment (camera). When these two independents are used in measurement equipment, there is definitely a need for integration. When the equipment integration is not fulfilled, an integration of data is crucial at least. Current TLS equipment and some of the sub-software satisfy some parts of this integration; however this integration is not used completely independent from operators. The processes and difficulties are determined in this chapter, which is made for a kind of data integration, and an approach to solve these issues is presented [18].

Generaly, made on terrestrial laser scanner coordinate system does not use affine transformation. Thus, point cloud data listed disorderly by Cyclone 5.2 software are classified in a surface-based way, so it passes through a filter.

Considering the core software as the result of the research, a general advantage-disadvantage list is shown below for the functional model:

Advantages;

- It can differentiate the surfaces with two distinct depth by the accuracy of 90%
- It is a function designed for the MatLAB environment.
- It has an object-oriented structure that can be developed under Windows.
- It can give faster and more accurate results for limited data intensity and for quick, small scale experiments.
- It provides hardware data integration.
- The extremes of the filter function can be defined and always are positive because of the exponential filtering technique.

Disadvantages;

- When the data density is increased, time of the operation gets longer.
- The credibility is an issue of question for various types of data and in terms of data intensity.
- It is sensible to the limitations deriving from the hardware and the processor.
- It does not have the structure of an object-oriented function.
- It has not yet been applied to wide range photogrammetric studies.

This approach for planar segmentation is valid for monuments with planar surfaces. It drastically reduces the huge amount of laser data and adds information about the nature of the data. Thus, this data can be treated easily with most of the existing CAD software for 3D modeling or other applications. Some precautions in data acquisition should be taken for future works in surveying engineering. First, the location of the stations should ensure both the complete covering of the object and the most possible homogeneity along each single scan in surveying. Another precaution is to perform the image rectification in the close range photogrammetric and surveying applications at the end.

Consequently, it is seen that TLS data can be used in photogrammetric, three dimensional modeling, in reducing the work steps that depend on operator and in point analysis research with the combination of software and hardware integration methods.

Abbreviation list

TLS : Terrestrial Laser Scanner
CRP : Close Range Photogrammetry
RGB : Red, Green, Blue
T : Filtering Matrix
MS : Microsoft

Biography of author

Bahadır Ergün, born in 1974. He graduated his bachelor's degree in 1996 in Istanbul Technical University in Turkey. He graduated in 1999 as master of science in Geodesy and Photogrammetry Engineering and obtained his doctorate in 2003 in Photogrammetry Engineering, all of from Istanbul Technical University. He worked as Research Assistant between 1997-2003 in Istanbul Technical University Photogrammetry Department. He has been since 2003 Associate Professor in Gebze Institute of Technology, Geodesy and Photogrammetry Department in Turkey. He has held research assignments in TU Berlin Photogrammetry Department in 2001.

5. References

[1] Kostamovaara, J., Maatta, K., Myllyla, R., 1991, Pulsed time of flight laser range techniques for industrial applications, SPIE Proceedings, Vol. 1614, pp.283-295
[2] Boehler W., Marbs A., 2002, 3D Scanning Instruments, University of Applied Sciences, Mainz, Germany,
[3] Heinz, G., 2002, Pharaoh Pepi I.: documentation of the oldest known life-size metal sculpture using laser scanning and photogrammetry. Proc. of the CIPA WG6 Int. Workshop on scanning for cultural heritage recording.

[4] Staiger, R., 2003, Terrestrial laser scanning technology, systems and applications. In: Proceedings. of 2nd FIG Regional Conference, Marrakech, Morocco.
[5] Briese, Ch., Pfeifer, N., Haring, A., 2003, Laserscanning and photogrammetry for the modelling of the statue Marc Anton. The ISPRS International Archives of the Photogrammetry, Remote Sensing and Spatial Information Sciences, vol. XXXIV-5/C15, pp. 528-533.
[6] Boehler, W., Bordas Vicent, M., Kanke, K., Marbs, A., 2003, Documentation of german emperor Maximilian I's tomb. The ISPRS International Archives of the Photogrammetry, Remote Sensing and Spatial Information Sciences, vol. XXXIV-5/C15, pp. 474-479.
[7] Drap, P., Sgrenzaroli, M., Canciani, M., Cannata, G.,Seienturier, J., 2003, Laser scanning and close range photogrammetry: towards a single measuring tool dedicated to architecture and archaeology. The ISPRS International Archives of the Photogrammetry, Remote Sensing and Spatial Information Sciences, vol. XXXIV-5/C15, pp. 629-634.
[8] Bornaz, L., Rinaudo, F., Roggero, M., 2003, Object segmentation in cultural heritage. The ISPRS International Archives of the Photogrammetry, Remote Sensing and Spatial Information Sciences, vol. XXXIV-5/C15, pp. 490-495.
[9] Forkuo K., King B., 2004, Automatic Fusion Of Photogrammetric Imagery And Laser Scanner Point Clouds, XXth ISPRS Congress, Istanbul, pp.475.
[10] Lichti D. D., Gordon S. J., 2004, Error Propagation in Directly Georeferenced Terrestrial Laser Scanner Point Clouds for Cultural Heritage Recording, FIG Working Week 2004, Athens
[11] Biosca Taronger, J.M., 2005, Filtrado y segmentación de datos obtenidos mediante láser escáner 3D. Universidad Politécnica de Valencia, Valencia.
[12] Lichti, D.D. and J. Franke, 2005, Self-calibration of the iQsun 880 laser scanner. In: Optical 3-D Measurement Techniques VII, Vienna, Austria, Vol. I, pp. 112-121.
[13] Lerma J.L., Biosca J.M., 2005, Segmentation And Filtering Of Laser Scanner Data For Cultural Heritage, CIPA XX International Symposium Proceedings, pp. 896-901.
[14] Lichti, D.D. and Licht, M. G., 2006, Experience with terrestrial laser scanner modelling and accuracy assessment. In: The International Archives of the Photogrammetry, Remote Sensing and Spatial Information Sciences, Vol. 36 (Part 5), pp. 155-160.
[15] Reshetyuk, Y., 2006, Calibration of Terrestrial laser scanners Callidus 1.1, Leica HDS 3000 and Leica HDS 2500. In:Survey Review 38 (302), 703-713.
[16] Luhmann T., S. Robson, S. Kyle and I. Harley, 2007, Close Range Photogrammetry, Whittles Publishing, England.
[17] Lichti D. D., 2007. Error modelling, calibration and analysis of an AM–CW terrestrial laser scanner system, In: ISPRS Journal of Photogrammetry and Remote Sensing, 61 (5), pp. 307-324.
[18] Ergun, B., 2010, A novel 3D geometric object filtering function for application in indoor area with terrestrial laser scanning data, Optics and Laser Technology, pp. 799-804
[19] FARO,http://www.faro.com/content.aspx?ct=uk&content=pro&item=5&subitem=55, 2007.
[20] LEICA, http://leica-geosystem.com/hds/en/lgs_5570.htm, 2007.
[21] OPTECH, http://www.optech.ca , 2007.

[22]RIEGL,http://www.reigl.com/terrestrial_scanners/terrestrial_scanner_overview_/terr_scanner_menu_all.htm, 2007
[23] ZOLLER+FROHLICH, http://www.zf-laser.com, 2007

26

Design, Calibration and Application of a Seafloor Laser Scanner

Chau-Chang Wang[1], Da-jung Tang[2] and Todd Hefner[3]
[1]*Institute of Applied Marine Physics and Undersea Technology*
National Sun Yat-sen University
[2,3]*Applied Physics Lab, Washington University*
[1]*Taiwan*
[2,3]*USA*

1. Introduction

The surface of a seabed is influenced by wave motions, current disturbance, sediment transportation and biological activities. The scale of the features on the seabed varies from meters for large sandwaves to less than one millimeter for prints left by benthic activities. These features are important information for the sea bottom related studies. For example, marine biologists are interested in the evolution of prints on a seabed because it reveals the quantity and activity of benthos. For underwater archaeology study, accurate in-situ measurement of the site is a crucial step for the site preservation. The roughness of the seabed is one of the key parameters in modeling high-frequency underwater acoustic wave scattering. Different tools, optical-, acoustic- or electric-based, have been developed to provide measurement under different environment conditions in different scales.

Photography is a popular method used as a monitoring or surveying tool for underwater research. It provides qualitative information for archaeological site documentation and coral reef habitat monitoring (Klimley & Brown, 1983; Broadwater, 1988). However, unlike the application in the air, underwater image quality is strictly limited by illumination, range and environmental factors such as turbidity. If the target is several meters away from the camera, even under the best water conditions, photos cannot give satisfactory results. For example, Fig. 1 is a photo taken during an underwater safety inspection of pier structures. The gray area is a patch of barnacle settlement, and the dark area is a rust pot. Unless we know what we are looking at in the photo, in general the undersea world lacks references to determine the nature and the dimension of features.

An popular method called *structured light* can overcome the aforementioned problem. The idea is to project a light source with know pattern onto the scene to enhance the signal-to-noise ratio. The dimension of the feature is inferred from the distortion of the structured light in the image or from the time-of-flight of the light pulse (Myers, 1980; Rocher & Keissling, 1975; K.S. Fu & Lee, 1987). Active light source is used, so it is classified as *active vision*. Laser scanning is a typical case of active vision. They are widely used in many applications of various scales, such as terrain survey and mapping (large scale), reverse engineering and prototyping (middle scale), and microscopic-particle image velocimetry (micro scale)

Fig. 1. An sample underwater photo taken during a pier inspection.

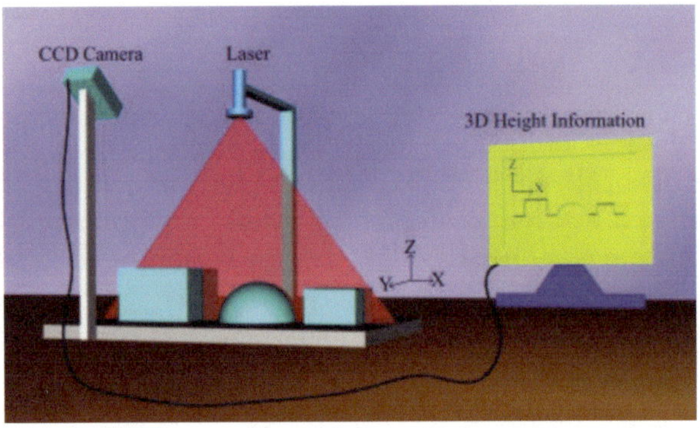

Fig. 2. Concept of structured light scanning: A laser light stripe projected on a target, and the reflection is captured by a camera to infer the dimension of the feature.

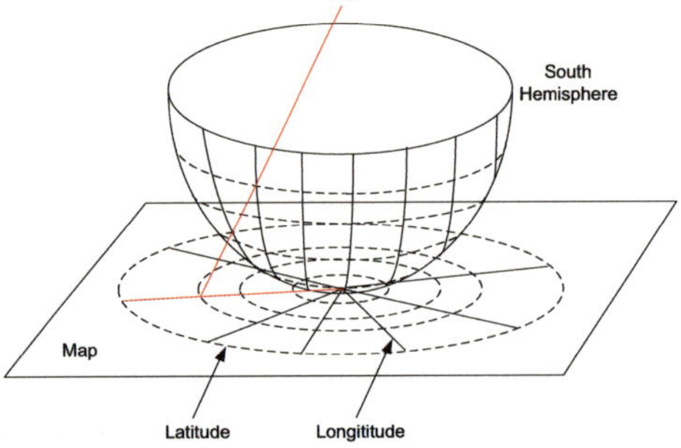

Fig. 3. Schematic of map projection: a light source projects the terrain of the earth on a plane sheet to make a map. The location of a point can be referred from the longitudes and latitudes.

(Campbell, 2006; Son & Lee, 2002; Melling, 1997; Raffel et al., 2007). As illustrated in Fig. 2, the light stripe projected on the target blocks leaves a trace on the surface which is the cross-section of the block at that location. A camera can be placed at a proper distance with an oblique perspective angle to observe the deformed light stripe. The laser stripe seen in the camera resembles the silhouette of the target. The high contrast of the laser scan line in the image can be extracted by proper thresholding. It should be noted that the offset of the laser line is not necessarily proportional to the height of the target because the camera is pointed at an oblique angle to the target such that the image generally has some distortion. Therefore a calibration is needed to convert the laser scan line described in pixel coordinates to the actual dimension of the profile (Li et al., 1989). Applying this idea in the water, Harbor Branch Oceanographic Institution installed a laser line scanning system in their Johnson-Sea-Link submersible to locate possible seafloor rock outcrops (Kocak et al., 1999; Ogawa, 2000). Crawford and Hay constructed a laser scanning system to record the seabed elevation variation and suspension events (Crawford & Hay, 1998).

To have a full 3-D measurement of the environment, the scanning head needs to be mounted on a linear track or rotating shaft to provide a third degree of freedom. As a result, for such a line scanning system, full 3D modeling of the camera is not necessary. The camera calibration can be simplified as finding a 2D-to-2D mapping, i.e. points in pixel coordinates to the coordinates on the scanning plane. Chen and Kak took this idea and simplified the optical model (Chen & Kak, 1989). They used a 4×3 transformation matrix to characterize this mapping. Theoretically, only four object points at known locations are required to determine this matrix. This method is attractive because it does not involve any low-level details of the camera and the projection light model parameters. On the other hand, a single constant transformation matrix cannot account for the nonlinear optics of the camera over the entire field of view. Another approach is to use multiple regression for relating a group of grid points on the laser scanning plane to the their images in the pixel coordinates (Gujarati, 2003). This approach is simple; however, for uneven distorted images, it is difficult to find appropriate polynomial functions to accurately characterize the nonlinearity.

In stead of tackling the camera calibration as a problem of finding intrinsic and extrinsic parameters for the camera model, we propose an empirical method in which the view seen by the camera is divided into smaller regions and local linear maps are built for each region. Ideas analogous to longitude and latitude lines of a map are adopted to refine the accuracy of laser line scanning. We utilize the properties of the structured-light-oriented geometry to simplify the process, while the desired accuracy is maintained. The goal is to develop a laser scanner to measure underwater features with millimeter resolution in general.

In this chapter, (1) a map-projection-based camera calibration for structural light scanning is introduced; (2) step-by-step analyses of the measurement error bounds are presented; (3) precise point-to-point comparisons between test pieces fabricated by a CNC (computer numerical control) milling machine and the scanning results are provided to verify the efficacy of our approach; (4) a roughness-known model surface is scanned to decide the reliable roughness spectrum measurement bounds of the scanner; (5) field experiments at water depth of eighty meters are reported.

2. Plane-based camera calibration

Before explaining the calibration method, let us first review the concept of map projection. Consider a transparent hemisphere painted with longitude and latitude lines on its surface

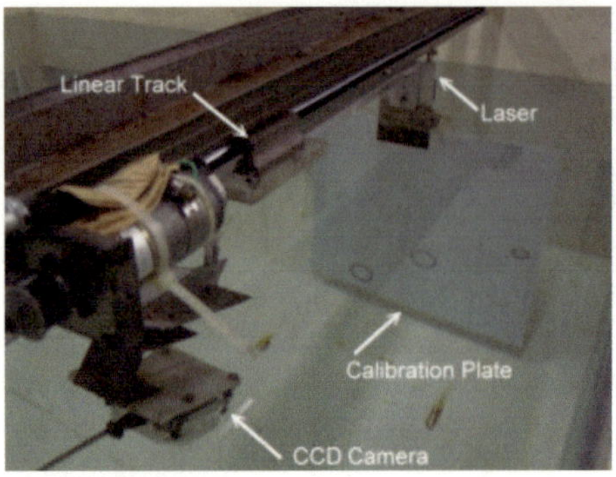

Fig. 4. Experimental Setup.

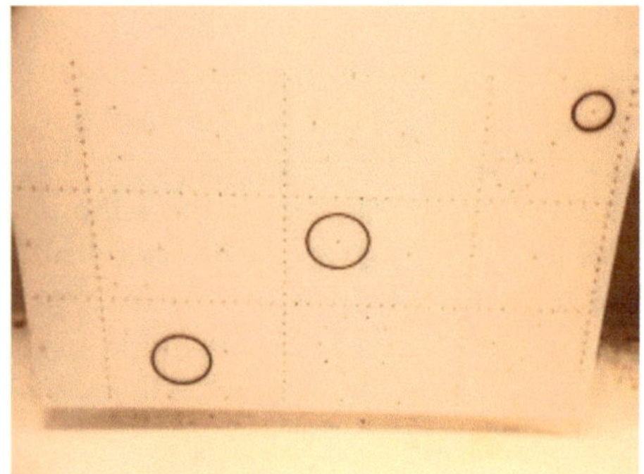

Fig. 5. The calibration board seen from the CCD camera.

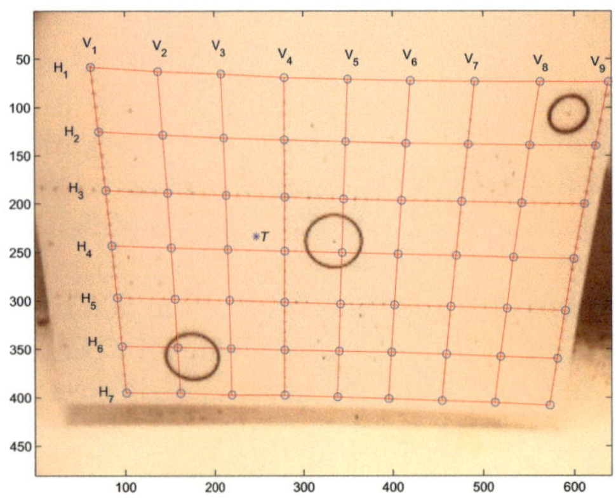

Fig. 6. Longitude and latitude lines created by curve-fitting the grid dots.

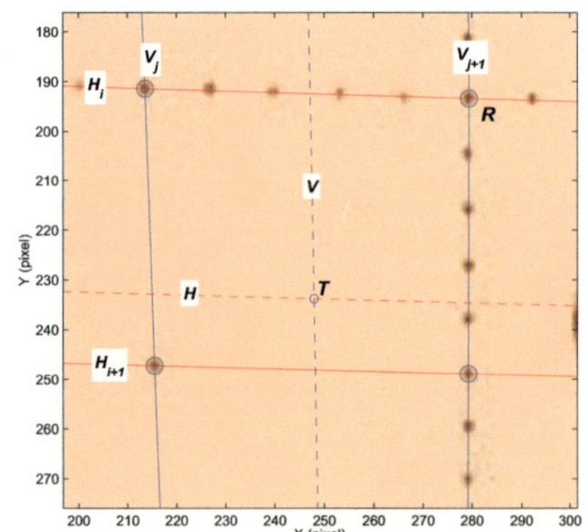

Fig. 7. Close-up of a target point T on the calibration board.

as shown in Fig. 3. To make a map, we put a light source at the top of the hemisphere and a screen tangent to the south pole. The light rays project the terrain and the longitude and latitude lines onto the screen to create the map. The longitude and latitude lines serve as references to relate a point on the map with its location on the hemisphere. For a location which is not right at the intersection of a longitude line and a latitude line, we can estimate its coordinates from its distances to the neighboring longitude/latitude lines. Bearing this idea in mind, the camera calibration for measuring features on a plane can be analogized to the projection problem: replace the projection screen with a camera, and the transparent hemisphere with a board with known graticules placed coplanarly with the measurement plane. To decide the coordinates of a point in the image is like reading a map to find out the coordinates of a town.

For our calibration setup, a laser line source and a camera were fixed on a linear stage and submerged into water (as shown in Fig. 4). Grid points, drilled with 50 mm separation by a CNC milling machine, are laid on an acrylic board as shown in Fig. 5. This board, called the *calibration board* hereafter, is carefully aligned with the plane of the laser scanning sheet. In order to cover a larger view, a camera with a wide-angle lens is placed roughly one meter away from the board. The optical axis of the camera is pointed downward approximately 30° from the horizon. The points seen in the camera are shown in Fig. 5. It is obvious that the upper half of the image is larger than the lower half. The formations of the grid points are not

straight and parallel lines but curvilinear near the boundary of the image frame. However, this distortion is mild and varies smoothly from the center to the sides. We curve-fit the grid points to create longitude lines, V_j, and latitude lines, H_i, as shown in Fig. 6, such that the scope is divided into smaller quadrangular regions. Within a quadrangle, we can assume that the distortion of the image is smooth and linear. Therefore, given a target point, T, shown in Fig. 6 or the close-up in Fig. 7, its location on the calibration board can be estimated from the distances (in pixels) between itself and its four neighboring sides V_j, V_{j+1}, H_i and H_{i+1}. A rough estimate of T is approximately one half of a spacing from V_j and three quarters of a spacing from H_i. In other words, its location is roughly 25 mm right from the longitude line V_j and 37.5 mm beneath the latitude line H_i. Since the grid points were laid with 50 mm spacing, the coordinates of T can be inferred from the closest known point, e.g. R. For example, if R's coordinates are (100 mm, 250 mm) in the calibration board coordinate system, T will be around (125 mm, 212.5 mm).
In stead of *estimation*, $T's$ position can be obtained with respect to a corner of the quadrangle with following procedures. Interpolate from the surrounding longitude and latitude lines a set of vertical and horizontal lines which intersect at the target point[1]. For the example we can have a linear combination of H_i and H_{i+1} with some proper weightings to construct a horizontal line H which passes through T as

$$H = w_i H_i + w_{i+1} H_{i+1}, \tag{1}$$

where w_i, $w_{i+1} \geq 0$ and $w_i + w_{i+1} = 1$. If the weighting factors ratio is $w_i : w_{i+1} = 0.26 : 0.74$, we find that T is 50 mm$\times$0.74 =37 mm beneath H_i , or 50 mm$\times$0.26 =13 mm above H_{i+1}. The horizontal component can be obtained likewise.
In summary, this method does not explicitly involve any intrinsic or extrinsic parameters of the camera or the relative position of the laser scanning sheet. All non-linearities are lumped together and linearized separately for each quadrangle in the map. Therefore when a laser scanning sheet is projected on the target, the relative position of the laser line segment in the calibration board coordinate system can be calculated with the same idea. In the following section, we will present the procedures of constructing such a projection map for the laser scanning system.

3. Implementation

First of all, we define the following terms used in explaining the calibration and measurement procedures.

A Control Points

Grid dots with 50 mm separation, called control points, are laid on the surface of an acrylic board as shown in Fig. 5. These dots were drilled with a CNC milling machine, so their positions are assumed to be accurate to the CNC machine accuracy which is 0.01 mm for our case. Each dot is of finite size, and its intensity-weight center is used as its nominal location. As shown in Fig. 8, the distances between control points in the image are not fixed but varying mildly from point to point. An algorithm was developed to search for the control points adaptively. As a first step, we manually select the upper-left dot (bounded in

[1] With the functional forms of the neighboring longitude lines and latitude lines, we can use bisection to decide a set of vertical and horizontal lines which intersect at the target point.

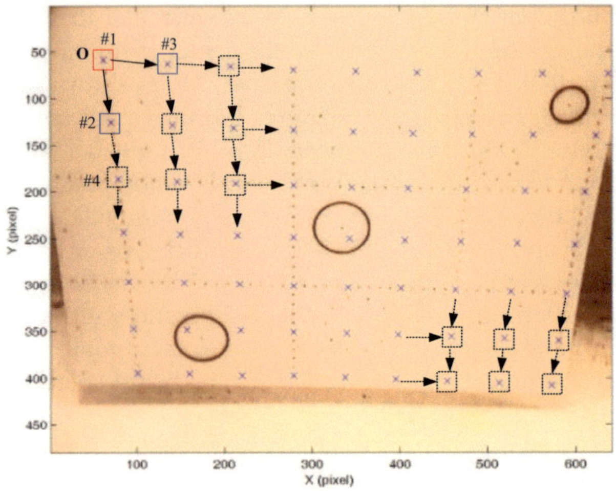

Fig. 8. The idea of adaptive searching for control points on the calibration board.

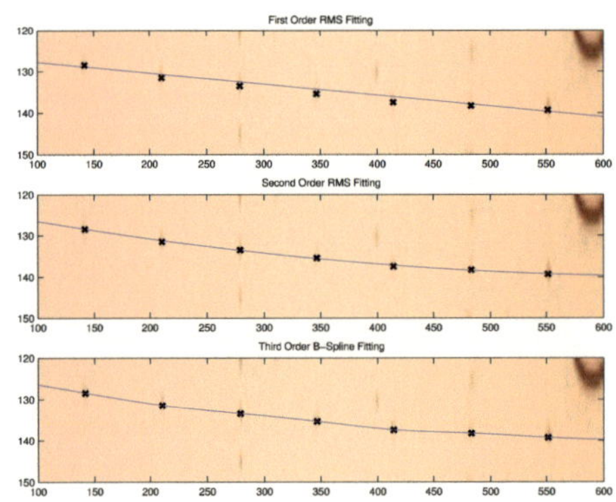

Fig. 9. Comparison of different fitting methods for grid points.

box #1), and the dot beneath (bounded in box #2) and right next to it (bounded in box #3). We use the displacement vector from dot #1 to #2 as a reference to estimate the approximate location of the control point beneath #2 (denoted as #4). A bounding box is placed at the tip of the displacement vector to search for the dot. Once dot #4 is found, a new displacement vector from #2 to #4 can be used to search downward further. In the same manner, we can search toward the right-hand side. This process is repeated until all the control points are located. We mark all the control points found with an "$\times$" as shown in Fig. 8. The control point at the upper left corner labeled as "O" is selected as the origin of the calibration board coordinate system.

B Longitude and Latitude Lines

All the control points are laid with equal spacing on the board. Therefore the trend of the local image distortion is characterized by the relative displacement of the control points and the overall distortion can be characterized by the longitude and latitude lines which are obtained by curve-fitting the control points (in pixel coordinates). To achieve better fitting results, the latitude lines are described as $Y = Y(X)$, while the longitude lines are described as $X = X(Y)$. In Fig. 9, we show the image of horizontal control points overlaid with curves obtained by three different fitting functions. From top to bottom, they are first

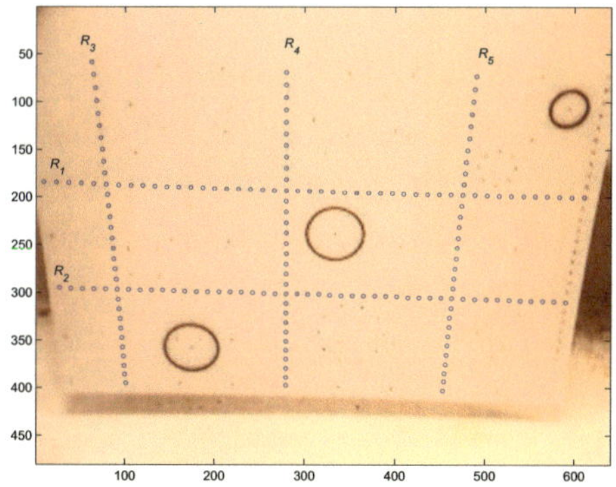

Fig. 10. Five auxiliary rulers on the calibration board.

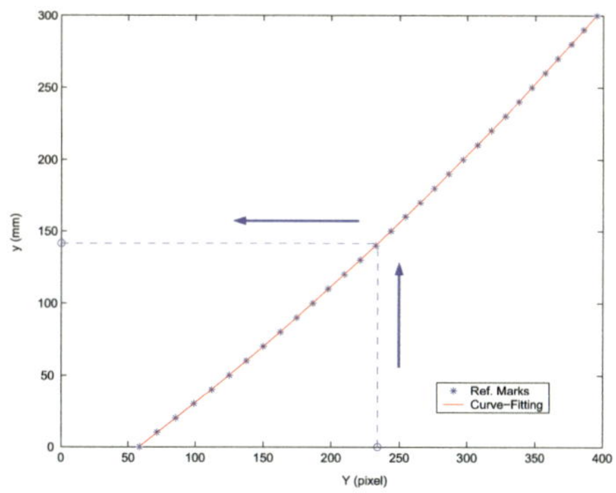

Fig. 11. Pixel coordinate of reference marks versus their calibration board coordinate for ruler R_4.

order RMS (Root Mean Square), second order RMS and cubic B-spline. In the case of first order RMS, the straight line can only pass through the neighborhood of control points, not right through them. As for second order RMS, the polynomial curve does not pass through the control points either, but it traces the control points closely. Mathematically, cubic B-spline curve is forced to pass all the control points. Because these lines are used as bases to determine the location of a target point, the correctness of the curve fit for the control points strongly influences the accuracy of the measurement. According to our analysis, the second order RMS gives the most stable performance which is a property preferred by most measurement systems (Wang & Cheng, 2007).

C Auxiliary Rulers

The longitude and latitude lines give us a rough idea of the location of a target point. The overall accuracy of the system can improved by adding denser control points on the calibration board. However, doing so would give us a crowded and confusing image. An alternative is to place marks of smaller spacing at selected locations for reference. For example, in Fig. 10 we place reference marks of 10 mm spacing on the calibration board labeled as R_1 to R_5. We give these lines another name, the *auxiliary rulers*. By the name *ruler* we mean that these reference lines help us to estimate where the actual coordinates are in a more accurate manner because the uncertainty is reduced from 50 mm to 10 mm.

We show again the closeup of T in Fig. 7. The interpolated latitude line H intersects with the closest vertical auxiliary ruler R_4 at T_4. T and T_4 are share the same y coordinate which

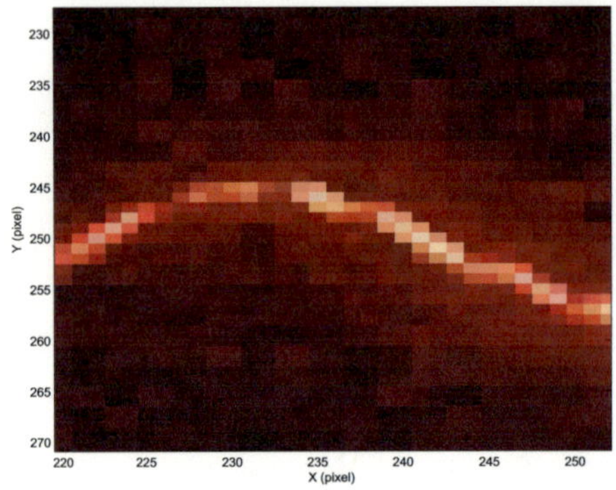

Fig. 12. A close-up image of a laser scan line casting on a sand ripple. It is obvious that the width of the laser line is more than one pixel.

can be obtained by interpolating between the two closest bounding reference marks on R_4. In Fig. 11, we plot the vertical pixel coordinate Y of each reference mark on R_4 versus the corresponding vertical coordinate y on the calibration board in millimeter. With this plot, the vertical component of T_4 in pixels is then used to look up its corresponding y coordinate in millimeter on the laser scanning plane. For example, according to Fig. 11, T_4's Y is 234.5 pixels which is converted to 142.3 mm. Likewise, T_1, the intersection of V with the near-by horizontal auxiliary ruler R_1, provides the x coordinate of T in millimeter.
We summarize the procedures as the following steps:

1 Choose two longitude lines and two latitude lines which surround the target point.

2 Iterate the weights (between one and zero) of the latitude lines until an interpolated curve passes through the target point.

3 Find the intersection of the interpolated latitude line with the closest vertical auxiliary ruler.

4 Substitute the vertical component of the coordinated in pixels found in step 3 into the ruler conversion curve to obtain the y coordinate of the point.

5 Repeat steps 2, 3 and 4 for the horizontal component.

4. Performance tests

The performance of the system depends on the longitude and latitude lines characterization, and also the effectiveness of the laser scan line image processing. Ideally, we would like to have a laser line as narrow as possible (one pixel) in width to reduce the uncertainty[2]. In reality, the scan line is several pixels wide (see the bright area in Fig. 12). In other words, depending on how we choose the location of the center of the scan line, the result might be off by a couple of pixels. From the look-up curve shown in Fig. 11 we know that a pixel in the image frame is roughly equivalent to one millimeter in linear dimension on the calibration board. This ambiguity contributes to the error in the measurement. To understand the nature of these errors, it is necessary to carry out separate experiments to identify the source of the error and its extent of influence.

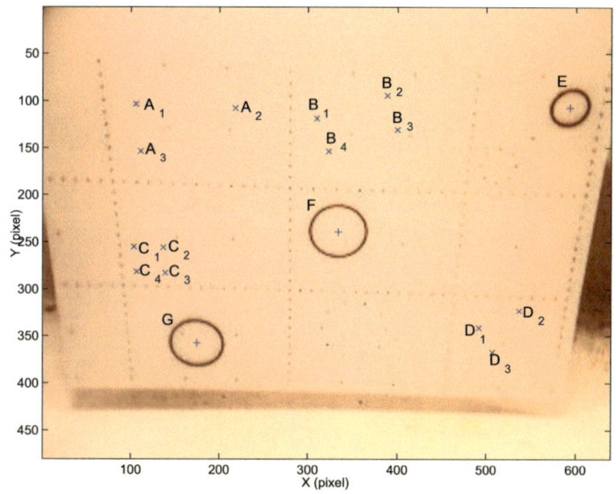

Fig. 13. Features on the calibration board for verification.

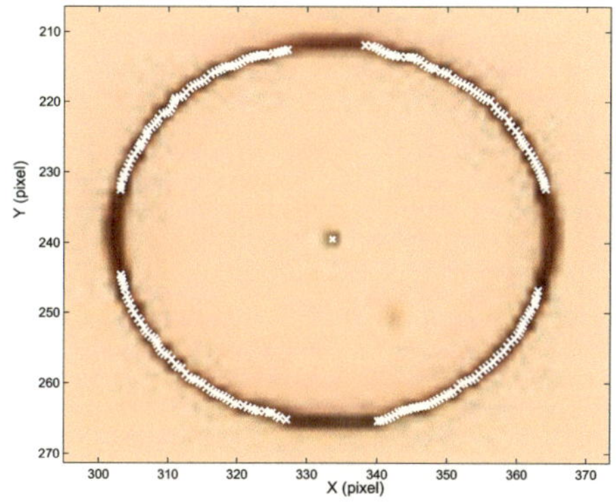

Fig. 14. Manually selected points (denoted as ×) on the circumference for estimating the diameter of circle E.

4.1 On-board verification

When making the calibration board, in addition to the control points and the auxiliary rulers, we also made several dots and shapes with known location and dimension. Measurement of the distance between two dots or the diameter of circles can be used to verify that the curve-fitting achieves the required accuracy.

A Point to Point Distance

There are several dots on the board, labeled as A_1 to A_4, B_1 to B_4, C_1 to C_4, and D_1 to D_4 in Fig. 13. We use the distance between points to estimate the error in our measurement. Because our approach is a point-based measurement, we expect to see that the absolute error remains bounded while the relative error tends to get smaller as the distance (denominator) increases. These results are summarized in Table 1.

B Diameter of Circles

Three circles, labeled as E, F and G, are laid at the upper, middle and lower part of the board. The circles are milled with a ϕ 0.8 end cutter and filled with the same black clay

[2] One-pixel width also poses problems if we want to have sub-pixel resolution.

Features	A_1A_2(80.00mm)			A_2A_3(89.44mm)			A_3A_1(40.00mm)		
Fitting Method	Measured	Err.	%	Measured	Err.	%	Measured	Err.	%
1st RMS	80.19	0.19	0.24	89.68	0.24	0.27	39.66	-0.34	-0.86
2nd RMS	80.01	0.01	0.01	89.43	-0.02	-0.02	39.84	-0.16	-0.39
Cubic B-Spline	86.82	5.82	7.28	94.73	5.28	5.91	40.10	0.10	0.26
Features	B_1B_2(60.00mm)			B_2B_3(30.00mm)			B_3B_4(60.00mm)		
Fitting Method	Measured	Err.	%	Measured	Err.	%	Measured	Err.	%
1st RMS	59.86	-0.14	-0.24	29.69	-0.31	-1.02	59.85	-0.15	-0.26
2nd RMS	59.80	-0.20	-0.33	29.90	-0.11	-0.35	59.84	-0.16	-0.27
Cubic B-Spline	63.77	3.77	6.28	30.03	0.03	0.09	63.89	3.89	6.48
Features	B_4B_1(30.00mm)			C_1C_2(25.00mm)			C_2C_3(25.00mm)		
Fitting Method	Measured	Err.	%	Measured	Err.	%	Measured	Err.	%
1st RMS	29.90	-0.10	-0.35	24.88	-0.12	-0.47	24.94	-0.06	-0.23
2nd RMS	29.88	-0.12	-0.39	24.91	-0.09	-0.37	25.09	0.09	0.35
Cubic B-Spline	30.16	0.16	0.55	25.62	0.62	2.49	25.09	0.09	0.37
Features	C_3C_4(25.00mm)			C_4C_1(25.00mm)			D_1D_2(40.00mm)		
Fitting Method	Measured	Err.	%	Measured	Err.	%	Measured	Err.	%
1st RMS	24.91	-0.09	-0.34	24.89	-0.11	-0.44	39.64	-0.36	-0.91
2nd RMS	24.91	-0.09	-0.34	24.95	-0.05	-0.20	39.82	-0.18	-0.46
Cubic B-Spline	25.58	0.58	2.33	24.90	-0.10	-0.42	40.45	0.45	1.13
Features	D_2D_3(50.00mm)			D_3D_1(30.00mm)					
Fitting Method	Measured	Err.	%	Measured	Err.	%			
1st RMS	49.66	-0.34	-0.68	30.45	0.45	1.51			
2nd RMS	49.72	-0.28	-0.57	30.11	0.11	0.38			
Cubic B-Spline	49.71	-0.29	-0.58	30.28	0.28	0.95			

Table 1. Distance measurement between known points. (The number in the parentheses is the nominal dimension given to the CNC milling machine.)

Features	E(28.40mm)			F(49.20mm)		
Fitting Method	Measure	Err.	Err.(%)	Measure	Err.	Err.(%)
1st RMS	28.39	-0.01	-0.04	49.04	-0.16	-0.32
2nd RMS	28.22	-0.18	-0.64	49.09	-0.11	-0.22
Cubic Spline	29.17	0.77	2.70	50.39	1.19	2.42
Features	G(49.20mm)					
Fitting Method	Measure	Err.	Err.(%)			
1st RMS	49.04	-0.16	-0.32			
2nd RMS	49.07	-0.13	-0.26			
Cubic Spline	49.63	0.43	0.88			

Table 2. Results of on-plane circles measurements. (The number in the parentheses is the nominal diameter of the circles.)

as used for control points. To obtain the diameter of the circle, we enlarge the image and manually pick out points on the circumference as shown in Fig. 14. These points are used to best-fit a circle. The diameters obtained for the circles are listed in Table 2. The results are accurate and are all well under one percent except for some measurements converted with cubic B-spline fitted longitude/latitude lines.

Here the error is defined as the measurement minus the nominal dimension, i.e. the value given to the CNC milling machine. Overall, first order and second order RMS curve-fitting give good measurement results. Almost all the relative errors of the measurement are less than 1% and the second order result is better than that of first order. Surprisingly, the cubic

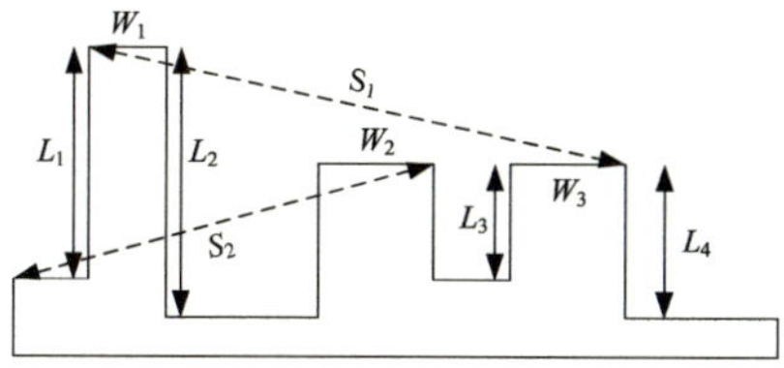

Fig. 15. The Dimensions of the grooved surface test piece.

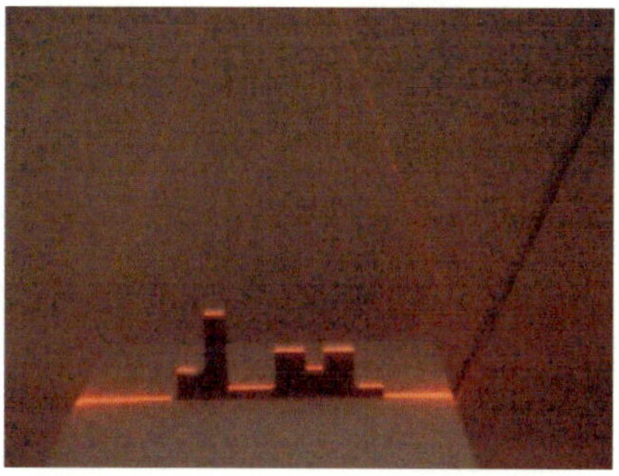
Fig. 16. A snap shot of the grooved surface test piece scanning.

B-Spline curve-fitting does not perform well; its performance fluctuates. It ranges from 0.09% for $\overline{B_2B_3}$ to 7.28% for $\overline{A_1A_2}$. This performance fluctuation results from the sensitive behavior of B-Spline fitting with respect to the minor difference of control points (Wang & Cheng, 2007). For the following experiments, only second order RMS is used to fit the longitude/latitude lines.

4.2 Grooved surface

The preceding experiments address the performance of the optical calibration alone. The final measurement result also depends on the reflection of the laser line from the target. For example a mirror-like surface may reflect the incoming beam perfectly. If the camera is not pointing along the path of the reflection, it will not see the laser profile at all. Therefore the texture of the target is a crucial factor. To determine whether the calibration remains accurate for an actual surface, we use a CNC milling machine to make a test piece which has grooves of different widths and heights as shown in Fig. 15. The laser profile seen from the CCD camera consists of line segments as shown in Fig. 16. The discontinuity of the segment is the location where the height changes. We measure the widths, the heights and the distances between edges and compare them with the nominal dimension. For this experiment, the widths are defined by the end points of the laser segments. To determine the height, we cannot treat it as the distance between the edge of a top and the corner of a groove because the corner at the bottom of a groove is a shadow zone where the laser cannot reach. Alternatively we fit the points of the laser segment with a straight line and infer the height from the distance between the two lines. We find that the absolute errors, summarized in Table 3, are all less than 0.6 mm. The measurement is accurate to 1% of the length of the features if the dimension is larger than 20 mm.

4.3 Sandwave

One major goal is to use the system to measure the roughness of a seafloor. To simulate a sandwave, we use a CNC milling machine to create a 3D surface defined by the following

Features	W_1(19.96mm)			W_2(29.96mm)			W_3(29.96mm)		
Fitting Method	Measured	Err.	%	Measured	Err.	%	Measured	Err.	%
2nd RMS	20.03	0.07	0.34	29.76	-0.20	-0.67	30.12	0.16	0.53
Features	H_1(60.00mm)			H_2(80.06mm)			H_3(20.04mm)		
Fitting Method	Measured	Err.	%	Measured	Err.	%	Measured	Err.	%
2nd RMS	59.45	-0.55	-0.91	79.40	-0.66	-0.83	19.97	-0.07	-0.36
Features	H_4(40.02mm)			S_1(154.96mm)			S_2(131.04mm)		
Fitting Method	Measured	Err.	%	Measured	Err.	%	Measured	Err.	%
2nd RMS	39.67	-0.35	-0.89	154.56	-0.40	-0.26	130.73	-0.31	-0.24

Table 3. Measurement of the grooved surface. (The numbers in the parenthesis are the dimensions measured with a caliper.)

equation:

$$z(x,y) = (A + By)\sin(Cx + D\sin(Ey)), \tag{2}$$

where x, y and z are all in millimeters and A, B, C, D and E are constants. There are two sine functions in this equation. The first one varies on the yz plane that creates ripples while the second one, appearing in the argument of the first, undulates the ripples along the y direction. The value of A determines the mean elevation of the surface, B decides how fast the amplitude of the ripple increases along the y axis, C controls the wavelength of the ripples, and D and E defines the undulating amplitude and frequency respectively. The constants are chosen such that four ripples were accommodated in a 210 mm $\times$ 210 mm $\times$ 40 mm block[3]. The ripples meander and change direction twice in the model while the amplitude increases linearly along the y direction from 1.5 mm to 8 mm. To simulate the granular texture of the seabed, a thin layer sea sand is attached to the test piece surface with adhesive. With this model, we can determine the lower limit of the amplitude which the system can detect. The CAD model and the workpiece is shown in Fig. 17 and 18 respectively.

A linear track carries the scanning head, moving at 0.5mm increments, to profile the sandwave workpiece. Altogether 400 images like the one shown in Fig. 19 are captured and processed to reconstruct the surface of the sandwave shown in Fig. 20. We plot all the data points and find that even the smallest amplitude (1.5 mm) of the sine wave is still visible on the plot. However the wave form is tilted because the base-plane of the test piece is not aligned with the xy plane of the scanning head. Several rotations and translations are carried out to align dataset with CAD model such that quantitative comparison can be done (Wang & Cheng, 2007).

We use the interpolated values of the calibrated scanning data at given locations for comparison. Four transects denoted as S_1, S_2, S_3 and S_D in Fig. 21 are selected for analysis. These locations are selected to verify the accuracy of measuring different amplitudes. The amplitude of the sine wave decreases from 7.5 mm for S_3, to 2 mm for S_1. S_D is even more challenging because it goes through different amplitudes, and it also indicates whether the phase shift between different ripples is captured correctly. The results are presented in Fig. 22 to Fig. 25. From the results we find that most of the errors are bounded between -1 mm to 1.5 mm except for the upper-left segment of the slice S_D. With this spatial resolution and accuracy, conservatively speaking, the laser scanning system should be able provide interface roughness information needed for modeling acoustic bottom interaction for 100 kHz (wavelength 15 mm) or lower.

[3] $A = 1.5$, $B = 0.043$, $C = 0.167$, $D = 2.0$ and $E = 0.042$.

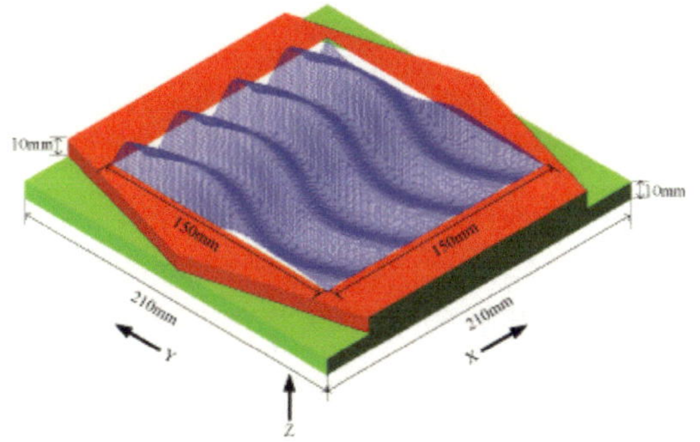

Fig. 17. The CAD model of the sandwave surface.

Fig. 18. The CNC machined workpiece of the simulated sandwave.

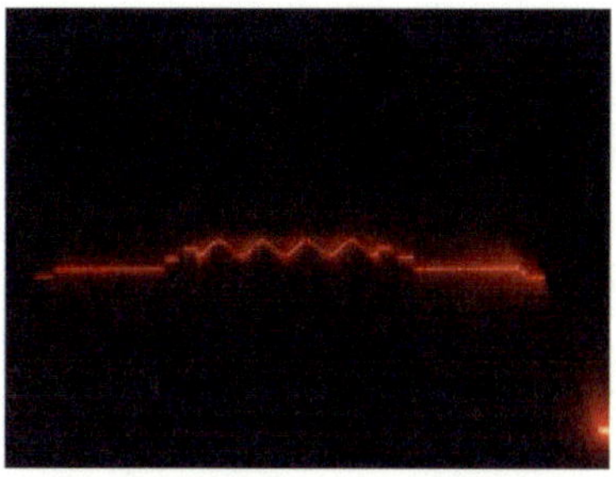

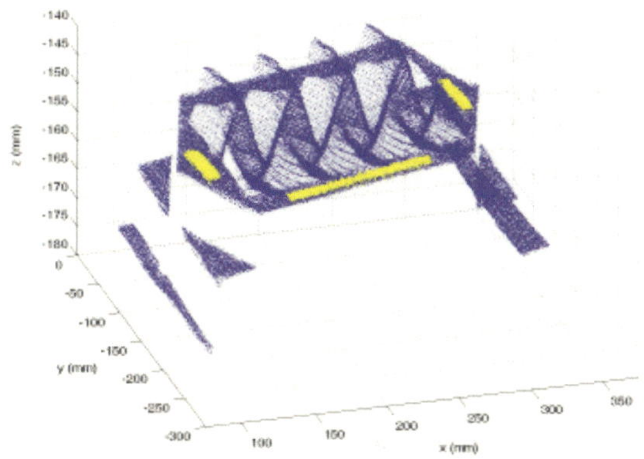

Fig. 19. A sample image of the sandwave scanning.

Fig. 20. The raw data of the sandwave scanned.

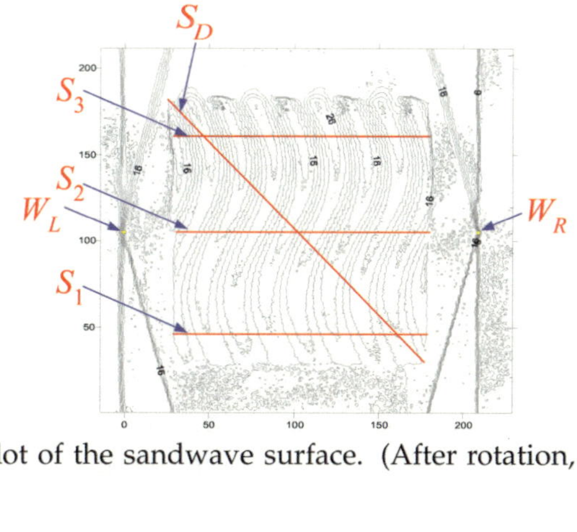

Fig. 21. The contour plot of the sandwave surface. (After rotation, translation and yawing calibration)

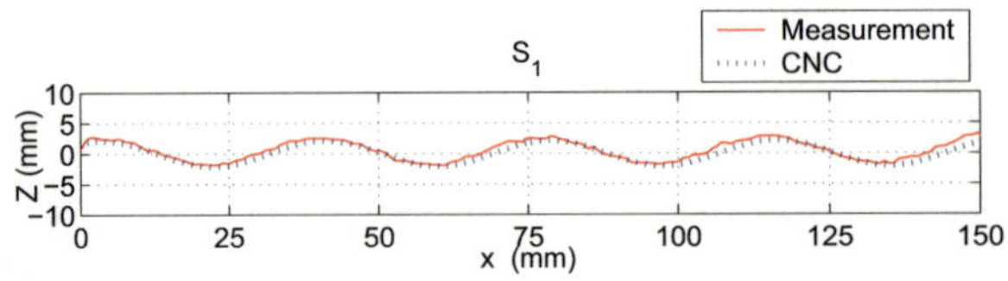

Fig. 22. Measurements (solid curves) v.s. CAD model theoretical values (dotted curves) at S_1 on the CNC machined workpiece of the simulated sandwave.

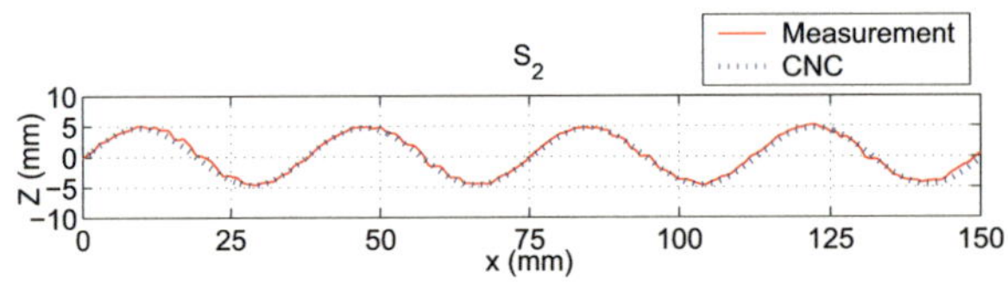

Fig. 23. Measurements (solid curves) v.s. CAD model theoretical values (dotted curves) at S_2 on the CNC machined workpiece of the simulated sandwave.

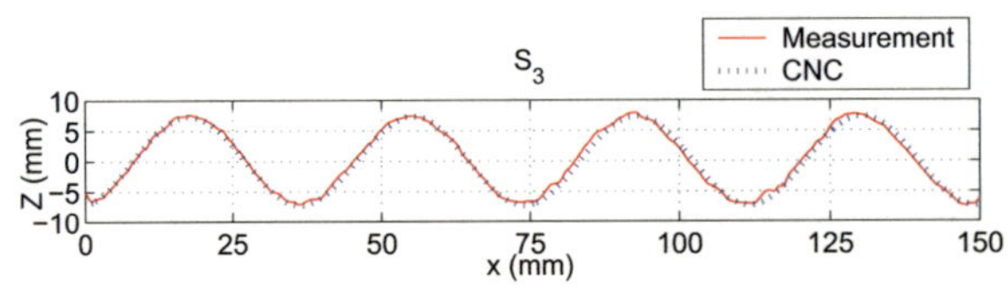

Fig. 24. Measurements (solid curves) v.s. CAD model theoretical values (dotted curves) at S_3 on the CNC machined workpiece of the simulated sandwave.

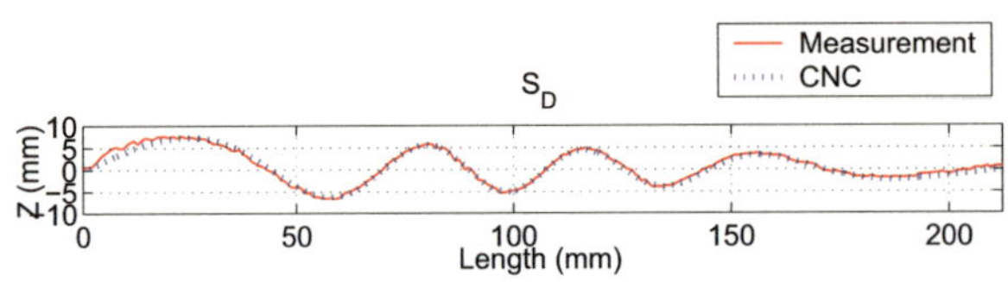

Fig. 25. Measurements (solid curves) v.s. CAD model theoretical values (dotted curves) at S_D on the CNC machined workpiece of the simulated sandwave.

5. Rough surface measurement

For studying high-frequency acoustic scattering, the most important input is the roughness power spectrum of the interface in the frequency band of interest (Williams et al., 2004). A proper estimation of the errors in this spectrum is essential for the validation and application of acoustic scattering models. It is necessary to establish a procedure to quantitatively assess the performance of the proposed laser scanner when measuring bottom interface roughness. However, conventionally, laser scanning performance assessment usually uses objects which are smooth and regular in shape. These test objects tend not to reflect the variation in surface height typically observed on sediment interfaces. The error bound obtained with these test objects is not sufficient to infer the system performance in measuring actual sediment roughness or the statistics of the roughness. In order to establish a subjective assessment of the scanner's performance in acquiring roughness spectrum, a model-realistic surface

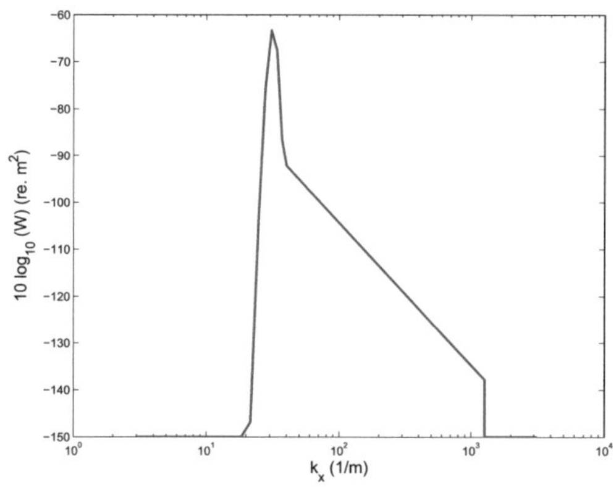

Fig. 26. Slice of the two-dimensional power spectrum along the positive k_x-axis. The spectrum has a Gaussian anisotropic (ripple) component and an isotropic power-law component.

was developed. Creating a realistic rough surface with known topography is challenging and is rarely done. With CNC milling machines, several researchers have constructed rough surfaces for use in studying the scattering of acoustic (Mellema, 1999; Summers et al., 2005) and electromagnetic waves (Kuga & Phu, 1996) from rough surfaces. These surfaces typically implement an isotropic roughness with a power-law spectrum and a surface size that is compatible with the acoustic or electromagnetic wavelength of interest. The surface topography we chose was to reflect the isotropic roughness and the anisotropic surface ripple observed during the high-frequency Sediment Acoustics Experiment (SAX99) (Thorsos et al., 2001; Briggs et al., 2004).

5.1 Power spectrum of the model surface

Certain ripple fields in near shore waters have the following properties: When the ripples are absent, the bottom roughness can be adequately described by an isotropic power-law power spectrum (Briggs, 1989) and when ripples are present, the bottom roughness can be considered as a superposition of an isotropic power-law component and a ripple component (Thorsos et al., 2000). To represent a realistic sediment interface, the design of the model surface was based on the SAX99 power spectrum and is composed of the two spectral components, $W = W_g + W_p$, where W_g is the shifted Gaussian power spectrum component, which generates ripples, and W_p is the isotropic power-law component.
The Gaussian power spectrum, W_g is expressed as

$$W_g = \frac{h^2}{4\pi\sigma^2}\left[\exp\left(-\frac{(k_x-k_0)^2+k_y^2}{2\sigma^2}\right)+\exp\left(-\frac{(k_x+k_0)^2+k_y^2}{2\sigma^2}\right)\right] \quad (3)$$

where (k_x,k_y) is the horizontal wave vector, h is the RMS height, $k_0 = 2\pi/\lambda_0$ is the center wavenumber with λ_0 being the dominant ripple wavelength, and σ determines the spread of the dominant ripple component. To make the size of the model surface manageable, we scaled the ripple size measured during SAX99 by a factor of 2/5, which applied to both the ripple wavelength and the ripple height. This choice ensures that shadowing during optical measurements behind ripples is the same for the field and model scenarios. As a result, the following set of parameters were used for the Gaussian spectrum: $h = 0.005$ m, $\lambda_0 = 0.2$ m, and $\sigma = 0.05$ m^{-1}. This spectrum produces ripples running predominantly along the x-axis.

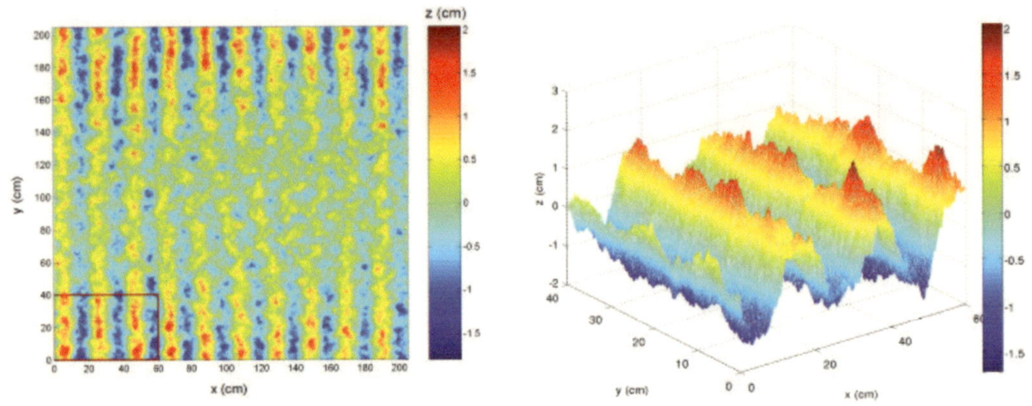

(a) (b)

Fig. 27. (a) The realization of the rough surface over a 200 × 200 cm^2 area. A smaller portion of this surface, a 40 × 60 cm^2 section on the lower left corner, was used for machining. (b) Magnification of the 40 × 60 cm^2 section used for the sample seafloor.

The isotropic portion of the power spectrum, W_p is written as a truncated power-law

$$W_p = \begin{cases} \frac{w_2}{k_r^{\gamma_2}}, & (k_1 \le k_r \le k_2) \\ -150 \text{ dB, otherwise} \end{cases} \quad (4)$$

where $k_r = \sqrt{k_x^2 + k_y^2}$, w_2 is the spectral strength, k_2 is the fine-scale or high-wavenumber cutoff, k_1 is the large-scale or low-wavenumber cutoff, which occurs just before the onset of the ripple component to make the observed ripples dominant at large scales. For this component, $k_1 = 1.1k_0$, $k_2 = 2\pi/0.005$ m, $\gamma_2 = 3.04$ and $w_2 = 0.0036$ cm$^{4-\gamma_2} = 4.331 \times 10^{-5}$ m$^{4-\gamma_2}$, as in SAX99.

The slice of the 2D spectrum along the positive k_x-axis is shown in Fig. 26 and a 200 cm × 200 cm rough surface realization based on the above spectrum is shown in Fig. 27(a). The surface resembles sand wave ripples with a 20 cm wavelength along the x-direction and has a superimposed power-law roughness as in Eq. (4). Limited by the available workspace of our CNC milling machine, we selected a region from the simulated surface (the lower left section labeled with a red box, 60 cm × 40 cm) to make the model seafloor since it has the most pronounced ripple features. This portion of the surface is shown in detail in Fig. 27(b). While the RMS amplitude of the entire realization is 5 mm, the RMS amplitude of the surface portion to be machined is actually 7.3 mm.

The seafloor model was designed specifically to be rough, not smooth. It has a continuously varying distribution of surface heights. Hence, it was not suitable to adopt the conventional coarse and fine machining process since they do not share the same the tool path. In theory, a simple approach to create the rough surface is to use a very small spherical cutter throughout the machining process. However it would take a tremendous amount of machining time to complete the project and the cutter would likely break during the long machining process. We used a flat end-cutter to do the coarse machining. The flat end-cutter was used to *peel* off the material above the design surface layer by layer. The cutter was commanded to move at a constant height such that the full cutting edge can be used to remove maximum amount

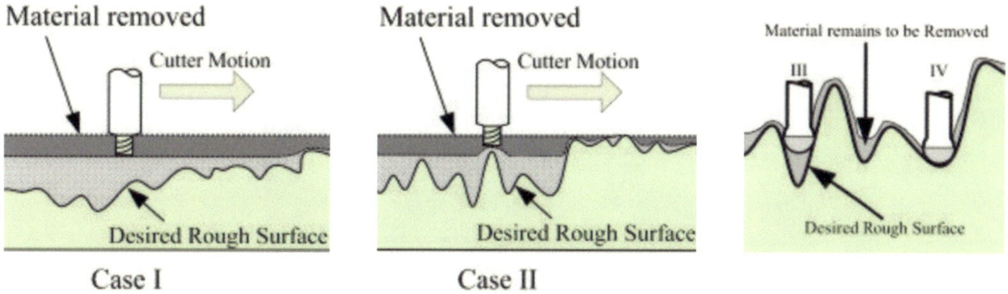

Fig. 28. (a) Coarse machining process for the rough surface. For Case I, the full cutting edge is used to remove material; for Case II, the cutter is lifted upward 1 mm to avoid overcutting the surface. (b) Fine machining process for the rough surface. For Case III, the radius of the cutter is larger than the curvature of the valley so an undercut is encountered, as opposed to Case IV where the cutter reaches the desired surface.

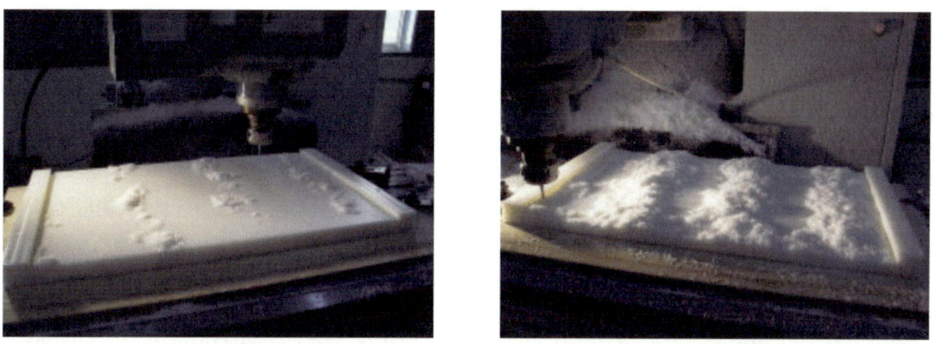

(a) (b)

Fig. 29. Coarse machining of (a) the first layer and (b) the second layer.

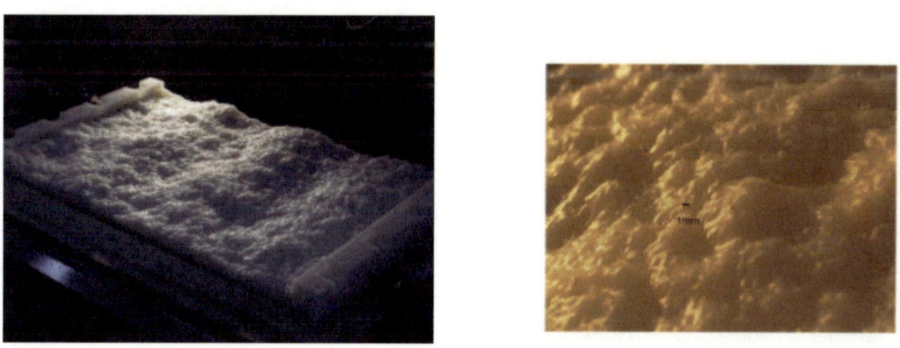

(a) (b)

Fig. 30. (a) Finished surface. (b) Close-up of the finished surface showing the small scale roughness features and the machining marks.

Fig. 31. IUT-NSYSU underwater laser line scanner and linear track.

of material while a one-mm clearance above the design surface was kept to preserve the material for fine machining later on. This process was repeated several times until any further feed along vertical direction would overcut the design surface. With the help of a machining simulation program, we decided to choose a ϕ 4 mm end cutter to do the coarse machining, and a ϕ 1 mm spherical cutter to do the finishing. These machining procedures are explained in Fig. 28. The model surface, corresponding to the input to the CNC milling machine, is shown in Fig. 29 and 30(a). As seen in the closeup of the finished surface shown in Fig. 30(b), the machining marks are much smaller than the relevant surface features.

5.2 Model surface measurement

The measurement is carried out by our underwater laser scanner called *Seafloor Laser Scanner* (SLS) as shown in Fig. 31. It consists of a water-proof laser line scanning head assembly and a one-meter servo linear stage. The scanning head assembly maintains the relative position and orientation between the camera and the laser source while the linear stage advances. The camera is a Basler A102f camera with a resolution 1388 $\times$ 1038 pixels CCD (charge coupled device). The wavelength of the laser is 650 nm, and a cylindrical lens is placed at its tip to generate a 60° fan angle. The scanning head assembly was kept roughly 75 cm above the model surface to have an effective scanning swath about 40 cm. The camera was tilted down to look at the laser reflection with a 30° grazing angle. The stage was commanded to profile the model surface with 2.3 mm/sec while the camera acquires about 5 to 6 images per second. With this frame rate, the scanning resolution is about 0.5 mm along track. For simplicity of the design, there is no synchronization signal between the motion of the linear stage and the camera acquisition. Assuming that the linear stage moved with constant speed, the distance relative to the first image frame along the track was estimated based on the time stamp of the image. The measured ripple surface is shown in Fig. 32(b).

5.3 Comparisons

Point-to-point comparisons of the measurement acquired by the laser scanning to the model surface is difficult because the measurement are not described in the CAD coordinate system. With that in mind, we display the surfaces with color coding of the height to have an overall idea of the similarity. We used the same color scale to plot the surfaces as shown in Fig. 32. Several features on the surface, denoted as **A** to **E**, are indicated for visual comparison. At **A**, a cross shape groove shown in the model surface is also found in the reconstructed surface. This is also the case at **B** for an isolated summit, at **C** for a kidney shape valley, at **D** for a T shape valley, and at **E** for two consecutive valleys. These similarities provide evidence that SLS works well, but the results do not provide a quantitative characterization of similarity.

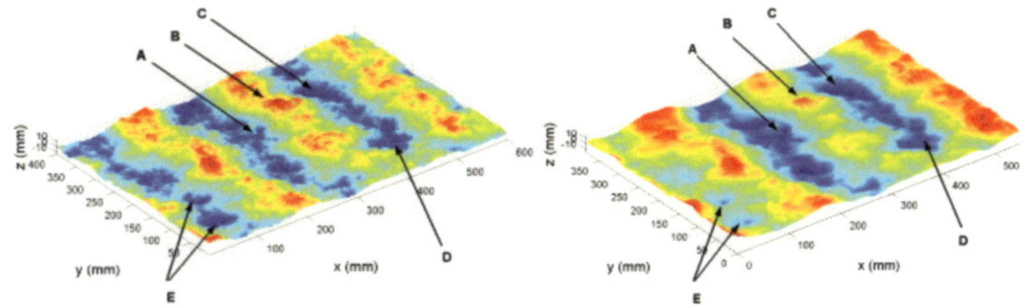

(a) (b)

Fig. 32. Reconstructed surface from (a) the CAD model and (b) the laser scan.

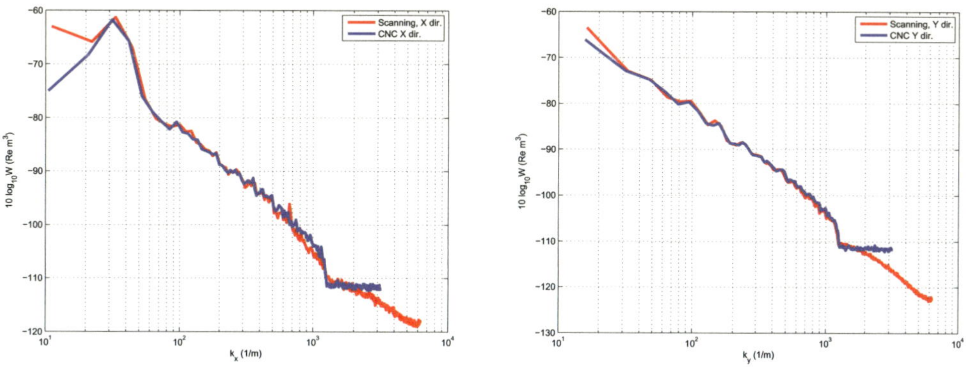

(a) (b)

Fig. 33. Comparison of the marginal spectrum (a) along the x-direction (perpendicular to the ripple crests) and (b) along the y-direction (parallel to the ripples) between CNC ripple model (blue-solid line) and laser scan (red-solid line).

While it was not possible to compare the measurement on a point-by-point basis, quantitative comparisons of the marginal spectra can be performed since they are mean 1-D spectra along the x- or y-axes. By taking the integral of the 2-D spectrum, the resulting marginal spectra are only weakly dependent on the orientation and alignment of the surface. These effects can further be mitigated by applying a 2D Hanning window to each the measurement prior to the determination of the power spectrum. This also reduces the effects of the wrap-around discontinuity of the boundary on the power spectrum of the reconstructed surface. The 1-D marginal spectra of the measurements and the model surface are shown in Fig. 33(a) and 33(b). For the marginal spectrum along the x-direction, SLS detects the existence of the ripples with the correct wavenumber and intensity. And it follows closely to the model surface spectra in the wavenumber range k_1 to k_2. The high frequency cut-off was also detected correctly. Wavenumber beyond the high frequency cut-off, SLS measurement starts to deviate from the CAD model. The RMS difference between the SLS results to the the model surface spectrum in the wavenumber range k_1 to k_2 is 1 dB in the x-direction, and 0.4 dB in the y-direction (Fig. 34). This indicates that SLS is accurately measuring the surface spectra in the wavenumber range k_1 to k_2. It should be noted that computation of the roughness spectra involves nearly two hundred thousand measurement points. It is the statistical characteristics of the large

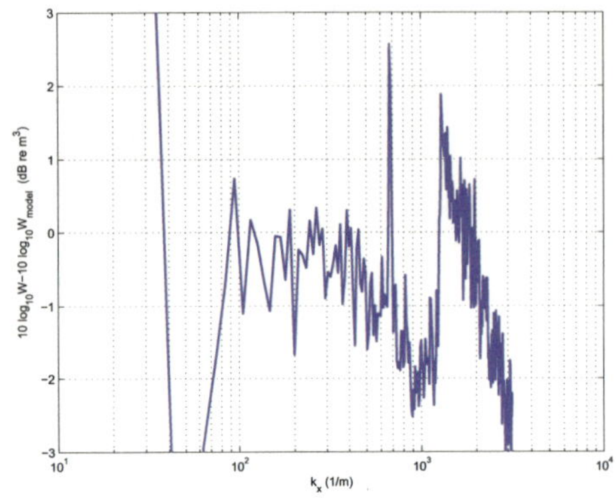

(a)

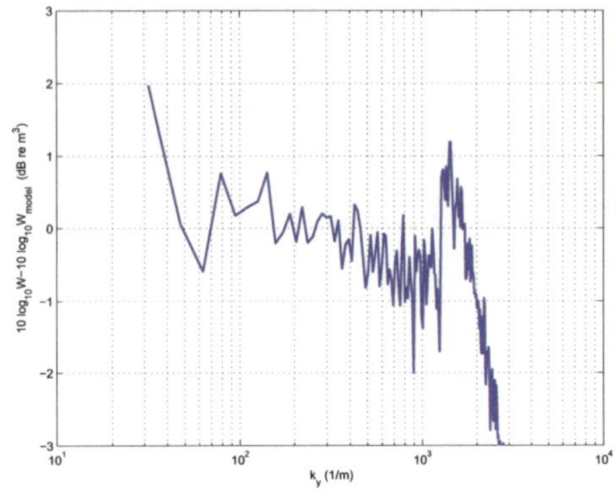

(b)

Fig. 34. Difference of the marginal spectra determined from the measurement and the CAD model surface along (a) the x-direction and (b) the y-direction. The range of k in each figure is from k_1 to k_2 which are defined in Subsection 5.1.

ensemble. In other words, it is not possible to affect trend of the spectra by removing or changing a subset of the data. It is accurate or inaccurate as it is.

6. Field experiment

In the summer of 2006, the US Office of Naval Research sponsored an experiment, Shallow Water 2006 (SW06), off the New Jersey coast. SLS was mounted on an existing sediment conductivity probe (In situ Measurement of Porosity, second generation - IMP2) and deployed jointly to measure bottom roughness spectra (Tang, 2004; Wang & Tang, 2009). IMP2 has been deployed in several acoustic experiments and provides reliable 1D bottom roughness along a 4-meter-long transect with a 10-mm horizontal sample interval and 1-mm vertical resolution. Its working principle is to measure the conductivity, so the system is robust to turbidity of the water but it provide only one transect per run. The details of its specifications and performance can be found in (Tang, 2004). On the other hand, SLS is limited by the water quality but it provides both 1D and 2D roughness spectra at a much higher spatial sampling density. Integrating these two methods offers an opportunity to compare the measured spectra from both methods at the same location in the field.

6.1 Field operation and seafloor microbathymetry

The SLS scanning head assembly, mounted on the horizontal linear stage of IMP2, was kept roughly 75 cm above the seafloor to achieve an effective scanning swath of 30 cm. The CCD camera was tilted downward to face the laser reflection at a 30° grazing angle. With this configuration, the optical resolution on the scanning plane (across track) is 0.3 mm.

The integrated system was deployed on August 14 (39°01.3559′N, 73°02.2294′W) and 17 (39°01.5506′N, 73°02.7994′W), 2006, in 80-m water depth. The measurement started with the conductivity probing from the home position of the linear stage out to 4 m followed by the laser scanning during the return trip of the linear stage to its home position. SLS was normally facing forward along the motion of the linear stage of IMP2. In this configuration, the IMP2 10-cm probing marks would not be in the view of SLS. To make direct comparison between the data from IMP2 and SLS, we turned the camera axis 30° to the side to "see" the IMP2 marks

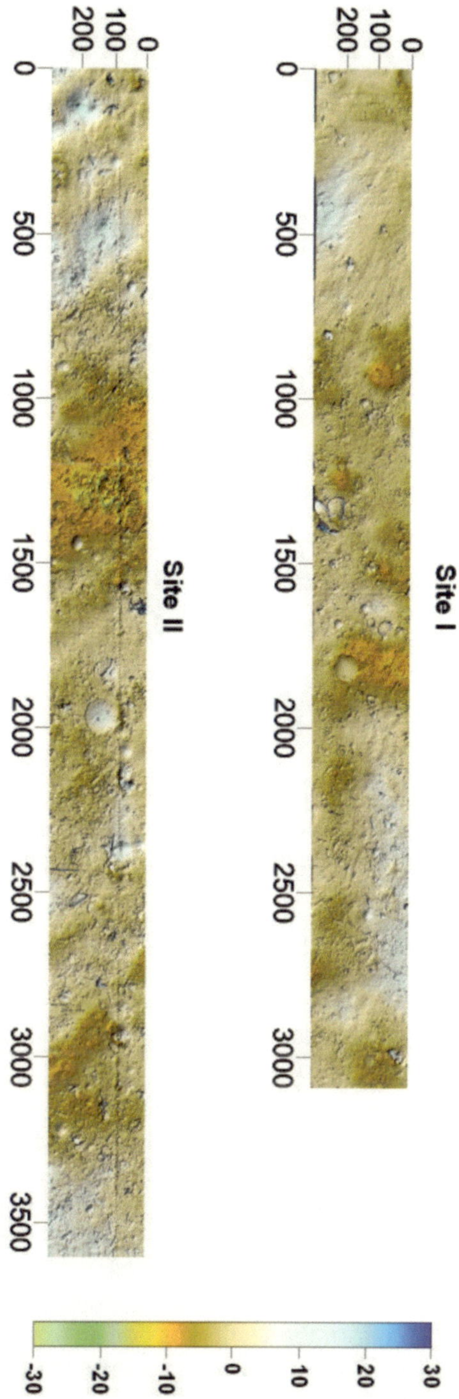

Fig. 35. SLS color-shaded relief maps for Site I and Site II. (All units are in mm). Note the IMP2 marks for Site II show as a stitch line near 100 mm on the vertical axis.

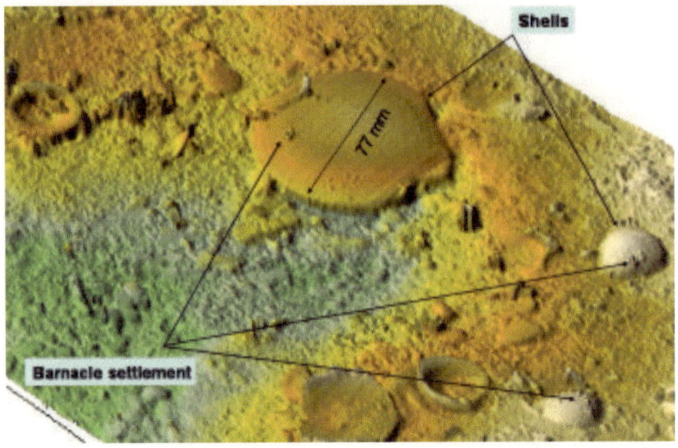

Fig. 36. A close-up view of Fig. 35 for Site I. Fine features depicted by the laser scanning: barnacle encrustations on shell surfaces.

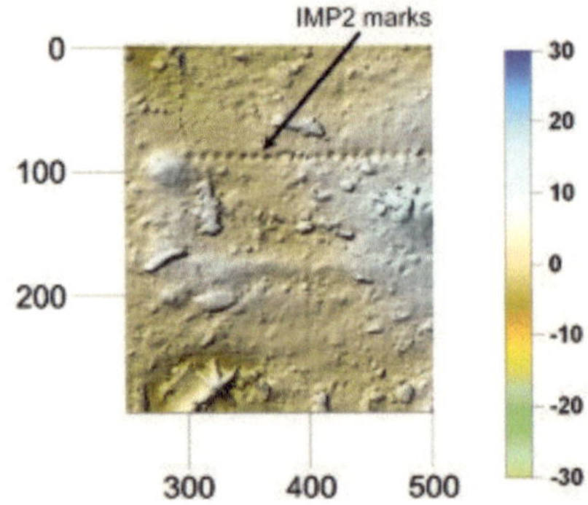

Fig. 37. A close-up view of the laser scanning that shows craters created by IMP2. The center-to-center distance between probing marks is 1 cm. (All units are in mm.)

in one of the two deployments. In either configuration, the overall area sampled is about 30 cm × 360 cm with sub-mm resolution in both vertical and horizontal directions. The sampled seafloor data points provided by these two apparatuses were roughly 400 for IMP2 and over 1,000,000 for SLS.

The rough sea bottom acquired by SLS presented in color-shaded relief maps for these two deployments are shown in Figs. 35. The shaded relief maps show that the seafloor was covered with abundant features such as shell fragments and sediment mounds and pits. Neither site shows the obvious presence of ripple fields. Site I showed no signs of disturbance by surface waves, and no apparent ripple field was found. At Site II, also full of benthic features like Site I, there were several weak crests, barely visible, lying along the upper-left to lower-right direction. According to Horikawa & Watanabe (1967), at 80 m depth, the seafloor is not influenced by the surface wave motion under normal conditions. The roughness features are most likely due to bottom currents and biological activities. The SLS's ability to portray the details of the features on the seafloor is demonstrated by a close-up of the SLS image from the first site (Fig. 36). Barnacle encrusting the surface of three shells are clearly visible in the figure. The dimension of the encrustations is 1 - 2 millimeters. Probing marks made during IMP2 measurements on the seafloor are also visible as a "stitch line" for Site II in Fig. 35. Center-to-center distance between the marks was 1 cm, which was the sampling

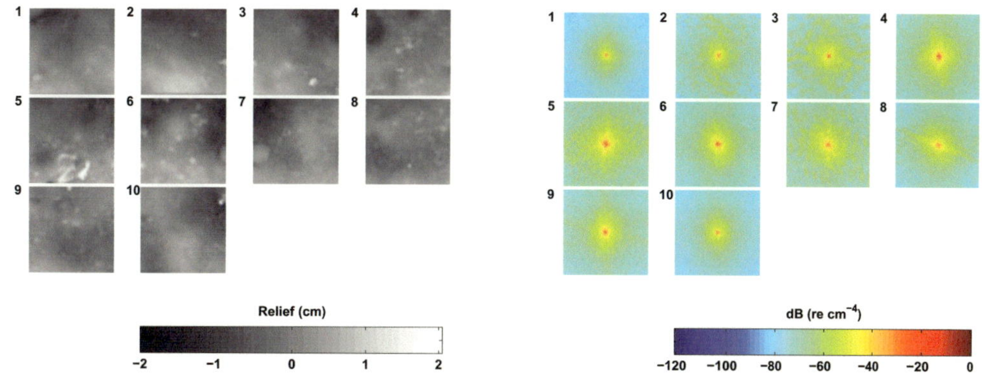

Fig. 38. Subdivided relief patches on the left and their 2D Spectra on the right for Site I. The size of the patch is 30 cm × 30 cm, and the wavenumbers k_x and k_y on the 2D spectra plots are from -31.5 (rad/cm) to to 31.5 (rad/cm) in linear scale. The number on the upper left corner of each relief map represents its sequence from left to right with respect to the original large map. The corresponding 2D spectrum is given with the same number on the right panel.

interval of IMP2. With a close-up look of the line as shown in Fig. 37, when the IMP2 probe was lowered into the sediment, it created a small pit with a diameter about 4 mm. As the probe retrieved, the sediment around the probe was piled up to form a mound around the pit like a crater. Individual 'stitch mark' was about 8 to 9 mm in diameter and a couple of millimeters in height. These marks make point-to-point comparison between the IMP2 and the SLS measurement impossible. However, the comparisons can be made on roughness spectra level.

6.2 Roughness spectrum

The area scanned by SLS is a slender rectangle, roughly 360 cm long and 30 cm wide. For the 2D spectral analysis, we divided the rectangle into 30-cm × 30-cm patches for Site I, and 26.4-cm × 26.4-cm for Site II; no overlap was taken between patches. A 2D Hanning tapering window was applied to each of the square patches to reduce spectral leakage. The relief maps of the square patches and their corresponding 2D spectra for Site I are shown in Fig. 38 as examples. The overall roughness intensity from patch to patch were about the same except for patch No. 1. On this patch, for wavenumbers greater than 20 rad/cm, the intensity is lower by 20 dB than the other patches. On its relief map, there are less shell fragments than the other patches. Patch No. 5 has brighter intensity than others. On the corresponding relief map, much more shell fragments with irregular shapes were found than the other patches. The averaged 2D spectra of the two sites show no apparent anisotropy, but there is a substantial difference in spectral level between the them (Fig. 39). In summary, no ripple field is observed at either site, and both sites show isotropic roughness spectra. To make roughness spectra comparisons between IMP2 and SLS, we use full-length, along-track SLS data. A 1D Hanning window is applied to the IMP2 roughness profiles before estimating their roughness spectra. IMP2 only provided one realization for each site, whereas the SLS yielded multiple transects. First, we compare the SLS spectrum using only those data neighboring the IMP2 transects. A 1-cm-wide stripe which covers the IMP2 probing marks is defined as the IMP2 probing zone where the seafloor was disturbed by the IMP2 probe. Two 1.5-cm-wide stripes that are 1.5 cm away from either side of the probing zone were selected. As shown in Fig. 40,

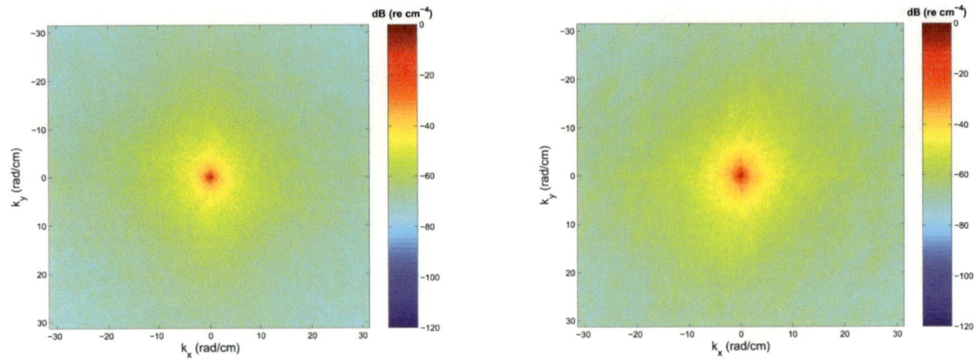

Fig. 39. Average 2D spectrum for Site I (left) and Site II (right).

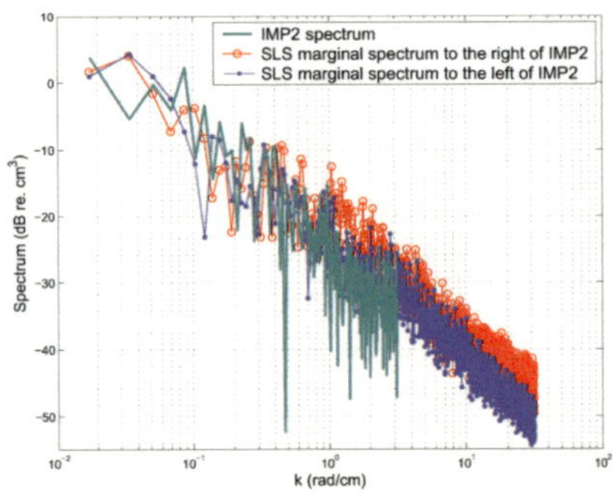

Fig. 40. Spectral comparison between data from IMP2 and data from its neighboring areas imaged with SLS.

the IMP2 spectrum agrees well with that of its left neighbor by SLS but is slightly different from that of its right neighbor for the overlapping wavenumbers. This comparison of spectra indicates that Site II roughness is variable on a small scale (centimeters). Because the SLS spectrum is a result of averaging over all transects across the 30-cm width, it is expected to be much smoother and having less statistical uncertainty than IMP2 (Figs. 41). In the figure, the IMP2 spectrum varies about the mean of the SLS spectrum for Site I; but for Site II the two measurements show visible difference for wavenumbers greater than 1 rad/cm. As discussed previously, the spectral difference is likely due to the small-scale spatial heterogeneity of the roughness. Spatial heterogeneity of bottom roughness is a major concern for modeling acoustic backscatter. The observed heterogeneity suggests that closely spaced multiple areas need to be scanned in order to get statistically representative estimates of the roughness.
The SLS's ability to detect minute roughness spectral components on the seafloor is shown in Fig. 42 as a close-up view of the Site II spectrum. A prominent spectral peak exists at wavenumber K=6.028 rad/cm, or spatial wavelength λ=1.033 cm, close to the sampling interval of 1 cm that IMP2 was configured. Also shown for comparison is the spectrum calculated from data excluding those from the IMP2 probing zone, and this spectrum does not have the spectral peak. The probing zone occupied about 3% of the totally scanning area. In other words, SLS in very keen on detecting the minute change of the seafloor.

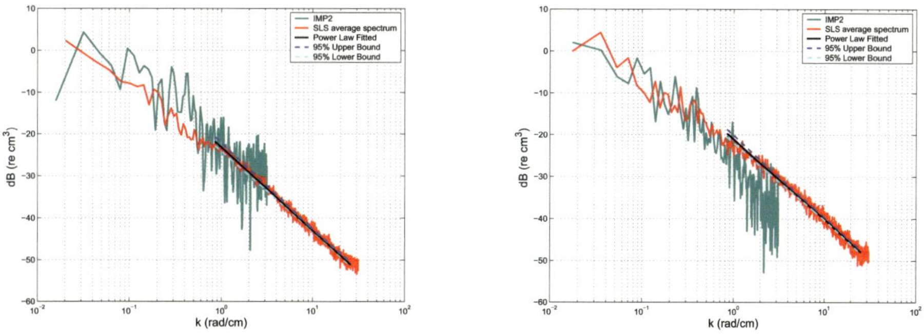

(a) Site I

(b) Site II

Fig. 41. Spectral comparison between IMP2 and SLS (the average of the 1-cm-wide stripes of the scanned area).

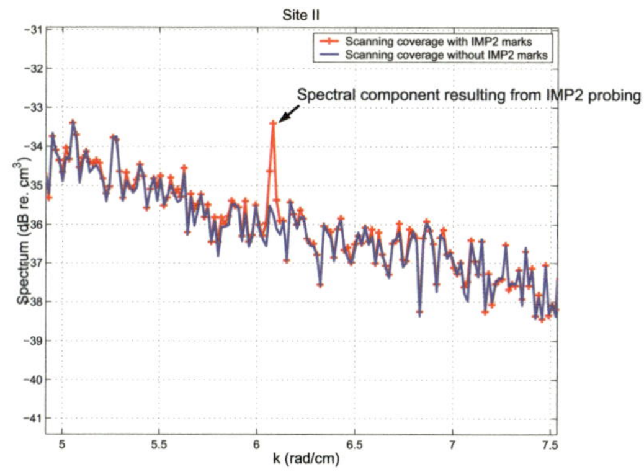

Fig. 42. Spectral component resulting from the IMP2 probing marks. (IMP2 sampling interval 1 cm is equivalent to wavenumber 6.28 rad/cm.)

7. Summary

We present a convenient method for calibrating the CCD camera used in an underwater laser scanning system, Seafloor Laser Scanner. We let the CCD camera capture the image of a calibration board on which grid points of equal span are laid. Longitude and latitude lines obtained with curve-fitting of the grid points are used to infer the world coordinates of target points. Because the method only uses a board for implementing the calibration scheme, it is easier than other approaches that need a rigid control frame. Only one image of the calibration board is needed, and it can be done before or after the measurement as long as the scanning head is not disassembled yet. Both on-plane measurement and laser scan line profiling give results with percentage error less than 1%. A simulated sandwave model test piece fabricated with CNC milling machine is scanned in the lab tank. The data points are adjusted to reconstruct the surface of a model sandwave. Because the mathematical equation of the sandwave is known, we compare the measurement results with the theoretical values at specific locations. The experiment shows that our approach can measure sandwave of amplitude as small as 2. mm from one meter away with error bound less than 1.5 mm.

Our ultimate goal is to use SLS to obtain the 2D roughness spectrum of seafloor, which is an important boundary condition for modeling high frequency acoustic scattering problem. Therefore, we fabricated a roughness-known surface model to evaluate the performance of SLS in measuring rough spectra. Our target spectrum, inheriting the characteristics of the seafloor at the SAX99 experiment site, consists of the superposition of a shifted Gaussian spectrum and an isotropic power law spectrum. The former generates ripples along the dominant direction and the latter produces fine scale rough features on the surface. A multi-stage CNC machining strategy, coarse machining and fine machining, was used to fabricate this model surface. In coarse machining, a large end-cutter is used to remove the relief depth down to a thin layer of clearance above the final desired surface. In fine machining, a small spherical cutter with dense tool paths was used to remove the remaining material to create the desired surface. Surface data measured by SLS were converted to marginal spectra parallel and perpendicular to the ripples, and compared to that of the model surface to assess its performance. SLS measured the model surface spectrum up to $k = 1256.6/m$ with an RMS error of less than 1 dB. The results of these assessments provide a means to set the reliable bounds for the roughness spectrum measurement of SLS. This surface is a useful tool for the evaluation of optically-based roughness measurements in both a laboratory setting and for *in situ* measurements.

SLS was designed to be compact and self-contained such that it can be mounted on IMP2 as an add-on module. This integrated system was deployed to acquire data on fine-scale seafloor roughness during the Shallow Water 2006 Experiment. The 3D bathymetry provided by the laser scanner revealed a seafloor full of shell hash. The effectiveness of the laser scanner for imaging the details of the seafloor was demonstrated by discerning fine features such as barnacles encrusting the surface of shells and small probing marks left by IMP2 in the imagery. The microbathymetry of the seafloor was divided into small patches, and the corresponding 2D spectra were estimated. No evidence of ripple fields was found and the seafloor roughness spectra were found to be azimuthally isotropic. The 1D spectra calculated from microbathymetry obtained by IMP2 and the laser scanner were compared over the same wavenumbers and exact locations. Individual 1D spectral lines from SLS, when compared with the IMP2 spectra, show good agreement at Site I, but different by several decibels at site II due to small-scale variability. Approximately a 4 dB difference was found in roughness spectral level between the two SW06 sites 900 m apart. A study of spatial heterogeneity of bottom roughness would benefit greatly by data that cover significantly greater areas of seafloor than so far attempted. One possibility is to mount an imaging system on a moving platform such as an AUV.

8. Acknowledge

This work was supported in part by National Science Council of Taiwan through contract NSC92-2611-E-110-005, and in part by the Asian Pacific Ocean Research Center, National Sun Yat-sen University. Field trials were supported by the Office of Naval Research Ocean Acoustics Program. We thank Min-Shine Chen, Po-Chi Chen and Min-Wei Hung at the Institute of Undersea Technology, Sun Yat-sen University for their technical contributions, and APL's technical staff for the deployment support.

9. References

Briggs, K. B. (1989). Microtopographical roughness of shallow-water continental shelves, *IEEE Journal of Oceanic Engineering*. 14: 360–367.

Briggs, K. B., Tang, D. & Williams, K. L. (2004). Characterization of interface roughness of rippled sand off Fort Walton Beach, Florida, *IEEE J. Oceanic Eng*. 29(4): 505–514.

Broadwater, J. (1988). Supporting Underwater Archeology with Ocean Technology, *OCEANS '88*, pp. 837 – 839.

Campbell, J. B. (2006). *Introduction to remote sensing*, 4th edition, edn, Guildford Press, New York.

Chen, C. & Kak, A. (1989). Modeling and calibration of a structured light scanner for 3-D robot vision, *Proceedings of 1987 International Conference on Robotics and Automation*, Vol. 4, pp. 807–815.

Crawford, A. & Hay, A. (1998). A Simple System for Laser-Illuminated Video Imaging of Sediment Suspension and Bed Topography, *IEEE Journal of Oceanic Engineering* 23(1): 12–19.

Gujarati, D. (2003). *Basic Econometrics*, McGraw-Hill, New York.

Horikawa, K. & Watanabe, A. (1967). A study on sand movement due to wave action, *Coastal Eng. in Japan* 10: 39–57.

Klimley, A. & Brown, S. (1983). Stereophotography for Field Biologist: Measurement of Length and Three Dimensional Positions of Free Swimming Sharks, *Marine Biology* pp. 175–185.

Kocak, D., Caimi, F., Das, P. & Karson, J. (1999). A 3-D Laser Line Scanner for Outcrop Scale Studies of Seafloor Features, *Proc. of MTS/IEEE Oceans '99*, Vol. 3, pp. 1105–1114.

K.S. Fu, R. G. & Lee, C. (1987). *Robotics: Control, Sensing, Vision and Intelligence*, McGRAW Hill, New York.

Kuga, Y. & Phu, P. (1996). Experimental studies of millimeter-wave scattering in discrete random media and from random surfaces, *Prog. Electromag. Res. (PIER)* 14: 37–88.

Li, Y., Young, T. & Huang, C. (1989). Noncontact Measurement Using Line-Scan Cameras: Analysis of Positioning Error, *IEEE Transactions on Industrial Electronics* 36(4): 545–551.

Mellema, G. R. (1999). *Subcritical acoustic scattering across a rough fluid-solid interface*, PhD thesis, Department of Electrical Engineering, University of Washington.

Melling, A. (1997). Tracer particles and seeding for particle image velocimetry, *Measurement Science and Technology* 8(12): 1406–1416.

Myers, W. (1980). Industry Begins to Use Visual Pattern Recognition, *Computer* 13(5): 21–31.

Ogawa, Y. (2000). Image Processing for Wet Welding in Turbid Condition, *Proc. of the 2000 International Symposium on Underwater Technology*, Tokyo, Japan, pp. 457–462.

Raffel, M., Willert, C., Wereley, S. & Kompenhans, J. (2007). *Particle Image Velocimetry: A Practical Guide (Experimental Fluid Mechanics)*, 2nd edition, edn, Springer, Berlin.

Rocher, L. & Keissling, A. (1975). Methods for Analyzing Three-Dimensional Scenes, *Proc. 4th Intl. Joint Conf. Artificial Intelligence*, pp. 669–673.

Son, S. & Lee, K. H. (2002). Automated laser scanning system for reverse engineering and inspection, *International Journal of Machine Tools & Manufacture* 42: 889–897.

Summers, J. E., Soukup, R. J. & Gragg, R. F. (2005). Characterization and fabrication of synthetic rough surfaces for acoustical scale-model experiments, *Technical report*, Naval Research Laboratory, Code 7140.

Tang, D. (2004). Fine-scale measurements of sediment roughness and subbottom variability, *IEEE J. Oceanic Eng.* 29(4): 929–939.

Thorsos, E. I., Williams, K. L., Chotiros, N. P., Christoff, J. T., Commander, K. W., Greenlaw, C. F., Holliday, D. V., Jackson, D. R., Lopes, J. L., McGehee, D. E., Richardson, M. D., Piper, J. E. & Tang, D. (2001). An overview of SAX99: Acoustic measurements, *IEEE Journal of Oceanic Engineering.* 26: 4–25.

Thorsos, E. I., Williams, K. L., Jackson, D. R. & Tang, D. (2000). High-frequency sound interaction in ocean sediments, *in* M. Zakharia, P. Chevret & P. Dubail (eds), *Proceedings of the fifth European Conference on Underwater Acoustics*, Vol. 1, Office for Official Publications of the European Communities, Luxemburg, pp. 275–280.

Wang, C. C. & Cheng, M. S. (2007). Nonmetric camera calibration for underwater laser scanning system, *IEEE Journal of Oceanic Engineering.* 32: 383–399.

Wang, C. C. & Tang, D. (2009). Seafloor roughness measured by a laser line scanner and a conductivity probe, *IEEE Journal of Oceanic Engineering.* 34: 459–465.

Williams, K. L., Jackson, D. R., Thorsos, E. I., Tang, D. & Briggs, K. B. (2004). Acoustic backscattering experiments in a well characterized sand sediment: Data/model comparisons using sediment fluid and Biot models, *IEEE Journal of Oceanic Engineering.* 29(4): 412–428.

Scanning and Image Reconstruction Techniques in Confocal Laser Scanning Microscopy

Peng Xi, Yujia Liu and Qiushi Ren
Peking University and Shanghai Jiao Tong University
China

1. Introduction

Taking advantage of the laser scanning methodology, confocal microscopy has been widely applied in cutting-edge biological research for its three-dimensional, high-resolution imaging capability (Pawley 1995). The principles of confocal were put forward by Marvin Minsky in 1955 (Minsky 1988), yet the experimental demonstration was not accomplished until 20 years later by Cremer brother (Cremer, Cremer et al. 1978), and Brakenhoff et al. (Brakenhoff, Blom et al. 1979). In confocal microscopy, one or more focal spots are scanned relative to the specimen, to extract the signal from the focal spots. The 3-D image can then be reconstructed digitally. Contrary to mechanically scanning the specimen, which is relatively slow as the specimen has to be moved, the laser provides fast scanning and imaging speed for confocal microscopy and therefore is routinely used in commercial state-of-the-art confocal systems. Several laser scanning mechanisms have been employed in confocal microscopy:

1. single spot laser scan with galvanometers;
2. single spot laser scan for real-time imaging and endoscopy;
3. line scan confocal with a line detector or a 2-D detector;
4. multiple spot scanning with 2-D detectors.

The single spot galvanometric scan is the most straightforward laser scanning mechanism, which provides an imaging speed of ~1 frame(s) per second (fps) for a typical 512 x 512 image, and is widely used in many confocal laser scanning systems. However, the relatively slow scanning speed of the galvanometer commonly used in a confocal laser scanning microscopy can dramatically limit the system performance in scanning speed and image quality if the data collection is simply synchronized to the galvanometric scanning. In Section 2 we discuss the command and data processing techniques for galvanometric scanning, for example, pixel delay and interlace line switching can be applied on bi-directional scans to cancel aliasing; pixel binning and/or image average can be used to improve the signal-to-noise level; Acquire-On-Fly scanning can be applied when the signal command is a limiting factor for the imaging speed.

Due to the limitations placed on maximum galvanometer scanning speed, real-time confocal microscopy relies on the fast scanning mechanism on its fast axis, such as a resonant scanner or rotational polygon mirror. This is discussed in Section 3. The resonant scanner runs at a fixed frequency with variable amplitude, and scans a temporal sinusoidal pattern.

Therefore, the pixel reconstruction has to be processed to correct the sinusoidal scan waveform and the bi-directional scan. Contrary to the resonant scan, a rotational polygon mirror provides a uni-directional, constant velocity scan, although the scanning angle is fixed the number of sides on the polygon. Fast scanning can be found in several versions of novel commercial confocal laser scanning microscopes. To miniaturize the size of the conventional scanning head, MEMS-based scanning mirrors are employed as well as vibrating optical fibers for fast endoscopic scanning.

In Section 4 we describe line scan confocal microscopy, in which a cylindrical lens is used to focus the beam to a line on the specimen. The line is scanned with a galvanometer to reach video-rate imaging. Signals from the specimen can be detected by a line detector, or be re-scanned and viewed by naked eye or a 2-D detector. In such a system, the signal collected from the line detector has to be synchronized with the scan to form a confocal image.

Multiple spot scanning is an alternative scan geometry that employs multiple focal spots to scan across the specimen simultaneously. As discussed in Section 5, spinning disk scanning is the most widely used multiple spot scan mechanism. To cover the specimen, a spiral geometry is used in the spinning disk to arrange the microlens array, which form focal spots, and the pinholes. Unlike other confocal laser scanning modalities, the 2-D CCD detector system can be used directly in multiple spot spinning disk scanning.

2. Galvanometric single spot laser scanning

The principle of confocal microscopy is illustrated in Fig. 1. In traditional wide-field imaging microscopy, the sample is illuminated uniformly and a 2-D detector is employed, such as a film, a CCD or CMOS sensor, or the retina of the eye. This imaging mode lets us see clearly when the target falls within the depth of field, while the targets out of the depth of field are blurred but still contribute intensity to the detector (see Fig. 1a). As a result, we cannot get 3-D information from this configuration. In contrast to wide-field imaging, confocal microscopy is able to achieve this from point detection. From Fig. 1b we can see that only

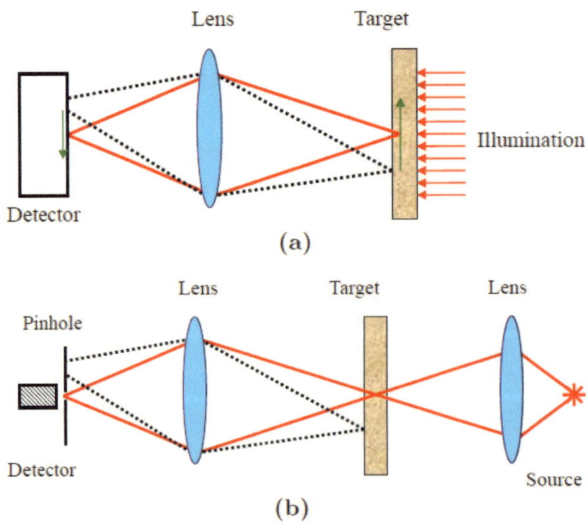

Fig. 1. Different types of microscopy: (a) wide-field microscopy; (b) confocal microscopy.(Xi, Rajwa et al. 2007)

one point is illuminated and the signal from this point is exclusively collected. Thus, through the relative movement of the detecting point and the target, 3-D information can be collected. The axial scan can be achieved by moving the objective or sample, while the lateral scan can be done by either the movement of the sample (stage scan), or the deflection of the angle of the laser incidence (laser scan). In this chapter, we will focus only on the lateral scan with laser scanning mechanisms.

Historically, a mirror galvanometer is used as mechanical meter that senses electric current and moves the direction of the mirror. The mirror reflects a beam of light, which projects onto a meter, and acts as a pointer to read out the electric current. In CLSM, the galvanometer is used in reverse, that is, to deflect the light to the specific angle driven by a known current. Galvanometric scanning is the most widely used scanning mechanism, and takes advantage of its accurate scanning across the sample. The most straightforward way to control the galvanometer is that a position (X,Y) is sent to the galvanometer, usually in a raster mode, and the image intensity data is read from the detector point by point.

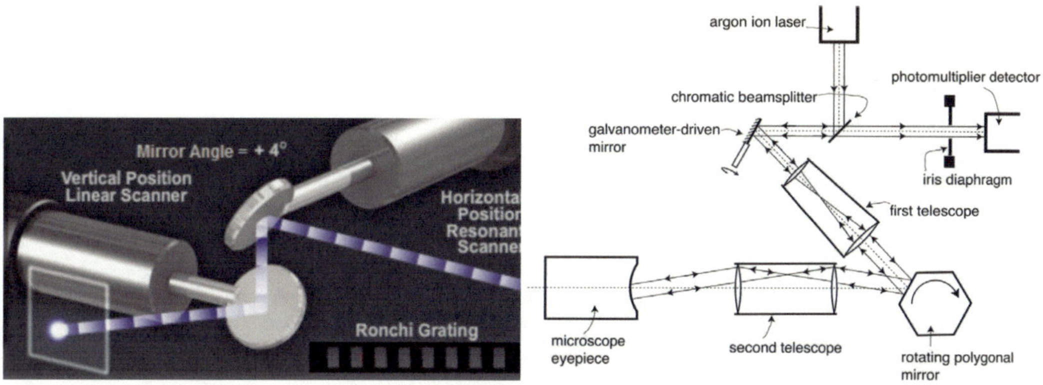

Fig. 2. Geometry of the scaning mirror configuration: Left: perpendicular galvanometers placed close to each other (http://www.microscopyu.com/tutorials/flash/resonantscanning/confocalresonantscanning/index.html); Right: telescope conjugated scanners (Amos and White 2003) Image courtesy of Elsevier.

Fig. 3 schematically illustrates the configuration of a CLSM. A scanning lens is employed to form a telescope; placing the scanning galvanometer at the entrance pupil will efficiently guide the light through the exit pupil of the telescope, where the objective is located. The scanning raster is generated at the back focal plane of the objective-tube lens. The scattered or fluorescent light signal from the objective goes back to the galvanometer and gets descanned, then is detected by a point detector, such as a photo-multiplier tube (PMT).

To realize a XY plane frame scan, the spot is generally scanned in a raster mode similar to that of an analoge television. There are two methods to deliver the light: (1) using two opposing mirrors rotating in perpendicular directions, one for a fast (X) line scan and the other for a slow (Y) frame scan. The two scanning mirrors are located close to each other, and a telescope is used to deliver the scanned beam to the back aperture of the objective. The conjugating plane of the objective aperture is usually chosen to lie in the middle of the scan axes. Because the beam is scanning across the objective, the beam size should be larger than the aperture of the objective so as to always completely overlfill it during scanning. An alternative scheme (2) is to use a 1:1 telescope between the two scanning mirrors located at

the entrance and exit pupil of the telescope, and then use another telescope to deliver the beam to the back aperture of the objective. In this scheme, the light can always pass the back aperture during scanning, therefore excessive beam expansion is not needed.
In the raster scan method, two scanning modes are generally employed: (1) a bi-directional scan; and (2) a unidirectional scan. Each scanning mechanism has its own pros and cons, as described in the following sections (Chen, Feng et al. 2010).

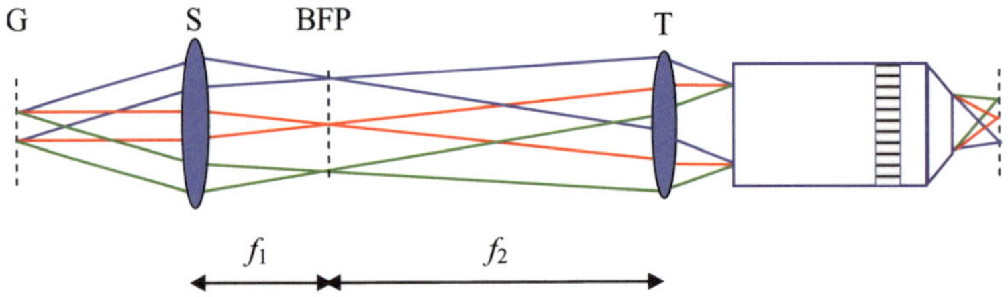

Fig. 3. The configuration of a confocal laser scanning microscopy system. G: Galvanometer; S: scan lens; BFP: back focal plane; T: tube lens; f_1: focal length of S; f_2: focal length of T.

2.1 Bi-directional scanning

In bi-directional scanning the galvanometer scans forward and then backwards at the same speed, and data is collected from both scan directions. The scan control scheme is shown in Fig. 4.

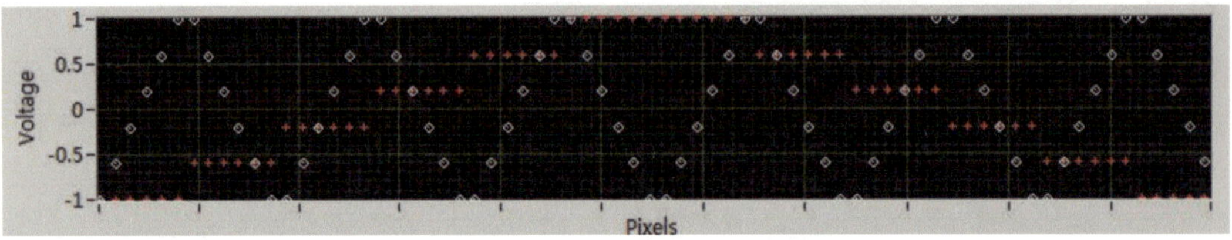

Fig. 4. Bi-direction scanning scheme. White (o): voltage applied onto X axis (fast) versus time (pixel sequence); red (+): voltage applied onto Y axis (slow) versus time (pixel sequence). The data collection of each pixel is synchronized with the scan control (Chen, Feng et al. 2010).

Unfortunately, one of the problems we found when bi-directionally scanning was that the speeds of the forward and backward scans are not simple symmetric functions but are attenuated by the acceleration of the galvanometric control due to the fast scan acceleration. This makes a line shift distortion apparent in the image. To correct the line shift, the pixels in the even (back) line of the forward scan are switched with the odd (forward) scan line of the back scan, so that in each obtained image all the scan lines have the same scan direction (see Fig. 5). It should be noted that the temporal sequence for the image lines is re-arranged, so this is not suitable for tracking particle movement.
A mouse kidney sample was imaged with two-photon fluorescence excitation to demonstrate the effect of line switching (the experimental setup is described in (Xi, Andegeko et al. 2008; Xi, Andegeko et al. 2009). It can be seen that when line switching is applied, the structures labelled with lectin wheat germ agglutin and the filamentous actin in the mouse kidney glumeruli are seen clearly (Fig. 5d); whereas, in the non-corrected image,

these structures are blurred (Fig. 5c). Naturally, the position of the blur is the same for all the three dyes, because the scanning direction is the same when collecting the images for the three channels. An alternative way to cancel the blur is to shrink the image to 25% of its total size; by doing so all the odd (or even) lines are removed; however the pixel number is dramatically decreased, and this can affect either the resolution (if the initial image contains proper sampling) or FOV (if the initial image is taken with twice excessive sampling and then shrinked to half). Note, that for uni-directional scanning (Fig. 9a) the line shift is not presented, simply because the scan directions are constant for all the lines on the image.

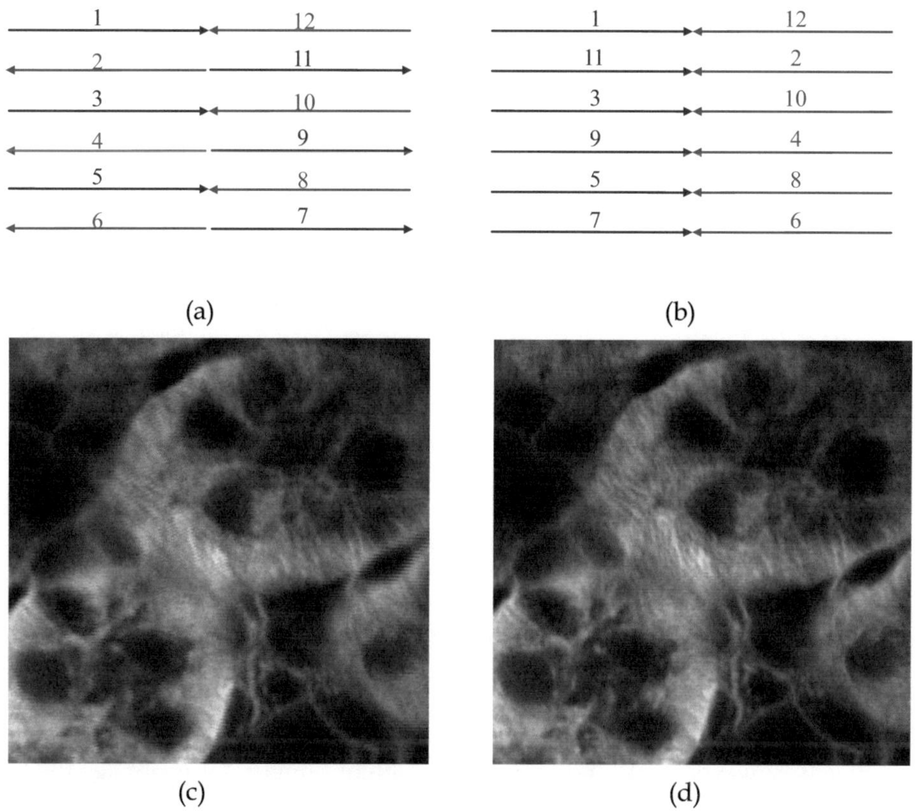

Fig. 5. Bi-directional scanning scheme: (a) before and (b) after line switching; the resulting images of a mouse kidney are shown in (c) before and (d) after line switching (Chen, Feng et al. 2010).

2.2 Uni-directional scan

One of the problems encountred in bi-directional scans is the fact that the torque for the galvanometer is generally not the same between backward and forward scans, which results in image aliasing. Moreover, the lack of constant scan speed results in a slight variation at the edge of the image, where the acceleration of the galvanometer reaches a maximum. As a result, the bi-direction scan yields two images that describe the structure of the biological specimen, but they are slightly different, which makes image averaging inapplicable. In uni-directional scanning, the backward scan is omitted, therefore, the scan direction on the reconstructed image is always the same (see Fig. 6). The proportion of each full-frame scanning interval that is utilized to actually scan the specimen is referred to as the duty cycle of the system. Due to limitations set by the speed of the galvanometer, the duty cycle of a confocal system is usually 66% at a high frame rate, or 1/2 of the time spent on the forward imaging scan. Because uni-directional scanning can minimize speed differences during forward and backward galvanometer scanning, it is commonly applied in many commercial confocal microscopes, in which a fast backward scan is employed, often followed by a shutter to avoid excessive photobleaching of the sample.

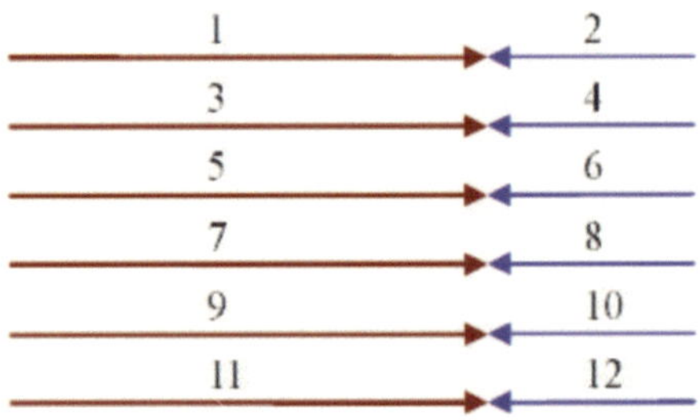

Fig. 6. Uni-directional scan: the brown lines indicate the forward imaging scan, and the blue lines indicate the backward scan, which is not used in imaging (Chen, Feng et al. 2010).

2.3 Pixel delay

Because of the response time of the galvanometer, the commands of the user cannot immediately generate galvanometer movement. Instead, the galvanometer has a fixed response delay if the data acquisition is at the same time as the scan command, and if not properly addressed, leads to a delay in acquiring data and a distortion of the image during image reconstruction. There exists several versions of ImageJ plug-ins, for example ``Correct X Shift'' by Krempp (http://rsbweb.nih.gov/ij/plugins/correct-shift.htm}) that shifts the X distortion by pixels on the odd lines so that the image width gets smaller when shifted, and ``X Shifter'' by Xi (http://rsbweb.nih.gov/ij/plugins/x-shifter.html) that shifts the pixels to the following line so there is a two pixel difference between the neighboring line on each pixel shift. These non-real time image processing tools can only be used to process the collected image, however, real time processing is highly important enabling the confocal user to visualize the structure of the sample properly.

The scheme of shifting pixels is illustrated in Fig. 7. In our program, the image data was collected at the same time as the scan command. To compensate for the response delay, a number of pixels at the front of the data collection were rolled back to the end of the data, resulting in pixel delay of the image (Fig. 8). This corresponds to the delay in data acquisition of the DAQ board and achieves the synchronization of the galvanometer scan and data acquisition. The number of shifted pixels is dependent on the galvanometer hardware. In our system, a response delay of 326.8 μs is applied, which corresponds to 21 pixels for a 256 x 256 image acquisition at 1 fps. It should be noted that due to the pixel shift, several of the last pixels do not contain the real information in the first image. However, as most biological information is located in the central part of the image, this does not create any noticeable effect in the image.

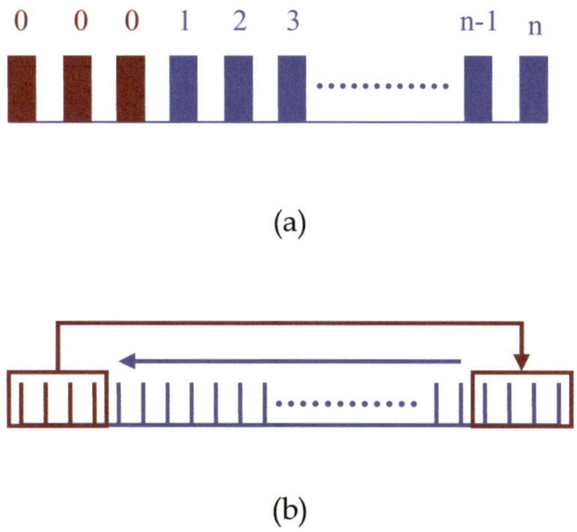

Fig. 7. Response delay of the galvanometer (a) and pixel shift correction method (b). The blue pixels contain imaging data from the sample, while the brown pixels indicate the data collection before the galvanometer responds and thus contain no information (Chen, Feng et al. 2010).

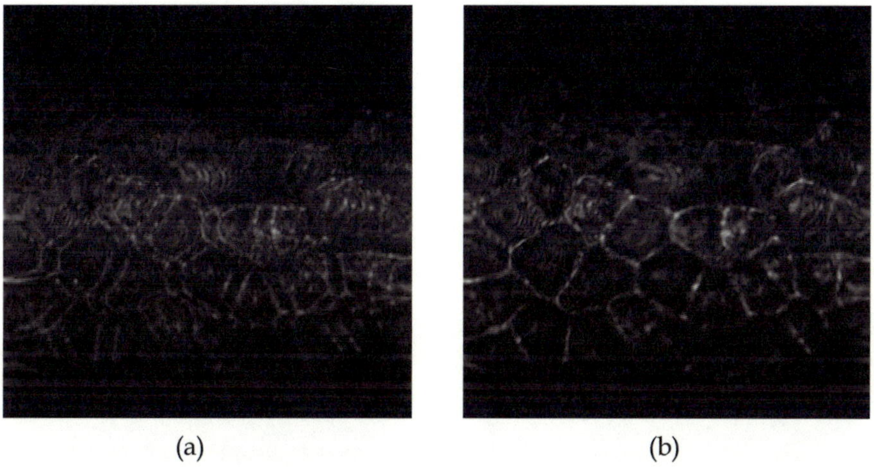

Fig. 8. Effect of pixel delay: before pixel delay correction (a) and after correction (b). Image size: 150 μm (Chen, Feng et al. 2010).

2.4 Data binning

As the data acquisition speed can be much higher than the scan command, multiple data acquisition can be performed with data binning to optimize the SNR. When the image is reconstructed, the software uses a simple arithmetic average to bin the data and applies the average value as the intensity of the pixel, as shown in Fig. 9. This method greatly improves the efficient use of the DAQ board and cancels the noise of the image.

In this system, because there are multiple input channels, the acquisition speed of the PCI-6251 is 1 MS/s. The galvanometer has a maximum scanning speed that is limited by the scan angle. To reach an optimal FOV, the galvanometer should work at ± 1 degree with a maximum scanning speed of ~500 Hz for uni-direction scanning with a filling factor of 66%. When the image is collected at a speed of 500 x 400 at 1 fps, the data collection command is sent at a speed of 800 KS/s. Thus, a 4-point data binning can be performed to match the speed difference between the galvanometer and data acquisition.

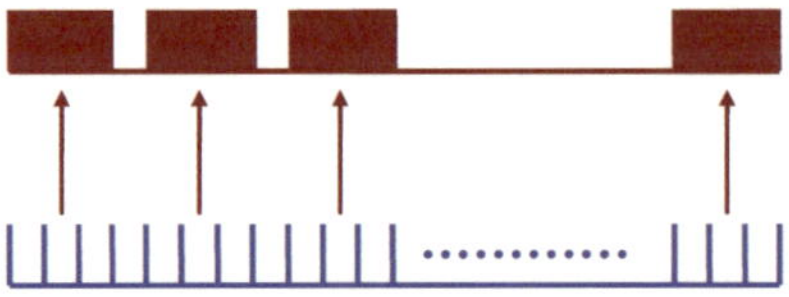

Fig. 9. The data binning algorithm with 4 data points collected and averaged to form one pixel (Chen, Feng et al. 2010).

Compared to the algorithm for frame averaging (Chou, Bower et al. 2005; Nguyen, Tsai et al. 2006), data binning is more robust with respect to sample drift, as each pixel is collected in the neighboring time window. For instance, for the same digitization speed, one can mathematically compare:

1. For frame averaging with an imaging speed of 500 x 500 at 1 fps and an average of four images, the temporal distance between each average pixel is one second, therefore the distortion was averaged one second away from the previous data pixel. The advantage of this mode is that with a sufficient signal (which is usually not the case when frame averaging or data binning is performed), one can observe the drifting movement of the sample from the collected images;
2. For data binning, the same imaging speed requires 2000 x 500 data points in four seconds, and bins every four points. Therefore, the time between each pixel average is four microseconds, a value six orders of magnitude shorter than the frame average.

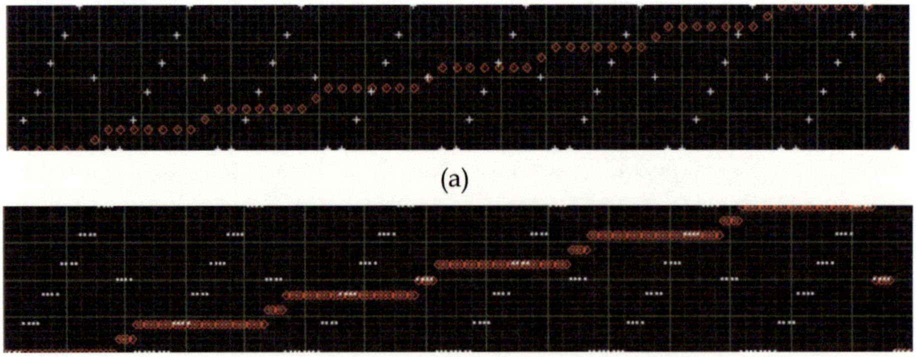

(a)

(b)

Fig. 10. Comparison of different scan modes: (a) Uni-directional Scan; (b) Uni-directional scan with data binning (Chen, Feng et al. 2010).

2.5 Acquire-on-fly

In CLSM the scan command pixel number and the pixel data collection number are the same (data binning is actually a mathematical method that sacrifices image pixel number for a better SNR). However, due to the fact that the speeds of the galvanometer and data acquisition are different, alternative methods to get around this problem only command several interval spots at one speed (this could be the fastest speed of the galvanometer), whereas data collection runs at another speed (usually much faster than the galvanometric scan for data acquisition techniques currently available). As illustrated in Fig. 11, because the galvanometer's velocity is rather constant when sufficient interval spots are commanded, an image with increased pixel number can be obtained through sequentially remapping the pixels to their respective location (rather than binning the pixels). The conceptual difference between data binning and AOF, relies on the increase of data collection speed in data binning, whereas AOFrelies on a decrease in scanning command speed, relative to the pixel clock.

Fig. 12 illustrates the effect of employing the AOF technique. The application of AOF with a 4 x interval pixel sampling rate between each scan command interval, can improve the image sampling frequency by four times. By adjusting the image aspect ratio accordingly, a square image can be readily collected. Note that in Fig. 12a the horizontal resolution is limited as insufficient pixels are sampled, and the 4x AOF apllied to the image shown in Fig. 12b has effectively increased the resolution. Therefore, for different imaging applications, the user can choose to either acquire a higher SNR image by data binning, or obtain a "larger" image using the AOF technique, depending on the relationships between the SNR, FOV, and Shannon-Nyquist sampling criteria. For example, if the FOV is set to 100 μm x 100 μm, and the optical resolution is 0.5 μm, then to obtain optimal resolution (to meet Shannon-Nyquist sampling criteria) we need at least 400 x 400 pixels. We may use the data binning and AOF together to command the galvanometer to run at 100 x 400 at 1 fps (where the fast axis runs at only 40 K intervals at 400 Hz for a uni-directional scan; the filling factor is

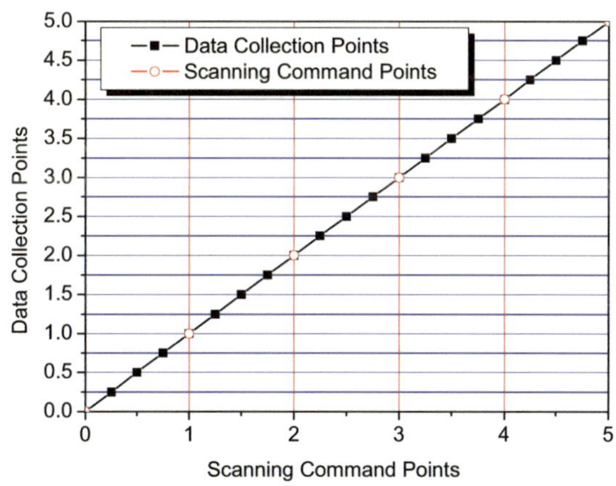

Fig. 11. The mismatch of scanning command speed and data collection speed. The line shows the trace of the galvanometric mirror response. By using data binning and/or Acquire-On-Fly technique, the mismatch can be compensated. (Chen, Feng et al. 2010)

omitted for simplicity) and collect data at 1600 x 400 at 1 fps (where the digitization speed is 640 KS/s). With four times data binning the SNR is improved; and with 4x AOF the image is rescaled to 400 x 400. Of course, the actual resolution is also affected by the SNR of the image, so in practice it is always a trade-off between AOF and data binning, with a fixed digitization speed.

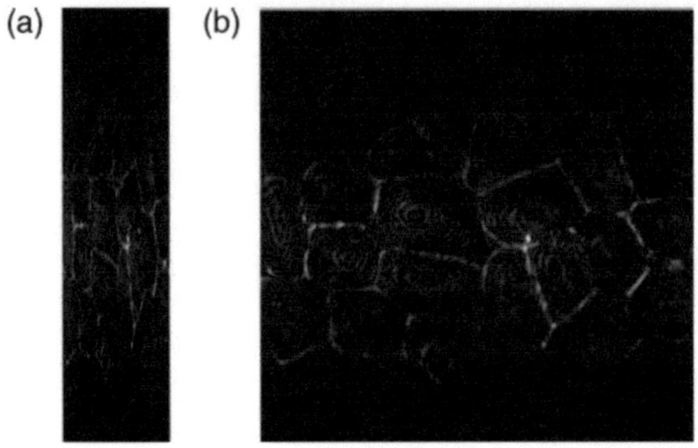

Fig. 12. Confocal images of a bifoliate bush leaf with: (a) conventional synchronized scanning and sampling mode (pixel ratio of X:Y=1:4); (b) Acquire-on-Fly mode with 4x the sampling speed resulting in a square-frame image with 4x the pixel numbers. Image size: 150 μm (Chen, Feng et al. 2010).

3. Real-time imaging with a single spot laser scan

3.1 Fast scan with a resonant scanner

The disadvantage of galvanometric scanning lies in the fact that galvanometer is accurate but slow to scan. To work around this problem, a resonant scanner running at a fixed frequency can be used. To maximize the velocity of scanning, the velocity of the resonant scanner is often a sinusoidal pattern, in which the specimen is scanned at the highest speeds in the central region of the image, where as the velocity progressively decreases as the scan reaches the edges. As a result, when the image data flow derived from a resonant scanner is acquired by a frame grabber clocked at a constant pixel rate (which assumes the beamscans linearly), the images appear stretched at the edges (see Fig. 13).

The key to restoring the distorted image is to generate a proper pixel clock that is syonchronized with the scan velocity. The image can be collected and then post-processed mathematically (Callamaras and Parker 1999). CRS Corp. has developped an electronic pixel clock generation circuit to work with their resonant scanner to provide a pixel clock output corresponding to the scanning velocity (CRS Application Notes 1994). A Ronchi grating can be utilized to optically generate a pixel clock when the beam scans across the grating groove (Leybaert, De Meyer et al. 2005).

3.2 Fast scanning with a polygon mirror

Another way to generate fast angle deflection is to use a rotating polygon mirror. Unlike the resonant scanner, which runs bi-directionally and scans with a sinusoidal velocity, the angle deflection provided by the polygon mirror is uni-directional and temporally uniform. Therefore, for resonant scanning post-processing of the pixel clock is avoided.

However, the disadvantage of a polygon mirror scanner is that spinning velocity is not constant, and fluctuates. A second a split photo-diode is generally used to monitor the velocity fluctuation and generate a corresponding pixel clock (Veilleux, Spencer et al. 2008).

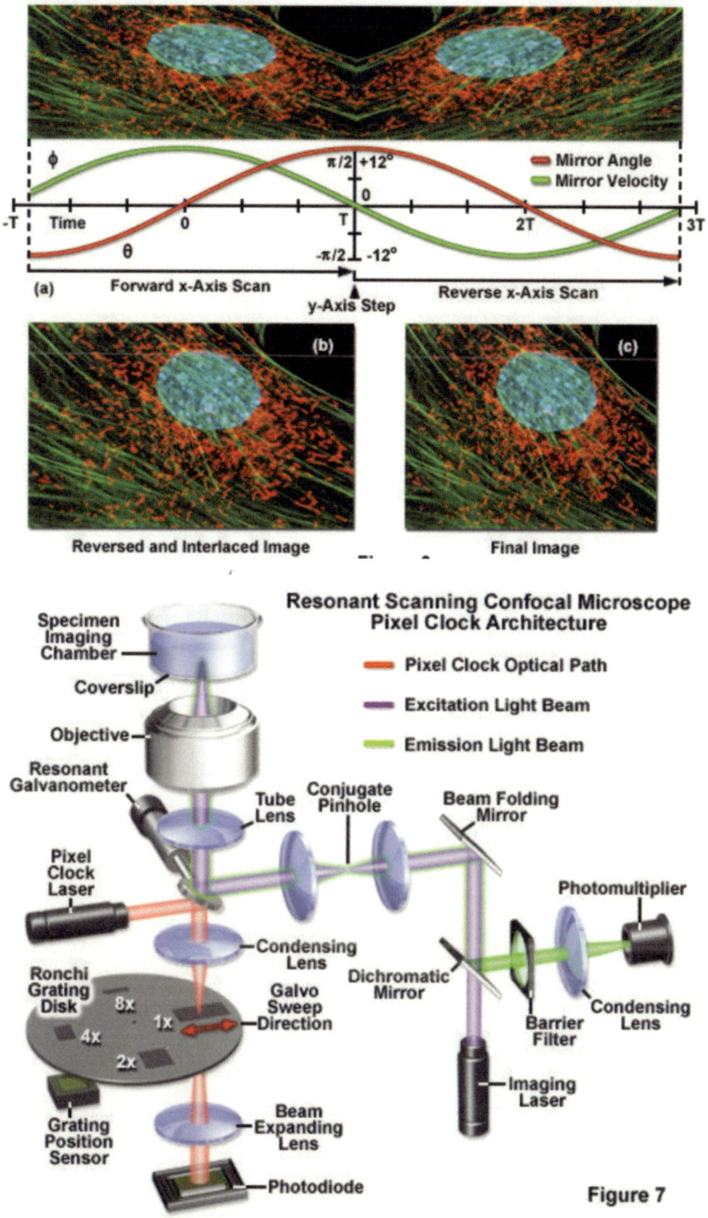

Fig. 13. The resulting distorted and corrected images are shown using the mirror angle and velocity of a sinusoidal resonant scanner; schematic diagram illustrates the optical pixel clock generation architecture (http://www.microscopyu.com/articles/confocal/resonantscanning.html).

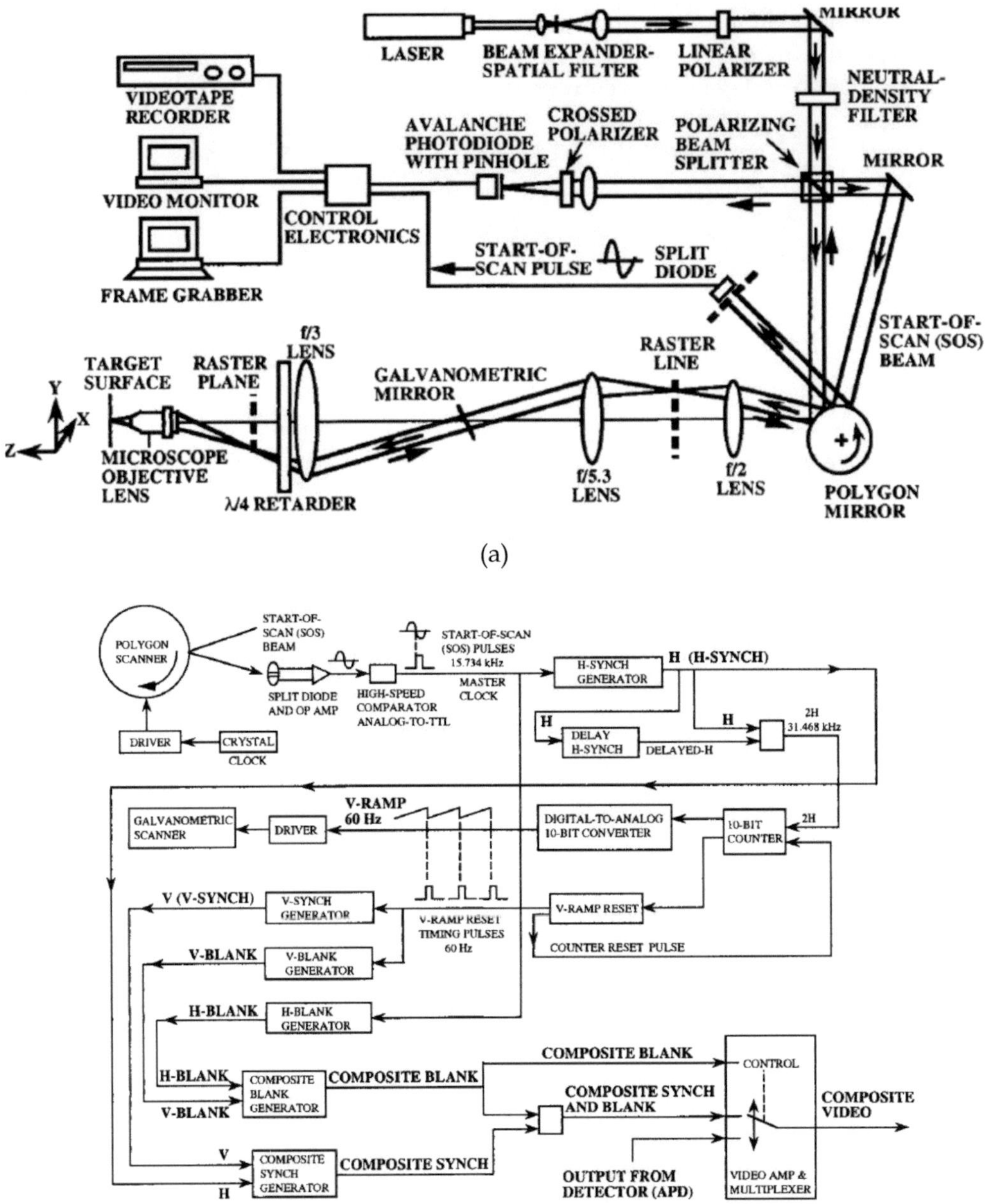

Fig. 14. (a) Schematic diagram of the polygon mirror based reflective confocal laser scanning microscopy; (b) electronic controlling diagram of the system(Veilleux, Spencer et al. 2008) Image courtesy of IEEE.

3.3 Endoscopic fast scan with MEMS

Taking advantage of cutting-edge MEMS fabrication techniques, a miniaturazed MEMS scanning mirror can be employed to dramatically decrease the size of conventional scanning schemes. Dikensheets and Kino (Dickensheets and Kino 1998) reported the instrumentation of a MEMS based confocal laser scaning system, the schematics of which are shown in Fig. 15. The two scanning MEMS mirrors are fabricated from the same silicon substrate.

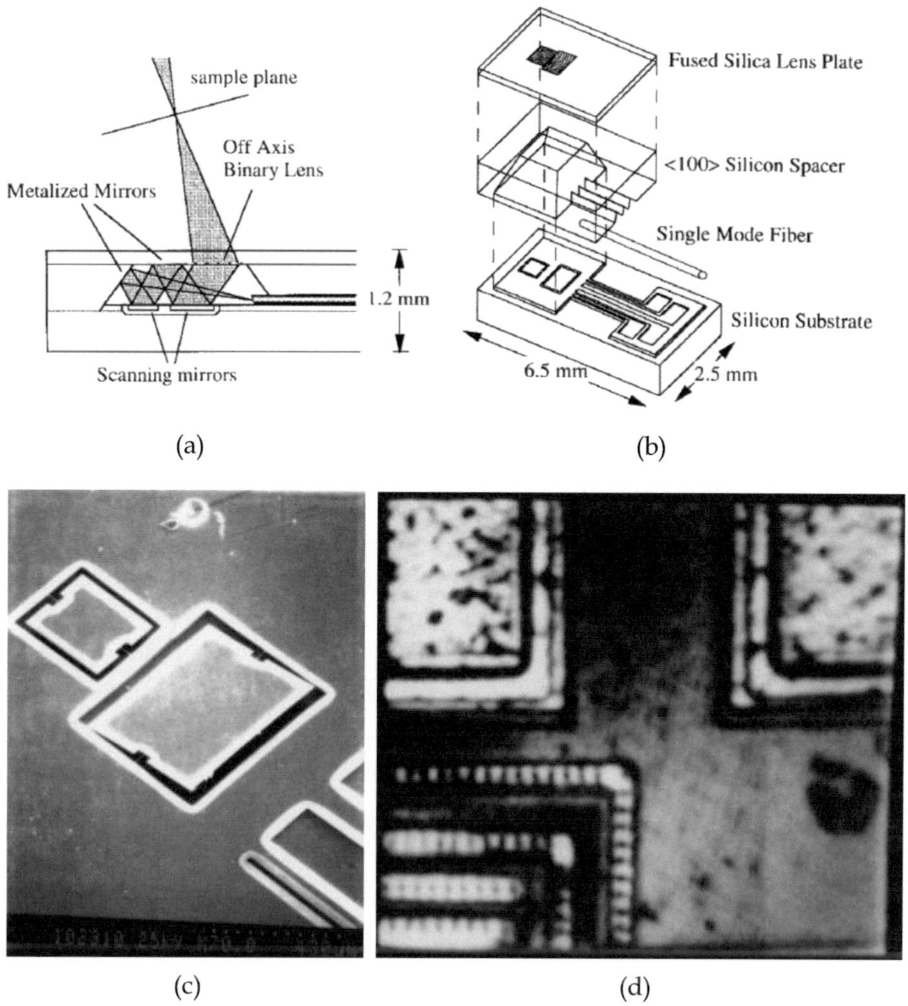

Fig. 15. (a) Schematic diagram of MEMS based CLSM; (b) the diagram of the structure of the MEMS-CLSM system; (c) a SEM image of the MEMS scanning mirrors; and (d) the image result of an IC chip (Dickensheets and Kino 1998) Image courtesy of IEEE.

3.4 Endoscopic fast scanning with fiber vibration

In endoscopy where the size of the probe is limited, optical fiber plays an important role in signal transferring. To obtain a confocal image *in vivo*, a mimiaturized CLSM system for endoscopy can be achieved with fiber vibrating at 2-D (Thong, Olivo et al. 2007). The handheld rigid probe consists of a handle with a focusing knob and an 8 mm diameter metal tube that houses a miniaturized tuning fork and the output end of the single-mode fiber is attached to one of the arms of the tuning fork. A magnetic coil driver induces a 700 Hz resonant oscillation in the tuning fork, providing a fast x scan, while a second coil causes the entire tuning fork to pivot in the y direction at a slower scanning rate of 1 to 2 Hz. Typically, a 1024 x 512 image can be obtained at 1.6 fps (Goetz, Fottner et al. 2007). The small tip of the single-mode fiber plays acts as both the illuminator and the detection pinhole of the confocal system. The fiber tip can also be driven to scan with a Lissajoul pattern (Helmchen, Fee et al. 2001) or spiral pattern for lateral imaging (Myaing, MacDonald et al. 2006).

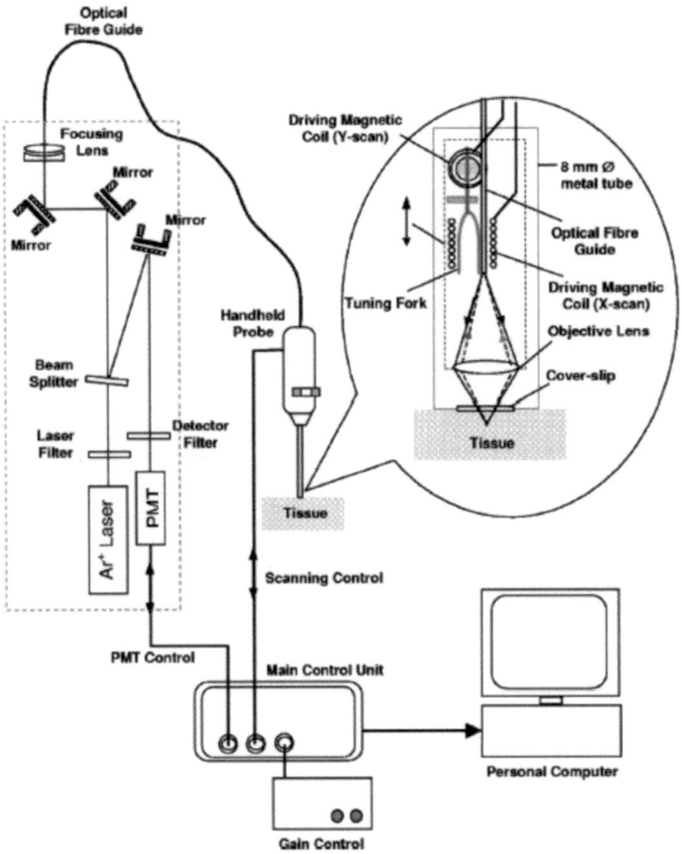

Fig. 16. Diagram of a confocal endomicroscope, which scans using a vibrating optical fiber with two tuning forks (Thong, Olivo et al. 2007) Image courtesy of SPIE.

4. Line scanning confocal microscopy

The imaging speed of a single spot scanning system is physically limited by the fast scanning unit. To reach a higher frame rate, one naturally thinks of using parallel spots or line scans. In line scan confocal microscopy, a cylindrical lens or mirror reshapes the light to a line at the focal plane of the objective where the specimen located. Since a galvanometer can reach speeds of ~500 Hz, it is used to scan the line and achieve a high imaging frame rate. The signal from the specimen is de-scanned by the galvanometer and detected by a line detector that is synchronized with the scanning command, thus forming a confocal image. One example of this setup is the Zeiss LSM5 Live system.

However, if such a line signal is re-scanned for 2-D imaging, it can be directly observed by the experimentor, or recorded by a CCD camera. Meridian released such a line scan system in the 1980's (see Fig. 17). In this system, the fluorescence signal from the sample is first de-scanned, then passes through a confocal slit located at the position of the mirror and focal

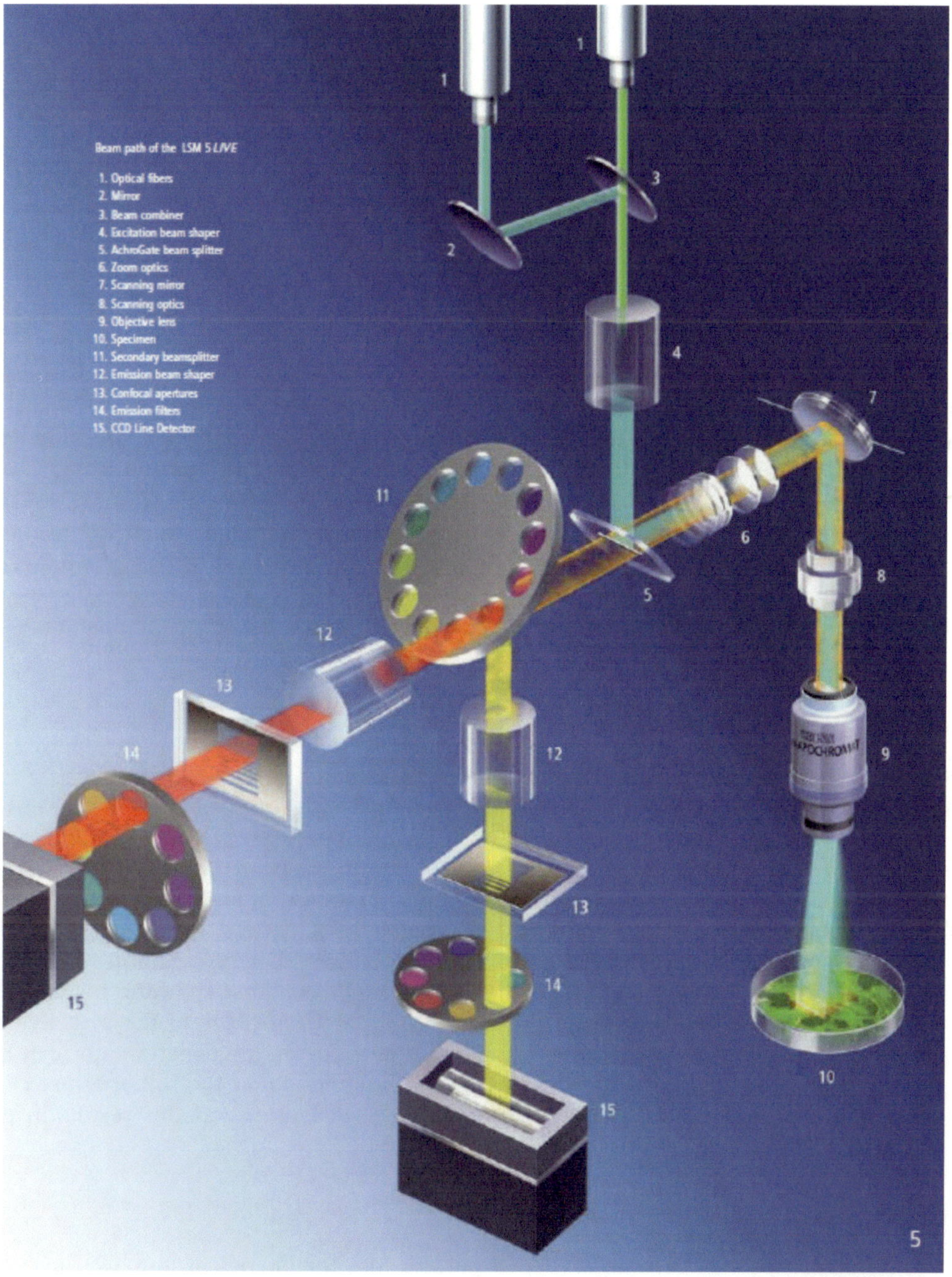

(a)

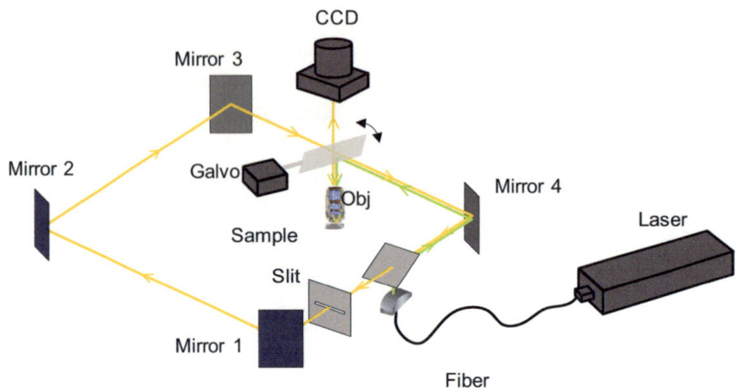

(b)

Fig. 17. Schematic diagrams of (a) a Leica LSM Live 5 (image from the manual); (b) a Meridian line scan confocal imaging system.

plane of the cylindrical lens. The back of the galvanometer is then used to re-scan the signal for CCD collection or the image viewed directly by the observer.
Theta-line scanning is an alternative method to that using in line scanning confocal imaging (Dwyer, DiMarzio et al. 2007). In theta line scanning, a D-shaped detection pupil is used with a f = 200 mm cylindrical lens and a line CMOS camera used for data collection. A non-resonant galvanometric scanner was used to scan across the sample. Similar to the control unit designed by Callamaras and Parker, as outlined in the previous section, the data from the CMOS detector is sent in conjunction with the PCLK, H-SYNC and V-SYNC signals, to a video frame grabber for signal processing. A 1143 x 1024, 4.86 fps image acquisition speed was acheived when using a 1 x 1 mm field-of-view.

5. Simultaneous multiple spot scanning

As described in Section 1, the application of a pinhoe in confocal microscopy can effectively block the stray light from the neighbourhood, attaining both improved resolution and the capability to reconstruct sections in 3-D. Multiple pinholes located at distant positions, make it is possible to simultaneously scan the sample with multiple beam foci. The Nipkow disk confocal scanning system utilizes this concept and is illustrated in Fig. 19. The Nipkow disk consists of two concentric disks spinning at the same velocity, and where the top disk is filled with a spinal shaped microlens array and the bottom disk has pinholes that are in correspondence with the lenslets of the upper disk. The illuminated spinning Nipkow disk generates multiple foci on the specimen postioned at the focal plane of the objective and when the disk is spinning fast enough a 2-D CCD detector or the experimentor can then detect the confocal image. One drawback in such a scheme is that the full aperture of the objective is not completely filled by the beam from each lenslet and this results in poor optical section resolution.

6. Conclusion

Confocal microscopy has become a very important tool for biological research as well as clinical diagnosis. Because laser scanning can provide faster imaging speeds compared to

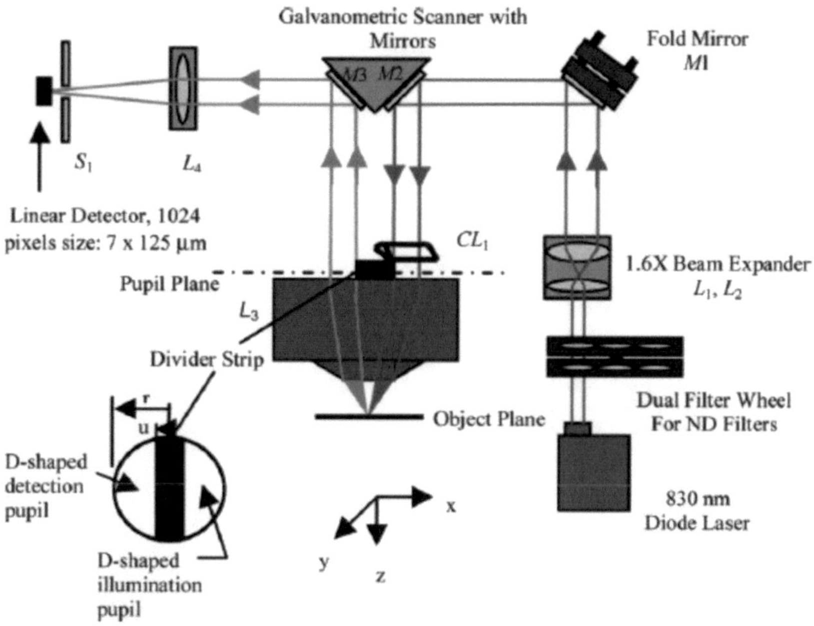

(a)

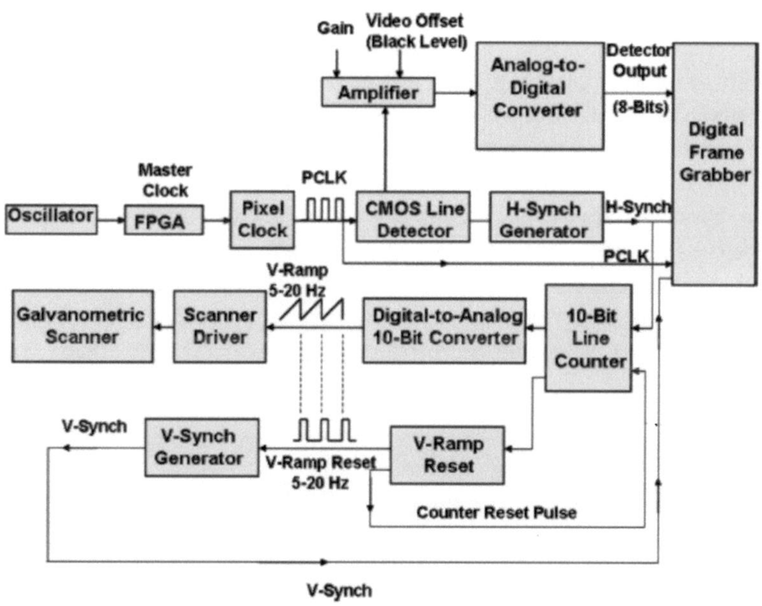

(b)

Fig. 18. Schematic diagram (a) and control circuit (b) of the theta-line scanning system. A 1024-element CMOS line camera was used for data collection (Dwyer, DiMarzio et al. 2007) Image courtesy of OSA.

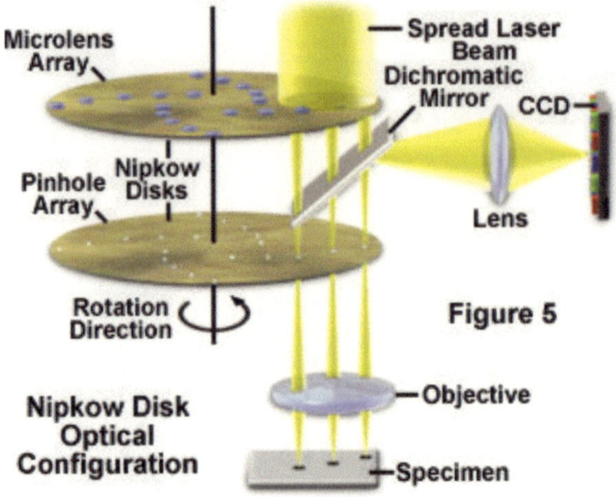

Fig. 19. Schematic diagram of the Nipkow disk scaning system (http://www.microscopyu.com/articles/confocal/confocalintrobasics.html).

conventional stage scanning movements, most confocal microscopic systems employ laser scanning as a means of lateral plane imaging. The most straightforward scanning mechanism can be achieved by using a pair of galvanometers and indeed, such a galvanometric configuration still plays a significant role in commercial CLSM systems. However, the speed of the galvanometer is generally limited to ~500 Hz, which in turn limits the imaging speed of a galvanometric based confocal system to ~1 Hz for a 500 x 500 image.

To obtain faster scanning speeds, a resonant scanner or a rotating polygon mirror can be used to generate a fast scan. Both these methods have their own pros and cons with respect to scanning control and data reconstruction. To minimize the size of the scanning head for confocal endo-microscopy, MEMS techniques can be applied, or through the use of optical fiber vibration. Since the speed of a galvanometric confocal system is limited by the fast line scan, a cylindrical lens can be used to generate a line focus and together with a slit, the confocal image can be detected. The image speed of a flying spot confocal system is limited by to scanning one spot at a time, however, multiple spots can be generated and scanned across the sample with a Nipkow spinning disk confocal system obtain confocal images with 2-D detectors.

7. Acknowledgements

We acknoledge the financial support from the National Basic Research Program of China ("973" Program, 2005CB724302, 2010CB933901), the National Natural Science Foundation of China (60588101, 60808029), the National High Technology Research and Development Program of China ("863" Program, 2008AA030118), Shanghai Commission of Science and Technology (08PJ14062, 064119540, 05DZ22318, 05DZ22325), Shanghai Jiao Tong University Med-Engineering Fund (YG2009ZD201), Georgia Institute of Technology- Emory University-Peking University Coulter Biomedical Engineering Seed Grant, and Peking University Med-Engineering Fund.

8. References

Amos, W. and J. White (2003). "How the Confocal Laser Scanning Microscope entered Biological Research." *Biology of the cell,* Vol 95, No. 6, 335-342, ISSN: 0248-4900.

Brakenhoff, G. J., P. Blom, et al. (1979). "Confocal scanning microscopy with high-aperture lenses." *Journal of Microscopy,* Vol. 117, No. 1, 219-232, ISSN:0022-2720.

Callamaras, N. and I. Parker (1999). "Construction of a confocal microscope for real-time xy and xz imaging." *Cell Calcium,* Vol. 26, No. 6, 271-279, ISSN:0143-4160.

Chen, S., X. Feng, et al. (2010). "Software controlling algorithms for the system performance optimization of confocal laser scanning microscope." *Biomedical Signal Processing and Control,* Vol. 5, No. 3, 223-228, ISSN:1746-8094.

Chou, D., B. Bower, et al. (2005). "Low-cost, scalable laser scanning module for real-time reflectance and fluorescence confocal microscopy." *Applied Optics,* Vol. 44, No. 11, 2013-2018, ISSN:0003-6935.

Cremer, C., T. Cremer, et al. (1978). "Considerations on a laser-scanning-microscope with high resolution and depth of field." *Microscopica Acta,* Vol. 81, No. 1, 31-44, ISSN:0044-376X.

Dickensheets, D. and G. Kino (1998). "Silicon-micromachined scanning confocal optical microscope." *Journal of Microelectromechanical Systems,* Vol. 7, No. 1, 38-47, ISSN:1057-7157.

Dwyer, P., C. DiMarzio, et al. (2007). "Confocal theta line-scanning microscope for imaging human tissues." *Applied optics,* Vol. 46, No. 10, 1843-1851,

Goetz, M., C. Fottner, et al. (2007). "In-vivo confocal real-time mini-microscopy in animal models of human inflammatory and neoplastic diseases." *Endoscopy,* Vol. 39, No. 4, 350-356, ISSN:0013-726X.

Helmchen, F., M. Fee, et al. (2001). "A Miniature Head-Mounted Two-Photon Microscope:: High-Resolution Brain Imaging in Freely Moving Animals." *Neuron,* Vol. 31, No. 6, 903-912, ISSN:0896-6273.

Leybaert, L., A. De Meyer, et al. (2005). "A simple and practical method to acquire geometrically correct images with resonant scanning-based line scanning in a custom-built video-rate laser scanning microscope." *Journal of Microscopy,* Vol. 219, No. 3, 133-140, ISSN:0022-2720.

Minsky, M. (1988). "Memoir on inventing the confocal scanning microscope." *Scanning,* Vol. 10, No. 4, 128-138, ISSN:0161-0457.

Myaing, M., D. MacDonald, et al. (2006). "Fiber-optic scanning two-photon fluorescence endoscope." *Optics Letters,* Vol. 31, No. 8, 1076-1078, ISSN:0146-9592.

Nguyen, Q., P. Tsai, et al. (2006). "MPScope: a versatile software suite for multiphoton microscopy." *Journal of Neuroscience Methods,* Vol. 156, No. 1-2, 351-359, ISSN:0165-0270.

CRS Application Notes (1994) "Pixel clock provides velocity correction."

Pawley, J. (1995). Handbook of biological confocal microscopy. New York, *Plenum Press.*

Thong, P., M. Olivo, et al. (2007). "Laser confocal endomicroscopy as a novel technique for fluorescence diagnostic imaging of the oral cavity." *Journal of Biomedical Optics,* Vol. 12, No. 014007, ISSN:1083-3668.

Veilleux, I., J. Spencer, et *al. (2008). "In vivo cell tracking with video rate multimodality laser scanning microscopy." IEEE Journal of Selected Topics in Quantum Electronics,* Vol. 14, No. 1, 10-18, ISSN:1077-260X.

Xi, P., Y. Andegeko, et al. (2009). "Two-photon imaging using ada*ptive phase compensated ultrashort laser pulses." Journal of Biomedical Optics,* Vol. 14, No. 1, 014002, ISSN:1083-3668.

Xi, P., Y. Andegeko, et al. (2008). "Greater signal, increased depth, and less photobleaching in two-photon microscopy with 10 fs pulses." *Optics Communications,* Vol. 281, No. 7, 1841-1849, ISSN:0030-4018.

Xi, P., B. Rajwa, et al. (2007). "The design and construction of a cost-efficient confocal laser scanning microscope." *American Journal of Physics,* Vol. 75, No. 3, 203-207, ISSN:0002-9505.

28

A New Laser Scanning System for Computed Radiography

Qibo Feng, Meng Zheng, Shuangyun Shao and Zhan Gao
School of science, Beijing Jiaotong University, Beijing 100044
P. R. China

1. Introduction

X-ray inspection is widely used in aviation, aerospace, weapons, nuclear energy, automobile and other fields for non-destructive testing and non-destructive evaluation. It plays an important role in engineering quality supervision and quality control, and it has become a basic technology in industry and high-tech industry.

With the strength of visuality, high sensitivity and high resolution, X-ray screen-film radiography is widely used in industrial non-destructive testing, medical imaging and other fields. However, it has some disadvantages of low efficiency and films being not reusable. Due to this, there have been some new methods of digital imaging for radiographic inspection such as the analog imaging based on image intensifier, the large array detector imaging based on Thin Film Transistor (TFT), and the computed radiography (CR) imaging based on the photo-stimulated luminescence (PSL). As the imaging plate of a CR system has some advantages of large dynamic range, erasibleness, high sensitivity, high spatial resolution, flexibility, etc. It has become a key research field (Brandt, 1993; Cowen et al., 2007; Dragusin et al., 2006; Satoshi et al., 1999; Schaetzing et al., 2002). Some commercial CR systems, especially for medical imaging, have also emerged.

With higher production standards for long-distance oil and gas pipeline systems, X-ray inspection of pipe ends and patching has become a key means for inspecting harmful defects in major domestic and international long-distance pipelines in order to guarantee their quality. Having shared the merits of industrial film radiography of large ray dosage absorption, flexibility, and higher spatial resolution, CR systems also enjoy new edges such as digital imaging, large image dynamic range and re-usable image plate, and it will play a key role in industry applications such as pipe quality testing and controlling. However, the high cost of CR systems has hindered their development in the field of industrial testing which is not so profitable. Therefore, to sharpen the competitive edge of CR systems in industrial testing, it is necessary not only to improve the inspection efficiency and image quality of the image plate, but also to significantly reduce the costs of CR systems.

Readout technology and equipment are the key technology and core part of CR systems and it also play a key role in the cost of the CR system. Theretofore, how to improve the resolution and scanning speed of readers and how to efficiently collect weak fluorescent signals in a simpler way and lower cost have been a focus for studies in this field.

2. The point-scanning methods for CR systems

A CR system mainly consists of three parts: information recording, completed by the imaging plate; image scanning and reading, done by the CR scanner; image processing and storing, undertaken by the computer and memory devices. Taking the industrial X-ray inspection of welding defects as an example, the specific work procedure of a CR system is shown in Fig.1.

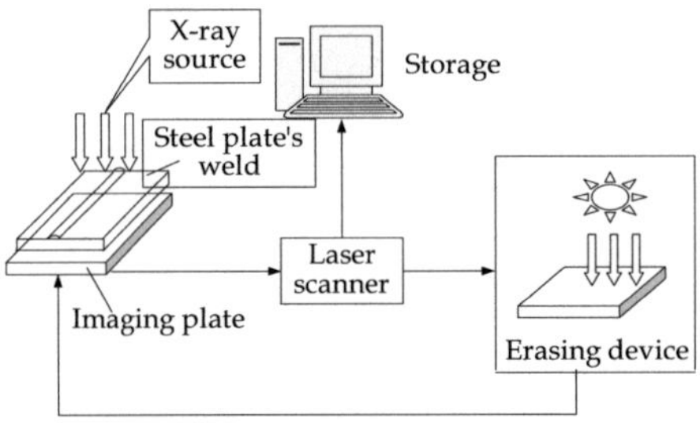

Fig. 1. CR system work flow

As shown in Fig.1, the CR system replace the conventional X-ray screen-film with a reusable imaging plate (IP) coated with a photo-stimulating storage phosphor, which retains latent image information when the IP is exposed to X-ray and the information of the X-ray energy is stored. Then the IP will be put through the laser scanner that reads and digitizes the image, and the image is stored in memory. When the laser scanner has collected all information on the IP, the IP will be taken to erasing device to clear the recorded information so that it can be re-used next time.

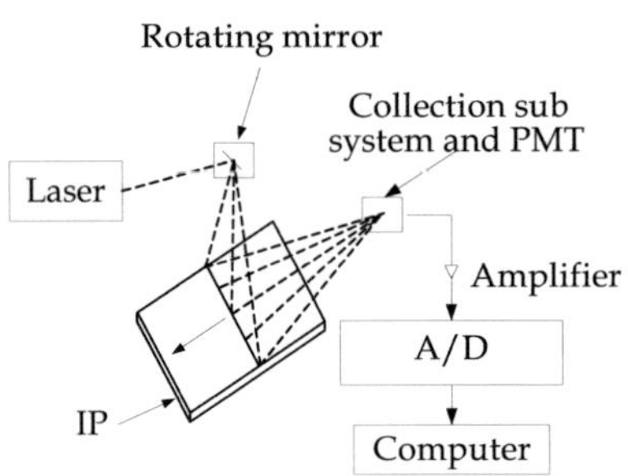

Fig. 2. Readout procedure of IP

The CR scanner is the core part of the CR system. Currently, there are mainly two kinds of scanning methods that are point-scanning and line-scanning. The procedure of the point-scanning and imaging is shown in Fig. 2. As the IP is exposed, the latent image is retained on fluorescent substances. Through the optical scanner such as a rotating mirror, the laser beam coming from the laser source is projected toward the IP to form a line scanning. As the IP moves along the direction of vertical scanning beam, a column scanning is formed. The weak luminescence signal excitated by the laser beam on the IP will be gathered by a photomultiplier tube through a luminescence optical collection device, and the signal from the photomultiplier tube will be amplified and put into a computer for processing through the A/D converter. Finally, the intensity value on one point of the IP can be obtained, and the whole high-quality digital images of the IP are produced through point by point scanning.

Compared with the line-scanning, the point-scanning has the advantages of simple configurations, low requirements for the laser source and the photo-detector and low cost, it has been widely adopted in CR systems. Currently, there still are different kinds of optical point-scanning in designing the CR scanner, which commonly works in the following ways:
(1) **Scanning Using a Rotary Polyhedron and a** $F\theta$ **Lens**. As shown in Fig. 3(Shan B Zh et al., 2005), in the scanning optical path of the IP, a rotary polyhedron is used for the scanning and a $F\theta$ lens is used for focusing the laser beam. The use of the $F\theta$ lens is to make sure that scanning spot maintain a linear movement on the IP in order to eliminate the distortion of scanning. However, the rotary polyhedron prism scanning will inevitably cause the discontinuity of scanning beams between two adjacent faces of the polyhedron prism, and there is a signal vacuum during the discontinuity time. Meanwhile, the rotary polyhedron has also caused some problems such as the sway and displacement error of the rotating shaft, the division error of the rotating mirror and the tilt error of reflector, all these errors directly influence the quality of the scanning and they are difficult to get rid of. In addition, it is difficult to for the laser beam to keep a linear movement by using a $F\theta$ lens, and using a $F\theta$ lens makes the CR system much more complicated and costs more money.

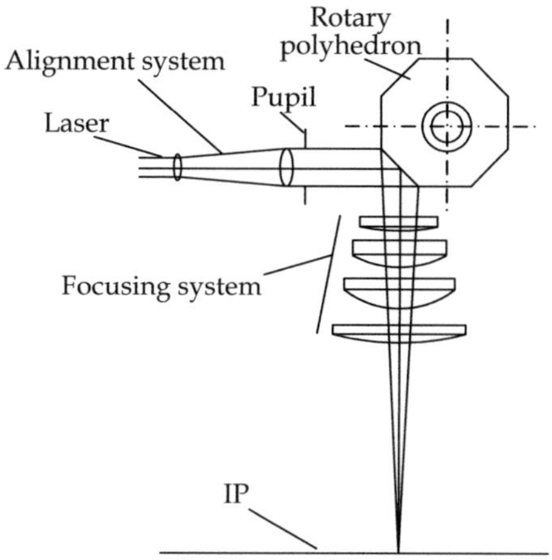

Fig. 3. Scanning with a $F\theta$ Lens

(2) **Scanning Using a Pentagonal Prism and a** $F\theta$ **Lens.** As shown in Fig. 4(Zhang J P et al., 2008), two pentagonal prisms are used for scanning the laser beam, and a $F\theta$ lens is used for focusing the laser beam. The collimated laser beam is first scanned by rotating the pentagonal prism, and then the laser beam is projected to the $F\theta$ lens, through which the laser beam is focused to a point on the IP that is in the focal plane of the $F\theta$ lens, a line scanning is formed by rotating the pentagonal prism, and whole digital image of the IP can be obtained when the IP move along the direction that is perpendicular to the scanning line. Two pentagonal prisms are used to enhance the efficiency of the scanning and to reduce the scanning time. This type of point-scanning can overcome the problem of discontinuity as shown in Fig.3, however, it has the same disadvantage of the complicated optical system and high cost as a complicated $F\theta$ lens is needed.

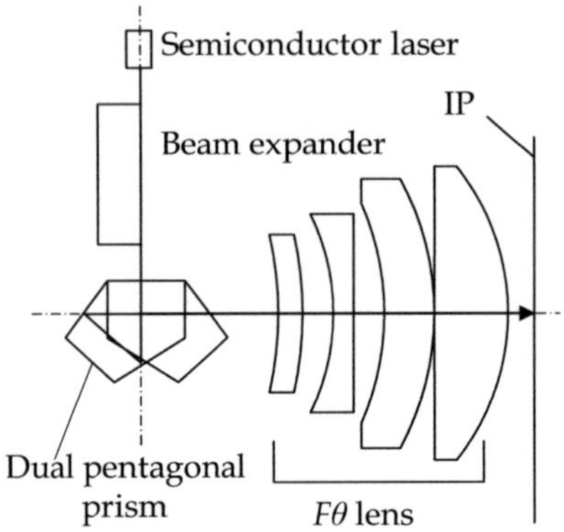

Fig. 4. Scanning with $F\theta$ and pentagonal prism

(3) **Scanning Using a Single Piece of Tilting Reflecting Mirror**. As shown in Fig.5 (Chen H Q, 2003), an objective lens and a single mirror are used to focus the laser beam to a point on the IP. Theoretically, a line scanning can be formed if the mirror is rotated around the axis that is also the axis of both the objective lens and the circular IP. It can be seen that the incident light beam occupies the whole surface of the mirror during the scanning and only one piece of mirror is required for the scanning. There is no transition of beam between mirrors and consequently no discontinuity during the laser beam scanning. Therefore, the receiving field is arc-shaped, the focused laser beam moves at a constant speed in the circular IP and a linear scanning can be gotten. This type of point-scanning has the advantages of a simpler optical system and lower cost of the scanning system; it has been adopted by some companies in their commercial CR systems. However, as there is inevitably a tilt of θ during the rotating of the mirror, the direction of the laser beam reflected by the rotating mirror will change 2θ, causing a position change of the focused laser beam point in the IP and worsening the quality of the image from the IP.

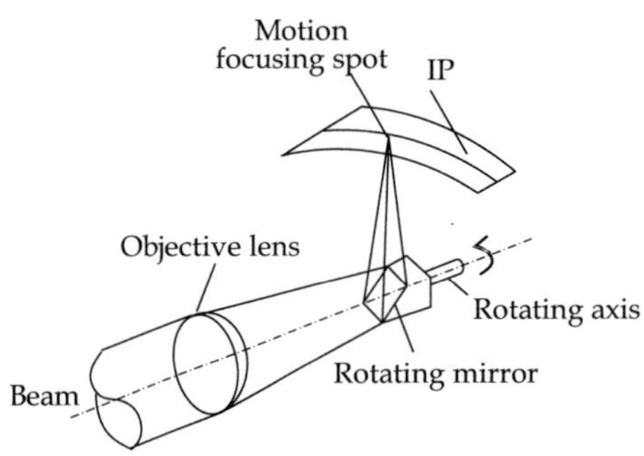

Fig. 5. Scanning with a single mirror

(4) **Scanning Using a Pyramid-shaped Rotating Mirror.** A pyramid-shaped multi-faceted prism is an equilateral prism, the surface of which tilts to form an angle with the rotating axis, placing itself under the full coverage of the beam, as shown in Fig.6 (Chen H Q, 2003). The major characteristic of the scanning with such a pyramid-shaped rotating mirror is that in order to make sure that the beam spot on the arc-shaped orbit for focusing moves with a uniform angular velocity, the focus of objective lens must be designed to be located on the rotation axis of the rotating mirror.

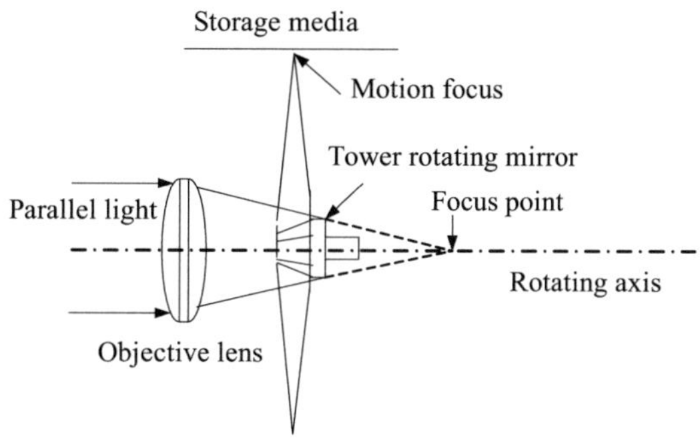

Fig. 6. Scanning by a pyramid-shaped mirror

In order to get stable scanning region, that is to say to eliminate scanning intermittence as much as possible, there must be at least two adjacent prism planes that are fully lightened in the effective scanning sector. The shortcoming of this kind of point-scanning is that it is difficult to manufacture a pyramid-shaped mirror, moreover, there will be inevitably a tilt of θ during the rotating of the mirror, and this will influence the quality of the image from the IP mentioned above in Fig.5. With the mirror plane inclining towards the axis, high-speed rotation will impose an uneven dynamic stress on the prism, which will in turn cause the shape of the prism to change.

In all four-kinds of point-scanning optical methods of the CR systems just mentioned above, a special separate optical collection system is needed in order to gather the weak fluorescence signals excitated by the scanning laser beam.

In summary, the current point- scanning optical systems are mainly faced with the following problems: (1) Polyhedron prism scanners will cause the discontinuity of scanning beams, resulting discontinuity in information transmission; (2) The $F\theta$ lens as focusing device would bloat the overall structure of the system and result a complicated and costly optical scanning system; (3) The scanning optical system and the luminescence collection optical system are usually separated from each other, which calls for the designing and manufacturing of a separate luminescence collection optical system, resulting in a more complex optical system of CR systems.

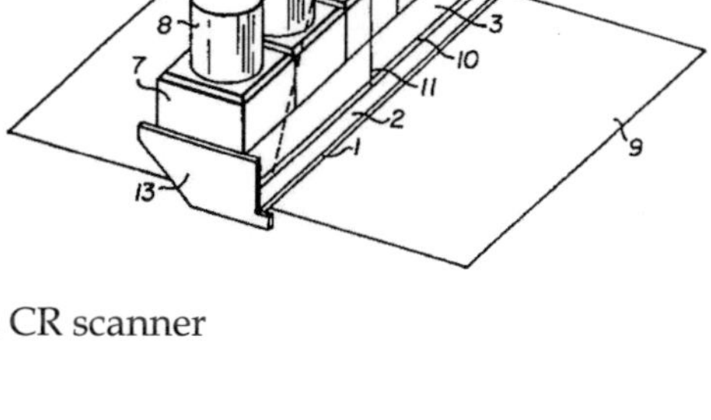

Fig. 7. Optical system of a CR scanner

A laser scanning system for computed radiography is shown in Fig.7 (Brandt et al., 1996). After being put through the beam expander 15, the laser beam emitted from a laser source 16 is turned into parallel light, and it is reflected by the laser beam scanner 14 and is focused on the IP 9 by the $F\theta$ lens 17. The luminescent are simulated by the scanning beam 11 and gathered by the special collection optical system, which uses the plane mirrors and 1,2,3,13 stands for the four pieces of these plane mirrors, and then the luminescent are sent to the PMT 8 for amplifying and further processing. The stray light can be removed by using a filter 7.

3. The new laser scanning system for CR

3.1 Principles of scanning and imaging

To overcome these three problems of the existing CR scanning systems mentioned above, a new laser scanning optical system has been designed, as shown in Fig.8. After being put through the filter and the beam expander, the laser beam emitted from a laser source is turned into parallel light, which will go through a dichroic mirror and is subsequently reflected by a pentagonal prism and is then focused by a scanning objective lens onto the IP. Thus luminescent signals are produced by the simulation of phosphor, and photo-stimulated luminescence becomes parallel light after being collected by the scanning objective lens. Having gone through the pentagonal prism and been reflected totally by the dichroic mirror, the parallel light will be focused by the condenser, filtered by a filter to eliminate the stray light, and then sent into the photomultiplier tube, where it will go through A/D converter and be transmitted into the computer to acquire the luminescence gray value.

As shown in Fig.8, the pentagonal prism and the doublet objective lens are used to form a scanning arm for both stimulating and collecting luminescence on the IP. Only a dichroic mirror is adopted to separate the laser beam and the photo-stimulated luminescence, which simplifies the optical system. The dichroic mirror can transmit almost all laser beams and reflect all the luminescence, thus improving the luminescence collection efficiency.

While at work, the scanning arm is driven to rotate at a high-speed by the scanning motor to start a horizontal scan of the IP; the plate feeding structure is pulsed by the step motor forward to feed the IP vertically, thus scanning the whole IP point by point can be finished and the digitized image of the IP can be obtained.

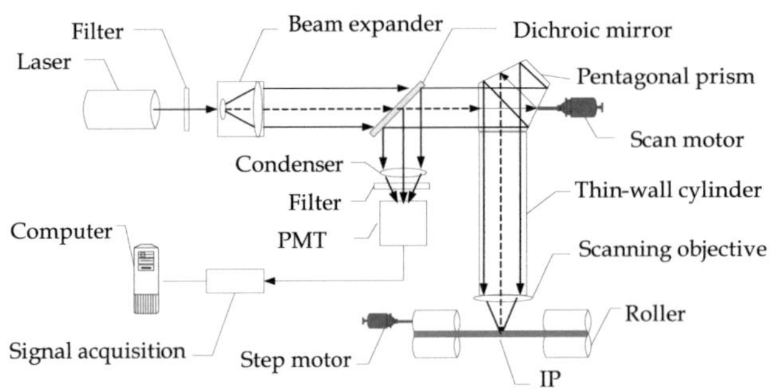

Fig. 8. Principle diagram of the CR scanner

To realize the rotation scanning, the IP has to be placed in the shape of a semi-circular arc and the circle central axis of the scaning arm shaft should coincide with the axis of the semicircular arc.

The configuration of the scanning optical system is shown in Fig.9. The advantages of such a scanning system is as follows: (1) The scanning optical system can guarantee that scanning laser beam is projected always vertically onto the IP and the focused spot of the laser beam on the arc keeps a continuous linear movement, there is no needs of using a $F\theta$ lens, resulting in a simpler and a lower price optical scanning system; (2) No need to design and manufacture a separate collection optical system as the scanning arm also can be used to collect the photo-stimulated luminescence, which further simplifies the scanning system.

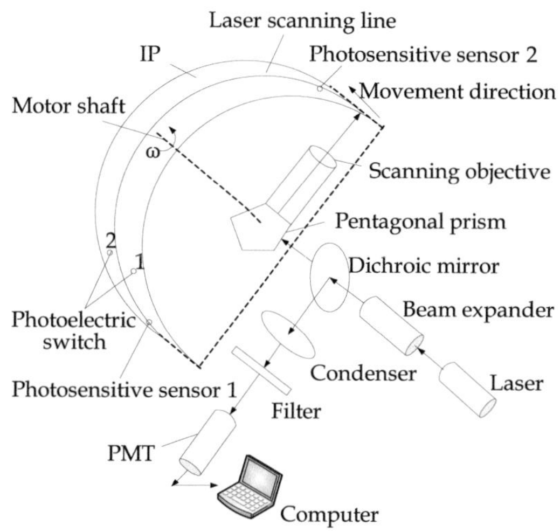

Fig. 9. Structure diagram of the CR scanner

3.2 Selections of key components

The laser source, the laser scanning objective lens and other components are quite vital to the performance of the scanning system. The following is a brief description of the selection strategy of these major components.

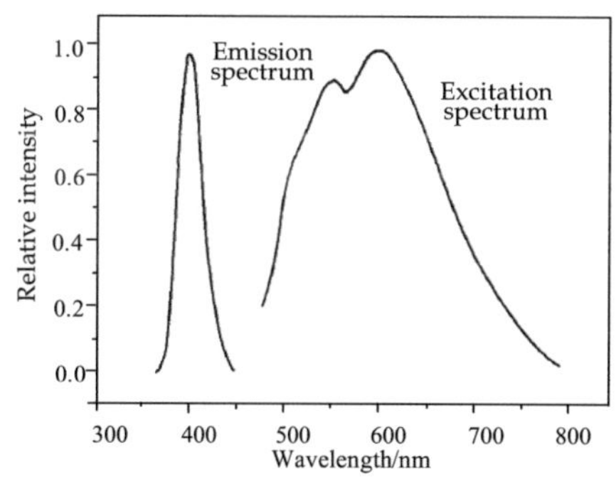

Fig. 10. Emission and excitation spectrum

(a) Laser

The emission spectrum and excitation spectrum curves of the IP are shown in Fig.10 (Zhou X L et al., 1993). The IP luminescence in this system has an emission peak wavelength of 390nm with its bandwidth of 20nm. The selected excitation laser beam has a wavelength of

638nm with its beam diameter about 0.8mm. To improve the quality of the laser spot, a single-mode coupled laser diode module has been adopted.

Within a certain range, the intensity of the luminescence stimulated on the IP is proportionate to both the X-ray energy stored in the IP and the exciting beam intensity.

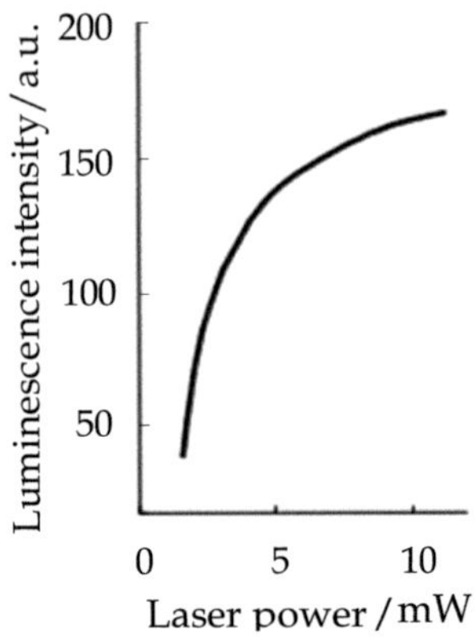

Fig. 11. Relationship between excitation flight intensity and laser power

Although excessively strong laser excitation beam can release more trapped electrons, the increased depth of the laser in the phosphor layer will enhance the excited luminescence diffusion and lower the probability of the stimulated luminescence to escape from the IP, resulting in lower collection efficiency. The relationship between the excited luminescence intensity and the power of the laser beam is shown in Fig.11 (Zhou X L et al., 1993), and it can be seen that a 10mw laser source is a good choice.

(b) Scan Lens

The scanning objective lens is an important component. For the excitation optical system, it must have a relatively small lens aberration to form a smaller laser scanning spot and to get a high spatial resolution; As far as the luminescence collection is concerned, the objective lens must have a greater numerical aperture in order to improve luminescence collection efficiency. The relationship between the excited luminescence on the IP and its emission direction can be expressed as follows

$$I = I_0 \cos\theta \tag{1}$$

Where I_0 is the luminescence intensity in the normal direction of the IP and θ the included angle with the normal direction.

The distribution of stimulated fluorescence from the IP is shown in Fig.12. The luminescence in the normal direction has the maximum intensity. With the increasing scattered angle, the stimulated fluorescence intensity is becoming lower. From Eq. (1), it can be gotten that $I = 0.866 I_0$ if $\theta = 30^\circ$. Across the scope of luminescence emission, a relatively strong luminescence can be collected within 30-degree solid angles formed by the launched luminescence with the normal direction. Therefore, the vertical excitation produces a concentrated energy distribution of the luminescence, and with an appropriate objective lens, relatively strong luminescence can be collected within the solid angle as shown in the Fig.12.

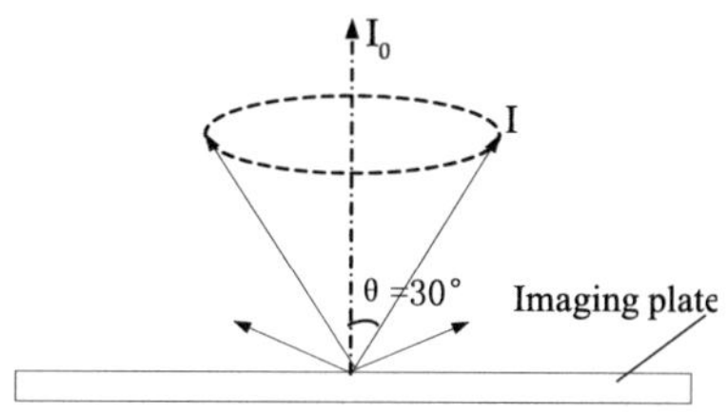

Fig. 12. Luminescence Distribution

A doublet lens is utilized as the scanning objective lens with the following parameters: the focal length $f = 25mm$, the back focal length $f_b = 15.6mm$, the diameter $\Phi = 25mm$, the effective aperture $CA = 22.5mm$ and thickness $t_c = 19mm$. For the luminescence gathering, the excited luminescence can fill the space of the lens aperture. For the laser beam excitation, as the laser beam has small diameter after collimation, it is approximately a paraxial beam, and the doublet lens can perfectly eliminate spherical aberration and coma. The calculated spot diagram is shown in Fig.13, which has a geometric aberration of a smaller size than the diameter of Airy disk and can meet the requirements for the scanning system.

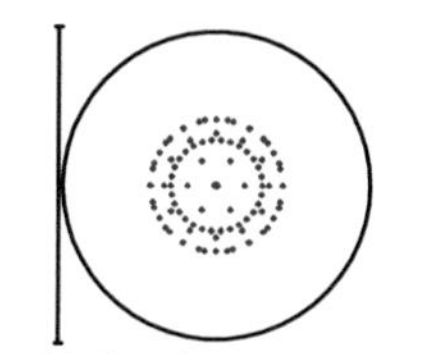

Fig. 13. Spots diagram of the doublet lens

(c) Dichroic Mirror

As the luminescence excitation optical system and the collection optical system are common-path in this system, effective measures must be taken to separate the laser beam and the stimulated luminescence after the luminescence has been collected. To this end, a special dichroic mirror which can almost transmit the laser excitation beam with the wavelength of 638nm and reflect nearly all the luminescence with the wavelength of 390nm

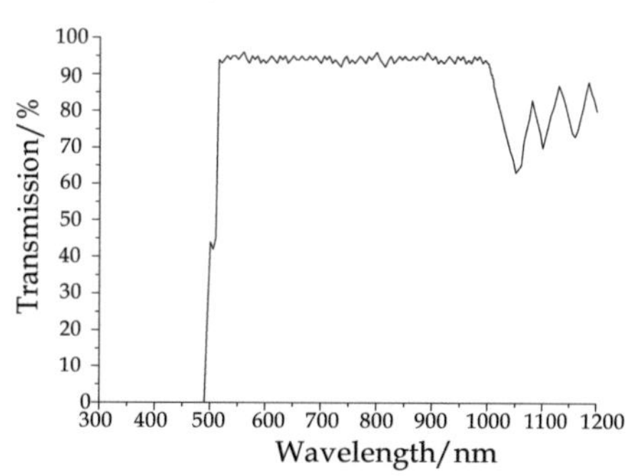

Fig. 14. Spectral curve of the dichroic mirror

was designed. Its properties are shown in Fig.14, and it can be seen that the dichroic mirror has the reflective band of 327-488nm with reflectance close to 100%, the transmission band of 515-850nm with the transmission rate of about 95%, thus contributing to the successful separation of the laser beam and the stimulated luminescence.

3.3 Computer signal acquisition and control

A control sub-system is mainly responsible for signal acquisition, motion control of the plate feeding sector, and synchronization between these two processes. The control scheme of the scanning system is shown in Fig.15.

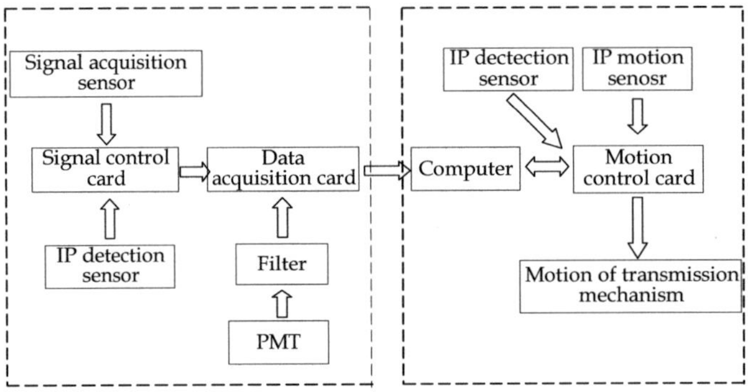

Fig. 15. Control scheme of the scanning system

As shown in Fig.9, the four sensors including two photoelectric switches and two photosensitive sensors are used to realize the synchronization and controlling of scanning, longitudinal movement of the IP and signal collection. Two photoelectric switches are longitudinally located on both sides of the laser scanning line, and can detect the entrance and exit of the IP through signals behind the NAND door. Two photosensitive sensors, which are triggered by the scanning laser beam, are located on the laser scanning line. When the scanning laser beam passes through the photosensitive sensor 1, the signal from the sensor 1 triggers the controlling circuit and the fluorescence signal is to be collected, at this time, the IP does not move; when the scanning laser beam passes through the photosensitive sensor 2, the circuit controls the IP move a step longitudinally and stop the fluorescence signal collection. Thus in the positive semi-circle of the scanning arm rotation path, the effective signal acquisition is under way. In the negative semi-circle, the stepper motor drives the IP to move one step forward. In this way, the scanning and imaging of the entire IP are completed.

The timing of overall control system is shown in Fig. 16.

When the scanner is powered up, the scanning motor and the laser start to work. When the scanning motor and the laser gains stability in their work, the IP, pushed by hand, has gone through the first optical switch, the IP detection output is a high level signal. Then stepper motor starts to work to move the IP quickly through a certain distance to reach the location of the scanning beam. Thereafter, the stepper motor regains normal speed with each step covering 50μm. The acquisition card goes into the normal acquisition state. When the scanning laser beam goes through photosensitive detector 1, signal acquisition is initiated. When it goes through the photosensitive detector 2, the stepper motor moves one step further. This process will be repeated until a complete image of the IP is gathered. When the

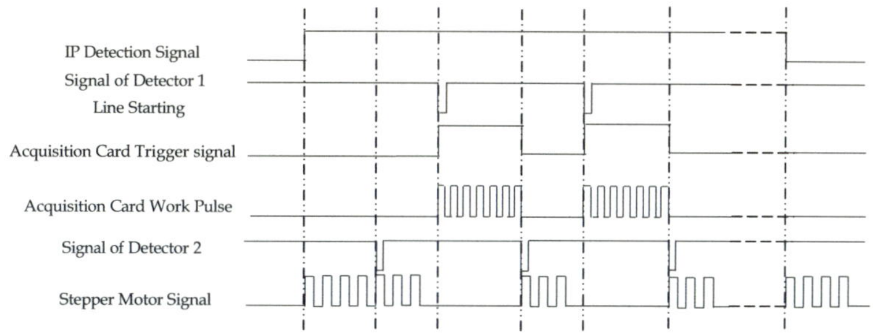

Fig. 16. Control system timing diagram

IP moves away from the two photoelectric switches, the IP detection signal output is at a low level and the acquisition card stops working. The stepper motor fast movement so that the IP can move fast out of the scanner and the whole scanning procedure is finished.

3.4 Mechanical system

On the one hand, the mechanical system should ensure the relative positions of the various components of the scanning system and their stability; on the other hand, it enables the rotation of the scanning arm and the movement of the plate driving mechanism as well as guarantees the accuracy of movement. The mechanical structure of the scanner designed according to these functions is shown in Fig.17. The photomultiplier tube lies below the excitation light path and its receiving window directly faces the luminescence emitting mouth of the dichroic mirror. While in the actual application, the photomultiplier tube is sealed in a tin box. The internal and external surfaces of the box are black anodized so that the photomultiplier tube remains in a shade environment. The scanning servo motor is fixed on a support, and the principal axis of the scanning motor is connected with the scanning arm and coincides with the optical axis of the scanning optical system.

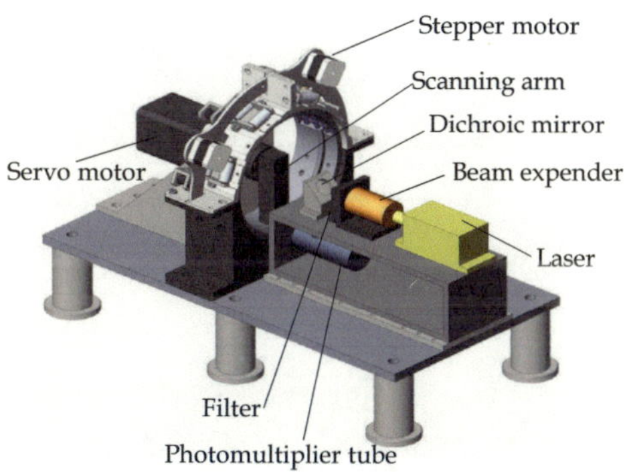

Fig. 17. Scheme diagram of the mechanical structure

3.5 Configuration of software

The software of the CR system is critical in keeping the system work properly. It is not only a program of human-computer interaction but also the center for controlling the system work and image acquisition and processing. It undertakes the tasks of deciding when to start work, displaying data that has been obtained after the IP is scanned, and image processing and weld defect identification.

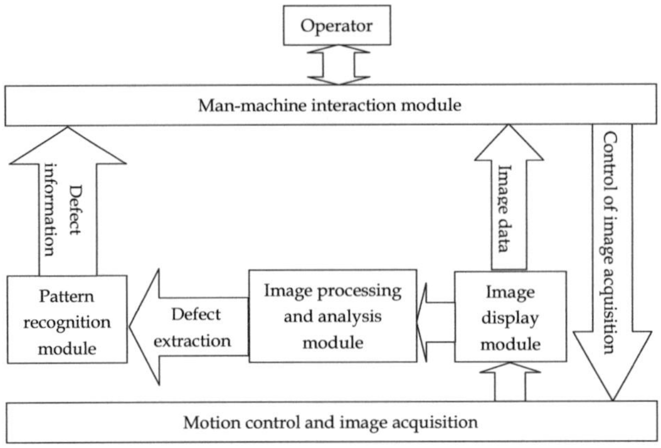

Fig. 18. Structure scheme of the software

As shown in Fig.18, the software of the scanner consists of four modules which are a human-machine interaction module that is responsible for defects identification and image gathering control; an image display module that is used for the conversion and storage of image data and the transmission of image data to the screen; an image processing and analysis module that is a systematic and modularized software for various image processing algorithms according to the inspection requirements, including a variety of basic image manipulation such as histogram transformation, image rotation, image reduction, basic filtering methods, and the geometric locations of the defects obtained through processing and analysis; a pattern recognition module that is used for the extraction of weld defects, marking, identification and classification.

The prototype of a computed radiography laser scanning system developed accordingly is shown in Fig.19.

Fig. 19. Prototype of the CR system

4. An analysis of the major factors influencing the performance of the system

The specifications that reflect the performance of the laser scanning systems mainly include the spatial resolution, the image stability and the speed of scanning and other parameters. The following is a brief analysis on these major factors so as to provide a basis for further improving such scanning systems.

4.1 The major factors affecting resolution of the system and related analysis

The spatial resolution of the CR image is related to properties of the IP, the laser spot size and the sampling frequency. In the scanner, the image is obtained by scanning the laser spot point by point on the IP, and the diameter of the laser spot is directly related to the quality of the image that has been read. Generally speaking, the smaller the diameter of the laser spot of the scanner is, the smaller the impact that it will have on the adjacent un-scanned IP latent image. And the image quality will be better and the resolution will be higher. Therefore, the laser spot size is an important factor that influences the spatial resolution of the system.

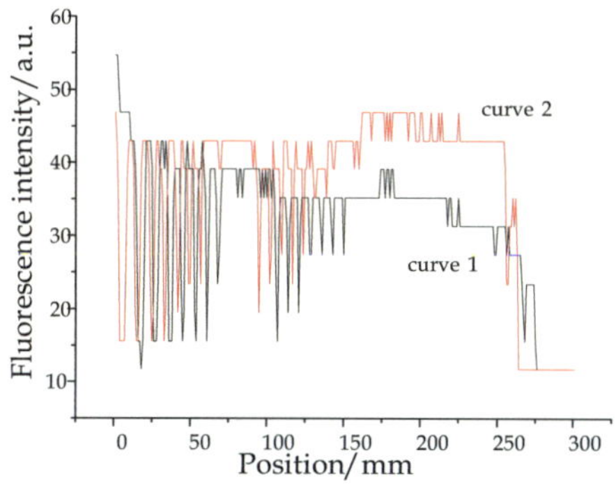

Fig. 20. Comparison with two laser spot size

The scanning curves by two different laser spot sizes are shown in Fig.20. In the experiment, 21 lead wires whose diameters are decreased in sequence are exposed to X-ray on the IP. The laser scanning system shown in Fig.19 was applied to scan the IP and a graph indicating the relationship between the luminescence intensity and its location was generated. In the graph, curve 1 represents the luminescence corresponding to the beam spot of 0.5mm in diameter and curve 2, the beam spot of 0.8mm diameter. The downward peaks show the wires that have been scanned. Curve 1 suggests that 16 wires can be distinguished, whereas curve 2 can only make it to 13. Therefore, the smaller the light spot size is, the higher the resolution will be.

The diameter of the laser spot should be as small as possible in order to obtain high spatial resolution. But the scanning arm deformation, the spherical aberration of the scanning objective lens, temperature deformation errors and installation errors will all cause the changes of the beam spot sizes of the laser beam on the IP, affecting the system's spatial resolution. Here is a simplified analysis:

(a) Scanning Arm Deformation

The mechanical structure of the scanning arm is shown in Fig.21. When the scanning arm begins to rotate at a high-speed, the centrifugal force will lead to the radial deformation of the scan arm, which will alter the distance between scanning objective lens and the IP, thus making the IP be not in the focus plane of the scanning objective lens and changing the diameter of the focusing spot.

A physical model of the scanning arm obtained by the three-dimensional mapping software is shown in Fig. 22. The scanning arm is made of the aluminium alloy 2024. With hexahedral element meshing, the result of the finite element analysis is that when the scanning arm rotates 4000 rounds per minute, the radial elongation is only 5μm, having little impact on the laser spot size and thus barely affecting spatial resolution of the scanning system.

Fig. 21. Mechanical structure of the scanning arm

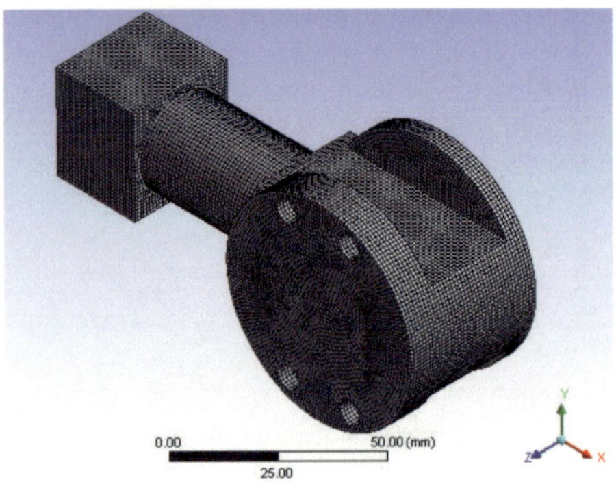

Fig. 22. Finite element mesh of scanning arm

(b) Spherical Aberration of the Scanning Objective Lens

The spherical aberration of the scanning objective lens will modify the beam waist in the image space, which means that the size of the focused spot changes and the spatial resolution of the system will reduce. In the laser scanning system, the selected laser source has following parameters: the incident laser wavelength $\lambda = 638nm$, beam waist radius $2\omega_0 = 2.4mm$. Since the image of this scanning system is mainly formed by axial laser beams, only the spherical aberration is considered here.

Theoretical derivation leads to the following formula expressing the relationship between the variant $\Delta\omega_0'$ representing the image space beam waist radius and the variation $\Delta R(z')$ representing the radius of curvature of the outgoing spherical wave (i.e., wave-front aberration)

$$\Delta\omega_0' = \frac{\pi^2\omega^5(z)}{\lambda^2}\left\{1+[\frac{\pi\omega^2(z)}{\lambda}]^2[\frac{1}{R(z)}-\frac{1}{f'}]^2\right\}^{-3/2}[\frac{1}{R(z)}-\frac{1}{f'}]^3\Delta R(z') \tag{2}$$

Where $\omega(z)$ stands for the radius of the light spot that the incident beam projects on the lens, $R(z)$ for the radius of curvature of the incident spherical wave, and f' for the focal length of the lens.

The relationship between the wave-front aberration of the optical system and the longitudinal spherical aberration can be gotten by

$$OPD = -\frac{n'}{2}\int LA'du'^2 \tag{3}$$

Where n' is the image space refractive index, LA' the longitudinal spherical aberration and u' the image space aperture angle and the exit surface divergence angle in the laser transmission.
The spherical aberration is dominated by the third order spherical aberration in most optical systems. As far as the third order spherical aberration alone is concerned, the following formula will hold

$$\Delta R(z') = OPD = -\frac{1}{4}n'\sin u'^2(a_2 y^2) = -\frac{1}{4}n'\sin u'^2 LA' = -\frac{1}{4}n'u'^2 LA' \quad (4)$$

Formula (2) minus formula (4) represents the relation between the variant $\Delta\omega_0'$ for the image space beam waist radius and the variant LA' for the spherical aberration of lens.
In the actual system, the distance from the incident light beam waist to the lens stands at $z = 200mm$. With the application of a commercial optical design software system, the spherical aberration of the doublet lens adopted by the system can be obtained. When formula (2) - formula (4) is applied, the wave-front aberration of lens can be obtained $OPD = 0.11\mu m$ and the variation of beam waist spot diameter is 0.44 ‰, which can be neglected.
It shows that the spherical aberration of the doublet lens utilized by the system is properly adjusted and has little impact on the size of focused laser beam spot and spatial resolution. However, if the scanning objective lens has a larger spherical aberration, it will result in an expanded diameter of the scanning laser spot and reduce the spatial resolution of the scanning system.

(c) Temperature Changes

In the most cases, the laser scanning system is used for the in-site applications, the temperatures of the environment are subjected to drastic changes. Since the scanner contains both optical and mechanical systems, changes in temperature will affect these systems and may eventually lead to enlarging the diameter of the laser spot at the IP, in which case the system's spatial resolution will reduce. However, through theoretical analysis, it can be discovered that in the temperature range from -10 °C to 50 °C, the system resolution is almost free from the impact of the environmental temperature.

(d) Installation Error

If there are some misalignments in adjusting the excitation optical system, especially in adjusting the laser source, as shown in Fig.23, the incident beam will not be perpendicular to the incidence surface of the pentagonal prism, in which case the axis of the scanning beam will not be perpendicular to the IP either. If there is an installation error in assembling the scanning arm, especially assembling the pentagonal prism, as shown in Fig.24, making the axis of the scanning beam be not perpendicular to the IP, the axis of the scanning beam will not be perpendicular to the IP either. In both cases, the size and shape of the scanning spot on the IP will change and the efficiency of luminescence collection will also be undermined.
The experimental results of the relationship between the incident angle of the scanning laser beam and the intensity of the luminescence collected after having excited IP is shown in Fig.25. It can be observed that when the incident angle enlarges gradually, the intensity of the luminescence collected will decrease and the collection efficiency will drop; when the axis of the scanning laser beam stimulates the IP perpendicularly, the luminescence intensity

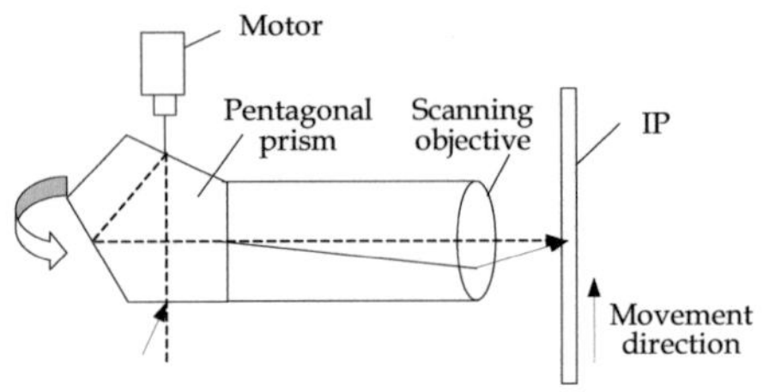

Fig. 23. The incident light be not perpendicular to incident plane of the pentagonal prism

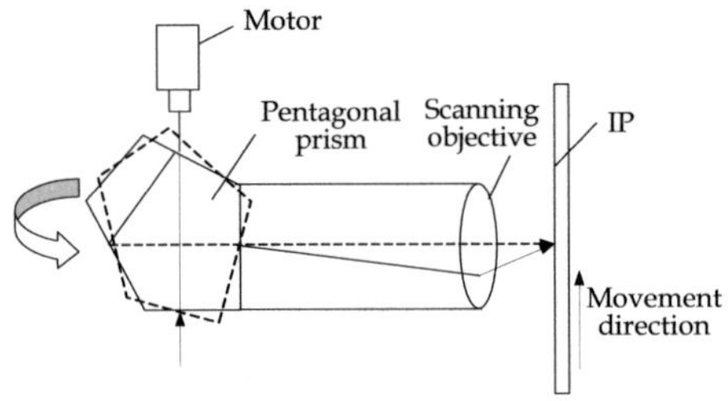

Fig. 24. Pentagonal prism with installation error

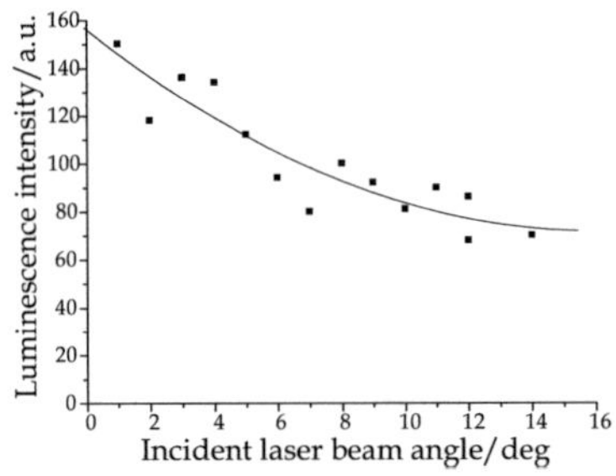

Fig. 25. Relation between angle & luminescence intensity

is the highest and collection is the most efficient. Therefore, when assembling the system, the optical system should be carefully aligned so that the incident laser beam and the IP can be as perpendicular to each other as possible.

It is difficult to ensure that the shaft of the scanning motor and the spindle of the circular plate feeding sector completely be coaxial during the assembling process of the scanning system, and this deviation will cause the IP on the plate feeding sector to be away from the focal plane of the scanning objective lens to somewhere before or behind it, creating the IP positioning error, as shown in Fig.26. In theory, the IP should be at the focal plane of the scanning objective lens, as in such a condition the laser spot diameter is the smallest and the spatial resolution of the system is the highest; when the IP has a positioning error, the spot diameter will become larger and the system's spatial resolution will reduce.

Considering all the types of installation errors mentioned above, i.e., the laser misalignment, the pentagonal prism assembling error, the IP position error generated by the off-axis between the shaft of rotating parts and the spindle of the circular plate feeding body, the formula for the transmission of decentred Gaussian beams through the optical systems is derived (Abdul-Azeez et al., 1995; Palma, 1997), and the law governing the influence of decentred Gaussian beams cases on the intensity of focusing light can be obtained. The transmission of decentred Gaussian beams through the focusing system is shown in Fig. 27.

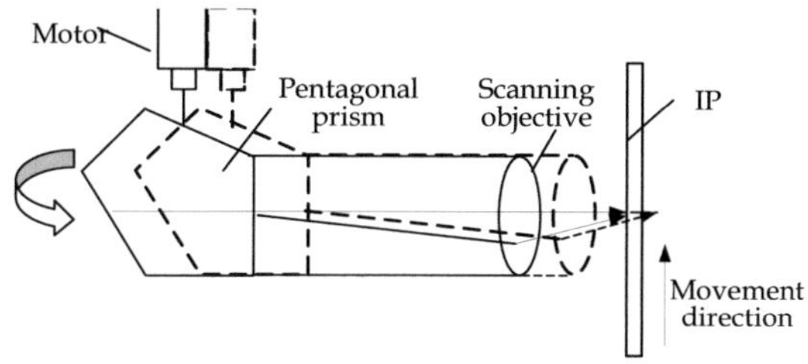

Fig. 26. Shaft of rotating parts and the principal axis of circular plate feeding sector being not coaxial

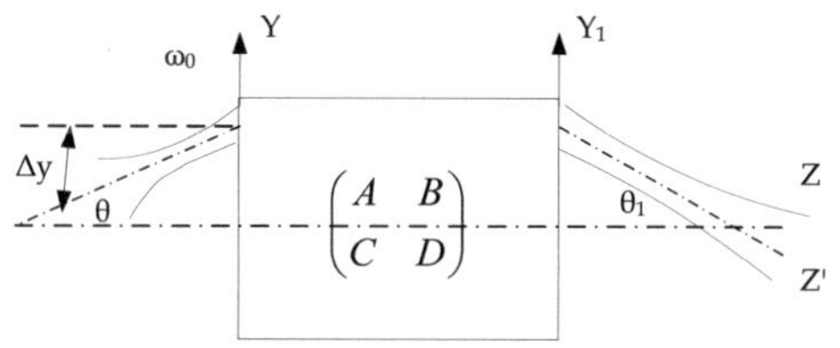

Fig. 27. Transmission of decentred Gaussian beams through the optical system

As shown in Fig.27, the input reference plane is at the waist of the incident beam and the output reference plane could be anywhere behind the objective lens. Z stands for the direction of the main axis of the objective lens, θ for the angle formed by the Gaussian beam and the axis Z, Δy for the displacement of the central ray of the entering light beam at the entering reference plane in the Y direction relative to the main axis of the optical system, ω_0 for the radius of the incident beam waist. To simplify the analysis, only one dimension is considered. The optical field distribution of the incident Gaussian beam at the plane of its waist is as follows

$$E_0(y_0,0)=\exp\left[-\frac{(y_0-\Delta y)^2\cos^2\theta}{{\omega_0}^2}\right] \tag{5}$$

Where y_0 stands for the coordinate value of the laser beam in the cross section. When the Gaussian beam goes through the optical system, its field distribution on the output reference surface can be known from the Collins formula

$$E_1(y_1,z)=\sqrt{\frac{ik}{2\pi B}}\int_{-\infty}^{+\infty}E_0(y_0,0)\exp\left[-\frac{ik}{2B}(A{y_0}^2-2y_0y_1+D{y_1}^2)\right]dy_0 \tag{6}$$

Where λ is the wavelength, k the wave number, and A, B, C and D the transformation matrix elements for the optical system. The phase factor $\exp(-ikl)$ is omitted here.
Substituting Eq. (5) into Eq. (6) and applying integral Eq. (7)

$$\int_{-\infty}^{+\infty}\exp(-p^2x^2\pm qx)dx=\frac{\sqrt{\pi}}{p}\exp(\frac{q^2}{4p^2})\ (p>1) \tag{7}$$

The field distribution of the output reference plane can be expressed as

$$E_1(y_1,z)=\sqrt{\frac{ik}{2\pi B}}\cdot\exp\left(-\frac{ikDy_1^{\ 2}}{2B}\right)\cdot\exp\left(-\frac{\Delta y^2\cos^2\theta}{\omega_0^{\ 2}}\right)\cdot\frac{\sqrt{\pi}}{p}\cdot\exp\left[-\frac{k^2(y_1-q)^2}{4B^2p^2}\right] \tag{8}$$

Where $q=\dfrac{2i\Delta yB\cos^2\theta}{k\omega_0^{\ 2}}$ and $p^2=\dfrac{\cos^2\theta}{\omega_0^{\ 2}}+\dfrac{ikA}{2B}$

The light intensity of Gaussian beam on the output reference plane can be written as

$$I_1(y_1,z)=E_1(y_1,z)\cdot E_1^{\ *}(y_1,z) \tag{9}$$

$E_1(y_1,z)$ can be obtained from Eq. (8), and $E_1^{\ *}(y_1,z)$ represents the conjugate of $E_1(y_1,z)$. Taking Eq. (8) into Eq. (9) with adjustment will lead to

$$I_1(y_1,z)=\sqrt{\frac{k^2}{4\pi^2B^2}}\cdot\exp\left(-\frac{2\Delta y^2\cos^2\theta}{\omega_0^{\ 2}}\right)\cdot\frac{\pi}{\sqrt{\dfrac{\cos^4\theta}{\omega_0^{\ 4}}+\dfrac{k^2A^2}{4B^2}}}\cdot\exp\left(-\frac{k^2}{4B^2}\cdot\frac{q_1}{p_1}\right) \tag{10}$$

Where $q_1=2y_1^{\ 2}\cos^2\theta-\dfrac{8\Delta y^2B^2\cos^6\theta}{k^2\omega_0^{\ 4}}-4y_1\Delta yA\cos^2\theta$ and $p_1=\dfrac{\cos^4\theta}{\omega_0^{\ 2}}+\left(\dfrac{kA\omega_0}{2B}\right)^2$

Since the parameters and relative positions of the pentagonal prism and the doublet lens are already known, the transmission matrix of the Gaussian beam M from the incident surface to the exit surface can be determined, which is

$$M=\begin{bmatrix} A & B \\ C & D \end{bmatrix} \tag{11}$$

If the parameters A, B, C and D in Eq. (11) are taken into the Eq. (10), Gaussian beam intensity of anywhere behind the doublet objective lens can be gained. When the beam is near to the axis($\theta\neq 0$) and $\Delta y=2mm$, the light intensity distribution on the reference plane at three positions, i.e., 2mm before the focal plane, in the local plane, and 3mm behind the focal plane, is shown in Fig.28. In front of the focal plane, central line of the maximum light intensity is shifting to the positive direction of y_1 axis and behind the focal plane, it moves toward the negative direction.

As shown in the Fig.28, the intensity of the scanning laser on the focal plane is at the maximum level and the spot diameter is the smallest. When the IP has a positioning error, the intensity of the Gaussian beam decreases and the diameter of spot rapidly expands, producing a great influence on the resolution of the system. At the same time, when the intensity of the laser goes down, the amount of excited luminescence will decrease, resulting in a bad image of the IP.

Therefore, it is necessary that the manufacturing and assembling accuracy for the mechanical parts be enhanced and the installation errors be reduced in order to improve the spatial resolution of the system.

After the previous theoretical analysis and simulation calculations, Table 1 is a comprehensive presentation of the influence that various factors will have on the spatial resolution of the scanning system.

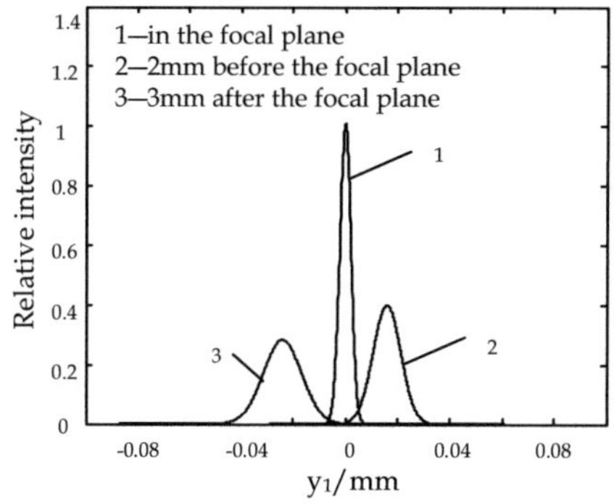

Fig. 28. Intensity distribution of different locations

	scanning arm deformation	Spherical aberration of scanning objective lens	temperature	Installation
resolution	Almost none	positive	Almost none	Huge influence

Table 1. The impact of various factors on the resolution

From Table 1, it can be seen that installation errors are the most important factors affecting the spatial resolution of a scanning system and should be given emphasis in the design, manufacturing and assembly of the laser scanning system.

4.2 An analysis of imaging stability

Since the scanning arm of the CR scanner rotates at a high speed while at work, an extra weight is needed at the opposite side of the scanning arm, which helps to keep the gravity center of the scanning sector on the motor shaft. Otherwise, if there is an offset between the gravity center and the geometric center of the scanning arm, the vibration of the scanning system will be inevitably induced, and it will make the scanning system work unstable. The model that is the equivalent of the mechanical structure when the scanning arm is rotating at a high-speed is shown Fig. 29.

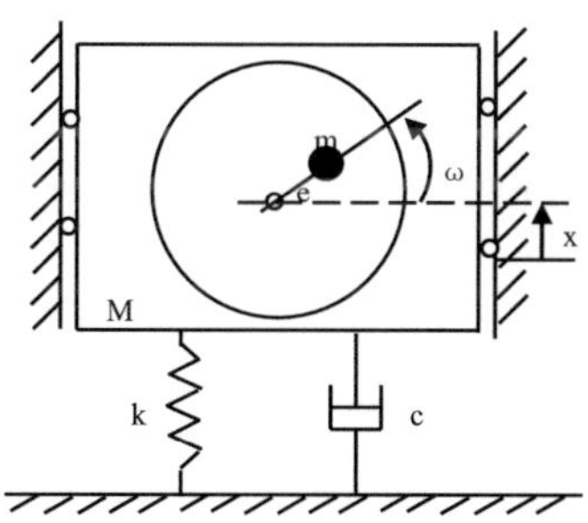

Fig. 29. Mechanical model of structure

The total mass of the rotating machinery is M, the eccentric mass of the rotor is m, the eccentricity is e, the rotation angular velocity of the rotor is ω, the base can be taken as a spring with stiffness k, and the resistance encountered in the process of vibration is simplified to the force on the mechanical structure by a damper that has been in a parallel connection with the spring and has a damping coefficient of c.

It is supposed that coordinate x represents the vertical displacement of the structure away from the equilibrium position, while the vertical displacement of the eccentric mass is $x + e\sin\omega t$, and the differential equation showing the vertical motion of the system obtained according to the D'Alembert principle is as follows

$$(M-m)\ddot{x} + m\frac{d^2}{dt^2}(x + e\sin\omega t) + c\dot{x} + kx = 0 \tag{12}$$

It can be obtained through

$$M\ddot{x} + c\dot{x} + k = me\omega^2 \sin\omega t \tag{13}$$

Where me is called amount of unbalance, and $me\omega^2$ is the centrifugal force caused by amount of unbalance. The steady state response of the system is as follows

$$x(t) = B\sin(\omega t - \phi) \tag{14}$$

The amplitude of vibration of the steady state response is as follows

$$B = \frac{me}{M}\frac{\lambda^2}{\sqrt{(1-\lambda^2)^2 + (2\xi\lambda)^2}} \tag{15}$$

The phase difference between the displacement of the vibrating object and the exciting force is as follows

$$\phi = arctg\frac{2\xi\lambda}{1-\lambda^2} \tag{16}$$

Where $\omega_n = \sqrt{\frac{K}{M}}$ is the natural frequency, $\lambda = \frac{\omega}{\omega_n}$ is the frequency ratio, and $\xi = \frac{c}{2\omega_n M}$ is the damping coefficient.

It can known from Eq. (15) and Eq. (16) that when the frequency is low ($\omega << \omega_n$), the amplitude of vibration is nearly zero; when the frequency is high ($\omega >> \omega_n$), the amplitude of vibration approaches a constant of me/M that is almost unrelated to the frequency and the damping coefficient of the system; when resonance happens ($\omega = \omega_n$), the amplitude of vibration is $B = me/2\xi M$ and smaller damping coefficient will lead to greater vibration. When $\omega \approx \omega_n$, the phase difference will undergo a sudden anti-phase change of 180° and the resonance will take place.

For rotating machinery, the centrifugal force caused by the deviation of the center of gravity is one of the major motivation sources of vibration. When the frequency of motivation force, namely, the angular velocity ω of the rotor is close to the natural frequency ω_n of the system, there will be resonance in the system. The motor speed at this moment is called the critical speed.

A simplified model of the mechanical structure of the scanner through three-dimension mapping software is shown in Fig.30, and the results of the finite element analysis are shown in Fig.31. There are total of 245041 grid nodes and 72149 units after finite element model meshing. The first six natural frequencies of the system and the corresponding speed of the motor when system resonates are shown in Table 2.

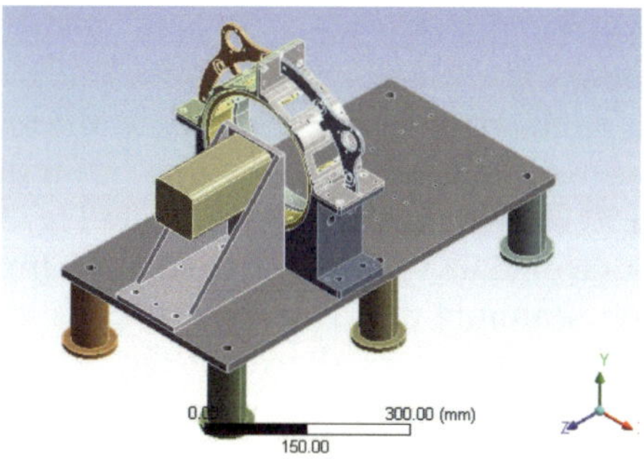

Fig. 30. Mechanical structure of scanner

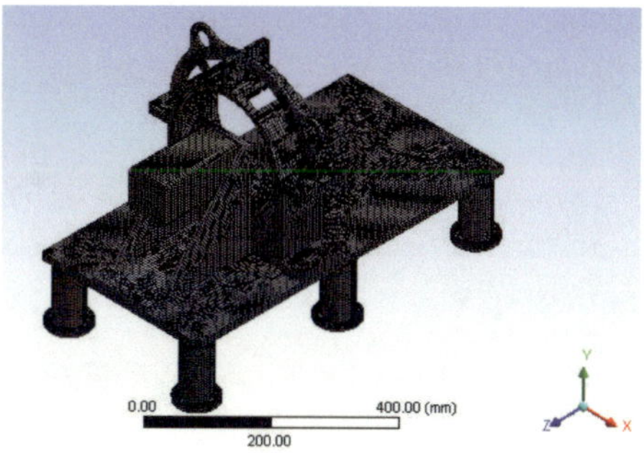

Fig. 31. Finite element mesh of the structure

From the Table2, it can be concluded that when the rotating parts of the scanner have the mass eccentricity, the system will vibrate, the amplitude of vibration will become large if the eccentricity increases; When the angular velocity is equal to the natural frequency of the system, the resonance will be generated, which will not only considerably deform the structure of the system and will also destabilize the system; The motor speed is associated to the systematic sampling frequency, and a high sampling frequency is needed if the motor rotates at a high speed. Therefore, after having determined the sampling frequency, the motor speed should be to calculated and also be kept away from the natural frequency of the mechanical system.

Mode order	1	2	3	4	5	6
Natural frequency(Hz)	220.58	289.68	379.33	413.63	444.51	463.75
motor rotation speed(rpm)	2106.4	2759.7	3622.3	3949.9	4244.8	4428.5

Table 2. Systematic natural frequency and related motor speed

5. Experimental results and analysis

Currently in China, the spatial resolution required in industrial radiographic testing is generally no less than 3Lp/mm. The German-made 20Lp/mm resolution testing card is applied to test this scanning system and the testing results show that the spatial resolution

of 7Lp/mm can be obtained by the developed system, which can meet the demand of China's industrial X-ray testing.
Experiments are carried out on the developed prototype in the testing of the steel pipe weld. The steel pipe used in the experiment of weld imaging has a diameter of 160 mm and its wall is 8mm thick, with the IP bent in the pipe as shown in Fig. 32. Experimental conditions include the X-ray tube voltage of 120KV, the tube current of 4mA, the focal length 700mm, the exposure time 40s. After scanning and reading with the laser scanner developed, the original image gotten by this system is as shown in Fig. 33, in which even a small weld defect can be seen clearly without any image processing.

Fig. 32. The experiment of radiography

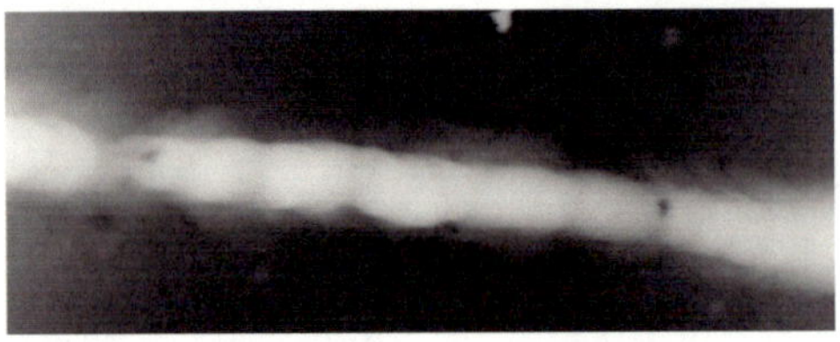

Fig. 33. Original image of scanning

6. Conclusions

A new laser scanning method and technology for the CR system is proposed here with designs accomplished for the optical system, the signal collection and control system, the mechanical system, and the software as well as the development of a CR scanner prototype; An analysis has been made on the a various factors that influence the system's spatial resolution and other performances, the results of which show that the positioning error of the IP is the key one. When the IP moves away from the focal plane of the scanning objective lens, the laser has a smaller intensity after focusing, the diameter of the spot becomes larger, and the spatial resolution decreases rapidly; the finite element simulation has been conducted on the mechanical structure to justify the effort to avoid resonance of the system. Some experiments have been carried out on the prototype that has been developed and the results of these experiments show that the quality of imaging of the scanning system developed is able to meet the demand of China's industrial X-ray imaging inspection.
Despite some of the good research results achieved during our work, much remains to be improved.

1. For the readout device that has been designed and developed, only half scanning is used to collect the luminescence. With regard to this problem, measures could be taken on the originally designed readout device to make the laser be focused on the IP from

two symmetrical directions at the same time. In this way, it amounts to the effect of having two scanning arms work together during the circular scanning of the IP, excluding the need for adding more counter weight. The scanning speed can be enhanced.

2. The prototype of the laser scanner leaves much to be desired, especially the mechanical part, including the fixing and support structure of the optical device and the transmission sector for the IP movement mechanism. The manufacturing and assembling accuracy is not so high, which causes the actual resolution to be lower than what has been designed. It is still quite necessary to further optimize and improve the mechanical structure and reduce the errors in process of the manufacturing and assembling.

7. Acknowledgments

This work was supported by the National Scientific and Technical Supporting Program of China under Grant NO. 2006BAK02B01. The Authors would like to express thanks to all the members of the research team. The following team members provided valuable research assistance: Keqin DING, Jianying CUI, Donglin HUANG, Xu Zhang , Na ZHAO, Sijin WU, Yan GAO, Daijun WU and Guang CHEN. Our hard works and efforts are the only things that have ever really made the project done.

8. References

AR Cowen; AG Davies & SM Kengyelics(2007). Advances in computed radiography systems and their physical imaging characteristics. *Clinical Radiology*, vol.62, No. 12, pp. 1132-1141, ISSN 0009-9260.

Abdul-Azeez R; Al-Rashed & Bahaa EA Saleh(1995). Decentered Gaussian beams. *Applied Optics*, Vol. 34, No.30,pp.6819-6825, ISSN 0003-6935.

Brandt et al. (1996). Light collector having optically coated acrylic substrate. *US. PAT.* NO. 5541421.

Claudio Palma(1997). Decentered Gaussian beams, ray bundles, and Bessel-Gauss beams. *Applied Optics*, Vol.36 No.6, pp.1116-1120, ISSN 0003-6935.

Chen H Q(2003).*Modern Practical Optical System*. Huazhong University of Science &Technology Press, ISBN：756092939, Wuhan

Michael Brandt (1993). The design and performance characteristics of a collector for high efficiency detection of photostimulable phosphor emissions. *Proceedings of SPIE - The International Society for Optical Engineering,* pp.289-295, ISBN 0819411302, Newport Beach CA USA, Febrary 1992, Publ by Society of Photo-Optical Instrumentation Engineers, Bellingham WA USA.

O. Dragusin et al. (2006). Characterization of a new generation of Computed Radiography system based on line scanning and phosphor needles. *Progress in Biomedical Optics and Imaging - Proceedings of SPIE*, pp.2113-2122,ISBN 0819464252, San Diego CA USA, Febrary 2006, SPIE, USA.

Satoshi Arakawa et al.(1999) Novel computed radiography system with improved image quality by detection of emissions from both sides of an imaging plate. *Proceedings of SPIE - The International Society for Optical Engineering*, pp.572-581, ISSN 0277786X, San Diego, CA, USA, Febrary 1999, SPIE, USA.

Schaetzing R et al. (2002). New high-speed scanning technique for computed radiography. *Proceedings of SPIE - The International Society for Optical Engineering*, pp.511-520, ISSN 0277786X, San Diego, CA, USA, Febrary 2002, SPIE-Int. Soc. Opt. Eng, USA.

Shan B Zh; Guo B P& Wang Sh Y(2005)Laser Scanning System of Large Field. *Shenzhen Science & Technology*, No.z1, pp2-7, ISSN 1009-5985.

Zhang J P; Feng X H & Huang L Q(2008). Design of laser scanning optical system for computed radiography. *Optics and Precision Engineering*, vol.16, No. 2, pp202-207, ISSN 1004-924x.

Zhou X L et al. (1993). The development of CR-IP system. *Chinese Journal of Medical instrumentation*, vol.17.No.2, pp63-68, ISSN 1671-7104